T0130832

Troubleshooting and Repairing Major Appliances

Third Edition

Eric Kleinert

New York Chicago San Francisco
Lisbon London Madrid Mexico City
Milan New Delhi San Juan
Seoul Singapore Sydney Toronto

Library of Congress Cataloging-in-Publication Data

Kleinert, Eric.
   Troubleshooting and repairing major appliances / Eric Kleinert. — Third
Edition.
      pages   cm
   ISBN 978-0-07-177018-7 (alk. paper)
   1. Household appliances, Electric—Maintenance and repair.   I. Title.
   TK7018.K56   2012
   683'.80288—dc23   2012032805

McGraw-Hill books are available at special quantity discounts to use as premiums and sales promotions, or for use in corporate training programs. To contact a representative, please e-mail us at bulksales@mcgraw-hill.com.

Troubleshooting and Repairing Major Appliances, Third Edition

17 18 19 20 21 LBC 28 27 26 25 24

ISBN   978-0-07-177018-7
MHID   0-07-177018-6

Sponsoring Editor Roger Stewart	**Acquisitions Coordinator** Molly Wyand	**Production Supervisor** Jean Bodeaux
Editorial Supervisor Patty Mon	**Copy Editor** Lisa McCoy	**Composition** Cenveo Publisher Services
Project Manager Anupriya Tyagi, Cenveo Publisher Services	**Proofreader** Amy Rodriguez **Indexer** Jack Lewis	**Illustration** Cenveo Publisher Services **Art Director, Cover** Jeff Weeks

I dedicate this third edition to my granddaughter Simone.
I know one day she will look at this book for guidance in repairing her own appliances.

About the Author

Eric Kleinert is a professional with 40 years of experience in commercial and domestic major appliances; refrigeration; and HVAC sales, service, and installation. He has owned and operated a number of major appliance and air conditioning sales and service corporations. As an instructor for a preeminent technical college, he taught adults aspects of preventative and diagnostic services and techniques, which enhanced their ability to better evaluate products and services necessary to maintain commercial and residential climate-control systems, refrigeration, and major appliances.

Contents

Acknowledgments

I would like to convey my utmost appreciation to the following companies for their valuable contributions to this improved third edition:

- BrandsMart USA
- Dwyer Instruments, Inc.
- Klein Tools, Inc.
- National Appliance Service Technicians Certification (NASTeC)
- Tecumseh Compressor Company

I would also like to thank Roger Stewart and the staff of McGraw-Hill for giving me the opportunity to write this third edition.

I would like to thank my wife, Eileen, for putting up with me while writing this book for the third time. I love you, honey.

I would like to thank my daughter, Brandi, and my son-in-law, Matt, for their support.

Introduction

In 2009, my granddaughter, Simone was born. At the age of one she was playing with her kitchen appliance toys and wondering how they work. One day in the not-too-distant future, Simone will open this book, just like her mother did, to repair her own real appliances. The future will have more advanced and efficient appliances and air conditioners. Some examples of newer products on the market today are condenser dryers, refrigerated ovens, and hybrid electric heat pump water heaters. In the not-too-distant future, we will see more smart appliances and air conditioners that can be managed and monitored from long distances. Trained service technicians will be able to troubleshoot mechanical issues over the phone and through computers, limiting service calls and in-home visits. Maybe in the future appliance and air conditioner manufacturers will come out with appliances and air conditioners that operate on a different type of fuel other than electricity, solar, or gas. The future has endless possibilities for service technicians in the major appliance and air conditioner field.

The principles and basics will never change; only the appliance products will change for the better. New products, energy conservation, and electronics have made the industry what it is today, and it will make it even better for my granddaughter's generation. All that's needed is the ability to think clearly, use hand tools, and use test meters.

I can only expose you to the knowledge that's needed for preventative, diagnostic, and service techniques in the major appliance and air conditioner industry. This will help you evaluate these products and services. If you are a homeowner, this book can help you decide if it's a problem you can fix yourself or if it's one that's best left for the experts. For the beginning technician, what you learn from this book will give you the keys for a rewarding service career. And for the experienced professional, this book will serve as a good resource and refresher on the latest techniques and technologies.

As a novice servicer, you may have to look up terms in the glossary or consult the component descriptions and diagnostic charts. However, you'll be amply rewarded for this effort when a major appliance or air conditioner is fixed for the first time. This book will also be informative and useful for the experienced servicer and repair technician.

This third edition is composed of six parts: general, types of fuels, electrical theory, refrigeration theory, parts, and specific information. Part I, "The Fundamentals of Service," identifies the problems of selecting, purchasing, and installing new major appliances and air conditioners. Also included is information on manufacturer warranties, use and care manuals, and where consumers can get help if needed.

Chapters 2 through 5 map out a complete guide for repairing major appliances and air conditioners. We'll discuss the safety precautions for installation, operation, and repair of major appliances and air conditioners, as well as the right tools needed for the job and the basic approach before beginning to repair these products. Included is a new chapter for those individuals who want to enter the service field.

Part II, "Electricity, Electronics, and Gas," covers everything you ever needed to know about electricity, gas, and electronics in major appliances and air conditioners. I will teach you how to use test meters, read circuit diagrams, test electronic parts, gas appliance venting, and much more.

In Part III, "Refrigeration and Air Conditioning," you'll learn about the refrigeration cycle and its components.

In Chapters 10, 11, and 12, we'll focus on the description, operation, and locations of the parts in today's major appliances and air conditioners. Chapter 13 will focus on the fault and error codes of today's modern electronic appliances and air conditioners.

In Part IV, "Parts," we'll explore in depth the different types of appliances and air conditioners you'll be expected to fix. Chapters 14 through 30 cover domestic dishwashers, garbage disposals, electric and gas water heaters, automatic washers, automatic electric and gas dryers, electric and gas ranges/ovens, microwave ovens, refrigerators, freezers, automatic ice makers, air conditioners, and dehumidifiers. We'll go over step-by-step procedures for testing and replacing parts, chart troubleshooting, preventative maintenance procedures, and appliance and air conditioner installation procedures.

To present a broad overview of service techniques, I've provided general troubleshooting information about the various types of major home appliances and air conditioners, rather than specific models that may become outdated. The pictures and illustrations are for demonstration purposes only to clarify the description of how to service them; they do not reflect a particular product's reliability.

Troubleshooting and Repairing Major Appliances aims to provide you with a basic understanding of the operation and common problems of major appliances and air conditioners as you become your own technician. Soon, you'll have the know-how to fix practically any appliance in the home that breaks down.

Fundamentals of Service

PART

I

Selecting, Purchasing, and Installing Major Home Appliances and Air Conditioners

Both the technician and the consumer will appreciate this chapter because it addresses some of the problems of selecting, acquiring, and installing major home appliances and air conditioners. It is also pertinent that technicians familiarize themselves with the features and functions of the appliances and air conditioners, which are also stated in every new appliance and air conditioner use and care manual. This will be essential when diagnosing problems with the diverse types of appliances and air conditioners.

This information will provide proper planning techniques and a better understanding of major home appliances and air conditioners, including appliance and air conditioner (A/C) warranties and where to get help when it is needed.

In today's market, major home appliances and air conditioners are manufactured to meet the needs of the average person. Remember that price should not be the most influential factor when choosing an appliance or air conditioner. Physical and mental limitations should also be considered when selecting the product that will be juxtaposed with the consumer's needs.

Electric and Gas Ranges, Cooktops, and Ovens

The domestic range was designed as a multipurpose cooking appliance. It consists of a surface area with heating elements on the top to cook the food. The oven cavity is used for baking food at a set temperature. Within the same oven cavity, the broiling of food is also incorporated.

Domestic ranges are available in either electric or gas, in sizes ranging from 20 inches to 48 inches in width, and with a wide selection of configurations and colors. The following sections describe a few of the common configurations available to the consumer.

Freestanding Ranges

The freestanding range stands between two base cabinets, or sometimes at the end of a cabinet line. The consumer has a choice of gas or electric. These ranges are available in 20- , 24- , 30- , 36- , 40- , and 48-inch widths (Figure 1-1). Some 40- and 48-inch models have two ovens. Designs include gas burners and gas ovens, or standard electric heating elements with electric ovens, and a glass cooktop with concealed electric elements underneath. Also available is a combination of electric and gas cooking. The controls might be located on the rear console or across the front.

Oven cleaning systems include self cleaning,[1] continuous cleaning,[2] and standard manual cleaning. Freestanding ranges have the following advantages and disadvantages:

Advantages

- Freestanding ranges are generally less expensive than other types. Prices vary with features.
- Freestanding ranges can be moved when the family moves.
- Most models have some center space for placing utensils.
- Front controls can be reached easily from a seated position.
- Bottom drawer adds to kitchen storage space.

Disadvantages

- Rear console controls are virtually impossible for a seated person to reach.
- Low broiler in gas ranges is less accessible from a seated position than the oven broiler in electric ranges.

Figure 1-1 The freestanding electric or gas range is available in 20- , 24- , 30- , 36- , 40- , and 48-inch widths and a variety of colors.

FIGURE 1-2
The slide-in range with a 30-inch width is available in a variety of colors.

Slide-In Ranges and Drop-In Ranges

Slide-in ranges and drop-in ranges might not always be available in all sizes: 30-inch width is the most common (Figure 1-2). However, drop-in ranges are also available in a 27-inch width. Designs include gas burners, standard electric heating elements, and a glass cooktop with concealed electric elements underneath. The oven and cooktop controls are usually located across the front of the range; however, some models have the cooktop controls along the side. They are an excellent choice for an island or peninsula-shaped counter because they are flush with the surrounding counter. These ranges can also be installed overlapping the adjacent countertop edges, thereby eliminating dirt-catching gaps. A drop-in range either hangs from a countertop or sits on a low cabinet base; a slide-in range sits on the floor.

Slide-in ranges and drop-in ranges (Figure 1-3) have the following advantages and disadvantages:

FIGURE 1-3
The drop-in range is most commonly available with a 30-inch width. It is available in a variety of colors.

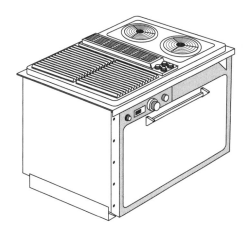

Advantages

- Can be installed a few inches higher or lower than a freestanding range.
- Controls can be reached by most cooks.

Disadvantages

- Built into kitchen and generally cannot be moved when the family relocates.
- Requires installation by a carpenter and an electrician.

Built-In Cooktops

Built-in cooktops are set into a countertop and are made in various sizes, from 15 to 48 inches wide. Built-in cooktop designs include gas burners, standard electric heating elements, and a glass cooktop with concealed electric elements underneath (Figures 1-4 and 1-5). They might have side or front controls and might be of modular design. Special plug-in cooking accessories are also available. Built-in cooktops have the following advantages and disadvantages:

Advantages

- Can be installed at the most convenient height for the cook.
- Side or front controls are easily reached by most cooks.
- Counter installation provides open space below the cooktop.
- The slide-in range with a 30-inch width is available in a variety of colors.

Figure 1-4   Electric or gas built-in cooktops are available with two, four, or five heating elements. They are available in a variety of colors.

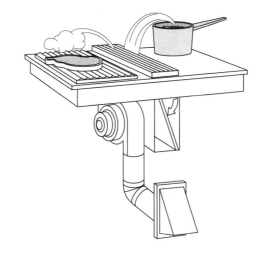

Disadvantages

- Built into kitchen and generally cannot be moved when the family relocates.
- Requires installation by a carpenter and an electrician.

Built-In Ovens

Built-in ovens usually have one oven cavity, but models with two oven cavities are also available. In the double oven cavity model, the second oven might be a conventional oven, a microwave oven, or a combination of both (Figure 1-6). Built-in ovens are available in 24-, 27-, and 30-inch widths. Their height varies, depending upon whether they are single- or double-oven units. Oven cleaning systems available include self-cleaning, continuous cleaning, and standard manual cleaning. Two-oven models also offer conventional ovens with two different cleaning systems. Built-in ovens with the microwave feature are available with either solid-state microcomputer or electromechanical controls.

Built-in ovens have the following advantages and disadvantages:

Advantages

- Can be installed at the most convenient height for the user, putting controls within reach for a standing or a seated cook.
- Automatic cleaning systems virtually eliminate the task of cleaning the oven manually.

Disadvantages

- Built into the kitchen and generally cannot be moved when the family relocates.
- Installation might involve structural and wiring changes, requiring a carpenter and an electrician.
- Two-oven combinations with microwave ovens often require learning some new cooking techniques.

FIGURE 1-6 The built-in oven is available in different configurations and in a variety of colors.

FIGURE 1-7 Countertop microwave oven.

Microwave Ovens

Do not go out and purchase a microwave oven until you do your homework. Brand, type, size, power, features, warranty, and location are just some of the features you need to consider. You can also go on the Internet and conduct research.

When comparing microwaves (Figure 1-7), fill out the following checklist. Check all that apply.

Measure the area available for the microwave oven:

_____width
_____depth
_____height

Before purchasing, read the installation instructions for the built-in or over-the-range model you are considering. These instructions will provide you with information such as cutout dimensions, venting, voltage, location, and more.

What is the voltage?
- ❑ 120 volts
- ❑ 240 volts

What type of microwave?
- ❑ Countertop
- ❑ Over the range
- ❑ Combination (range and microwave)
- ❑ Convection
- ❑ Under the counter
- ❑ Built-in
- ❑ Other (list the features)

How much room do I need to fit the microwave oven on the countertop?
- ❑ Large: 22 to 27 inches wide by 13 to16 inches high by 15 to 21 inches deep
- ❑ Medium: 20 to 24 inches wide by 16 to 16 inches high by 13 to 18 inches deep
- ❑ Small: 18 to 20 inches wide by 10 to 12 inches high by 11 to 12 inches deep
- ❑ Touchpad controls:
  - ❑ Are the numbers large and easy to read?
  - ❑ Is the oven easy to program?
  - ❑ Does the model display prompters that will guide you as you set the controls?
- ❑ Mechanical controls:
  - ❑ Does the mechanical timer have a wide time range setting?
    - ❑ Seconds?   ❑ Minutes?   ❑ Hours?
  - ❑ Is there an on/off switch to operate the oven after setting the time function?
  - ❑ Does the timer knob turn on the oven after setting the time function?

How many watts of cooking?
- ❑ 400 to 700 watts
- ❑ 600 to 800 watts
- ❑ 800 to 1000 watts
- ❑ 1000 to 1600 watts

What additional features do you want?
- ❑ Auto defrost
- ❑ Pre-programmed cooking

Does the oven have a light inside the cavity?

❑ yes   ❑ no

Is there a window in the door?

❑ yes   ❑ no

Does the oven have a rack?

❑ yes   ❑ no

Do you want a turntable for the food to rotate on?

❑ yes   ❑ no

Does the model have multi-stage cooking?

❑ yes   ❑ no

❑ Temperature probe

❑ Cooking sensors

Will my plates fit the oven cavity?

❑ yes   ❑ no

Other (list here)

Additional information on warranties, safety, recalls, and maintenance must be taken into account when purchasing a microwave oven. Ask the salesperson for a demonstration of the microwave you are considering.

Checklist for Cooking Products

Prior to selecting and purchasing an appliance, read this section. Fill out the following checklist. Check all that apply.

Measure the area available for the range:

_____width

_____depth

_____height

These measurements are the cut-out measurements, not the range measurements. Also, be sure that the range can fit through the doorways of the house.

The type of range desired:

❑ Freestanding with one oven   ❑ Double oven

❑ Built-in   ❑ Drop-in

❑ Slide-in   ❑ Electric

❑ Gas

❑ LP   ❑ Natural   ❑ Dual fuel (gas and electric)

Type of oven needed:

❑ Single   ❑ Double

❑ Conventional   ❑ Microwave

❑ Combination oven   ❑ Convection oven

Location desired:
- ❏ Below cooktop
- ❏ Separate built-in oven(s)
- ❏ One over, one under

Oven controls:
- ❏ On back console
- ❏ On hood
- ❏ Touch pads
- ❏ On range front console
- ❏ Dial type
- ❏ Automatic oven clock/timer

Cleaning system:
- ❏ Self-cleaning (pyrolytic)
- ❏ Manual
- ❏ Continuous

Broiler type:
- ❏ Top of oven
- ❏ Variable heat
- ❏ Low broiler

Cooktop style:
- ❏ Conventional
- ❏ Grill/griddle convertible
- ❏ Glass-ceramic
- ❏ Induction

Cooktop controls:
- ❏ Thermostatic control
- ❏ Eye-level controls on hood
- ❏ Controls at front of cooktop
- ❏ Eye-level controls
- ❏ Controls on backsplash
- ❏ Controls at side of cooktop

Accessories:
- ❏ Rotisserie
- ❏ Griddle
- ❏ Roast temperature probe

Venting system:
- ❏ Separate hood
- ❏ Vent-microwave oven combination
- ❏ Vent over regular cooktop
- ❏ Built-in down draft
- ❏ Hood attached to upper oven
- ❏ Vent over grill/griddle cooktop

The preferred color of the range _____

Who will use the range? _____

How many people are being cooked for? _____

Is the range design convenient for all family members? ❏ yes ❏ no

Price range: $_____

Warranty and service information:_____

Refrigerators and Freezers

As with other kitchen appliances, refrigerators and freezers come in a wide variety of styles, sizes, and colors (Figure 1-8). Some designs might meet the needs of a family member who has a physical or mental limitation better than others.

FIGURE 1-8   Refrigerators and freezers are available in a wide variety of sizes and colors. (*continued*)

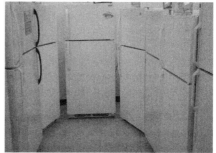

FIGURE 1-8 Refrigerators and freezers are available in a wide variety of sizes and colors.

Some questions to consider when choosing a refrigerator are:

- Does it have a true no-frost system to do away with manual defrosting?
- Does it have an automatic ice maker, which produces ice without trays to fill or empty?
- Does it have in-the-door dispensers to deliver ice and cold water without opening the door?
- Does it have shelves, bins, and drawers that pull out to make reachable those foods stored in the back?

Five basic types of refrigerators are on the market today:

- Compacts or portables: 1.2 to 6 cubic feet (12 inches to 24 inches wide)
- Single-door models: 9 to 14 cubic feet (23 inches to 30 inches wide)
- Top-mount refrigerator-freezer combination: 10 to 23 cubic feet (24 inches to 33 inches wide)
- Bottom-mount refrigerator-freezer: 18 to 20 cubic feet (32 to 36 inches wide)
- Side-by-side refrigerator-freezer: 17 to 30 cubic feet (30 to 48 inches wide)

Excluding compacts and portables, refrigerators range in height from 56 to 84 inches and in depth from 24 to 31 inches.

When selecting the capacity of a refrigerator, the following guidelines should be considered:

- Allow a minimum of 12 cubic feet for the first two persons in the household.
- Add 2 cubic feet for each additional member. The most popular size for an average family is 18 cubic feet.
- Subtract from this rule if many meals are eaten away from home, using the number of meals eaten outside the household as a basis.
- Add to the rule if the customer entertains often, if the family is expanding, if there is a vegetable garden growing, or if cooking is enjoyable.

Compact and Portable Refrigerators

Compact and portable refrigerators are often used as a supplementary model in family rooms, offices, dorms, vacation homes, campers, and other convenient places. Many fit on or under a countertop. They might be freestanding or built-in models, and they come in a variety of colors and finishes.

Compact and portable refrigerators have the following advantages and disadvantages:

Advantages

- Refrigerators of this type have a smaller capacity—less than 6 cubic feet. They can be installed at any height.
- Has a small freezer compartment for ice trays.
- Some models have an optional ice maker.

Disadvantages

- Frozen food storage is limited to a few days—a week at most—depending on the temperature.
- A few compact models have automatic defrosting. However, most are manually defrosted.
- Small size provides limited storage capacity.

Single-Door Refrigerators

Single-door refrigerators provide both fresh and frozen food storage. Frozen food compartments are located on top of the inside of the refrigerator and usually contain ice cube trays.

Single-door refrigerators have the following advantages and disadvantages:

Advantages

- Refrigerators of this type generally have a small capacity—less than 14 cubic feet. Most shelf areas are within reach of a seated person.
- Freezer compartments have side-opening doors, rather than drop-down doors, for easier accessibility.

Disadvantages

- Refrigerators of this type usually require manual defrosting, a difficult chore for disabled persons.
- Freezer compartments with drop-down doors are inaccessible from a seated position.
- Limited storage capacity.
- Freezer compartment can only be used for short-term storage of commercially frozen food and for making ice. High-sugar foods, such as ice cream, might not stay frozen.

Two-Door Refrigerators

Two-door refrigerators with top freezers provide storage for fresh and frozen food. The freezer maintains a temperature of 0 to 10 degrees Fahrenheit. These models come with and without automatic ice makers. Some models have an ice or water dispenser on the outside of the doors.

Two-door refrigerators with top freezers have the following advantages and disadvantages:

Advantages

- Provide proper storage conditions for both fresh and frozen foods.
- Keep ice cream frozen.

Disadvantages

- Top freezer is not accessible from a seated position.
- Foods stored near the rear of deep shelves might be difficult to reach without using special aids.

Two-Door Refrigerators with Bottom Freezers

Two-door refrigerators with bottom freezers provide storage for both fresh and frozen foods. The freezer maintains a temperature of 0 to 10 degrees Fahrenheit. Models with and without automatic ice makers are available.

Two-door refrigerators with bottom freezers have the following advantages and disadvantages:

Advantages

- Freezer shelf and basket slide out for easy accessibility.
- Lower shelves of the fresh food storage area are within easy reach from a seated position.

Disadvantages

- Although this design might meet the needs of some users with disabilities, the bottom-freezer refrigerator is generally not convenient for those in wheelchairs. The low freezer is also inconvenient for users who have trouble stooping or bending.

Side-by-Side Refrigerators

Side-by-side refrigerators have deep, vertical fresh and frozen food compartments. They require less room for the doors to open than other types.

Side-by-side refrigerators have the following advantages and disadvantages:

Advantages

- Provide universal access to the majority of shelves in both refrigerator and freezer compartments.
- Models with in-door dispensers give easy access to ice and ice water without opening the door.
- Models with a shallow, third-door compartment have the option of putting frequently used foods within easy reach without opening the main refrigerator door.
- Pull-out shelves, drawers, bins, and baskets provide easier reach for foods stored at the back.

Disadvantages

- Special features add to the total cost.
- Might require more space than available in existing kitchens due to a wider design.

Freezers

A compact, upright freezer will best meet the needs of the disabled person. Installing the freezer on a box or raised platform can help make its contents more easily accessible. Freezers are conveniences for people who do not frequent the supermarket. They are especially useful in homes with smaller refrigerators or refrigerators having only an ice cube tray compartment. Home freezers come in chest and upright models. Two designs of upright models are available on the market today: manual defrost and automatic defrost.

Home freezers are available with wire shelves and baskets, and with storage shelves on the doors in upright models.

Freezers have the following advantages and disadvantages:

Advantages

- Make it possible to keep a supply of all kinds of frozen food on hand.
- By stocking up on food at sale prices and storing them for later use, consumers can easily take advantage of price specials.
- Require fewer shopping trips.

Disadvantages

- Kitchen might not allow space for the freezer.

Checklist for Refrigerator and Freezer Products

Prior to selecting and purchasing a refrigerator or freezer, fill out the following checklist. Check all that apply.

Size

- How large is the space for the freezer?
  _____ width _____ depth _____ height
- How much room does the door need to swing open? _____
- Will there be enough room to open the doors completely so as to remove the storage bins?_____
- Direction of door swing: handle on the left side ❑ right side ❑
- How many people will be using the refrigerator?

  ❑ One to two people (need at least 14 cubic feet)

  ❑ Three to four people (need 16 to 18 cubic feet)

  ❑ More than four people (add 2 cubic feet per additional person). To accommodate for later expansion needs, plan for additional refrigerator space, especially if the family is growing larger or to accommodate peak loads.

- Is there a regular stock of cold beverages in the refrigerator?

  ❑ yes   ❑ no

- How often does the customer go food shopping?

  ❑ two to three times a week

  ❑ weekly   ❑ daily

- Does the consumer host large holiday dinners?

  ❑ yes   ❑ no

Each "yes" answer will add to the refrigerator size requirements, as specified earlier.

Model

- Are there any handicapped or disabled members in the household?

    ❑ yes   ❑ no

A side-by-side model allows easy access to both freezer and fresh food compartments for those who have limited activity requirements.

- Choose the model desired:
    ❑ Side-by-side   ❑ Top-mount freezer
    ❑ Bottom-mount freezer   ❑ Three door
    ❑ Compact   ❑ Single door (no long-term freezer compartment)

- Which features are important?
    ❑ Automatic defrost
    ❑ Cycle defrost (requires manual defrosting of freezer)
    ❑ Manual defrost (requires manual defrosting of both refrigerator and freezer)
    ❑ Reversible doors

- Automatic ice maker:
    ❑ Factory installed   ❑ Equipped for later installation

- Through-the-door dispenser:
    ❑ Ice cubes only   ❑ Cubes/crushed ice
    ❑ Cubes/crushed ice/chilled water

- Storage drawers:
    ❑ See-through   ❑ Adjustable temperature
    ❑ Adjustable humidity   ❑ Sealed snack pack for lunch meats/cheese

- Refrigerator shelves:
    ❑ Glass   ❑ Full width only
    ❑ Wire   ❑ Half width only
    ❑ Adjustable   ❑ Combination full and half width
    ❑ Nonadjustable

- Door storage:
    ❑ Egg compartment   ❑ Removable
    ❑ Covered

- Dairy compartment:
    ❑ Butter only   ❑ Butter and cheese
    ❑ Deep enough for liter-size bottles or six packs   ❑ Removable storage/servers

- Freezer compartment:
    - ❏ Needed only for one to two days' storage of frozen food
    - ❏ Needed for storage of ice cream, meats, and other frozen food over longer periods
- Freezer shelves:
    - ❏ No shelves needed
    - ❏ Nonadjustable shelves acceptable
    - ❏ Need adjustable shelves
- Convenience features:
    - ❏ Juice/can dispenser
    - ❏ Ice tray shelves
    - ❏ Ice cube bin
    - ❏ Wine bottle holder
- Price range: $_____
- Warranty and service information:_____

Dishwashers

During the past decades, dishwashers have proven their value and usefulness in reducing the cleanup tasks in the kitchen. They not only save time, energy, and labor, but they also deliver dishes cleaner than those washed by hand.

The most common type is the built-in dishwasher. However, other styles are available for special situations. When selecting a new dishwasher (Figure 1-9), consider the following:

- Is it a front-loading undercounter model? Portables are less convenient because they must be moved into position and hooked up to a faucet every time they are used.
- Does it have a self-cleaning filter, rather than one that has to be removed and cleaned?
- Does it have dispensers and silverware baskets in an easy-to-reach location?
- Are the silverware baskets on the door or in the bottom rack?
- Are the racks designed with flexibility for easy loading of tall or bulky dishes?

Built-In Dishwashers

Built-in dishwashers are designed to fit into a 24-inch-wide space between two kitchen cabinets and under the countertop. Also available are smaller-capacity models that fit into an 18-inch space. The consumer now has a choice between a drop-down door (standard) and a single or double slide-out drawer.

Convertible/Portable Dishwashers

Convertible/portable models are essentially the same as built-ins, but they have finished sides and tops, drain and fill hoses with a faucet connector, and casters for easy rolling to the sink. These can be installed later as built-ins, if desired.

Undersink Dishwashers

An undersink model is designed to fit under a special six-inch-deep sink in just 24 inches of space; or under a special double-bowl sink, with a disposer under the second bowl,

FIGURE 1-9 Undercounter dishwasher. Available in 18- or 24-inch width. Portable models are also available. Consumers will also have a choice of dish-drawers, either single or double stacked.

in 36 inches of space. These dishwashers were designed for small kitchens with limited cabinet space for storage.

Dishwasher-Sink Combination

A dishwasher-sink combination unit is also available. It includes a stainless steel sink with drainboard, an enameled metal undersink cabinet, and a dishwasher in 48 inches of space. Some "Pullman-type" combination units include a dishwasher as well as a sink, range, and/or, in some, a refrigerator—all in one unit. Dishwashers have the following advantages and disadvantages:

Advantages

- Can save the physical labor of washing dishes.
- Provide out-of-the-way storage for dirty or clean dishes.

Disadvantages

- Require an 18-inch or 24-inch space in the kitchen, so a cabinet might have to be removed.
- Are nearly always designed for use with standard 36-inch-high countertops.

This might require a two-level counter if the sink is installed at a 30-inch height for a wheelchair-bound person.

Checklist for Dishwashers

Before selecting and purchasing a dishwasher, complete the following checklist. Check all that apply.

- Model:
  - ❏ Built-in
  - ❏ Slide-out drawer
  - ❏ Dishwasher-sink combination
  - ❏ Standard drop-down door
  - ❏ Convertible/portable
- Loading convenience:
  - ❏ Racks designed to handle
  - ❏ Special baskets:_____
  - ❏ Adjustable racks large dishes
- Wash system:
  - ❏ Water temperature booster
  - ❏ Rinse aid dispenser
  - ❏ Soft food dispenser
  - ❏ Two- or three-level spray arms
  - ❏ Self-cleaning filter
- Control panel:
  - ❏ Dials
  - ❏ Touch pads
  - ❏ Energy level indicator
  - ❏ Push buttons
  - ❏ Cycle time indicator
  - ❏ System status

- Cycles:
  - ❏ Normal
  - ❏ Economy/water saver
  - ❏ No-heat drying
  - ❏ China and glass
  - ❏ Delay start
  - ❏ Rinse only
  - ❏ Super-heated rinse
  - ❏ Light duty
  - ❏ Pots and pans
- Finishes:

  Tub _____   Racks_____

  Color panels_____   Trim kits_____
- Preferred color_____
- Price range: $_____
- Warranty and service information:_____

Laundry Equipment

Today's laundry equipment, along with changes in fibers and fashions, has eliminated the need for hand laundering, clothesline drying, and routine ironing. There features can often be preprogrammed into the appliance for any type of laundry load. These all give excellent results. From diapers and jeans, to delicate silks and knits, today's laundry system is equipped for all fabric needs.

The typical laundry pair, a standard washer and dryer, will stand side by side in 4-1/2 to 5 feet of wall space, depending on the brand and model (Figure 1-10). Some questions to consider when choosing laundry appliances are as follows:

- Is there enough space available for laundry appliances?
- What control location will be best for the principal user? Some models offer front or rear controls.
- What capacity will best satisfy family needs?
- Will built-in dispensers for bleach and fabric softener increase the washer's utility?
- Will the dryer need to be vented outside?
- Which is preferred (gas or electric) for drying clothes?
- How many different washing cycles are needed?
- How many different drying cycles are needed?

Automatic Washers

Basically, all automatic washers will wash the clothes in the same manner, but there are some key differences in design and special features from model to model and manufacturer to manufacturer.

FIGURE 1-10   The automatic washer and automatic dryer. The dryer is available in electric or gas.

Top-Loading Automatic Washers

Top-loading models vary in width from about 24 to 30 inches (Figure 1-11). They are available in a variety of load capacities. Standard-capacity washers are built for the average two- to four-person household. However, a large-capacity model reduces the number of loads washed, saving time. Some models offer front panel controls, and many models have dispensers. Top-loading washers have the following advantages and disadvantages:

Advantages

- Provide a convenient, at-home way to do laundry.
- Models installed in a small space, only 24 inches wide, are available.
- Models with front controls can be reached and operated easily from a seated position.
- Provide a variety of designs and control positions to meet varying user needs.

Disadvantages

- Models with rear console controls are virtually impossible to operate from a seated position.
- Some designs might require special aids to remove loads, set controls, and clean filters from a seated position.
- Compact models—although they provide the greatest accessibility from a wheelchair—have a smaller load capacity than other designs.

FIGURE 1-11
The top-loading
washer with dryer built
in above is available
in a variety of colors.

Front-Loading Automatic Washers

Front-loading models might have drop-down or side-opening doors (Figures 1-12 and 1-13). In these models, the entire wash basket revolves. As clothes tumble, they are lifted by vanes on the sides of the basket. Front loaders use less water than top loaders, and they use high-efficiency detergent, but only certain models will handle very large loads because they must have empty space in the drum to tumble clothes. Front-loading washers have the following advantages and disadvantages:

Advantages

- Front controls can be reached and operated easily from a seated position.
- Front opening makes loading and unloading easier for users with limitations.
- No agitator needed.
- Consumer can add the pedestal unit to raise the washer higher by 18 inches.

Disadvantages

- Drop-down door might create wheelchair barrier.
- Door opening might be too low for some wheelchair users or those who cannot stoop or bend.
- Smaller front-load washers are unable to handle large wash loads.

FIGURE 1-12 Front-load automatic washer.

FIGURE 1-13 The front-loading washer can be stacked, placed under the counter, or installed in the laundry area. Available in a variety of colors.

Compact Automatic Washers

These compact, "apartment-sized" washers range from 24 to 27 inches wide to fit spatial needs. They are available in two forms: built-in or on casters so that they can be rolled to the kitchen sink for use. Matching dryers can be installed next to the washer, stacked on a special rack, wall-hung, or purchased as a one-piece unit with the washer (Figure 1-14).

FIGURE 1-14 Compact or portable washer.

Automatic Dryers

Automatic dryers perform in the same tumbling manner as front-loading automatic washing machines. However, there are some key differences in design and special features from model to model and manufacturer to manufacturer.

Dryers are available in electric or gas. They vary in load capacity. Some models offer front or rear controls and side-opening or drop-down doors (Figure 1-15). For optimum efficiency, an electric dryer should have a minimum rating of 4400 watts. Gas dryers require a 120-volt outlet for such features as the motor, lights, and ignition. The gas heater should have a rating of at least 20,000 BTU/hour for top performance. Dryers should be installed in an area that permits proper venting.

Compact dryers are electric (either 120- or 240-volt). The 120-volt dryer takes at least twice as long to dry clothes as the 240-volt model does. While venting is recommended for all dryers, some 120-volt models can be used without venting if they are not in an enclosed space. The 240-volt dryers must be vented to prevent damage from moisture buildup in the home. Compact dryers can be installed next to the washer, stacked above the washer on a special rack, wall hung, or purchased as a one-piece unit with the washer.

Automatic dryers have the following advantages and disadvantages:

Advantages

- Eliminate the difficulties inherent in line drying.
- Give modern fabrics proper care, practically eliminating the ironing chore.
- Designs available to meet the needs of most disabled persons.

Figure 1-15   The automatic dryer is available with a drop-down or side-opening door.

Disadvantages
- Models with rear controls are virtually impossible to operate from a seated position.
- Dryer door might be too low without a raised installation.
- As with washers, compact dryers have a smaller load capacity than other designs.

Checklist for Washers and Dryers

Before selecting and purchasing laundry equipment, fill out this checklist on washers and dryers. Check all that apply.

Washer
- What size is wanted?
  - ❏ Compact/portable
  - ❏ Large capacity
  - ❏ Standard capacity
  - ❏ One-piece washer/dryer combination
- Style:
  - ❏ Front-loading
  - ❏ Top-loading
- Cycle selections:
  - ❏ Permanent press
  - ❏ Knits
  - ❏ Soak
  - ❏ Delicate
  - ❏ Pre-wash
  - ❏ Extra-clean
- Options:
  - ❏ Variable water level
  - ❏ Extra rinse cycle
  - ❏ Water saver
  - ❏ Bleach dispenser
  - ❏ Detergent dispenser
  - ❏ Water temperature control
  - ❏ Electronic controls
  - ❏ Small load basket (to reuse wash water)
  - ❏ Fabric softener dispenser
  - ❏ Other_____
- Color_____
- Price range: $_____
- Warranty and service information:_____

Dryer
- Which type is preferred?
  - ❏ Electric
  - ❏ LP or natural
  - ❏ Gas
- What size is needed?
  - ❏ One-piece washer
  - ❏ Large capacity
  - ❏ Standard capacity dryer combination
- Cycle selections:
  - ❏ Permanent press/medium heat
  - ❏ Delicate/low heat

❏ No heat   ❏ Timed cycles

❏ Automatic drying   ❏ No-heat tumbling at end of the drying cycle

• Other options:

   ❏ Electronic controls   ❏ End-of-cycle signal

   ❏ Drying shelf   ❏ Side-opening door

   ❏ Drop-down door

• Color_____

• Drum (stainless or porcelain)_____

• Price range: $_____

• Warranty and service information:_____

Air Conditioners

Room air conditioners are self-contained units that can be installed in a window, through a wall, or moved around on wheels (Figure 1-16). The smaller air conditioners are designed to cool the immediate area of a room. However, manufacturers are designing room air conditioners from 4000 BTU/hour to over 30,000 BTU/hour. These larger-sized units will

FIGURE 1-16   Room air conditioners are available in a variety of sizes and configurations.

cool a larger area or cool multiple rooms in a home or office at one time. The standard features are manual or electronic controls, straight cool only, electric heat and cooling, or reverse-cycle air conditioning (heat pump). Air flow circulating from the air conditioners discharge grill might have fixed, directional, or motorized louvers. Some models are available with wireless remote control handheld units. The remote control makes it easier to control the on/off, fan speed, and temperature of the air conditioner from across the room. Room air conditioners are available in 120 volts or 230 volts. To properly install a room air conditioner, it is strongly recommended that you follow the manufacturer's recommendations. Factors that should be taken into account when purchasing an air conditioner include the following:

- Size
- Energy rating
- Location of air conditioner within the home or office
- Location of the electrical outlet and voltage needed to run the air conditioner

Room air conditioners have the following advantages and disadvantages:

Advantages
- Save money on electric bill by only cooling rooms needed.
- Easy installation.
- No ductwork needed.
- You can have different temperature settings for each room air conditioner in a home or office.

Disadvantages
- Room air conditioners are a little noisier than central air conditioning.
- Window air conditioners are drafty in winter time and most likely have to be removed.

Sizing Room Air Conditioners

To properly size an air conditioner that is needed for a particular room, you must determine the square footage of that room. If the room is rectangular or square (Figure 1-17), multiply the length by the width. This will provide you with the square footage of the area to be cooled. Some rooms may be oddly shaped, consisting of both a rectangular or square area and a triangular area (Figure 1-18). First determine the square footage of a triangle—measure the base of the triangle and the height of the triangle in feet. Now multiply both amounts, and divide that figure by two. The result will be the square footage of that triangle. Then determine the square footage of the rectangle or square as described previously. When you have determined the square footage of the area to be cooled, see Table 1-1 to find out what air conditioner capacity is needed.

FIGURE 1-17
In a square or rectangular room, multiply the length by the width in feet to determine the square footage.

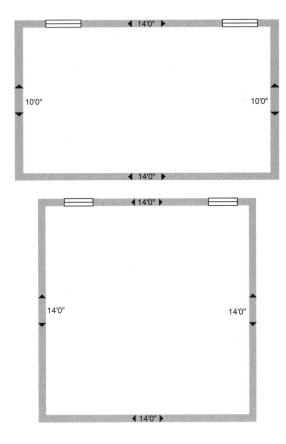

FIGURE 1-18
In oddly shaped room, determine the square footage of the rectangle. Then determine the square footage of the triangle. Subtract the square footage of the triangle from the total square footage of the rectangle to determine the square footage of the oddly shaped room.

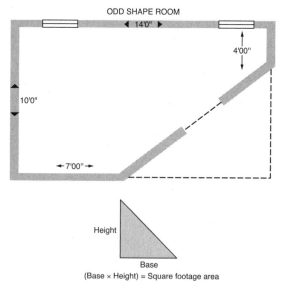

Area to Be Cooled (square feet)	Capacity Needed (BTUs per hour)
100 to 150	5000
150 to 250	6000
250 to 300	7000
300 to 350	8000
350 to 400	9000
400 to 450	10,000
450 to 550	12,000
550 to 700	14,000
700 to 1000	18,000
1000 to 1200	21,000
1200 to 1400	23,000
1400 to 1500	24,000
1500 to 2000	30,000
2000 to 2500	34,000

TABLE **1-1**   Room Air Conditioner Sizing Chart

Once you have determined what size air conditioner is needed for the room, there are other circumstances that have to be taken into account before you purchase an air conditioner:

- Most air conditioners are designed to cool a room based on two occupants. For every additional person in the room, you will have to add 600 BTUs to the basic cooling capacity.

- If the room to be cooled is not insulated, you will have to add 15 percent to the basic cooling capacity.

- You will have to add to the basic cooling capacity by 10 percent if the room has a west and/or southwest exposure.

- If the room to be cooled is facing north or northeast or is heavily shaded, the cooling capacity will be reduced by 10 percent.

- If you are only going to use this air conditioner at night, deduct 30 percent from the basic cooling capacity.

- If you have high ceilings, or if the ceilings are not insulated, increase the basic cooling capacity by 10 percent.

Figures are based on sizing an air conditioner in a moderate climate around 75 to 80 degrees Fahrenheit with an average humidity of 50 percent. If the climate is warmer and the humidity is higher, you might have to increase the basic cooling capacity by 10 to 20 percent.

How to Purchase the Best Air Conditioner or Appliance Value Using the EnergyGuide Label

Proper planning and evaluation before buying can save time, trouble, and money each step of the way. Take the time to determine the air conditioner or appliance's annual cost of operation.

Remember, while some energy-efficient products have higher purchase prices than less efficient ones, they will cost less in the long run because they require less electricity to operate. Calculate and evaluate the product's annual cost of operation and carefully read the EnergyGuide label, which appears on most appliances and air conditioners sold today, to get the best buy.

The EnergyGuide label (Figure 1-19) is required by the U.S. government on many home appliances and air conditioners, and gives information to help customers select and save.

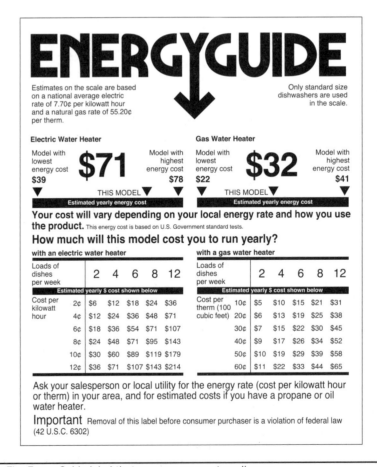

FIGURE 1-19   The EnergyGuide label that appears on most appliances.

The information gained from the EnergyGuide label is important because is helps determine the following:

- It will help you compare the estimated annual operating costs of one model versus another.
- It will give information about the size range of the models being compared.
- It will tell how each model compares in terms of its energy costs to other models in the same size range.

To read the EnergyGuide label, first look for the estimated annual energy cost in the center. To either side are energy costs of lower-rated and higher-rated models. These costs are derived from national average electricity rates, so knowledge of the local rate for electricity is helpful.

Major Appliance Consumer Action Panel (MACAP)

The Major Appliance Consumer Action Panel, or MACAP, was an independent, complaint-mediation group made up of professionals with expertise in textiles, equipment, consumer law, and engineering who volunteered their time. Unfortunately, MACAP went out of business in the last few years, but the information printed in this book is still valuable and worth reading for guidance. Panelists cannot be connected with the appliance industry. They received no financial remuneration other than the reimbursement of their travel and living expenses while attending meetings. MACAP received comments and complaints from appliance owners; excessive charges, delays in parts and service, and alleged unnecessary repairs are frequently mentioned. Other complaints are related to product performance, such as operating noise, temperature, maintenance, and running time. Nonresponsiveness of dealers and manufacturers, warranty coverage, food loss claims, imperfect finishes, improper installation, and purchasing dissatisfaction are also alleged.

MACAP also studied industry practices and advised industries of ways to improve their services to consumers. It recommended how to educate consumers on proper appliance purchase, use, and care. The panel developed and distributed educational publications and periodic news releases when its review of individual consumer complaints pinpointed information that would be useful to consumers. The panel was sponsored by the Association of Home Appliance Manufacturers.

The types of appliances represented include:

- Compactors
- Dehumidifiers
- Microwave ovens
- Refrigerators and freezers
- Room air conditioners
- Laundry equipment
- Ranges
- Dishwashers
- Disposers

After a complaint reached MACAP, the staff screened it to ascertain whether the consumer has already requested assistance from the local dealer and brand-name owner's headquarters office. If this had been done, the complaint was sent to the sponsoring association and thus entered MACAP's communications phase. This phase comprised the following actions:

- Copies of the consumer's correspondence or a summary of telephone comments were sent to the brand-name owner involved, requesting a report on proposed action within two weeks.

- A letter was sent to the consumer acknowledging receipt of the complaint, reporting action taken, and asking for any additional information, if necessary.

- When an answer was received from the brand-name owner, the staff wrote a letter to the consumer confirming any company action or information. The consumer was asked to return a card verifying this action.

- The file was then resolved, unless the consumer otherwise advised MACAP.

- If the complaint reached an impasse, the file moved to the study phase.

In the study phase, the panel discussed the file at a meeting. Preparation for this review included the gathering of an exact and detailed background of the complaint from the consumer and from the brand-name owner, if such information was not already on file. If conflicting reports were received, the panel could ask for an independent, on-site evaluation by a utility, extension, university home economist, or an engineer.

All information is included in a confidential summary prepared by the staff and discussed by MACAP. The panel might have made a recommendation to the company and/or consumer, have asked for additional information, or closed the file on the basis of the information presented.

If you are experiencing any problems with your product or service company, contact your local Better Business Bureau (BBB).

Purchasing Decisions

The purchase of a major appliance is one of the most important investments made for the home. MACAP's experience in handling consumer appliance complaints has shown that poor purchasing decisions lead to unhappy appliance owners.

Pointers from MACAP to help consumers make wise decisions include the following:

- Ask the dealer for specification sheets from several manufacturers of the appliance types you plan to purchase. Study them carefully and note the different features, designs, and capacities.

- Ask the dealer to see the warranty before purchasing the appliance. Does the warranty cover the entire product? Only certain parts? Is labor included? How long is the warranty coverage?

- Ask the dealer for the use and care manual. Read it carefully before you purchase the appliance. The dealer should have manuals available from the floor models on display. These manuals will help in asking pertinent questions, and they will explain how the product operates and what special care it needs.

- Decide what special features are essential. Consider the possibility of adding on features at a later date, such as an ice maker for a refrigerator.

- Check the space available for the appliance. Will it fit where it is planned? Is there adequate clearance space in the hallways and doors through which the appliance will have to pass before installation?
- Check the product design carefully prior to purchase. Compare the designs of different brands. If a combination microwave oven/range is being purchased, check the space between the units to be sure that everything will fit.
- Clearly establish the cost of delivery and installation. Are these costs included, or are they extra?
- Ask the dealer if he services the appliances he sells. If not, ask him where to go for authorized factory service.
- Compare price in relation to convenience and service. Both vary according to the model. As more features and conveniences are included, the price increases.
- Be sure the house has adequate electrical service for the appliance in order to avoid overloading circuits. Also, be sure it has adequately grounded three-hole receptacles.[3]

Appliance Warranties

MACAP urges consumers to compare warranties of different brands the way you would compare price, size, and features—it's a basic consumer responsibility! Yet the panel's experience in working with consumer appliance problems indicates that many consumers are not aware of the importance of warranty buymanship.

Warranty inspection is a legislated consumer right. A federal law, the Magnuson-Moss Warranty/Federal Trade Commission Improvement Act of 1975, requires warranty information to be available at the point of purchase for products costing $15 or more. The law does not, however, require manufacturers to provide warranties on their products.

The store must provide this information in one of the following ways:

- Displayed near the appliance.
- Shown on the package displayed with the appliance.
- In an indexed and updated binder that is prominently displayed, containing all warranties for products sold in the department.
- If ordering the appliance through a catalog, the catalog must include the warranty text or offer it upon request.[4]

Full vs. Limited Warranties

The full warranty offers more protection. Under a full warranty, at a minimum, the warrantor must remedy the problem within a reasonable time and without charge for as long as the warranty is in effect. In addition, the manufacturer might not limit the duration of any implied warranty. Any limitation (or exclusion) of consequential damages (for example, food loss or floor damage) must appear conspicuously on the warranty. If a reasonable number of attempts to correct an in-warranty problem fails, the consumer must be given the choice of a replacement or a refund.

Under a limited warranty, the protection is limited to what is outlined in the terms of the warranty. For example, the warranty might require that the consumer pay for diagnostic costs, labor costs, or other costs of that kind. Also, any implied warranty might be limited to the duration of the written warranty. A limited warranty should be studied carefully to determine exactly what the warrantor will provide and what costs the consumer must pay.[4]

Warranty Time Limitations

Some major appliances are hardly used over a year's time. The window air conditioner in Minnesota, the refrigerator at the lake cabin in Michigan, the range in a Florida vacation condominium—all of these have only seasonal use.

When problems with these appliances arise, some owners feel that four years of seasonal usage should constitute less than one year's worth of warranty coverage. The manufacturers, however, think differently. MACAP knows it is necessary for consumers to be aware of the contents of their appliance warranties, and also of the period covered. The coverage is stated for a specific period of time, not for how often the appliance is used.

The coverage period might vary with the product's components. For example, a window air conditioner warranty might provide free repairs of any part that breaks down in the first year, but only partially cover repair expenses of the sealed system components (compressor, evaporator, etc.) for an extra four years. Manufacturers can offer almost any type of warranty as long as the provisions are clearly stated and the warranty is available for review before buying.

Although seasonal usage might result in less wear and tear on the appliance, MACAP believes that there are at least two valid reasons for not basing warranties on usage:

- Basing a warranty on usage (rather than on appliance age) is unrealistic because it's impossible for a manufacturer to monitor actual consumer usage.

- Some appliances actually suffer more from extended periods of non-use than from continuous daily operation.

For example, refrigerator and air conditioner sealed systems stay vacuum-tight (generally for many years) if the refrigerant gas and oil they contain are kept moving. This way, the various seals are kept lubricated and soft. They might dry out if the mechanisms involved don't move for long periods. The dishwasher provides another example. Many consumers don't realize that the seals in the pump area of the dishwasher are meant to be immersed in water at all times. Water keeps the seal soft and pliable, which is necessary for proper operation.

MACAP urges consumers to carefully read the warranty that comes with each major appliance and use the appliance enough during the warranty period so that any defects have time to surface. It is important to use every feature and control on the new appliance soon after it is installed.[5]

Appliance Installation Instructions

The manufacturer supplies the installation instructions with every new appliance or air conditioner that is purchased. These instructions will help the installer to plan, locate, install, and secure the product for safe and proper operation.

Appliance Use and Care Manual

This manual contains information and suggestions to help customers get the best results from their appliances. The manual will disclose to the customer how to start the appliance, maintain it, and use all the features that come with it. Also included in the use and care manual are the following:

- Safety precautions
- What to do before using the appliance
- Maintenance instructions
- Vacation and moving care
- Warranty information
- Parts and features
- How to use the appliance
- Common problems and solutions
- Assistance

Where to Get Help

Keep careful records. Always put complaints in writing, and keep copies of all correspondence and service receipts. Be sure to ask for service receipts, even for no-charge, in-warranty calls. Note details: when the problem was first noticed, when it was reported, and the servicing history (who serviced the appliance or air conditioner, when, what was done, and how often service was required).

If there are complaints about the appliance or air conditioner, there are three steps to follow:

- Read the use and care manual that comes with the product. Also check the plug, as well as fuses, pilots, and controls.
- Call the service company authorized to fix the brand. They have the training and equipment to deal with appliance or air conditioner service problems.
- If not satisfied, contact the manufacturer's main customer relations office. The address and phone number are located in the use and care manual.

Product Recalls and the Internet

There are times when a product might have an electrical, gas, or mechanical issue that might be hazardous to the consumer. The U.S. Consumer Product Safety Commission (CPSC) was created to protect the consumer from serious injuries, unreasonable risks, death, and property damage.

The CPSC's website (http://www.cpsc.gov) includes information on appliance, air conditioner, and other product recalls. It also includes other information beneficial to the consumer. In addition, Table 1-2 provides a listing of websites for major appliance and air conditioner manufacturers. You can also look in the use and care manual that comes with the product for the manufacturer's website. Remember, safety and education must be considered at all times when operating or repairing any major appliance or air conditioner.

www.amana.com	www.jennair.com
www.americanwaterheater.com	www.kitchenaid.com
www.aosmith.com	www.kuppersbuschusa.com
www.avantiproducts.com	www.lge.com
www.boschappliance.com	www.lifeluxurymarvel.com
www.bradfordwhite.com	www.maytag.com
www.broan.com	www.miele.com
www.brockwaterheaters.com	www.northlandnka.com
www.carrier.com	www.panasonic.com
www.convaircoller.com	www.rheem.com
www.dacor.com	www.roperappliances.com
www.danby.com	www.samsung.com
www.delonghi.com	www.sears.com
www.dynastyrange.com	www.sharpusa.com
www.eemaxine.com	www.staber.com
www.electrolux.com	www.subzero.com
www.equatoronline.com	www.sunfrost.com
www.fedders.com	www.thermador.com
www.friedrich.com	www.thorappliances.com
www.frigidaire.com	www.u-line.com
www.gaggenau.com	www.vaughncorp.com
www.ge.com	www.veronaappliances.com
www.goldstarappliances.com	www.vikingrange.com
www.haieramerica.com	www.weathercraft.com
www.hatachi.com	www.whirlpool.com
www.hotwater.com	www.wolfappliance.com
www.insinkerator.com	www.zephyronline.com
www.jadeappliances.com	

TABLE 1-2 Major Appliance and Room Air Conditioner Manufacturers

National Appliance Service Technicians Certification (NASTeC)

The roots of NASTeC go back to the Certified Appliance Technician (CAT) certification issued by ISCET (International Society of Certified Electronics Technicians) in the mid-1970s, when the bulk of appliance servicing was done by retail appliance dealers, who serviced the products they sold. In many instances, these same dealers sold items such as televisions, recorders of various types, stereos, radios, etc. Since ISCET was already in contact with these dealers with respect to the other electrical appliances, it was natural that the major appliances known as "white goods" should be served as well.

The North American Retail Dealers Association (NARDA) attracted dealers and thus servicers to join their efforts at promoting the retail dealers' interests through their trade organization. This put NARDA in a position to better offer its members an appliance technician certification. NASTeC was then born as a separate entity from ISCET, but as a partner with equity in NASTeC.

As the "white goods" dealers became more separated from the televisions, recorders, etc., the servicing of these products became independent operations. This diminished the ability of NARDA to offer appliance certifications.

In 2002, ISCET secured the interests of NARDA in the NASTeC venture and began developing the NASTeC certifications to once again actively work with the appliance-servicing organizations to establish both certification and education in the appliance-servicing segment of the "white goods" industry.

Certification is offered in basic skills, laundry equipment, cooking equipment, and refrigeration equipment. Once a technician acquires all four certificates, he is considered a "universal technician" and is awarded that certificate. Each step is signified by a shoulder patch that identifies a person as a certified appliance technician. These technicians are known for their abilities and professional approach to both the customers and their work.[6] For more information, visit the NASTeC website (http://www.nastec.org).

Endnotes

1. Pyrolytic cleaning is the true self-cleaning system. It uses high heat during a special 1- to 3-hour cycle to decompose food, soil, and grease. During the cycle, which is clock-controlled, the oven door is latched and locked. It cannot be opened until the oven cools down. All of the oven walls, racks, and the door (except for a small area outside the door gasket) are completely cleaned. After cleaning, a small bit of white ash might be found, which can be easily wiped up.

2. Catalytic and continuous-cleaning ovens use a special porous coating on the oven walls that partially absorbs and disperses the soil. This process takes place during normal baking and keeps the oven presentably clean, but the racks and door parts must be cleaned by hand. Some manufacturers recommend occasionally operating an empty oven at 500 degrees Fahrenheit to remove any buildup of soil. This special oven coating cannot be cleaned with soap, detergent, or commercial oven cleaners without causing permanent damage.

3. Reprinted from *MACAP Consumer Bulletin*, Issue no. 8, December 1985.

4. Reprinted from *MACAP Consumer Bulletin*, Issue no. 1, December 1979.

5. Reprinted from *MACAP Consumer Bulletin*, Issue no. 5, February 1983.

6. Reprinted with permission from NASTeC, April 2006.

CHAPTER 2

Safety Precautions

S afety starts with accident prevention. Injuries are usually caused because learned safety precautions are not practiced. In this chapter are listed some tips to help the technician to correctly and safely install, operate, and repair major appliances.

Any person who cannot use basic tools should *not* attempt to install, maintain, or repair any major appliance or air conditioner. Any improper installation, preventative maintenance, or repairs will create a risk of personal injury, as well as property damage. Call the service manager if installation, preventative maintenance, or the repair procedure is not fully understood.

Every technician should carry a first aid kit and should know the location of the kit and how to properly use its contents. Technicians should carry a fire extinguisher in their service vehicles in case of an emergency. It is also recommended that they take a first aid course, such as those offered by the American Red Cross.

Safety Procedures

Individual, electrical, gas, chemical, appliance and air conditioner, operating, and installation safety precautions are generally the same for all major appliances and air conditioners. Carefully observe all safety cautions and warnings that are posted on the appliance or air conditioner being worked on. Understanding and following these safety tips can prevent accidents.

Individual Safety Precautions

Protecting yourself from injuries is vital. Before installing, maintaining, or servicing any major appliance or air conditioner, do the following:

- Wear gloves. Sharp edges on appliances or air conditioners can hurt hands.
- Wear safety shoes. Accidents are often caused when dropping heavy appliances, especially on feet that are not protected.
- Avoid loose clothing that could get caught in the appliance while it is operating.
- Remove all jewelry when working on appliances.

- Tie long hair back.
- Wear safety glasses to protect your eyes from flying debris.
- Use proper tools that are clean and in good working condition when repairing appliances and air conditioners.
- Have ample light in the work area.
- Be careful when handling access panels or any other components that may have sharp edges.
- Avoid placing hands in any area of the appliance or air conditioner that has not been visually inspected for sharp edges or pointed screws.
- Be sure that the work area is clean and dry from water and oils.
- When working with others, always communicate with each other.
- Always ask for help moving heavy objects.
- When lifting heavy appliances or air conditioners, always use your leg muscles and not your back muscles.

Electrical Safety Precautions

Know where and how to turn off the electricity to the appliance or air conditioner. For example: Know the location of plugs, fuses, circuit breakers, and cartridge fuses in the home. Label them. If a specific diagnostic check requires that voltage be applied, reconnect electricity only for the time required for such a check, and disconnect it immediately thereafter. During any such check, be sure that no other conductive parts come into contact with any exposed current-carrying metal parts. When replacing electrical parts or reassembling the appliance or air conditioner, always reinstall the wires back on their proper terminals according to the wiring diagram. Then check to be sure that the wires are not crossing any sharp areas, are not pinched in some way, are not between panels, and are not between moving parts that may cause an electrical problem.

These additional safety tips are also important to remember:

- Always use a separate, grounded electrical circuit for each major appliance or air conditioner.
- Never use an extension cord for major appliances.
- Be sure that the electricity is off before working on the appliance or air conditioner.
- Never remove the ground wire from a three-prong power cord or any other ground wires from the appliance or air conditioner.
- Never bypass or alter any appliance or air conditioner switch, component, or feature.
- Replace any damaged, pinched, or frayed wiring before repairing the appliance or air conditioner.
- Be sure that all electrical connections within the product are correctly and securely connected.

Gas Safety Precautions

As a technician, I have full respect for electricity and gas fuels. Like electricity, natural and liquefied petroleum (LP or LPG) gas can become dangerous if not handled properly. First, locate the gas shutoff valve by the appliance, and know where the main gas shutoff is located before beginning repairs. The fuel supply (natural or LP gas), supply lines, pressure regulator, and LP storage tank will need servicing at least once a year to make certain that there are no leaks. All gas appliances must be kept clean and free of soot.

The following are some additional safety procedures that a homeowner or technician must know before working on gas appliances:

- Never smoke or light a flame when working on gas appliances.
- Always follow the manufacturer's recommendations as stated in the use and care manual for gas appliances.
- Keep all combustible materials, such as gasoline, liquids, paint, paper materials, and rags, away from gas appliances.
- Do not allow pilot lights to go out. The gas fumes will seep into the room, causing a hazardous condition for anyone who lights a match, turns on a light switch, or uses the telephone.
- When working on gas appliances always have a fire extinguisher nearby.
- When using gas appliances, always have the proper ventilation according to the manufacturer's specifications.
- Keep ventilation systems clean of debris and obstructions.
- If you smell gas where the appliance is located, shut off the gas supply and vent the room. Warning: If you cannot turn the gas supply off, you must leave the area, go to a neighbor, and telephone the gas company and the fire department.
- When the flame turns from a blue color to a yellow color, or if soot forms, you could have a condition known as incomplete combustion. Warning: If not corrected immediately, you could have carbon monoxide buildup, which could be potentially fatal. Carbon monoxide is a poisonous nonirritating gas that is odorless, colorless, and has no taste. If you breathe in carbon monoxide, it will mix with your blood and prevent oxygen from entering into the blood supply, causing illness or death.
- All homes should have smoke and carbon monoxide detectors installed.

Chemical Safety Precautions

Chemicals are also dangerous. Knowledge of chemical safety precautions is essential at all times. The following tips are examples of important practices:

- Remove all hazardous materials from the work area.
- Always store hazardous materials in a safe place and out of the reach of children.
- Before turning on appliances that use water, run all of the hot water taps in the house for approximately five minutes. This clears out the hydrogen gas that can build up in the water heater and pipes if they have not been used for more than two weeks.

Appliance and Air Conditioner Safety

Call the service manager to check out the appliance or air conditioner if the safety of the product is in doubt.

Only use replacement parts of the same specifications, size, and capacity as the original part. If you have any questions, contact your local appliance and air conditioner parts dealer, your service manager, or the manufacturer of the product.

Check water connections for possible water leaks before reconnecting the power supply. Then completely reassemble the appliance, remembering to include all access panels.

Operating Safety

After repairing the appliance or air conditioner, do not attempt to operate it unless it has been properly reinstalled according to the use and care manual and to the installation instructions supplied by the manufacturer. If these instructions are not available, do not operate the appliance or air conditioner. Call the service manager to check out the reinstallation or ask for a copy of the installation instructions from the manufacturer.

Know where the water shutoff valves are located for the washer, dishwasher, ice maker, and water heater, as well as the house's main water shutoff valve. Label them. Following these additional safety tips can also prevent injuries:

- Do not allow children to play on or to operate appliances or air conditioners.

- Never allow anyone to operate an appliance or air conditioner if they are not familiar with its proper operation.

- When discarding an old appliance, remove all doors to prevent accidental entrapment and suffocation.

- The refrigerant in refrigerators, freezers, and air conditioners will have to be reclaimed before disposal of the product can begin.

- Instruct the customer to use the appliance or air conditioner only for the job that it was designed to do.

Installation Safety Precautions

The first step in assuring safety with major appliances and air conditioners is to be sure that they are installed correctly. Read the installation instructions and the use and care manual that comes with the appliance or air conditioner. Observe all local codes and ordinances for electrical, plumbing, and gas connections. Ask your local government agency about these codes. Additional safety tips include the following:

- Carefully observe all safety warnings that are contained in the installation instructions and in the use and care manual.

- The work area should be clear of unnecessary materials so that there is plenty of room to work on the appliance or air conditioner.

- Be sure that the appliances are installed and leveled on a floor that is strong enough to support their weight.

- The appliance or air conditioner should be protected from the weather, as well as from freezing or overheating.

- The appliance or air conditioner should be correctly connected to its electric, water, gas, drain, and/or exhaust system. It should also be electrically grounded.
- Be sure that the appliance has a properly installed anti-tip device, as in the case of kitchen ranges.

Grounding of Appliances

In 1913, the National Electrical Code (NEC) made grounding at the consumer's home mandatory. The NEC required that range frames be grounded to a neutral conductor in 1943. Later, in 1946, it required that receptacles in laundry areas be grounded. Soon after, the NEC required that the frames of automatic dryers be grounded to neutral conductors. Then, in 1959, NEC required that automatic washers, automatic dryers, automatic dishwashers, and certain motor-operated handheld appliances be grounded. All 15-amp and 20-amp branch circuits have grounding type receptacles, as specified in 1962. Finally, in 1968, the code required that refrigerators, freezers, and air conditioners be grounded.

The greatest importance of grounding appliances and air conditioners is that it prevents people from receiving shocks from them. However, the major problem associated with the adequate grounding of appliances and air conditioners is that many homes are not equipped with three-prong or four-prong grounded receptacles. To solve this problem, the consumer must install, or have installed, a properly grounded and polarized three-prong or four-prong receptacle. A qualified electrician should connect the wiring, and properly ground and polarize the receptacle.

Remember that safety is the paramount concern, especially when dealing with electricity. Both the technician and the consumer must be aware that it only takes about 100 milliamperes of current to cause death in one second. Here are some safety tips:

- Do not install or operate an appliance or air conditioner unless it is properly grounded.
- Do not cut off the grounding prong from the appliance or air conditioner plug.
- Where a two-prong wall receptacle is encountered, it must be replaced with a properly grounded and polarized three-prong receptacle.
- Call the service manager if you doubt your abilities. When dealing with electricity, there is no leeway for mistakes.

Checking Appliance and Air Conditioner Voltage

If it becomes necessary to test an appliance or air conditioner with the voltage turned on, observe the following precautions:

- The floor around the appliance or air conditioner must be dry. Water and dampness increase the probability of a shock hazard.
- When using a multimeter, always set the meter correctly for the voltage being checked.
- Handle only the insulated parts of the meter probes.

- Touch components, terminals, or wires with the meter probe tip only.
- Touch the meter probe tips only to the terminals being checked. Touching other components could damage good parts.
- Be sure that the appliances have properly installed anti-tip devices, as found on kitchen ranges.

What to Do If You Smell Gas

If there is a gas smell, the following precautions should be taken:

- Turn off the gas supply.
- Do not light matches.
- Do not turn the lights on or off.
- Do not use the telephone.
- Open the windows to ventilate the room.

Then call the gas company to come out and repair the gas leak.

Tools Needed for Installation and Repair

A basic knowledge of hand tools, electrical tools, and test meters is necessary to effectively complete most installations and repairs. This chapter will cover the basics of each tool. A working knowledge of these tools is a must for the installation and repair of appliances and air conditioners. Always follow safety precautions and manufacturer's recommendations and warnings when handling tools.

Before starting on any type of repairs, take the time to put together a toolkit with a selection of good quality hand tools. A partial list of common hand tools includes:

- **Screwdrivers**   A complete set of flat-blade screwdrivers, ranging from 1/8 inch to 5/16 inch. Handle sizes may vary with the blade dimension. Phillips-tip sizes also vary; the two most common are #1 and #2.

- **Nut drivers**   A complete set is recommended. The common sizes are: 3/16, 1/4, 5/16, 11/32, 3/8, and 1/2 inch.

- **Wrenches**
  - **Socket wrenches**   Either 6-point or 12-point, ranging in size from 5/32 to 1 inch.
  - **Box wrenches**   Common sizes range from 1/4 to 11/2 inch.
  - **Open-end wrenches**   Common sizes range from 1/4 to 15/8 inch.
  - **Adjustable wrenches**   The handle size indicates the general capacity. For example, a 4-inch size will take up to a 1/2-inch nut. A 16-inch handle will take up to a 17/8-inch nut.
  - **Allen wrenches**   Sizes range from 1/16 to about 1/2 inch.
- **Claw hammer**
- **Adjustable pliers**
- **Flashlight**
- **Drop-cloth**

Safety Precautions

Safety starts with accident prevention. Listed in this chapter are some tips to help the technician when using hand and power tools.

WARNING *Any person who cannot use basic tools should* not *attempt to install, maintain, or repair any major appliance or air conditioner. Any improper installation, preventative maintenance, or repair creates a risk of personal injury and property damage.*

Individual Safety Precautions

Injuries abound when using tools. To be protected from injuries when using hand tools and power tools, do the following:

- Wear gloves.
- Avoid wearing loose clothing when working with power tools.
- Wear safety glasses to protect the eyes from flying debris.
- Use tools according to manufacturer's specifications, and never alter their use.

Safety Precautions When Handling Tools

Regardless of which tool is being used, these same rules of care and safety apply:

- Keep tools clean and in good working order.
- Use the tool only for jobs for which it was designed.
- When using power tools, be certain that the power cord is kept away from the working end of the tool.
- If the tool has a shield or guard, be sure it is working properly and remember to use it.
- If an extension cord is used, be sure it is in good working order. Do not use it if bare wires are showing. Also, use a heavy-gauge wire extension cord to ensure adequate voltage for the tool being used.
- Be sure that the extension cord is properly grounded.
- Grip the tool firmly.
- Never use worn-out tools. A worn-out tool has more potential for causing injuries. For example, with a worn-out screwdriver, there is a greater possibility for slips, which could make medical attention necessary.
- If there is a problem with a power tool, never stick your fingers in the tool. Unplug it first, and then correct the tool's problem.
- Make sure to use insulated hand tools when working with electricity and electrical components.

Screwdrivers

A screwdriver is a hand tool used either to attach or remove screws. The two most common types are the flat-blade and Phillips. The flat-blade screwdriver is used on screws that have a slot in the screw head (Figures 3-1 and Figure 3-3). The flat-blade screwdriver is available in

FIGURE 3-1 Flat-blade screwdriver.

FIGURE 3-2 Phillips screwdriver.

many sizes and shapes. Always use the largest blade size that fits snugly into the slot on the screw head so that it will not slip off the screw. The screwdriver should never be used as a pry bar or a chisel; it was not designed for that purpose.

A Phillips screwdriver is used to attach or remove screws that have two slots crossing at right angles in the center of the screw head (Figures 3-2 and Figure 3-3).

When using a Phillips screwdriver, exert more pressure downward in order to keep the tool in the slots. Always use the largest Phillips size that fits snugly into the slots, just as with the flat-blade screwdriver.

Never use worn-out screwdrivers when working on appliances. A worn screwdriver may damage the head of the screw. It can also damage the product on which you are working.

Nut Drivers

Many manufacturers use metal screws with a hexagonal head. A nut driver is a hand tool similar to a screwdriver, except that the working end of the driver is hexagonal-shaped and fits over a hexagonal nut or a hexagonal bolt head. Each size nut requires a different sized driver (Figure 3-4).

Wrenches

Wrenches are the most frequently used tool. There are many different types and sizes of wrenches. Their purpose is to hold and turn nuts, bolts, cap screws, plugs, and various threaded parts. Wrenches are generally available in five different types (Figure 3-5):

- Socket wrenches
- Box wrenches

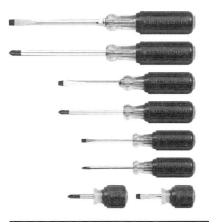

FIGURE 3-3 Combination screwdriver set.

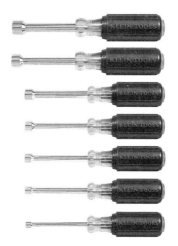

FIGURE 3-4 Hex-nut drivers.

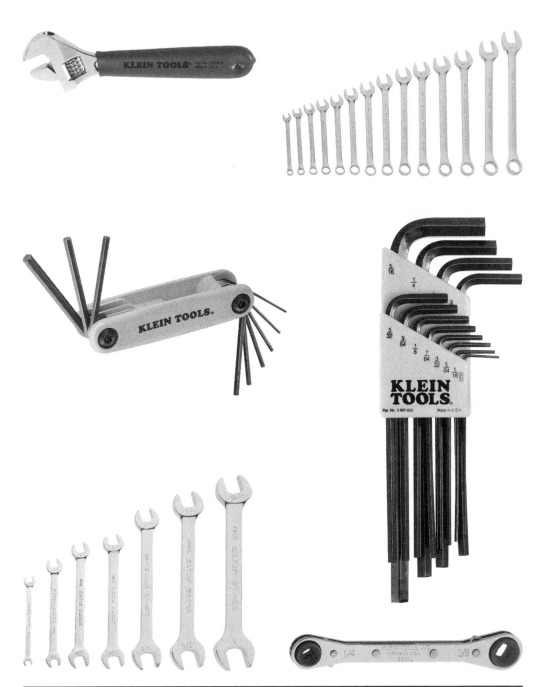

Figure 3-5 Wrenches are used to remove and fasten nuts and bolts. They are available in socket wrench, box, open-end, adjustable, and Allen types. (*continued*)

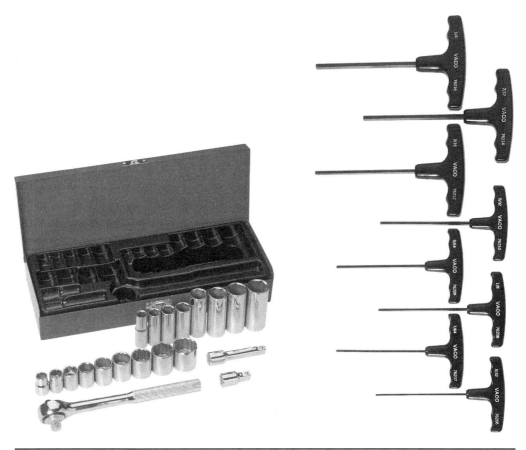

FIGURE 3-5 Wrenches are used to remove and fasten nuts and bolts. They are available in socket wrench, box, open-end, adjustable, and Allen types.

- Open-end wrenches
- Adjustable wrenches
- Allen, or hex, wrenches

Socket wrenches are used to slip over bolt heads, as opposed to other wrenches listed, which are used at right angles to the nut or bolt. This arrangement allows more leverage to be applied to loosen or tighten the nut or bolt (Figure 3-6). Select the size and type of socket to fit the nut with the proper drive size for the load. See Table 3-1 for the proper drive size loading recommendations.

Always choose the correct wrench for the job. Box wrenches should be used for heavy-duty jobs and in certain close-quarter situations. Open-ended wrenches are useful for

Hex Size	1/4" Drive	3/8" Drive	1/2" Drive	3/4" Drive	1" Drive
1/8" to 7/32"	USE	DO NOT USE	DO NOT USE	DO NOT USE	DO NOT USE
1/4" to 11/32"	USE	USE	DO NOT USE	DO NOT USE	DO NOT USE
3/8" to 9/16"	USE	USE	USE	DO NOT USE	DO NOT USE
19/32" to 11/16"	DO NOT USE	USE	USE	DO NOT USE	DO NOT USE
3/4" to 1"	DO NOT USE	DO NOT USE	USE	USE	DO NOT USE
1 1/16" to 1 1/4"	DO NOT USE	DO NOT USE	USE	USE	USE
1 1/2"	DO NOT USE	DO NOT USE	DO NOT USE	USE	USE
1 9/16" to 3 1/2"	DO NOT USE	DO NOT USE	DO NOT USE	DO NOT USE	USE

TABLE 3-1 Drive Size and Hex Size Loading Recommendations

FIGURE 3-6
(a) Socket wrench,
(b) adjustable wrench,
(c) open-end wrench.

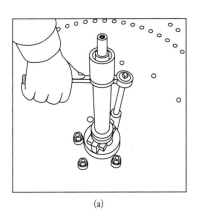

(a)

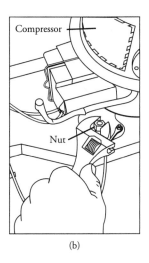

(b)

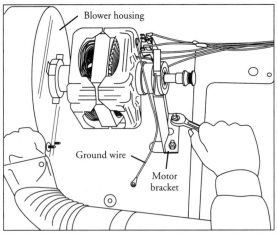

(c)

medium-duty work or situations where it is impossible to fit a socket or box wrench on a nut, bolt, or fitting from the top. Adjustable wrenches help with light-duty jobs, work well with odd-sized nuts and bolts, and are useful where a regular open-end wrench could be used. It is adjustable to fit any size object within its maximum opening. Allen wrenches are used for adjusting and removing fan blades or other components that are held in place by Allen set screws. The Allen wrench has a six-pointed flat face on either end.

Hammers

A hammer is a hitting tool. There are many sizes and styles of hammers available (Figure 3-7). The most common type used in appliance repairs is the claw hammer. The claw hammer can also be used for prying objects.

Prying Tools

Prying tools are available in many sizes and shapes. The most common are the crowbar, ripping bar, and the claw hammer. The claw hammer is basically used for light-duty work: removing nails and prying small objects. The ripping bar is used for medium-grade work, and the crowbar is used for heavier work.

Pliers

Pliers are one of the most frequently used tools. A pliers is a tool for holding or cutting, depending on the type. Generally, they are not made to tighten or unscrew heavy nuts and bolts. They are available in many sizes and shapes (Figure 3-8). Choose a pliers that fills a

FIGURE 3-7
While there are many different kinds of hammers, the claw hammer is the one most often used.

Short handle
sledge hammer

Claw
hammer

Ball-peen
hammer

particular need, being careful that it is the proper pliers for the job. Some of the most common types of pliers include:

- Slip-joint
- Slip-joint adjustable
- Vise grip
- Needle nose
- Diagonal cutting

Slip-joint pliers are pliers for everyday tasks (Figure 3-8a). The jaws can be adjusted into two different positions. Do not use them on nuts, bolts, or fittings. They can easily slip and injure both the technician and the device.

Pump pliers, also known as *slip-joint adjustable pliers,* are also used for general jobs (Figure 3-8i). They would be preferred over slip-joint pliers when working on a larger object. The jaws of slip-joint adjustable pliers can be moved into many different positions.

The *vise grip pliers* is actually four tools in one (Figure 3-8j): a clamp, a pipe wrench, a hand vise, and pliers. The lever holds the jaws in one position, allowing the vise to hold up to one ton of pressure.

FIGURE 3-8
Pliers are available in many sizes and shapes.

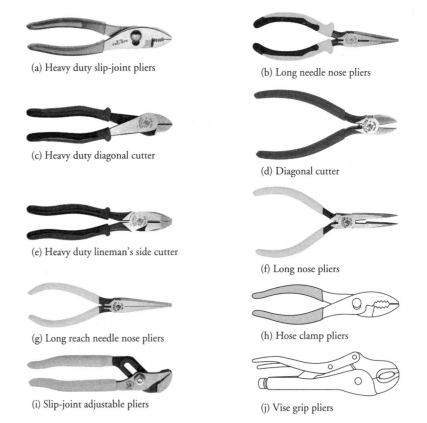

(a) Heavy duty slip-joint pliers

(b) Long needle nose pliers

(c) Heavy duty diagonal cutter

(d) Diagonal cutter

(e) Heavy duty lineman's side cutter

(f) Long nose pliers

(g) Long reach needle nose pliers

(h) Hose clamp pliers

(i) Slip-joint adjustable pliers

(j) Vise grip pliers

PART I

FIGURE 3-9
Chisels come in many different sizes and shapes, such as wood chisels, metal chisels, and concrete chisels.

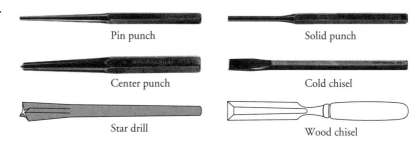

Pin punch

Solid punch

Center punch

Cold chisel

Star drill

Wood chisel

The *needle nose pliers* are mostly used with electronic, telephone, and electrical work (Figure 3-8b, Figure 3-8f, and Figure 3-8g). Other uses include in confined areas, to form wire loops, and to grip tiny pieces firmly. (The long nose is particularly useful for this latter task.) They are also available with side cutters.

Diagonal cutting pliers are used in electrical and electronic work (Figure 3-8c, Figure 3-8d, and Figure 3-8e). They are used for cutting wire and rope.

Cutting Tools

Many different types of tools are used for cutting. The key is to know which tool to use in each situation. Chisels are used for cutting metal and wood. They are made of high-carbon steel, which makes them hard enough to carve through metal (Figure 3-9). These should be used when removing rusted bolts and nuts.

Hacksaws are used for cutting metal (Figure 3-10). The hacksaw consists of a handle, frame, and a blade. The frame is adjustable, so it can accept any length of blade. The blades are available with different numbers of teeth per inch.

A file is also a cutting tool. It is used to remove excess material from objects. Files come in a variety of sizes and shapes (Figure 3-11).

Drill bits are also cutting tools. They are designed for cutting holes in metal, wood, and concrete (Figure 3-12).

FIGURE 3-10
Hacksaws are used for cutting metal, wood, etc.

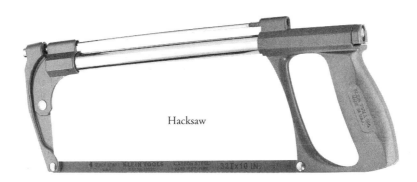

Hacksaw

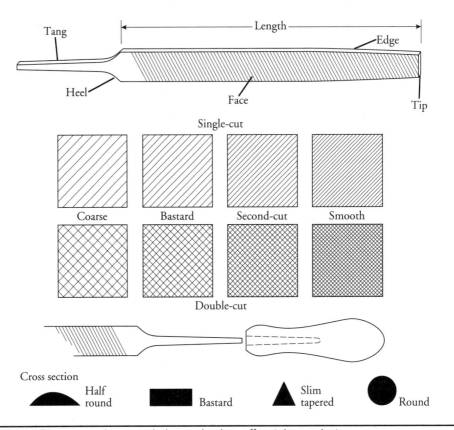

FIGURE 3-11   Files are used to smooth the rough edges off metals, wood, etc.

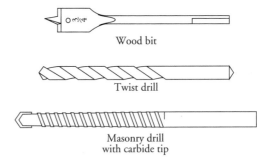

FIGURE 3-12
A general-purpose twist drill set will handle most of the technician's needs. The wood bit and masonry drill bit are used in installation work.

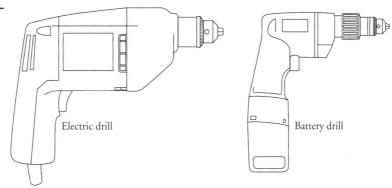

FIGURE 3-13
Power tools can have either a line- or battery-operated power supply.

Electric drill

Battery drill

Power Tools

Power tools do the same job as hand tools. However, they do the job faster. The most common power tools can be either electric or battery-powered (Figure 3-13).

When using power tools, there are some safety precautions that must be followed to prevent accidental injury. Always read the use and care manual that comes with each power tool.

Specialty Tools

These tools are specifically designed for a particular use and are used for in-depth servicing of the appliance. For example, special tools are used for the installation and removal of special screws and nuts (Figure 3-14); they are required to remove the bearings in washing machines; and they are also used for adjusting switch contacts. Figure 3-15 illustrates some of the many types of specialty tools. Other specialty tools and their uses will be discussed in later chapters of this book.

Test Meters

Test instruments are important tools used in assisting with the diagnosis of the various problems that arise with appliances. Varieties of test meters (Figure 3-16) include the following:

- The analog volt-ohm-milliammeter is used for testing the resistance, current, and voltage of the appliance or air conditioner. This type of meter moves a needle along a scale. It is also the most important test meter to have in the tool box.

- The digital multimeter displays a readout in numbers, usually on an LCD screen. This type of meter is similar in operation to the analog meter.

- An ammeter is a test instrument that is connected into a circuit to measure the current of the circuit without interrupting the electrical current.

- A wattmeter is a test instrument used to check the total wattage drawn by an appliance.

- A temperature tester is a test instrument used to measure the operating temperatures of the appliance.

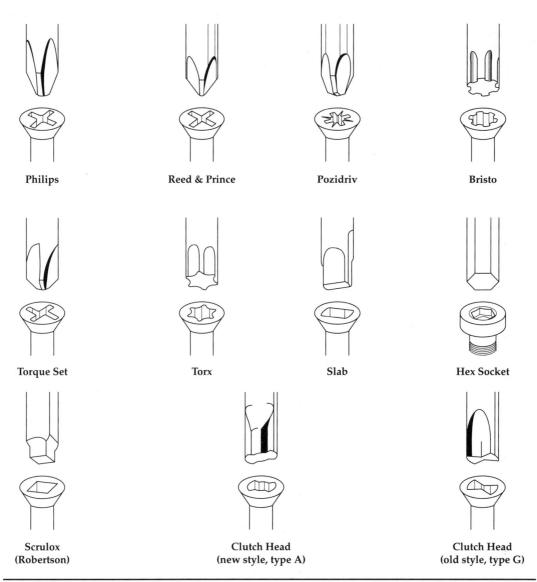

Philips   Reed & Prince   Pozidriv   Bristo

Torque Set   Torx   Slab   Hex Socket

Scrulox
(Robertson)

Clutch Head
(new style, type A)

Clutch Head
(old style, type G)

Figure 3-14   Here are just some of the different types of specialty screws and the types of drivers needed to remove them.

Advantages of Digital Meters

- A digital meter offers greater sensitivity, accuracy, and a better resolution in reading measurements.
- Digital meters have a faster readout as compared to analog meters.

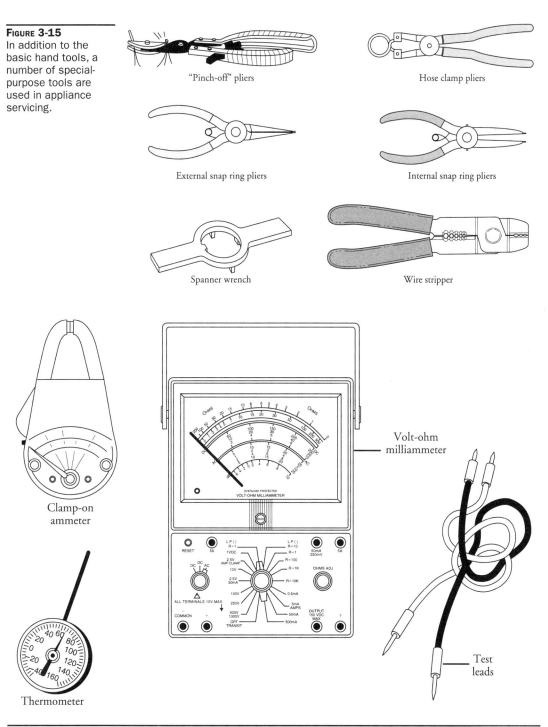

PART I

FIGURE 3-15
In addition to the basic hand tools, a number of special-purpose tools are used in appliance servicing.

"Pinch-off" pliers

Hose clamp pliers

External snap ring pliers

Internal snap ring pliers

Spanner wrench

Wire stripper

Clamp-on ammeter

Volt-ohm milliammeter

Test leads

Thermometer

FIGURE 3-16   Test meters are available in analog or digital readouts. For example, thermometers measure the temperature.

Becoming a Professional Technician

Today's service technician must be educated in major appliances and air conditioning theory and have the ability to diagnose problems with these products. The service technician also needs another important tool in their arsenal: customer service relations. The customer must be satisfied with the service call, the service company, and the service technician. When the service technician is in a customer's home, to the customer, the service technician represents the service company. While the service technician is in the customer's home, he or she must be able to accurately diagnose the problem and make efficient repairs. Repairing major appliances and air conditioners is only part of the job. The service technician must establish good customer relation skills in order to have good customer satisfaction. Service companies place service technicians in a unique position that affects customer satisfaction. The service technician will have to meet the customer face to face in the customer's home or place of business. Service technicians meet customers when their products have failed or are not performing to their expectations or when the customer has to pay for a service call and repairs when the product is out of warranty.

The service technician has the opportunity to restore customer confidence with the product when they meet face to face, when they are not satisfied with the product, with the service company, or with the service technician. In most cases, the service technician has no control over the customer's dissatisfaction of the appliance or air conditioner performance.

As a service technician, whether you are just starting out in your career or you have more experience, you will have to practice good customer relation skills to prevent negative experiences with customers.

The purpose of this chapter is to provide you with the tools needed to develop and provide excellent customer satisfaction and to avoid increased customer dissatisfaction. I will present the information needed to prevent customer relation problems and to promote customer satisfaction. Also, I will present the skills needed to deal with dissatisfied and angry customers.

What Is Customer Satisfaction?

Customer satisfaction is the customer's impression that the service company and the service technician have done something meaningful and in the way that the customer wants it done. This is not the same as having done a good job by providing good service. The service company you are working for defines these concepts, which are generally based on the technical aspects of the job. However, when the service technician provides good service for the customer, the customer may not be satisfied. When you are on a service call, the customer must believe that you provided a worthwhile repair and that you did it in a manner that pleases the customer. Just remember this: the customer has the final word, whether they are satisfied or not. As a service technician you must balance what is attainable in your field and customer satisfaction.

Customer Satisfaction Is Important for Business

From a business standpoint, when a customer is satisfied with the service company and the service technician, a bond of loyalty begins. When a customer is satisfied and happy, a great thing happens: they tell their friends and family of the experience. That is the best advertising for a service company. They do not have to pay for advertising and it brings repeat business and profits. When you have good customer satisfaction, the customers stay with that company; they do not have to look for another company to fill their needs. Customers place a high priority on good service and if they are not satisfied, they will go elsewhere for that satisfaction.

Why Is Customer Satisfaction Important to the Technician?

When the technician finds ways to increase customer satisfaction, they can make the service call and the job more interesting and rewarding for themselves and their company. Along with customer satisfaction, the service technician can expect to receive pay increases or even a promotion from the company. Once the technician has learned customer service techniques and treats the customer as a special person, it will carry over to other service calls. Going on service calls to happy customers is more rewarding and less stressful than going on service calls for customers who are not happy with you or the service company.

What Does the Customer Expect of the Service Technician?

If the customer does not recognize the quality of service being provided as superior, then the skills of the service technician are not enough. The quality of the service call is whatever the customer recognizes it to be. When the service technician can recognize what the customer perceives to be quality service, and knowing what the customer expects, then the technician will exceed the customer's expectations. The customer expects the service technician to solve their problems with their appliances or air conditioner. When the service technician cannot solve the customer's problems, or creates new problems, then the customer becomes dissatisfied with the technician and the service company.

The customer expects the service technician to observe and perform the following:

- Arrive at the customer's home or place of business on the day that the service call is scheduled.

- Arrive on the scheduled time that was provided to the customer.
- Identify themselves when they arrive at the service call.
- Wear a uniform or have an identification badge that identifies the service company.
- Be properly groomed, knowledgeable of the product being serviced, and courteous to the customer.
- Show a genuine concern for the customer's problems with the product.
- Listen to the customer's complaint and ask questions regarding the complaint. This will allow you to better diagnose the faults with the product and to efficiently perform repairs.
- Do not leave the customer in the dark about the repairs. Always explain to the customer what the repairs involve. The service technician must provide the customer an estimate and a breakdown of the cost for the repairs to the product before beginning any repairs. At this point in time the customer will have to provide you with an answer as to whether they want the repairs or if they will replace the product with a new one.
- When the customer tells you that they want the repairs, you will make sure that you have the necessary parts to complete the repairs. The technician must work quickly and efficiently to complete the job. Make sure that you have no leftover screws or parts that belong in the product. Double-check.
- When the repairs are completed, you will demonstrate to the customer that the repair has solved the customer's problem with the product. You will then offer to leave the old parts with the customer if they are out of warranty. Let the customer decide what to do with the old parts.
- Leave the work area clean and free of debris.
- When you are ready to leave the customer, do not forget to thank them for their business.

How Can a Service Technician Increase Customer Satisfaction Skills?

The service technician should be aware of specific procedures that will help them to increase customer satisfaction. By learning these procedures and practicing them so that they become automatic and natural, the technician should be able to demonstrate better customer skills. Remember that the overall objective is to demonstrate to the customer your ability to resolve their service problem, while providing them with great service that will encourage future business.

The following is a list of actions that the service technician can establish for good customer relations:

- While in the presence of the customer, always respect the customer.
- Arrive on time to the service call.
- When addressing the customer, always use their name.

At the customer's home or business, always respect their property:

- Park your service vehicle in the street, not on the customer's driveway.
- Wipe your feet before entering the customer's home or business.
- Use a protective covering for your tools to protect the customer's property from being damaged while servicing their product.
- When uninstalling an appliance or air conditioner, be careful not to damage anything.
- When the repairs are completed, clean up the mess.

Always show a concern for the customer's problem with the product:

- Listen to the concerns of the customer.
- Give assurances to the customer that the problem will be solved.

The service technician is building customer confidence by showing their ability to repair the product. You will maintain a professional appearance and have a great attitude. You will maintain a clean and properly labeled service vehicle. Most of all, as a technician you will demonstrate technical abilities in completing the repairs.

The service technician will have to rebuild customer confidence in the appliance or air conditioner in the following circumstances:

- When repairs are completed, you have to resell the product to the customer.
- Always avoid any negative comments about the product manufacturer, other service technicians, the service company staff, or any other service company.

Communication Skills

A service technician must have good communication skills to help them establish good customer relations. There are three important channels for communicating with the customer. They are verbal, nonverbal, and paraverbal communications. The reason these three channels are important is that they can send conflicting messages from each other since they convey information independently of the others. When the three types of communications are sending different messages, the customer gets confused, and your point can be lost. Inconsistency can also create a lack of trust and undermine the chance to build a good working relationship between you and the customer. If a technician sends a message with conflicting verbal, nonverbal, and paraverbal information, the nonverbal information tends to be believed.

Some other skills that you will need include:

- Customer empathy
- Listening to the customer
- Reflective listening skills (feedback)
- Paraphrasing
- Asking questions

Verbal Communication

As a service technician you must be able to communicate to customers effectively. Verbal communication is one way for people to communicate face to face. Speak clearly and know what you are going to say before you say it. Words that are critical, blaming, judgmental, or harsh tend to create an unwilling and defensive mindset that is not conducive to productive problem solving. When we speak, only 10 percent of the words we use get through to the customer. Spoken words are unlike written words, where a person can go over a passage several times to ensure understanding. It is the technician's responsibility to make sure that the message gets across to the customer the first time. Therefore, if you want your message to be understood, you must be careful of the words you use. Technicians who are well versed in their fields have the tendency to use industry jargon when speaking to customers. While this is comfortable for the technician, the customer feels lost in the conversation. Choose your words with the intent of making your message as clear as possible for the customer; keep it simple and not confusing.

Nonverbal Communication

Nonverbal communication is the unspoken messages that a technician sends to a customer by gestures, posture, facial expressions, and sounds. Of the two types of communication already mentioned, nonverbal communication is considered to be a truer reflection of a person's feelings and thoughts. Never underestimate the power of nonverbal communication; it accounts for 52 percent of what is perceived and understood by others. Our body language is always communicating to the customer whether we want to or not. Because of this, the customer is most likely to believe and trust the nonverbal communication over the verbal communication, especially when both types of communication are contradicting each other. If what you say to a customer is not supported by your body language, the customer might interpret your words as insincere. It also tells the customer how you actually feel and that you do not care about their problems with the appliance or air conditioner.

Paraverbal Communication

Paraverbal communication refers to how we say something, our tone, our volume, our pitch, and how we pace our speaking and articulation of the sentences we use when speaking with customers. Remember this: it is how we say something, not what we say. A paraverbal message accounts for 38 percent of what is communicated to the customer. When speaking to the customer, a sentence can convey entirely different meanings depending on the emphasis on words and tone of voice. Here are some points to remember about paraverbal communication:

- When you are angry or excited, your speech tends to become more rapid and higher pitched.
- When you are bored or feeling down, you speech tends to show and take on a monotone quality.
- When you are feeling defensive, your speech is often abrupt.

All of these points will significantly influence the customer's opinion of the service you will be providing. The customer must feel that you are treating them as a highly valued individual who is unique and important to you.

Customer Empathy

Empathy is the ability to place yourself in the shoes of your customer and see through their eyes, the customer's point of view. It's thinking about the customer's fears, anxiety, and trying to understand their previous service call experiences (good or bad) before they met you. Other times the customer is someone you've seen before and they are expecting to repeat a good experience with you. To effectively communicate empathy, you must have empathy. Your ability to view the situation from the customer's point of view depends upon the following:

- Observing and listening to the customer's problem
- Imagining yourself in the customer's situation
- Recalling how you felt in a similar situation

Once you put yourself in the customer's place, you can communicate that empathy. Remember, it's our privilege to create a long-term relationship with the customers. So, listen carefully and look into their hearts before responding. Then, answer with kindness, thoughtfulness, and understanding. You'll be rewarded with trust, friendship, and customer loyalty.

Listening to the Customer

Improve verbal communication by also listening diligently to what the customer has to say. If you dominate the conversation or don't pay attention to the responses you're receiving, you won't comprehend the customer's point of view or know if they truly understand yours. Listening to the customer will provide you with the following:

- You will obtain valuable information about the appliance or air conditioner, the customer's perception of the problem, and the customer's concerns.
- You will demonstrate your respect for the customer and your concern for their problem.

When you are in the customer's home or place of business, do not start to work on the appliance when the customer is speaking or look at your cell phone for messages without first listening to the customer's concerns. When we open our minds to new ideas, we have the opportunity to learn new things and to hear different perspectives. Give the customer your full attention. Allow the customer to finish speaking, or wait for a pause in the speaking before you begin to speak. A good listener will:

- Set aside your own thoughts and agendas.
- Step into the customer's shoes and try to see the world through their eyes.
- Suspend judgment, evaluation, and approval in an attempt to understand the customer.
- Not finish the sentence for the customer; let the customer finish speaking first.
- Share with the customer only positive comments.

- Move your body in response to the speaker, that is, appropriate head nodding and facial expressions.
- Always face the customer and show good eye contact.
- Maintain an open posture with arms and legs uncrossed.
- Not lean against anything; stand on your own feet. Show good posture.

Reflective Listening Skills (Feedback)

Technicians have a tendency to filter out parts of the message, or to fill in parts that have not been communicated by the customer. It is important to not only listen effectively, but to provide feedback to confirm what you think you heard from the customer. Effective feedback tells the customer that the technician has heard the message and indicates a response to the message. Feedback includes nodding your head up and down, a particular facial expression, direct eye contact, or posture. Feedback can also be communicated through words, like "yes" or "I understand." Feedback is the technician's response to the customer, so useful feedback can only be given when the technician has listened effectively.

Paraphrasing

Paraphrasing is a form of feedback to the customer. The paraphrase should be in the technician's own words rather than using the customer's words. In other words, do not be a parrot by repeating word for word. Paraphrasing not only indicates that you are listening, but it gives the customer the opportunity to verify your understanding of what was said. You do not want to interrupt the customer while they are stating their issues or explanations.

If you can state the customer's ideas in your own words, you have demonstrated that you have listened carefully and that you understand what was said. Even though you are paraphrasing, make sure you are using words that the customer understands. Good paraphrasing not only helps the technician understand what the customer is saying; it helps you to remember what was said.

Asking Questions

Asking questions indicates that the service technician does not understand, or wants to verify their understanding of the problem with the appliance. This helps focus the technician on the topic, encourages the technician to talk, and provides the technician with the opportunity to give feedback. Asking questions will help you to:

- Obtain more information that you will need to solve the customer's problem with the appliance or air conditioner.
- Gain an opportunity to learn what the customer feels about the appliance or air conditioner.

Questions will help you to discover the customer's actions that may have caused the problem with the appliance. By knowing what actions have occurred, it will give you the opportunity to give instructions that will prevent the problem from recurring. Not asking questions or not obtaining enough information can lead to a misdiagnosis of the appliance or air conditioner. It also reduces the opportunity to know how the customer feels about the appliance. Knowing how the customer feels about the product will help you know what to do to restore, or improve, customer satisfaction.

Preparing for a Service Call

Before a service technician arrives at a home or place of business, some important steps must take place first. When you prepare first, you minimize the conditions that can lead to customer dissatisfaction, while promoting customer satisfaction. Customers expect the service technician to be knowledgeable about their appliance or air conditioner. It is essential that you gather as much information as possible about servicing the appliances and/or air conditioners that you are likely to see in your area. One of the most important means of receiving information on these products is to attend an appliance and air conditioning school. While in school or working for a service company, attend any and all of the manufacturers' scheduled training sessions. There are also appliance and/or air conditioning parts distributors that have training sessions. They not only give you information presented about new product introductions and updates, but also give you the opportunity to discuss issues that you are seeing in the field day-to-day. Training sessions also give you the opportunity to interact with other technicians from your area. You may find they are experiencing the same difficulties, or that they have a remedy for a problem that you were experiencing. Manufacturer's service manuals or the service mini-manual (technical data sheet) that is attached to the appliance or air conditioner is another source of information. Most contain information on product operation, disassembly, diagnosis tips, and procedures for checking specific components. These manuals also contain error codes (on certain models only) that help the technician with the diagnosis. Trade magazines and periodicals often cover general and sometimes specific appliances or air conditioners and can be beneficial to the technician or the service company. The Internet is another source for technical and parts information. Yet another source will be salespeople who sell appliances and air conditioners to customers based on features that are intended to satisfy the customer's needs. These features apply mainly to high-end appliances and air conditioners that are advertised as being the "quietest," "most energy efficient," "largest," etc. Insights to these features will help you get an idea of what is important to the customer who purchased one of these products.

Prescreen the Service Call for Parts

Customers expect the technician to repair the appliance or air conditioner on the first call. When a service call is placed, customers are facing two major sources of inconvenience:

- The appliance or air conditioner is not functioning properly.
- They have to wait for the service technician to arrive to repair it.

When a return trip must be scheduled due to lack of parts to repair the product, it further adds to the customer's inconvenience, as well as to their dissatisfaction. Prescreening service calls so that you can attempt to have the proper parts to complete the call will greatly reduce the dissatisfaction and add to the customer's satisfaction. It also serves to reduce costs to your service company. It costs money to schedule a second trip, and the cost can be minimized, and customer satisfaction maximized, with just a little extra effort from you. Believe it or not, it's part of your job description.

Proper Grooming for a Service Technician

As a service technician you have only one chance to make an impression on a customer. Most customers will be more likely to overlook some flaws in the service call visit if they feel good about the technician. Customers will be more likely to call back the same company and ask for the same technician if they like you. First impressions make a technician's job a little easier. Make sure that you start your day with a clean uniform and shined shoes. While it's difficult to stay fresh all day, an extra shirt or two in the truck, with a shoe brush, will help you look your best. Shower, shave, (if applicable), and comb your hair (also if applicable). When you present the image that you care about yourself, the customer will feel that you care about them, and about their problem with the appliance and/or air conditioner.

When You Arrive at the Service Call

If a commitment has been made between the service company and/or service technician with the customer, you must arrive on the scheduled date and time—this is what the customer expects. In many cases, the customer has to make arrangements to be home for the technician to do the repair. This may include taking time off work, rescheduling other activities, or having someone else there to let you into the residence. Do not enter a residence, under any circumstances, where there is a minor child present, unless an adult is present. This is for your protection along with the minor's. If there is no adult present, use your cell phone to make contact with the customer or notify your office to call the customer to reschedule the service call. Be mindful of the fact that customers are not there for your convenience. You are there to help their situation. Be on time if a commitment has been made.

If circumstances make it impossible to keep an appointment, call the customer in advance to let them know that you will not be there. This "phoning ahead" would also include arriving prior to a prearranged time or date.

When arriving at the customer's home or place of business, park on the street. A customer will more than likely be dissatisfied with a service van that leaks oil, or other fluids, onto the driveway or in front of their place of business. By parking in the driveway, you may block family members or consumers from entering or leaving the residence/business.

Check your grooming before walking up to the door. Present a positive attitude. To the customer, you are the individual who will answer all of their questions and solve all of their problems. Stand up straight, walk to the door with confidence, and convey the image that you know what you are doing.

When the customer answers the door, greet them by saying, for example: "Good *morning/ afternoon* Mr. / Mrs. _____." This is not only good to establish a rapport with the customer, but also to identify that this person is the responsible party that you are there to see.

At the customer's residence or business, identify yourself, your company, and your intended purpose. For example: "I am _____ from WXYZ Service Company. I'm here to service your refrigerator." Always ask if there is an alternate entrance the customer would prefer you to use, such as a garage or a back door. Wipe your feet before entering, and, if necessary, wear protective shoe covers so as not to dirty or damage carpeting or floors.

If the appliance or air conditioner is out of warranty, explain any service call charges, diagnostic charges, and establish a method of payment before you begin diagnosing the problem. This will avoid problems later. If the appliance or air conditioner is under warranty, you must make sure that your company is authorized to do the repairs. Use protective mats and protective equipment on your tools and for the product being moved for repairs. Do not place your tools or other diagnostic equipment on other appliances, countertops, or furniture. Do not damage the customer's property.

Repairing the Appliance or Air Conditioner

Before you begin to take apart the product, ask the customer questions regarding the problem they were experiencing with the product and why they called. This involves actively listening to what the customer is saying. Do not presume that you've seen this problem before, so you know what it will take to repair it. To the customer, they are the only one having this experience, and many want to be certain that you completely understand their needs. By listening to the customer describe the perceived or real problem with the product, you can save valuable time by not checking or disassembling parts of the appliance or air conditioner that are not affected. Verify that the product has the problem that the customer described. Customers are not supposed to know about the technical aspects of their appliance or air conditioner. In most cases, customers can only relate the end result of a malfunction. Talking to the customer and asking questions may lead to a completely different condition than the one stated on your service invoice. There can be several reasons for obtaining an incorrect diagnosis:

- The customer's spouse may have made the service call and did not understand the nature of the problem.
- The person responsible for taking the service call did not get the exact reason for the call.
- The person taking the call used "shorthand" in describing the problem, or all of the necessary information did not make it into the description of the complaint.
- The customer thinks they have a problem with the product, but they do not.
- The customer may have forgotten how to use the product, or they forgot to read the use-and-care guide.

After diagnosing the problem, explain what repairs are needed, and give an estimated cost to the customer. On some occasions you might have to give a written estimate. Remember, show empathy for the customer's situation. When describing the nature of the repair or the product's condition, explain it to the customer in a way they'll understand. Never talk down to the customer, assuming they won't understand anyway. That may be the case, but they do not want to be made to feel unintelligent.

When the repair is completed, demonstrate to the customer that the appliance or air conditioner is operating properly. This gives the customer the assurance that the appliance or air conditioner has been repaired and that you have solved the problem. Demonstration of a repaired product will also reduce the possibility of callbacks, since you and the customer have witnessed that the product was operating according to manufacturer's specifications.

If the old part is not returnable for "dud" allowances, offer to leave it with the customer. This will give the customer the assurance that you have installed the parts as stated. Use the old part to demonstrate the reason that the product malfunctioned. This will help the customer to understand the nature of the failure and its repair and assure that you are a trustworthy professional service technician. If the customer does not want the old part, remove it and dispose of it when you return to your company.

Explain any warranty on the repair, and point the warranty out on the service invoice. Resell the appliance or air conditioner by reaffirming its features. Give the customer the satisfaction of feeling that they have selected a quality product and that they can depend on it. Do not make the customer feel that they are destined to live with recurring problems that cannot be resolved. If you diagnosed the problem correctly and tested the product according to manufacturer's specifications, and the product is working according to specifications, then there is no reason in your mind to doubt the product's longevity.

Leave a copy of a written invoice or receipt, signed by the customer, even if the product is under warranty. Last, clean up the work area (water, lubricants, dirt, or debris) left as a result of the repair. Thank the customer for their business. Invite future business by offering the customer a number to call if there are any questions that may arise. Close the service call by saying "Good-bye," "Have a nice day," etc.

Follow-Up Calls

Technicians try to complete all of the service calls within the first trip to the customer. Unfortunately, most service calls will not be completed on the first trip, but you should try to make the number of calls as few as possible. This will lead to some customer dissatisfaction, since they must wait to have their appliance or air conditioner operational, and may have to make arrangements to leave work early again so that you can complete the repair. How you handle the follow-up call can help minimize the customer's dissatisfaction with the delay.

If a part needs to be ordered, or you need to come back to complete the repair, explain the reasons to the customer. If possible, give the customer an approximation as close as possible on when you will return to complete the repair. If you cannot estimate the date of your return, advise the customer that you or someone from your company will be contacting them to give them an update on the status of the delay and when the service call can be rescheduled. It is absolutely critical that the customer is contacted within a day or two, even if the status is yet unclear, to keep them from wondering if anyone is actively working on the problem. Many customers do not realize that their appliance or air conditioner repair is one of many to your company. The customer should be made to feel that they are receiving your utmost attention, and that you are doing everything possible to complete the service repair.

Things to Avoid While on a Service Call

There are some things you'll want to avoid while on the service call that can turn a good impression into a bad one. The customer is a valued person and should be treated as one. Keep a positive attitude when dealing with the customer. Make every effort to keep your conversations with the customer on a professional level. It is okay to be friendly, but avoid becoming too familiar. Here are some other things to avoid while on a service call:

- Avoid placing blame on the appliance or air conditioner, its design, the person who installed it, the company that built it, other technicians and other service companies, etc. It serves no purpose to point fingers. As long as the problem is taken care of, the customer probably doesn't care who or what caused it.

- Avoid a negative attitude. You are representing your company, and the perception the customer forms about your company is based largely on you.

- Avoid getting too technical when explaining the repairs required. Don't talk down to the customer, but try to speak to them in a way they'll understand.

- Avoid profanity or joking around with the customer. There are limits to how friendly you should be with a customer. You are a professional and should act like one.

- Avoid causing damage to the appliance or air conditioner or to the customer's home. You are there to correct a problem, not cause a new one.

- Avoid making promises you can't keep. You need to maintain a balance between serving the customer and running a business. Some things just aren't doable. Be realistic in attempting to satisfy the customer. Making a promise and failing to follow through will make matters worse.

- Avoid leaving the customer without an explanation of the "next steps," if needed. If parts are required, or another service call will have to be scheduled, make it clear that you or someone from your office will be contacting them and give them an estimate of when. If the customer needs to make arrangements with someone else to resolve their problem, advise that they should call the dealer or electrician, plumber, etc., and give them the reason why they'll need to do it.

- Avoid using the word "can't." Most customers don't care about what you can't do. They are only interested in what you can do. Do not give the customer any reason to tell you to leave; they will call someone else to repair their product.

How to Turn Customer Dissatisfaction into Customer Satisfaction

There will be a time when your best efforts will not be enough and your customer will be dissatisfied. Unforeseen circumstances, a missed diagnosis, a problem where you couldn't keep an appointment are all events that can lead to customer dissatisfaction. While these are things that should be avoided, sometimes they are inevitable. There are steps, however, that can be taken to regain customer satisfaction:

- **Apologize sincerely**   If you are the reason for the dissatisfaction, apologize to the customer in a way that is sincere. This does not mean a quick "sorry" mumbled under your breath, but an apology that the customer will understand as being meaningful.

- **Fix the problem quickly**   Okay, you've made a mistake and you've apologized. Now, you must resolve the problem and resolve it in a timely manner. All of the apologies in the world will not help the situation if you don't take steps to fix it.

Most customers will understand that mistakes will happen. The sooner you fix the mistake, the more likely the customer will forgive the mistake.

- **Do something extra**   Having apologized and fixed the problem will not necessarily bring the customer to a feeling of satisfaction. They may no longer be dissatisfied, but there is more to be done to bring them back. It doesn't have to be anything extraordinary, but something that the customer would not expect. You could offer a free inspection of another appliance or air conditioner. Also, you could offer an extension of the labor warranty. This will depend on your company's policy and capabilities and should be checked prior to making the offer.

- **Finally, follow up**   After the service call has been completed, call the customer to check up on the appliance's performance. Follow up by sending a letter thanking them for their business and making sure everything is satisfactory. You must demonstrate to the customer that you are making a conscious attempt to regain their business and make them feel important and valuable.

A customer who experiences too many incidents that lead to dissatisfaction may never be brought back to being satisfied, no matter what you do or how many times you apologize. Obviously, the best way to avoid recovering customer satisfaction is to do the service call right the first time.

Angry Customers

Customers become angry when their dissatisfaction is poorly managed. By taking the appropriate steps in dealing with an angry customer, you can minimize their anger, resolve their problem, and end up with a satisfied customer. These steps, as they apply to the service technician, can be summarized as follows:

- The service technician sets the tone by being calm and using reason.
- The customer expresses feelings and facts about the problem.
- The service technician, after listening carefully to understand the feelings and the facts, responds to the customer's feelings by expressing understanding, and responds to the facts by suggesting a solution to the problem.
- The service technician ends the process by focusing on the positive.

Don't Lose Your Cool

As a trained professional, it is your responsibility to maintain your self control. The customer is frustrated by the situation, and has become angry. If you become angry as well, the customer is likely to respond with increased anger, and to focus the anger on you. While you cannot control the customer's response, you can influence it by remaining objective, reasonable, and calm. Remember, the customer is frustrated because things are not as they expected. Their anger may not be directed at you, but you may not be the cause of the anger, and should not take it personally. If you are the cause of the anger, apologize sincerely. Let the customer know that you acknowledge the anger, and that you will take actions to correct the cause. Let the customer express emotions and facts verbally, but remain calm. Do not panic, and do not respond with anger.

Encourage the Customer to Vent Their Emotions

When consumers' expectations are not met, they are likely to become angry. The anger is likely to increase if the customer feels powerless to correct the problem. To the customer, it may seem that the only way they can get the problem resolved is to express their anger. The result is increased frustration. Allowing the customer to vent their frustrations verbally will help them relieve some of the frustration, because expressing the anger is doing something about the problem. Good listening skills, such as providing feedback, will encourage customer expression.

Find Out the Facts

To correct the problem, you will need to get the facts. With the customer's expression of anger, some of the facts will surface. You will need to ask clarifying questions to get all of the facts you'll need. Avoid jumping to conclusions about the customer, or about the appliance or air conditioner problem. Many times you'll think you know what the customer is leading up to, but you may be incorrect in your assumption. Continue getting as many facts as you can about the problem. Ask questions to get clarification that will help get the facts you'll need to diagnose the problem. Be careful to ask pertinent questions. If a customer gets the impression you're asking frivolous questions that are just wasting their time, they may become even angrier. Paraphrase the information the customer has given you to check your understanding of the problem, and to demonstrate to the customer that you are actively listening and making a commitment to resolve their problem. The facts you are seeking are not only about the appliance or air conditioner, but about the customer as well. The problem may be with the customer's expectations, and not an appliance or air conditioner issue at all.

Understand Your Customer's Feelings

Use the skills you've learned in this chapter about empathy, listening, and providing feedback. As a service technician, you will have the technical knowledge and skills as well as a concern for your customer. You have to understand the customer's feelings and communicate that understanding to the customer. Acknowledge the customer's distress and let them know that you are willing to help. Recall how you felt in a similar situation. Be aware of the customer's nonverbal communication. Remember that the customer is aware of what you are communicating nonverbally. If you listened carefully and got the facts about the product and about the customer, you will be able to better understand the situation.

Suggest a Way to Fix the Problem

Using the skills learned in this chapter, explain your diagnosis to the customer in terms that they can understand. Do not talk down to the customer, and do not use terms that are too technical. This is sometimes the cause of their anger in the first place.

- State the problem.
- Show or describe where the problem occurred.
- Explain how the problem affects the overall performance of the product.
- Describe the likely cause of the problem.

- Explain what is required to correct the problem.
- Explain what can be done to prevent a recurrence of the problem.

If you cannot provide the solution the customer expects, provide alternative solutions, rather than no solution at all, or explain why a particular condition exists. Avoid using words like "no" or "can't." Give the customer a solution that uses the words "can" or "will." It is important that you keep in mind your limits. Do not make promises that you, or the product, will be unable to keep.

Many of your calls are likely to involve customer error. If the problem is with the customer rather than with the product, keep your suggestions focused on the product or on the process. Do not blame the customer for the problem. In most cases you may have to remind the customer how to set the controls properly or explain how the product functions. In your attempt to resolve a customer error and not an appliance or air conditioner problem, you may think that installing a part on the product will make the customer happy. In many cases, this will only prove to the customer that there is a problem with the product when there isn't. This may lead to more problems in the future.

End on a Positive Note

This last step helps ensure that a situation that began in anger ends with a customer who is pleased by your ability to solve their problem. Good service will help ensure repeat business, not only from that customer, but from people that the customer talks to about the service you provided. Believe me, customers talk to each other about service companies and service technicians. Here are some more ways to end on a positive note:

- Do more to solve the problem than a customer would ordinarily expect. Be willing to go the "extra mile" to demonstrate your willingness to resolve the customer's problem.

- Resell the product by pointing out its useful features to give the customer the feeling that they have selected a quality product.

- Let the customer know that if any further problems develop, you are trained and prepared to resolve the problem.

- Be careful not to make statements that may not be true. Do not promise that this appliance or air conditioner will last for years or never have another problem.

Never Stop Learning

In this rapidly changing field, you must keep up to date with the changes (Figure 4-1). This textbook offers you a broad basic background on the appliances and air conditioners that are already in the field. The one thing we all have in common today is the Internet, where you can keep abreast of the new developments, changes, and improvements in the appliance and air conditioner field. Also, there are publications serving the appliance and air conditioner fields that you can subscribe to. Manufacturers often offer training programs, and studying manufacturers' literature will help you to update your knowledge on appliances and air

FIGURE 4-1
Never stop learning.

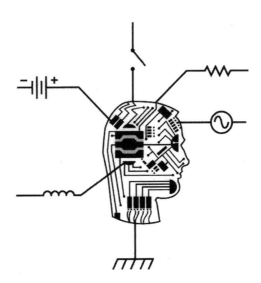

conditioners. By belonging to a trade organization and subscribing to its publication, you will have a source for detailed information on any given product. Technical knowledge is a never-ending process. The appliance and air conditioner service industry is a multibillion-dollar business that is still growing. The benefit of staying abreast of updates in the industry is job security and a better chance to stay in business, if you are in business, and do not forget your customer relation skills.

Basic Techniques

There are many different ways to diagnose a problem, but all of them use basically the same reasoning:

- Where does the consumer think the malfunction is located within the appliance or air conditioner?
- Where is the actual problem located within the appliance or air conditioner?
- Are there any related problems with the appliance or air conditioner?
- How can the problem with the appliance or air conditioner be solved?

For example, the consumer states that the dryer does not dry the clothes and believes that the heating element is bad. The actual problem might be a restricted exhaust vent, a clogged lint filter, bad heating elements, faulty operating thermostats or safety thermostat, or improper control settings.

When checking the dryer, you might notice that the control settings are set for air drying instead of heat drying. Thus the actual problem was that the control settings were not positioned correctly. The related problem is: "How did the control setting move to the air-dry position?" This leads to the question: "Does the consumer know how to operate the dryer?" To solve this problem, you will have to instruct the consumer in the proper operation of the dryer.

All appliances and air conditioners go through a certain sequence of events. Understanding the proper operation and this sequence as indicated in the use and care manual is beneficial when diagnosing the appliance or air conditioner.

Product Failure

Given the information about the product's problem, information and diagnostic charts from this book, and the information you have read in the use and care manual, as the servicer you will be able to perform the following steps in sequence to diagnose and correct a malfunction. The basic steps to follow when diagnosing an appliance or air conditioner problem are:

1. **Verify the complaint.** Ask the consumer what symptoms were caused by the problem with the appliance or air conditioner.

2. **Check for external factors.** For example: is the appliance or air conditioner installed properly, does the product have the correct voltage, etc.?

3. **Check for physical damage.** Look for internal and external physical damage. Any damage will prevent the appliance or air conditioner from functioning properly. Two examples are broken parts at the base of the washing machine or a damaged cabinet that prevents the doors from closing properly.

4. **Check the controls.** The controls must be set to the proper settings. If the controls are not set correctly, the appliance or air conditioner might not function properly or complete its cycle.

5. **Operate the product.** Operate the appliance or air conditioner, and let it run through its cycle. Check the cycle operation against the operational sequence of events that is listed in the use and care manual.

6. **Is the product operating properly?** If it is, explain to the consumer how to operate the appliance or air conditioner according to the manufacturer's specifications.

7. **The product is not operating properly.** If the appliance or air conditioner is not operating properly, proceed to locate which component has failed. Check the diagnostic charts that are listed in this book to assist you in the correct direction to take.

Diagnosis and Correction Procedure

When diagnosing a problem with an appliance or air conditioner, use your five senses to determine the condition of the product. This will help in analyzing and defining the problem:

- *Example #1:* When turning on the washing machine, there is a smell of something burning. You can track down the location of the burning smell and therefore discover which part has failed.

- *Example #2:* When turning on a dishwasher, unusual noises are heard coming from underneath the machine. Stop the dishwasher and attempt to track down where the noises are coming from.

Along with your hand tools, there are a variety of test meters that can assist you in analyzing and defining the problem. This is the sequence of events to follow when servicing an appliance or air conditioner:

- **Unplug the product** Change the range setting on the multimeter to voltage. Check the voltage from the appliance or air conditioner's receptacle. If there is an uncertainty, check the name plate rating for the correct voltage rating; this is located on the product. When diagnosing a component failure, there are three types of circuit failure—the open circuit, the grounded circuit, and the short circuit—all of which are thoroughly explained in Chapter 6.

- **Gain access** Only remove the panels and screws necessary to gain access to where the suspected component failure is located.

- **Isolate and/or remove the defective part** Using the multimeter, isolate and/or remove the part, set the range to ohms, and check for component failure. This will be further explained in Chapter 6 and then in Chapters 14 through 30.

- **Install the new component**   When you find a defective part, replace it with a new original part. Reconnect all the wires in their original places.
- **Reattach all panels and screws**   Close the appliance or air conditioner, and reattach all panels and screws.
- **Test the product for proper operation**   Plug in the appliance or air conditioner and test it.

Technicians Diagnostic Guide

Before a technician begins to service an appliance or air conditioner, he or she must check the following conditions before the service call begins, during the diagnostic phase of the service call, and after the service call is completed:

- Make sure that there is the correct voltage at the receptacle for the appliance or air conditioner to operate correctly.
- Check and see if a fuse has blown or a circuit breaker has tripped.
- Check and see if the appliance or air conditioner has been installed according to the manufacturer's instructions.
- All meter readings should be made with a multimeter (VOM or DVM) with a sensitivity of 20,000 ohms per volt DC or greater.
- Locate the technical data sheet in the product. It is usually located behind the control panel or tucked away on the bottom of the product behind an access panel. The technical data sheet contains the wiring diagram and other useful information needed to complete the repairs.
- On electronic models you will need the actual service manual for the model you are working on to properly diagnose the product. The service manual will assist you in properly placing the product in the service test mode for testing the product functions.
- During the diagnostic part of the service call, check all connections first before you replace any parts. Check for loose connections or burnt wires. During the testing phase you will have to disconnect and reconnect wires. Be careful not to pull on the wires; you might pull the wire off the terminal connector.
- Check all wire harness connectors first. Inspect each wire harness connector and make sure that there are no loose or broken wires on the connectors. Make sure that all wires are pressed into the connector properly.
- Resistance checks must be performed with the service cord removed from the receptacle.
- Voltage checks must be performed with the service cord plugged into the receptacle with the correct voltage present.
- When you complete the service call, make sure that the appliance or air conditioner operates according to the manufacturer's specifications.

PART

II

Electricity, Electronics, and Gas

Electricity

The technician must be knowledgeable in electrical theory to be able to diagnose and repair major appliances and air conditioners properly. Although this chapter cannot cover all there is to know about electricity, it provides the basics. In the field of major appliances and air conditioners, the greatest number of potential problems is in the electrical portions of the product.

Electrical Wiring

The flow of electricity from a power source to the home can be made easier to understand by comparing it to a road map. Electricity flows from a power source to a load. This is similar to a major highway that runs from one location to another.

High-voltage transformers are used to increase voltages for transmission over long distances. The power lines that go to different neighborhoods are like the smaller roads that turn off the major highway. The electricity then goes to a transformer that reduces the voltage going into the home. This is the intersection between the small roads and the medium-sized highways. The small road that goes into the neighborhood, and all the local streets, are like the wiring that goes inside the home.

When all the streets, roads, and highways are connected together, the city is accessible. This is similar to having electricity flowing from the power source to all of the outlets in the home.

Imagine driving down a road and coming to a drawbridge (in this case, the switch), and it opens up. This stops the flow of traffic (electricity). In order for traffic (or electricity) to flow again, the drawbridge must close.

What Is a Circuit?

A *circuit* is a complete path through which electricity can flow and then return to the power source. Figure 6-1 is an example of a complete circuit. To have a complete path (or *closed circuit*), the electricity must flow from Point A to Point B without interruption.

When there is a break in the circuit, the circuit is *open*. For example, a break in a circuit is when a switch is turned to its "off" position. This will interrupt the flow of electricity, or current, as shown in Figure 6-2. When a broken circuit is suspected, it is necessary to discover the location of the opening.

The complete circuit. Current flows from Point A, through the light bulb, and then back to Point B.

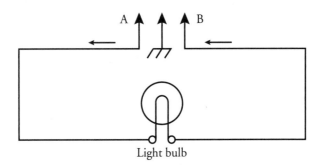

Light bulb

Circuit Components

In an appliance or air conditioner, an electric circuit has four important components:

- **Power source**   This source might be a battery or the electricity coming from the wall outlet. Without the applied voltage, current cannot flow.

- **Conductors**   A conductor will usually be a wire and sometimes the metal chassis (frame). The function of the wire conductor is to connect a voltage source to a load.

- **Loads**   These are the components that do the actual work in the appliance or air conditioner. A load is anything that uses up some of the electricity flowing through the circuit. For example, motors turn the belt, which turns the transmission. That, in turn, turns the agitator in a washing machine. Other examples are heating elements and solenoids.

- **Controls**   These control the flow of electricity to the loads. A control is a switch that is either manually operated by the user of the product or operated by the appliance or air conditioner itself.

Three Kinds of Circuits

You will come across three kinds of circuits: series circuits, parallel circuits, and series-parallel circuits (a combination of series and parallel circuits).

With the switch open, the current flow is interrupted.

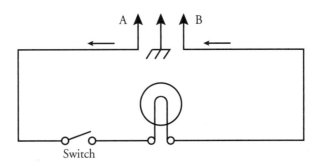

Switch

PART II

FIGURE **6-3**
A series circuit.

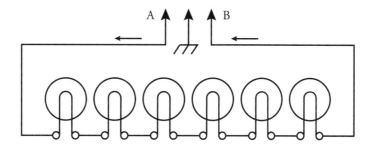

Series Circuit

The components of a series circuit are joined together in successive order, each with an end joined to the end of the next (Figure 6-3). There is only one path that electricity can follow. If a break occurs anywhere in the circuit, the electricity, or current flow, will be interrupted and the circuit will not function (Figure 6-4). Figure 6-5 shows some of the many different shapes of series circuits, all of which are used in wiring diagrams. In each series circuit, there is only one path that electricity can follow. There are no branches in these circuits where current can flow to take another path. Electricity only follows one path in a series circuit.

Parallel Circuits

The components of a parallel circuit are connected across one voltage source (Figure 6-6). The voltage to each of these branches is the same. The current will also flow through all the branches at the same time. The amount of current that will flow through each branch is determined by the load, or resistance, in that branch.

Figure 6-7 and Figure 6-8 show examples of parallel circuits. If any branch has a break in it, the current flow will only be interrupted in that branch. The rest of the circuits will continue to function.

Series-Parallel Circuits

A series-parallel circuit is a combination of series circuits and parallel circuits. In many circuits, some components are connected in series to have the same current, but others are in parallel for the same voltage (Figure 6-9). This type of circuit is used where it is necessary

FIGURE **6-4**
If there is a break in the wiring, all of the light bulbs will be off.

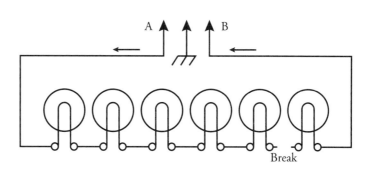

Break

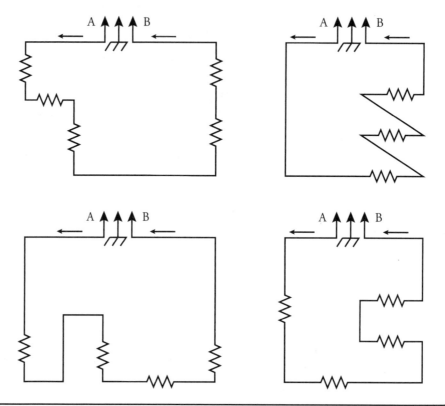

FIGURE 6-5 When you look at wiring diagrams, you will find series circuits in all sorts of shapes.

to provide different amounts of current and voltage from the main source of electricity that is supplied to that appliance.

Series and parallel rules apply to this type of circuit. For example, if there is a break in the series portion of the circuit (Figure 6-10), the current flow will be interrupted for the entire circuit. If the break is in the parallel portion of the circuit (Figure 6-11), the current will be interrupted for only that branch of the circuit. The rest of the circuits will still function.

FIGURE 6-6
Parallel circuit.

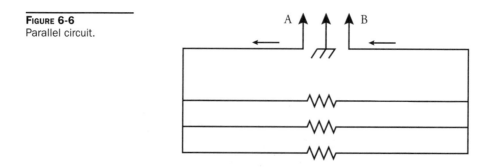

FIGURE 6-7
Another parallel
circuit.

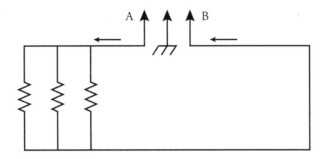

FIGURE 6-8
Notice that as in
series circuits, the
same parallel circuit
can be drawn in many
different ways.

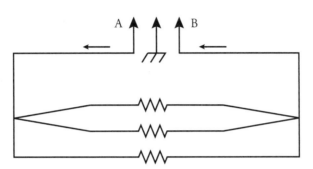

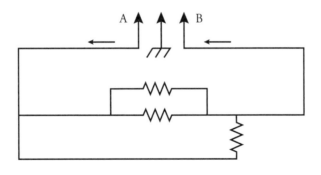

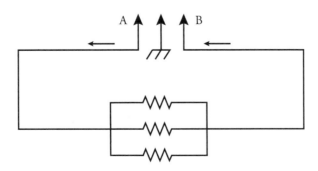

PART II

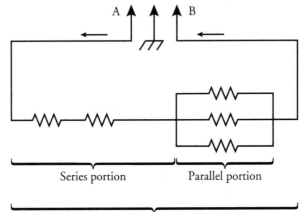

Figure 6-9
Series-parallel circuit.

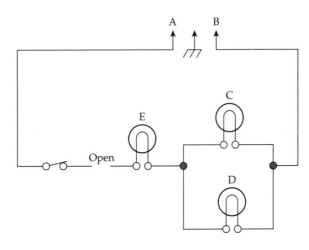

Figure 6-10
Series-parallel circuit
with open in the
series portion of the
circuit. No current flow
in the entire circuit.

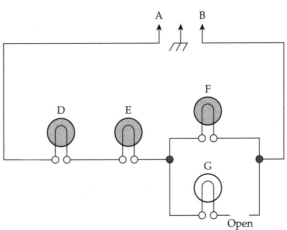

Figure 6-11
Series-parallel circuit
with open in the
parallel portion of the
circuit. Current flows
through D, E, and F,
but not through G.

Types of Shorts in a Circuit

When a short occurs in a series circuit, it means there is zero resistance across an electrical source. Figure 6-12 shows a wire across load C. This short, which is the path of least resistance, allows the current to flow through the wire instead of the load.

A short can be added to part of the series circuit to bypass one or more of the loads in the circuit. This type of short is referred to as a "shunt." Manufacturers can design this shunt into a circuit. The difference between a "short" and a "shunt" is as follows:

- With a short, there is no resistance to the flow of current in a circuit.
- A shunt will form a bypass around the load(s), but the circuit will still offer resistance to the flow of current.

In Figure 6-13, if a short is connected in the series circuit between C and D, the loads E, F, and G would by bypassed because they are shorted out of the circuit. Electricity takes the path of least resistance.

In Figure 6-14, if a permanent shunt was connected in the series circuit between points C and D, loads F and G would be bypassed because they were shorted out of the circuit.

In Figure 6-15, if the manufacturer designed a load to be turned on and off, they will install a switch to act as a shunt across a load. When switch X is closed, load G is turned off, allowing loads E and F to remain on.

In Figure 6-16, the parallel circuit will allow current to flow from point A through the loads C and D, and back to point B.

In the parallel circuit in Figure 6-17, a short was placed between points X and Y, shorting out loads C and D. The current will flow from point A through the loads and through the shorted wire back to point B. Remember, electricity takes the path of least resistance.

In the series-parallel circuit in Figure 6-18, the current will flow from point A, through the short beginning at point C, to point D, and back to point B, bypassing loads E, F, G, and H. Keep in mind that electricity takes the path of least resistance.

In Figure 6-19, current will flow through the series portion of the circuit from point A through the load C. In the parallel portion of the circuit, current will flow from load C through the wire (shunt) across load D, and back through point B. The three loads would be bypassed. Electricity takes the path of least resistance.

FIGURE 6-12
A wiring short in a series circuit.

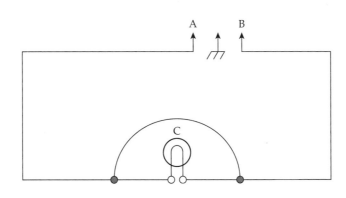

FIGURE 6-13
A series circuit with a short across loads E, F, and G.

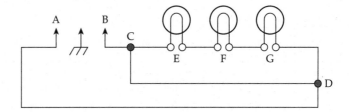

FIGURE 6-14
A series circuit with a permanent shunt across loads F and G.

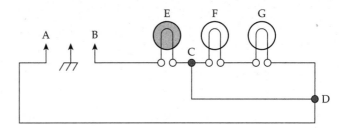

FIGURE 6-15
A series circuit with a switch acting as a temporary shunt across load G.

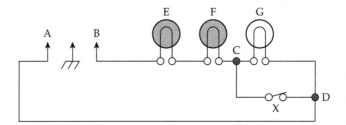

FIGURE 6-16
A parallel circuit with two loads.

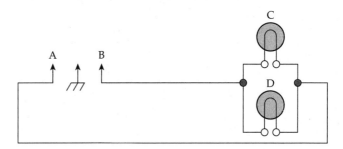

FIGURE 6-17
A parallel circuit with a short between loads C and D.

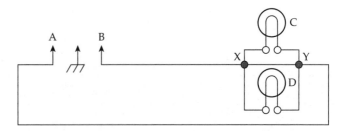

FIGURE 6-18
A series-parallel circuit
with a short.

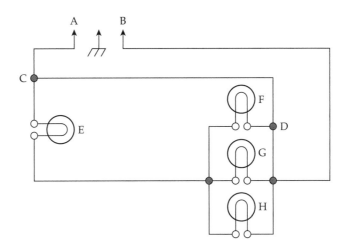

Strip Circuits

A strip circuit is a piece of the whole circuit; it is easy to use as a diagnostic tool. When you compare a strip circuit (Figure 6-20) to a complete standard wiring diagram, you will notice many different components. In the wiring diagram, it can be confusing to figure out the active circuits to diagnose. In the strip circuit in Figure 6-20, only the circuits needed for an active broil and bake circuit are shown. You read a strip circuit from left to right, which makes it easier to read and to diagnose the circuit.

FIGURE 6-19
A series-parallel circuit
with a shunt.

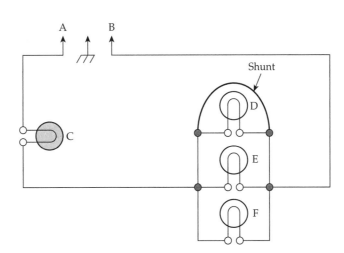

Broil Circuit

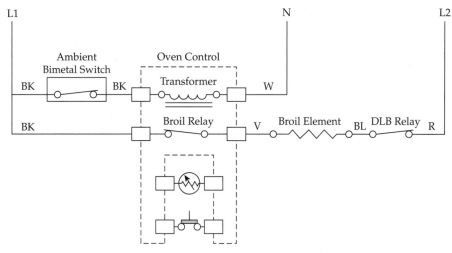

Bake Circuit

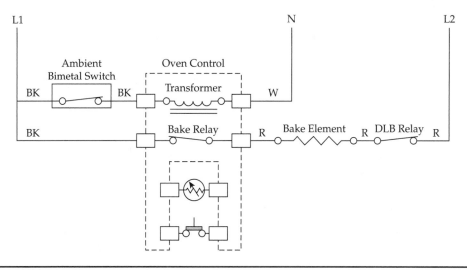

Figure 6-20 A broil and bake strip circuit.

Types of Electric Current

There are two types of electric current:

- Direct current (DC) flows continuously in the same direction (Figure 6-21).

- Alternating current (AC) flows in one direction and then reverses itself to flow in the opposite direction along the same wire. This change in direction occurs 60 times per second, which equals 60 Hz (Figure 6-22).

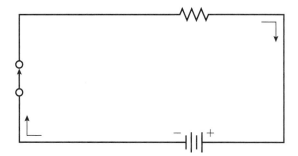

FIGURE 6-21
A simple DC electrical circuit. Current flows from the negative side of the battery through the switch and load, and back to the positive side of the battery.

Direct current (DC) is used in automobile lighting, flashlights, and cordless electric appliances (such as toothbrushes, shavers, and drills), and in some major appliances and air conditioners.

Alternating current (AC) is used in most homes. This current can be transmitted more economically over long distances than direct current can. Alternating current can also be easily transformed to higher or lower voltages.

Theory of Current Flow

When servicing electronic solid-state circuits in major appliances or air conditioners, the technician should be aware that explanation of the circuit is often given assuming conventional current flow as opposed to electron flow. Conventional current flow theory states that current flow is from positive to negative. Electron current flow theory states that current flows from negative to positive.

Ohm's Law

Ohm's Law states ($I = E/R$): The current that flows in a circuit is directly proportional to the applied voltage and inversely proportional to the resistance. So, an increase in the voltage will increase the current as long as the resistance is held constant. Alternately, if the resistance in a circuit is increased and the voltage does not change, the current will decrease.

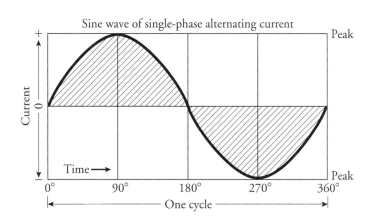

FIGURE 6-22
A waveform of a single-phase alternating current.

Sine wave of single-phase alternating current

FIGURE 6-23
An ohmmeter
connected to read
resistance.

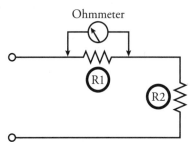

Ohmmeter

The second version states (E = I × R): It can be seen from this equation that if either the current or the resistance is increased in the circuit (while the other is unchanged), the voltage will also have to increase.

The third version states (R = E/I): If the current is held constant, an increase in voltage will result in an increase in resistance. Alternately, an increase in current while holding the voltage constant will result in a decrease in resistance.

In any of the three formulas, when two elements of the electric circuit are known, the unknown factor can be calculated.

Ohms

Resistance is measured in ohms and opposes the flow of electrons (current). An instrument that measures resistance is known as an ohmmeter. Figure 6-23 is a schematic showing an ohmmeter connected to read the resistance of R1. The resistance of any material depends on its type, size, and temperature. Even the best conductor offers some opposition to the flow of electrons. Figure 6-24 shows another type of meter, the digital multimeter, used for measuring

FIGURE 6-24
A digital multimeter
for measuring ohms.

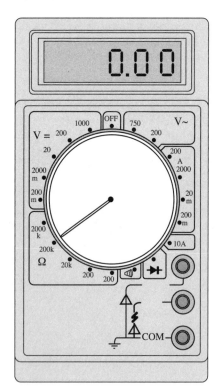

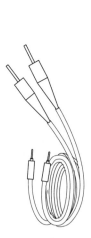

ohms. The fundamental law used to find resistance is stated as follows: The resistance (R) in ohms is equal to the potential difference measured in volts (V), divided by the current in amperes (A). The equation is: $R = V \div A$.

Amperes

Current is measured in amperes. The term *ampere* refers to the number of electrons passing a given point in one second. When the electrons are moving, there is current. The ammeter is calibrated in amperes, which we use to check for the amount of current in a circuit.

FIGURE 6-25
An ammeter connected in a circuit, measuring amperes.

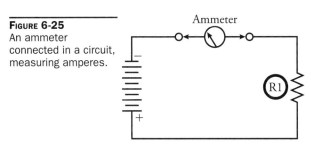

Ammeter

An instrument that measures amperes is known as an ammeter. Figure 6-25 is a schematic showing an ammeter connected in a circuit to measure the current in amperes. Figure 6-26 shows an ammeter that is used in diagnosing electrical problems with appliances. Current is the factor that does the work in the circuit (light the light, ring the buzzer). The fundamental law to find current is stated as follows: The current in amperes (A) is equal to the potential difference measured in volts (V), divided by the resistance in ohms (R). The equation is: $A = V \div R$.

Volts

Electromotive force is measured in volts. This is the amount of potential difference between two points in a circuit. It is this difference of potential that forces current to flow in a circuit. One volt (potential difference) is the electromotive force required to force one ampere of current through one ohm of resistance.

FIGURE 6-26
In the ammeter, the jaws clamp around a wire to measure the amperage of a circuit.

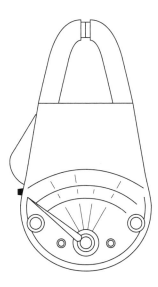

FIGURE 6-27
A voltmeter connected
in a circuit to measure
voltage.

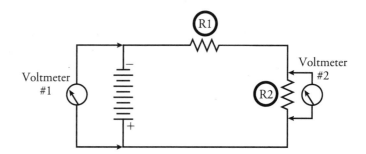

An instrument that measures voltage is known as a voltmeter. Figure 6-27 is a schematic showing a voltmeter connected in the circuit to measure the voltage. Voltmeter #1 is connected to read the applied (or source) voltage. Voltmeter #2 is connected to measure the voltage drop, or potential difference, across R2. Figure 6-28 shows an actual volt-ohm-milliammeter (VOM) that is used in measuring voltage.

The fundamental law to find voltage is stated as follows: The potential difference measured in volts (V) is equal to the current in amperes (A) multiplied by the resistance in ohms (R). The equation is: $V = A \times R$.

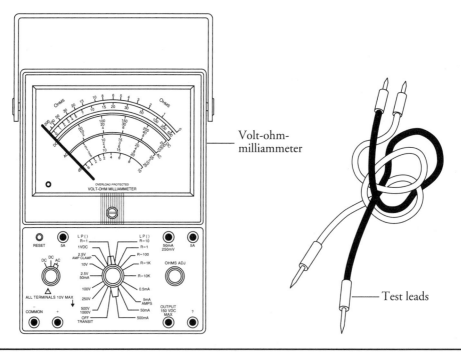

FIGURE 6-28 The volt-ohm-milliammeter with test leads.

FIGURE 6-29
The wattmeter is used
to measure watts.

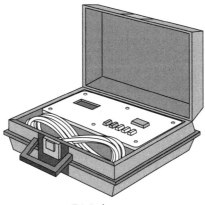

Digital wattmeter

Watts

Power is measured in watts, and an instrument that measures watts is known as a wattmeter (Figure 6-29). One watt of power equals the work done in one second, by one volt of potential difference, in moving one coulomb of charge. One coulomb per second is equal to one ampere. Therefore, the power in watts (W) equals the product of amperes (A) times volts (V). The equation is: $W = A \times V$.

Ohm's Law Equation Wheel

The Ohm's Law equation wheel in Figure 6-30 shows the equations for calculating any one of the basic factors of electricity. Figure 6-31 shows the cross-reference chart of formulas as used in this text. If you know any two of the factors (V = voltage, A = amperage, R = resistance, W = power), you can calculate a third. To obtain any value in the center of the equation wheel for direct or alternating current, perform the operation indicated in one segment of the adjacent outer circle.

FIGURE 6-30
The Ohm's Law
equation wheel.

Conversion chart for determining amperes, ohms, volts, or watts
(Amperes = A, Ohms = Ω, Volts = V, Watts = W)

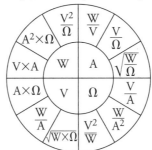

Term	Measured in	Referred to in formulas as	Identification used in this text
Amperage	Amperes (Amps)	I	A for amps
Current	Amperes (Amps)	I	A for amps
Resistance	Ohms	Ω or R	R for resistance
Voltage	Volts	V or E	V for volts
Electromotive force	Volts	V or E	V for volts
Power	Watts	W	W for watts

Figure 6-31   The cross-reference chart of formulas.

Example 1: A 2400-watt heating element is connected to a 240-volt circuit. How many amps does it draw?

When finding amperage, the formula will be found in the Amperes section of the wheel.

$$\frac{W \text{ (watts)}}{V \text{ (volts)}} = A \text{ (amps)}$$

Then, solving for amperage:

$$\frac{2400 \text{ (watts)}}{240 \text{ (volts)}} = 10 \text{ amps}$$

What is the resistance (ohms)?

$$\frac{V^2 \text{ (volts squared)}}{W \text{ (watts)}} = \text{ohms}$$

Then, solving for resistance:

$$\frac{240^2 \text{ (volts squared)}}{2400 \text{ (watts)}} = 24 \text{ ohms}$$

Example 2: What is the resistance of a 100-watt resister if the voltage is 120 volts and the current is 0.83 amps?

When finding resistance, the formula will be found in the Ohms section of the wheel.

$$\frac{V \text{ (volts)}}{A \text{ (Amps)}} = R \text{ (Resistance)}$$

$$\frac{120 \text{ (volts)}}{0.83 \text{ (amps)}} = R \text{ (Resistance)}$$

$$145 \text{ Ohms (resistance)}$$

Example 3: What is the voltage of a circuit if the resistance of the load is 48 Ohms and current is 5 amps?

$$A \text{ (Amps)} \times R \text{ (Ohms)} = V \text{ (Volts)}$$
$$5 \text{ (amps)} \times 48 \text{ Ohms} = 240 \text{ Volts}$$

Wiring Diagram Symbols

These symbols are commonly used in most wiring diagrams. Study each symbol so that you can identify it by sight (Table 6-1, Table 6-2, Table 6-3, and Table 6-4).

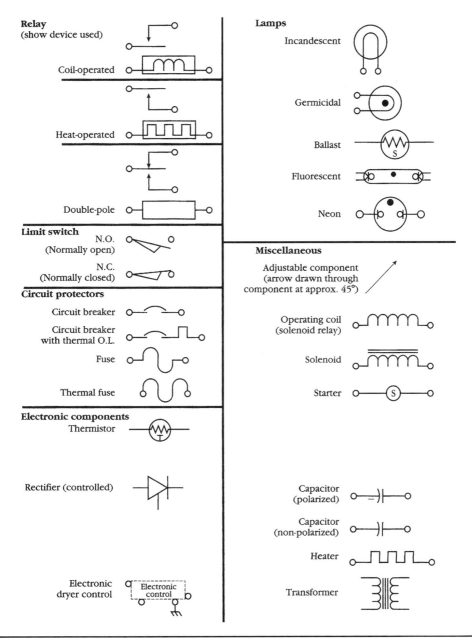

TABLE 6-1   Wiring Diagram Symbols

Temperature-actuated components

(Note: Symbols shown to be used for thermostats, bimetal switches, overload protectors, or other similar components, as required.

Temp. actuated
(close on heat rise)

Temp. actuated
(open on heat rise)

S.P.S.T.
(open on heat rise)

S.P.D.T.

S.P.D.T.

S.P.S.T.
(two contacts)

S.P.S.T. (adj.)
(close on heat rise)

S.P.D.T. (adj.)

S.P.S.T. (adj.)
(open on heat rise)

S.P.D.T. (adj.)
(with aux. "off" contacts)
(typical example)

S.P.S.T. (with internal heater)
(close on heat rise)

S.P.S.T. (with internal heater)
(open on heat rise)

Combination devices

Relay-magnetic
(arrangement of
contacts as necessary
to show operation)

Relay-thermal
(arrangement of
contacts as necessary
to show operation)

Timer (defrost)

Manual and mechanical switches

Normally closed (S.P.S.T.)
(single-pole, single-throw)

Normally open (S.P.S.T.)
(single-pole, single-throw)

Transfer (S.P.D.T.)
(single-pole, double-throw)

Multi position

Number of terminals

Timer switch

Automatic switch

N.O.
(normally open)

N.C.
(normally closed)

Integral switch
(timer, clock, etc.)

Push button switch

(momentary or spring return)

Circuit closing
N.O. (normally open)

Circuit opening
N.C. (normally closed)

Two circuit

SPDT
(single-pole, double-throw)

TABLE 6-2 Wiring Diagram Symbols

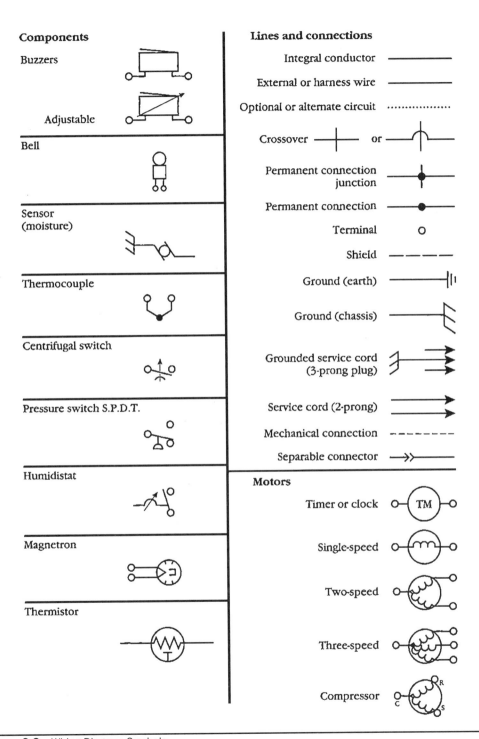

Components

Buzzers

Adjustable

Bell

Sensor (moisture)

Thermocouple

Centrifugal switch

Pressure switch S.P.D.T.

Humidistat

Magnetron

Thermistor

Lines and connections

Integral conductor

External or harness wire

Optional or alternate circuit

Crossover   or

Permanent connection junction

Permanent connection

Terminal

Shield

Ground (earth)

Ground (chassis)

Grounded service cord (3-prong plug)

Service cord (2-prong)

Mechanical connection

Separable connector

Motors

Timer or clock   TM

Single-speed

Two-speed

Three-speed

Compressor

TABLE 6-3 Wiring Diagram Symbols

Electronic wiring symbols

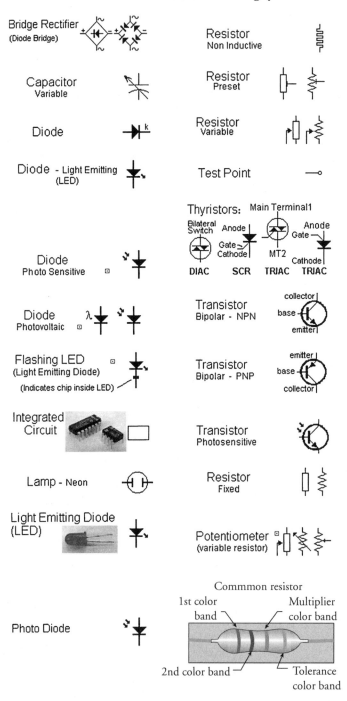

Bridge Rectifier (Diode Bridge)		Resistor Non Inductive	
Capacitor Variable		Resistor Preset	
Diode		Resistor Variable	
Diode - Light Emitting (LED)		Test Point	
Diode Photo Sensitive		Thyristors: Bilateral Switch / DIAC, SCR, TRIAC, TRIAC	
Diode Photovoltaic		Transistor Bipolar - NPN	
Flashing LED (Light Emitting Diode) (Indicates chip inside LED)		Transistor Bipolar - PNP	
Integrated Circuit		Transistor Photosensitive	
Lamp - Neon		Resistor Fixed	
Light Emitting Diode (LED)		Potentiometer (variable resistor)	
Photo Diode		Common resistor	

TABLE 6-4 Electronic Wiring Diagram Symbols

Terminal Codes

Terminal codes are found on all wiring diagrams. To help you identify the color codes, they are listed in Table 6-5.

Terminal Color Code	Harness Wire Color
BK	Black
BK-Y	Black with yellow tracer
BR	Brown
BR-O or BR-OR	Brown with orange tracer
BR-R	Brown with red tracer
BR-W	Brown with white tracer
BU or BL	Blue
BU-BK or BL-BK	Blue with black tracer
BU-G or BU-GN	Blue with green tracer
BU-O or BU-OR	Blue with orange tracer
BU-Y	Blue with yellow tracer
G or GN	Green
B-Y or GN-Y	Green with yellow tracer
G-BK	Green with black tracer
GY	Gray
GY-P or GY-PK	Gray with pink tracer
LBU	Light blue
O or OR	Orange
O-BK or OR-BK	Orange with black tracer
P or PUR	Purple
P-BK or PUR-BK	Purple with black tracer
P or PK	Pink
R	Red
R-BK	Red with black tracer
R-W	Red with white tracer
T or TN	Tan
T-R	Tan with red tracer
V	Violet
W	White
W-BK	White with black tracer
W-BL or W-BU	White with blue tracer
W-O or W-OR	White with orange tracer
W-R	White with red tracer
W-V	White with violet tracer
W-Y	White with yellow tracer
Y	Yellow
Y-BK	Yellow with black tracer
Y-G or Y-GN	Yellow with green tracer
Y-R	Yellow with red tracer

TABLE 6-5 Terminal Codes

Timer Sequence (Esterline) Charts

Figure 6-32 represents a sample timer sequence chart. Dishwashers, washers, and dryers have an Esterline Chart, which is also known in the appliance industry as a Timer Sequence Chart, and is part of the technical data sheet along with the wiring schematic, and other

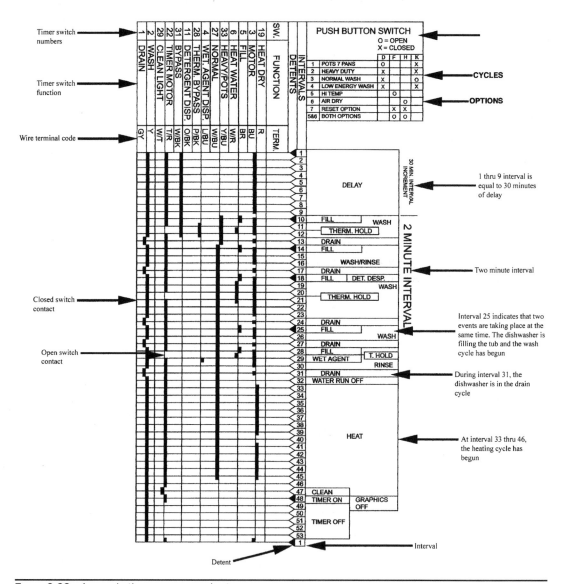

Figure 6-32   A sample timer sequence chart.

helpful information for the technician to use in diagnosing the appliance. This chart indicates the switch contacts that are open or closed at a specific time (interval) during the operating cycle. It is another diagnostic tool that is used in conjunction with the wiring schematic by technicians to assist in diagnosing problems with the appliance.

The timer sequence chart is divided into the following:

- Overview of all of the columns on the chart.

- Correspondence between the timer dial and the intervals on the timer sequence chart.

- The consumer selects the additional options through the push button switch that will affect the switch contacts opening and closing on the wiring schematic.

- At a certain interval in the operating cycle, the timer sequence chart will show the technician the position of the switch contacts that are open or closed on the wiring schematic.

- The timer sequence chart will indicate which components are energized at any time interval.

Wiring Diagrams

In this section, each of the four examples presented will take you through a step-by-step process in how to read wiring diagrams.

Example 1: Take a look at a simple wiring diagram for a refrigerator (Figure 6-33). Note the black wire on the diagram. This is the wire that goes to the temperature control. The circuit is

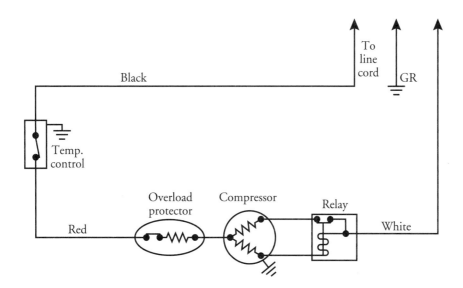

FIGURE 6-33 A simple wiring schematic of a refrigerator circuit.

not energized when the temperature control is in the "off" position. When the temperature control knob is turned to the "on" position (switch contacts closed) and the circuit is energized, current will flow through the temperature control, through the red wire, through the overload protector, through the compressor and the relay, and back through the white wire to the line cord.

Example 2: In the wiring diagram in Figure 6-34, trace the active circuits with your finger. Note switch number 1. This is the first switch in the line, and it is the main switch that supplies voltage to the timer. No circuits are energized when the timer dial is in the "off" position (as the diagram indicates). When the user selects the wash cycle with a warm wash and turns on the timer, switch contacts 1, 2, 3, 6, and 7 are closed. (Use a pencil to close the switches on the diagram.) Voltage is supplied to the water level switch and to the hot and cold water valves. Warm water is now entering the washing machine tub. When the water

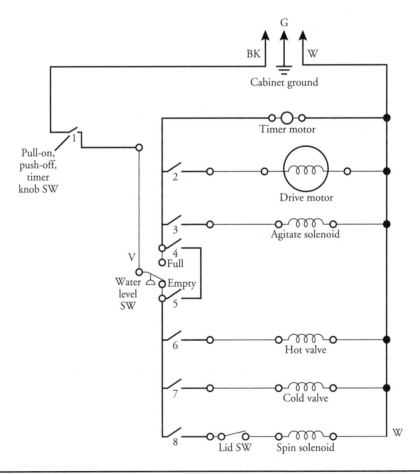

FIGURE 6-34 The wiring schematic of an automatic washer circuit.

level reaches the selected position, the water level switch contacts close from V to number 4, indicating that the water in the tub has reached the selected water level. When the water level switch is in this position, the water is turned off. Voltage is now supplied to the timer motor, drive motor, and the agitate solenoid. The washing machine is now agitating and cleaning the clothes. As the timer advances to the spin cycle, timer switch contacts 2 and 3 open, thus turning off the agitate solenoid and the drive motor. Switch contacts 8 and 2 now close, thus supplying voltage to the drive motor and the spin solenoid. The timer motor advances to the end of the cycle, and timer switch contacts 1, 2, and 8 will open. The washing machine is now off, and no circuits are energized.

Example 3: The wiring diagram in Figure 6-35 is for a refrigerator. Assume that the thermostat is calling for cooling and the compressor is running. With your finger, trace the active circuits. The thermostat in the wiring diagram for the refrigerator is closed. The evaporator fan motor and the condenser fan motor are running. Voltage is supplied through the overload to the relay. Current is flowing through the relay coil to the compressor-run-winding. Also notice that the door switch is open and the refrigerator light is off. When the temperature in the refrigerator satisfies the thermostat, the thermostat switch contacts will open, thus turning off the compressor, the evaporator fan motor, and the condenser fan motor.

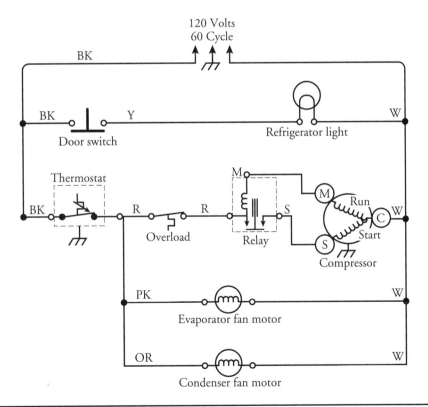

FIGURE 6-35   The wiring schematic of a refrigerator circuit.

Example 4: The wiring diagram in Figure 6-36 is for a no-frost refrigerator. Note the defrost timer in the lower-left part of the diagram. The defrost timer switch contact is closed to contact 4, the thermostat is calling for cooling, and the compressor is running. Trace the active circuits with your finger. Voltage is supplied to the defrost timer terminal 1. Current will flow through the defrost 2 timer motor to the white wire and then back to the line cord. At the same time, current flows through contact 4 in the defrost timer to the thermostat. At this point, the current passes through the thermostat to a junction and splits in two directions. Current will flow through the temperature control, through the overload protector, through the compressor and the relay, and back through the white wire to the line cord. At the same time, the compressor is running and current will flow through the evaporator and condenser fan motors.

When the defrost timer activates the defrost cycle, the defrost timer switch contact is closed to contact 2, and the compressor is not running. The evaporator and condenser fan

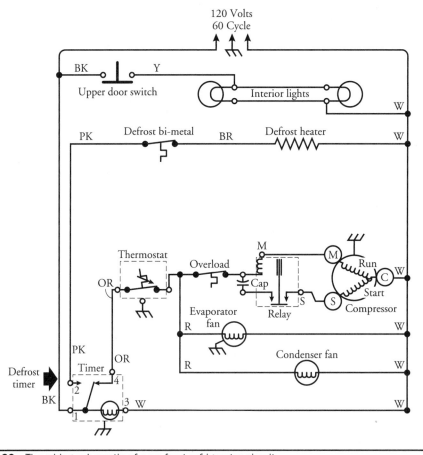

Figure 6-36 The wiring schematic of a no-frost refrigerator circuit.

motors will also stop running. The current will flow from terminal 2, through the defrost bi-metal, through the defrost heater, and back through the white wire to the line cord. With your finger, trace the active circuits.

Sample Wiring Diagrams

The diagram in Figure 6-37 is known as a ladder diagram. It is a simplified diagram using symbols for parts and control components attached to wires. The timer is represented by a dotted line running vertically through the switch contacts. To read and understand this type of diagram, assume that the complaint is that the dryer is not heating. Look at the section of the diagram marked HEAT. Starting on the left side of the diagram, trace the circuit with your finger. You will notice that L1 goes to one side of the timer switch. As your finger moves from left to right, you will pass over the cycling thermostat, high-limit thermostat, heater, motor centrifugal switch (CS), and on to L2. These are the components that make up the heating circuit. You must also include the motor and the door switch. If the door switch should fail, the motor will not run, thus opening the motor centrifugal switch (CS). If the motor fails, the high-limit thermostat will open, shutting off the heater element. If any of these components fail, the dryer will not dry the clothing.

The diagrams in Figure 6-38 and Figure 6-39 illustrate a pictorial diagram and a schematic diagram. The pictorial diagram shows the actual picture of the components, and the schematic diagram uses symbols for the components. Figure 6-40 shows a pictorial diagram of a refrigerator. In this type of diagram, you can see where the components are actually located.

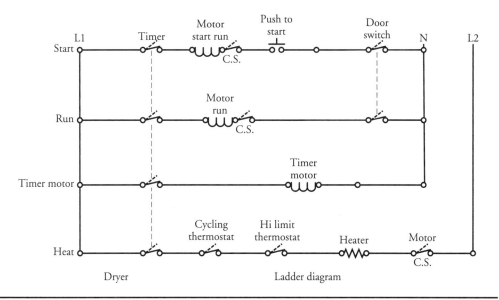

Figure 6-37 A simple ladder diagram.

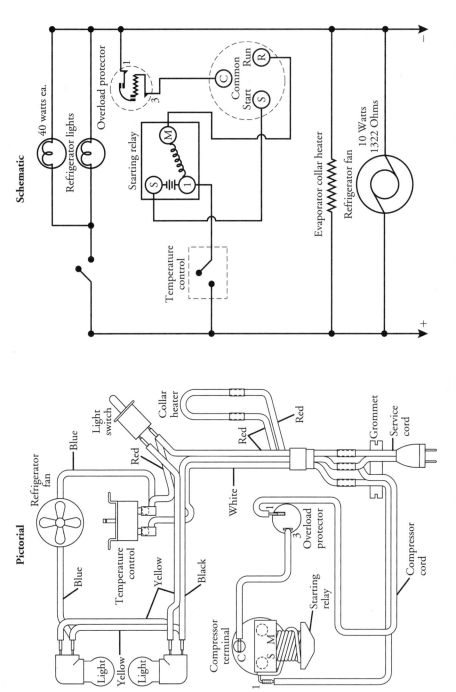

FIGURE 6-38 A pictorial diagram and a schematic diagram. Both show the same components.

Schematic   **Pictorial**

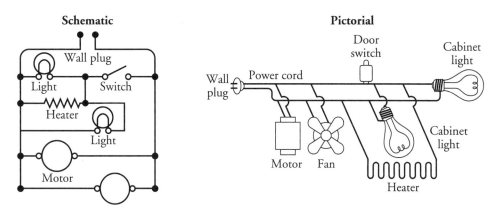

FIGURE 6-39   A pictorial diagram and schematic diagram.

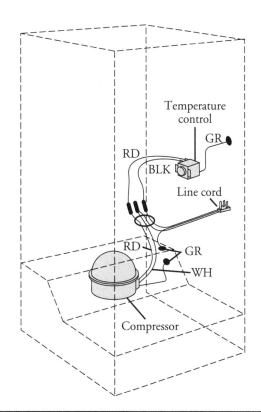

WIRE CODE	
COLOR	CODE
RED	RD
WHITE	WH
BLACK	BLK
ORANGE	OR
GREEN	GR
BLUE	BLU

FIGURE 6-40   A pictorial diagram of a refrigerator, showing where the components are located.

How to Read the Volt-Ohm-Milliammeter Test Instrument

The volt-ohm-milliammeter (VOM) is sometimes called a multimeter because it can perform more than one function. A typical VOM will allow you to measure voltage, resistance, and current:

- Voltage equals electromotive force.
- Resistance equals the amount of resistance holding back the flow of current (measured in ohms).
- Current equals the amount of electricity flowing through a wire or circuit component (measured in amperes). There are many different types and brands of VOMs (Figure 6-41). However, all VOMs are used in the same way to measure voltage, current, or resistance. Most VOMs will have the following:
  - **Test leads**   These are the wires coming from the meter to the part being tested.
  - **Meter scales and pointer (or a digital display on a digital meter)**   These show the amount of whatever value you are measuring.
  - **Function switch**   This allows you to select whether you will be measuring AC or DC voltage (volts), current (amps), or resistance (ohms).
  - **Range selector switch**   This allows you to select the range of values to be measured. On many meters, you can select functions and ranges with the one switch (as in the meter pictured in Figure 6-41a and Figure 6-41c).

Measuring Voltage

If you don't have the right voltage in an appliance or air conditioner, the product won't function properly. You can find out whether the product is getting the right voltage by measuring the voltage at the wall outlet (receptacle). If an appliance or air conditioner isn't getting the proper voltage, nothing else you do to fix it will help.

For your safety, before using any test instrument, it is your responsibility as a technician to read and understand the manufacturer's instructions on how the test instrument operates.

Making the Measurement

Making voltage measurements is easy once you know how to select and read the scales on your meter. When measuring voltage, you should perform the following steps:

- Attach the probes (another term for *test leads*) to the meter. Plug the black probe into the meter jack marked negative or common. Plug the red probe into the positive outlet.
- Set the function switch to AC VOLTS.
- Select a range that will include the voltage you are about to measure (higher than 125 Vac if you are measuring 120 volts; higher than 230 Vac if you are measuring 220 volts). If you don't know what voltage to expect, use the highest range, and then switch to a lower range if the voltage is within that amount.

FIGURE **6-41**
(a) Volt-ohm-
milliammeter, (b) VOM
and ammeter
multimeter
combination, (c) digital
multimeter.

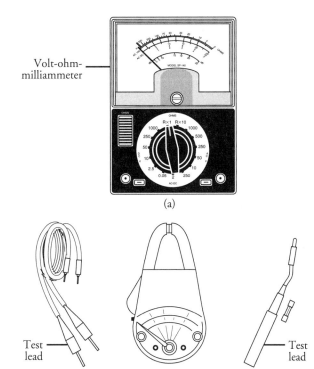

Volt-ohm-
milliammeter

(a)

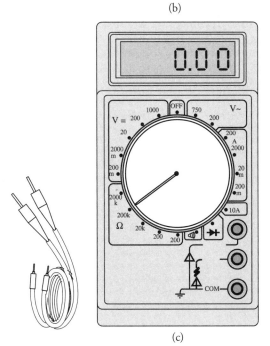

Test lead

Test lead

VOM and ammeter multimeter combination

(b)

(c)

- Touch the tips of the probes to the terminals of the part to be measured.
- Read the scale.
- Decide what the reading means.

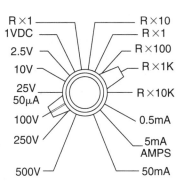

FIGURE 6-42
An example of scales used on some VOMs.

Selecting the Scales

To measure the voltage in appliances and air conditioners, use the AC/DC scales (Figure 6-42). These same scales are used for both AC and DC readings. The numbers on the right side of the AC/DC scales tell you what ranges are available to you. The meter face in Figure 6-43 has three scales for measuring voltage: one is marked 10, another 50, and the third is marked 250. Remember—the scale you read is determined by the position of the range switch.

EXAMPLE *If the range is set on 250 V, as in Figure 6-44, you read the 0- to 250-V scale.*

Reading the AC Voltage Scale

When reading the pointer position, be sure to read the line marked AC (Figure 6-43). The spaces on the voltage scales are always equally divided. When the pointer stops between the marks, just read the value of the nearest mark. In Figure 6-45, the pointer is between 115 and 120 volts on the 250 scale. Read it as 120 volts. With an analog meter, you're not gaining anything by trying to read the voltage exactly.

Measuring Line Voltage

Measuring line voltage is the first, and most important, part of checking out an appliance or air conditioner that does not operate. Line voltage is the voltage coming from the wall

FIGURE 6-43
The meter face of an analog meter.

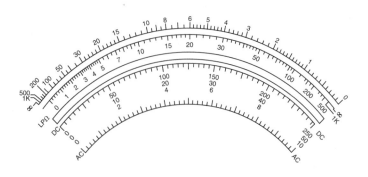

FIGURE 6-44
The range is set on
the 250-V scale.

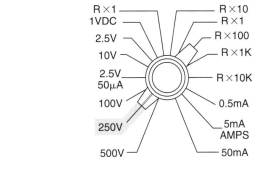

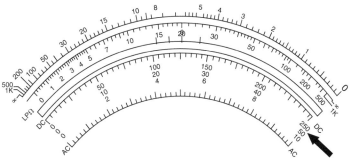

FIGURE 6-45
The range is set
on the 250-V scale,
and the pointer
reads 120V.

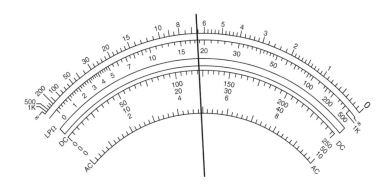

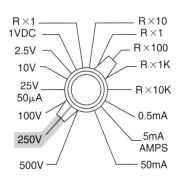

outlet. There should be approximately 120 volts AC at the outlet under "no-load" conditions. No-load means that no appliance or air conditioner is connected or that an appliance or air conditioner is connected but it is turned off.

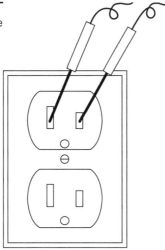

Figure 6-46
Measuring line voltage with no load.

To measure line voltage under no-load conditions at 120 volts (Figure 6-46):

1. Set the meter to measure AC volts.

2. Set the range selector to the range nearest to but higher than 120 volts.

3. Insert either test lead into one slot of an empty wall receptacle.

4. Insert the other test lead into the other slot of the same outlet. (Disregard the ground terminal for this test.) *Warning:* Do not touch or handle the test leads by the metal portion of the probe. Hold the probe by the plastic grips that are attached to the test leads to avoid electric shock.

5. Read the meter. The reading should be between 115 and 120 volts.

6. When testing for 240 volts, be sure that the range selector is set to the nearest range higher than 240 volts. *Note:* Most appliances and small BTU-sized air conditioners are rated at 120 volts, but will work on voltages ranging from 110 to 125 volts. If the voltage drops more than 10 percent, the appliance or air conditioner will not operate, and most likely will damage some electrical components if the appliance or air conditioner keeps running.

To measure line voltage under load at 120 volts (Figure 6-47):

1. Be sure that the appliance or air conditioner is plugged into one of the receptacles and that the product is turned on.

2. Follow steps 1–5 for no-load conditions, inserting the test leads into the empty receptacle next to the one into which the product is plugged.

3. Under load conditions (appliance or air conditioner is turned on), your reading will be slightly less than under no-load conditions.

4. When testing for 240 volts, be sure that the range selector is set to the nearest range higher than 240 volts. *Note:* Products with drive motors and compressors, such as automatic washers, dishwashers, trash compactors, refrigerators, air conditioners, and microwaves, should also be tested at the moment of start. If the voltage drops more than 10 percent of the supplied voltage when the motor is started, there is a problem with the electrical supply.

FIGURE **6-47**
Measuring line voltage under load.

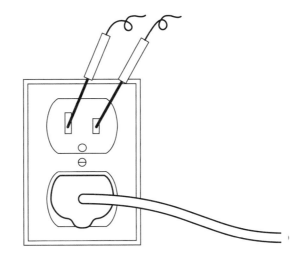

Testing for Ground and Polarity

When a component is grounded to a chassis, there is no voltage between the two. When a chassis is grounded to the earth, there is no voltage between the chassis and the earth. If there is voltage between a chassis and the earth, it's dangerous. If you stand on the earth (dirt, concrete slab, etc.) and touch the chassis, electricity will flow through your body. You could injure yourself, or even someone else, and death could occur. It is important to be sure that the electrical outlets in the home from which appliances or air conditioners are powered are correctly wired. That will protect the users from electrical shock.

If an outlet is wired "backward" (that is, the black or red "hot wire" is connected to the long slot of the outlet), the appliance or air conditioner connected to that outlet might be unsafe to operate, blow the fuse, or trip the circuit breaker. Appliances and air conditioners with solid-state controls will not function properly if the outlets are wired backward.

To test for ground:

1. First, test for line voltage. (If there is no voltage, you can't test for ground.)

2. Notice that the receptacle has a longer and a shorter slot (Figure 6-48). If the outlet has been mounted right-side up, the longer slot will be on the left.

3. Test for voltage between the short slot and the ground receptacle (the round hole), as shown in Figure 6-49. If there is line voltage between these two points, the receptacle is grounded.

4. If there is no round hole (ground prong receptacle), touch one of your probes to the screw that fastens the coverplate to the outlet (Figure 6-50).

To test for polarity:

1. Test to be sure that there is line voltage between the longer and shorter slots (Figure 6-51).

FIGURE 6-48
Receptacle
identification points.

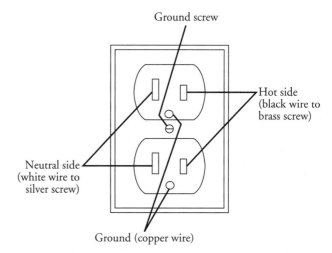

Ground screw

Hot side
(black wire to
brass screw)

Neutral side
(white wire to
silver screw)

Ground (copper wire)

2. Test to be sure that there is line voltage between the short slot and the center screw or the round hole (ground prong receptacle) (Figure 6-50 and Figure 6-51).

3. Test to be sure that there is no voltage between the longer slot and the center screw or the round hole (Figure 6-52 to Figure 6-55). *Note:* If these three tests don't test out this way, the outlet is incorrectly wired and should be corrected by a licensed electrician.

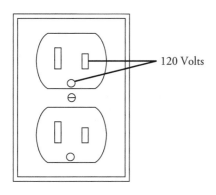

120 Volts

FIGURE 6-49 Testing for ground.

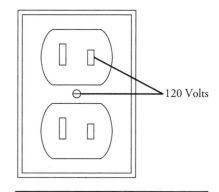

120 Volts

FIGURE 6-50 Testing for ground again.

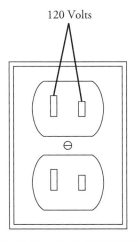

FIGURE **6-51**   Testing for polarity.

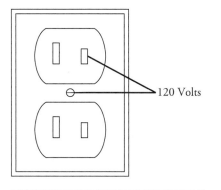

FIGURE **6-52**   Testing the polarity of a receptacle.

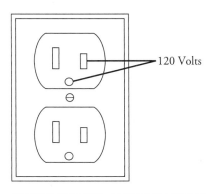

FIGURE **6-53**   Testing the polarity of a receptacle.

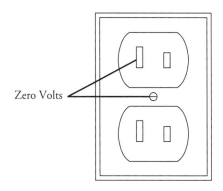

FIGURE **6-54**   Testing the polarity of a receptacle.

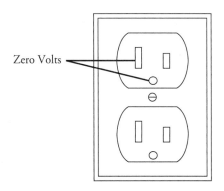

FIGURE **6-55**   Testing the polarity of a receptacle.

Measuring 240-Volt Receptacles

Warning: When testing a three-wire or four-wire receptacle for 240 volts, the technician must be extremely careful not to touch or handle the test leads by the metal portion of the probe. Hold the probe by the plastic grips that are attached to the test leads to avoid electric shock. The multimeter must be set on the AC scale to 240 volts or higher. Do not use the ohms scale—the voltage will damage your meter.

Testing a Three-Wire Receptacle for 240 Volts

To test the voltage on a three-wire receptacle with a no-load condition:

1. Select the correct AC range on your multimeter.

2. Insert the test leads into the receptacle as shown in Figure 6-56. You might have to move the probes around a little to make contact within the receptacle. Check the meter reading—there should be approximately 240 volts. If no voltage is present, check the fuse or circuit breaker.

3. Insert the test leads into the receptacle as shown in Figure 6-57. Check the meter reading—there should be approximately 120 volts. If no voltage is present, check the fuse or circuit breaker.

4. Next, insert the test leads into the receptacle as shown in Figure 6-58. Check the meter reading—there should be approximately 120 volts. If no voltage is present, check the fuse or circuit breaker.

Testing a Four-Wire Receptacle for 240 Volts

The red and black wires attached to the receptacle will each carry 120 volts. The white wire attached to the receptacle is the neutral wire, and the bare wire is the ground wire. This type of receptacle adds additional safety for the consumer and to the product.

FIGURE 6-56
Testing the voltage of a three-wire receptacle for 240 volts.

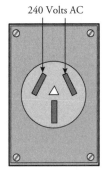 

240 Volts AC

FIGURE 6-57
Testing the voltage
of a three-wire
receptacle for
120 volts.

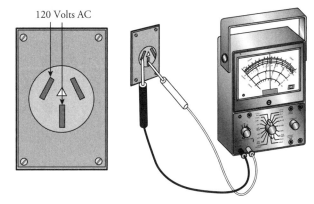

120 Volts AC

FIGURE 6-58
Testing the voltage of
a three-wire receptacle
for 120 volts.

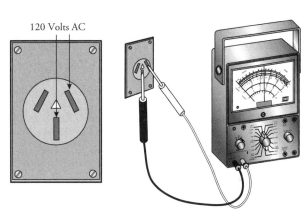

120 Volts AC

To test the voltage on a four-wire receptacle with a no-load condition:

1. Select the correct AC range on your multimeter.

2. Insert the test leads into the receptacle as shown in Figure 6-59. You might have to move the probes around a little to make contact within the receptacle. Check the meter reading—there should be approximately 120 volts. If no voltage is present, check the fuse or circuit breaker.

3. Insert the test leads into the receptacle as shown in Figure 6-60. Check the meter reading—there should be approximately 120 volts. If no voltage is present, check the fuse or circuit breaker.

4. Next, insert the test leads into the receptacle as shown in Figure 6-61. Check the meter reading—there should be approximately 240 volts. If no voltage is present, check the fuse or circuit breaker.

5. Now, insert the test leads as shown in Figure 6-62. The meter should read 0 volts.

FIGURE **6-59**
Testing the voltage of
a four-wire receptacle
for 120 volts.

120 Volts AC

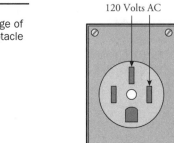

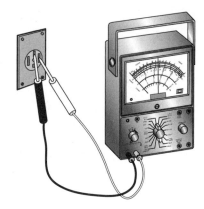

FIGURE **6-60**
Testing the voltage of
a four-wire receptacle
for 120 volts.

120 Volts AC

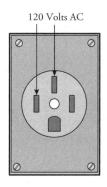

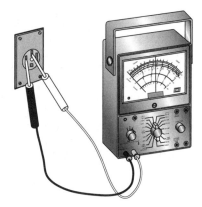

FIGURE **6-61**
Testing the voltage of
a four-wire receptacle
for 240 volts.

240 Volts AC

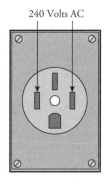

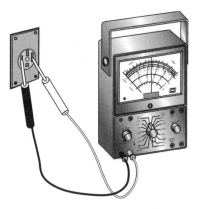

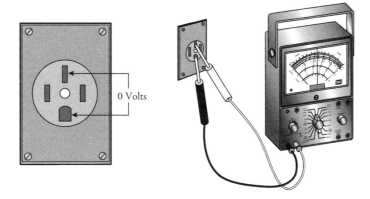

Figure 6-62
Testing the voltage of
a four-wire receptacle
for 0 volts.

0 Volts

Measuring Resistance (OHMS)

Electrical appliances and air conditioners need a complete path around which electricity can flow. If there is infinite resistance to the flow of electricity, you have an open circuit or infinite resistance between the two points being measured. When there is a complete path, you have continuity in the circuit. When you test to find out whether there is a break in the path, you say you are making a continuity check.

Continuity checks are made by measuring the amount of resistance there is to the flow of electricity. If there is so much resistance that it is too high to measure (called "infinite"), then you say that the circuit is open (there is no complete path for the electricity to follow). If there is some resistance, it means that there is continuity, but that there is also one (or more) load on the line—a light, a motor, etc. *Note:* A load is an electrical component that uses electricity to work (for example, a light bulb, a motor, or a heater coil).

If there is no resistance between the two points, it means that the electricity is flowing directly from one point to the other. If the electricity flowed directly from one point to the other by accident or error, then you say you have a short circuit or a "short."

Setting Up the Meter

Measuring resistance (ohms) is like measuring voltage, except that the measurements are made with the electricity turned off. The steps for measuring ohms are as follows:

1. Attach the leads to the meter. Plug the black lead into the negative outlet and the red lead into the positive outlet.

2. If your meter has a function switch, set the function switch to OHMS.

3. If your meter has a range switch, set the range. The range selector switch will have several ranges of resistance (ohms) measurements (Figure 6-63). The ranges are shown like this:

   - $R \times 1$: The actual resistance shown on the meter face times 1.

   - $R \times 10$: The resistance reading times 10 (add one zero to the reading).

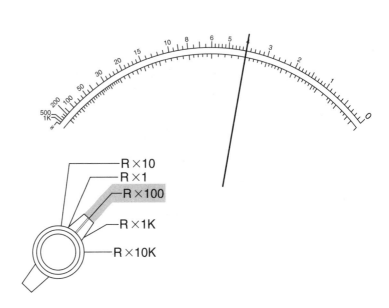

FIGURE 6-63
With the range set on
R × 100, the meter
reads 400 ohms.

- $R \times 100$: The resistance reading times 100 (add two zeros to the reading).
- $R \times 1K$: The resistance reading times 1000 (add three zeros to the reading).
- $R \times 10K$: The resistance reading times 10,000 (add four zeros to the reading).

4. Set the range so that it is higher than the resistance you expect. If you don't know what measurement to expect, use the highest setting and adjust downward to a reading of less than 50. The left side of the scale is too crowded for an accurate reading.

5. Zero the meter. You should do this each time you set it. To "zero the meter" means to adjust the pointer so that it reads 0 when the two test leads are touched together. Use the Ohms Adjust knob on the front of the VOM to line up the pointer over the zero on the ohms scale.

6. Attach the test leads to the component you are measuring.

7. Take the measurement. In Figure 6-63, the range selector switch is on $R \times 100$, and the measurement is 400 ohms.

Electrical Safety Precautions

Know where and how to turn off the electricity to your appliance or air conditioner—for example: plugs, fuses, circuit breakers, or cartridge fuses. Know their location in the home and label them. When replacing parts or reassembling the appliance or air conditioner, you should always install the wires on their proper terminals according to the wiring diagram. Then check to be sure that the wires are not crossing any sharp areas, are pinched in some way, or are between panels or moving parts that might cause an electrical problem.

These additional safety tips can help you and your family:

- Always use a separate, grounded electrical circuit for each major appliance or air conditioner.

- Never use an extension cord for major appliances.
- If you have to use an extension cord for a room air conditioner, make sure to use one that is properly rated for the size air conditioner installed.
- Be sure that the electricity is off before working on the appliance or air conditioner.
- Never remove the ground wire of a three-prong power cord, or any other ground wires from the appliance or air conditioner.
- Never bypass or alter any appliance or air conditioner switch, component, or feature.
- Replace any damaged, pinched, or frayed wiring that might be discovered when repairing the appliance or air conditioner.

Operating Appliances and Room Air Conditioners on a Generator

There are times when we have to seek alternate sources of electricity to run appliances and room air conditioners. Many people are installing auxiliary power generators as an alternative. In the 2004–2005 hurricane season, my family and neighbors were without electricity for up to three weeks per each hurricane. Luckily, we had a generator to power our home. When I purchased the generator, I had to take into account what appliances I intended to run off it.

When you consider running lights, appliances, and room air conditioners on a generator, you must calculate how much wattage is needed for each product, including the start-up wattage. In addition, the voltage (120 or 240) and cycles (60) are critical for proper operation of appliances and air conditioners. If the voltage is less than 10 percent of the operating voltage, refrigerators, freezers, and room air conditioners will not operate properly, which might cause damage to the components in the product. Any variation in the cycles could speed up the clocks or slow them down. It is also important when searching for a generator that it has the ability to regulate the voltage and has surge protection for use with appliances and room air conditioners. Table 6-6 provides a guide for the average wattage requirements for some products when choosing the appropriate size generator to purchase.

Product	Starting Wattage	Running Wattage
Dishwasher (cool dry only)	540	216
Electric range (8-inch element only)	2100	2100
Electric range (all functions)	10,000 to 12,000	10,000 to 12,000
Microwave	1000 to 14,000	1000 to 14,000
Refrigerator or freezer	1200	132 to 192
Automatic washer	1200	1200
Electric dryer	6750	5400
Room air conditioner (10,000 BTU)	2200	1500
Formula's Watt = 3.414 BTU/hour, BTU per hour × 0.293 = Watts, Volt × Amps = Watts		

TABLE 6-6   Wattage Requirement Chart

Flowcharts

A flowchart is a diagnostic tool that helps the technician to diagnose an appliance or air conditioner. It provides you with a step-by-step process to solve the problem. Choose the flowchart needed, start at the beginning, follow the arrows to the next box, and so on, until you solve the problem. The following are for diagnosing the electrical circuits of a product:

- DC voltage diagnostic flowchart (Figure 6-64 a and b)
- 24-volt AC diagnostic flowchart (Figure 6-65 a and b)
- 120-volt AC diagnostic flowchart (Figure 6-66 a and b)
- 240-volt AC diagnostic flowchart (Figure 6-67 a and b)

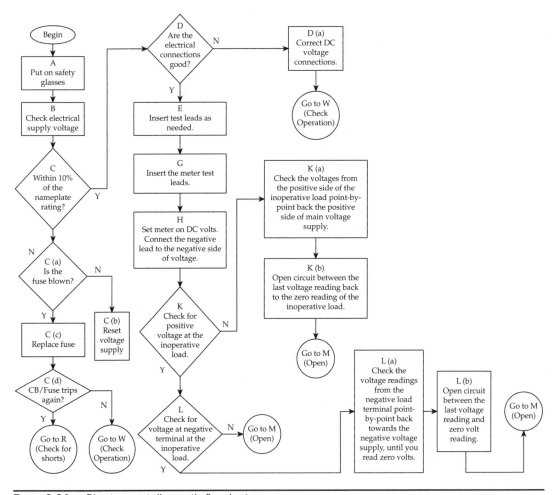

Figure 6-64a   Direct current diagnostic flowchart.

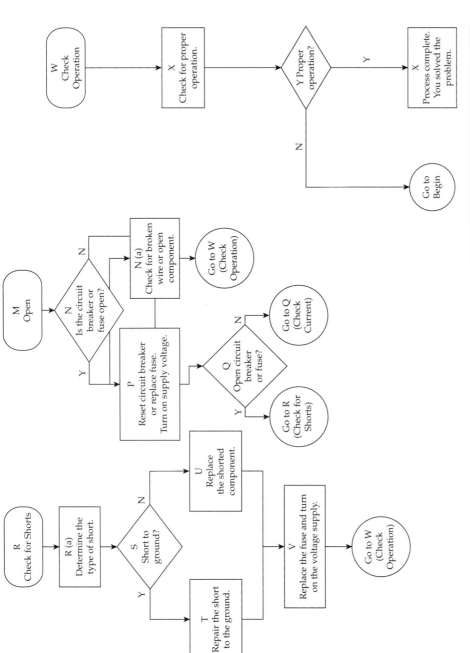

FIGURE 6-64b   Direct current diagnostic flowchart.

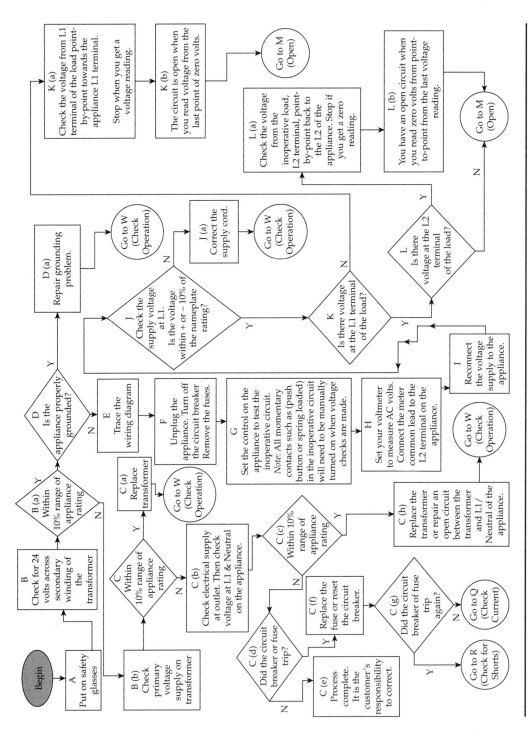

Figure 6-65a 24-volt AC diagnostic flowchart.

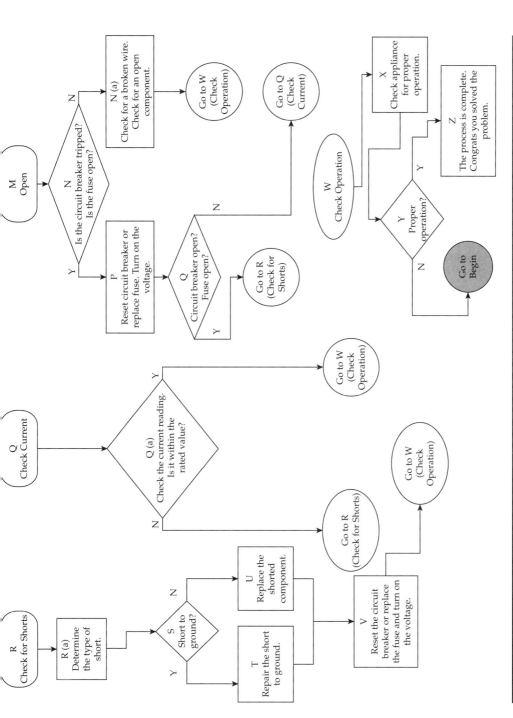

FIGURE 6-65b   24-volt AC diagnostic flowchart.

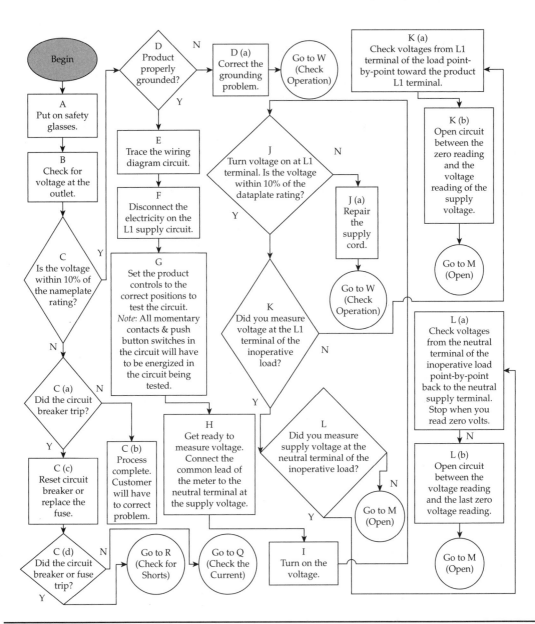

Figure 6-66a   120-volt AC diagnostic flowchart.

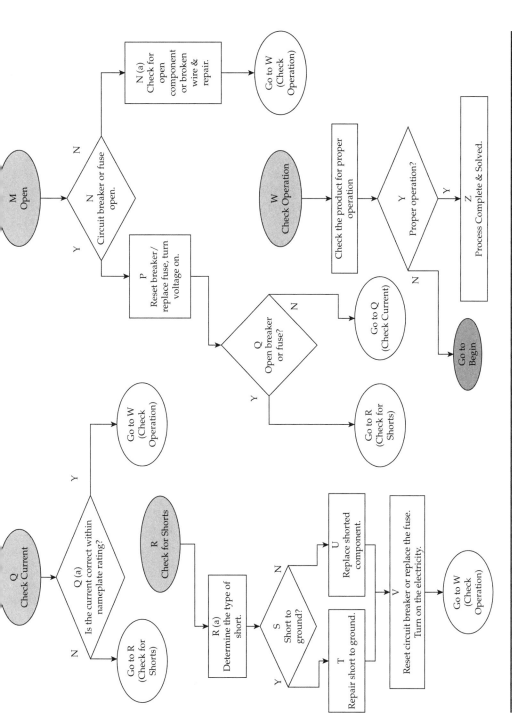

Figure 6-66b 120-volt AC diagnostic flowchart.

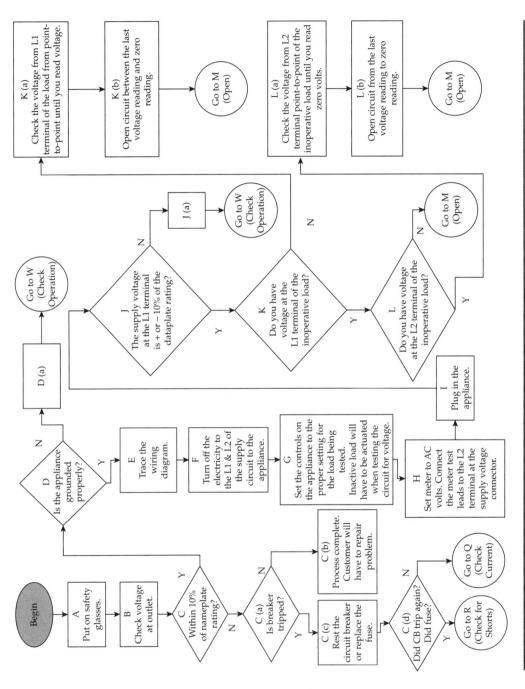

FIGURE 6-67a 240-volt AC diagnostic flowchart.

PART II

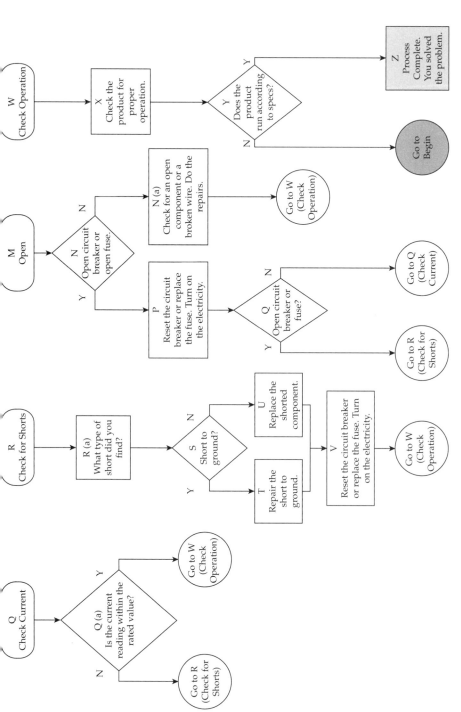

FIGURE 6-67b   240-volt AC diagnostic flowchart.

CHAPTER

Electronics

When servicing major appliances and air conditioners, you will also encounter electronic parts and/or integrated circuit boards within them. Manufacturers introduced electronic parts for a number of reasons, such as reliability, miniaturization, standardization, and maintainability. Here are some more basic reasons for electronic integrated circuit boards:

- Electronic integrated circuit boards will cost less to produce than their mechanical counterparts.
- You can program the electronic integrated circuit board to save the customer money during operation.
- The customer can choose more programs that come with the component.
- To make the appliance more efficient, the electronic control can use the feedback circuits.
- Electronic integrated circuit board displays are easier to read and set than the old-fashioned mechanical timers.
- During power outages, some electronic integrated circuit boards have a battery backup to keep track of what was programmed into the board, including the time.
- Electronic integrated circuit boards have built-in self-diagnostics and error codes.

When troubleshooting these electronic parts or integrated circuit boards, the technician will determine that an integrated circuit board has failed and will change the entire circuit board rather than repair the electronic components on the board. Although this chapter cannot cover all there is to know about electronics, it provides the basics. With the influx of electronic components into appliances and air conditioners, it is imperative that service technicians be able to troubleshoot them.

In order to troubleshoot electronic circuits, the technician will need the following:

- The technical data sheet or wiring diagram.
- An understanding of the appliance's or air conditioner's operation.
- Knowledge of how the controls were set and usage of the product prior to your arrival.

- Knowledge of how to access and use the diagnostic service mode. Each product has a different way to access the service mode.
- An understanding of how to read wiring diagrams and other circuit charts.

In order to properly diagnose a problem in an appliance or air conditioner, you will have to determine if the problem is any, or all of the following:

- Electrical
- Mechanical
- Customer related

Before you can replace an electronic component, you must rule out everything else.

Simplified Circuits

When compared to mechanical timers, the electronic integrated circuit board, also known as the electronic control board, has wiring circuits that are more simplified and easier to read. Figure 7-1 shows two wiring strip circuits of a range set on bake. The top circuit illustrates a mechanical timer and a mechanical thermostat. The bottom circuit illustrates the electronic control board. You can see in both strip circuits that the bottom circuit is easier to read and easier to diagnose.

Electronic Controls

In an appliance or air conditioner, in theory, the electronic control system could use a low-voltage control to run a high-voltage control. There are three main areas in the basic electronic control system (Figure 7-2). These areas are

- The low-voltage control board
- The high-voltage relay board
- The touchpad (customer interface)

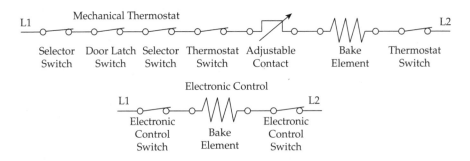

Figure 7-1 Wiring strip circuit for a range with the bake mode on. Mechanical components vs. electronic control components.

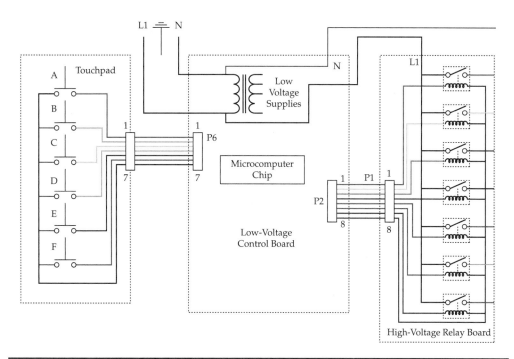

FIGURE 7-2 Basic electronic control system.

Low-Voltage Board

The low-voltage board (Figure 7-3a) consists of the following components:

- Transformer
- Microcomputer chip (the brain)
- Sensor inputs
- Control outputs
- Ribbon connectors

The electronic control board has a 120-volt step-down transformer attached to it. It converts the high voltage to smaller voltages consisting of 24 volts, 10 volts, 5 volts, 3 volts, etc. The smaller voltages will operate the circuits on the control board, including other circuits in the appliance or air conditioner. These other circuits include input from sensors and touchpad circuits. The microcomputer chip on the integrated circuit board is the brain that controls and interprets the inputs and the outputs of the circuit, and decides what components should be operating according to the preprogrammed instructions from the customer.

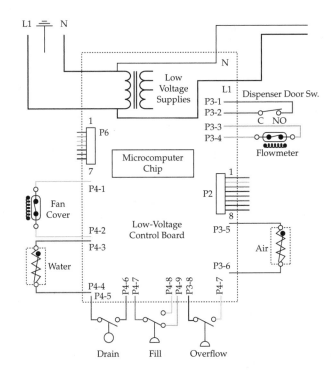

Figure 7-3a
The low-voltage board, sometimes called the main computer board.

High-Voltage Relay Board

The components and functions of the high-voltage relay board are

- Inputs from the power supply
- The relays
- Component/load outputs

On the high-voltage relay board (Figure 7-3b), there are two different types of inputs:

- The main voltage supply to the relay board, usually 120 volts.
- The control voltage. This is a lower-voltage AC or DC originating from the electronic control board. These lower voltages operate the relay coils, closing the switch contacts, allowing the load to operate. These loads can be motors, water valves, or even heating elements.

Touchpad (Customer Interface)

The customer interface, also known as the touchpad (Figure 7-3c), consists of the following items:

- The touchpad itself
- Input
- Output
- Ribbon connector

FIGURE 7-3b
The high-voltage relay
board.

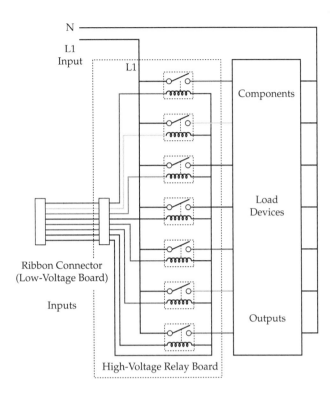

In order for the electronic control board to operate, it must receive inputs (commands) from the touchpad. This device contains switches in a plastic overlay. Every time the customer touches a keypad, it closes a switch temporarily. When the customer touches the keypad, they are programming the microcomputer on the electronic control board by sending low-voltage signals back and forth from the low-voltage board through the ribbon cable that is connected between them.

FIGURE 7-3c
The touchpad
assembly.

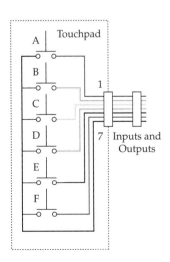

Three-Board Electronic Control System

A three-board control system consists of the touchpad, the electronic control board, and the relay board (Figure 7-4). Here is how the system operates. When electricity is applied to the circuit, 120 volts goes to the low-voltage transformer (Figure 7-4, circle #1). The transformer will step down (lower) the secondary voltage to be used as a control voltage for the other circuit boards. Also, the transformer supplies different voltage levels to the other areas of the control circuits.

When the touchpad (Figure 7-4, circle #2) is programmed by the customer, a signal is transmitted to the microcomputer chip on the electronic control board. The microcomputer chip will then check with the sensor circuits (Figure 7-4 circle, #3) for feedback. If the microcomputer receives a signal from the sensors that a load needs to be turned on, it will then send a low-voltage supply to the relay coil assigned to that load, a switch will then close (Figure 7-4, circle #4), and the load will operate. On the electronic control board, the microcomputer chip continues to monitor the sensing circuit (Figure 7-4, circle #5); when the preprogrammed instructions are met, the low-voltage supply to the relay coil is turned off, causing the switch to open up and turning off the load.

Troubleshooting Circuits

Troubleshooting electronic control system circuitry will depend on the type of method that is used to test the areas of the circuit, which needs to be checked (Figure 7-5). High-voltage circuitry can be tested by using the same procedures as for nonelectronic circuits and components. For example, Figure 7-6 represents a simple range circuit, where voltage testing will be performed. The switches on the electronic control board will be tested the same way as any other standard mechanical switch. The load devices attached to the three-board system will be tested in the same manner as the loads in a mechanical controlled system. The method used to test the low-voltage circuits will depend on what type of circuit is involved. For example, in Figure 7-7, low-voltage checks can be performed on the relay coil if the circuit goes to the high-voltage relay board. How the sensor input circuits are tested will depend on what type of circuit is involved. For example, in Figure 7-8, for a sensor to measure the correct temperature, you will need to perform a resistance check on the sensor. Then you will compare the resistance reading to the corresponding temperature reading on the chart that is on the technical data sheet that comes with the appliance or air conditioner. When the sensor acts as a switch in a circuit (Figure 7-9) you can test the component using resistance readings or voltage checks. The touchpad (Figure 7-10) that programs the microcomputer on the electronic control board is a set of switches. When a keypad is pressed, signals are transmitted. Test the touchpad using resistance checks. In order to do this, you will need the technical data sheet.

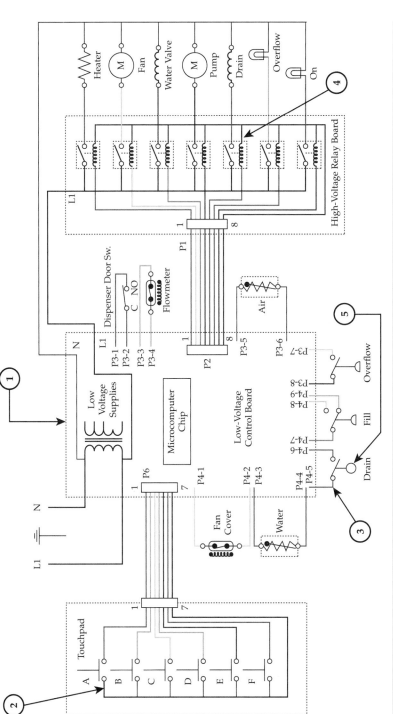

Figure 7-4 The three-board electronic control system, consisting of a touchpad, low-voltage board, and the relay board: (1) transformer, (2) touchpad, (3) sensor circuits, (4) relay assigned to a load, (5) sensor circuits still monitoring.

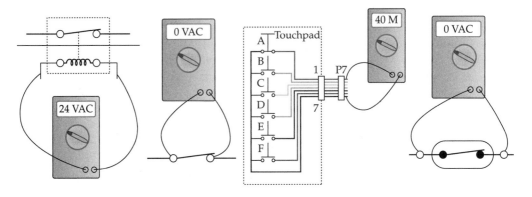

Figure 7-5 Testing the following components: (1) relay coil, (2) switch, (3) touchpad circuit, (4) sensor circuit.

Figure 7-6
A simple range circuit, also known as a strip circuit, for the bake element.

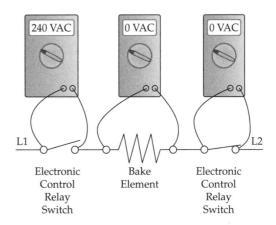

Figure 7-7
Testing the low-voltage control circuit on the high-voltage board.

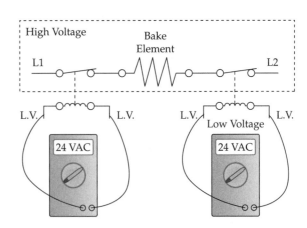

FIGURE 7-8
Temperature-to-resistance chart. For instance, if the thermistor is supposed to be reading 70 degrees Fahrenheit, your ohmmeter reading for the thermistor should be 11.9 k-ohms.

Temperature	Resistance
50 degrees F.	19.9 K ohm
60 degrees F.	15.3 K ohm
70 degrees F.	11.9 K ohm
80 degrees F.	9.2 K ohm
90 degrees F.	7.4 K ohm

FIGURE 7-9
This is the symbol for a sensor that acts like a switch.

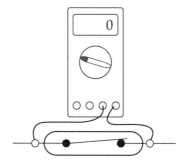

FIGURE 7-10
Using the ohmmeter to test the keypad switches on the touchpad assembly.

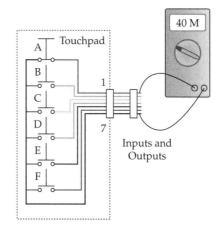

Two-Board Electronic Control System

Some manufacturers' products use a two-board control system (Figure 7-11) instead of a three-board control system. The two types of control systems may be different, but the theory of operation is the same. They simply combine some of the components on a single board, which drives down the production cost. The components forming the low-voltage

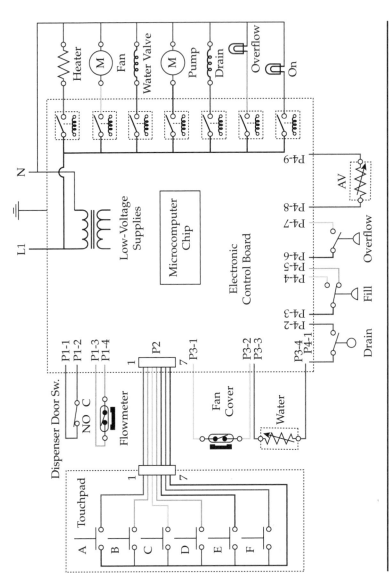

FIGURE 7-11 A two-board electronic control system.

control board and the high-voltage relay board are combined into one board (Figures 7-12 and 7-13), and include the following:

- Low-voltage transformer
- Microcomputer chip
- Inputs from the sensors
- Inputs from the power supply
- Outputs from the loads
- Relays

All of the sensors and load devices operate the same on both systems. The second board will be the touchpad.

When troubleshooting the two- or the three-board electronic control systems, there are fewer test points (see Table 7-1).

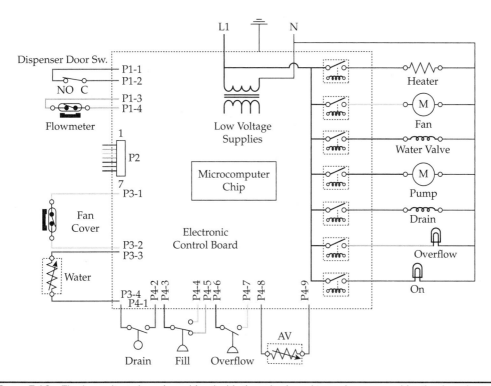

Figure 7-12 The low-voltage board combined with the relay board to make one combination board.

Type of Circuits	Two-Board Electronic Control System	Three-Board Electronic Control System
Sensor circuits	Same checks	Same checks
Touchpad circuits	Same checks	Same checks
Load components	Same checks	Same checks
High-voltage load circuits	Same checks	Same checks
Power supply inputs	Only one to electronic control board	Checks to both low-voltage control board and high-voltage relay board
Low-voltage control voltages	Not required	Outputs from low-voltage control board and input to high-voltage relay board

TABLE 7-1   Compare the Testing of a Two-Board Electronic Control System vs. Three-Board Electronic Control System

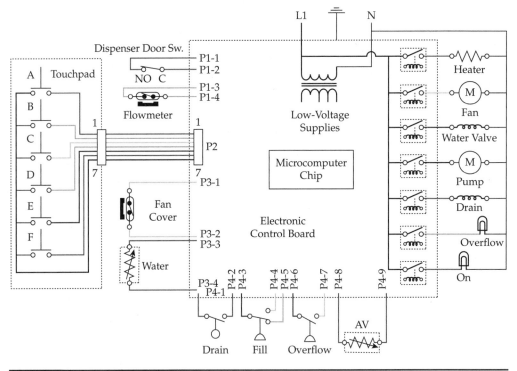

FIGURE 7-13   Another type of two-board electronic control system. This combination board includes a touchpad assembly.

Motor Board Electronic Control System

If you take a motor control board system and add it to a two-board control system, you will have another version of a three-board electronic control system. The motor control board is used to run a variable-speed motor. This type of control system (Figure 7-14) incorporates a touchpad, an electronic control board, and a sensor circuit.

Figure 7-15 illustrates how voltage flows through one type of a variation of a two-board electronic control system with a motor control board. When voltage is applied to the product and turned on, voltage is supplied to the transformer (Figure 7-15, circle #1) on the electronic control board, and at the same time voltage is supplied to the motor control board (Figure 7-15, circle #4). The microchip on the electronic control board receives a signal from the touchpad (Figure 7-15, circle #2). The microchip monitors the reed switch in the sensor

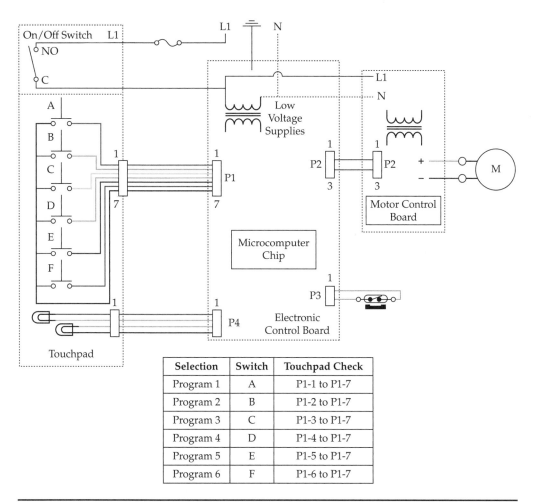

Selection	Switch	Touchpad Check
Program 1	A	P1-1 to P1-7
Program 2	B	P1-2 to P1-7
Program 3	C	P1-3 to P1-7
Program 4	D	P1-4 to P1-7
Program 5	E	P1-5 to P1-7
Program 6	F	P1-6 to P1-7

FIGURE 7-14 Another version of a two-board electronic control system that incorporates a motor control board.

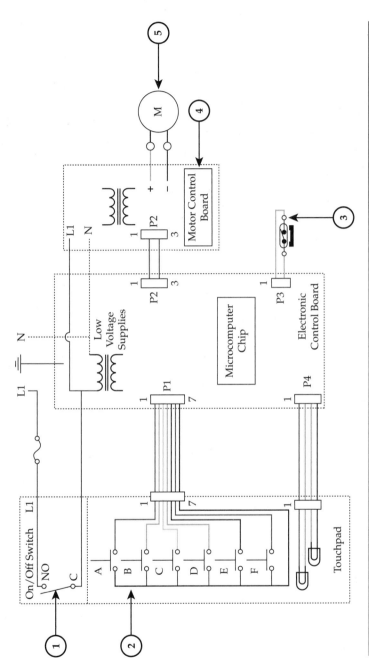

Figure 7-15 Two-board electronic control system with a motor control board: (1) on/off switch, (2) touchpad, (3) sensor circuit, (4) motor control board, (5) the motor (load).

circuit (Figure 7-15, circle #3). In order to turn on the motor (Figure 7-15, circle #5), the microchip sends a signal to the motor control board (Figure 7-15, circle #4). The microchip on the electronic control board monitors the sensor circuit and tells the motor control board to adjust the motor speed to go faster or slower. The motor control board will adjust the supply voltage to the motor according to the type of signal received from the microchip. There are many different types of motor control systems in use and manufactured according to the appliance, brand, and model. Each control system has its own diagnostic codes built into the microchip and troubleshooting procedures. These service procedures and diagnostic codes can be found on the technical data sheet that is attached to all appliances and air conditioners.

In many of these control systems, the signal voltage and the voltage supplied to the motor will not only vary in voltage, but also in frequency. The frequency is the number of complete cycles per second of alternating current. These voltages and frequencies will also vary according to the load placed on the product. You cannot successfully measure these voltages and frequencies using a digital meter. When you read through the technical data sheet, it will provide you with the necessary information needed to take resistance measurements, and other techniques used for troubleshooting the problem for each model.

There are many troubleshooting checkpoints in an electronic control system using a motor control board for the service technician to test. There are voltage inputs and outputs to both the electronic control board and to the motor control board. Some other checkpoints are as follows:

- High-voltage circuit
  - 120-volt source circuits
  - The fuse
  - The on/off switch
- Electronic control/signal circuits
  - The connection on the electronic control board for the wiring between the electronic control board
  - The connection on the motor control board for the wiring between the electronic control board and the motor control board
- Motor/motor control circuits
  - The connection on the motor control board for the wiring between the electronic control board and the motor control board
  - The motor and the connection on the motor for the wiring between the motor control board and the motor
- Touchpad
  - The line side and neutral side wiring to a switch
  - The wiring for the on/off light
- Sensor
  - The reed switch (check for continuity)
  - The wiring to and from the reed switch

Resistors

A resistor, when installed into an electrical circuit, will add resistance, which will produce a specific voltage drop, or a reduction in current. Resistors can be either fixed or variable. The most common problem found with resistors when testing them with a milliammeter is that they read as open (infinite ohms). To determine the resistance value in ohms of a resistor, you have to isolate it from the circuit and then use your ohmmeter to read its value. On larger resistors, the resistance value is stamped on the body of the resistor.

A visual inspection of the resistor should also be performed to rule out any physical flaws in the resistor and the surrounding components. If you touch a resistor and it begins to flake apart, most likely it is bad and it needs to be replaced. If the technician smells a burning smell coming from the resistor, most likely this component is bad and it must be replaced.

Resistance Color Bands

Because of their size, most axial resistors (Figure 7-16a) are color coded, usually with four bands to indicate their resistance value. Reading the resistor bands from left to right, the first band closest to the end indicates the first numerical digit of resistance. The second band

FIGURE 7-16
(a) A common resistor.
(b) Resistor color-code chart.

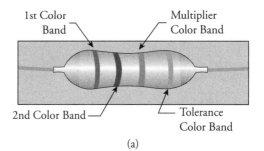

1st Color Band   Multiplier Color Band

2nd Color Band   Tolerance Color Band

(a)

Color	1st Band	2nd Band	Multiplier	
Black	0	0	× 1	
Brown	1	1	× 10	
Red	2	2	× 100	
Orange	3	3	× 1000	
Yellow	4	4	× 10,000	
Green	5	5	× 100,000	
Blue	6	6	× 1,000,000	
Violet	7	7	× 10,000,000	
Gray	8	8	× 100,000,000	
White	9	9	× 1,000,000,000	
4th Band = Tolerance				
Gold	5%	Silver	10%	No Color	20%

(b)

indicates the next numerical digit of resistance, and the third band indicates the numerical multiplier in zeros. The fourth band on a resistor is the resistance tolerance measured in a percentage (Figure 7-16b). For example, let's say the color bands are brown, red, orange, and silver:

Brown = 1
Red = 2
Orange = 1000
Silver = + or −10 percent

The resistor is rated at 12,000 ohms with a ±10 percent tolerance. The tolerance is the safe operating percentage difference of the color-coded resistance value. For instance, a 12,000-ohm resistor with a 10 percent tolerance will operate safely at 10 percent below or 10 percent above the 12,000-ohm valve.

Thermistor

Thermistors are constructed from a semiconductor material with a specified temperature range. This electronic device exhibits a change in resistance with a change in temperature. There are two types of thermistors: negative temperature coefficient (NTC) and positive temperature coefficient (PTC), with the most common being NTC. The resistance of the NTC thermistor decreases with an increase in temperature, whereas the resistance of the PTC thermistor increases with an increase in temperature. There are many applications for thermistors, including their use as temperature sensors, resettable fuses, current limiters, and power indicators. They are also used in applications where interchangeability without recalibration is required.

One type of test performed on thermistors to check if they are good or bad is the resistance-versus-temperature test. To perform this test, the technician will have to locate the technical data sheet that comes with the appliance. The data sheet will provide the resistance (ohms) value at the test temperature listed. Another way to test the thermistor is by testing it with an ohmmeter for an open or a short.

Diodes

A diode is an electrical device that allows current to flow in one direction only. If you connect a positive voltage to the anode (positive side of the diode) and a negative voltage to the cathode side (negative side of the diode), the diode becomes forward-biased (Figure 7-17a). When the diode is forward-biased, current will flow. When a positive voltage is connected to the cathode (negative side of the diode) and a negative voltage is connected to the anode (positive side of the diode), the diode becomes reverse-biased (Figure 7-17b). Then the diode will not allow current to flow. A diode may be thought of as a switch "closed" when forward-biased and "open" when reverse-biased. There are many types of uses for diodes besides rectification. These include capacitance that varies with the amount of voltage applied to the diode and photoelectric effects. The wiring diagram symbol for a diode is shown in Figure 7-18.

When a diode is connected in a circuit, there must be a means to identify the cathode and the anode (Figure 7-19). Diodes are manufactured in different case styles. On large

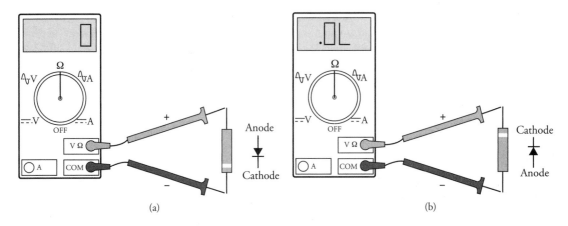

Figure 7-17 (a) When testing the diode with an ohmmeter, the forward bias indicates zero ohms of resistance. (b) When testing the diode with an ohmmeter, the reversed bias indicates infinite resistance.

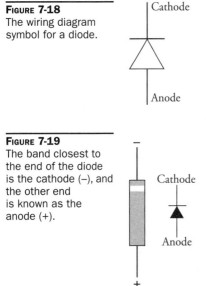

Figure 7-18
The wiring diagram symbol for a diode.

Figure 7-19
The band closest to the end of the diode is the cathode (–), and the other end is known as the anode (+).

diodes, there is a diode symbol stamped on the side to indicate to the technician the cathode and anode. On smaller diodes, there is a band around one end to identify the cathode.

Testing a Diode

A diode can be tested with an ohmmeter. To test a diode, you must first disconnect one side of the diode to isolate it from the remainder of the circuit. Set your ohmmeter to the ohms scale, and connect the leads to the diode. The meter may or may not read continuity. Then reverse the leads on the diode and check for continuity. You should read continuity in one direction only. If continuity is not indicated in either direction, the diode is open (Figure 7-20a). If continuity is indicated in both directions, the diode is shorted (Figure 7-20b).

With some microwave oven high-voltage diodes, you will not be able to read continuity in any direction. Not all multimeters can perform this function. You will then attach a 9-volt battery in series with the diode (removed from the microwave oven circuit) and take voltage readings. Read the voltage through the diode in one direction, and then reverse the meter leads, and you should not have a voltage reading in the other direction.

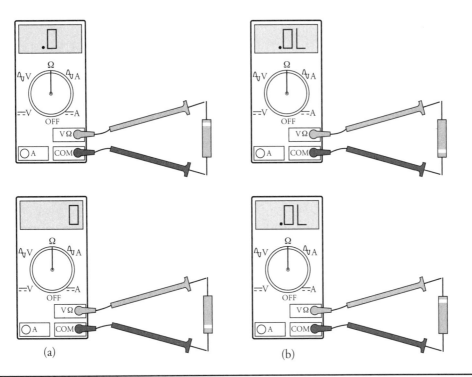

Figure 7-20 (a) The ohmmeter indicates no continuity in either direction. The diode is open. (b) The ohmmeter reads continuity in both directions, indicating a shorted diode.

Bridge Rectifier

The bridge rectifier (Figure 7-21) consists of four diodes connected together in a bridge configuration on the circuit board. On electronic control boards, a bridge rectifier is used to convert alternating current into direct current for the low-voltage circuitry.

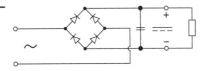

Figure 7-21
A bridge rectifier circuit converting AC to DC.

The technician cannot test the bridge rectifier directly because it is wired directly into the circuitry on the integrated circuit board. You can only test the AC input voltage and DC output voltages. If you're reading on your voltmeter AC input voltage to the electronic board and no DC output voltage from the board, the board is bad and will need to be replaced. If you're reading on your voltmeter no AC input voltage and no DC output voltage on the electronic board, check the AC voltage supply to the board. If you're reading on your voltmeter AC voltage input and DC voltage output on the electronic board and the load is not operating, check the load and the connecting wiring from the load to the relay on the circuit board.

Triac

A triac (Figure 7-22) is a three-terminal electronic device that is similar to a diode, except that it allows current to flow in both directions, as with alternating current. There is no anode or cathode in the triac, and it acts as a high-voltage switch on an electronic control board that will turn loads on or off in the circuit. A signal pulse to the gate lead on the triac will trigger the current to flow through the device. The gate loses all control of the triac until the current falls below the manufacturer's specifications. A triac can supply more or less power to a load depending on the trigger point. They are available with current ratings up to 40 amps and voltage ratings up to 600 volts.

Testing a Triac

Care must be taken if you are testing a triac in a circuit. In a microwave, for example, unplug the appliance and discharge the capacitor. Locate the triac (Figure 7-22) within the circuitry

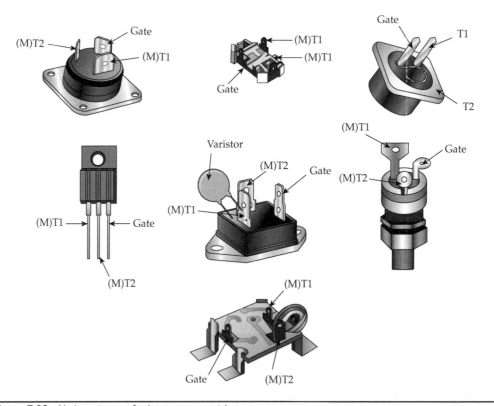

Figure 7-22   Various types of microwave oven triacs.

and identify the triac terminals. The three terminals are generally designated as G (Gate), T1, and T2. Do the testing as follows:

- Remove the wires from the triac terminals.
- Set your multimeter on the ohms scale.
- Measure from the T1 to the G (gate) terminal, note the meter reading, then reverse the leads and note the meter reading again.
- The reading in each measurement should be between 10 and 200 ohms, depending on the make and model.
- Next, measure each of the following readings; you should read infinity:

  (a) From T1 to T2

  (b) From T2 to G (gate)

  (c) From each terminal to chassis ground

NOTE *These readings are approximate and may vary from manufacturer to manufacturer, but generally speaking, any results that are significantly different would indicate a defective triac.*

There is a second test. Turn on the triac in the following manner:

- Remove the wires from the triac terminals.
- Set your multimeter on the ohms scale.
- Attach your negative lead from your meter to the T1 terminal and the positive lead to the T2 terminal.
- Using a screwdriver, create a momentary short between the T2 terminal and the G (gate) terminal. This action should turn on the triac, and the meter reading will between 15 and 50 ohms.
- Next, disconnect one of the meter leads and then reconnect it to the terminal. The meter reading should indicate infinity.
- Finally, reverse the meter test leads and repeat the tests. The results should be the same.

NOTE *Any abnormal tests would suggest a defective triac.*

Transistor

A transistor is a three-element, electronic, solid-state component that is used in a circuit to control the flow of current or voltage. It opens and closes a circuit just like a switch (Figure 7-23a). Other uses for transistors include amplification and oscillation. Transistors are located on circuit boards. Figure 7-23b indicates the wiring diagram symbol for a transistor.

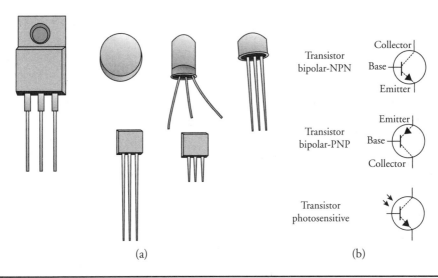

Figure 7-23   (a) A sample of the various types of transistors used in circuits. (b) The wiring symbol for an NPN, PNP, and photosensitive transistor.

Integrated Circuits and Circuit Boards

An integrated circuit (IC), shown in Figure 7-24, is a miniature electric circuit consisting of transistors, diodes, resistors, capacitors, and all the connecting wiring—all of it manufactured on a single semiconductor chip. Integrated circuits are found in products like computers, televisions, cell phones, and major appliances and air conditioners. IC chips are not repairable.

A printed circuit board (PCB) is a layout of conductive pathways (an electric circuit) secured to a board. Electronic components are soldered on to the board for a particular function(s) within an appliance or air conditioner. Only a small percentage of printed circuit boards are repairable. The remaining percentage of PCBs is replaced with new boards.

Figure 7-24
An integrated circuit
(IC) chip.

LED Multiple Segment Displays

LED segment displays are available in 7, 9, 14, and 16 segment displays (Figure 7-25). Each segment will light up to produce an alphanumeric character. Applications today include displays fitted to ranges, microwaves, automatic washers, automatic dryers, dishwashers, refrigerators, and air conditioners (Figure 7-26). Light-emitting diodes (LEDs) are a semiconductor light source, and they are available in a variety of shapes, sizes, and colors.

FIGURE 7-25
(a) A seven-segment LED display. (b) Alphanumeric characters. (c) An LED. (d) and (e) 16-segment LED displays used on major appliances and A/Cs. (f) Light-emitting diodes (LEDs) are a semiconductor light source.

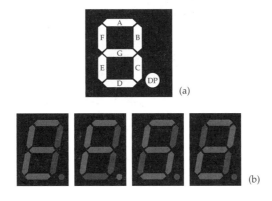
(a)

(b)

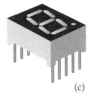
(c)

(d)   (e)

(f)

FIGURE 7-26
An LED display in a microwave oven. It tells a customer the time of day and how much time is left to cook the food.

LEDs are used as a lamp indicator in ranges and ovens. They are used as lamps in other products also. The LED was first introduced for electronics in 1962.

Electrostatic Discharge

The electronic parts in an appliance or air conditioner are sensitive to electrostatic discharge (ESD). An example of electrostatic discharge occurs when you rub your feet on a carpet and then touch a metal doorknob. The little spark that you generated by touching the metal doorknob is ESD. This type of electrostatic discharge will damage the electronics in the appliance or air conditioner at the microscopic level. When servicing an appliance or air conditioner or installing a new electronic part, ESD may cause failure in the future due to stress from these occurrences.

To prevent ESD from damaging expensive electronic components, follow these steps:

- Turn off the electricity to the appliance or air conditioner before servicing any electronic component.
- Before servicing the electronics in an appliance or air conditioner, discharge the static electricity from your body by touching your finger repeatedly to an unpainted surface on the appliance or air conditioner. Another way to discharge the static electricity from your body is to touch your finger repeatedly to the green ground connection on that product.
- The safest way to prevent ESD is to wear an antistatic wrist strap.
- When replacing a defective electronic part with a new one, touch the antistatic package that the part comes in to the unpainted surface of the appliance or air conditioner, or to the green ground connection of the appliance or air conditioner.
- Always avoid touching the electronic parts or metal contacts on an integrated board.
- Always handle integrated boards by the edges.

Gas

A s a technician or novice servicer, this chapter on gas is very important to the safety and technical knowledge that is needed to repair major gas appliances. In addition, consumers should be knowledgeable about safety procedures and basic characteristics of natural and liquefied petroleum gas (LP or LPG). Any person who cannot use basic tools or follow written instructions should not attempt to install, maintain, or repair gas appliances.

If you do not fully understand the procedures in this chapter, or if you doubt your ability to complete the task on your gas appliance, please call your service manager.

Currently, gas is used in millions of homes for heating, cooling, cooking, drying laundry, and water heating. Understanding gas theory, gas conversion, ignition systems, and the different types of gases used is something that every technician needs in his arsenal to service gas appliances. In addition, the technician needs to understand two other important elements when servicing gas appliances: combustion and ventilation. Some appliances use only gas as the main fuel, while others use gas and electricity. Although this chapter cannot cover all there is to know about gas, it provides the basics.

Gas Safety

The following are a few safety tips to help you in handling major gas appliances in your home:

- Always follow the manufacturer's use and care manual for the gas appliance.
- Always keep combustible products away from gas appliances.
- Keep your gas appliance clean from soot, grease, and food spillages.
- Teach your children not to play near or with gas appliances.
- Always have a fire extinguisher nearby just in case of mishaps that might lead to a fire.
- Have a smoke detector and a carbon monoxide detector installed in the home, and check the batteries yearly.
- Never use a gas range to heat the home.
- Make sure that gas appliances have proper venting according to the manufacturers' recommendations.

Types of Gas

The two most common gases that are used in homes today are natural gas and liquefied petroleum gas (LP or LPG). Gas is a form of chemical energy, and when it is converted by combustion, it becomes heat energy. This type of heat energy is used for cooking, drying, heating, cooling, and lighting.

Natural Gas

Natural gas is a naturally occurring product made up of hydrocarbon and non-hydrocarbon gases. The main ingredient found in natural gas is methane (70 to 90 percent), with the remainder of the ingredients being nitrogen, ethane, butane, carbon dioxide, oxygen, hydrogen sulphide, and propane. These gases are located beneath the earth and can be removed through constructed wells. Another method for producing and harvesting natural gas is through landfills. The methane gas that is produced by the decomposition of materials can be harvested and added to the natural gas supply.

The heating value of natural gas is between 900 and 1200 BTUs per cubic foot. The air we breathe has a specific gravity of 1.00, and natural gas is lighter than air, with a specific gravity varying from 0.58 to 0.79. Natural gases are odorless and colorless—gas companies add an odor agent to warn for leaks.

The pressure of natural gas that is supplied to a residence will vary, between a 5- and 9-inch water column. A gas pressure regulator that is connected to the appliance will further reduce the pressure. For example, on some gas ranges the pressure will be reduced to a 4-inch water column, while on other models the pressure might be a 6-inch water column. On some water heaters manufactured today, the pressure is between a 4- and 7-inch water column. To determine the correct pressure rating, the technician must refer to the manufacturer's specifications or the installation instructions for that product.

Liquefied Petroleum Gas

Liquefied petroleum gas (LP or LPG) is obtained from natural gas sources or as a by-product of refining oil. LP gas for domestic use is usually propane, butane, or a mixture of the two. This type of gas is compressed and stored in storage tanks under pressure in a liquid state at approximately 250 pounds per square inch. The pressure in an LP tank will vary according to the surrounding temperatures and altitude. LP gas tanks can be transported to areas that are not supplied by natural gas supply lines.

The heating value of propane gas is 2500 BTUs per cubic foot, with a specific gravity of 1.53. The heating value of butane is much higher—about 3200 BTUs per cubic foot, with a specific gravity of 2.0. Liquefied petroleum gas is heavier than air and will accumulate in low-lying areas on the floor, in an enclosure, or in pockets beneath the ground, creating a hazard if it encounters an open flame. As mentioned, these gases are odorless and colorless, and gas companies add an odor agent to warn of leaks.

LP or LPG gas pressure for residential appliances, as established by the gas industry, will be between a 9- and 11-inch water column. To determine the correct pressure rating, the technician must refer to the manufacturer's specifications or the installation instructions for that product.

Combustion

Combustion is a rapid chemical reaction (burning) of LP or natural gas and air to produce heat energy and light. To sustain combustion in a gas appliance, an ignition source—such as that produced by a flame or by electrical means—is used to ignite the gas vapors. Three elements are needed to produce a flame when burning gas vapors: fuel, heat, and oxygen. If any one element is missing, flame or burning of gas vapors will not exist (Figure 8-1).

It takes 1 cubic foot of gas mixed with 10 cubic feet of air to have complete combustion of natural gas. The process produces approximately 11 cubic feet of combustion product, consisting of approximately 2 cubic feet of water vapor, 1 cubic foot of carbon dioxide, 8 cubic feet of nitrogen, and the excess air from the gas appliance. These combustion products must be properly vented or discharged safely from the gas appliance.

It takes 1 cubic foot of gas mixed with 24 cubic feet of air for complete combustion to happen with propane gas. The process produces approximately 25 cubic feet of combustion product. With butane gas, it takes 1 cubic foot of gas mixed with 31 cubic feet of air and produces 32 cubic feet of combustion product (Figure 8-2). With the proper mixture of air, gas, and flame, complete combustion will take place. The combustion product that is produced from complete combustion will be carbon dioxide and water vapor.

Inadequate venting of a gas appliance will restrict the flow of air into the gas appliance. This lack of proper ventilation will reduce the amount of oxygen within the air for complete combustion to occur. The air that we breathe is a mixture of gases containing nitrogen, oxygen, argon, carbon dioxide, water vapor, and other trace gases (Figure 8-3). Incomplete combustion will cause the reduction of oxygen levels in the air supply within a room. Proper ventilation—adequate fresh air into the room—is important and cannot be stressed enough for the proper operation of a gas appliance, as well as for the safety of human life within the home.

Carbon Monoxide

Carbon monoxide (CO) is a toxic gas that can cause death if inhaled in copious amounts. It is odorless, colorless, and has no taste. The human body cannot detect carbon monoxide with its senses. When carbon monoxide is inhaled, it is absorbed into the bloodstream and stays there longer, preventing oxygenated blood from performing its job in the body. Two factors affect the amount of carbon monoxide absorbed into the bloodstream: the amount of

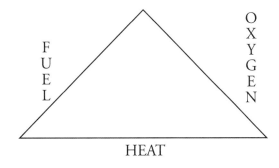

FIGURE 8-1
The combustion triangle.

Properties of		Natural gas	Propane	Butane
Chemical formula		CH_4	C_3H_8	C_4H_{10}
Boiling point of liquid at atmospheric pressure	°F	−258.7	−44	32
Specific gravity of vapor (air = 1)		0.6	1.53	2.00
Specific gravity of liquid (water = 1)		0.6	0.51	0.58
Calorific value @ 60°F	BTU/cubic foot	1012	2516	3280
	BTU/gallon		91,690	102,032
	BTU/lb		21,591	21,221
Latent heat of vaporization	BTU/gallon	712	785.0	808.0
Liquid weight	lbs/gallon	2.5	4.24	4.81
Vapor volume from 1 gallon of liquid at 60°F	Cubic foot		36.39	31.26
Vapor volume from 1 lb of liquid at 60°F	Cubic foot		8.547	6.506
Combustible limits	% of gas in air	5–15	2.4–9.6	1.9–8.6
Amount of air required to burn 1 cubic foot of gas	Cubic foot	9.53	23.86	31.02
Ignition temperature in air	°F	1200	920–1020	900–1000
Maximum flame temperature in air	°F	3568	3595	3615
Octane number		100	Over 100	92
All data are approximate.				

FIGURE 8-2 Properties of utility gases.

carbon monoxide in a room and the length of exposure. Lower levels of CO inhalation can cause flu-like symptoms, including headaches, dizziness, disorientation, fatigue, and nausea. Other exposure effects can vary depending on the age and health of the individual.

Carbon monoxide in homes without gas appliances varies between 0.5 to 5 parts per million (ppm). CO levels in homes with properly maintained gas appliances will vary from 5 to 15 parts per million. For those gas appliances that are not maintained properly, carbon monoxide levels may be 30 ppm or even higher.

Testing for Carbon Monoxide

Consumer products for detecting carbon monoxide in a home have been on the market for years. Carbon monoxide detectors should have an alarm that alerts consumers before they are exposed to hazardous levels of carbon monoxide. In order to prevent false alarms, CO detectors must be able to distinguish carbon monoxide gases from other types of gases, such as butane, heptane, alcohol, methane, and ethyl acetate. Two manufacturers that you can visit on the Internet to view the different types of carbon monoxide detectors available are www.kidde.com and www.firstalert.com.

Technicians who test for carbon monoxide use a special handheld meter to check the levels of carbon monoxide in a room or home. For a handheld carbon monoxide test meter, visit www.fluke.com.

FIGURE 8-3
Composition of air.
The air we breathe is
made up of 99.99
percent of nitrogen,
oxygen, carbon
dioxide, and argon.

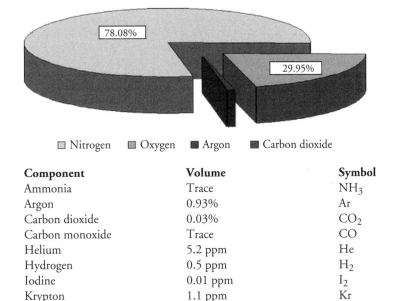

Component	Volume	Symbol
Ammonia	Trace	NH_3
Argon	0.93%	Ar
Carbon dioxide	0.03%	CO_2
Carbon monoxide	Trace	CO
Helium	5.2 ppm	He
Hydrogen	0.5 ppm	H_2
Iodine	0.01 ppm	I_2
Krypton	1.1 ppm	Kr
Methane	2.0 ppm	CH_4
Neon	18.2 ppm	Ne
Nitrogen	78.08%	N_2
Nitrous oxide	0.5 ppm	N_2O
Oxygen	20.95%	O_2
Sulfur dioxide	1.0 ppm	SO_2
Water vapor	0 to 5%	H_2O

When to check for carbon monoxide in a home:

- When the consumer complains of headaches or nausea
- Houseplants are dying
- Unknown chronic odors from unknown sources
- Condensation on cool surfaces that might lead to flue gas products in the home

Figure 8-4 illustrates the locations in a home to test for carbon monoxide gas. These tests should be conducted near gas appliances, gas heating systems, heating ducts, and the atmosphere in a room, approximately 6 feet above the floor.

When testing for CO in a gas appliance that has not been running for a while, you should follow these steps:

1. Test the air near the appliance and the surrounding air in the room before you turn on the appliance.
2. Test the air after you turn on the appliance.
3. Test the air near the appliance after the appliance has been running for 15 minutes.

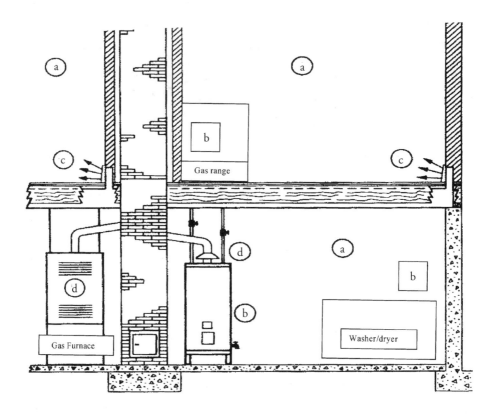

(a) Test in the atmosphere about 5 to 6 feet off the ground
(b) Test near gas appliances
(c) Test near heating ducts
(d) Test near appliance diverters and fire-doors on appliances in the basement or
utility room

FIGURE 8-4 Typical test locations for carbon monoxide.

Flue Gases

Flue gases are the by-products of combustion that are exhausted through a chimney or flue to the outside of the home. These gases consist of the following:

- **Nitrogen (colorless, odorless, and tasteless)** The air we breathe is made up of 79 percent nitrogen.

- **Carbon dioxide (colorless and odorless)** The human respiration process produces carbon dioxide.

- **Oxygen (colorless, odorless, and tasteless)** Oxygen is the main ingredient for combustion.

- **Carbon monoxide (colorless, odorless, and tasteless)**
- **Nitrogen oxide (colorless and odorless)**   Nitrogen oxide forms in the combustion process of gas fuels.
- **Sulfur dioxide (colorless and smells like burnt matches)**   Sulfur dioxide is irritating to the lungs.
- **Hydrocarbons (colorless, odorless, and tasteless)**   Hydrocarbons are found in natural and liquefied petroleum gases.
- **Water vapor**   Approximately 10 percent will be vented through the flue or chimney.
- **Soot**   The remains of incomplete combustion.

The "Flame"

Prior to the invention of the Bunsen burner in 1842, gas burners would produce a yellow flame for light and heat. The process allowed the gas to enter into a tube, expel through a port, and, when lit with an ignition source, the gas would burn without pre-mixing the air before it left the burner (Figure 8-5). As the flame burned, the gas temperature began to rise

FIGURE 8-5
Gas lamp.

within the flame. Without the presence of pre-mixed air and gas, carbon particles began to pass through the flame, causing it to turn yellow. As more air was introduced into the combustion process, the flame would turn blue or blue with yellow tips.

Appliance manufacturers had the freedom to tailor the flame patterns from the Bunsen burner design to design gas appliances. The Bunsen burner worked on the principle of introducing air into the mixture of the gas before it left the burner port. An orifice was used to regulate the gas flow within the burner body (Figure 8-6). Attached to the burner body is an adjustable shutter to control the primary air mixture that enters into it. The flame that is

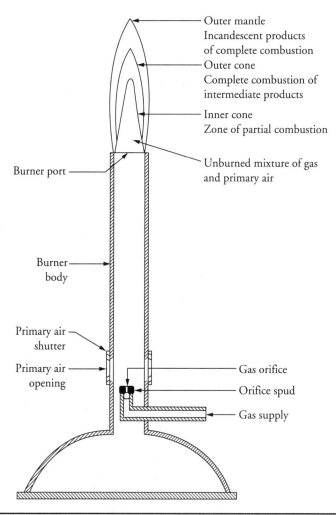

Outer mantle
Incandescent products
of complete combustion

Outer cone
Complete combustion of
intermediate products

Inner cone
Zone of partial combustion

Burner port

Unburned mixture of gas
and primary air

Burner body

Primary air shutter

Primary air opening

Gas orifice

Orifice spud

Gas supply

Figure 8-6 Bunsen burner.

produced by the Bunsen burner has multiple colors within it. Each color within the flame marks a stage of the burning process of gas (Figure 8-7). The stages are as follows:

- The inner cone is the first stage of the burning process. Within this cone the gas is burned, which forms by-products, such as aldehydes, alcohols, carbon monoxide, and hydrogen.
- The outer cone surrounds the inner cone, and as air diffuses into the flame, it continues the burning process. If enough air is present during the burning process, the by-products from the inner cone will burn up in the outer cone. The by-products that are produced in the outer cone are carbon dioxide and water vapor.
- The outer mantle that is produced by the Bunsen burner is nearly invisible, with all the gas burned up. The reason it glows is due to the high temperature produced by the complete combustion of the gas products from the outer cone.

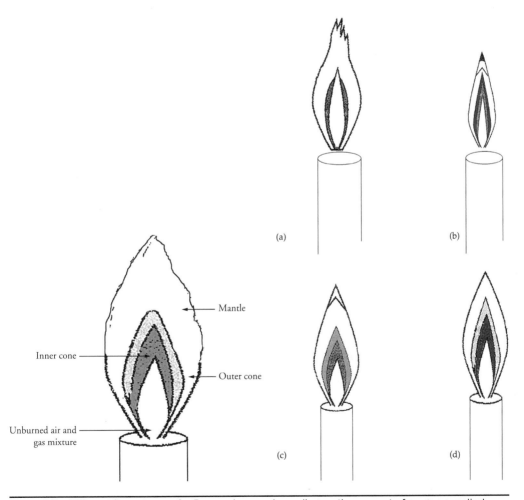

FIGURE 8-7 Different flame types of a Bunsen burner, depending on the amount of oxygen supplied.

The temperature of a flame can vary, depending on what type of gas is used and how much air is pre-mixed with the gas. Temperatures within the flame will vary; the inner cone is cooler than the outer cone, and the hottest flame temperature is just above the outer cone.

Air Supply

When air is mixed with gas before it leaves the burner port, it is known as the primary air supply. Under ideal conditions, most burners only use 50 percent of the primary air supply to burn the gas, and the remaining 50 percent of the air is supplied by the secondary air supply from around the flames. In reality, additional excess air is needed to ensure that enough air is present to completely burn off the gas. The total air supply is accomplished by adding the sum of the primary, secondary, and excess air supplies. The total air supply needed for combustion is measured in percentages of air needed for complete combustion. Every manufacturer has its own specifications in the installation instructions or service manual for the total air percentages needed for the appliance model that you are servicing.

Flame Appearance

The appearance and stability of a flame is influenced by the amount of the primary air supply, the gas-burning speed, and the burner port. When the primary air supply is at 100 percent, under ideal conditions, the speed of burning is at its maximum. The speed of burning is also affected by the type of gas being burned. The flow velocity of the gas depends on the size of the orifice—the smaller the orifice size, the greater the flow velocity. Flow velocity from a burner port will also vary, depending on the size of the opening, with the greatest flow from the center of the port (Figure 8-8).

After the air-gas mixture leaves the port, the flow velocity slows down, and the flame begins to stabilize when the flow velocity equals the burning speed. At the same time, the flame cone begins to take shape when the flow velocity of the air-gas mixture levels off when it leaves the burner port. The burning speed increases and the flame temperature

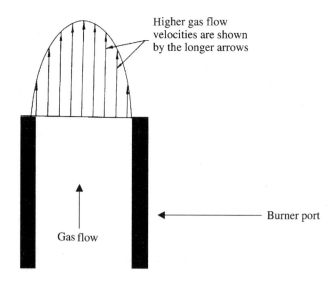

FIGURE 8-8
Flow velocities of gas.

Higher gas flow velocities are shown by the longer arrows

Burner port

Gas flow

increases near the top of the flame cone. Therefore, the shape of the flame cone is rounded off at its tip. The inner cone is determined by the effects of the velocity away from the centerline of the port for each layer of the flame burning.

Appliance manufacturers design gas appliances to have a stable burner flame by port loading. Port loading is expressed as BTU per hour per square inch of open burner port area. It is obtained by dividing the gas input rate by the total area of the port opening.

Flame Lifting

When enough primary air is introduced into a burner, a stabilized flame begins to form, and the flame burns quietly. If too much air is introduced, or if the port loadings are increased, the flame on the burner will have the tendency to lift (Figure 8-9). When the flames begin to lift (blowing) off the burner, they become very noisy. If this condition persists, the flame cones begin to rupture and complete combustion will not take place. In addition, the efficiency of the gas burner begins to drop along with the BTUs and the appliance begins to lose its heat content. If this condition is not corrected, aldehydes and carbon monoxide will begin filling the room.

Additional flame-lifting possibilities include:

- Flame lifting is more likely to occur with natural gas appliances.
- Flame lifting will also occur if any operating or design factors cause the burner to decrease the burning speed.
- If any operating or design factors increase the flow velocity from the ports, flame lifting will occur.
- The port size and depth will affect flame lifting.
- Overrating the burners is a primary cause of flame lifting.
- A cold burner will cause the flame to lift, but it should settle down after a while.

FIGURE 8-9
Too much air supply will cause the flames to lift off the burner.

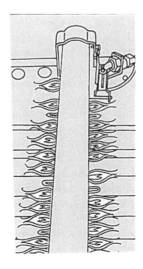

Flashback

When the flame begins to ignite from within the burner head, a condition known as flashback occurs. This is caused when the reduced flow velocity of the air-gas mixture is less than the burning speed near the burner port. In other words, the inner flame cone becomes inverted and ignites the gas from within the burner port. If the burner is properly adjusted, flashback will not occur during normal operation of the gas appliance.

Additional flashback possibilities include:

- Flashback occurs in faster-burning gases.
- Flashback could occur when primary air is increased, which will increase the burning speed.
- Flashback occurs with underrated burners with incorrect orifice size.
- Incorrect gas pressure will cause flashback.
- Decreased flow velocities from a burner port will cause flashback.
- Leaking burner valves cause flashback.
- Flashback can occur when a burner is first turned on if the air-gas mixture is too rich or too lean.

When a burner is turned off, a condition known as "extinction pop," or flashback on extinction, could occur. This could happen immediately, or it may take a few seconds. The extinction pop is caused by excessive air entering the burner when the gas burner valve is turned off. The reduced flow rate of the normal air-gas mixture in a burner is replaced with all air. In addition, it is possible for the flame speed to exceed the flow velocity of the gas, causing a flashback. This condition is not hazardous—it is just annoying to the appliance owner. With a burner that is properly designed, maintained, and adjusted, flashback will not occur under normal operating conditions.

Yellow Flame Tips

In gas appliances, if the primary air supply is reduced, the inner cones of the flames will begin to lengthen, and eventually they will disappear, turning the blue flame tips to yellow. If the primary air supply is cut off completely, the flames will turn yellow (Figure 8-10). When these yellow flames and glowing carbon particles impinge on a cool surface, they begin to quench the complete combustion process and the carbon particles will not burn off. In addition, carbon monoxide, soot, or both will begin to form when the yellow flames impinge on a cool surface.

Dust particles in the air will begin to glow as they pass through a flame (Figure 8-11); the flame's appearance begins to have color streaks. These streaks will appear orange-yellow and will not have an effect on complete combustion. The technician should not confuse dust particles with yellow flame tips. True yellow flame tips will be a pale yellow color and they can be eliminated by increasing the primary air supply. Depending on the air conditions of the surrounding area near the gas appliance, when you increase the primary air supply, you might also increase the amount of dust particles entering into the flame. Another way to distinguish yellow flame tips from dust particles is to put on a pair of brazing goggles and look into the flame. True yellow flame tips will not disappear.

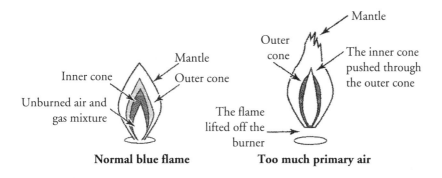

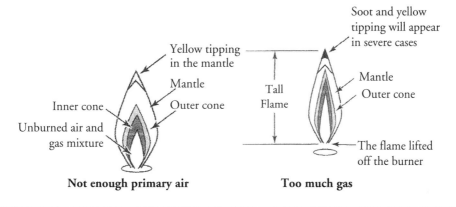

FIGURE 8-10   Characteristics of gas flames.

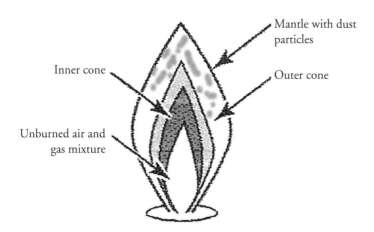

FIGURE 8-11   Dust particles passing through a normal flame.

Appliance Burner Components

Over the years, appliance manufacturers have come out with many types of burner designs (see Figure 1-1 in Chapter 1 and Figure 8-12). In fact, appliance burners must have the following features:

- The burner design must be able to provide complete combustion of the gas.
- The burner design must provide for rapid ignition of the gas and be able to carry over the flame across the entire burner.
- The burner must operate reasonably quietly during ignition, burning, and extinction of the flame.
- The burner design must not allow for excessive flame lifting, flashback, or flame outage.
- The burner must provide uniform heat over the heated area.
- Most important to the consumer and manufacturer, the burner must have a long service life.

Gas Orifice

The gas orifice is not part of the burner, but it plays an important role in its operation (Figure 8-13). In the service field, the gas orifice goes by many names, such as orifice hood, spud, hood, or cap. The purpose of the gas orifice is to regulate or limit the flow of gas into the burner. The size of the hole in the orifice will depend on what type of gas is in use

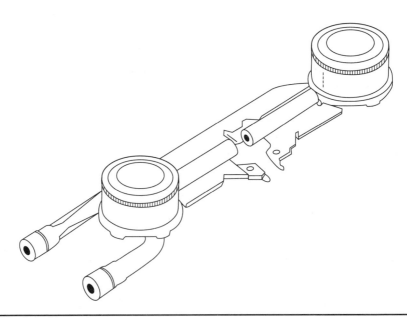

FIGURE 8-12 A double gas burner.

PART II

FIGURE 8-13
Gas orifices are used
to limit gas flow to
burners.

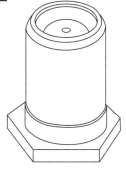

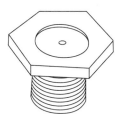

and what constant pressure is needed to light the burner evenly. Another reason for use of an orifice is to force primary air into the burner. Three types of orifices that are in use are:

- The fixed orifice has a predetermined opening in the orifice hood to allow a certain rate of gas to flow (Figure 8-14).

- The adjustable orifice is mainly used in ranges. This type of orifice will allow you to control the gas rate, from zero to its maximum rated flow (Figure 8-15).

- The universal orifice is also mainly used on ranges. This type of orifice was designed to allow the range to be operated on natural or liquefied petroleum gas (Figure 8-16).

Most gas appliances that operate below the elevation of 2000 feet may not need to have the orifices replaced. If you have to replace the orifice, or if you have to check to see if the correct orifice size is installed, refer to Tables 8-1, 8-2, 8-3, and 8-4.

NOTE *Gas appliances shipped from the manufacturer are set up for natural gas. If the home has LP gas, then the appliance will have to be converted for LP use.*

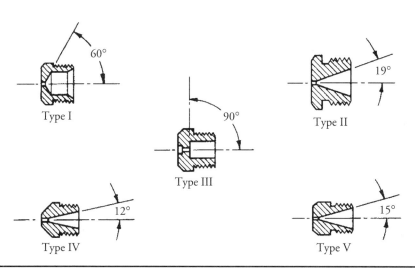

FIGURE 8-14   Different types of fixed orifices.

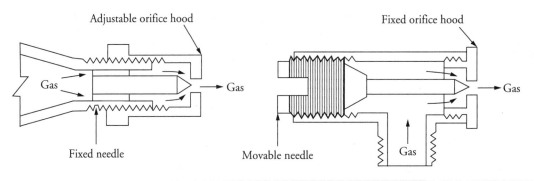

Figure 8-15 Adjustable orifices.

Air Shutter

Most gas appliances have an adjustable primary air shutter to adjust the air intake, and they come in many designs. In addition, manufacturers have designed the air shutter to withstand rust and corrosion and to be secured in any position (Figure 8-17).

Venturi Throat

When gas passes through an orifice, it enters the throat of the mixing tube (Figure 8-18). Called the Venturi throat, this was designed to provide a more constant flow of primary air into the mixing tube of the burner.

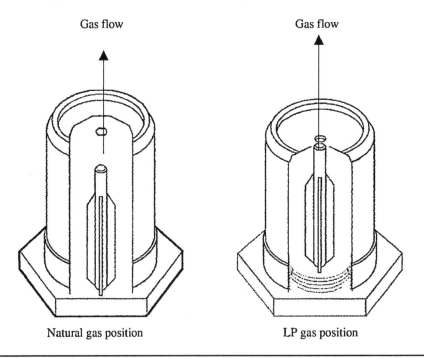

Figure 8-16 Universal orifices.

Orifice Size Drill Size (Decimal or DMS)	Gas Pressure at Orifice—Inches Water Column								
	3	3.5	4	5	6	7	8	9	10
0.008	0.17	0.18	0.19	0.23	0.24	0.26	0.28	0.29	0.30
0.009	0.21	0.23	0.25	0.28	0.30	0.33	0.35	0.37	0.39
0.010	0.27	0.29	0.30	0.35	0.37	0.41	0.43	0.46	0.48
0.011	0.33	0.35	0.37	0.42	0.45	0.48	0.52	0.55	0.59
0.012	0.38	0.41	0.44	0.50	0.54	0.57	0.62	0.65	0.70
80	0.48	0.52	0.55	0.63	0.69	0.73	0.79	0.83	0.88
79	0.55	0.59	0.64	0.72	0.80	0.84	0.90	0.97	1.01
78	0.70	0.76	0.78	0.88	0.97	1.04	1.10	1.17	1.24
77	0.88	0.95	0.99	1.11	1.23	1.31	1.38	1.47	1.55
76	1.05	1.13	1.21	1.37	1.52	1.61	1.72	1.83	1.92
75	1.16	1.25	1.34	1.52	1.64	1.79	1.91	2.04	2.14
74	1.3	1.44	1.55	1.74	1.91	2.05	2.18	2.32	2.44
73	1.51	1.63	1.76	1.99	2.17	2.32	2.48	2.64	2.78
72	1.64	1.77	1.90	2.15	2.40	2.52	2.69	2.86	3.00
71	1.82	1.97	2.06	2.33	2.54	2.73	2.91	3.11	3.26
70	2.06	2.22	2.39	2.70	2.97	3.16	3.38	3.59	3.78
69	2.25	2.43	2.61	2.96	3.23	3.47	3.68	3.94	4.14
68	2.52	2.72	2.93	3.26	3.58	3.88	4.14	4.41	4.64
67	2.69	2.91	3.12	3.52	3.87	4.13	4.41	4.69	4.94
66	2.86	3.09	3.32	3.75	4.11	4.39	4.68	4.98	5.24
65	3.14	3.39	3.72	4.28	4.62	4.84	5.16	5.50	5.78
64	3.41	3.68	4.14	4.48	4.91	5.23	5.59	5.95	6.26
63	3.63	3.92	4.19	4.75	5.19	5.55	5.92	6.30	6.63
62	3.78	4.08	4.39	4.96	5.42	5.81	6.20	6.59	6.94
61	4.02	4.34	4.66	5.27	5.77	6.15	6.57	7.00	7.37
60	4.21	4.55	4.89	5.52	5.95	6.47	6.91	7.35	7.74
59	4.41	4.76	5.11	5.78	6.35	6.78	7.25	7.71	8.11
58	4.66	5.03	5.39	6.10	6.68	7.13	7.62	8.11	8.53
57	4.84	5.23	5.63	6.36	6.96	7.44	7.94	8.46	8.90
56	5.68	6.13	6.58	7.35	8.03	8.73	9.32	9.92	10.44
55	7.11	7.68	8.22	9.30	10.18	10.85	11.59	12.34	12.98
54	7.95	8.59	9.23	10.45	11.39	12.25	13.08	13.93	14.65
53	9.30	10.04	10.80	12.20	13.32	14.29	15.27	16.25	17.09
52	10.61	11.46	12.31	13.86	15.26	16.34	17.44	18.57	19.53
51	11.82	12.77	13.69	15.47	16.97	18.16	19.40	20.64	21.71
50	12.89	13.92	14.94	16.86	18.48	19.77	21.12	22.48	23.65
49	14.07	15.20	16.28	18.37	20.20	21.60	23.06	24.56	25.83
48	15.15	16.36	17.62	19.88	21.81	23.31	24.90	26.51	27.89
47	16.22	17.52	18.80	21.27	23.21	24.93	26.62	28.34	29.81

TABLE 8-1   Orifice Gas Pressure (*continued*)

Orifice Size Drill Size (Decimal or DMS)	Gas Pressure at Orifice—Inches Water Column								
	3	3.5	4	5	6	7	8	9	10
46	17.19	18.57	19.98	22.57	24.72	26.43	28.23	30.05	31.61
45	17.73	19.15	20.52	23.10	25.36	27.18	29.03	30.90	32.51
44	18.45	21.01	22.57	25.57	27.93	29.87	31.89	33.96	35.72
43	20.73	22.39	24.18	27.29	29.87	32.02	34.19	36.41	38.30
42	23.10	24.95	26.50	29.50	32.50	35.24	37.63	40.07	42.14
41	24.06	25.98	28.15	31.69	34.81	37.17	39.70	42.27	44.46
40	25.03	27.03	29.23	33.09	36.20	38.79	41.42	44.10	46.38
39	26.11	28.20	30.20	34.05	37.38	39.97	42.68	45.44	47.80
38	27.08	29.25	31.38	35.46	38.89	41.58	44.40	47.27	49.73
37	28.36	30.63	32.99	37.07	40.83	43.62	46.59	49.60	52.17
36	29.76	32.14	34.59	39.11	42.76	45.77	48.88	52.04	54.74
35	32.36	34.95	36.86	41.68	45.66	48.78	52.10	55.46	58.34
34	32.45	35.05	37.50	42.44	46.52	49.75	53.12	56.55	59.49
33	33.41	36.08	38.79	43.83	48.03	51.46	54.96	58.62	61.55
32	35.46	38.30	40.94	46.52	50.82	54.26	57.95	61.70	64.89
31	37.82	40.85	43.83	49.64	54.36	58.01	61.96	65.97	69.39
30	43.40	46.87	50.39	57.05	62.09	66.72	71.22	75.86	79.80
29	48.45	52.33	56.19	63.61	69.62	74.45	79.52	84.66	89.04
28	51.78	55.92	59.50	67.00	73.50	79.50	84.92	90.39	95.09
27	54.47	58.83	63.17	71.55	78.32	83.59	89.27	95.04	99.97
26	56.73	61.27	65.86	74.57	81.65	87.24	93.17	99.19	104.57
25	58.87	63.58	68.22	77.14	84.67	90.36	96.50	102.74	108.07
24	60.81	65.67	70.58	79.83	87.56	93.47	99.83	106.28	111.79
23	62.10	67.07	72.20	81.65	89.39	94.55	100.98	107.49	113.07
22	64.89	70.08	75.21	85.10	93.25	99.60	106.39	113.24	119.12
21	66.51	71.83	77.14	87.35	95.63	102.29	109.24	116.29	122.33
20	68.22	73.68	79.08	89.49	97.99	104.75	111.87	119.10	125.28
19	72.20	77.98	83.69	94.76	103.89	110.67	118.55	125.82	132.36
18	75.53	81.57	87.56	97.50	108.52	116.03	123.92	131.93	138.78
17	78.54	84.82	91.10	103.14	112.81	120.33	128.52	136.82	143.91
16	82.19	88.77	95.40	107.98	118.18	126.78	135.39	144.15	151.63
15	85.20	92.02	98.84	111.74	122.48	131.07	139.98	149.03	156.77
14	87.10	94.40	100.78	114.21	124.44	133.22	142.28	151.47	159.33
13	89.92	97.11	104.32	118.18	128.93	138.60	148.02	157.58	165.76
12	93.90	101.41	108.52	123.56	135.37	143.97	153.75	163.69	172.13
11	95.94	103.62	111.31	126.02	137.52	147.20	157.20	167.36	176.03
10	98.30	106.16	114.21	129.25	141.82	151.50	161.81	172.26	181.13
9	100.99	109.07	117.11	132.58	145.05	154.71	165.23	175.91	185.03
8	103.89	112.20	120.65	136.44	149.33	160.08	170.96	182.00	191.44
7	105.93	114.40	123.01	139.23	152.56	163.31	174.38	185.68	195.30

TABLE 8-1   Orifice Gas Pressure (*continued*)

Orifice Size Drill Size (Decimal or DMS)	Gas Pressure at Orifice—Inches Water Column								
	3	3.5	4	5	6	7	8	9	10
6	109.15	117.88	126.78	142.88	156.83	167.51	178.88	190.46	200.36
5	111.08	119.97	128.93	145.79	160.08	170.82	182.48	194.22	204.30
4	114.75	123.93	133.22	150.41	164.36	176.18	188.16	200.23	210.71
3	119.25	128.79	137.52	156.26	170.78	182.64	195.08	207.66	218.44
2	128.48	138.76	148.61	168.64	184.79	197.66	211.05	224.74	235.58
1	136.35	147.26	158.25	179.33	194.63	209.48	223.65	238.16	250.54
A	145.34	155.48	165.62	189.28	206.18	219.70	236.60	250.12	263.63
B	150.36	160.85	171.33	195.81	213.29	227.25	244.76	258.74	272.73
C	155.45	166.30	177.14	202.44	220.52	234.98	253.06	267.51	281.97
D	160.63	171.84	183.04	209.19	227.87	242.81	261.48	276.44	291.38
E	165.89	177.47	189.05	216.05	235.34	250.77	270.06	285.49	300.93
F	175.43	187.61	199.78	227.75	248.70	265.02	285.40	301.70	318.02
G	180.81	193.43	206.04	235.49	256.50	273.33	294.35	311.18	327.99
H	187.81	200.91	214.01	244.59	266.42	283.89	305.74	323.20	340.67
I	196.38	210.08	223.77	255.75	278.58	296.85	319.68	337.95	356.20
J	203.66	217.87	232.08	265.24	288.92	307.87	331.55	350.49	396.44
K	209.59	224.22	238.84	272.95	297.33	316.82	341.19	360.69	380.18
L	223.23	240.49	257.75	290.71	316.68	337.44	363.40	384.17	404.92
M	231.00	247.12	263.23	300.83	327.69	349.18	376.03	397.52	419.01
N	242.09	258.98	275.06	315.27	343.42	365.94	394.09	416.61	439.13
O	265.05	283.54	302.03	345.18	376.00	400.66	431.47	456.13	480.79
P	276.92	296.24	315.56	360.64	392.84	418.60	450.80	476.56	502.32
Q	292.57	312.98	333.39	381.03	415.05	442.26	476.28	503.49	530.71
R	305.03	326.33	347.62	397.26	432.73	461.10	496.59	524.95	553.32
S	321.45	344.01	366.57	418.62	456.01	485.91	523.28	553.19	583.09
T	340.19	363.92	387.65	443.04	482.59	514.24	553.79	585.44	617.09
U	359.46	383.54	409.61	468.14	509.93	543.36	585.17	618.60	652.04
V	377.26	403.58	429.90	491.31	535.17	570.27	614.14	649.23	684.33
W	395.48	423.07	450.66	515.05	561.04	597.83	643.82	680.60	717.39
X	418.34	447.53	476.72	544.82	593.47	632.39	681.03	719.94	758.86
Y	433.23	463.45	493.67	564.20	614.58	654.87	705.25	745.56	785.86
Z	452.75	484.33	515.91	589.64	642.26	684.38	737.02	779.14	821.26

UTILITY GASES

(Cubic feet per hour at sea level)

Specific gravity = 0.60

Orifice coefficient = 0.9

For utility gases of another specific gravity, select factor from Table 8-3

For altitudes above 2000 feet, first select the equivalent orifice size at sea level from Table 8-4

TABLE 8-1   Orifice Gas Pressure

Orifice Drill Size (Decimal or DMS)	Gas Input, BTU Per Hour For:	
	Propane	Butane or Butane-Propane Mixture
0.008	500	554
0.009	641	709
0.01	791	875
0.011	951	1053
0.012	1130	1250
80	1430	1590
79	1655	1830
78	2015	2230
77	2545	2815
76	3140	3480
75	3465	3840
74	3985	4410
73	4525	5010
72	4920	5450
71	5320	5900
70	6180	6830
69	6710	7430
68	7560	8370
67	8040	8910
66	8550	9470
65	9630	10,670
64	10,200	11,300
63	10,800	11,900
62	11,360	12,530
61	11,930	13,280
60	12,570	13,840
59	13,220	14,630
58	13,840	15,300
57	14,550	16,090
56	16,990	18,790
55	21,200	23,510
54	23,850	26,300
53	27,790	30,830
52	31,730	35,100
51	35,330	39,400
50	38,500	42,800
49	41,850	45,350
48	45,450	50,300

TABLE 8-2 LP Orifice Drill Size (*continued*)

Orifice Drill Size (Decimal or DMS)	Gas Input, BTU Per Hour For:	
	Propane	Butane or Butane-Propane Mixture
47	48,400	53,550
46	51,500	57,000
45	52,900	58,500
44	58,050	64,350
43	62,200	69,000
42	68,700	76,200
41	72,450	80,200
40	75,400	83,500
39	77,850	86,200
38	81,000	89,550
37	85,000	94,000
36	89,200	98,800
35	95,000	105,300
34	97,000	107,200
33	101,000	111,900
32	105,800	117,000
31	113,200	125,400
30	129,700	143,600
29	145,700	163,400
28	154,700	171,600
27	163,100	180,000
26	169,900	187,900
25	175,500	194,600
24	181,700	201,600
23	186,800	206,400
22	193,500	214,500
21	198,600	220,200
20	203,700	225,000
19	217,100	241,900
18	225,600	249,800

Liquefied Petroleum Gas (LP or LPG)

(BTU per hour at sea level)

	Propane	Butane
BTU per cubic foot	2500	3175
Specific gravity	1.53	2.00
Pressure at orifice measured in inches water column	11	11
Orifice coefficient	0.9	0.9

For altitudes above 2000 feet, first select the equivalent orifice size at sea level from Table 8-4

TABLE 8-2 LP Orifice Drill Size

Specific Gravity	Factor		Specific Gravity	Factor
0.45	1.155		0.95	0.795
0.50	1.095		1.00	0.775
0.55	1.045		1.05	0.756
0.60	1.000		1.10	0.739
0.65	0.961		1.15	0.722
0.70	0.926		1.20	0.707
0.75	0.894		1.25	0.693
0.80	0.866		1.30	0.679
0.85	0.840		1.35	0.667
0.90	0.817		1.40	0.655

To select an orifice size to provide a desired gas flow rate:

(1) Obtain the factor for the known specific gravity of the gas from above.
(2) Divide the desired gas flow rate by this factor, which gives the equivalent flow rate with a 0.60 specific gravity gas.
(3) Calculate the gas flow from a gas meter using the following formula:

$$\text{Gas flow measured in cubic feet per hour} = \frac{(\text{cubic feet per revolution of gas meter dial}) \times 3600}{\text{Time one revolution of the gas meter dial in seconds}}$$

The number 3600 converts seconds to hours.

Example: If the gas meter measures ½ cubic foot per revolution and the amount of time for that revolution is found to be 15 seconds, then the gas flow will calculate as follows:

$$\text{Gas flow rate} = \frac{0.5 \times 3600}{15} = 120 \text{ cubic feet per hour}$$

(4) Use Table 8-1 to select orifice size.

To estimate gas flow rate for a given orifice size and gas pressure when it is not possible to meter gas flow:

(1) Use Table 8-1 to obtain equivalent gas flow rate with a 0.60 specific gravity gas.
(2) Obtain the factor from Table 8-3 for the specific gravity of the gas being burned.
(3) Multiply the rate obtained in (1) by the factor from (2) to estimate gas flow rate to the burner.

TABLE 8-3   Factors for Utility Gases of Another Specific Gravity

Orifice Size at Sea level	Orifice Size Required at Other Elevations								
	2000	3000	4000	5000	6000	7000	8000	9000	10,000
1	2	2	3	3	4	5	7	8	10
2	3	3	4	5	6	7	9	10	12
3	4	5	7	8	9	10	12	13	15
4	6	7	8	9	11	12	13	14	16
5	7	8	9	10	12	13	14	15	17
6	8	9	10	11	12	13	14	16	17
7	9	10	11	12	13	14	15	16	18
8	10	11	12	13	13	15	16	17	18
9	11	12	12	13	14	16	17	18	19
10	12	13	13	14	15	16	17	18	19
11	13	13	14	15	16	17	18	19	20
12	13	14	15	16	17	17	18	19	20
13	15	15	16	17	18	18	19	20	22
14	16	16	17	18	18	19	20	21	23
15	16	17	17	18	19	20	20	22	24
16	17	18	18	19	19	20	22	23	25
17	18	19	19	20	21	22	23	24	26
18	19	19	20	21	22	23	24	26	27
19	20	20	21	22	23	25	26	27	28
20	22	22	23	24	25	26	27	28	29
21	23	23	24	25	26	27	28	28	29
22	23	24	25	26	27	27	28	29	29
23	25	25	26	27	27	28	29	29	30
24	25	26	27	27	28	28	29	29	30
25	26	27	27	28	28	29	29	30	30
26	27	28	28	28	29	29	30	30	30
27	28	28	29	29	29	30	30	30	31
28	29	29	29	30	30	30	30	31	31
29	29	30	30	30	30	31	31	31	32
30	30	31	31	31	31	32	32	33	35
31	32	32	32	33	34	35	36	37	38
32	33	34	35	35	36	36	37	38	40
33	35	35	36	36	37	38	38	40	41
34	35	36	36	37	37	38	39	40	42
35	36	36	37	37	38	39	40	41	42
36	37	38	38	39	40	41	41	42	43
37	38	39	39	40	41	42	42	43	43
38	39	40	41	41	42	42	43	43	44
39	40	41	41	42	42	43	43	44	44
40	41	42	42	42	43	43	44	44	45
41	42	42	42	43	43	44	44	45	46
42	42	43	43	43	44	44	45	46	47

TABLE 8-4   Equivalent Orifice Sizes at High Altitudes (*continued*)

Orifice Size at Sea level	Orifice Size Required at Other Elevations								
	2000	3000	4000	5000	6000	7000	8000	9000	10,000
43	44	44	44	45	45	46	47	47	48
44	45	45	45	46	47	47	48	48	49
45	46	47	47	47	48	48	49	49	50
46	47	47	47	48	48	49	49	50	50
47	48	48	49	49	49	50	50	51	51
48	49	49	49	50	50	50	51	51	52
49	50	50	50	51	51	51	52	52	52
50	51	51	51	51	52	52	52	53	53
51	51	52	52	52	52	53	53	53	54
52	52	53	53	53	53	53	54	54	54
53	54	54	54	54	54	54	55	55	55
54	54	55	55	55	55	55	56	56	56
55	55	55	55	56	56	56	56	56	57
56	56	56	57	57	57	58	59	59	60
57	58	59	59	60	60	61	62	63	63
58	59	60	60	61	62	62	63	63	64
59	60	61	61	62	62	63	64	64	65
60	61	61	62	63	63	64	64	65	65
61	62	62	63	63	64	65	65	66	66
62	63	63	64	64	65	65	66	66	67
63	64	64	65	65	65	66	66	67	68
64	65	65	65	66	66	66	67	67	68
65	65	66	66	66	67	67	68	68	69
66	67	67	68	68	68	69	69	69	70
67	68	68	68	69	69	69	70	70	70
68	68	69	69	69	70	70	70	71	71
69	70	70	70	70	71	71	71	72	72
70	70	71	71	71	71	72	72	73	73
71	72	72	72	73	73	73	74	74	74
72	73	73	73	73	74	74	74	74	75
73	73	74	74	74	74	75	75	75	76
74	74	75	75	75	75	76	76	76	76
75	75	76	76	76	76	77	77	77	77
76	76	76	77	77	77	77	77	77	77
77	77	777	77	78	78	78	78	78	78
78	78	78	78	79	79	79	79	80	80
79	79	80	80	80	80	0.013	0.012	0.012	0.012
80	80	0.013	0.013	0.013	0.012	0.012	0.012	0.012	0.011

Equivalent orifice sizes at high altitudes
(Includes 4% input reduction for each 1000 feet)

TABLE 8-4   Equivalent Orifice Sizes at High Altitudes

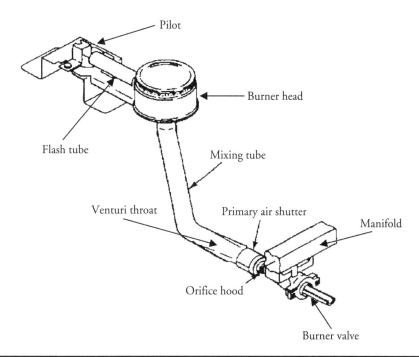

FIGURE 8-17   The primary air shutter is located at the end of the mixing tube. The burner head is designed with ports around the burner to allow for even burning and a blue flame.

Mixing Tube

The mixing tube (see Figure 8-17) is designed to mix the air and gas uniformly after it leaves the Venturi throat but before entering the burner head.

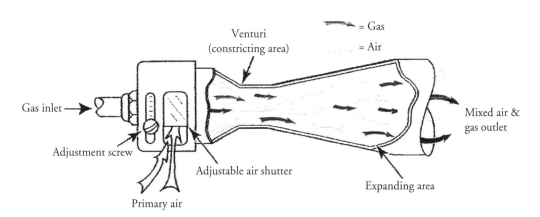

FIGURE 8-18   The Venturi throat is incorporated into the mixing tube of the burner.

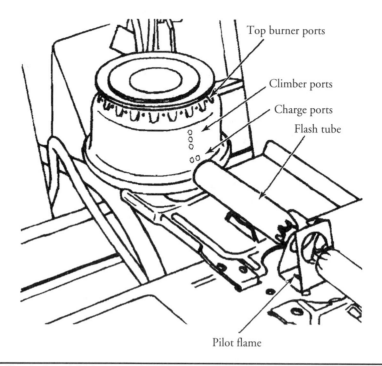

Top burner ports

Climber ports

Charge ports

Flash tube

Pilot flame

FIGURE 8-19   The design of the burner ports and the location of the burner portholes.

Burner Head

When gas leaves the mixing tube, it enters the burner head, which contains the burner ports. The air-gas mixture is then distributed evenly among the burner ports. As the air-gas mixture emerges from the burner ports, it is lit with an ignition source. The size and shape of the burner head will allow even heat transfer to occur (see Figure 8-17).

Burner Ports

The burner ports (Figure 8-19) serve to distribute the flame evenly, provide even heat transfer, and provide a stable blue flame. In addition, when the flames are distributed throughout the burner ports, they will then pick up secondary air to assist in complete combustion.

Burner Operation

When the gas is turned on, whether by a knob or automatically by a control device, gas will flow from the supply line through a gas pressure regulator, gas manifold, burner valves, and orifice into the Venturi throat and mixing tube. The air-gas mixture then enters the burner head to be ignited. As the air-gas mixture leaves the burner head, when ignited, the flame begins to burn evenly around the burner head. Figure 8-20 illustrates an exploded view of the gas range's components.

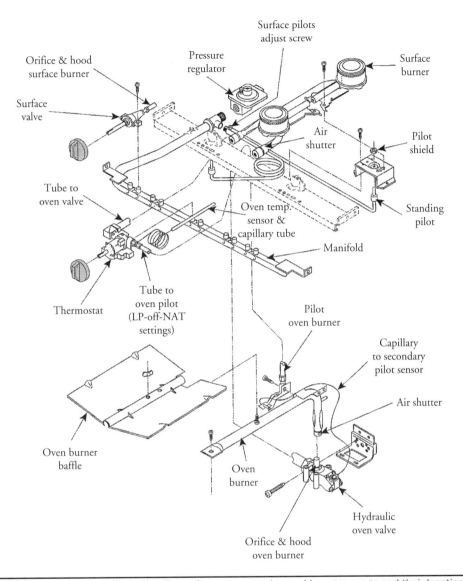

FIGURE 8-20   Exploded view illustrating the various components used in a gas range and their location.

Ventilation

When you properly install a ventilation system, it will serve the following important purposes:

- The by-products of hot combustion gases will be transported outside the home.
- The home will be protected from fire hazards.

- A ventilation system will provide good air circulation and adequate oxygen supply for the gas appliance and the occupants of the residence.

- It will remove the water vapors produced when burning gas.

A wide variety of venting options are available from vent manufacturers, and every gas appliance manufacturer will provide venting instructions for properly installing its products. It is strongly recommended that you follow these instructions. You must use the proper venting materials as described in the installation instructions for the model being installed, and install the venting system according to your local building codes. For additional information on proper ventilation, visit www.epa.gov/iaq/homes/hip-combustion.html.

Troubleshooting Ventilation Problems

The most common problems with ventilation systems are:

- Improper maintenance of the ventilation system

- Incorrect sizing and installation of the ventilation system

- Inadequate air supply

If you suspect a problem with the ventilation system, check the following:

- Make sure that you have the correct vent sizing.

- Check if there are too many elbows or if the length of the venting system exceeds the manufacturer's recommendations or local building codes.

- Inspect the entire ventilation system for faults, such as a disconnected or crushed pipe.

- Inspect for an obstruction, such as a clogged vent cap.

- Inspect all air openings for obstructions.

Measuring Gas Pressure

When installing or repairing gas appliances, it is often necessary to measure the pressure of the gas supply to the appliance. There also may be times when the technician will have to test the gas pressures at the manifold or orifices. The two types of test instruments used today to test the gas pressures are the manometer and the magnehelic gauge.

The manometer is a U-shaped tube equipped with a scale to measure gas pressure in water column inches (Figure 8-21). Before you check the gas pressure on an appliance, locate the model and serial number nameplate. This nameplate will have the gas rating stamped on it. Now turn off the gas supply to the appliance. Before you can begin to test the gas pressure, you must first set up the manometer by adding water to the U-shaped tube until both columns read zero on the scale (Figure 8-22a). If reading the scale is difficult, you can add a food coloring to the water to enhance the color so that you can read the scale better. When measuring gas pressure in an appliance, the tubing from the manometer is attached to the orifice, manifold, or gas supply line, and the other end of the tube on the manometer is left open to the room atmosphere. To test the gas pressure at the burner orifice, remove the burner (Figure 8-23). Next, attach the long hose from the manometer to the burner orifice. Now turn on the gas supply to the appliance. In addition, turn on the burner valve being tested, and turn on one other burner to serve as the load. When the gas pressure is applied to the manometer, it pushes down on the water column and the water column rises on the other side of the manometer (Figure 8-22b). To read the manometer scale, observe where the water column stops. In Figure 8-22b, the incoming gas has pushed the water in the manometer down 2 inches below zero, and the water column on the other side is pushed up 2 inches above zero. The pressure reading is obtained by adding the two readings together. Figure 8-22b shows that the total change equals a 4-inch water column. The reading should be within the rating on the model and serial nameplate.

The magnehelic gauge shown in Figure 8-24 also measures gas pressure. This gauge can measure gas pressure faster than the manometer can. To use this type of gauge, just follow the same procedures for the manometer. The only difference is that you read the gas pressure directly from the gauge scale. Some models have different scales on the gauge dial—just read the gauge that indicates water column pressure.

Many companies manufacture test instruments for checking gas pressures. Websites to check out include:

- http://www.uniweld.com/catalog/gauges _thermometers/gas_pressure_test_kit.htm
- http://www.dwyer-inst.com
- http://www.marshbellofram.com

Testing for Gas Leaks

When testing for a gas leak, do not use a lighter or a lit match. That is not the safe or correct way to test for gas leaks. All gas appliances, when they are installed or repaired, must be tested for leaks before placing the appliance in operation. Remember, LP or LPG gas is heavier than air and settles in low-lying areas. If you do not have a combustible gas detector (and I recommend that you purchase one), use your nose to smell for a gas leak all around the appliance at floor level. A safe and effective way to test for gas leaks is to use a chloride-free soap-and-water solution to check all connections and fittings. The solution can be sprayed, applied directly, or an applicator can be used to spread the solution. If a gas leak is detected by the soap solution it will begin to form bubbles.

Sometimes, air might be present in the gas lines from installation or repairs, and it could prevent the pilot or burner from lighting on initial startup. The gas

FIGURE 8-21
A manometer. This instrument is used to test gas pressure.

PART II

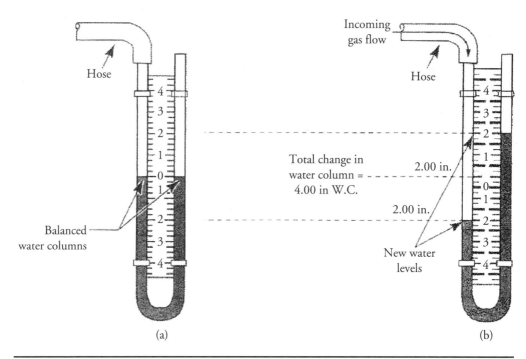

Total change in
water column =
4.00 in W.C.

2.00 in.

2.00 in.

Figure 8-22 (a) Before using a manometer, check to see if the columns have equal amounts of water. (b) When reading a manometer, you add the change in both columns; the sum equals the pressure of the gas.

Figure 8-23 Removing the burner to test the gas pressure at the gas valve.

FIGURE 8-24
A magnehelic gauge.
This instrument is
used to test gas
pressure.

lines should be purged of air in a well-vented area. Before proceeding with purging the gas lines, inspect the work area for any other sources of ignition, such as flames, burning candles, running electrical appliances, etc. With a safe working environment, you can begin to purge the gas lines. On gas ranges/ovens, just turn on the burners until you smell the gas or until the burners are lit properly. With gas dryers and gas water heaters, turn off the gas supply valve. Remove the gas line from the appliance, and slowly begin to turn on the gas supply valve. When you begin to smell gas, turn off the main supply valve. Next, reconnect the gas line to the appliance, open the gas supply valve, and test for gas leaks. If there are no gas leaks, run the appliance.

Use an approved pipe joint compound for connecting gas piping and fittings to the appliance. In addition, never use white Teflon tape on gas lines. White Teflon tape does not guarantee a leak-free connection. You can use a PTFE yellow Teflon tape on gas connections and fittings. This type of approved Teflon tape is a specially designed thread-sealing tape for use on gas appliances. To attach PTFE yellow Teflon tape to pipe thread, wrap the tape in the direction with the thread, as shown in Figure 8-25.

FIGURE 8-25
Wrap tape in the
thread direction on a
pipe fitting.

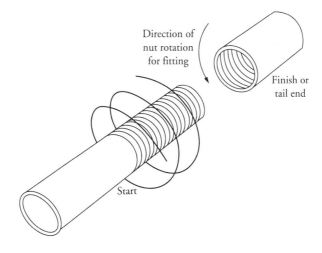

Direction of
nut rotation
for fitting

Finish or
tail end

Start

Combustible Gas Detector

Combustible gas detectors are another general-purpose tool that the service technician could use to detect gas leaks (Figure 8-26). This instrument uses an audible alarm that gets louder when you approach a gas leak. This device has the capabilities of detecting LP (LPG) or natural gas and other hydrocarbons and halogenated hydrocarbons, depending on the manufacturer. The sensitivity of a combustible gas detector can be between 50 and 1500 ppm of gas detection, with an instantaneous response.

One manufacturer of combustible gas detectors can be viewed on the Web at www .aprobe.com.

Gas Conversion Procedures

Before installing a new gas appliance, look at the model and serial dataplate to determine that the appliance is made for the type of gas supplied in the home. Do not install an LP or LPG gas appliance when natural gas is supplied to the home. Conversely, do not install a natural gas appliance when LP or LPG gas is supplied to the home.

Appliance manufacturers do make gas appliances that can be installed with either LP or natural gas. All you have to do is follow the manufacturer's recommendations for converting the appliance from one type of gas to the other.

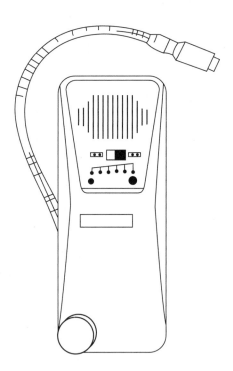

FIGURE 8-26
A combustible gas detector. This meter is used for detecting gas leaks.

Gas Pressure Regulator

To convert a gas pressure regulator from one type of gas to the other, you must first shut off the main gas supply to the appliance (Figure 8-27). Also, if the gas range has an electrical connection, turn off the electricity. Locate the gas pressure regulator valve. On some models, the gas pressure regulator valve is located in the lower rear of the range (Figure 8-28a) or under the cooktop (Figure 8-28b). To convert the gas pressure regulator valve, you do not have to remove it from the range. Just gain access to it. As you can see in Figure 8-29, remove the plastic cover from the regulator cap. Next, unscrew the gas pressure regulator cap with a wrench; turn it counterclockwise to remove the cap and leave the spring in the regulator.

NOTE *In Figure 8-29, the regulator cap has a gas designation stamped on it.*

For LP gas operation, the "<-LP" indicator must face in the direction of the gas pressure regulator, and the hollow end faces away from the regulator. For natural gas operation, the "<-NG" indicator must face in the direction of the regulator, with the dimple end facing away from the valve. Reinstall the cap and tighten with a wrench in the clockwise direction. Do not overtighten the cap, as you might damage it. Finally, replace the plastic cover on the regulator cap.

Converting Surface Burners

To convert the surface burners from one type of gas to the other, you must first shut off the main gas supply to the appliance (see Figure 8-27). Also, if the gas range has an electrical connection, turn off the electricity. Remove the burner grate and burner cap (Figure 8-30). Remove the screws from the burner base, and reinstall one screw to hold the orifice holder (Figure 8-31) secured to the cooktop. This will steady the orifice holder so that you can remove the orifice without damaging the orifice holder. To remove the orifice, use a 5/16 nut driver with masking tape applied to the end of it. The masking tape will hold the orifice in the nut driver when you remove it. Turn the nut driver counterclockwise to remove the orifice. Locate the extra orifices that came with the range. Most likely, they will be attached

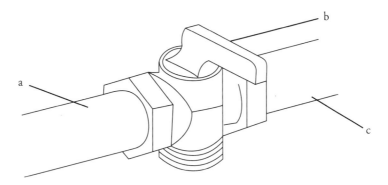

FIGURE 8-27
The main gas supply
shut-off valve.

a. The gas supply line from the shut-off valve to the range
b. Gas supply-line shut-off valve
c. The main gas supply entering the home to the shut-off valve

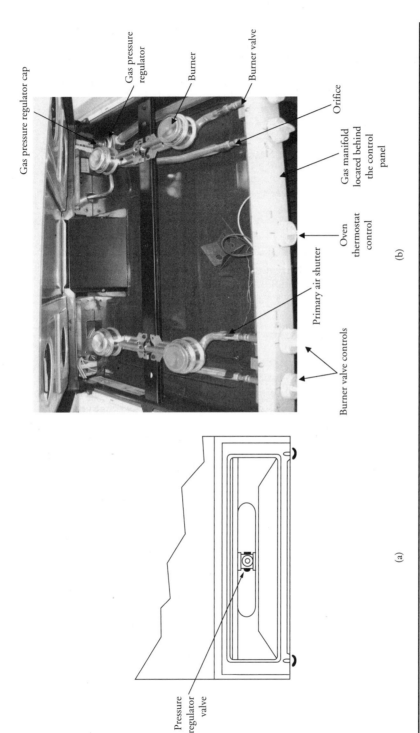

FIGURE 8-28 (a) On some models, the gas pressure regulator is located in the rear of the range behind the lower drawer. (b) The location of gas components under the range cooktop.

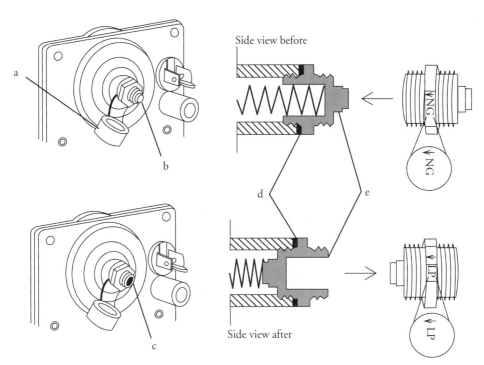

Side view before

Side view after

a. Plastic cover
b. Pressure regulator cap with the solid end facing toward you
c. Pressure regulator cap with the hollow end facing toward you
d. Washer
e. Pressure regulator cap

FIGURE 8-29 (a) Plastic cover. (b) Pressure regulator cap with the solid end facing outward for natural gas. (c) Pressure regulator cap with the hollow end facing outwards for LPG gas. (d) Washer. (e) Gas pressure regulator cap.

somewhere near the main gas inlet in a holder or small bag. To choose the correct size orifice for the burner that you are converting, refer to the model and serial dataplate. Orifices are stamped with a number and are color-coded, with a groove in the center side of the hex head (Figure 8-32). Use Table 8-5 for the proper sizing of the orifices. With the correct orifice size in hand, just reverse the order of disassembly, and reassemble the burner.

If the range you are servicing resembles the one in Figure 8-28, convert the burner by first removing the screw that secures it to the burner bar. Remove the burner by lifting it up and out of the range. The orifice is now ready to be converted. To convert the orifice from natural to LP gas, use a wrench around the orifice head and turn the orifice clockwise to seat the orifice. Do not overtighten; you might damage the orifice opening. On some models, the orifice hood will be replaced entirely with the correct orifice size for the type of gas used. With the correct orifice adjustment completed, just reverse the order of disassembly, and reassemble the burner.

Burner grate

Burner cap

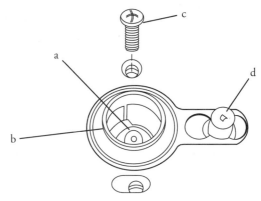

c

a

d

b

a. Orifice
b. Orifice holder
c. Screw
d. Spark electrode

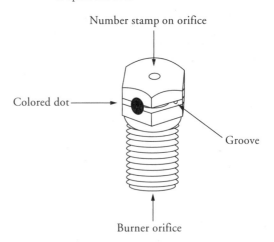

Number stamp on orifice

Colored dot

Groove

Burner orifice

LP or LPG Gas Orifice Sizing Chart		
Burner BTU Rating	**Color**	**Orifice Size**
12,000	Green/Magenta	1.03 mm
11,000	Green/Yellow	0.99 mm
8000	Green/Light blue	0.80 mm
5000	Green/Brown	0.65 mm
Natural Gas Orifice Sizing Chart		
Burner BTU Rating	**Color**	**Orifice Size**
13,500	Blue/Orange	1.86 mm
12,000/12,500	Blue/Red	1.76 mm
9500	Blue/Brass	1.51 mm
8000	Blue/Black	1.39 mm
5000	Blue/White	1.12 mm

Note: To choose the correct size orifice for the burner that you are converting, refer to the model and serial dataplate for the proper sizing of the orifices for each burner location.

TABLE 8-5   LP or LPG Gas Orifice Sizing Chart

Converting the Oven Bake Burner

To convert the oven bake burner from one type of gas to the other, you must first shut off the main gas supply to the appliance (see Figure 8-27). Also, if the gas range has an electrical connection, turn off the electricity. When converting the oven bake burner, remove the racks and gain access to the oven burner (Figure 8-33). Place a wrench on the orifice, and turn the wrench clockwise for the LP gas setting. Turn the wrench in a counterclockwise direction for natural gas. Only turn the wrench about two or two and a half turns. Do not overtighten the orifice hood; you might damage the orifice pin. With the correct orifice adjustment completed, just reverse the order of disassembly, and reassemble the oven.

Converting the Oven Broil Burner

To convert the oven broil burner from one type of gas to the other, you must first shut off the main gas supply to the appliance (see Figure 8-27). Also, if the gas range has an electrical connection, turn off the electricity. When converting the oven broil burner, gain access to the burner (Figure 8-34). Use a wrench to turn the orifice hood counterclockwise for natural gas or clockwise for LP or LPG gas. Only turn the wrench about two or two and a half turns. Do not overtighten the orifice hood; you might damage the orifice pin. With the correct orifice adjustment completed, just reverse the order of disassembly, and reassemble the oven.

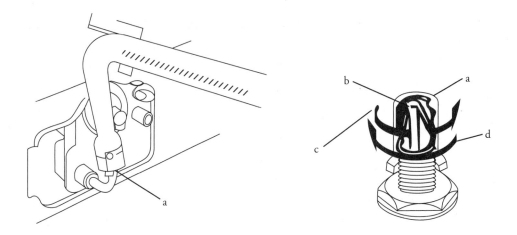

a. Orifice
b. Orifice pin
c. Turn in this direction for natural gas
d. Turn in this direction for LP or LPG gas

FIGURE 8-33   The oven burner is located underneath the bottom oven cover. Remove the racks and lift out the bottom cover for access.

FIGURE 8-34
The broiler burner
location.

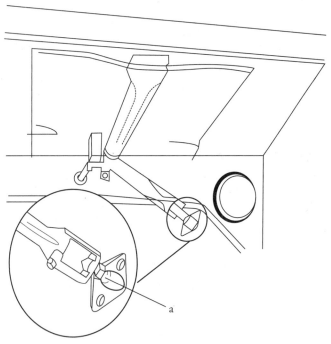

a. Orifice

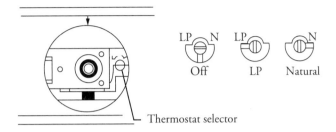

FIGURE 8-35
Oven pilot conversion.
The pilot control screw
is located behind the
thermostat knob.

Thermostat selector

Converting the Bake/Broil Thermostat

Gas ranges with oven pilot lights will have to be converted from one type of gas to the other. Remove the oven thermostat knob to expose the oven thermostat selector. Insert a small screwdriver into the pilot control screw, turn the screw clockwise for LP or LPG gas or counterclockwise for natural gas (Figure 8-35). Replace the oven thermostat knob. Gas ranges with electronic spark ignition for pilot control will be discussed later in the book.

Gas Appliance Maintenance

To maintain gas appliances, always follow the manufacturer's recommendations for periodic maintenance as stated in the use and care manual. The range, oven, or cooktop can be cleaned with warm water, mild detergent, and a soft cloth on all cleanable parts, as recommended in the use and care manual. Also, never use abrasive cleaners that are not recommended by the manufacturer.

Do not allow grease spillovers to accumulate after cooking on top of the range; it will become a fire hazard. When cleaning the burners, always make sure that all of the portholes are free of debris. If, for any reason, the burner portholes are blocked, the flame appearance will be different. Blocked portholes will reduce gas flow, and the heating value of the burner will be reduced. Maintenance procedures for gas water heaters and gas dryers will be discussed in later chapters.

PART III

PART

Refrigeration and Air Conditioning

CHAPTER 9

Principles of Air Conditioning and Refrigeration

Many excellent books have been written on the subject of air conditioning and refrigeration, and this chapter will only cover the basics needed to diagnose and repair the electrical portion of domestic refrigerators and room air conditioners (RACs). The chapter will cover the refrigeration process, but will not cover the replacement procedures of any sealed system components. This is a specialized area, and only an EPA-certified technician can repair the sealed system of a refrigerator or air conditioner (AC). The HVAC/R-certified technician has the proper training, tools, and equipment required to make the necessary repairs to the sealed system. An individual who is not certified to service the sealed system could raise the risk of personal injury, as well as property damage. The uncertified individual could void the manufacturer's warranty on the product if he or she attempts entry into the sealed system.

Introduction to Air Conditioning and Refrigeration

Before the 1900s and mechanical refrigeration, people cooled their foods with ice and snow. The ice and snow were transported long distances and stored in cellars below ground or in icehouses. Refrigeration is the process of removing heat from an enclosed area, thus lowering the temperature in it. Air conditioning and refrigeration began to come to life in the early 1900s. The first practical refrigerating machine was invented in 1834, and it was improved a decade later.

The two most important figures involved with refrigeration and air conditioning were John Gorrie and Willis Carrier. John Gorrie, an early pioneer of refrigeration and air conditioning, was granted a patent in 1851 for his invention of the first commercial refrigeration machine to produce ice. Improvements were made to Gorrie's invention in the late 1880s, and the reciprocating compressor was used commercially in meat and fish plants and for ice production. In 1902, Willis Carrier designed and invented the first modern air conditioning system, and in 1922, Carrier invented the first centrifugal refrigeration machine. Willis Carrier went on to found the company we know today as Carrier Corporation. Toward the end of the 1920s, the first self-contained room air conditioner was introduced.

Refrigerators from the late 1880s until 1929 used ammonia, methyl chloride, or sulfur dioxide as refrigerants. These refrigerants were highly toxic and deadly if a leak occurred. In 1931, the DuPont Company began producing commercial amounts of "R-12," also known by its trade name "Freon." With safer refrigerants being introduced, the refrigerator and air conditioner market began to grow. Only half a century later did people realize that these chlorofluorocarbons endangered the earth's ozone layer. Today, refrigerant companies are working harder to produce even safer refrigerants that will not endanger the earth's ozone layer.

Saving the Ozone Layer

High above the earth is a layer of ozone gas that encircles the planet. The purpose of the gas is to block out most of the damaging ultraviolet rays from the sun. Such compounds as CFCs, HCFCs, and halons have depleted the ozone layer, allowing more ultraviolet (UV) radiation to penetrate to the earth's surface.

In 1987, the United States, the European Economic Community, and 23 other nations signed the Montreal Protocol on Substances that Deplete the Ozone Layer. The purpose of this agreement was to reduce the use of CFCs throughout the world. To strengthen the original provisions of this protocol, 55 nations signed an agreement in London on June 29, 1990. At this second meeting, they passed amendments that called for a full phaseout of CFCs and halons by the year 2000. Also passed at that meeting was the phaseout of HCFCs by the year 2020, if feasible, and no later than the year 2040, in any case.

On November 15, 1990, President George H.W. Bush signed the 1990 Amendment to the Clean Air Act, which established the National Recycling and Emissions Reduction Program. This program minimizes the use of CFCs and other substances harmful to the environment, while calling for the capture and recycling of these substances. The provisions of the Clean Air Act are more stringent than those contained in the Montreal Protocol as revised in 1990.

Beginning on July 1, 1992, the Environmental Protection Agency (EPA) developed regulations under Section 608 of the Clean Air Act (the Act) that limit emissions of ozone-depleting compounds. Some of these compounds are known as *chlorofluorocarbons* (CFCs) and *hydrochlorofluorocarbons* (HCFCs). The Act also prohibits releasing refrigerant into the atmosphere while maintaining, servicing, repairing, or disposing of refrigeration and air conditioning equipment. These regulations also require technician certification programs. A sales restriction on refrigerant is also included, whereby only certified technicians will legally be authorized to purchase such refrigerant. In addition, the penalties and fines for violating these regulations can be rather severe.

In 1993, Section 605 of the Clean Air Act established the phaseout framework. This is an accelerated phaseout of class II controlled substances (including R-22). It also limited production and consumption of R-22 between 2010 and 2020 to the servicing of equipment manufactured prior to January 1, 2010.

By January 1, 2015, The Montreal Protocol requires the United States to reduce its consumption of HCFCs by 90 percent below the U.S. baseline, and by January 1, 2020, the United States needs to reduce its consumption of HCFCs by 99.5 percent below the U.S. baseline.

Matter

In order to understand how refrigeration and air conditioners work, it is necessary to understand several basic laws of matter. There are currently five states of matter: solid, liquid, gas, plasma, and Bose-Einstein condensate.

In our discussion of matter, only three of the five states apply to air conditioning and refrigeration (Figure 9-1):

- Matter in the form of a solid will retain its shape and volume without a container. The molecules within a solid are compressed and bound together, and under normal conditions will not move at all.

- Matter in the form of a liquid will take the shape of its container, without losing any volume if not under pressure. Light-density fluids, such as water, will eventually lose some of their volume due to evaporation. The molecules within a liquid are spaced apart from each other and are constantly on the move.

- Matter in the form of a gas will take the shape and volume of its container and will expand to fill the container. The molecules within a gas are spaced far apart from each other.

Change of State

There are five principles regarding how matter can change from one form to another. When matter is heated, cooled, or an increase or decrease of pressure occurs, matter will be transformed to another state:

- Liquefaction occurs when matter changes from a solid to a liquid. For example, an ice cube will begin to change state when heat is applied.

- Solidification occurs when matter changes from a liquid to a solid. For example, when the temperature of water reaches its freezing point, it will freeze and become a solid.

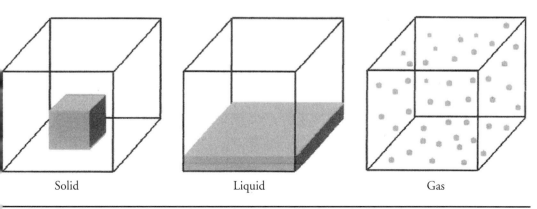

Solid   Liquid   Gas

FIGURE 9-1   The three states of matter that pertain to air conditioning and refrigeration.

- Vaporization occurs when a liquid matter like water is heated to its boiling point and transforms into a vapor.

- Condensation occurs when a vapor transforms into a liquid. For example, when water vapor is boiled off into steam in a closed vessel and then allowed to cool down, it will begin to turn back into a liquid.

- Sublimation can occur when a solid matter transforms into a vapor without passing through the liquid state. For example, carbon dioxide in a solid form (dry ice) at atmospheric pressure will transform into a vapor state at normal temperatures, thus bypassing the liquid state.

Law of Thermodynamics

Energy exists in many forms, such as heat, light, mechanical, chemical, and electrical energy. The first law of thermodynamics is the study of that energy. Energy itself is the ability to do work, and heat is only one form of energy—ultimately, all forms of energy end up as heat energy. The first law of thermodynamics states that energy cannot be created or destroyed; it can only be converted from one form to another.

The second law of thermodynamics as it pertains to air conditioning and refrigeration deals with heat energy travel. The law states that heat energy can only travel in one direction—from hot to cold. When two objects are placed together, heat will travel from the warmer object to the cooler object until the temperatures of both objects are equal; only then will heat transfer stop. The rate of travel will depend on the temperature difference between the objects. The greater the temperature difference between the objects, the faster heat will travel to the cooler object.

It's All About Heat

Heat is a form of energy that:

- Causes temperatures to rise in an object
- Has the capacity to do work
- Has the ability to flow from a warmer substance to a cooler one
- Can be converted to other forms of energy

Heat energy can be measured with a thermometer. The scales used are Fahrenheit (F) and Celsius (C). Heat intensity has three reference points: freezing point (32 degrees F or 0 degrees C), boiling point (212 degrees F and 100 degrees C), and absolute zero (–470 degrees F or –273 degrees C). This third reference point is believed to be where all molecular action ceases.

Another way to measure heat energy is in units called British Thermal Units (BTUs). A BTU represents the amount of heat required to raise the temperature of 1 pound of water 1 degree Fahrenheit at sea level. All refrigeration equipment and air conditioning equipment are rated in terms of how many BTUs of cooling or heating they can remove or put into a given area.

FIGURE 9-2
The transfer of heat from a warmer object to a cooler object is known as conduction.

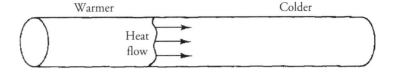

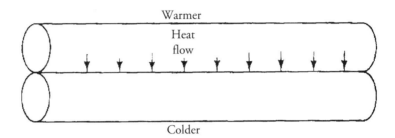

On larger room air conditioners and central air conditioning, it is necessary to calculate the heat load and heat gain of a given area to properly size the unit. The same is required for large refrigeration equipment and commercial refrigeration.

Methods of Heat Transfer

Heat energy is an important concern in refrigeration and air conditioning. The room air conditioner and domestic refrigerator must be able to remove the heat from a given area and transport it to an area outside of the cabinet or room. Heat transfer will take place using one or more of the following basic transfer methods:

- Conduction occurs when two objects are in contact with each other and heat transfers from one object to another (Figure 9-2). The rate of heat transfer will depend on the makeup of the materials used. Conduction will also occur in a single object. For example, if one end of a solid metal rod is heated, the heat will travel toward the cooler end of the rod. Copper and aluminum are good conductors of heat and are used mostly in air conditioners and refrigerators.

- Convection occurs when heat transfers are caused by air movement or fluid movement, whether naturally or by the means of forced air movement (Figures 9-3 and 9-7). Using the refrigerator as an example, the forced air moving over the cold evaporator coil will cause the air to become colder and denser, and it begins to fall to the bottom of the cabinet. In doing so, the air will absorb the heat from the food and air within the refrigerator cabinet. With the heat absorbed in the colder air, it will begin to expand and become lighter; it will rise and, once again, it will be exposed to the colder evaporator coil temperatures, where heat will transfer to the cold coils. This cycle will continue until the temperature requirements are met. Convection also occurs in a liquid when heat transfers in and out of the liquid, as in a refrigerant used in the sealed system of a refrigerator or air conditioner.

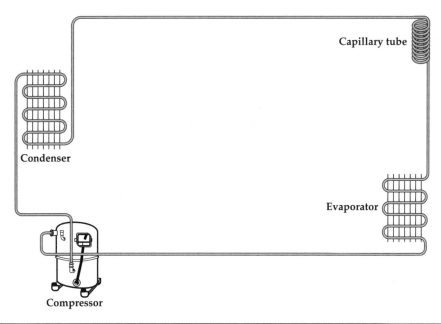

The refrigeration cycle.

- Radiation is the travel of heat energy through the atmosphere by means of radiant waves (light or radio frequency). It travels in a straight path, does not heat the air, and heats only a solid object. For example, if you are standing outside directly in the sunlight, it feels much hotter than if you were to stand in the shade outside. This type of heat transfer comes into play when the refrigerator door is opened or a window covering is opened in an air-conditioned room, allowing heat to enter by radiation. The amount of heat transferred into a refrigerator cabinet or air-conditioned room depends on the temperature difference between the atmosphere and the object.

Refrigerants

Refrigeration involves the process of lowering the temperature of a substance or cooling a designated area. Manufacturers must consider safety, reliability, environmental acceptability, performance, and economics of each refrigerant that is used in refrigeration and air conditioning. The product that absorbs the heat from a substance to be cooled is known as a refrigerant. Refrigerants are classified into two main groups. The first group will absorb the heat by having the refrigerant go through a change of state; the second group involves the absorption of heat without changing the state of the refrigerant. A secondary refrigerant is a refrigerant that will transport the "cold" or "hot" from one location to another (water, for example). The most widely used method of refrigeration is known as the vapor compression cycle (refrigeration cycle using a compressor), which relies on the evaporation of a liquid. This type of method is widely used in domestic refrigeration and air conditioning.

Refrigerants are made up of chemical compounds that are used in a sealed refrigeration system of an air conditioner or refrigerator. They absorb the heat by the process of evaporation or boiling the refrigerant, thereby changing it from a liquid state to a gaseous state. Refrigerants have a much lower boiling point than water when it begins to change state. To remove the heat from the refrigerants, they must be able to change back into a liquid state by the process known as condensation.

Refrigerants are classified into three groups according to their flammability:

- **Class 1**   Nonflammable refrigerant
- **Class 2**   Moderately flammable refrigerant
- **Class 3**   Highly flammable refrigerant

A refrigeration system's reliability will also depend on the chemical stability of the refrigerant, its compatibility with the components within the refrigeration circuit, and the type of lubricant used. At all times, the refrigerant should be miscible at various temperatures with the oil used in the compressor to ensure that the oil will circulate throughout the refrigeration circuit and return to the compressor, where it belongs.

The refrigerant used in a domestic refrigerator or freezer is R-134a. The refrigerant used in room air conditioners is R-22, which will be phased out and replaced with R410a. Other substitute refrigerants are being used today as well.

Special note: Before charging a sealed system the technician must verify the type of refrigerant being used in the product.

Remember: You must be certified to repair sealed refrigeration systems.

When a sealed system has to be repaired, the acceptable method for recharging the product is the weighed-in method. The weight of the refrigerant, usually in ounces, is weighed; this is the most accurate method of refilling the sealed system. Also, these small refrigeration systems are hermetically sealed from the factory, and the manufacturer requires that the products be hermetically resealed again after the repairs are made.

Tables 9-1, 9-2, and 9-3 illustrate the different types of refrigerants used in modern refrigeration and air conditioning equipment.

Beginning technicians or do-it-yourselfers should check out this website before attempting any sealed system repair: http://www.epa.gov/Ozone/title6/608/general/index.html.

PART III

Room Air Conditioner vs. Refrigerator/Freezer

A room air conditioner (RAC) consists of a cabinet, a sealed refrigeration system (see Figure 1-16 and Figure 9-3), and the electrical circuitry, including the fan motor and other components. The RAC will circulate air (Figure 9-4), remove the humidity in a room, filter out the dust particles, and maintain a desired temperature within the room. Some RAC models are equipped with electric heating elements or reverse-cycle heating. Room air conditioners are less expensive than central air conditioning, and they are easier to install. The RAC refrigeration cycle is similar to the refrigeration cycle of a refrigerator/freezer. The only physical difference is in the packaging of the appliance (Figure 9-5).

A refrigerator/freezer consists of a cabinet, a sealed refrigeration system (Figure 9-6a, b, c, and d), and the electrical circuitry. It circulates the air (Figure 9-7), removes humidity, and

Refrigerant ASHRAE #	Chemical Name of Refrigerant	Color of Container
R11	Trichlorofluoromethane	Orange
R12	Dichlorodifluoromethane	White
R13	Chlorotrifluoromethane	Light Blue
R113	Trichlorotrifluoroethane	Dark Purple
R114	Dichlorotetrafluoroethane	Navy Blue
R12/114	Dichlorodifluoromethane, Dichlorotetrafluoroethane	Light Gray
R13B1	Bromotrifluoromethane	Pinkish-Red
R22	Chlorodifluoromethane	Light Green
R23	Trifluoromethane	Light Blue Gray
R123	Dichlorotrifluoroethane	Light Blue Gray
R124	Chlorotetrafluoroethane	DOT Green
R134A	Tetrafluoroethane	Light Blue
R401A	Chlorodifluoromethane, Difluoroethane, Chlorotetrafluoroethane	Pinkish-Red
R401B	Chlorodifluoromethane, Difluoroethane, Chlorotetrafluoroethane	Yellow-Brown
R402A	Chlorodifluoromethane, Pentafluoroethane, Propane	Light Brown
R402B	Chlorodifluoromethane, Pentafluoroethane, Propane	Green Brown
R403B	Chlorodifluoromethane, Octafluoropropane, Propane	Light-Gray
R404A	Pentafluoroethane, Trifluoroethane, Tetrafluoroethane	Orange
R407C	Difluoromethane, Pentafluoroethane, Tetrafluoroethane	Brown
R408A	Chlorodifluoromethane, Trifluoroethane, Pentafluoroethane	Medium Purple
R409A	Chlorodifluoromethane, Chlorotetrafluoroethane, Chlorodifluoroethane	Medium Brown
R410A	Difluoromethane, Pentafluoroethane	Rose
R414B	Chlorodifluoromethane, Chlorotetrafluoroethane, Chlorodifluoroethane, Isobutane	Medium Blue
R416A	Tetrafluoroethane, Chlorotetrafluoroethane, Butane	Yellow-Green
R417A	Pentafluoroethane, Tetrafluoroethane, Isobutane	Green
R500	Dichlorodifluoromethane, Difluoroethane	Yellow
R502	Chlorodifluoromethane, Chloropentafluoroethane	Light Purple
R503	Chlorotrifluoromethane, Trifluoromethane	Blue-Green
R507	Pentafluoroethane, Trifluoroethane	Aqua Blue
R508B	Trifluoromethane, Hexafluoroethane	Dark Blue

TABLE 9-1 Refrigerants

OTHER REFRIGERANTS	
Refrigerant	**Chemical Name of Refrigerant**
R14	Tetrafluoromethane
R32	Difluoromethane
R41	Fluoromethane
R50	Methane
R125	Pentafluoroethane
R141B	Dichlorofluoroethane
R152A	Difluoroethane
R290	Propane
R600	Butane
R600A	Isobutane
R610	Ethyl Ether
R611	Methyl Formate
R630	Methyl Amine
R631	Ethyl Amine
R702	Hydrogen
R704	Helium
R717	Ammonia
R718	Water
R720	Neon
R728	Nitrogen
R732	Oxygen
R740	Argon
R744	Carbon Dioxide
R744A	Nitrous Oxide
R764	Sulfur Dioxide
Refrigerant Categories	
CFC	Chlorofluorocarbons: Being phased out due to Montreal Protocol agreement
HCH	Hydrocarbons
HCFC	Hydrochlorofluorocarbons: Soon to be phased out; production has ceased
HFC	Hydrofluorocarbons: Relatively new for vapor compression systems
Azeotropes	A mixture of two or more refrigerants that retains the same composition in vapor form as in liquid form with similar boiling points for a given temperature and pressure and act as one refrigerant
Zeotropes	A mixture of two or more refrigerants that shift in composition during the boiling or condensing process

TABLE 9-2   Other Types of Refrigerants

Refrigerant ASHRAE #	Trade Name	Refrigerant Components	Components Weight %	Type of Refrigerant	Type of Lubricant
R123		Pure	100%	HCFC	Alkyl benzene or mineral oil
R124		Pure	100%	HCFC	Alkyl benzene or mineral oil
R134A		Pure	100%	HFC	Polyolester oil
R22		Pure	100%	HCFC	Alkyl benzene or mineral oil
R23		Pure	100%	HFC	Polyolester oil
R401A	MP39	R22 / R152A / R124	53% / 13% / 34%	HCFC blended	Alkyl benzene or MO/AB mixture
R401B	MP66	R22 / R152A / R124	61% / 11% / 28%	HCFC blended	Alkyl benzene or MO/AB mixture
R402A	HP80	R125 / R290 / R22	60% / 2% / 38%	HCFC blended	Alkyl benzene or MO/AB mixture
R402B	HP81	R125 / R290 / R22	38% / 2% / 60%	HCFC blended	Alkyl benzene or MO/AB mixture
R403B	ISCEON 69L	R290 / R22 / R218	5% / 56% / 39%	HCFC blended	Alkyl benzene or mineral oil
R404A	HP62, FX70	R125 / R143A / R134A	44% / 52% / 4%	HFC blended	Polyolester oil
R407C	SUVA 9000	R32 / R125 / R134A	23% / 25% / 52%	HFC blended	Polyolester oil
R408A	FX10	R125 / R143A / R22	7% / 46% / 47%	HCFC blended	Alkyl benzene or mineral oil
R409A	FX56	R22 / R124 / R142B	60% / 25% / 15%	HCFC blended	Alkyl benzene or mineral oil
R410A	AZ20, PURON	R32 / R125	50% / 50%	HFC blended	Polyolester oil
R414B	HOT SHOT	R22 / R600A / R124 / R142B	50% / 1.5% / 39% / 9.5%	HCFC blended	Alkyl benzene or mineral oil
R416A	FRIGC FR12	R134A / R124 / R600	59% / 39% / 2%	HCFC blended	Polyolester oil
R417A	NU-22	R125 / R134A / R600A	46.6% / 50% / 3.4%	HFC blended	Alkyl benzene or mineral oil or polyolester oil
R507	AZ50	R125 / R143A	50% / 50%	HFC blended	Polyolester oil
R508B	SUVA 95	R23 / R116	46% / 54%	HFC blended	Polyolester oil

TABLE 9-3 Refrigerant Components, Types, and Lubricants

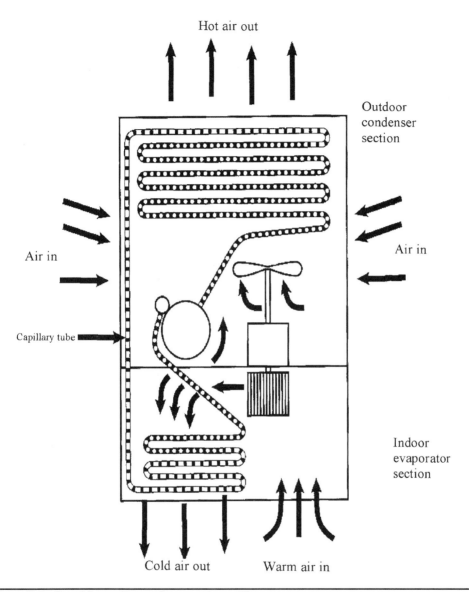

Hot air out

Outdoor
condenser
section

Air in

Air in

Capillary tube

Indoor
evaporator
section

Cold air out   Warm air in

FIGURE 9-4   The airflow patterns of a room air conditioner.

controls the temperature within a closed cabinet that prevents food from spoiling. The refrigerator/freezer refrigeration cycle is similar to the refrigeration cycle of a room air conditioner.

The temperature range for a refrigerator or freezer will range from –20 degrees Fahrenheit to 45 degrees Fahrenheit. A room air conditioner's temperature range varies from 60 degrees to 86 degrees Fahrenheit.

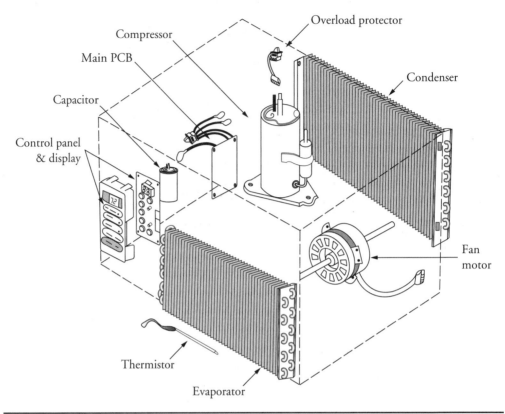

Compressor

Main PCB

Capacitor

Control panel
& display

Overload protector

Condenser

Fan
motor

Thermistor

Evaporator

FIGURE 9-5 Typical component locations in a room air conditioner (RAC).

FIGURE 9-6
Typical components of
a sealed refrigeration
system. (*continued*)

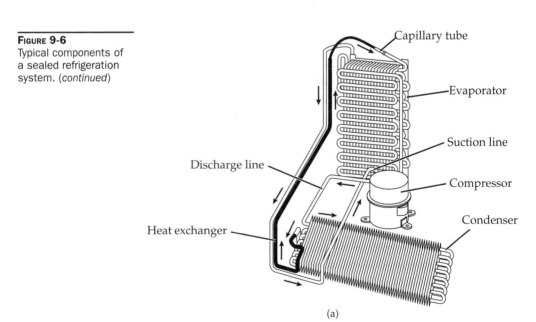

Capillary tube

Evaporator

Suction line

Compressor

Condenser

Discharge line

Heat exchanger

(a)

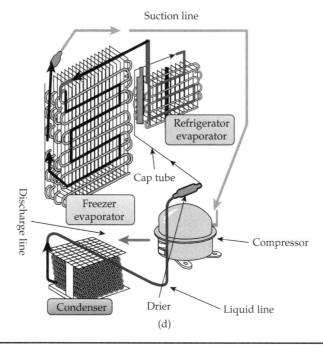

Evaporator

Heat
exchanger

Discharge
line

Drier -
strainer

Suction
line

Accumulator

Compressor

Condenser

(b)

Evaporator

Heat
exchanger

Second discharge
line

Drier -
strainer

Out
Suction
line

In

Accumulator

First discharge
line

Precooler

Condenser   Compressor   Precooler
discharge
return

(c)

Suction line

Refrigerator
evaporator

Cap tube

Freezer
evaporator

Discharge line

Compressor

Condenser   Drier   Liquid line

(d)

Figure 9-6 Typical components of a sealed refrigeration system.

PART III

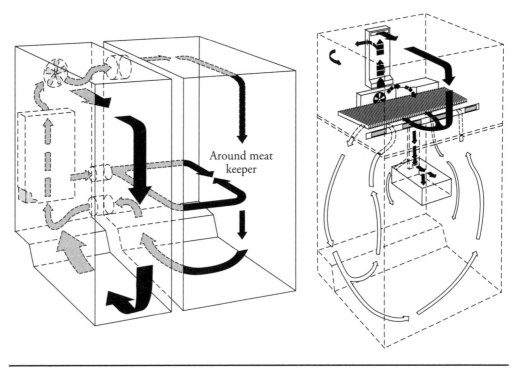

FIGURE 9-7 The cold airflow pattern in side-by-side and two-door refrigerators.

The Basic Refrigeration Cycle

A service technician needs a good understanding of the refrigeration cycle in order to diagnose a refrigerator, freezer, or air conditioner. Without this understanding, the technician will not be able to diagnose a sealed system problem correctly. Mechanical refrigeration is accomplished by continuously circulating, evaporating, and condensing a fixed supply of refrigerant in a closed system. Evaporation occurs at a low temperature and low pressure, while condensation occurs at a high temperature and high pressure. Thus, it is possible to transfer heat from an area of low temperature (that is, the refrigerator cabinet) to an area of high temperature (that is, the kitchen area).

There are five components that make up the refrigeration cycle (Figures 9-3 and 9-6a, b, c, and d). They are

1. Compressor
2. Condenser coil
3. Metering device
4. Evaporator coil
5. Refrigerant

Let's start the refrigeration cycle at the compressor. The compressor, also known as a vapor pump, is the heart of the refrigeration cycle. Once the compressor has started, the suction side of the compressor will draw in a superheated, low-pressure, low-temperature vapor refrigerant from the evaporator. This vapor is then compressed by the compressor piston into a high-pressure, high-temperature, and superheated vapor. When the refrigerant exits from the compressor discharge side and enters into the discharge line, it is a high-pressure, superheated vapor. The refrigerant will then travel to and enter the condenser coil, giving up some of its heat upon entry.

The beginning portion of the condenser coil is often referred to as the desuperheating section and the refrigerant is in the 100 percent vapor state. Around the middle of the condenser coil, the refrigerant gives up latent heat and begins to condense. At this point, the refrigerant's state is 50 percent liquid and 50 percent vapor. As the refrigerant continues to travel and condense in the condenser coil, the liquid refrigerant increases and the vapor refrigerant decreases. During this time the refrigerant temperature remains stable. At the bottom of the condenser coil, the refrigerant will become 100 percent subcooled liquid, and the temperature of the refrigerant is also reduced. The subcooled refrigerant will prevent the refrigerant from boiling. The boiling of the refrigerant at the bottom of the condenser coil is known as flash gas. If this flash gas happens, the capacity of the sealed system would be reduced. To prevent this from happening, we must subcool the refrigerant before it reaches the metering device.

The condenser coil's main purpose is to allow the refrigerant to give up its heat. As the refrigerant circulates through the condenser coil, the condenser fan motor will circulate the surrounding air through the condenser coil. As the air passes over the condenser coil, heat will transfer from the refrigerant to the surrounding air.

Inside the condenser coil, the temperature of the high-pressure vapor refrigerant will determine the temperature at which condensation begins. Condensation usually begins when the refrigerant is approximately 30 degrees Fahrenheit higher than the surrounding air temperature. At that point, heat will transfer from the refrigerant to the surrounding air. The state of the refrigerant when it enters the liquid line is 100 percent liquid.

The liquid refrigerant now enters the metering device. The flow of refrigerant into the evaporator is controlled by the pressure differential across the metering device. The metering device for refrigerators, freezers, residential ice machines, and room air conditioners is called the capillary tube. The capillary tube has a very small diameter opening at both ends and is constructed out of copper tubing, with a predetermined length. This type of metering device will control and measure the flow of refrigerant into the evaporator coil. The location of the metering device is at the end of the liquid line before the inlet of the evaporator coil.

When the refrigerant leaves the metering device, the refrigerant is 75 percent liquid and 25 percent vapor. As the high-pressure liquid refrigerant enters the evaporator coil, it is subjected to a much lower pressure due to the suction of the compressor and the pressure drop across the metering device. Thus the refrigerant tends to expand and evaporate. In order to evaporate, the liquid must absorb heat from the air passing over the evaporator coil. The liquid refrigerant in the evaporator coil will begin to boil and vaporize. This is called latent heat of vaporization. About midway in the evaporator coil the state of the refrigerant is 50 percent liquid and 50 percent vapor.

As the refrigerant moves through the evaporator coil, it goes from a saturated liquid to a more saturated vapor. When the refrigerant leaves the evaporator coil, it is superheated vapor. This superheated vapor returns to the compressor to begin the cycle over again.

PART III

Eventually, the desired air temperature is reached and the thermostat, or cold control, will break the electrical circuit to the compressor motor and stop the compressor. As the temperature of the air running through the evaporator rises, the thermostat reestablishes the electrical circuit. The compressor starts, and the refrigeration cycle continues.

It is extremely important to analyze every system completely and understand the intended function of each component before attempting to determine the cause of a malfunction or failure.

Refrigeration Components

As a new technician, you may understand the process of a refrigeration cycle, but you must also understand how each component functions in the refrigeration cycle. Each component has a specific job to assist in maintaining the objective of removing heat from a given area.

Compressor

The compressor (see Figures 9-8a, b, and c) is the heart of the vapor compression system, also known as the pump. All refrigerant circulating within the sealed system must pass through the compressor. This component has two functions within the refrigeration cycle:

- Retrieve the refrigerant vapor from the evaporator, thereby maintaining a constant pressure and temperature

- Increase the refrigerant vapor pressure and temperature to allow the refrigerant to give up its heat in the condenser coil

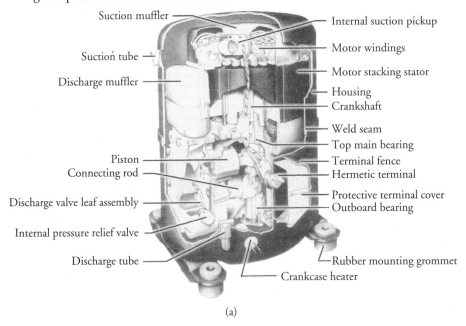

(a)

FIGURE 9-8a The internal view and identification of the internal components in a hermetically sealed air conditioner compressor.

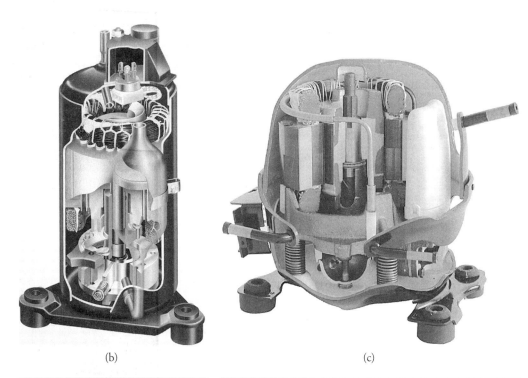

(b)

(c)

FIGURE 9-8b An internal view of a rotary compressor.

FIGURE 9-8c An internal view of a compressor.

Compressors used for domestic refrigeration and room air conditioners are either reciprocating or rotary in design and are hermetically sealed. A hermetically sealed compressor cannot be repaired in the field; it can only be replaced as a complete unit by a certified technician. The electric motor within the compressor can be tested using a multimeter to determine if the windings of the motor are good or bad.

Compressor Safety

In the interest of promoting safety in the refrigeration and air conditioning industry, Tecumseh Products Company has prepared the following information to assist service personnel in safely installing and servicing equipment. This section covers a number of topics related to safety. However, it is not designed to be comprehensive or to replace the training required for professional service personnel.

Trained Personnel Only

Refrigeration and air conditioning devices are extremely complicated by nature. Servicing, repairing, and troubleshooting these products should be done only by those with the necessary knowledge, training, equipment, and certification.

Terminal Venting and Electrocution

Improperly servicing, repairing, or troubleshooting a compressor can lead to electrocution or fire due to terminal venting with ignition. The following precautions can help avoid serious injury or death from electrocution or terminal venting with ignition.

A. Fire Hazard from Terminal Venting with Ignition

Oil and refrigerant can spray out of the compressor if one of the terminal pins is ejected from the hermetic terminal. This "terminal venting" can occur as a result of a grounded fault (also known as a short circuit to ground) in the compressor. The oil and refrigerant spray from terminal venting can be ignited by electricity and produce flames that can lead to serious burns or even death. When spray from terminal venting is ignited, this is called "terminal venting with ignition." (See Figures 9-9a, b, and c for details.)

B. Terminal Venting and Electrocution Precautions

To reduce the risk of electrocution or serious burns or death from terminal venting with ignition:

- Be alert for sounds of arcing (sizzling, sputtering, or popping) inside the compressor. IMMEDIATELY GET AWAY if you hear these sounds.
- Disconnect ALL electrical power before removing the protective terminal cover. Make sure that all power legs are open. (*Note*: The system may have more than one power supply.)
- Never energize the system unless: (1) the protective terminal cover is securely fastened and (2) the compressor is properly connected to ground. Figures 9-9d, 9-9e, and 9-9f illustrate the different means of fastening protective terminal covers.
- Never reset a breaker or replace a fuse without first checking for a ground fault (a short circuit to ground). An open fuse or tripped circuit breaker is a strong indication of a ground fault. To check for a ground fault, use the procedure outlined in "Identifying Compressor Electrical Problems" on page 47 in the Tecumseh Hermetic Compressor Service Handbook (located at www.tecumseh.com).
- Always disconnect power before servicing, unless it is required for a specific troubleshooting technique. In these situations, use extreme caution to avoid electric shock.

Compressor Motor Starting Relays

A hermetic motor starting relay is an automatic switching device to disconnect the motor start capacitor and/or start winding after the motor has reached running speed. There are two types of motor starting relays used in refrigeration and air conditioning applications: the current responsive type and the potential (voltage) responsive type.

Never select a replacement relay solely by horsepower or other generalized rating. Select the correct relay as specified in the Tecumseh Electrical Service Parts Guide Book.

Current Type Relay

When power is applied to a compressor motor, the relay solenoid coil attracts the relay armature upward, causing the bridging contact and stationary contact to engage. This energizes the motor start winding.

(a)

(b)

(c)

(d)

(e)

(f)

Figure 9-9 (a) Compressor with (1) protective terminal cover and (2) bale strap removed to show (3) hermetic terminals. (b) Close-up view of hermetic terminal showing individual terminal pins with power leads removed. (c) Close-up view of hermetic terminal after it has vented. (d) Compressor with (1) protective terminal cover held in place by (2) metal bale strap. (e) Compressor with (1) protective terminal cover held in place by (2) nut. (f) Compressor with (1) snap-in protective terminal cover.

When the compressor motor attains running speed, the motor main winding current is such that the relay solenoid coil de-energizes, allowing the relay contacts to drop open and disconnecting the motor start winding.

The relay (Figure 9-10a) must be mounted in true vertical position so that armature and bridging contacts will drop free when the relay solenoid is de-energized.

PTC Type Relay

Solid-state technology has made available another type of current sensitive relay: a PTC starting switch (see Figure 9-10b). Certain ceramic materials have the unique property of greatly increasing their resistance as they heat up from current passing through them. A PTC solid-state starting device is placed in series with the start winding and normally

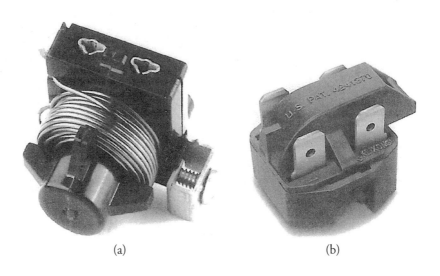

(a)   (b)

FIGURE 9-10   (a) Current type relay. (b) PTC type relay.

has a very low resistance. Upon startup, as current starts to flow to the start winding, the resistance rapidly rises to a very high value, thus reducing the start winding current to a trickle and effectively taking that winding out of operation.

Usage is generally limited to domestic refrigeration and freezers. Because it takes 3 to 10 minutes to cool down between operating cycles, it is not feasible for short-cycling commercial applications.

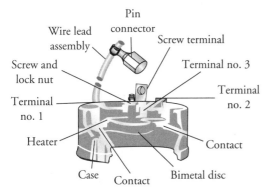

External thermal protector (Models AE, TP, TH, AK, AJ, CAJ, AZ, RK, RG, TW, and some CL)

(a)

FIGURE 9-11a   The compressor overload protector.

Compressor Thermal Protectors (Overloads)

All hermetic compressors have either an internal or an external protector (overload) to protect the motor from overheating. The compressor overload in Figure 9-11a is attached to the compressor in Figures 9-11c and 9-11d. The overload is designed

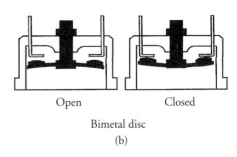

Open   Closed

Bimetal disc

(b)

FIGURE 9-11b   An internal view of the compressor overload protector.

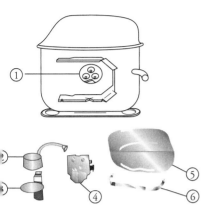

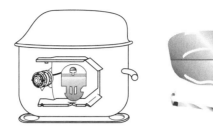

AE refrigeration compressor with the thermal protector and relay assembled.

(d)

AE refrigeration compressor showing (1) hermetic terminal, (2) thermal protector, (3) thermal protector clip, (4) push-on relay, (5) protective terminal cover, and (6) bale strap.

(c)

FIGURE 9-11c An illustration of the components before they are attached to the exterior of the compressor.

FIGURE 9-11d An illustration of the components after they are attached to the exterior of the compressor.

to react quickly by means of a bimetal disk to an increase in temperature or an increase in current.

The bimetal disk (Figure 9-11b) consists of two dissimilar metals combined together. Any change in temperature or current draw will cause it to deflect, actuating the switch contacts. When the bimetal cools, the reverse action takes place. The external protector is a non-serviceable part, and it should be replaced with a duplicate of the original. If the compressor has an internal overload and is diagnosed as being defective, the compressor will have to be replaced by a certified technician.

Condenser Coil

A condenser coil is manufactured out of copper, aluminum, or steel tubing, with metal fins attached to the tubing to assist in dissipating the heat (Figures 9-12a and b). The main purpose of the condenser coil is to dissipate the heat from the refrigerant along with the heat from the compression stage of the compressor. Condenser coils in refrigerators, freezers, and room air conditioners usually have a fan motor and fan blade circulating the air across the condenser coil to assist in removing the heat from the refrigerant. On some models of refrigerators and freezers, the condenser coils are static-cooled whereby the heat from the condenser coil will rise and dissipate into the room. In some models of freezers, the condenser coil is attached to the inside of the outer cabinet so that the cabinet is the heat transfer medium.

Evaporator Coil

The evaporator coil, manufactured in the same manner as the condenser coil, absorbs the heat from the food product or from the air in a room to be cooled (Figures 9-5, 9-12a, and 9-12b). Manufacturers design evaporator coils in different shapes and designs to meet the needs of specific products. Most refrigerator models and room air conditioners use a fan motor and fan blade to circulate the air across the evaporator coil, which absorbs the heat.

PART III

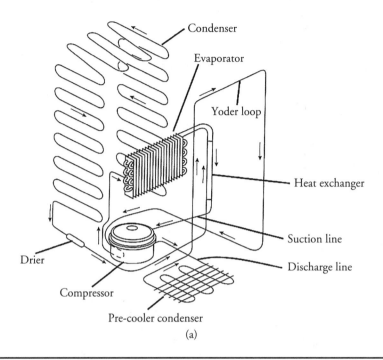

(a)

FIGURE 9-12a The condenser coil in a refrigerator's sealed refrigeration system.

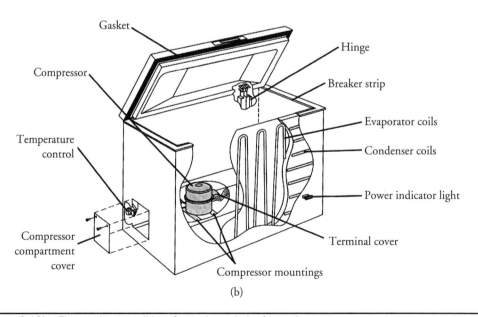

(b)

FIGURE 9-12b The condenser coil in a freezer's sealed refrigeration system.

Other refrigerator/freezer models have rows of tubing placed between two plates, which are welded together to form an interior cabinet in the freezer compartment or shelving.

Metering Device

The most commonly used metering device in a domestic refrigerator and room air conditioner is a capillary tube, also called a cap tube (Figures 9-3, 9-4, and 9-6). This metering device is manufactured out of copper tubing, with a very small-diameter opening at both ends, and is premeasured at a specific length for maximum heat load efficiency for the product being cooled or refrigerated. The small-diameter opening in the cap tube will restrict the flow of refrigerant, causing a pressure drop from the high-pressure side (condenser) of the sealed system to the low-pressure side (evaporator). This will allow the refrigerant to enter the evaporator at the rate needed to remove the unwanted heat from the product being cooled or refrigerated. There are no moving parts in a cap tube, and it does not require any service. The capillary tube is located between the condenser coil and the evaporator coil at the end of the liquid line.

Other Components Used in the Sealed Refrigeration System

The refrigeration sealed system in an appliance or air conditioner can have additional components added to the sealed system. These additional components added are usually from manufacturers' designs of the product. These additional components will enhance the operation of the product. Some additional components are:

- **Refrigeration drier**   The function of the refrigerant drier (Figure 9-12a) is to trap any moisture, contaminants, and large particulate matter within the sealed refrigeration system. In a refrigerator or air conditioner, the drier is installed within the liquid line before the metering device. The drier can be either a unidirectional flow or bidirectional flow drier. The drier must be compatible with the oils and refrigerants that are used in today's products.

- **Accumulator**   Some models of refrigerators and freezers have accumulators (Figure 9-3b, c, and d). An accumulator is a device located at the end of the evaporator coil. Its appearance makes the tubing size look much larger. It looks like a small elongated storage tank attached to the evaporator coil. The accumulator allows liquid refrigerant to be collected in the bottom of the accumulator and stay there. Only vapor refrigerant will be drawn from the accumulator and returned to the compressor. On air conditioners, the accumulator is located on the suction line returning to the compressor. The compressor and accumulator are one component.

- **Heat exchanger**   The heat exchanger (Figure 9-6 and Figure 9-12a) consists of the suction line soldered to the capillary tube. The reason for this is to make sure that there is good contact between the two lines. These two tubes are located inside a refrigerator or freezer. There are two purposes for the heat exchanger:

  1. The capillary tube will give up some of its heat to the cooler suction line, thus allowing the refrigeration system to run more efficiently.

  2. No liquid refrigerant will be allowed to return to the compressor. The warmer refrigerant in the capillary tube will boil off and vaporize any remaining liquid refrigerant in the suction line.

Diagnosing Sealed Systems

When servicing the refrigerator/freezer or room air conditioner, do not overlook the simple things that might be causing the problem. Before you begin testing and servicing an appliance or air conditioner, check for the most common faults. Inspect all wiring connections for broken or loose wires. Check and see if all components are running at the proper times during the cycle. Test the temperatures in the refrigerator/freezer cabinet or air conditioner inlet and outlet grills.

The following charts (Tables 9-4 and 9-5) is intended to serve as an aid to assist the technician in determining if there is a sealed system malfunction, providing that all other components are functioning correctly within the appliance or conditions that mimic a sealed system failure. You must rule out everything else before you enter the sealed system.

NOTE *Before you enter a sealed system, you must be certified to do so.*

Sealed System Condition	Suction Line Pressure (TECHNICIANS ONLY)	Liquid Line Pressure Line Pressure (TECHNICIANS ONLY)	Suction Line to Compressor	Compressor Discharge Line	Condenser Coil	Capillary Tube	Evaporator Coil	Frost Line	Amperage or Wattage	Pressure Equalization Rate
Normal Operation	Normal pressure readings	Normal pressure readings	Slightly below room temperature	**WARNING** Very hot to the touch	**WARNING** Very hot to the touch	Warm	Cold coil and maintaining proper temperatures	Suction line from inside box not frozen	Normal meter readings	Normal
Sealed System Overcharged	Higher than normal pressure readings	Higher than normal pressure readings	Heavily frosted; may be very cold to the touch	Between slightly warm to hot	Between hot to warm to the touch	Cool; below room temperature	Cold coil and possibly not maintaing temperatures	All the way back to the suction line	Higher than normal meter readings	Normal to slightly longer
Sealed System Undercharged	Pressure readings are lower than normal	Pressure readings are lower than normal reading	Warm to the touch; possibly near room temperature	**WARNING** Hot to the touch	The entire coil feels warm	Warm	Inlet to coil feels extremely cold while the outlet from the coil will be below room temperature	Partial	Lower than normal meter readings	Normal
Partial Restriction Within Sealed System	Lower than normal pressure readings, possibly in a vacuum	Intermittent lower than normal reading	Warm to the touch; possibly near room temperature	**WARNING** Very hot to the touch	The top passes in the coil are warm and the lower passes are cool, near to room temperature	Feels like room temperature between cool to colder	Inlet to coil feels extremely cold while the outlet from the coil will be below room temperature	Intermittent / Frost line will begin to grow in length	Lower than normal meter readings	Intermittent
Complete Restriction Within Sealed System	Pressure readings are in a deep vacuum	Ambient readings	Feels the same as room temperature	Feels the same as room temperature	Feels the same as room temperature	Feels the same as room temperature	No refrigeration or air conditioning	None	Lower than normal meter readings	No equalization
Out of Refrigerant Possible Leak in System	The pressure reading will be from 0 PSIG to 30" vacuum	Atmospheric reading	Feels the same as room temperature	Can feel like cool to hot	Feels the same as room temperature	Feels the same as room temperature	No refrigeration or air conditioning	Non existent	Lower than normal meter readings	Normal
Low Capacity Compressor	Higher than normal pressure readings	Lower than normal readings	Cool to room temperature	Cooler than normal	Low	Warm to box temperature	Partial or half of the evaporator frost pattern	Partial to non-existent	Lower than normal	Quicker than normal

TABLE 9-4 Refrigeration Sealed-System Diagnosis Chart

225

Conditions	Amperage or Wattage	Condenser Coil Temperature	Frost Line	Compressor Discharge Line Temperature	Low Side Pressure (for Service Technicians Only)	High Side Pressure (for Service Technicians Only)	Fresh Food Compartment Temperature	Freezer Compartment Temperature
Plugged Condenser Coil	Higher than normal	Higher than normal	Full	Higher than normal	Higher than normal	Higher than normal	Warmer than normal readings	Warmer than normal readings
Blocked Condenser Fan Assembly	Higher than normal	Higher than normal	Full	Higher than normal	Higher than normal	Higher than normal	Warmer than normal readings	Warmer than normal readings
Blocked Evaporator Fan Assembly	Lower than normal	Lower than normal	Frost back to compressor	Lower than normal	Lower than normal	Lower than normal	Warmer than normal readings	Warmer than normal readings
Evaporator Coil Iced Up (Defrost Failure)	Lower than normal	Lower than normal	Frost back to compressor	Lower than normal	Lower than normal	Lower than normal	Warmer than normal readings	Warmer than normal readings
High Head Load	Higher than normal	Higher than normal	Full	Higher than normal	Higher than normal	Higher than normal	Warmer than normal readings	Warmer than normal readings
High Ambients	Higher than normal	Higher than normal	Full	Higher than normal	Higher than normal	Higher than normal	Warmer than normal readings	Warmer than normal readings
Damper Failed Closed	Lower than normal	Lower than normal	Full	Lower than normal	Lower than normal	Lower than normal	Warmer than normal readings	Cooler than normal readings
Damper Failed Open	Slightly higher than normal	Slightly higher	Full	Normal	Slightly higher than normal	Normal	Cooler than normal	Normal to slightly warmer readings

Conditions That Mimic Sealed System Failures

TABLE 9-5   Conditions that Mimic a Sealed System Failure

IV PART

Parts

Electric Appliance and Air Conditioner Parts

Servicing a highly complex electromechanical appliance or room air conditioner is not as hard as might be expected. Just keep in mind that an appliance or air conditioner is simply a collection of parts, located inside a cabinet, coordinated to perform a specific function (Figures 10-1a, 10-1b, 10-1c, 10-1d, and Figure 9-4). Before servicing an appliance or air conditioner, you must know what these parts are and how they function.

ᴇ Switch

The switch is a mechanical device used for directing and controlling the flow of current in a circuit. Simply put, the switch can be used for turning a component on or off (Figure 10-2). Internally, the switch has a set of contacts that close, allowing the current to pass; when opened, current is unable to flow through it. Built into the switch, a linkage mechanism actuates these contacts inside of the closed housing (Figure 10-3).

Switches come in a wide range of sizes and shapes, and can be used in many different types of applications (Figure 10-4). The voltage and amperage rating is marked on the switch or on the mounting bracket for the type of service the individual switch was designed to do. The switch housing is usually marked with the terminal identification numbers that correspond to the wiring diagram. These identify the contacts by number: normally open (NO) contacts, normally closed (NC) contacts, or common (COM) contacts. Internally, the switch can house many contact points for controlling more than one circuit.

When a switch failure is suspected, remember that there are only three problems that can happen to a switch:

- The contacts of the switch might not make contact. This is known as an *open switch*.

- The switch's contacts might not open, causing a *shorted switch*.

- The mechanism that actuates the contacts might fail. This is a *defective switch*.

When these problems arise, the switches are not repairable, and they should be replaced with a duplicate of the original.

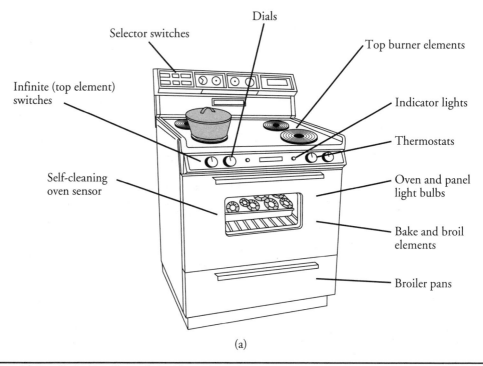

Selector switches

Dials

Top burner elements

Infinite (top element) switches

Indicator lights

Thermostats

Self-cleaning oven sensor

Oven and panel light bulbs

Bake and broil elements

Broiler pans

(a)

FIGURE 10-1a Typical locations of electric range parts.

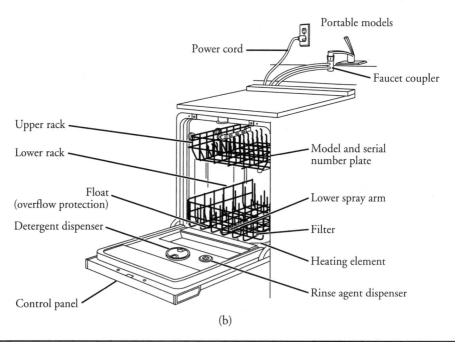

Portable models

Power cord

Faucet coupler

Upper rack

Model and serial number plate

Lower rack

Float (overflow protection)

Lower spray arm

Detergent dispenser

Filter

Heating element

Rinse agent dispenser

Control panel

(b)

FIGURE 10-1b Typical locations of dishwasher parts.

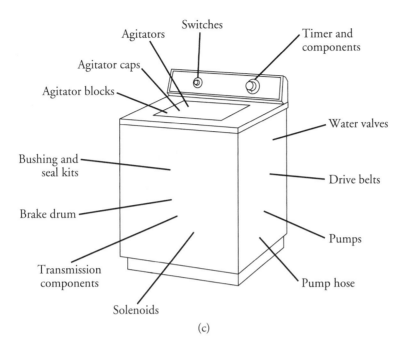

(c)

FIGURE 10-1c   Typical locations of automatic washer parts.

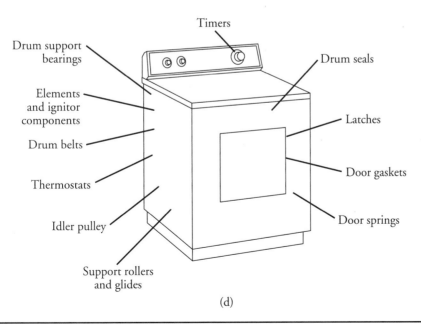

(d)

FIGURE 10-1d   Typical locations of automatic dryer parts.

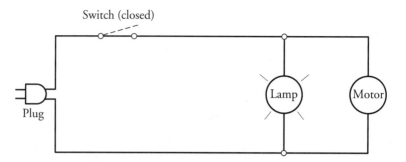

FIGURE 10-2 The wiring diagram illustrates a switch in the closed position. If the switch is closed, the light and motor are on. If the switch is open, the light and motor are off.

Pressure Switch

The *pressure switch* is a specialty switch, with a similar operation to those mentioned previously, but with one important exception: The pressure switch is actuated by a diaphragm that is responsive to pressure changes (Figure 10-5). This switch can be found in washing machines and in dishwashers, and it operates as a water level control. Other uses include furnaces, gas heaters, computers, vending machines, sump pumps, and other low-pressure applications. The pressure switch is not serviceable, and should be replaced with a duplicate of the original.

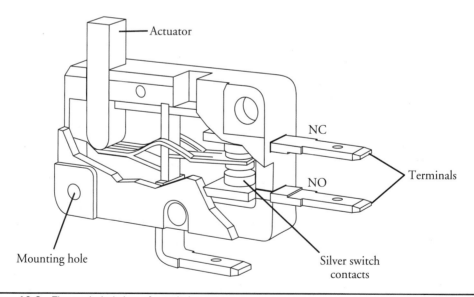

FIGURE 10-3 The exploded view of a switch.

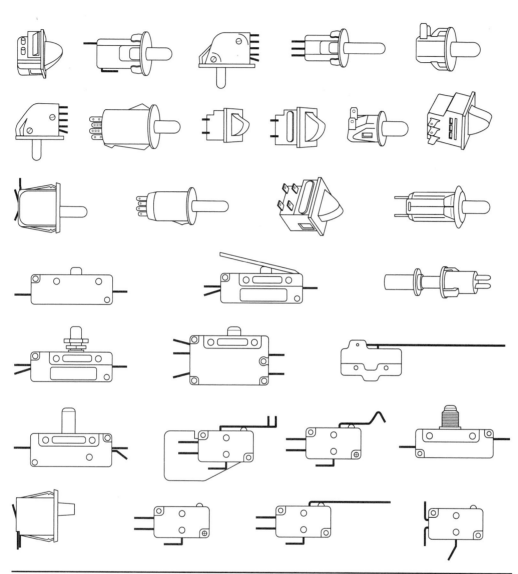

FIGURE 10-4 This is a sample of the many types of switches used in major appliances.

Thermostat

The *thermostat* operates a switch. It is actuated by a change in temperature. The two most common heat-sensing methods used in appliances and air conditioners are the bimetal and the expansion thermostats (Figure 10-6).

The *bimetal thermostat* (Figure 10-7) consists of two dissimilar metals combined together. Any change in temperature will cause it to deflect, actuating the switch contacts. When the bimetal cools, the reverse action takes place.

FIGURE 10-5
Construction of a
pressure switch.

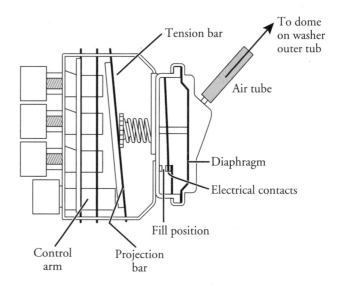

The *expansion (temperature control) thermostat* (Figure 10-8) uses a liquid in a tube that is attached to bellows. The liquid converts to a gas when heated and travels up the tube to the bellows. This causes the bellows to expand, thus actuating the switch contacts. When the gas cools, the reverse action occurs.

Thermostats are used in applications as diverse as gas and electric ranges, automatic dryers, room air conditioners, irons, waterbeds, spas, and in heating and refrigeration units.

Electromechanical Timer

Although the timer is the most complex component in the appliance, don't assume that it is the malfunctioning part. Check all of the other components associated with the symptoms as described by the customer.

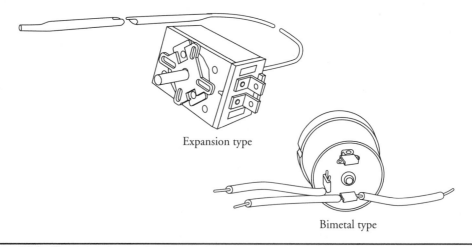

FIGURE 10-6   Bimetal and expansion thermostats.

FIGURE 10-7
Construction of a
bimetal thermostat.

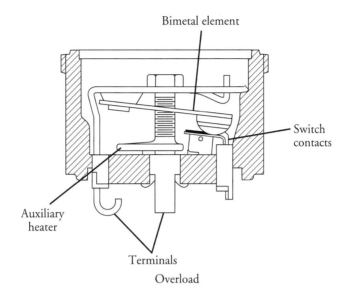

The timer assembly is driven by a synchronous motor in incremental advances. It controls
and sequences the numerous steps and functions involved in each cycle of an appliance
(Figure 10-9). The timer directs the on and off times of the components in an electrical circuit.
It consists of three components assembled into one unit: the motor, the escapement, and the
cam switches (Figure 10-10).

Electromechanical timers are utilized for controlling performance in automatic washers,
automatic dryers, and dishwashers. Most of these timers are not serviceable and should be
replaced with a duplicate of the original.

FIGURE 10-8
Construction of
an expansion
(temperature-control)
thermostat.

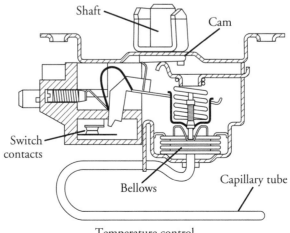

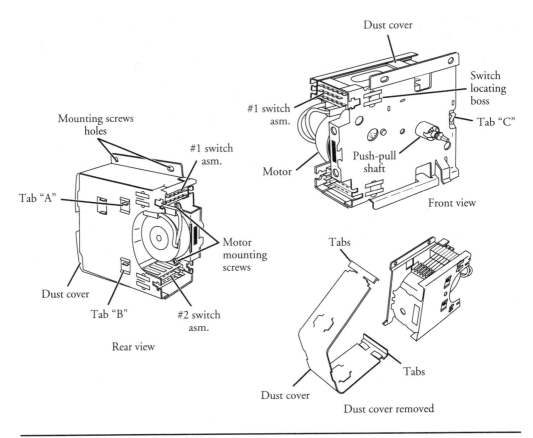

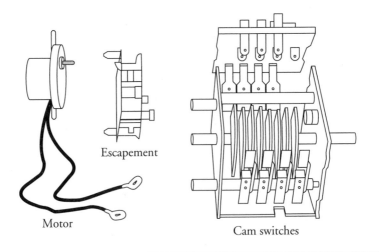

FIGURE 10-9 An electromechanical timer.

FIGURE 10-10 The timer components: motor, escapement, and cam switches.

Figure 10-11
The construction of
the timer cam and
switch contact.

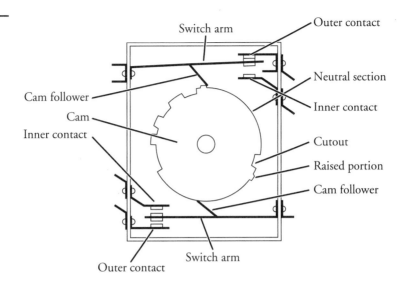

Figure 10-11
The construction of
the timer cam and
switch contact.

The Motor

The motor is a *synchronous motor*, geared to drive the escapement. In Figures 10-9 and 10-10, the motor is mounted on the timer assembly. These are specially designed motors whose speed is controlled by the 60-cycle period of the current, rather than by the fluctuating supply voltage.

The Cam Switches

The number of switch contacts and cams varies with the number of functions an appliance performs. Some of the switches perform two or more functions (Figure 10-11). The particular shape of the cam varies with the number of switch contacts that it controls and with the length of time each switch contact opens and closes (Figure 10-12). A single metal strip, called a cam follower (see Figure 10-11), is anchored to each cam switch arm. As the cam turns, this metal strip follows the contour of the circumference of the cam, causing the cam switch to open or close at the proper time.

Figure 10-12
How a switch contact
opens and closes on
the cam.

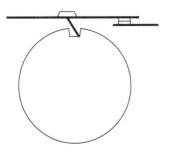
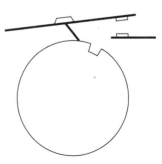

The Escapement

The escapement is a spring-controlled mechanism that limits the cam shaft rotation to a set number of degrees in each increment (see Figure 10-10). The cam follower moves rapidly to ensure a snap action of the switch contacts and also to prevent any arcing of the points. Not all timers have an escapement. For example, some dryer timers, and all defrost and electronic timers, have no escapement.

Relays

A relay is an electrically operated switch. It contains an electromagnet with a fixed coil and a movable armature, which actuates a set of contacts to open and close electric circuits (Figure 10-13). Relays have heavy-duty switching contacts, and they are used to operate larger components, such as motors and compressors. The most common failures of relays are:

- Open relay coil
- Burned switch contacts
- Armature not actuating the contacts

A broken relay is not serviceable and should be replaced with a duplicate of the original.

Solenoids

A solenoid is a device used to convert electrical energy into mechanical energy (Figure 10-14). When a solenoid is energized, it acts like an electromagnet and is positioned to move a pre-designated metal object. The work performed by the moving plunger makes the solenoid coil useful in appliances. Some solenoids are equipped with a free-moving armature or plunger. Common failures of a solenoid are:

- Open coil
- Shorted coil
- Jammed armature

FIGURE 10-13
Construction of
a relay.

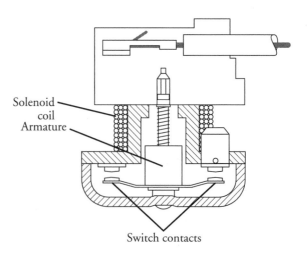

Solenoid
coil
Armature

Switch contacts

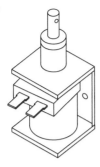

FIGURE 10-14
The solenoid coil and plunger. When the coil is activated, the plunger will be drawn to the center of the magnetic field.

A solenoid is not a serviceable part and should be replaced with a duplicate of the original. These devices are manufactured in a variety of designs for various load force and operational requirements. Solenoids are found in automatic washers and dryers, gas and electric ranges, automatic dishwashers, refrigerators, freezers, automatic ice machines, and in heating and air conditioning units.

Water Valves

The water inlet valve controls the flow of water into an appliance, and is solenoid-operated (Figure 10-15). When it is energized, water in the supply line will pass through the valve body and into the appliance. Some of the different types of water inlet valves that are used on appliances include:

- **Single water inlet valve (Figure 10-16)** Used on dishwashers, ice makers, refrigerators, undercounter appliances and ice machines.
- **Dual water inlet valve (Figure 10-17)** Used on washing machines, refrigerators, and ice makers. Some dishwasher models also use dual water inlet valves. The inlet side of the valve has a fine mesh screen to prevent foreign matter from entering the valve. Some water valves also have a "water hammer" suppression feature built into them.

FIGURE 10-15
Construction of a water valve.

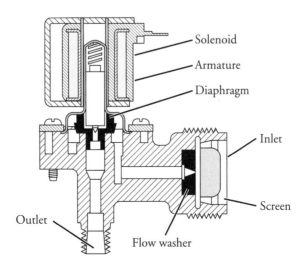

Solenoid

Armature

Diaphragm

Inlet

Screen

Outlet

Flow washer

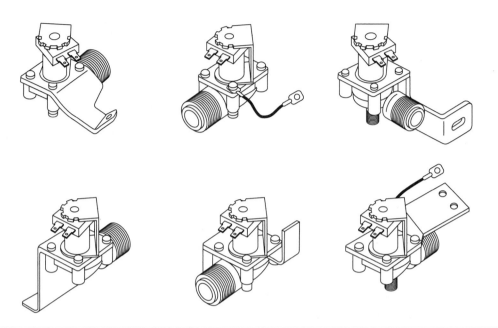

Figure 10-16 The single water valves are just some of the different types available that are used in major appliances.

Drain valves (Figure 10-18) are used on some dishwasher and washing machine models to control the drainage of the water in the tub and its expulsion into the sewage system of the residence.

The water valve should not be serviced. Replace it with a duplicate of the original.

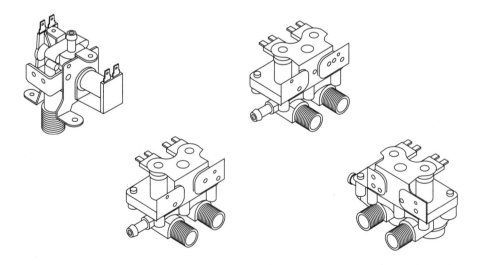

Figure 10-17 These dual water valves are designed for dual water inlet connections.

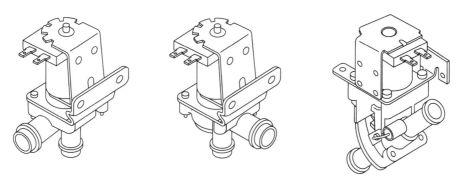

FIGURE 10-18   The drain water valves.

Motors

The two major assemblies that form an electric motor are the rotor and the stator (Figure 10-19). The rotor is made up of the shaft, rotor core, and (usually) a fan. The stator is formed from steel laminations, stacked and fastened together so that the notches form a continuous lengthwise slot on the inside diameter. Insulation is placed so as to line the slots; and then coils, wound with many turns of wire, are inserted into the slots to form a circuit. The wound stator laminations are pressed into, or otherwise assembled within, a cylindrical steel frame to form the stator (Figure 10-20). The end bells, or covers, are then placed on each end of the motor. One important function of the end bells is to center the rotor or armature accurately within the stator to maintain a constant air gap between the stationary and moving cores (Figures 10-19 and 10-21).

These coils of wire are wound in a variety of designs, depending upon the electrical makeup of the motor. They provide two or more paths for current to flow through the stator windings. When the coils have two centers, they form a two-pole motor; when they have

PART IV

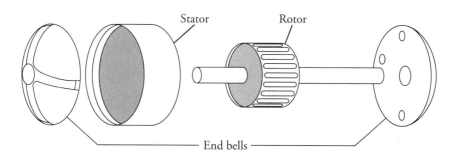

FIGURE 10-19   The stator and rotor.

Figure 10-20
Stator and rotor
construction.

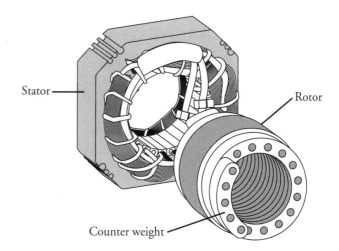

Stator —

Rotor

Counter weight —

four centers, they form a four-pole motor. In short, the number of coil centers determines
the number of poles that a motor has (Figure 10-22).

Thermal protection in a motor is provided by a temperature-sensitive element, which
activates a switch. This switch will stop the motor if it reaches the pre-set temperature limit.
The thermal protector in a motor is a non-replaceable part, and the motor will have to be
replaced as a complete component. There are two types of thermal protection switches:

- **Automatic reset**   It automatically resets the switch when the temperature has been
  reduced.

- **Manual reset**   It has a small reset button on the motor on the opposite end from
  the shaft.

Figure 10-21
End bells position the
motor shaft in the
center of the stator.

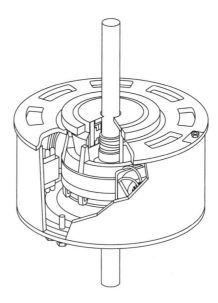

FIGURE 10-22
Two-pole motor and
four-pole motor.

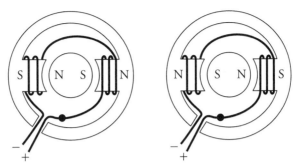

Two-pole stator motor

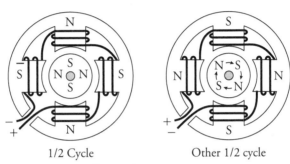

1/2 Cycle   Other 1/2 cycle

Four-pole stator motor

Several types of motors that are used for different types of applications include the following:

- Synchronous motors are permanent magnet-timing motors, often used in automatic ice cube makers, water softeners, and humidifiers. In addition, they are integral to timers for automatic washers, automatic dryers, and dishwashers.

- Shaded pole motors are used as continuous duty motors, with limited or adjustable speeds. They are used for small fans and clocks.

- Split phase motors are used as continuous duty motors, with fixed speeds. They are often used in automatic washer and dryer drive motors.

- Capacitor start motors are similar to the split-phase motors, and they are used in hard-to-start applications, such as compressors and pumps.

- Permanent split capacitor motors are used in a variety of direct-drive air-moving applications—for example, air conditioner fans.

- Three-phase motors are used in industrial or large commercial applications where three-phase power is available.

- Multispeed, split-phase motors are used in fans, automatic dryers, automatic washers, and many other appliances.

- Induction three-phase AC motor, 120 volt, used in some domestic washer models.

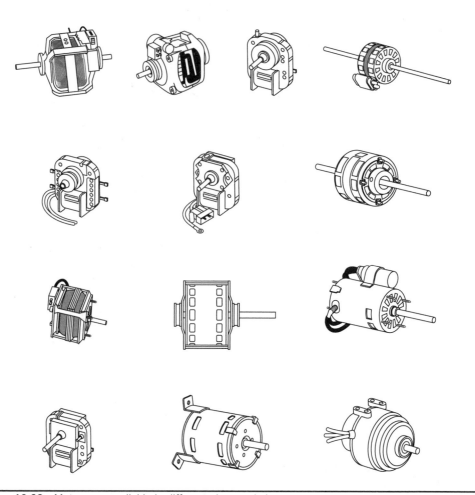

Figure 10-23   Motors are available in different sizes and shapes.

- Variable-speed, reversible, three-phase induction DC motor, used in some domestic washer models.
- Direct current (DC) motors, used in refrigerators, washers, dryers, ranges, microwave models.

Figure 10-23 illustrates some of these motors. Appliance motors are not repairable, and they should be replaced with a duplicate of the original.

Compressors

The compressor is an electric motor that drives a mechanical compression pump designed to compress the refrigerant vapors and to circulate the refrigerant within a sealed system. Domestic refrigeration and room air conditioners use a hermetic compressor. The electric

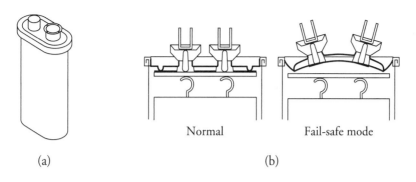

Normal   Fail-safe mode

(a)   (b)

FIGURE 10-24   (a) The capacitor is rated by voltage and by capacitance (in microfarads). (b) This built-in disconnect device is also known as a fail-safe.

motor and mechanical compression pump are sealed within the same housing (Figure 9-7), and it is a non-serviceable part. If the compressor fails, it must be replaced with a duplicate of the original by a certified technician. Two types of compressors are used in domestic refrigeration and room air conditioners: reciprocating and rotary. For more information on compressor construction, visit the following compressor manufacturer's websites:

- www.copeland-corp.com
- www.compressors.danfoss.com
- www.embraco.com
- www.tecumseh.com

Capacitors

A capacitor is a device that stores electricity to provide an electrical boost for motor starting (Figure 10-24). Most high-torque motors need a capacitor connected in series with the start winding circuit to produce the desired rotation under a heavy starting load.

There are two types of capacitors:

- **Start capacitor**   This type of capacitor is usually connected to the circuit between the start relay and the start winding terminal of the motor. Start capacitors are used for intermittent (on and off) operation.

- **Run capacitor**   The run capacitor is also in the start winding circuit, but it stays in operation while the motor is running (continuous operation). The purpose of the run capacitor is to improve motor efficiency during operation.

Capacitors are rated by voltage and by their capacitance value in microfarads (μF). This rating is stamped on the side of the capacitor. A capacitor must be accurately sized to the motor and the motor load. Always replace a capacitor with one having the same voltage rating and the same (or up to 10 percent greater) microfarad rating. On larger capacitors, the rating is stamped on the side. Also, watch out for the decimal point on some capacitors.

PART IV

The rating might read .50 μF instead of 50 μF. Small capacitors in electronic circuits are rated by numbering or are color-coded.

Capacitors are used in electrical circuits to perform the following:

- An electrical voltage boost in a circuit
- Control timing in a computerized circuit
- Reduce voltage disruptions and allow voltage to maintain a constant flow
- Block the flow of direct current when fully charged and allow alternating current to pass in a circuit

Both run and start capacitors can be tested by means of an ohmmeter or a capacitor tester.

Testing a Capacitor

Before testing a capacitor, disconnect the electricity. This can be done by pulling the plug from the electrical outlet. Be sure that you only remove the plug for the product you are working on. Or, you can disconnect the electricity at the fuse panel or at the circuit breaker panel.

Some appliance or air conditioner models have the capacitor mounted on the motor, and some are mounted to the cabinet interior in the rear of the machine. Access might be achieved through the front or rear panel, depending on which model you are working on. Do not touch the capacitor until it's discharged.

WARNING *A capacitor will hold a charge indefinitely, even when it is not currently in use. A charged capacitor is extremely dangerous. Discharge all capacitors immediately any time that work is being conducted in their vicinity. Redischarge after repowering the equipment if further work must be done.*

Many capacitors are internally fused. If you are not sure, you can use a 20,000-ohm, 2-watt resistor to discharge the capacitor. Do not use a screwdriver to short out the capacitor. By doing so, you will blow out the fuse in the capacitor and the capacitor will not work. Safely use an insulated pair of pliers to remove the wires from the capacitor, and place the resistor across the capacitor terminals. Set the ohmmeter on the highest scale, and place one probe on one terminal and the other probe on the other terminal (Figure 10-25). Observe the

FIGURE 10-25
Placing ohmmeter test leads on the capacitor terminals.

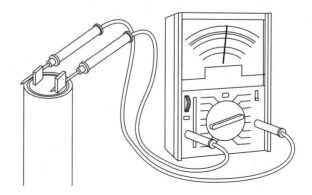

meter action. While the capacitor is charging, the ohmmeter will read nearly zero ohms for a short period of time. Then the ohmmeter reading will slowly begin to return toward infinity. If the ohmmeter reading deflects to zero and does not return to infinity, the capacitor is shorted and should be replaced. If the ohmmeter reading remains at infinity and does not dip toward zero, the capacitor is open and should be replaced.

When using a capacitor analyzer to test capacitors, it will show whether the capacitor is "open" or "shorted." It will tell whether the capacitor is within its microfarads rating, and it will show whether the capacitor is operating at the proper power-factor percentage. The instrument will automatically discharge the capacitor when the test switch is released.

eating Elements

Most heating elements are made with a nickel-chromium wire, having both tensile strength and high resistance to current flow. The resistance and voltage can be measured with a multimeter to verify if the element is functioning properly. Heating elements are available in many sizes and shapes (Figure 10-26). They are used for

- Cooking food
- Heating air for drying clothes
- Heating water to wash clothes, dishes, etc.
- Environmental heating

Heating elements are not repairable, and they should be replaced with a duplicate of the original.

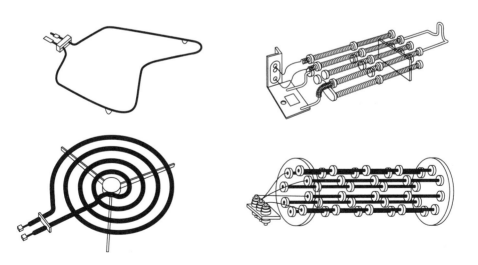

FIGURE 10-26   Heating elements.

Mechanical Linkages

The mechanical linkages are those devices (connecting rods, gears, cams, belts, levers, pulleys, etc.) that are used on appliances and air conditioners in order to transfer mechanical energy from one point to another. Figure 10-27, the automatic ice maker, is an excellent example of this. Some other examples are:

- In the automatic dryer, the motor is turning a pulley, which moves a belt, which turns the drum.

- In the automatic washing machine, the motor turns a pulley, which moves the belt, which turns the transmission gears, which performs the agitation or spin cycle.

- In the automatic ice maker, the timer gear turns the drive gear, which moves the cam, which actuates the switches and rotates the ice ejector.

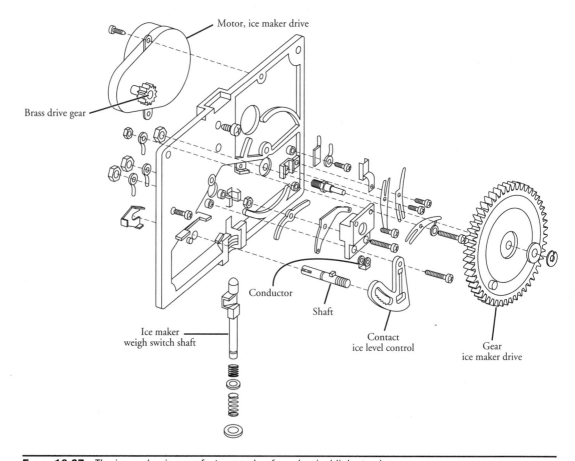

FIGURE 10-27   The ice maker is a perfect example of mechanical linkages in use.

Wires

The wiring, which connects the different components in an appliance or air conditioner, is the highway that allows current to flow from point A to point B. Copper and aluminum are the most common types of wires that are used in appliances. They are available as solid or stranded. Wires are enclosed in an insulating sleeve, which might be rubber, cotton, or one of the many plastics. Wires are joined together or to the components by:

- Solderless wire connectors
- Solderless wire terminal connectors
- Solderless multiple-pin plug connectors
- Soldering

Never join copper and aluminum wires together, because the two dissimilar metals will corrode and interrupt the flow of current. The standard wire-gauge sizes for copper wire are listed in Table 10-1. As the gauge size increases from 1 to 20, the diameter decreases and the amperage capacity (ampacity) will decrease also (see Table 10-2).

Gauge No.	Diameter, Mil	Circular-Mil Area	Ohms Per 1000 ft. of Copper Wire at 25°C*
1	289.3	83,690	0.1264
2	257.6	66,370	0.1593
3	229.4	52,640	0.2009
4	204.3	41,740	0.2533
5	181.9	33,100	0.3195
6	162.0	26,250	0.4028
7	144.3	20,820	0.5080
8	128.5	16,510	0.6405
9	114.4	13,090	0.8077
10	101.9	10,380	1.018
11	90.74	8234	1.284
12	80.81	6530	1.619
13	71.96	5178	2.042
14	64.08	4107	2.575
15	57.07	3257	3.247
16	50.82	2583	4.094
17	45.26	2048	5.163
18	40.30	1624	6.510
19	35.89	1288	8.210
20	31.96	1022	10.35

TABLE 10-1   Standard Wire-Gauge Sizes for Copper Wire

TABLE **10-2** Wire Size and Ampacity

Size (AWG)	Ampacity
18	6
16	8
14	17
12	23
10	28

How to Strip, Splice, Solder, and Install Solderless and Terminal Connectors, and How to Use Wire Nuts on Wires

To strip the insulation off the wire, there are certain steps you need to follow. First, you must have a good wire stripper (Figure 10-28). Now, place the wire in the proper sized slot in the wire stripper and work the stripper back and forth until a cut is made in the entire insulation. Do not damage any of the strands in a stranded wire or put a knick in a solid wire; this will cause a weakness in the wire that may cause a break in the circuit in the not-too-distant future. To remove the insulation, hold the wire tight with one hand and use the other hand to gently move the insulation back and forth until the cut breaks clean and the unwanted insulation can be pulled off the wire (Figure 10-29). Next, taper the insulation with a knife to increase the wire's flexibility, because a straight cut in the insulation will create a force that can cause a wire to break prematurely (Figure 10-30). Figures 10-31 and 10-32 illustrate the different methods of splicing single and stranded wires together.

To connect wires to screw terminals, the wire being attached at a screw terminal should be connected so that the loop lies in the direction the screw turns (Figure 10-33). The wire should loop the screw a little less than one full turn, but excessive loops around the screw terminal are not recommended, as this could cause wire damage.

FIGURE **10-28**
Always use a wire stripper to remove the insulation from the wire.

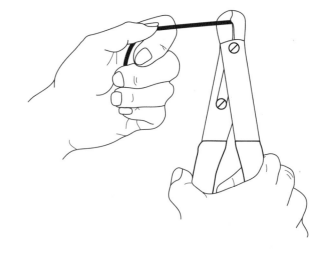

Figure 10-29
Move insulation back and forth to remove insulation from wire.

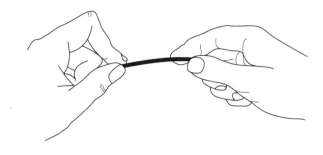

Figure 10-30
Taper the insulation on the wire with a knife to increase the wire's flexibility.

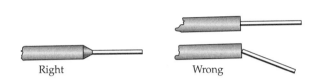

Right   Wrong

Figure 10-31
Different ways to splice stranded and solid wire together.

Types of Splices

A.   Simple Splice

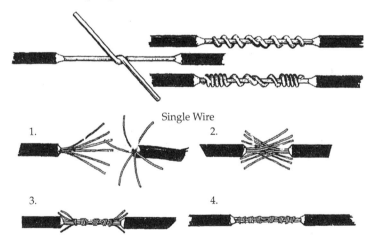

Single Wire

1.   2.

3.   4.

Stranded Wire

B.   Simple Tap

Single Wire

FIGURE 10-32
Different ways to
correctly splice wire
together.

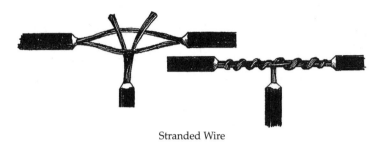

Stranded Wire

C.   Pigtall Splice

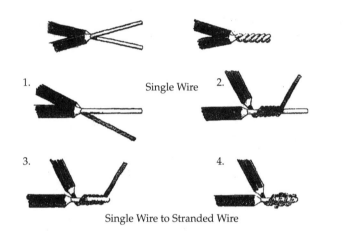

Single Wire

Single Wire to Stranded Wire

D.   Hook Splice

Here is a good guideline that you should practice for soldering wire splices (Figures 10-34, 10-35, and 10-36). First, the wire being soldered together should be bright and clean at the point of connection. The connection point should be tight so that the solder can flow between the joint and solidify without any wire movement. When soldering wires together, the wires should be coated with an electric soldering paste of flux, and soldered so that the solder melts and flows into every crevice of the spliced joint. After the soldered joint cools, the entire splice area should be covered with a waterproof plastic tape or heat-shrink covering to protect the joint from shorting out against the cabinet of the product.

FIGURE 10-33
Connecting a wire
correctly to a screw
terminal.

Right

Wrong

When you have to attach a solderless connector or wiring terminal to a wire, you should follow these guidelines. Solderless connectors should be used according to their color codes, and a connector for a smaller-gauge wire should never be used on a heavier-gauge wire. The wire connector might burn off the wire and break the circuit. A screw-on wire connector (also known as a wire nut) works well on pigtail splices (Figure 10-37).

When installing a crimp-on connector (Figure 10-38), there should not be a gap between the insulation and the terminal connector, and if there is a gap, a plastic sleeve should be added to cover the bare wire, or reinstall a new connector if needed. Always prevent wires from shorting out and breaking the circuit, causing you to have to return to the service call.

Right   Wrong

Figure 10-34   Preparing to solder wires together. Connect the two ends of the wires as shown.

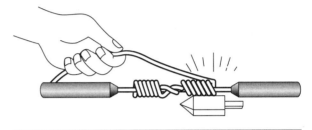

Figure 10-35   Having a tight wire connection and clean wires will allow the solder to flow through the joint.

Figure 10-36   When the solder cools off, tape the soldered splice to protect the wire from shorting out.

Circuit Protection Devices

Circuit protection devices are important for appliances and air conditioners. These devices will protect the electrical circuits and components from damage from too much current flow. Each fuse (Table 10-3) or circuit breaker (Figures 10-39, 10-40, and 10-41) must be rated for voltage and current. Never replace a fuse or circuit breaker with one that is not correctly rated for the product. Fuses and circuit breakers must be able to do the following:

- Sense a short in the circuit
- Sense an overloaded circuit (too much current)

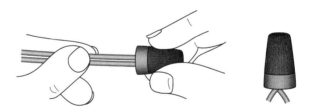

Figure 10-37   Wire connector, also known as a wire nut. Twist wires in a clockwise direction; screw the wire nut cap in a clockwise direction onto the connected wires as shown.

Right   Wrong

Figure 10-38   When attaching terminal connectors, try not to have any gaps from the end of the insulation to the end of the wire connector.

PART IV

Cartridge Fuse Classification	Fuse Characteristics
L	Time delay
RK1	Time delay; fast acting
RK5	Time delay
T	Fast acting
J	Time delay; fast acting
CC	Time delay; fast acting
CD	Time delay
G	Time delay
K5	Fast acting
H	Renewable fuse; fast acting

TABLE 10-3   Classifications of Cartridge Fuses According to Underwriters Laboratories (UL)

- Not open a circuit under normal operating current conditions
- Open a circuit before electrical damage to components or product
- Not exceed the current carrying capacity of the circuit

Other circuit protection devices include:

- **Thermal overload**   Operated by heat; usually resettable; can also be one-time use (Figure 10-7).
- **Thermal fuse**   Encased in an insulated case; one-time use; turns circuit off (Figure 10-42).
- **Bimetal thermostats**   Resettable switches; rated by temperature change; turns circuit on or off (Figure 10-43).
- **Bimetal switches**   Turns a component on or off by design temperature rating (Figure 10-6).
- **Fusible links**   Has a one-time thermal limit; turns the circuit off (Figure 10-44).

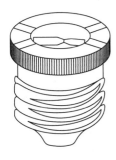

FIGURE 10-39   Two types of screw-in fuses.

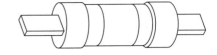

FIGURE 10-40   A cartridge fuse.

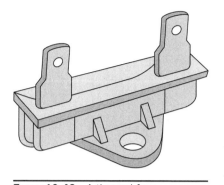

On

Tripped

OFF

Off

Reset

FIGURE 10-41   A circuit breaker.

FIGURE 10-42   A thermal fuse.

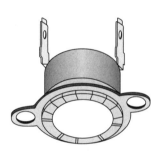

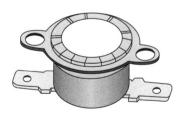

FIGURE 10-43   A bimetal thermostat.

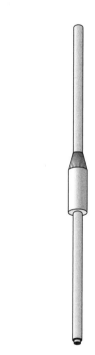

FIGURE 10-44   A fusible link (also known as a fuselink).

Electronic Parts

Did you ever wonder what happens when you touch a function on the touch panel keypad on a microwave, range, washer, or room air conditioner? In this chapter, I will explain the sequence of events from behind the control panel (Figure 11-1) that occurs when you turn on the appliance or room air conditioner for the first time. In addition, I will discuss the service techniques needed to service electronic components in appliances and air conditioners.

With the introduction of electronic components in appliances and room air conditioners, there are consumers, first-time repairers, and even some technicians who do not have a clue as to the operation of the sequence of events that takes place after the product is programmed.

On standard appliances and room air conditioners, consumers will turn knobs and press buttons to set the functions, and sometimes they will have to manually turn their appliances on or off. Appliances and room air conditioners with electronic touch panels (see Figure 11-1) can now be programmed to perform a single event or multiple events and to automatically turn on or off.

Electronic Components in General

Much of the information in this chapter covers electronic components in general, rather than specific models, in order to present a broad overview of operation and service techniques. The pictures and illustrations that are used in this chapter are for demonstration purposes to clarify the description of how to service these appliances and room air conditioners. They in no way reflect a particular brand's reliability.

How Electronic Appliances and Air Conditioners Operate

Beginning with the electronic touch panel on the product, the individual will place his or her finger on a function or number to begin the process of programming the product or telling it what action to perform. The electronic touch panel is made up of a thin membrane with a matrix configuration of pressure-sensitive resistive elements that are sealed. When you touch any key pad, you are closing a circuit in the touch panel membrane to be transmitted to the printed circuit board (PCB), called a display board (Figure 11-2). The display board consists of LEDs (light-emitting diodes) or an LCD (liquid crystal display) that shows the consumer

FIGURE 11-1
A typical oven control panel with manual and electronic controls.

what functions have been stored. In Figure 11-3, the schematic illustrates the matrix configuration of the touch panel membrane. A technician can test the individual key pad functions with the electricity off to the product. For example, if you hold down the cook time key pad and place the ohmmeter probes on pins 11 and 13 on the ribbon connector, you should measure a resistance from 50 to 200 ohms. If you measure zero ohms, the touch panel membrane is faulty and must be replaced. Depending on what model you are servicing, the display board may be part of the main processor PCB or it may be a separate PCB entirely. This PCB is powered by a low-voltage transformer that is either mounted on the PCB or mounted somewhere within the appliance. The touch panel and display board are connected

FIGURE 11-2
An exploded view of the touch pad membrane, bracket, and the display board.

Touch pad membrane

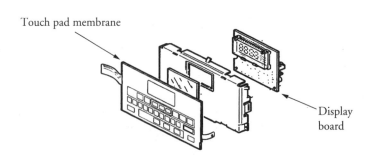

Display board

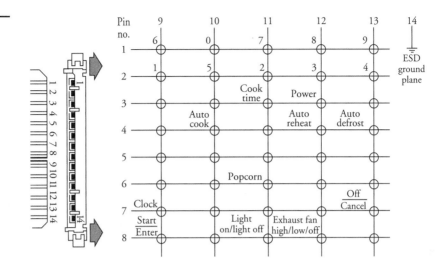

FIGURE 11-3
A technician can test the matrix configuration of the touch panel membrane at the ribbon connector.

to the main processing board (Figure 11-4) by a ribbon cable that is plugged into the main processor board. On some models, all of the components discussed so far may be assembled on to one printed circuit board. After the commands have been entered into the touch panel, the signal is transferred to the main processing PCB to be stored in the main microcomputer chip (CPU) that is mounted on the board (see Figure 11-4). When the user presses the start button, a signal goes to the main CPU chip on the main PCB to initiate the start cycle. The main CPU chip will then select the correct relays to turn on or off the functions that were programmed into it. On some models, the main processor board will send a signal to other PCBs within the appliance to initiate the cycle of events.

Electrostatic Discharge

To prevent electrostatic discharge (ESD) from damaging expensive electronic components, simply follow these steps:

- Turn off the electricity to the appliance or air conditioner before servicing any electronic component.
- Before servicing the electronics in an appliance or air conditioner, discharge the static electricity from your body by touching your finger repeatedly to an unpainted surface on the appliance or air conditioner. Another way to discharge the static electricity from your body is to touch your finger repeatedly to the green ground connection on that product.
- The safest way to prevent ESD is to wear an antistatic wrist strap.
- When replacing a defective electronic part with a new one, touch the antistatic package that the part comes in to the unpainted surface of the appliance or air conditioner or to the green ground connection of the appliance or air conditioner.
- Always avoid touching the electronic parts or metal contacts on an integrated board.
- Always handle integrated circuit boards by the edges.

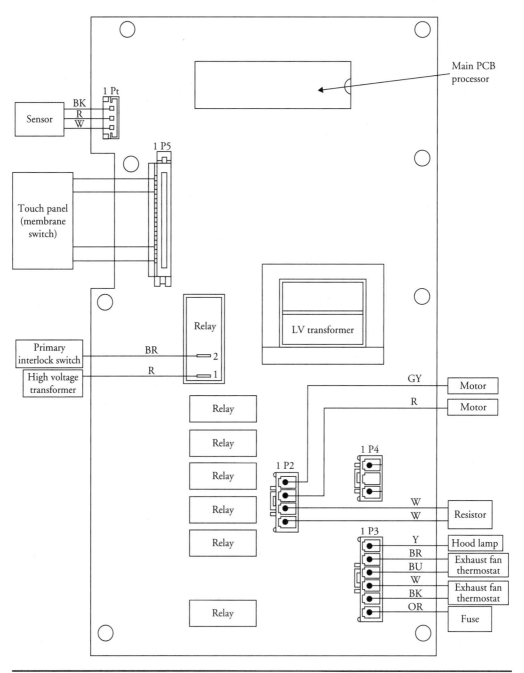

FIGURE 11-4 A pictorial diagram of a main processor printed circuit board.

﹖sting Printed Circuit Boards

Before disassembling or testing can begin, look for the technical data sheets. These are typically attached to the outer cabinet under the appliance, behind the control panel, or behind an access panel. These data sheets will provide you with a lot of helpful information when diagnosing and testing procedures for the appliance or room air conditioner. Most technical data sheets will provide you with a self-diagnostic test sequence that can be programmed through the touch panel. On some models, you can isolate and operate the components through the electronic control to see if they operate. When you initiate the diagnostic test, the main PCB will respond, in most cases, with a code that will indicate where the problem lies. On other models, the fault code appears when a malfunction occurs. In addition, the technical data sheet provides other important information, such as the position of the switch contacts, color-coding of wires, performance data tests, a wiring schematic, and other information that might be helpful to the technician.

The most common problem with electronic components in appliances is loose plug connections and corrosion. Before you begin to replace any component, it is recommended that you disconnect the plug connections from the circuit boards and reconnect them. This process will eliminate any corrosion buildup on the plug connectors or pin connections on the circuit board. In addition, if any plug connections were loose, they will be reattached when you plug them back into the circuit board. Most printed circuit boards have fuses soldered to the circuit board. These fuses must be tested first before condemning the component.

Touch Panel Membrane

Before condemning the touch panel, you need to perform certain inspections. The following is a list of procedures to follow:

- Examine the touch panel membrane (see Figure 11-1) for dents or scratches in the panel. This might cause a short in one or more of the touch pads.
- Inspect the ribbon cable from the touch panel to the display board. Look for evidence of corrosion, tarnishing, or wear on the cable.
- Test all of the keypads and check to see if all functions are working properly.
- If you have to press hard on the touch panel to activate a function, the touch panel will have to be replaced.
- If you press a number and the display shows a different number, the touch panel may have to be replaced.

Transformer

A transformer is an electrical device that can increase (step up) or decrease (step down) the voltage and current. It works on the principle of transferring electrical energy from one circuit to another by electromagnetic induction (Figure 11-5). The primary side of the transformer is the high-voltage side, with the voltage ranging from 120 volts AC to 240 volts AC. On the secondary, or low-voltage, side, the voltage will range from 5 volts AC to 24 volts AC, depending on the amount of voltage and current needed to operate the circuit boards. Some circuit boards require DC voltage to operate, depending on the manufacturer's requirements for the product.

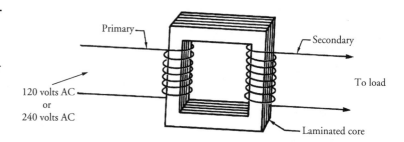

FIGURE 11-5
A transformer. Most appliances and room air conditioners use a step-down transformer to supply a low voltage to electronic PCBs.

Circuit Board

When diagnosing the circuit board, the wiring schematic for the appliance will be helpful in diagnosing, understanding wire color codes, and reading the correct voltages. For example, in Figure 11-6, the main PCB controls the on/off functions and the temperature for the air

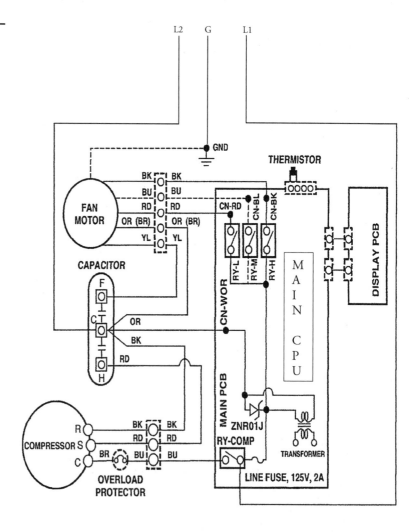

FIGURE 11-6
A sample RAC wiring schematic.

conditioner. To determine if the main PCB is defective, you would check for the correct supply voltage coming into the PCB. In this case, the voltage should be 120 V AC. In addition, you can also check the voltage at the primary winding of the transformer mounted on the PCB for 120 V AC. Next, test the secondary side of the transformer for output voltage. There is a line fuse on this printed circuit board. Turn off the electricity and check the fuse for continuity. If all checks out, test the relays on the PCB for voltage to the relay coils, or test if the switch contacts on the relays are opening and closing.

A good rule to remember when testing any printed circuit board is that there must be voltage supplied to the board and there must be voltage leaving the board to turn a function on.

Integrated Circuit Chip

An integrated circuit (IC), shown in Figure 7-24, is a miniature electric circuit consisting of transistors, diodes, resistors, capacitors, and all the connecting wiring—all of it manufactured on a single semiconductor chip.

Resistors

A resistor (Figure 7-16a), when installed into an electrical circuit, will add resistance, which will produce a specific voltage drop, or a reduction in current. Resistors can be either fixed or variable.

Sensors

A sensor (examples in Figure 12-17 and Figure 14-55a) is a device that produces a measurable response to a change in a physical condition, such as temperature or humidity, and converts it into a signal that can be read by the microcomputer chip on the printed circuit board. Sensors are used to measure basic physical phenomena, including:

- Acceleration
- Angular/linear position
- Chemical/gas concentration
- Humidity
- Flow rate
- Force
- Magnetic fields
- Pressure
- Proximity
- Temperature
- Velocity

Temperature Detectors

Thermocouples, thermistors, and resistance temperature detectors (RTDs) are devices that sense and measure temperature. Thermocouples are useful in applications where a wide temperature operating range is anticipated. Thermistors are recommended for applications with a specified temperature range in mind. RTDs are recommended for applications where accuracy and repeatability are important.

Thermistor

A thermistor is a thermally sensitive resistor that exhibits a change in electrical resistance with a change in its temperature. They are a semiconductor composed of metallic oxides such as manganese, nickel, cobalt, copper, iron, and titanium. Thermistors can be various shapes. There are two types of thermistors: negative temperature coefficient (NTC) and positive temperature coefficient (PTC), with the most common being NTC.

Thermistors are used in the following products:

- Automatic dryers
- Automatic washers
- Refrigerators
- Ovens

Thermocouple

A thermocouple (Figure 12-16) is a measuring device manufactured by joining two dissimilar metals at one end. A voltage is generated when a temperature gradient exists between the wire junction and a reference junction. This measurable change of electric potential is the basis of the thermocouple method. Thermocouple junctions are manufactured in three forms: exposed, grounded, and ungrounded. The exposed junction was designed for a faster response.

Resistance Temperature Detector

An RTD (Figure 12-15) is a resistance temperature detector. The RTD's function is similar to the thermistor. It is a device that provides a useable change in resistance to a specified temperature change. Unlike thermocouples, RTDs are not self-powered. A current must be passed through the RTD, the same as with thermistors, and the change of voltage with temperature is measured.

Thermopile

A thermopile is a thermoelectric device that consists of an array of thermocouple junction pairs connected electrically in series. This device does not measure temperature, but generates an output voltage proportional to the temperature difference or temperature gradient where the device is installed.

Transducer

A transducer is a sensing device that converts one type of energy to another and sends information to the microcomputer chip on the electronic control board.

Diode

A diode (Figures 6-24 and 6-25) is an electrical device that allows current to flow in one direction only. There are many types of uses for diodes besides rectification. These include capacitance that varies with the amount of voltage applied to the diode and photoelectric effects.

LED

Light-emitting diodes, or LEDs as they are usually called, generate light when a current is passed through them. LEDs are used in appliances to indicate if a control is on or off.

Bridge Rectifier

The bridge rectifier (Figure 7-21), consists of four diodes connected together in a bridge configuration on the circuit board. On electronic control boards, a bridge rectifier is used to convert alternating current into direct current for low-voltage circuitry.

Triac

A triac (Figure 7-22), is a three-terminal electronic device that is similar to a diode, except that it allows current to flow in both directions, as with alternating current. There is no anode or cathode in the triac, and it acts as a high-voltage switch on an electronic control board that will turn loads on or off in the circuit.

Transistor

A transistor is a three-element, electronic, solid-state component that is used in a circuit to control the flow of current or voltage. It opens or closes a circuit just like a switch (Figure 7-23a).

Inverter Board

Inverter boards (Figure 11-7) are used on refrigerators, microwaves, and automatic washers. They convert 120-volt, single-phase, 60-Hertz alternating current into three-phase alternating current, either 230-volt alternating current with frequency variations from 57 Hertz to 104 Hertz, or into a specified direct current voltage, single- or three-phase, with varying frequencies.

Figure 11-7
A microwave inverter
board.

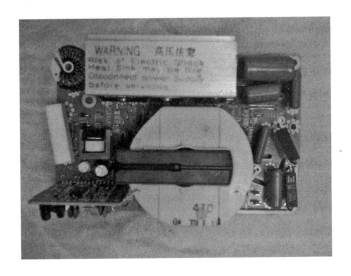

Piezoelectric Ignitor

A piezoelectric ignitor (Figure 11-8) can generate voltages sufficient to spark across an electrode gap, and thus can be used as ignitors in gas water heaters, gas ranges, and gas ovens. Piezoelectric ignition systems are small and simple, and are made from crystalline minerals.

Figure 11-8
A piezoelectric ignitor
used to light the gas
flame in a gas water
heater.

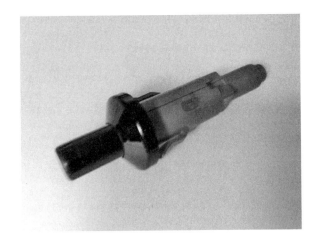

Gas Appliance Parts

This chapter explains how to identify, locate, and understand the operation of gas appliance parts. In addition to electrical parts, the gas components play an important role in the proper operation and safety of gas appliances. Figures 12-1, 12-2, and 12-3 will help you to identify and locate the parts in a gas range, gas dryer, and gas water heater, respectively. Gas parts are divided into the following groups:

- Control parts: Manual and automatic controls used in gas appliances to turn the gas supply on or off or to regulate the flow of gas in the appliance.
- Safety parts: Gas controls that prevent a hazardous condition.
- Combination parts: Gas controls that act as both control parts and safety parts.
- Sensing parts: Sensing devices that are used to activate or deactivate a control.
- Ignition parts: Gas appliances require an ignition source to ignite the burners.

Gas appliance parts are factory-set upon installation and manufacture of the product. These settings should not be tampered with, unless it is determined that the setting was improperly set. It is recommended that you adjust the factory setting according to the manufacturer's recommendation.

ontrols

Manual and automatic controls are the two types of controls used in major appliances. Manual controls are operated by the consumer and are adjusted by eyesight. For example, a consumer will manually turn on a gas burner and adjust the flame height with the burner knob. Automatic controls require three elements to control the gas flow:

- A device to sense the operating conditions
- A device to regulate the flow of gas
- A means to actuate the control

Over the years, controls have evolved from simple controls to complex electronic systems using microprocessors that provide integrated control over all of the components in an appliance.

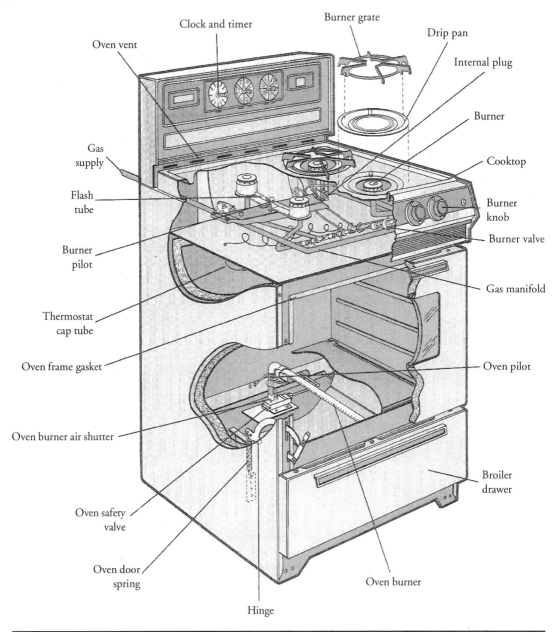

FIGURE 12-1 Typical gas range parts identification.

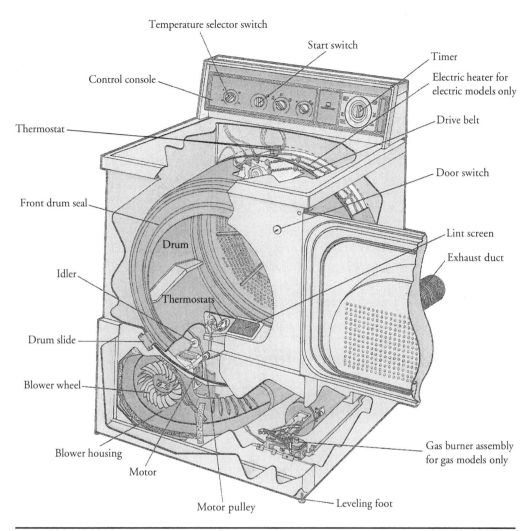

Temperature selector switch

Start switch

Timer

Electric heater for electric models only

Control console

Thermostat

Drive belt

Door switch

Front drum seal

Drum

Lint screen

Exhaust duct

Idler

Thermostats

Drum slide

Blower wheel

Blower housing

Motor

Gas burner assembly for gas models only

Motor pulley

Leveling foot

Figure 12-2 Typical gas dryer parts identification.

Pressure Regulator Controls

The pressure regulator (Figures 12-4, 12-5, and 12-6) is either a mechanical or electric control that regulates and maintains gas flow. This device reduces the incoming gas pressure to a level that is desired for a particular application. It is recommended that a main shutoff valve be installed between the pressure regulator valve and the main gas supply entering the appliance (Figure 12-7). With the shutoff valve located near the appliance, the technician will be able to shutoff the gas supply to the product before beginning repairs.

PART IV

FIGURE 12-3
FIGURE 12-3
Typical gas water
heater parts
identification.

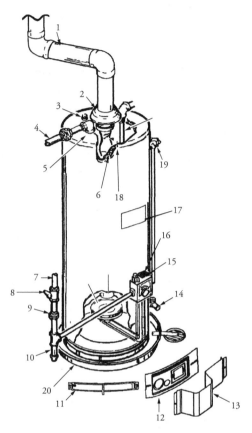

1	Vent pipe	5	Outlet	9	Ground joint union	13	Outer door	17	Name tag
2	Drafthood	6	Insulation	10	Sediment trap	14	Drain valve	18	Flue baffle
3	Anode	7	Gas supply	11	Air intake screen	15	Thermostat	19	TPR valve
4	Hot water outlet	8	Gas shutoff valve	12	Inner door	16	Gas igniter	20	Drain pan

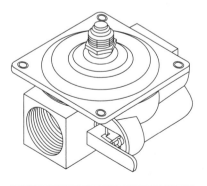

FIGURE 12-4   A gas pressure
regulator valve.

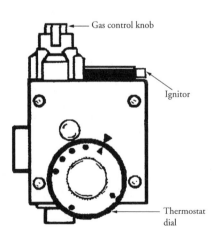

FIGURE 12-5
A water heater
combination control.
This control includes a
gas pressure regulator
valve, thermostat, and
ignitor.

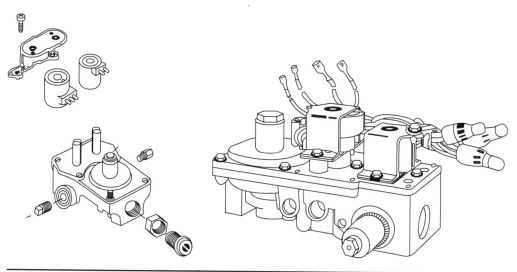

FIGURE 12-6   Two types of dryer gas valves.

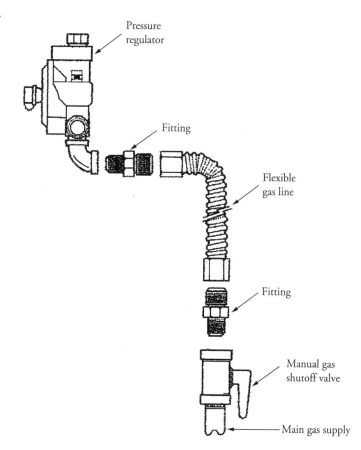

FIGURE 12-7   The main gas supply line to an appliance should include a manual gas shutoff valve. If you intend to use a flexible gas line, you must check local building codes first.

Pressure regulator

Fitting

Flexible gas line

Fitting

Manual gas shutoff valve

Main gas supply

PART IV

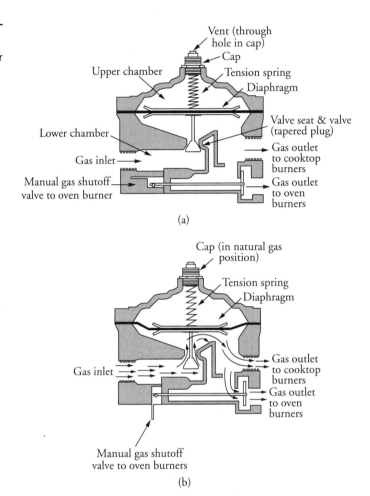

Figure 12-8
(a) An illustration of a gas pressure regulator valve in the closed position. (b) An illustration of a gas pressure regulator valve in the open position.

Vent (through hole in cap)

Cap

Upper chamber

Tension spring

Diaphragm

Valve seat & valve (tapered plug)

Lower chamber

Gas outlet to cooktop burners

Gas inlet

Manual gas shutoff valve to oven burner

Gas outlet to oven burners

(a)

Cap (in natural gas position)

Tension spring

Diaphragm

Gas outlet to cooktop burners

Gas inlet

Gas outlet to oven burners

Manual gas shutoff valve to oven burners

(b)

As the gas enters the pressure regulator valve (see Figure 12-4 and Figure 12-8a), the gas pressure pushes against the spring-loaded diaphragm, forcing the valve to close and shutting off the supply of gas to the appliance. When the consumer turns on the appliance, or when the appliance itself is calling for more gas, the pressure within the pressure regulator valve (Figure 12-8b) decreases, allowing the spring tension to push down on the diaphragm and forcing the valve to open, allowing more gas to the burner(s). The design of the tapered plug and diaphragm allows for metering and maintaining a constant pressure of gas to the burner(s). Another feature incorporated into the pressure regulator is an air vent in the upper chamber. The main purpose of this air vent is to allow air to enter and leave the upper chamber during the operation of the pressure regulator. As a secondary feature, the vent will allow gas to escape at a predetermined amount if the diaphragm ever ruptures.

Dryer gas valves (see Figure 12-6) contain a pressure regulator and two solenoid-operated gas valves. During normal operation, both solenoid valves are energized simultaneously to allow gas to flow to the burner.

When diagnosing a pressure regulator failure, common causes to consider include:

- The valve portion within the regulator may have worn out or may be broken.
- Accumulation of dirt and debris around the valve seat can cause erratic operation or a complete shutdown of the regulator valve.
- The air vent might be plugged or restricted.
- The diaphragm has ruptured and gas is venting into the atmosphere.
- With LP gas, corrosion can occur within the regulator valve if water enters the gas supply.
- An electrical component may have failed.

Pressure regulator valves and dryer gas valves are not serviceable and should be replaced with a duplicate of the original if they fail.

Water Heater Thermostat/Regulator Combination Control

Water heaters use a combination control that incorporates a thermostat and a gas pressure regulator in one control (see Figure 12-9). In addition, the control has a gas cutoff device incorporated into the control in the event that the thermostat fails to shut off the gas supply. The combination valve is activated by a thermocouple that opens the gas inlet to the pressure regulator. The temperature probe will actuate a lever from within the pressure regulator to open or close the valve to the main burner. On top of the control is a knob that you will depress or turn, depending on the type of control, to begin the process to light the pilot light. If any part of this control fails, it is not serviceable and should be replaced with a duplicate of the original.

FIGURE 12-9
A water heater
combination control
valve.

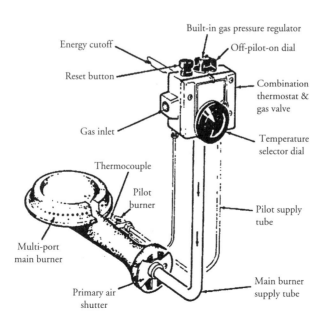

Built-in gas pressure regulator

Energy cutoff

Off-pilot-on dial

Reset button

Combination
thermostat &
gas valve

Gas inlet

Temperature
selector dial

Thermocouple

Pilot
burner

Pilot supply
tube

Multi-port
main burner

Main burner
supply tube

Primary air
shutter

FIGURE 12-10
(a) The single gas
safety valve in a
standing pilot system
in the open position.
(b) The single gas
safety valve in a
standing pilot system
in the closed position.

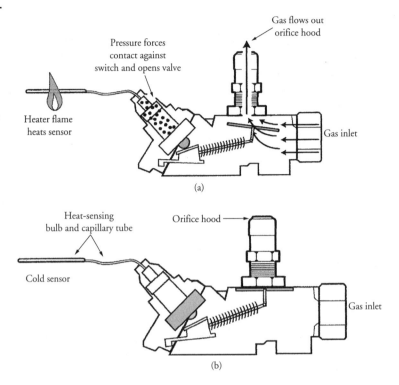

FIGURE 12-10
(a) The single gas
safety valve in a
standing pilot system
in the open position.
(b) The single gas
safety valve in a
standing pilot system
in the closed position.

Safety Valve

Ovens with a standing pilot-light ignition system have a safety valve (Figure 12-10) that controls the gas flow. The safety valve's main function is to allow the gas coming from the thermostat to enter the oven burner. In Figure 12-10a, as the pilot flame heats up the safety valve sensor, the mercury-filled sensor expands and forces the switch to open the safety valve, allowing the gas to enter the oven burner. When the temperature in the oven is satisfied, the sensor begins to cool down, closing the safety valve (Figure 12-10b), stopping the gas flow to the oven burner. The safety valve and the oven thermostat must work together to operate the oven burner correctly. Further discussion on the thermostat and safety valve operation will be covered in a later chapter.

Ovens that have a glow-bar ignition system use a bimetal-operated safety valve (Figure 12-11). This type of valve has one gas inlet and one gas outlet. It is used for the bake burner and the broil burner combination. At the outlet end of the safety valve, there is an electrically operated bimetal strip with a rubber seat that covers the outlet, preventing the flow of gas at room temperature. When current is applied to the bimetal strip, it will warp, allowing the safety valve to open. Gas ranges with the self-cleaning feature in a single oven cavity have a dual safety valve (Figure 12-12). This valve will allow the gas to flow to the bake and broil burners separately when needed. It will not operate both burners at the same time. The operation of the dual safety valve is similar to the single safety valve.

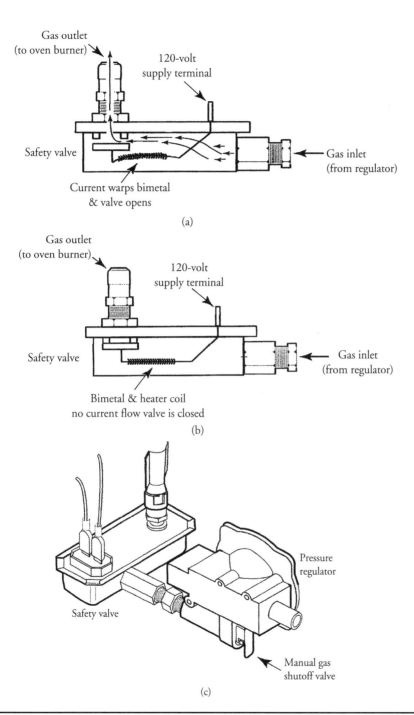

Figure 12-11 (a) The single gas safety valve in a glow-bar ignition system in the open position. (b) The single gas safety valve in a glow-bar ignition system in the closed position. (c) A bimetal single safety gas valve and gas regulator connected together in an automatic ignition system.

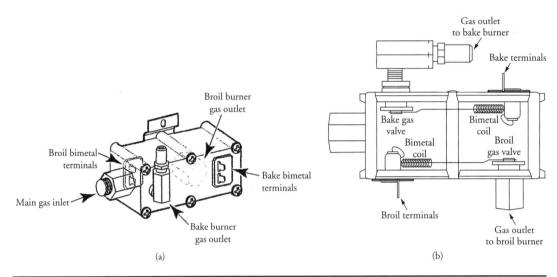

FIGURE 12-12 (a) A dual safety gas valve. (b) An internal view of a dual safety gas valve.

When diagnosing a safety valve failure, common causes to consider include:

- A broken capillary tube
- Loss of voltage to the safety valve
- Bimetal and heater coil failure from within the safety valve
- Debris buildup around orifice
- Mechanical failure

Safety valves are not serviceable and should be replaced with a duplicate of the original if they fail.

Dryer Gas Valve

The dryer gas valve in Figure 12-13 is a combination control consisting of a pressure regulator and dual shutoff valves, housed in one body to regulate the gas flow when the thermostats call for more heat. The solenoid coils (Figure 12-14) will activate by means of electrical power and open the gas valves by electromagnetism, allowing gas to flow to the burner. When the temperature is satisfied, the electrical power is turned off and the solenoid coils deactivate, allowing the internal spring pressure to close the valve. Dryer gas valves are not serviceable; only the solenoid coils are serviceable. The gas valve body should be replaced with a duplicate of the original if it fails.

Sensing Devices

A sensing device can be a temperature-responsive or pressure-responsive device that transmits a signal or motion to activate or deactivate a control device. In electrical control circuits, resistive coils, resistance temperature detectors (RTD), and thermistors are used in a circuit to activate or deactivate the controls. The electrical resistance of these devices varies by temperature change to control current flow.

PART IV

FIGURE 12-13
Two different designs of a dryer gas valve combination.

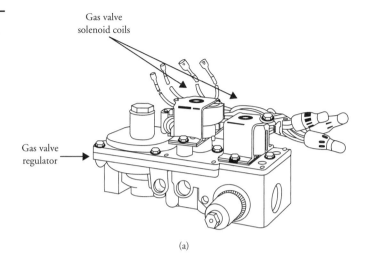

Gas valve solenoid coils

Gas valve regulator

(a)

Gas valve solenoid coils

Gas valve regulator

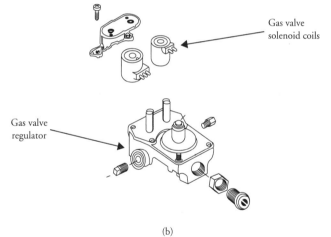

(b)

FIGURE 12-14
(a) A de-energized solenoid coil in a dryer gas valve assembly indicating no gas flow. (b) An energized solenoid coil in a dryer gas valve assembly indicating the flow of gas.

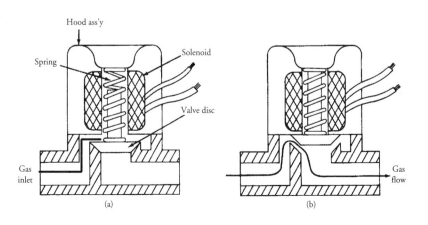

Hood ass'y

Spring

Solenoid

Valve disc

Gas inlet

Gas flow

(a)　　　　(b)

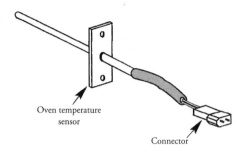

FIGURE 12-15 A resistance temperature detector (RTD).

Resistance Temperature Detector

The resistance temperature detector (RTD) sensor operates on the principle that as the temperature increases, the resistance in the metal increases. With a constant voltage, the current through the metal will drop off as the temperature increases. Ovens with electronic control circuits use an oven temperature sensor (Figure 12-15) to activate or deactivate the bake, broil, and self-clean functions. This sensor is an RTD composed of a stainless steel tube coated with platinum at one end, and two wires connected to a connector that plugs into the electronic circuitry. The location of the sensor is in the upper corners of the oven cavity. This device is neither adjustable nor repairable, and should be replaced with a duplicate of the original if it fails.

Thermocouple

A thermocouple (Figure 12-16) is a measuring device consisting of two dissimilar metals, which produces a low DC voltage when heated by a gas pilot flame that is measured in millivolts. This thermoelectric device is commonly used in gas appliances to power automatic-pilot safety devices. The average output voltage for a single thermocouple is between 20 to 30 millivolts. If the thermocouple voltage drops below 5 millivolts, which can vary in design from manufacturer to manufacturer, the pilot safety device will shut off the gas supply to the burner. This device is neither adjustable nor repairable, and should be replaced with a duplicate of the original if it fails.

Flame Sensors

Flame sensors (Figure 12-17) are used in gas appliances to detect the presence of a pilot flame or the main burner flame. For safety reasons, in an automatic ignition system, it is required that a flame sensor be installed to detect the presence of a flame in the gas pilot or the main gas burner. Before the gas valve can open in an automatic pilot system, the flame sensor must detect the presence of the pilot flame. In an electronic ignition system (pilotless ignition), the flame sensor must be mounted over a window cut out in the burner tube to ensure that the burner flame is present or it will not allow the gas valve to open. The switch will open within 15 to 90 seconds if a flame is detected. Also, the ignitor temperature must be within 1800 to 2500 degrees Fahrenheit to open the gas valve. This device is neither adjustable nor repairable, and should be replaced with a duplicate of the original if it fails.

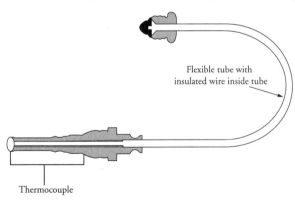

FIGURE 12-16 A thermocouple.

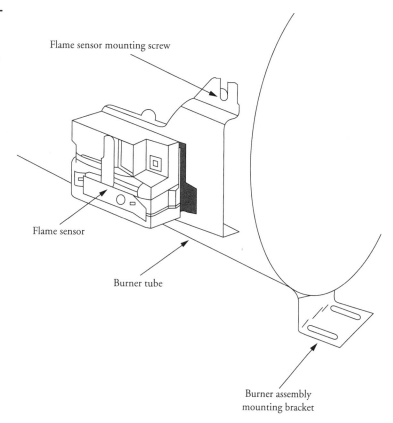

FIGURE 12-17
A flame sensor is a thermostatically controlled single pole, single throw, normally closed switch.

Flame sensor mounting screw

Flame sensor

Burner tube

Burner assembly mounting bracket

Ignition

There are two ways to ignite a gas burner: using matches or using an automatic ignition source. Many appliances manufactured today have some type of automatic ignition source. This automatic system can be continuous, intermittent, interrupted, or a combination of these things.

Glow-Bar Ignitor

To achieve direct ignition, a silicon carbide glow-bar device (Figure 12-18) is positioned in the path of the burner flame. The reason for this positioning is to achieve the best performance for ignition and flame sensing. Line voltage is applied to the ignitor. When it reaches a temperature between 1800 and 2500 degrees Fahrenheit, in about 15 to 100 seconds (depending on design), a signal is sent to open the gas valve, allowing gas to flow to the burner, and gas ignition occurs. When using the glow-bar as a sensor, if the

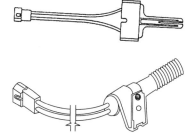

FIGURE 12-18
Two types of a silicon carbide glow-bar device (ignitor) used in gas appliances.

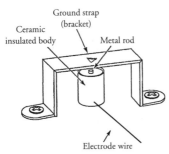

FIGURE 12-19
Type of ignitor used to light the gas burner(s) on a gas range.

temperature of the glow-bar begins to drop below the ignition temperature, the gas valve will close, shutting off the gas supply to the burner.

There are times when the glow-bar ignitor will appear to glow properly and be reddish in color, but the gas burner will not light. In addition, there are times when the ignitor will not light at all. If this happens, you will need to perform a visual inspection and test the ignitor with a clamp-on multimeter. This ignitor is neither adjustable nor repairable, and should be replaced with a duplicate of the original if it fails.

Spark Electrode Ignitor

The spark electrode ignitor replaces the standing pilot flame system with electrodes and a spark module. The ignitor (Figure 12-19) consists of a metal rod embedded into a ceramic insulating body that is wired to a spark module located in the gas appliance. The spark module will send a number of pulses to the spark electrode ignitor, which will begin to arc between the metal rod and the grounding strap bracket. This device is neither adjustable nor repairable, and should be replaced with a duplicate of the original if it fails.

Spark Module

The spark module (Figure 12-20) is an electronic device that delivers a high-voltage pulse to the spark electrode ignitor. These pulses are delivered by a repeatable timing sequence from within the module every few seconds (depending on design) and operate at very low amperage. When a flame is detected, the spark module will stop transmitting pulses to the ignitor. Some gas appliance models are designed with automatic flame recovery and/or automatic lockout of the gas valve. This device is neither adjustable nor repairable, and should be replaced with a duplicate of the original if it fails.

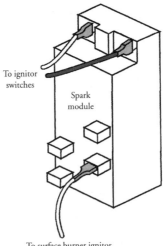

FIGURE 12-20 A spark module.

V

PART

Error/Fault or Function Codes

Error/Fault or Function Codes

rror/Fault or Function Codes

Over the years, appliance manufacturers have incorporated fault codes, diagnostic and test information, and appliance usage information into their appliances. All appliances with electronic displays and electronic components have fault codes programmed into the microprocessor on the main printed circuit board. These codes act as a service aid and starting point for the technician when diagnosing a fault with the appliance. An error/fault code will appear on the display when the appliance encounters some type of error detection. This can be in the form of a series of blinking lights or an actual alphanumeric code.

Before you begin diagnostics, locate the technical data sheet and the manufacturer's use and care guide for the product you are about to service. The technical data sheet will have the fault codes and the information on how to get into the appliance test mode. The test mode is another diagnostic tool that allows the technician to advance to a specific function, or the technician can turn on or off a single component. The use and care guide will provide you with the usage information and some fault codes as well. There are times when an error/fault code will appear that is not on the technical data sheet or in the use and care guide. If this occurs, go online to the manufacturer's website or contact the appliance manufacturer's customer service or technical department for the code description.

Do not assume that a service problem is directly caused by the main printed circuit board. The problem might exist with the voltage supply to the product or a line voltage component (including the power cord and the appliance wiring) that has shorted, grounded, opened, or otherwise malfunctioned.

What follows is a sampling of some of the appliances that have error/fault and function codes. Not all of the codes listed in this chapter will cover every appliance manufacturer or model. Rely on the technical data sheet for most of the information for the product you are servicing.

Admiral Oven and Range Error/Fault or Function Codes

The error/fault or function codes listed here will vary from model to model. The codes on the model you are servicing might be different from what is listed. If this is the case, check the technical data sheet, the use and care guide, or the manufacturer's website, or call the manufacturer for the description of the error/fault code you are experiencing.

Sometimes, by disconnecting the electricity to the product and waiting for two minutes and then turning on the electricity, this may reset the error/fault or function code, allowing you to reprogram and test the operation of the appliance.

For your safety and the safety of others, all testing and physical inspections of the product must be performed with the electricity turned off. On occasion, you will have to test a component with the electricity on. Extreme care must be taken to prevent electrical mishaps.

F0   *Description*   One or more of the function keys are stuck.

             Solution   Replace the touchpad assembly. If the touchpad is part of the clock (ERC), replace the clock.

F1   *Description*   Defective touchpad or clock.

             Solution   Replace the touchpad and/or clock (ERC).

F2   *Description*   The oven temperature is too hot.

             Solution   If there is a relay board, replace it. Replace oven sensor (RTD). Also, check the sensor wiring harness.

F3 or F4   *Description*   Oven sensor is open or shorted.

             Solution   Check the sensor harness and connector between the oven control and the oven sensor. If the wiring checks out okay, replace the oven sensor.

F5   *Description*   Main PCB and hardware conflict.

             Solution   Replace the ERC (clock).

F7   *Description*   One or more of the function keys are stuck or shorted.

             Solution   Replace the touchpad assembly. If the touchpad is part of the clock (ERC), replace the clock.

F8   *Description*   Analog/digital problem.

             Solution   Replace the ERC (clock).

F9   *Description*   Door latch problem

             Solution   Check door latch assembly.

ASKO Dishwasher Error/Fault or Function Codes

The error/fault or function codes listed here will vary from model to model. The codes on the model you are servicing might be different from what is listed. If this is the case, check the technical data sheet, the use and care guide, the manufacturer's website, or call the manufacturer for the description of the error/fault code you are experiencing. Sometimes, by disconnecting the electricity to the product and waiting for two minutes and then turning on the electricity, this may reset the error/fault or function code, allowing you to reprogram and test the operation of the appliance.

For your safety and the safety of others, all testing and physical inspections of the product must be performed with the electricity turned off. On occasion, you will have to test a component with the electricity on. Extreme care must be taken to prevent electrical mishaps.

F1   *Description*   A problem has occurred in the water heating cycle. The water temperature in the dishwasher did not reach the predetermined temperature within 60 minutes.

      Solution   Check the water temperature at the sink next to the dishwasher. It should be around 120–140 degrees Fahrenheit. Check the water heater temperature also. Check the dishwasher heater and circuit components. Do not take showers or baths or run the automatic washer before running the dishwasher—this will deplete the hot water supply in the water heater.

F2   *Description*   Dishwasher overfilled or float stuck and/or float switch failure or clogged sump or drain.

      Solution   Check for debris in the sump area of the tub. Check the water float and switch assembly. Check the water inlet valve.

F3   *Description*   Thermistor malfunction.

      Solution   Check the thermistor and connecting wiring circuit.

F4   *Description*   Water fill problem. Dishwasher models with level-controlled fill only. *Note*: If the control lights are blinking, the pump is running, and the controls are dead, the fault will indicate an overfill problem.

      Solution   Check for water in the base pan. The dishwasher will not operate if water is in the pan. Remove the water, reset the dishwasher, and inspect for leaks. Check float assembly.

F5   *Description*   Valve leakage fault code.

      Solution   Inspect valve for leaks.

F7   *Description*   Blocked drain fault code.

      Solution   The drain pump can be cleaned out. Refer to the use and care guide for the cleanout procedure.

F8   *Description*   Clogged filter fault code. One or all three filters are clogged.

      Solution   Before beginning to disassemble the filters, refer to the use and care guide for proper removal and cleaning instructions for the filters.

Some ASKO dishwasher models use flashing indicator lights as error/fault or function codes. See Table 13-1.

ASKO Washer Error/Fault or Function Codes

The error/fault or function codes listed here will vary from model to model. The codes on the model you are servicing might be different from what is listed. If this is the case, check the technical data sheet, the use and care guide, the manufacturer's website, or call the manufacturer for the description of the error/fault code you are experiencing. Sometimes, by disconnecting the electricity to the product and waiting for two minutes and then turning

Flashing Indicator Light(s)		ASKO Dishwasher Error/Fault or Function Code
⊖	Heavy Wash	Dishwasher overfilled. Check float assembly and float switch. Check for clogged filters or clogged drain.
⊻	Delicate Wash	No water to dishwasher. Check water supply and water inlet valve.
⊻	Quick Wash	Valve leakage fault. Check valves for leaks.
⊖-⊖	Pots and Pans and Heavy Wash	The drain is blocked. Refer to the use and care guide for cleanout procedure.
⊖⊻	Pots and Pans and Normal Wash	The filter is clogged. Refer to the use and care guide for proper removal and cleaning instructions for the filters.

TABLE 13-1   ASKO Dishwasher Error/Fault or Function Codes

on the electricity, this may reset the error/fault or function code, allowing you to reprogram and test the operation of the appliance.

For your safety and the safety of others, all testing and physical inspections of the product must be performed with the electricity turned off. On occasion, you will have to test a component with the electricity on. Extreme care must be taken to prevent electrical mishaps.

F1   *Description*   An overfill condition was detected.

Solution   Check the water inlet valve. Replace if necessary.

F3   *Description*   The washer has a draining problem.

Solution   Check drain filter, pump, and drain hose. Replace defective part.

F4   *Description*   A water inlet error was detected.

Solution   Make sure that the water is turned on to the washer. Check water inlet valve. Replace defective part.

F6   *Description*   The control detected a door-open condition.

Solution   Check door latch assembly. Check wiring connectors and wiring.

ASKO Dryer Error/Fault or Function Codes

The error/fault or function codes listed here will vary from model to model. The codes on the model you are servicing might be different from what is listed. If this is the case, check the technical data sheet, the use and care guide, the manufacturer's website, or call the manufacturer for the description of the error/fault code you are experiencing. Sometimes, by disconnecting the electricity to the product and waiting for two minutes and then turning on the electricity, this may reset the error/fault or function code, allowing you to reprogram and test the operation of the appliance.

For your safety and the safety of others, all testing and physical inspections of the product must be performed with the electricity turned off. On occasion, you will have to test a component with the electricity on. Extreme care must be taken to prevent electrical mishaps.

F1   *Description*   The condenser drain hose has detected a blockage.

      Solution   Clear the blockage and restart.

F2   *Description*   Dryer takes too long to dry.

      Solution   Check dryer installation, dryer venting system, and heater. Replace defective part if necessary.

Dacor Oven and Range Error/Fault or Function Codes

The error/fault or function codes listed here will vary from model to model. The codes on the model you are servicing might be different from what is listed. If this is the case, check the technical data sheet, the use and care guide, or the manufacturer's website, or call the manufacturer for the description of the error/fault code you are experiencing. Sometimes, by disconnecting the electricity to the product and waiting for two minutes and then turning on the electricity, this may reset the error/fault or function code, allowing you to reprogram and test the operation of the appliance.

For your safety and the safety of others, all testing and physical inspections of the product must be performed with the electricity turned off. On occasion, you will have to test a component with the electricity on. Extreme care must be taken to prevent electrical mishaps.

F0   *Description*   The function key is stuck.

      Solution   Replace the touchpad assembly. If the touchpad is part of the clock (ERC), replace the clock .

F1   *Description*   PCB or main board problems.

      Solution   Press the "Cancel" button to turn off the beeping sound. If the sound continues, replace the ERC.

F2   *Description*   The temperature in the oven is too hot.

      Solution   Check the oven temperature sensor, wiring, and for a stuck relay. Test self-clean cycle.

F3   *Description*   The oven temperature sensor is shorted.

      Solution   Check the sensor harness and connector between the oven control and the oven sensor. If the wiring checks out okay, replace the oven sensor.

F4   *Description*   The oven temperature sensor is open.

      Solution   Check the sensor harness and connector between the oven control and the oven sensor. If the wiring checks out okay, replace the oven sensor.

F6   *Description*   Main computer board (ERC) problem.

      Solution   Replace the ERC.

PART V

F7   *Description*   Oven door lock enabled above temperature.

   Solution   Check the door lock switches and wiring.

F8   *Description*   Door locking switch malfunction.

   Solution   Check and replace, if necessary, the door lock switches and wiring.

Estate Oven and Range Error/Fault or Function Codes

F0   *Description*   Main board failure.

   Solution   Replace the main board or ERC.

F1   *Description*   Main board failure.

   Solution   Replace the main board or ERC.

F2   *Description*   Open oven temperature sensor.

   Solution   (1) Check the bake and broil relays on the ERC; if contacts are bad, replace the ERC.
(2) For the self-clean models, check the door lock assembly for proper operation.
(3) Check the oven temperature sensor.

F3 or   *Description*   Check for open oven temperature sensor.
F4
   Solution   Replace the oven temperature sensor.

F5   *Description*   Main board failure (ERC).

   Solution   Replace the main board (ERC).

F7   *Description*   Touchpad key stuck on ERC.

   Solution   Replace the ERC.

F8   *Description*   Main board failure.

   Solution   Replace the ERC.

F9   *Description*   There is an oven door lock failure.

   Solution   Check oven door lock switches, wiring harness, and connectors.

Friedrich Room Air Conditioner Error/Fault or Function Codes

The error/fault or function codes listed here will vary from model to model. The codes on the model you are servicing might be different from what is listed. If this is the case, check the technical data sheet, the use and care guide, the manufacturer's website, or call the manufacturer for the description of the error/fault code that you are experiencing. Sometimes, by disconnecting the electricity to the product and waiting for 10 minutes and then turning on the electricity, this may reset the error/fault or function code, allowing you to reprogram and test the operation of the appliance.

For your safety and the safety of others, all testing and physical inspections of the product must be performed with the electricity turned off. On occasion, you will have to test a component with the electricity on. Extreme care must be taken to prevent electrical mishaps.

E1   *Description*   The electronic control board detected a short-cycle malfunction.

          Solution   Check for conditions that caused the short-cycle situation. Check thermostat setting, air flow conditions, refrigeration cycle, and fan motor.

E2   *Description*   The electronic control board detected a touch keypad error.

          Solution   Reset the air conditioner controls and test. Check for stuck keys. If error code returns, replace the touch keypad assembly.

E3   *Description*   The electronic control detected the frost probe is open.

          Solution   Test the valve of the frost probe. If the probe tested bad, replace the frost probe. If good, replace the electronic control board.

E4   *Description*   The electronic control board detected a short in the frost probe.

          Solution   Replace the electronic control board.

E5 or E6   *Description*   The electronic control board detected an open or shorted indoor temperature sensor.

          Solution   Replace the electronic control board.

CHECK FILTER INDICATOR LAMP   *Description*   The check filter indicator lamp will come on after 250 hours of operation. It is a reminder for the consumer to clean the air conditioner filter. This code can be reset to start the countdown to the next filter cleaning.

Frigidaire/Electrolux Dishwasher Error/Fault or Function Codes

The error/fault or function codes listed here will vary from model to model. The codes on the model you are servicing might be different from what is listed. If this is the case, check the technical data sheet, the use and care guide, the manufacturer's website, or call the manufacturer for the description of the error/fault code you are experiencing. Sometimes, by disconnecting the electricity to the product and waiting for two minutes and then turning on the electricity, this may reset the error/fault or function code, allowing you to reprogram and test the operation of the appliance.

For your safety and the safety of others, all testing and physical inspections of the product must be performed with the electricity turned off. On occasion, you will have to test a component with the electricity on. Extreme care must be taken to prevent electrical mishaps.

* or Lo or status LEDs   *Description*   Wetting agent level is low.

          Solution   Check the wetting agent level in the dispenser. After five wash cycles, the indicator light will turn off when the dispenser is not filled up.

PF   *Description*   A power failure occurred. The high-temp wash and the no-heat dry/power-dry off lights will begin to flash when the power resumes.

          Solution   Press the start/cancel pad and reset the controls. If the code persists, turn the electricity off for five minutes. Check all electrical connections to the electronic control panel.

HO   *Description*   Delay in the water heating cycle.

            Solution   This is a normal condition until the water reaches the predetermined temperature. Check the water temperature at the sink next to the dishwasher—it should be around 120 to 140 degrees Fahrenheit. Check the water heater temperature also. Check dishwasher heater and circuit components. Do not take showers or baths or run the automatic washer before running the dishwasher—this will deplete the hot water supply in the water heater.

CL   *Description*   The door is open.

            Solution   Close and latch the dishwasher door.

01-24   *Description*   On some models, this is the amount of time in hours left before the dishwasher begins the cycle.

01-25 or   *Description*   On some models, this is the amount of time in hours left
01-10   before the dishwasher begins the cycle.

SENSING   *Description*   On some models, a thermistor called the turbidity sensor is checking the condition of the wash/rinse water.

            Solution   Check thermistor and electronic control board.

Frigidaire/Electrolux Washer Error/Fault or Function Codes

The error/fault or function codes listed here will vary from model to model. The codes on the model you are servicing might be different from what is listed. If this is the case, check the technical data sheet, the use and care guide, the manufacturer's website, or call the manufacturer for the description of the error/fault code you are experiencing. Sometimes, by disconnecting the electricity to the product and waiting for two minutes and then turning on the electricity, this may reset the error/fault or function code, allowing you to reprogram and test the operation of the appliance.

For your safety and the safety of others, all testing and physical inspections of the product must be performed with the electricity turned off. On occasion, you will have to test a component with the electricity on. Extreme care must be taken to prevent electrical mishaps.

E11   *Description*   Washer taking too long to fill up.

            Solution   Check water inlet supply. Check the water pressure—it should be at or above 30 PSI. Check the washer fill valve, pressure switch or pressure sensor, and the control board. If the washer fills in the off mode and the electricity is turned off, replace the water inlet valve.

E13   *Description*   The electronic control detected a water leak in the tub or an air leak in the air bell.

            Solution   Check for water leaks. Check the air bell system, pressure switch, or pressure sensor. Check the control board. Repair any water leaks. Replace any defective part.

E21

Description   The water in the washer is not pumping out fast enough.

Solution   Check washer installation. Make sure that the drain hose is not restricted. Check the drain pump for 120 volts in the drain cycle. If the voltage reading is zero, replace the control board. If the voltage is correct, check the drain pump for a restriction or blockage. If none is found, replace the pump.

E23 or E24

Description   The electronic control detected that the drain pump relay malfunctioned.

Solution   Test the drain pump relay on the electronic control board. If the test failed, replace the electronic control board. Check all wiring connections and wiring to the drain pump.

E31

Description   A communication error was detected between the pressure sensor and the electronic control board.

Solution   Check the wiring connections and wiring for the pressure sensor to the electronic control board. Test the pressure sensor valve and replace if necessary.

E35

Description   The pressure sensor has detected a water overfill problem.

Solution   Check to see if the water level is above the manufacturer's recommendations. Check the water inlet valve for proper operation. Check the wiring connections and the wiring from the electronic control to the pressure sensor and to the water inlet valve. Check the electronic control board for proper operation. Replace any defective part.

E36, E43, E44, E45, or E46

Description   The electronic control board has detected a malfunction.

Solution   Check all the wiring connections to the electronic control board. If okay, replace the electronic control board.

E41

Description   The electronic control board has detected a door-open code.

Solution   Check that the door closes properly. Check the door strike and door switch assembly. Check the electronic control. Replace any defective part.

E47 or E48

Description   The electronic control board detected a problem in the door thermistor circuit.

Solution   Check the door lock assembly and measure the resistance of the thermistor. If you detect a short or an open, replace the door lock assembly. If your measurement of the thermistor is at least 1500 ohms, replace the electronic control board.

E52

Description   The electronic control board received a bad signal from the tacho generator.

Solution   Test the drive motor for 105 to 130 ohms at pins 4 and 5 on the motor only. If the readings are other than 105 to 130 ohms, replace the speed control board.

PART V

E56, E57, or E58	*Description* The electronic control board detected a motor problem.
	Solution Check the following components: motor, tub bearings, speed control. After removing the belt from the motor, spin the motor—it should turn freely. If it does not, replace the motor. With the belt off, spin the tub—it should spin freely. If it does not, replace the tub bearings. Test the motor windings and test pins 1 and 2, pins 1 and 3, and pins 2 and 3 on the motor. It should read about 6 ohms. If this is not the case, replace the motor.
E59	*Description* The tacho generator lost a signal.
	Solution Check the following components: motor, tub bearings, speed control. After removing the belt from the motor, spin the motor—it should turn freely. If it does not, replace the motor. With the belt off, spin the tub—it should spin freely. If it does not, replace the tub bearings. Test the motor windings and test pins 1 and 2, pins 1 and 3, and pins 2 and 3 on the motor. It should read about 6 ohms. Test the motor for 105 to 130 ohms at pins 4 and 5 on the motor only. If the readings are other than 105 to 130 ohms, replace the speed control board.
E5A, E5B, or E5C	*Description* The heat sink overheated on high temperature.
	Solution Test the heat sink. If it is bad, replace the speed control board.
E5D, E5E, or E5F	*Description* A communication error has occurred between the electronic control board and the speed control board.
	Solution Check the wiring connections and the wiring between the electronic control board and the speed control board. Check the electronic control board and the speed control board. Replace the defective part.
E66 or E68	*Description* The electronic control detected a heater problem.
	Solution Check the wiring connections and the wiring for the heater circuit. Check the heater element for 14 ohms. Check for a grounded heater. Replace the heater if the readings are incorrect.
E67	*Description* The electronic control detected the wrong input voltage to the microprocessor.
	Solution Check the wiring connections and wiring. Check the tub thermistor for 4.8 ohms. If incorrect, replace the heating element.
E71	*Description* Wash thermistor (tub heater) malfunction.
	Solution Check the wiring. If okay, replace the electronic control board.
E74	*Description* The wash water will not heat up.
	Solution Place the thermistor in the correct position.
E75	*Description* Water temperature sensor detection.
	Solution Check the thermistor for resistance—it should be 50 ohms. If it is not, replace the water inlet valve assembly. If the resistor measured 50 ohms, replace the electronic control board.

E76   *Description*   The thermistor temperature is off for the cold water inlet valve.

               Solution   The water fill hoses need to be installed correctly.

E82, E83, E93,   *Description*   The electronic control board has a communications problem.
E94, E95, E97,
E98, EBE, or   *Solution*   Check all wiring connections and wiring. If okay, replace the
EBF   electronic control board.

E91   *Description*   There is a communication error between the electronic control board and the UI board.

               Solution   Check all wiring connections and wiring.

EB1   *Description*   Incoming voltage frequency problem.

               Solution   Have your local electric company come out and check the input voltage frequency to the house. If okay, replace the electronic control board.

EB2   *Description*   Incoming voltage too high.

               Solution   If the voltage at the washer receptacle is below 130 volts, replace the electronic control board.

EB3   *Description*   Incoming voltage too low.

               Solution   If the voltage at the washer receptacle is above 90 volts, replace the electronic control board.

EF1   *Description*   The drain pump is clogged.

               Solution   Clear the blockage from the drain pump.

EF2   *Description*   The electronic control board detected too much soap in the washer.

               Solution   Advise the consumer on proper soap usage.

EF5   *Description*   The thermistor temperature is off for the hot water inlet valve.

               Solution   The water fill hoses need to be installed correctly.

1 beep/E10   *Description*   The washer is not filling up with water or water is leaking onto the floor.

               Solution   Make sure that the water is turned on to the washer. Check the water pressure, fill hoses for kinks, and the drain hose installation. If you found water leaking, check the household drain for a blockage, and check for too much soap usage.

2 beeps/E20   *Description*   The water will not drain out.

               Solution   Check the drain hose installation and check for a drain blockage in the household drain.

3 beeps/E30 *Description*   The electronic control has detected that the drum is overfilled with water.

Solution   Check the water inlet valve and pressure switch for proper operation. Check the electronic control board. Replace the defective part. Drain the water from the washer and reset the controls.

4 beeps/E40 *Description*   The electronic control detected that the door is open.

Solution   Close the door and reset the controls. The washer will not operate when the door is open.

5 beeps/E50 *Description*   The electronic control detected a motor overheat problem.

Solution   Let the motor cool down. Check the wiring connections and wiring. Reset the controls and restart the washer.

7 beeps/E70 *Description*   The cold water is not entering the washer.

Solution   Make sure that the cold water is turned on all the way. Check the water fill hose installation. Check the water inlet valve.

15 beeps/EF0 *Description*   The electronic control detected an oversudsing condition or no hot water entering the washer.

Solution   Check for too much detergent added to the load. Advise the consumer on proper detergent usage. Read the use and care guide. Check for hard water conditions, water temperature, and the size of the load. Make corrections if necessary. For no hot water, check and see if the hot water is turned on all the way. Also check the water fill hose installation.

Frigidaire/Electrolux Dryer Error/Fault or Function Codes

The error/fault or function codes listed here will vary from model to model. The codes on the model you are servicing might be different from what is listed. If this is the case, check the technical data sheet, the use and care guide, the manufacturer's website, or call the manufacturer for the description of the error/fault code you are experiencing. Sometimes, by disconnecting the electricity to the product and waiting for two minutes and then turning on the electricity, this may reset the error/fault or function code, allowing you to reprogram and test the operation of the appliance.

For your safety and the safety of others, all testing and physical inspections of the product must be performed with the electricity turned off. On occasion, you will have to test a component with the electricity on. Extreme care must be taken to prevent electrical mishaps.

E10, E11, *Description*   There is a communication error in the electronic control
or E12   microprocessor.

Solution   Cancel the cycle by pressing cancel. Reset the controls and test the cycle. If the problem persists, replace the electronic control.

E24   *Description*   Thermistor sensor malfunction.

Solution   Check the thermistor sensor for a short. Check all wiring between the thermistor and the electronic control. Test the electronic control for proper operation.

E25   *Description*   Thermistor sensor problem.

     Solution   Test the thermistor sensor. Check the wiring from the thermistor to the electronic control. Test the electronic control for proper operation.

E4A   *Description*   The drying cycle was interrupted because the drying time exceeded the time that was programmed.

     Solution   Inspect the clothes. If they are still wet, check for proper installation of the dryer. Check vent system, blower assembly, and motor. Check the moisture sensor bars for dirt. Check and test the heater. Check all the wiring and wiring connections. Check the electronic control for proper operation.

E5B   *Description*   The thermistor is sensing no heat and the cycle is stalled.

     Solution   Check and test the entire heating circuitry, including the heater. Check the electronic control for proper operation. See the technical data sheet for the specific procedure for the model you are working on.

E68   *Description*   A stuck button fault code.

     Solution   Inspect the buttons and run the test mode to determine which button is stuck in the active mode. Check the electronic control for proper operation.

E8C   *Description*   The high limit safety thermostat has tripped.

     Solution   Check for the proper installation of the dryer. Check the entire exhaust system for proper installation or blockage. Check the blower assembly and motor for proper operation. Check the drum seals. Check the lint filter. Check the door seal.

EAF   *Description*   The microprocessor in the electronic control has been reset by the watchdog timer.

     Solution   Clear the code and reset the controls. Test the dryer. If the problem persists, replace the electronic control.

PF   *Description*   A power failure occurred.

     Solution   Correct the problem by pressing the cancel button and reset the controls. If the code persists, turn the electricity off for five minutes. Check all electrical connections to the electronic control.

ERR   *Description*   When programming the control, a mistake was made.

     Solution   Press cancel and reprogram the controls.

LOC   *Description*   The control lock feature is activated. This is a function display code during normal operation.

PAU   *Description*   The drying cycle has been interrupted.

     Solution   The customer might have opened the door in the middle of the cycle. If not, make repairs to the door and latch.

"ad"   *Description*   This is a function display code indicating that the dryer is in the auto-dry cycle.

"dn"   *Description*   This is a function display code indicating that the dryer has completed the auto-dry cycle.

PART V

Frigidaire/Electrolux Refrigerator Error/Fault or Function Codes

ALARM INDICATOR LIGHT

Description When the light is on, the alarm system is activated.

Solution This is a normal condition. If the refrigerator does not reach the set temperature setting, an alarm will sound. Check refrigeration cycle.

DOOR AJAR INDICATOR LIGHT

Description The refrigerator door was left open for over five minutes. The indicator light will be flashing and the audible alarm will sound.

Solution Close the doors. Rest the alarm. If problem persists, check wiring connectors and wiring.

HIGH TEMP INDICATOR LIGHT

Description The refrigerator temperature has been above the temperature setting for over an hour. The indicator light will begin to flash amber and an audible alarm will sound.

Solution Reset the alarm, and take a temperature reading to confirm. Check the refrigeration cycle, the defrost cycle, and components for proper operation.

POWER FAILURE INDICATOR LIGHT

Description A power failure was detected, and the amber indicator light will begin to flash.

Solution Press the reset button and check the food storage temperature.

General Electric and Hotpoint Room Air Conditioner Error/Fault or Function Codes

The error/fault or function codes listed here will vary from model to model. The codes on the model you are servicing might be different from what is listed. If this is the case, check the technical data sheet, the use and care guide, the manufacturer's website, or call the manufacturer for the description of the error/fault code you are experiencing. Sometimes, by disconnecting the electricity to the product and waiting for two minutes and then turning on the electricity, this may reset the error/fault or function code, allowing you to reprogram and test the operation of the appliance.

For your safety and the safety of others, all testing and physical inspections of the product must be performed with the electricity turned off. On occasion, you will have to test a component with the electricity on. Extreme care must be taken to prevent electrical mishaps.

E1

Description The electronic control detected that the filter needs cleaning.

Solution Clean the filter.

ER

Description The electronic control detected a temperature sensor malfunction.

Solution Check the value of the temperature sensor. Check for debris buildup on the probe. Check the sensor for proper positioning. Replace if necessary.

General Electric and Hotpoint Dishwasher Error/Fault or Function Codes

The error/fault or function codes listed here will vary from model to model. The codes on the model you are servicing might be different from what is listed. If this is the case, check the technical data sheet, the use and care guide, the manufacturer's website, or call the manufacturer for the description of the error/fault code you are experiencing. Sometimes, by disconnecting the electricity to the product and waiting for two minutes and then turning on the electricity, this may reset the error/fault or function code, allowing you to reprogram and test the operation of the appliance.

For your safety and the safety of others, all testing and physical inspections of the product must be performed with the electricity turned off. On occasion, you will have to test a component with the electricity on. Extreme care must be taken to prevent electrical mishaps.

PF	*Description* Power failure.
	Solution Reset the electronic control by pressing the cleaning options and start key pad. If this does not correct the problem, press the cancel button and reset the control panel. If the code persists, turn the electricity off for five minutes. Check all electrical connections to the electronic control.
C1	*Description* The drain cycle exceeded two minutes.
	Solution Check for obstruction in the drain line to the sink. If there is a garbage disposer, run it and clean it out. If clear, check the drain line valve.
C2	*Description* The motor/pump did not completely drain the water from the tub within seven minutes.
	Solution Press the reset button to halt the alarm. Check the drain valve, drain hose, and garbage disposer for obstruction.
C3	*Description* The motor/pump will not drain the water from the tub.
	Solution Disconnect the electricity to the dishwasher for two minutes and then reset the electronic control. Go into the test mode to run the drain cycle over and check the motor/pump assembly.
C4	*Description* Dishwasher overfilled or float stuck; float switch failure or clogged sump or drain.
	Solution Reset the electronic control by touching the start key pad and then the cancel/reset key pad.
	Sometimes, after a power failure, the dishwasher will fill the tub again. Check for debris in the sump area of the tub. Check the water float and switch assembly.
C5	*Description* Dishwasher drain cycle is too short.
	Solution Reset the electronic control and restart the cycle. Check water supply and float/switch assembly.

PART V

C6   *Description*   Low water temperature in the tub.

Solution   Check the water temperature at the sink next to the dishwasher—it should be around 120 to 140 degrees Fahrenheit. Check the water heater temperature also. Check dishwasher heater and circuit components. Do not take showers or baths or run the automatic washer before running the dishwasher. This will deplete the hot water supply in the water heater.

C7   *Description*   A fault has occurred in the water temperature sensor.

Solution   Turn the electricity off to the dishwasher for two minutes and then reset the electronic control. Check the water temperature sensor.

C8   *Description*   Detergent cup malfunction or blocked.

Solution   Remove the obstruction, close the dispenser, and reset the electronic control.

CUP OPEN   *Description*   The detergent dispenser cup was left open.

Solution   Close the detergent dispenser cup and press start.

● Sensing LED off   *Description*   Clean sensor error.

Solution   Check control circuit board and wiring harness. Replace sensor if necessary. The clean sensor must come on during the normal and pots and pans cycles.

☀ Washing LED flashing   *Description*   The start/reset touchpad has been pressed.

Solution   Let the dishwasher drain for 90 seconds before resetting it.

☀ Rinsing LED flashing and beeping sound   *Description*   Control board error.

Solution   To turn off the beeper, press the start/reset pad. The rinse LED will continue to flash. Turn off the power to the dishwasher to reset. If the flashing continues, check sequence switch, control circuit board, and wiring harness; replace defective part.

☀ Drying LED Flashing and beeping sound   *Description*   Control board error. The sequence switch is not reaching its target within 30 seconds.

Solution   Restart the dishwasher. If the beeping continues, check the sequence switch, wiring harness, and control circuit board. Replace defective part.

☀ Start/Reset   *Description*   Someone has pressed the start/reset pad.

Solution   Before selecting a new cycle, allow the dishwasher to drain for 90 seconds.

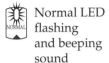 Clean LED
flashing and
beeping sound

Description   Control board error.

Solution   To turn off the beeper, press the start/reset pad. The clean LED will continue to flash. Turn off the power to the dishwasher to reset. If the flashing continues, check sequence switch, control circuit board, and wiring harness; replace defective part.

Normal LED
flashing
and beeping
sound

Description   Control board error. The sequence switch is not reaching its target within 30 seconds.

Solution   Restart the dishwasher. If the beeping continues, check the sequence switch, wiring harness, and control circuit board. Replace defective part.

General Electric and Hotpoint Microwave Error/Fault or Function Codes

The error/fault or function codes listed here will vary from model to model. The codes on the model you are servicing might be different from what is listed. If this is the case, check the technical data sheet, the use and care guide, the manufacturer's website, or call the manufacturer for the description of the error/fault code you are experiencing. Sometimes, by disconnecting the electricity to the product and waiting for two minutes and then turning on the electricity, this may reset the error/fault or function code, allowing you to reprogram and test the operation of the appliance.

For your safety and the safety of others, all testing and physical inspections of the product must be performed with the electricity turned off. On occasion, you will have to test a component with the electricity on. Extreme care must be taken to prevent electrical mishaps.

PF or RESET   *Description*   A power failure occurred.

Solution   Correct the problem by pressing the cancel button and reset the control. If the code persists, turn the electricity off for five minutes. Check all electrical connections to the electronic control.

F1   *Description*   Thermal sensor problem; excessive heat from within the microwave cavity.

Solution   Test the thermal sensor. Check the fan motor and air circulation pattern for a blockage (dirty filter).

F2   *Description*   Thermal sensor problem.

Solution   Test the thermal sensor for a short. Check all circuits and components for proper operation.

F3   *Description*   The touch panel may have a short.

Solution   Inspect and test the touch panel and connectors.

F4   *Description*   Humidity sensor problem.

Solution   Test the humidity sensor. Check the fan motor and air circulation pattern for a blockage (dirty filter).

PART V

F5   *Description*   Humidity sensor problem.

                      Solution   Test the humidity sensor for a short. Check all circuits and components for proper operation.

F6 or PROBE *Description*   Temperature probe problem.

                      Solution   Test the temperature probe value. Check the probe receptacle wiring.

F10   *Description*   Touch screen problem.

                      Solution   Test touch screen for a short. Check connecting wiring harness.

General Electric and Hotpoint Range Error/Fault or Function Codes

The error/fault or function codes listed here will vary from model to model. The codes on the model you are servicing might be different from what is listed. If this is the case, check the technical data sheet, the use and care guide, the manufacturer's website, or call the manufacturer for the description of the error/fault code you are experiencing. Sometimes, by disconnecting the electricity to the product and waiting for two minutes and then turning on the electricity, this may reset the error/fault or function code, allowing you to reprogram and test the operation of the appliance.

For your safety and the safety of others, all testing and physical inspections of the product must be performed with the electricity turned off. On occasion, you will have to test a component with the electricity on. Extreme care must be taken to prevent electrical mishaps.

F0 or F1   *Description*   Key panel problem.

                      Solution   Check key panel and wiring. Replace key panel.

F2   *Description*   Oven temperature malfunction.

                      Solution   Test the oven temperature during operation for preset temperature. Check temperature sensor.

F3 or F4   *Description*   Oven sensor problem.

                      Solution   Test oven sensor value and replace.

F5   *Description*   Electronic control problem.

                      Solution   Check the wiring connections to the electronic control. Replace the electronic control.

F6 or F7   *Description*   Key panel problem.

                      Solution   Check key panel and wiring. Replace key panel.

F8   *Description*   Electronic control problem.

                      Solution   Check the wiring connections to the electronic control. Replace the electronic control.

F9 or FC   *Description*   Door lock component malfunction.

                      Solution   Check the door latch assembly and switches. Check wiring connections. Replace failed component.

FD	*Description* Probe and/or receptacle malfunction.
	Solution Test the probe value and replace if necessary. Check the receptacle and make repairs or replace the receptacle.
FF	*Description* Electronic control problem.
	Solution Check the wiring connections to the electronic control. Replace the electronic control.
BAD LINE	*Description* This indicates that the main electric supply to the oven is wired incorrectly.
	Solution Check the voltage supply to the product. Make sure that the product is wired correctly, and normal operation should resume.
UNLOCK DOOR	*Description* Door lock malfunction at the end of the cleaning cycle.
	Solution Check the door latch assembly and circuitry. Read the use and care guide and check the GE website for the steps necessary to open the door lock.
LOC	*Description* A safety feature added to the programming to prevent children from turning on the product.
	Solution Read the use and care guide to learn about the programming for each model.
ERR	*Description* When programming the control, a mistake was made.
	Solution Press cancel and reprogram the controls.
	Description A holiday program function.
	Solution A backwards "C" means that your oven was programmed for the Sabbath Mode. Read the use and care guide for correction, or go to the GE website for instructions.

JennAir Oven and Range Error/Fault or Function Codes

The error/fault or function codes listed here will vary from model to model. The codes on the model you are servicing might be different from what is listed. If this is the case, check the technical data sheet, the use and care guide, or the manufacturer's website, or call the manufacturer for the description of the error/fault code you are experiencing. Sometimes, by disconnecting the electricity to the product and waiting for two minutes and then turning on the electricity, this may reset the error/fault or function code, allowing you to reprogram and test the operation of the appliance.

For your safety and the safety of others, all testing and physical inspections of the product must be performed with the electricity turned off. On occasion, you will have to test a component with the electricity on. Extreme care must be taken to prevent electrical mishaps.

PART V

F0 or F1   *Description*   A function key is stuck.

Solution   Replace the touchpad. If the touchpad is part of the ERC, then replace the ERC too.

F2   *Description*   Oven temperature is too hot.

Solution   Replace the oven temperature sensor. If there is a relay board, check it and the wire harness.

F3 or F4   *Description*   Open or shorted oven temperature sensor.

Solution   Check the wiring harness and connectors; if okay, replace the sensor.

F5   *Description*   Main computer board problems.

Solution   Replace the ERC.

F7   *Description*   The function key is stuck.

Solution   Replace the touchpad. If the touchpad is part of the ERC, then replace the ERC too.

F8   *Description*   Analog/digital problem.

Solution   Replace the ERC.

F9   *Description*   Door latch problem.

Solution   Check the door latch assembly and switches.

LG Dishwasher Error/Fault or Function Codes

The error/fault or function codes listed here will vary from model to model. The codes on the model you are servicing might be different from what is listed. If this is the case, check the technical data sheet, the use and care guide, or the manufacturer's website, or call the manufacturer for the description of the error/fault code you are experiencing. Sometimes, by disconnecting the electricity to the product and waiting for two minutes and then turning on the electricity, this may reset the error/fault or function code, allowing you to reprogram and test the operation of the appliance.

For your safety and the safety of others, all testing and physical inspections of the product must be performed with the electricity turned off. On occasion, you will have to test a component with the electricity on. Extreme care must be taken to prevent electrical mishaps.

Code	Description	Solution
1E	Water supply problem	Check water supply, inlet water line, water pressure, and hall sensor. Check air braker. Replace defective part.
OE	Drain problem	Check drain hose, drain outlet on sump, and drain pump motor. Check circuit and wiring harness. Repair and/or replace defective part.
E1	Leakage problem	Check hose connections and sump assembly. Check drain hose installation. Check impeller on washing pump. Repair and/or replace defective part.

Code	Description	Solution
FE	Excess water problem	Check inlet water valve and controller. Repair and/or replace defective part.
_tE_	Thermistor problem	Check inlet water temperature. Check wiring harness. Test the thermistor. Check wiring harness and impeller on washing pump. Check motor and blade. Repair and/or replace defective part.
LE	Motor problem	Check wiring harness and impeller on washing pump. Check motor and blade. Repair and/or replace defective part.

LG Side X Side Refrigerator Error/Fault or Function Codes

The error/fault or function codes listed here will vary from model to model. The codes on the model you are servicing might be different from what is listed. If this is the case, check the technical data sheet, the use and care guide, or the manufacturer's website, or call the manufacturer for the description of the error/fault code you are experiencing. Sometimes, by disconnecting the electricity to the product and waiting for two minutes and then turning on the electricity, this may reset the error/fault or function code, allowing you to reprogram and test the operation of the appliance.

For your safety and the safety of others, all testing and physical inspections of the product must be performed with the electricity turned off. On occasion, you will have to test a component with the electricity on. Extreme care must be taken to prevent electrical mishaps.

Code	Description	Solution
Er/FS	Freezer sensor problem	Check freezer sensor for a short; replace if necessary.
Er/rS	Refrigerator sensor #1 problem	Check refrigerator sensor #1 for a short and replace if necessary.
Normal display*	Refrigerator sensor #2 problem	Check refrigerator sensor #2 for a short and replace if necessary.
Er/dS	Defrost sensor problem	Check sensor and wiring for a short. Repair or replace part if necessary.
Er/dh	Defrosting problem	Check defrost heater, temperature fuse, and wiring harness. Repair or replace part if necessary.
Er/FF	BLDC motor freezing problem	Check motor. Check wiring harness. Check circuit for an open or short. Repair or replace part if necessary.
Er/CF	BLDC motor cooling problem	
Er/CO	Communication problem	Check wiring harness for an open or short between the main PCB and the display PCB. Repair or replace part if necessary.

Code	Description	Solution
Normal display **	Ambient sensor problem	Check ambient sensor for a short. Replace part if necessary.
Normal display *	Optichill sensor problem	Check optichill sensor for a short. Replace part if necessary.

* When the refrigerator is in the failure mode, pressing the buttons has no effect on the operation.
** While the refrigerator is in operation, the error code indicates the failure mode.

LG French Door Refrigerator Error/Fault or Function Codes

The error/fault or function codes listed here will vary from model to model. The codes on the model you are servicing might be different from what is listed. If this is the case, check the technical data sheet, the use and care guide, or the manufacturer's website, or call the manufacturer for the description of the error/fault code you are experiencing. Sometimes, by disconnecting the electricity to the product and waiting for two minutes and then turning on the electricity, this may reset the error/fault or function code, allowing you to reprogram and test the operation of the appliance.

For your safety and the safety of others, all testing and physical inspections of the product must be performed with the electricity turned off. On occasion, you will have to test a component with the electricity on. Extreme care must be taken to prevent electrical mishaps.

Code	Description	Solution
Er/F5	Freezer sensor problem	Test sensor. Check wiring harness for an open or short.
Er/rS	Refrigerator sensor problem	Test sensor. Check wiring harness for an open or short.
Er/d5	Defrost sensor problem	Test sensor. Check wiring harness for an open or short.
Er/rt	Temperature sensor problem	Test sensor. Check wiring harness for an open or short.
Er/dh	Defrost problem	Check wiring harness. Check defrost heater, temperature fuse, and connectors. Replace defective part.
Er/FF	BLDC freezer fan motor problem	Check wiring harness.

LG Washer Error/Fault or Function Codes

The error/fault or function codes listed here will vary from model to model. The codes on the model you are servicing might be different from what is listed. If this is the case, check the technical data sheet, the use and care guide, or the manufacturer's website, or call the manufacturer for the description of the error/fault code you are experiencing. Sometimes, by disconnecting the electricity to the product and waiting for two minutes and then turning on the electricity, this may reset the error/fault or function code, allowing you to reprogram and test the operation of the appliance.

For your safety and the safety of others, all testing and physical inspections of the product must be performed with the electricity turned off. On occasion, you will have to test a component with the electricity on. Extreme care must be taken to prevent electrical mishaps.

Code	Description	Solution
1E	Inlet water problem	Check water supply and water valve.
UE	Imbalance problem	Check installation, load size, and laundry collected on one side.
OE	Drain problem	Check drain hose and pump.
FE	Overflow problem	Check drain hose and pump. Check wiring harness.
PE	Pressure sensor problem	Check the pressure sensor switch assembly. Replace if necessary.
dE	Door open problem	Check door and switch assembly. Check wiring harness. Check PWB assembly. Replace if necessary.
tE	Heating problem	Check thermistor.
CE	Over current problem	Replace motor and main PWB assembly.
LE	Motor locked rotor problem	Check wiring harness. Check main PWB assembly. Check hall sensor. Replace defective part.
bE	Ball sensor problem	Check wiring harness. Check ball sensor.
EE	EEPROM error problem	EEPROM is defective. Replace part.
PF	Power failure	The washer experienced a power failure.

LG Frontload Washer Error/Fault or Function Codes

The error/fault or function codes listed here will vary from model to model. The codes on the model you are servicing might be different from what is listed. If this is the case, check the technical data sheet, the use and care guide, or the manufacturer's website, or call the manufacturer for the description of the error/fault code you are experiencing. Sometimes, by disconnecting the electricity to the product and waiting for two minutes and then turning on the electricity, this may reset the error/fault or function code, allowing you to reprogram and test the operation of the appliance.

For your safety and the safety of others, all testing and physical inspections of the product must be performed with the electricity turned off. On occasion, you will have to test a component with the electricity on. Extreme care must be taken to prevent electrical mishaps.

PART V

Code	Description	Solution
1E	Inlet water problem	Check water supply and water valve.
UE	Imbalance problem	Check installation, load size, and laundry collected on one side.
OE	Drain problem	Check drain hose and pump.
FE	Overflow problem	Check drain hose and pump. Check wiring harness.

Code	Description	Solution
PE	Sensor pressure switch problem	Check sensor pressure switch and replace if necessary.
dE	Door open problem	Check door switch and replace if necessary.
tE	Heating problem	Check thermistor and replace if necessary.
SE	Sensor problem	Check wiring harness, hall sensor, main PWB assembly, motor, and controller. Replace defective part.
CE	Current problem	Check main PWB assembly and motor. Replace if necessary.
LE	Lock problem	Check wiring harness, motor, main PWB assembly, and hall sensor. Replace defective part.

Maytag Laundry Center Error/Fault or Function Codes

The error/fault or function codes listed here will vary from model to model. The codes on the model you are servicing might be different from what is listed. If this is the case, check the technical data sheet, the use and care guide, the manufacturer's website, or call the manufacturer for the description of the error/fault code you are experiencing. Sometimes, by disconnecting the electricity to the product and waiting for two minutes and then turning on the electricity, this may reset the error/fault or function code, allowing you to reprogram and test the operation of the appliance.

For your safety and the safety of others, all testing and physical inspections of the product must be performed with the electricity turned off. On occasion, you will have to test a component with the electricity on. Extreme care must be taken to prevent electrical mishaps.

L or 1E   *Description*   A lid lock error was detected.

Solution   Check the lock assembly for proper operation. Check all wiring connections and wiring. Check the electronic control board. Replace defective part if necessary.

U   *Description*   An unbalanced load was detected.

Solution   Check the unbalance switch, wiring, and electronic control board. Advise the consumer on loading the washer. Replace defective part if necessary.

6E or 7E   *Description*   The electronic control detected a temperature error in the dryer.

Solution   Check the temperature sensor, thermostats, electronic control board, and the wiring connections and wiring. Check the installation of the appliance. Replace part if necessary.

8E   *Description*   An error was detected in the motor and circuitry in the dryer.

Solution   Check the motor, relay, electronic control board, and wiring connections and wiring. Replace defective part if necessary.

9E   *Description*   The electronic control board detected a door switch failure in the dryer.

Solution   Check the door switch, electronic control board, and the wiring connections and wiring. Replace defective part if necessary.

Maytag Washer Error/Fault or Function Codes

The error/fault or function codes listed here will vary from model to model. The codes on the model you are servicing might be different from what is listed. If this is the case, check the technical data sheet, the use and care guide, the manufacturer's website, or call the manufacturer for the description of the error/fault code you are experiencing. Sometimes, by disconnecting the electricity to the product and waiting for two minutes and then turning on the electricity, this may reset the error/fault or function code, allowing you to reprogram and test the operation of the appliance.

For your safety and the safety of others, all testing and physical inspections of the product must be performed with the electricity turned off. On occasion, you will have to test a component with the electricity on. Extreme care must be taken to prevent electrical mishaps.

Do   *Description*   Door lock problem detected.

      Solution   Make sure that the door closes properly. Check door lock assembly.

FL   *Description*   The control detected that the washer did not lock the door.

      Solution   Make sure that the door closes properly. Check door lock assembly.

Sd   *Description*   The washer detected an oversudsing problem.

      Solution   The suds recovery cycle should correct the problem. Advise the consumer on proper detergent usage.

Pf   *Description*   A power failure has occurred.

      Solution   Restart the washer.

Nd   *Description*   The washer did not drain the water out of the machine.

      Solution   Check the drain hose for proper installation. Check the household drain for blockage. Check the pump for blockage. Replace defective part if necessary.

LO   *Description*   The door lock will not open.

      Solution   Make sure the door closes properly. Push the off button and wait a few minutes for the door to unlock. Disconnect the electricity to the washer for a few minutes and try to unlock the door. Check door lock assembly. Replace defective part if necessary.

lr   *Description*   The control detected a motor problem.

      Solution   Check the motor windings. Check the wiring connections and wiring. Replace defective part if necessary.

nf   *Description*   Water will not fill the washer.

      Solution   Make sure that the water is turned on all the way. Check for kinked fill hoses. Check water inlet valve. Replace defective part if necessary.

od   *Description*   This is a safety feature that will not allow the washer to start on another cycle unless the door is opened first.

Maytag Dryer Error/Fault or Function Codes

The error/fault or function codes listed here will vary from model to model. The codes on the model you are servicing might be different from what is listed. If this is the case, check the technical data sheet, the use and care guide, the manufacturer's website, or call the manufacturer for the description of the error/fault code you are experiencing. Sometimes, by disconnecting the electricity to the product and waiting for two minutes and then turning on the electricity, this may reset the error/fault or function code, allowing you to reprogram and test the operation of the appliance.

For your safety and the safety of others, all testing and physical inspections of the product must be performed with the electricity turned off. On occasion, you will have to test a component with the electricity on. Extreme care must be taken to prevent electrical mishaps.

PF   *Description*   A power failure has occurred.

            Solution   Reset the control.

Hp or hl   *Description*   The temperature sensor has detected an error.

            Solution   Check all the components in the heat circuit. Replace defective part if necessary.

hll   *Description*   The hanging rod in the dryer cabinet has a problem.

            Solution   Check all the components in the dryer cabinet for proper operation.

Maytag/Jenn-Air Oven Error/Fault or Function Codes

The error/fault or function codes listed here will vary from model to model. The codes on the model you are servicing might be different from what is listed. If this is the case, check the technical data sheet, the use and care guide, the manufacturer's website, or call the manufacturer for the description of the error/fault code you are experiencing. Sometimes, by disconnecting the electricity to the product and waiting for two minutes and then turning on the electricity, this may reset the error/fault or function code, allowing you to reprogram and test the operation of the appliance.

For your safety and the safety of others, all testing and physical inspections of the product must be performed with the electricity turned off. On occasion, you will have to test a component with the electricity on. Extreme care must be taken to prevent electrical mishaps.

F0, F1,   *Description*   The electronic control detected a stuck or shorted key on the
or F7   touch panel.

            Solution   Check the touch key pad for proper operation. Check all wiring connections and wiring. Replace touch pad.

F2   *Description*   The electronic control has detected a temperature problem.

            Solution   Check the sensor, relay board, and wiring connections and wiring. Replace defective part if necessary.

F3 or F4   *Description*   The electronic control has detected a sensor problem.

            Solution   Check the value of the sensor. Check the wiring connections and wiring. Replace the sensor.

F5   *Description*   A communication error was detected.

            Solution   Replace the clock.

F6   *Description*   A voltage problem has been detected in the oven.

            Solution   Check the voltage to the oven.

F8   *Description*   An analog/digital error occurred.

            Solution   Replace the clock.

F9   *Description*   The electronic control detected a door lock problem.

            Solution   Check the door latch assembly. Check the wiring connections and wiring. Replace the clock if necessary.

KitchenAid/Roper/Whirlpool Gas and Electric Range Error/Fault or Function Codes

The error/fault or function codes listed here will vary from model to model. The codes on the model you are servicing might be different from what is listed. If this is the case, check the technical data sheet, the use and care guide, the manufacturer's website, or call the manufacturer for the description of the error/fault code you are experiencing. Sometimes, by disconnecting the electricity to the product and waiting for two minutes and then turning on the electricity, this may reset the error/fault or function code, allowing you to reprogram and test the operation of the appliance.

Listed are the code designations for a four-digit display (**/** or ****) and for a three-digit display (***).

For your safety and the safety of others, all testing and physical inspections of the product must be performed with the electricity turned off. On occasion, you will have to test a component with the electricity on. Extreme care must be taken to prevent electrical mishaps.

F0/E0, F1/E0, or   *Description*   The microprocessor EEPROM on the PCB received a
E0 or F1   communication error and the heating LED will begin flashing.

                     Solution   Turn off the electricity to the appliance and wait for two minutes before turning it back on. If the code reappears, replace the electronic control.

F1/E1, F1/E2,   *Description*   An electronic control error occurred and the heating
F1/E4, or F1   LED will begin flashing.

                     Solution   Turn off the electricity to the appliance and wait for two minutes before turning it back on. If the code reappears, replace the electronic control.

F1/E5   *Description*   The electronic control detected a calibration error.

                     Solution   Check operating temperatures. Turn off the electricity to the appliance and wait for two minutes before turning it back on. If the code reappears, replace the electronic control.

F1/E9

Description   A stack overflow error was detected by the electronic control.

Solution   Turn off the electricity to the appliance and wait for two minutes before turning it back on. If the code reappears, replace the electronic control.

F1/E6

Description   The electronic control received a latch error.

Solution   Turn off the electricity to the appliance and wait for two minutes before turning it back on. If the code reappears, replace the electronic control.

F2/E0, F2/E1, F2/E5, or F1

Description   The touch key pad has possibly shorted out.

Solution   Turn off the electricity to the appliance and wait for two minutes before turning it back on. If the code reappears, replace the touch key pad. Check the touch pad connector to the electronic control PCB.

F3/E0 or F3

Description   The oven sensor might be open in the top oven. The oven will be on and the heating LEDs will begin to flash continually.

Solution   Test the oven sensor valve. Turn off the electricity to the appliance and wait for two minutes before turning it back on. If the code reappears, replace the oven sensor.

F3/E1 or F3

Description   The oven sensor might be shorted out in the top oven. The oven will be on and the heating LEDs will begin to flash continually.

Solution   Test the oven sensor valve. Turn off the electricity to the appliance and wait for two minutes before turning it back on. If the code reappears, replace the oven sensor.

F3/E2 or F3

Description   The oven sensor will not maintain temperature in the bake cycle. The oven will be on and the heating LEDs will begin to flash continually.

Solution   Test the oven sensor valve. Turn off the electricity to the appliance and wait for two minutes before turning it back on. If the code reappears, replace the oven sensor.

F3/E3 or F3

Description   The oven sensor will not maintain temperature in the clean cycle. The oven will be on and the heating LEDs will begin to flash continually.

Solution   Test the oven sensor valve. Turn off the electricity to the appliance and wait for two minutes before turning it back on. If the code reappears, replace the oven sensor.

F3/E4

Description   The oven sensor might be defective in the bottom oven.

Solution   Test the oven sensor valve. Turn off the electricity to the appliance and wait for two minutes before turning it back on. If the code reappears, replace the oven sensor.

F3/E5

Description   The oven sensor might be shorted out in the bottom oven.

Solution   Test the oven sensor valve. Turn off the electricity to the appliance and wait for two minutes before turning it back on. If the code reappears, replace the oven sensor.

F3, E6, or E7

Description   The warming drawer sensor might be defective or shorted out.

Solution   Check and test the warming drawer sensor and switches. Check the door alignment.

Turn off the electricity to the appliance and wait for two minutes before turning it back on. If the code reappears, replace the control.

F5/E0, F5/E1, F5/E2, or F5

Description   In the clean cycle, an error occurred in the door latch assembly. The door will be locked and cleaning LEDs will begin to flash.

Solution   Test the electronic control. Check the oven latch assembly, switch actuation, and the clearance between the latch and door. Check the wiring and connectors. Check the door alignment. Turn off the electricity to the appliance and wait for two minutes before turning it back on. If the code reappears, test and/or replace the control.

F5/E5

Description   The electronic control detected the self-clean temperature was not reached in time.

Solution   Check oven temperatures, thermostat, latch assembly, and latch switch actuation. Turn off the electricity to the appliance and wait for two minutes before turning it back on. If the code reappears, test and/or replace the control.

F5/E7

Description   The electronic control detected that the door latch will not unlock.

Solution   Test the electronic control. Check the oven latch assembly, switch actuation, and the clearance between the latch and door. Check the wiring and connectors. Check the door alignment. Turn off the electricity to the appliance and wait for two minutes before turning it back on. If the code reappears, test and/or replace the control.

F8/E0

Description   A refrigeration problem was detected.

Solution   Check all wiring connectors and wiring. Check electronic control. Check the refrigeration unit.

PF

Description   A power failure occurred.

Solution   Reset the controls and test. Check wiring connections.

OVEN LED ON AND FLASHING

Description   The oven knob was not in the "off " position when the oven went into cycle.

Solution   Turn the knob to the "off" position and then reset the temperature.

PART V

DOOR LOCKED/ CLEANING AND HEATING LEDs FLASHING

Description There is a voltage failure.

Solution Check voltages and all connectors and wiring. Turn off the electricity to the appliance and wait for two minutes before turning it back on. If the code reappears, replace the control.

Samsung Freezer Error/Fault or Function Codes

The error/fault or function codes listed here will vary from model to model. The codes on the model you are servicing might be different from what is listed. If this is the case, check the technical data sheet, the use and care guide, or the manufacturer's website, or call the manufacturer for the description of the error/fault code you are experiencing. Sometimes, by disconnecting the electricity to the product and waiting for two minutes and then turning on the electricity, this may reset the error/fault or function code, allowing you to reprogram and test the operation of the appliance.

For your safety and the safety of others, all testing and physical inspections of the product must be performed with the electricity turned off. On occasion, you will have to test a component with the electricity on. Extreme care must be taken to prevent electrical mishaps.

LED Display segment number

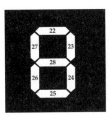 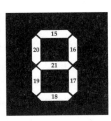

*Refer to the LED segments above for the correct segment number location on the refrigerator LED display.

15
Description Ambient temperature sensor problem.

Solution Check wiring harness. Check for shorted circuit. Check for contact failure. Test the sensor.

16
Description Freezer sensor problem.

Solution Check wiring harness. Check for shorted circuit. Check for contact failure. Test the sensor.

17
Description Freezer room defrost sensor problem.

Solution Check wiring harness. Check for shorted circuit. Check for contact failure. Test the sensor.

18
Description Freezer fan motor problem.

Solution The fan motor is not spinning at the correct RPM. Check the fan feedback line; it might be open.

19
Description Condenser fan motor problem.

Solution The fan motor is not spinning at the correct RPM. Check the fan feedback line; it might be open.

20   *Description*   CF sensor problem.

            Solution   Check wiring harness. Check for shorted circuit. Check for contact failure. Test the sensor.

20   *Description*   French door ice room sensor problem.

            Solution   Check wiring harness. Check for shorted circuit. Check for contact failure. Test the sensor.

21   *Description*   Defrost heater problem.

            Solution   Check wiring harness. Check for contact failure. Check for a shorted circuit. Check freezer sensor for correct readings.

22   *Description*   Evaporator fan motor problem. CF-compartment fan four-door.

            Solution   The fan motor is not spinning at the correct RPM. Check the fan feedback line; it might be open.

23   *Description*   Evaporator fan motor problem. CR-compartment fan four-door.

            Solution   The fan motor is not spinning at the correct RPM. Check the fan feedback line; it might be open.

23   *Description*   Evaporator fan motor problem. French door model.

            Solution   The fan motor is not spinning at the correct RPM. Check the fan feedback line; it might be open.

25   *Description*   Ice maker fill pipe problem.

            Solution   Check the ice maker fill pipe heater for an open or short.

26   *Description*   Communication error.

            Solution   Ignore during diagnostic testing.

27   *Description*   Communication error.

            Solution   Check wiring harness. Replace main PCB board if necessary.

28   *Description*   Lost communication between touchpad and PCB.

            Solution   Check wiring harness. Replace defective board if necessary.

Samsung Refrigerator Error/Fault or Function Codes

The error/fault or function codes listed here will vary from model to model. The codes on the model you are servicing might be different from what is listed. If this is the case, check the technical data sheet, the use and care guide, or the manufacturer's website, or call the manufacturer for the description of the error/fault code you are experiencing. Sometimes, by disconnecting the electricity to the product and waiting for two minutes and then turning on the electricity, this may reset the error/fault or function code, allowing you to reprogram and test the operation of the appliance.

For your safety and the safety of others, all testing and physical inspections of the product must be performed with the electricity turned off. On occasion, you will have to test a component with the electricity on. Extreme care must be taken to prevent electrical mishaps.

LED display
segment
number

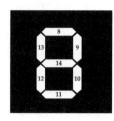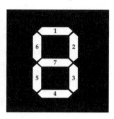

*Refer to the LED segments above for the correct segment number location on the refrigerator LED display.

1   *Description*   Ice maker sensor problem.

Solution   Check wiring harness. Check for contact failure. Check sensor for correct readings.

2   *Description*   Refrigerator sensor problem.

Solution   Check wiring harness. Check for contact failure. Check for a shorted circuit. Check sensor for correct readings.

3   *Description*   Refrigerator defrost sensor problem.

Solution   Check wiring harness. Check for contact failure. Check for a shorted circuit. Check refrigerator sensor for correct readings.

4   *Description*   Refrigerator fan motor problem.

Solution   The fan motor is not spinning at the correct RPM. Check the fan feedback line; it might be open.

5   *Description*   Ice maker problem.

Solution   The ice maker tray did not return to its level position after the third attempt. Check ice maker assembly and wiring harness. Check for external factors.

6   *Description*   Cool select zone sensor problem.

Solution   Check wiring harness. Check for a shorted circuit. Check zone sensor for correct readings.

7   *Description*   Refrigerator defrost problem.

Solution   Check wiring harness. Check for a short circuit. Check for contact failure. Check for correct temperature readings.

8   *Description*   Pantry-damper problem.

Solution   Check sensor and wiring harness.

9   *Description*   CR-sensor four-door problem.

Solution   Check wiring harness. Check for a short circuit. Check for contact failure. Test the sensor.

9   *Description*   Pantry sensor problem.

Solution   Check wiring harness. Check for a short circuit. Check for contact failure. Test the sensor.

10   *Description*   Defrost sensor problem of CR room, four-door.

        Solution   Check wiring harness. Check for a short circuit. Check for contact failure. Test the sensor.

11   *Description*   Defrost sensor problem of CF room, four-door.

        Solution   Check wiring harness. Check for a short circuit. Check for contact failure. Test the sensor.

12   *Description*   Defrost heater problem. CR room, four-door. Defrost over 80 minutes.

        Solution   Check for shorted circuit. Check for contact failure. Check sensor. Check defrost bimetal.

13   *Description*   Defrost heater problem. CF room, four-door. Defrost over 80 minutes.

        Solution   Check for shorted circuit. Check for contact failure. Check sensor. Check defrost bimetal.

14   *Description*   Water heater tank problem.

        Solution   Check for an open or shorted heater and test the circuit.

Samsung Refrigerator Single Evaporator Error/Fault or Function Codes

The error/fault or function codes listed here will vary from model to model. The codes on the model you are servicing might be different from what is listed. If this is the case, check the technical data sheet, the use and care guide, or the manufacturer's website, or call the manufacturer for the description of the error/fault code you are experiencing. Sometimes, by disconnecting the electricity to the product and waiting for two minutes and then turning on the electricity, this may reset the error/fault or function code, allowing you to reprogram and test the operation of the appliance.

For your safety and the safety of others, all testing and physical inspections of the product must be performed with the electricity turned off. On occasion, you will have to test a component with the electricity on. Extreme care must be taken to prevent electrical mishaps.

No.	Error Code	LED Display	Solution
1	Fridge sensor	Fridge "Mid"	Check wiring harness. Check for a short circuit. Check for contact failure. Test the sensor.
2	Ambient temperature sensor	Fridge "Mid"	Check wiring harness. Check for a short circuit. Check for contact failure. Test the sensor.
3	Freezer temperature sensor	Freezer "Max"	Check wiring harness. Check for a short circuit. Check for contact failure. Test the sensor.
4	Freezer defrost sensor	Freezer "Mid"	Check wiring harness. Check for a short circuit. Check for contact failure. Test the sensor.
5	Freezer defrost heater	Freezer "Min"	Check for shorted circuit. Check for contact failure. Check sensor. Check defrost bimetal.

No.	Error Code	LED Display	Solution
6	Ice maker problem	No ice	The ice maker tray did not return to its level position after the third attempt. Check ice maker assembly and wiring harness. Check for external factors.
7	Ice maker sensor problem	Cubed ice	Check wiring harness. Check for a short circuit. Check for contact failure. Test the sensor.

Samsung RB Series Refrigerator Error/Fault or Function Codes

The error/fault or function codes listed here will vary from model to model. The codes on the model you are servicing might be different from what is listed. If this is the case, check the technical data sheet, the use and care guide, or the manufacturer's website, or call the manufacturer for the description of the error/fault code you are experiencing. Sometimes, by disconnecting the electricity to the product and waiting for two minutes and then turning on the electricity, this may reset the error/fault or function code, allowing you to reprogram and test the operation of the appliance.

For your safety and the safety of others, all testing and physical inspections of the product must be performed with the electricity turned off. On occasion, you will have to test a component with the electricity on. Extreme care must be taken to prevent electrical mishaps.

No.	Error Code	LED Display	Solution
1	Refrigerator sensor problem	Fridge 5	Check wiring harness. Check for a short circuit. Check for contact failure. Test the sensor.
2	Defrost sensor problem	Fridge d	Check wiring harness. Check for a short circuit. Check for contact failure. Test the sensor.
3	Ambient temperature sensor problem	Freezer E5	Check wiring harness. Check for a short circuit. Check for contact failure. Test the sensor.
4	Freezer temperature sensor problem	Freezer F5	Check wiring harness. Check for a short circuit. Check for contact failure. Test the sensor
5	Defrost heater problem	Freezer d5	Check for shorted circuit. Check for contact failure. Check sensor. Check defrost bimetal.

Samsung Washer Error/Fault or Function Codes

The error/fault or function codes listed here will vary from model to model. The codes on the model you are servicing might be different from what is listed. If this is the case, check the technical data sheet, the use and care guide, or the manufacturer's website, or call the manufacturer for the description of the error/fault code you are experiencing. Sometimes, by disconnecting the electricity to the product and waiting for two minutes and then turning on the electricity, this may reset the error/fault or function code, allowing you to reprogram and test the operation of the appliance.

For your safety and the safety of others, all testing and physical inspections of the product must be performed with the electricity turned off. On occasion, you will have to test a component with the electricity on. Extreme care must be taken to prevent electrical mishaps.

No	Error Displayed	Code	Description	Solution
1	nd	1	Within 15 minutes, the water level will not drop below the water reset level position.	Check for blockage in hoses and water pump. Check wiring harness. Check to see if the drain hose is not frozen.
2	LO	2	After three attempts the door fails to unlock.	Check door lock assembly. Check door switch. Check wiring harness. Check lock mechanism alignment. Check main PCB.
3	nf	3	After 16 minutes, the water fill continues, with no change in water level in the tub for 3 minutes.	Check water supply, hose installation, water valve harness, and water temperature. Check the relays for proper operation. Replace the main PCB.
4	FL	4	After three attempts the door fails to lock.	Check door lock assembly, door switch, and wiring harness. Check lock mechanism alignment. Check main PCB.
5	nF1	5	Hot and cold water problem.	Check water hoses to see if they are installed correctly.
6	LeE or LE	8	Water level sensor problem. Beeping and the cycle has stopped.	Check the sensor and hose. Check wiring harness. Check main PCB.
7	OE	E	Water level sensor problem.	Check water level sensor. Check for stuck water valves.
8	dc	10	Unbalanced load problem.	Customer education. Press "pause," open door, and reposition clothes or remove some clothes.
9	-	11	EEPROM error.	If the display shows "fail" in the "EEPROM clear mode," then replace main PCB.
10	E2	15	Jammed or stuck key problem.	A signal is being detected for more than 30 seconds. Reset everything.
11	dL	18	When the motor is running, the door is detected as open.	Check door lock assembly, door switch, and wiring harness. Check lock mechanism alignment. Check main PCB.

PART V

No	Error Displayed	Code	Description	Solution
12	dS	22	Door lock problem on startup.	Check door lock assembly, door switch, and wiring harness. Check lock mechanism alignment. Check main PCB.
13	bE	25	Hall sensor problem.	Replace machine control board.
14	*tE*	29	Thermal sensor problem.	Check wiring harness. Check water temperature. Check thermistor and PCB. Replace if necessary.
15	E3	2E	No hall sensor signal.	Check wiring harness and test the motor.
16	Sr	34	Min relay failure.	Replace the PCB.
17	Hr	36	Heater relay failure.	Replace PCB.
18	3E	3E	Motor will not run.	Check wiring harness. Test motor.
19	2E	91	Too much voltage detected.	Check line input voltage. Replace PCB.
20	2E	92	Not enough voltage detected.	Check line input voltage. Replace PCB.
21	8E	8E	MEMS sensor failure.	Check main PCB, MEMS PCB, and the wiring harness.
22	7E	7E	Silver care PCB failure.	Check main PCB, silver care PCB, and the wiring harness.
23	PF	-	Power failure.	This occurs when voltage is interrupted and turned back on. Normal operation. It notifies the customer of a power failure.
24	SUdS or Sd	-	Suds detected.	Advise customer to reduce detergent usage.
25	9E1		Line filter noise.	Line noise triggered error code. Check EMC filter.
26	AE		Communication problem.	Check the main and sub-PCB and wiring harness. Replace main PCB if necessary.
27	SF1, SF2, SF3		System error.	Replace PCB.

Samsung Dryer Error/Fault or Function Codes

The error/fault or function codes listed here will vary from model to model. The codes on the model you are servicing might be different from what is listed. If this is the case, check the technical data sheet, the use and care guide, or the manufacturer's website, or call the manufacturer for the description of the error/fault code you are experiencing. Sometimes, by disconnecting the electricity to the product and waiting for two minutes and then turning on the electricity, this may reset the error/fault or function code, allowing you to reprogram and test the operation of the appliance.

For your safety and the safety of others, all testing and physical inspections of the product must be performed with the electricity turned off. On occasion, you will have to test a component with the electricity on. Extreme care must be taken to prevent electrical mishaps.

Error Displayed	Description	Solution
88	Possible shorted thermistor	Check for a clogged screen and vent system. Check thermistor resistance.
88	Possible open thermistor	Check for a clogged screen and vent system. Check thermistor resistance.
do	Open door	Check wiring harness. Check door sense circuit.
FE	Power source frequency problem	Check for nonutility power supply.
df	Door circuit problem	Check wiring harness. Check door sense circuit.
hE or HE	Heater problem	Check for a clogged screen and vent system. Check thermistor resistance. Check heater.
bE	Button problem	Check for stuck button on the display.
od	Excessive dry time	Check and inspect sensor bars.
88	EEPROM communication error	Replace the main PCB.

Samsung Range Error/Fault or Function Codes

The error/fault or function codes listed here will vary from model to model. The codes on the model you are servicing might be different from what is listed. If this is the case, check the technical data sheet, the use and care guide, or the manufacturer's website, or call the manufacturer for the description of the error/fault code you are experiencing. Sometimes, by disconnecting the electricity to the product and waiting for two minutes and then turning on the electricity, this may reset the error/fault or function code, allowing you to reprogram and test the operation of the appliance.

For your safety and the safety of others, all testing and physical inspections of the product must be performed with the electricity turned off. On occasion, you will have to test a component with the electricity on. Extreme care must be taken to prevent electrical mishaps.

Code	Description	Solution
E27	Open oven sensor	Check sensor resistance. Check wiring harness and terminals. Test sensor at PCB. Replace defective part.
E28	Shorted oven sensor	
E08	Oven not heating	Test sensor; replace if necessary.
EOA	Oven overheating	

Code	Description	Solution
SE	Shorted key	Check keypad cable connections. Check for short between the main PCB and the keypad. Replace main PCB if necessary.
EOE	Door locking problem	Check wiring harness, door lock switch, and motor. Check voltage while operating the door lockout.

Samsung Dishwasher Error/Fault or Function Codes

The error/fault or function codes listed here will vary from model to model. The codes on the model you are servicing might be different from what is listed. If this is the case, check the technical data sheet, the use and care guide, or the manufacturer's website, or call the manufacturer for the description of the error/fault code you are experiencing. Sometimes, by disconnecting the electricity to the product and waiting for two minutes and then turning on the electricity, this may reset the error/fault or function code, allowing you to reprogram and test the operation of the appliance.

For your safety and the safety of others, all testing and physical inspections of the product must be performed with the electricity turned off. On occasion, you will have to test a component with the electricity on. Extreme care must be taken to prevent electrical mishaps.

Code	Description	Solution
4E	Water supply problem	Check water valve. Check wiring harness. Replace main PBA if necessary.
4E1	Water supply problem	Check thermistor. Check water temperature. Replace main PBA if necessary.
5E	Drain problem	Check drain pump. Check level sensor. Check wiring harness. Check for clogged drain hose. Replace main PBA if necessary.
oE	Overflow problem	Check water valve. Check leakage sensor. Check flow meter and replace if necessary. Replace main PBA if necessary.
LE	Leakage problem	Check hose connections. Check the tub and sump area for leaks. Check drain pump and replace if necessary. Check main PBA and replace if necessary.
HE	Heater problem	Check wiring harness. Check heater, heater relay, and main PBA; replace if necessary.
HE1	Heater problem	Check heater relay, thermistor, and main PBA and replace if necessary.
tE1	Temperature sensor problem	Check thermistor connections. Check thermistor and main PBA; replace if necessary.
9E	Low water level problem	Check wiring harness. Check low level sensor and main PBA; replace if necessary.
bE2	Button problem	Check sub-PBA and the main PBA for defects and replace if necessary.
PE	Half-load function problem	Check microswitch and wiring. Check distributor motor and main PBA, and replace if necessary.

Sears Appliances

The error/fault or function codes for Sears appliances will vary from model to model. Sears is a department store and does not manufacture appliances. See Table 13-2 for appliance manufacturers. Then check the technical data sheet, the use and care guide, the manufacturer's website, or call the manufacturer for the description of the error/fault code you are experiencing. Sometimes, by disconnecting the electricity to the product and waiting for two minutes and then turning on the electricity, this may reset the error/fault or function code, allowing you to reprogram and test the operation of the appliance.

Manufacturer	Model Number							
Absocold	126							
Admiral (Inglis)	C646	C880						
Amana	335	596						
Bosch	630							
Broan	233							
Brown Stoveworks	196							
Caloric	174	960						
Camco	C362	C363	C978					
Carrier	416							
D&M	587							
Defiance	789							
Electrolux	336	790						
Fedders	484							
Freidrich	840							
Frigidaire	119	790	970					
General Electric	362	363	464					
Gibson	253							
Goldstar (LG Electronics)	580	721	795					
Haier	183							
Hitachi	934							
Hoover	473							
Inglis	C106	C110	C646					
InSinkErator	175							
JennAir	629							
KeepRite	867	C938						
Kelvinator	417	622	628	662	C933	C970		
KitchenAid	666							

BLE 13-2 Manufacturers of Appliances for Sears (*continued*)

Manufacturer	Model Number								
Lennox	292								
Litton	747								
Maytag	925								
Moffat (Camco)	C968								
Monarch/Simpson	143								
Panasonic	568	586							
Phico Italy	683								
Preway	155	850							
Ranney/Marvel, Imperial	757								
RCA	274								
Rheem/Rudd	879								
Roper	103	155	278	647	835	911	917		
Royal Chef	119								
Samsung	401								
Sanyo/Fisher	143	564	565	566					
Sharp	575								
Speed Queen	651								
State	153								
Tappen	719	791							
Toshiba	562								
Trane	144								
Whirlpool	106	110	154	198	484	562	664	665	850
Woods	C675								
York	342								

Sears is not a manufacturer of appliances. It is a department store. Sears appliances are manufactured for Sears and are sold under the Sears/Kenmore label. To locate which manufacturer built the appliance for Sears, locate the first three digits of the model number and match them with the list above.

TABLE 13-2   Manufacturers of Appliances for Sears

For your safety and the safety of others, all testing and physical inspections of the product must be performed with the electricity turned off. On occasion, you will have to test a component with the electricity on. Extreme care must be taken to prevent electrical mishaps.

Thermador Oven and Range Error/Fault or Function Codes

The error/fault or function codes listed here will vary from model to model. The codes on the model you are servicing might be different from what is listed. If this is the case, check the technical data sheet, the use and care guide, or the manufacturer's website, or call the manufacturer for the description of the error/fault code you are experiencing. Sometimes, by disconnecting the electricity to the product and waiting for two minutes and then turning on the electricity, this may reset the error/fault or function code, allowing you to reprogram and test the operation of the air conditioner.

For your safety and the safety of others, all testing and physical inspections of the product must be performed with the electricity turned off. On occasion, you will have to test a component with the electricity on. Extreme care must be taken to prevent electrical mishaps.

F0 or F1   *Description* Electronic oven control failure (ERC).

            Solution Replace ERC.

F2   *Description* Oven temperature hot.

            Solution Replace oven temperature sensor. Check and replace relay board if necessary. Replace ERC.

F3   *Description* Oven temperature sensor (RTD) problem.

            Solution Check the wiring and connectors between the oven temperature sensor and the ERC. Replace ERC if necessary.

F4   *Description* Oven temperature shorted.

            Solution Check wiring and connectors; replace RTD if necessary.

F7   *Description* Function key stuck.

            Solution Check touchpad keys. Replace ERC if other components check okay.

F8   *Description* ERC problems.

            Solution Check wiring between oven temperature sensor and ERC. Replace oven temperature sensor and ERC if wiring checks okay.

F9   *Description* ERC problems.

            Solution Check the door latch assembly and switches; replace if necessary. Replace ERC if everything else checks okay.

Fd   *Description* Bake/broil relay board problems.

            Solution Test the relay board and replace if necessary. Replace the ERC if defective.

E1   *Description* ERC problem.

            Solution Replace ERC.

E2   *Description* Oven temperature problem.

            Solution Check the oven temperature sensor and replace if necessary. Replace ERC if necessary.

E3   *Description* Oven temperature problem.

            Solution Check for open oven temperature sensor and replace if necessary.

PART V

E4   *Description*   Oven temperature problem.

     Solution   Check for a shorted oven temperature sensor and replace if necessary.

E5, E6, E7,   *Description*   ERC problems.
E8,   *Solution*   Replace ERC.

E9   *Description*   Problems with the latch switch assembly.

     Solution   Check the latch assembly and switches, and replace if necessary.

E10   *Description*   ERC problems.

     Solution   Replace ERC if necessary.

E11 or E12   *Description*   Problems with the latch switch assembly.

     Solution   Check the latch assembly and switches, and replace if necessary.

E13   *Description*   Latch assembly problems or ERC problems.

     Solution   Check for voltage to the latch motor. ERC not converted.

E14   *Description*   Latch switch assembly problem.

     Solution   Check the latch assembly and switches, and replace if necessary.

E15   *Description*   ERC problems.

     Solution   Replace ERC.

Whirlpool Room Air Conditioner Error/Fault or Function Codes

The error/fault or function codes listed here will vary from model to model. The codes on the model you are servicing might be different from what is listed. If this is the case, check the technical data sheet, the use and care guide, the manufacturer's website, or call the manufacturer for the description of the error/fault code you are experiencing. Sometimes, by disconnecting the electricity to the product and waiting for two minutes and then turning on the electricity, this may reset the error/fault or function code, allowing you to reprogram and test the operation of the air conditioner.

For your safety and the safety of others, all testing and physical inspections of the product must be performed with the electricity turned off. On occasion, you will have to test a component with the electricity on. Extreme care must be taken to prevent electrical mishaps.

FF   *Description*   The electronic control has detected an open thermistor or a temperature above 140 degrees Fahrenheit.

     Solution   Test the thermistor for an open circuit. Also check the wiring connections and wiring and the location of the thermistor-sensing bulb. Replace if necessary.

00   *Description*   The electronic control has detected a shorted thermistor or a temperature below 32 degrees Fahrenheit.

     Solution   Test the thermistor for a short. Also check the wiring connections and wiring and the location of the thermistor-sensing bulb. Replace if necessary.

Whirlpool and KitchenAid Microwave Error/Fault or Function Codes

The error/fault or function codes listed here will vary from model to model. The codes on the model you are servicing might be different from what is listed. If this is the case, check the technical data sheet, the use and care guide, the manufacturer's website, or call the manufacturer for the description of the error/fault code you are experiencing. Sometimes, by disconnecting the electricity to the product and waiting for two minutes and then turning on the electricity, this may reset the error/fault or function code, allowing you to reprogram and test the operation of the appliance.

Listed are the code designations for a four-digit display (**/** or ****) and for a three-digit display (***).

For your safety and the safety of others, all testing and physical inspections of the product must be performed with the electricity turned off. On occasion, you will have to test a component with the electricity on. Extreme care must be taken to prevent electrical mishaps.

F1/E1	*Description* The relay control board is faulty.
	Solution Replace the relay control board and test the microwave.
F2	*Description* The electronic control detected a touch pad error.
	Solution Test the touch pad assembly and the electronic control board.
F2/H1 or F2/H2	*Description* A problem occurred in the hood light switch and keys.
	Solution Test the hood light switch and replace if necessary.
F2/Q1	*Description* A problem occurred in the touch control key pad.
	Solution Check the user interface board and the door control assembly for a malfunction. Turn off the electricity to the microwave and wait for two minutes before turning it back on. If the code reappears, replace the defective part.
F3/H1 or F3H	*Description* There is a problem with the humidity sensor.
	Solution Test the value of the humidity sensor. Check all connections from the humidity sensor to the relay control board. Test the relay control board. Turn off the electricity to the microwave and wait for two minutes before turning it back on. If the code reappears, replace the defective part.
F3T	*Description* The electronic control detected a temperature sensor failure.
	Solution Test the microwave oven cavity temperature. The sensor will not operate below 41 degrees Fahrenheit or above 140 degrees Fahrenheit. If tests are okay, replace the electronic control board.
F3/T1	*Description* A user interface board thermistor error occurred.
	Solution Check all wiring connections and wiring from the user interface board, relay control board, and relay board. Test the user interface board and thermistor. Turn off the electricity to the microwave and wait for two minutes before turning it back on. If the code reappears, replace the defective part.

PART V

F3/T2   *Description*   The FC thermistor detected a problem.

               Solution   Check the wiring connections and wiring from the FC thermistor to the relay control board. Test the value of the FC thermistor. Test the relay control board. Turn off the electricity to the microwave and wait for two minutes before turning it back on. If the code reappears, replace the defective part.

F3/T4   *Description*   The magnetron thermistor detected a problem.

               Solution   Check the wiring connections and wiring from the magnetron to the relay control board. Turn off the electricity to the microwave and wait for two minutes before turning it back on. If the code reappears, replace the magnetron thermistor.

F6   *Description*   The electronic control detected a microwave relay problem.

               Solution   Check all wiring connections and wiring to the inverter and the relay control board. Test the relay board. Turn off the electricity to the microwave and wait for two minutes before turning it back on. If the code reappears, replace the relay control board.

F7   *Description*   The magnetron stopped operating due to voltage interruption.

               Solution   Check the cooling fans, magnetron, inverter, relay control board, and interlock switches. Check all wiring connections and wiring between these components. Replace the defective component and retest the microwave.

F9   *Description*   The electronic control received a communication error from the relay control board.

               Solution   Check all wiring connections and wiring between the relay control board and the relay board. Turn off the electricity to the microwave and wait for two minutes before turning it back on. If the code reappears, replace the relay board and the relay control board.

F9/Q   *Description*   A communication error was detected at the touch panel.

               Solution   Check all wiring connections and wiring from the user interface board to the touch panel. Turn off the electricity to the microwave and wait for two minutes before turning it back on. If the code reappears, replace the door control assembly.

Whirlpool Front-Load Washer Error/Fault or Function Codes

The error/fault or function codes listed here will vary from model to model. The codes on the model you are servicing might be different from what is listed. If this is the case, check the technical data sheet, the use and care guide, the manufacturer's website, or call the manufacturer for the description of the error/fault code you are experiencing. Sometimes, by disconnecting the electricity to the product and waiting for two minutes and then turning on the electricity, this may reset the error/fault or function code, allowing you to reprogram and test the operation of the appliance.

For your safety and the safety of others, all testing and physical inspections of the product must be performed with the electricity turned off. On occasion, you will have to test a component with the electricity on. Extreme care must be taken to prevent electrical mishaps.

F/H
Description The electronic control detected no water entering the washer, or the pressure switch did not detect the water.

Solution Verify if the washer has water in the tub. If it does, check the following components for proper operation: drain pump, pressure switch and hose connection, flow meter, and electronic control board. In addition, check the washer installation, drain hose installation, and all wiring connections and wiring. If the tub has no water in it, check the water inlet valve and screen, make sure that the water is turned on, and check the inlet hoses for kinks. To clear the error code, press the pause/cancel button two times.

F/02
Description The drain time exceeded the eight-minute time limit.

Solution Check the drain hose and the drain pump filter for obstructions. Check the electrical connections and wiring from the drain pump to the electronic control board. If no problems were found, then replace the drain pump.

F/05
Description The electronic control has detected a water temperature fault.

Solution Check the heating element, relay, and the water temperature sensor. Check the wiring connections and wiring.

F/06
Description The drive motor tachometer, which is mounted on the motor, has detected a problem with the speed of the motor.

Solution Check all of the wiring connections and wiring between the motor, motor control board, and the electronic control board. Test the drive motor for proper operation. Check the electronic control board and the motor control board.

F/07
Description The electronic control board has detected a possible short in the circuitry with the motor control board.

Solution Check the wiring connections and wiring between the motor control board, drive motor, and the electronic control board. Check the following components for proper operation: drive motor, drive motor control board, and electronic control board.

F/09
Description A water overflow error has been detected.

Solution Check the wiring connections and wiring between the electronic control board, drain pump, and the pressure switch. Check the drain hose installation for blockages or a kinked hose. Clean the drain pump filter and check the drain pump for proper operation. Check the water inlet valve for proper operation. Check the pressure switch for proper operation. Replace any defective part found.

PART V

F/10   *Description*   The electronic control board received a signal from the motor control board that the thermal heat sink has tripped.

   Solution   Check washer installation for proper ventilation. Check the wiring connections and wiring between the electronic control board and the motor control board. Check for worn or failed components in the drive mechanism. Check the electronic control board and the drive motor for proper operation.

F/11   *Description*   A communication error has been detected between the electronic control board and the motor control board.

   Solution   Check all wiring connections and wiring. Check for worn or failed components in the drive mechanism. Check the electronic control board, motor control board, and the motor for proper operation.

F/13   *Description*   The electronic control board has detected a problem in the dispenser circuitry.

   Solution   Check the dispenser motor and the mechanical linkage from the motor to the dispenser. Check the electronic control board for proper operation.

F/14   *Description*   The electronic control board has detected an EEPROM error from the microprocessor chip on the board.

   Solution   Disconnect the electricity to the washer for a few minutes and try to reprogram the electronic control. There might have been a power surge to the washer. Check the electronic control board for proper operation if the error code returns.

F/15   *Description*   The electronic control board detected a problem with the motor control board.

   Solution   Check the wiring connections and wiring from the electronic control board, motor control board, and the motor. Check the following components: drive belt, motor control board, electronic control board, and motor for proper operation.

Sud   *Description*   A suds lock error was detected.

   Solution   The motor control board has detected a suds lock condition by analyzing the current draw on the motor. Check the pump operation. In addition, check for a heavy wash load, too much detergent, or over-sudsing condition.

F/dt   *Description*   A door lock error occurred after six attempts to lock the door.

   Solution   Check the door lock switch and assembly. Check the wiring connections and wiring. Check the electronic control board. Repair or replace the defective part.

F/du   *Description*   The electronic control board has detected a door unlock error after six attempts to open the door lock.

   Solution   Check the door switch/lock assembly for proper operation and for foreign objects. Check the wiring connections and wiring from the door lock assembly to the electronic control board. Check the electronic control board.

Whirlpool Front-Load Dryer Error/Fault or Function Codes

The error/fault or function codes listed here will vary from model to model. The codes on the model you are servicing might be different from what is listed. If this is the case, check the technical data sheet, the use and care guide, the manufacturer's website, or call the manufacturer for the description of the error/fault code you are experiencing. Sometimes, by disconnecting the electricity to the product and waiting for two minutes and then turning on the electricity, this may reset the error/fault or function code, allowing you to reprogram and test the operation of the appliance.

For your safety and the safety of others, all testing and physical inspections of the product must be performed with the electricity turned off. On occasion, you will have to test a component with the electricity on. Extreme care must be taken to prevent electrical mishaps.

PF   *Description*   The electronic control board has detected a power failure.

      Solution   Press start and the dryer will continue with the cycle, or press cancel to reprogram the dryer.

E1   *Description*   The electronic control has detected a faulty thermistor.

      Solution   Test the thermistor and the electronic control board, and check the wiring connections and wiring.

E2   *Description*   The electronic control has detected a short in the thermistor.

      Solution   Test the thermistor and the electronic control board, and check the wiring connections and wiring.

E3   *Description*   The electronic control has detected a communication or software error, a defective electronic control board, or a defective component in the control panel.

      Solution   Check the electronic control board and check all of the components in the control panel. Check all wiring connections and wiring.

Who Makes What?

In the marketplace today there are only a few appliance manufacturers left. Over the past decades the larger manufacturers have bought out the smaller ones and merged those products into their product lines. Today the main manufacturers are producing the products under the various brand names they own.

To help the technician better understand the comparison of appliances and air conditioners, I compiled this list (see Table 13-3). Cosmetically, the product looks different, but internally, the mechanical components making up the product are similar. The manufacturer's warranty may be different between the various brands and models. After servicing these products for a while, the technician will see the similarities.

Manufacturer	Brand Name
Amana *(now owned by Whirlpool)*	Amana Caloric Dansby *(laundry equipment and gas ranges)* Econowash Glenwood Imperial *(microwaves, refrigerators)* Litton Maytag *(certain washer models)* Menu Master Modern Maid Speed Queen *(older domestic models)* Sunray Viking *(U.S. refrigerators except built-in models)*
BSH Home Appliances *(BSH Bosch and Siemans Home Appliances Group)*	Bosch Siemans Thermador Gaggenau
Carrier	*(air conditioners and air conditioning equipment for residential and commercial uses)*
Dacor	Dacor
Dynamic Cooking Systems (DCS)	Dynamic Cooking Systems (DCS)
Fedders	Airtemp Climette *(current models)* Comfort Aire *(window a/c's)* Crosley *(some window a/c's)* Emerson Quiet Kool *(a/c's)* Fedders Hampton Bay *(some a/c's)* Maytag *(window a/c's and dehumidifiers)* Microsonic
Fisher & Paykel	Dynamic Cooking Systems (DCS) *(cooking, refrigeration, dishwashing, and laundry products)*

TABLE 13-3 Manufacturers of Brand-Name Appliances and Air Conditioners *(continued)*

Manufacturer	Brand Name
Frigidaire/Electrolux (non-U.S. brands below)	Affinity Airdryer (dehumidifiers) Bosch (cooking appliances) Capehart (freezers and dehumidifiers) Citation (freezers) Dometic (Some model microwaves) Design Manufacturing (D&M) Electrolux Euroflair (unique product line) Frigidaire Gallery General General Freezer Gibson Harvard Logic (dehumidifiers and microwaves) Kelvinator Kenmore (see Sears below) Icon Leonard O'Keefe & Merritt Polaris (freezers and dehumidifiers) Roy Tappean Vesta White Westinghouse (major appliances only)
Frigidaire/Electrolux (commercial products)	Arctic Air (refrigerators and freezers distributed by Broich Enterprises) Edina (refrigerators) Fedpak (freezers converted to soft ice cream freezers) Imperial (freezers – distributed by Broich Enterprises; sold through Blue Ribbon Foods, Colorado Prime, Dutterer's, Guaranteed Foods, Service Foods, Town & Country, and Farm & Home) Venex (freezers converted to vending machines)
Game Keeper Cooler	(walk-in coolers, various styles and models)
General Electric	Amana (older dishwasher models) Americana Beau*Mark (Most) Beefeater Camco Concept II Eterna GE General Electric Hotpoint Kenmore (see Sears below) McClary Moffat Monogram (unique product line) Profile RCA

TABLE 13-3   Manufacturers of Brand-Name Appliances and Air Conditioners (continued)

Manufacturer	Brand Name
InSinkErator	Ace Hardware (garbage disposers and hot water dispensers only) Badger Crosley (garbage disposers and hot water dispensers only) Dayton (garbage disposers and hot water dispensers only) Emerson (garbage disposers and hot water dispensers only) Franke (garbage disposers and hot water dispensers only) Frigidaire (garbage disposers and hot water dispensers only; current models) ISE (InSinkErator) Kenmore (garbage disposers and hot water dispensers only; most models) KitchenAid (garbage disposers and hot water dispensers only; current models) Master Plumber (garbage disposers and hot water dispensers only) Maytag (garbage disposers and hot water dispensers only) True Value (garbage disposers and hot water dispensers only) Whirlpool (garbage disposers and hot water dispensers only) Wolverine (garbage disposers and hot water dispensers only)
LG Electronics	Arco Aire (mini-split a/c) Carrier (recent a/c's and dehumidifiers) Comfort Aire (recent a/c's and dehumidifiers) Comfort Maker (mini-split a/c) Crosley (recent a/c's) Fast (mini-split a/c's) Freidrich (recent a/c's and dehumidifiers) GE (recent a/c's and dehumidifiers) Glacier Breeze Goldstar Hampton Bay Heat Controller (mini-split a/c's) Heil (mini-split a/c's) I-City (mini-split a/c's) Kenmore (see Sears below) LG Panasonic Quasar Ritetemp Tempstar (mini-split a/c's) Texas Furnace (mini-split a/c's) Whirlpool (recent a/c's) Zenith

TABLE 13-3 Manufacturers of Brand-Name Appliances and Air Conditioners (continued)

Manufacturer	Brand Name
Maytag Corporation (now owned by Whirlpool)	Admiral (USA) Amana Atlantis Crosley (except a/c's; also see Whirlpool) Dynasty Gaffers & Sattier Gemini (unique product line) Hardwick Jade Jade Range Jenn-Air (Unique product line) Magic Chef (Major appliances only) Maytag (Performa brand models) Neptune (unique product line) Norge
Samsung	(refrigerators, washers, dryers, microwaves, ranges, and dishwashers)
Sears	Capri Coldspot Galaxy Oasis Sears does not manufacture any of their products; instead, they are all produced by the other leading manufacturers, often with added features. They are then rebranded with the Kenmore (or other) brand name. See Table 13-2
Sub-Zero/Wolf	Sub-Zero (refrigeration products) Wolf (cooking products and ventilation products)
Techtronics	Hoover
U-Line	(refrigeration products)
Viking Range	(cooking products)
Whirlpool	Admiral (Canada) Chambers Coovert (a/c's) Crosley (newer refrigerators, a/c's, and washers) Danby (Canadian full-size refrigerators and some electric ranges) Estate Inglis Ikea Kenmore (see Sears) KitchenAid (dishwasher with stainless tub and redesigned wash system and pump) Kirkland Maytag Epic Roper Speed Queen (Canada only) Sub-Zero (undercounter ice makers) Whirlpool

Manufacturer	Brand Name
WC Wood *(factory purchased by Whirlpool in 2010)*	Amana *(older freezer models)* Artic Aire Country Squire Crosley *(freezers)* Danby *(freezers and some dehumidifiers)* Edison Electrohome *(range hoods, humidifiers, and dehumidifiers)* Estate *(freezers)* Frost Queen KitchenAid *(freezers)* Maytag *(freezers)* Miami Carey *(range hoods)* Quickfreeze Quickfrez Roper *(freezers)* Sahara Whirlpool *(freezers)* Woods
Electrolux *(non-North American)*	ACEC AEG Alfatec Arthur Martin Arctis Atlas Bendix Buderus Castor Elektro Helios Elektra Electrolux Faure Flymo Husqvarna Ibelsa Juno Lehel Marynen Moffat Nestor Martin Parkinson Cowan Progress Rex Rosenlew Simpson Therma Tornado Tricity Bendix Volta Voss Zanker Zanussi Zoppas

TABLE 13-3 Manufacturers of Brand-Name Appliances and Air Conditioners *(continued)*

Manufacturer	Brand Name
Nordyne (HVAC products)	*(HVAC products only listed below)* Frigidaire Gibson Grandaire Intertherm Kelvinator Mammoth Maytag Miller Phico Tappen Westinghouse

TABLE 13-3   Manufacturers of Brand-Name Appliances and Air Conditioners

Where to Locate the Model and Serial Numbers on the Appliances

Sometimes a technician or the consumer cannot locate the model or serial numbers on the appliance or air conditioner. Figure 13-1 illustrates the location of the data plate on the appliances. This information is needed so that the technician or consumer can purchase the correct parts needed to repair the product.

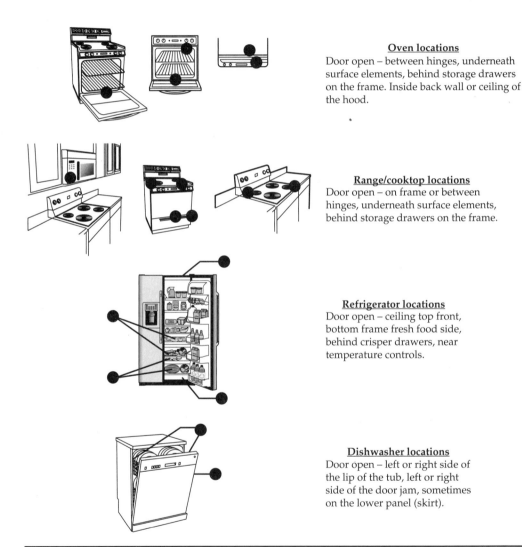

Oven locations
Door open – between hinges, underneath surface elements, behind storage drawers on the frame. Inside back wall or ceiling of the hood.

Range/cooktop locations
Door open – on frame or between hinges, underneath surface elements, behind storage drawers on the frame.

Refrigerator locations
Door open – ceiling top front, bottom frame fresh food side, behind crisper drawers, near temperature controls.

Dishwasher locations
Door open – left or right side of the lip of the tub, left or right side of the door jam, sometimes on the lower panel (skirt).

Figure 13-1 This illustration contains important information regarding the location of the model and serial data plate. (*continued*)

Washer locations
Either side of the cabinet, back of the control panel, under lid, down by the kickplate.

Dryer locations
Door open – around door recess at the top, right center, either lower side of outer cabinet, back of the control panel.

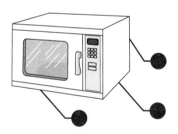

Microwave locations
Door open – inside cavity on the left or right side of the wall, on the back of the microwave, side of the control panel, on the bottom near the rear of the microwave.

Garbage disposer locations
Around the bottom front edge.

FIGURE 13-1 This illustration contains important information regarding the location of the model and serial data plate.

PART VI

Appliance Service, Installation, and Preventive Maintenance Procedures

Automatic Dishwashers

When a dishwasher is properly used, it will provide satisfactory results. There are times, however, when the dishwasher is blamed for poor performance. Perhaps the customer does not know how to load the dishwasher properly or the wrong amount of detergent was used—in some cases, the dishwasher might not run at all. Whatever the case might be, it is up to the technician to either repair the dishwasher or instruct the owner in its proper usage.

This chapter provides the technician with the basic skills needed to diagnose and repair automatic dishwashers. The actual construction and features might vary, depending on what brand and model you are servicing.

Principles of Operation

After placing the dishes properly in the dishwasher, the detergent is placed in the dispenser, and the rinse conditioner is checked for the proper level. The door is closed, and the type of wash cycle is selected. The door latch holds the door closed and activates the door latch switch. This will complete the electrical circuit for the dishwasher to operate. If the door is opened during the cycle, this will cause all operations to cease.

The timer will energize the water inlet valve, and water will begin to enter the tub. The dishwasher does not fill with water like a washing machine. It is designed so that the tub does not have more than two gallons of water in it at any one time. Should the timer switch contacts fail to open during the fill cycle, a float switch assembly, located inside the tub, will open the electrical circuit to the water inlet valve at a preset level.

The fill safety switch is part of the float assembly. Should the timer fail to open its switch contacts, water will keep entering the tub until the float, located inside the tub, rises and engages the float switch to shut off the water. *Note:* The float switch will not protect against a mechanical failure of the water inlet valve.

During the wash and rinse portion of a cycle, the heater element heats the water (on some models) to at least 140 degrees Fahrenheit. This feature is built into the dishwasher and is designed to save the customer money on the operating cost. Also, the customer does not have to raise the water temperature of the water heater to 140 degrees Fahrenheit. He or she only needs to set the water heater temperature at 120 degrees Fahrenheit. This will prevent any member of the household from getting burned.

Front view of
dishwasher with
water circulating
in wash cycle

Side view of
dishwasher with
water circulating
in wash cycle

The water is repeatedly pumped through the lower and upper spray arms and onto the dishes (Figure 14-1). As the water runs off the dishes and back to the pump, it flows through a filtering system. On some models, the filter is designed to separate most food particles from the water so that they aren't sprayed back onto the dishes (Figure 14-2).

At the end of a wash or rinse cycle, the water is pumped out of the dishwasher, flushing the filter of any small food particles. The larger pieces of food are trapped on the pump

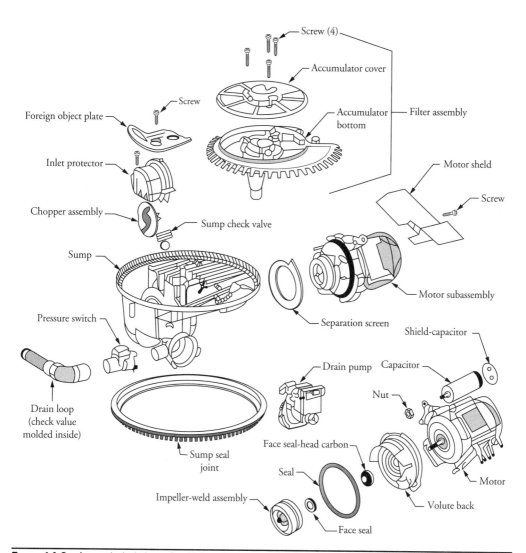

FIGURE 14-2   An exploded view of a dishwasher pump, motor, and filter assembly.

guard, which must be cleaned out before the next use. On some models, the pump screen removes food particles from the water, stores them, and then grinds them up as they are washed down the drain. During grinding, some sounds will be heard. All dishwasher models have some type of mechanism to dispense a rinse aid agent during the designated rinse cycle. The rinse aid agent will not allow the water to stay on the dishes. This will improve the drying time and prevent spotting and filming on the dishes.

At the end of the cycle, the heater element (Figure 14-3a) comes on (if selected) and helps dry the dishes. Certain models have a fan that circulates the air to speed up the drying cycle,

FIGURE 14-3a
Dishwasher
component locations.

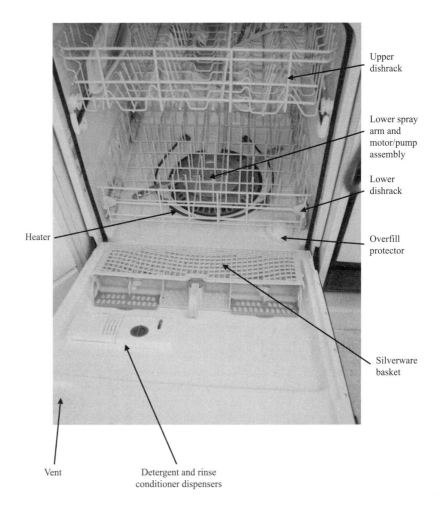

Upper
dishrack

Lower spray
arm and
motor/pump
assembly

Lower
dishrack

Overfill
protector

Heater

Silverware
basket

Vent

Detergent and rinse
conditioner dispensers

thus making sure that the dishes dry evenly. Some models have a cool-dry cycle. This allows
the dishes to be dried without the heater element operating. Combined with the heated air
within the tub (from the wash cycle) and the dishwasher door opened a little bit, it will cause
the water that remains on the dishes to condense and roll off them. Figure 14-3a and b
illustrates component locations within the automatic dishwasher (under-the-counter,
portable, and dish drawer).

Safety First

Any person who cannot use basic tools or follow written instructions should *not* attempt to
install, maintain, or repair an automatic dishwasher. Any improper installation, preventive
maintenance, or repairs could create a risk of personal injury or property damage.

URE 14-3b
hwasher
nponent locations.

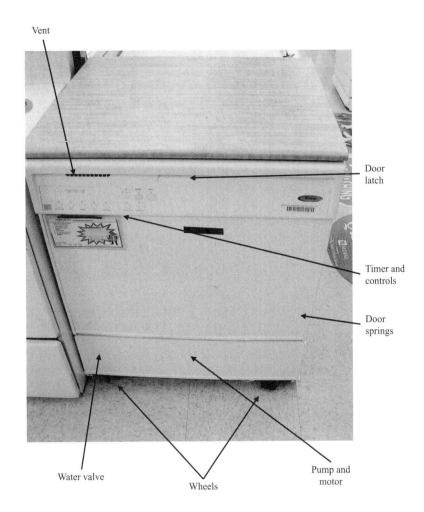

Vent

Door
latch

Timer and
controls

Door
springs

Water valve

Wheels

Pump and
motor

If you do not fully understand the installation, the preventive maintenance, or the repair procedures in this chapter, or if you doubt your ability to successfully complete the task on the automatic dishwasher, call your service manager.

Before continuing, take a moment to refresh your memory on the safety procedures in Chapter 2.

ishwashers in General

Much of the troubleshooting information in this chapter covers the various types of dishwashers in general, rather than specific models, in order to present a broad overview of service techniques. The pictures and illustrations that are used in this chapter are for demonstration purposes only—they clarify the description of how to service an appliance, and they in no way reflect any particular brand's reliability.

FIGURE 14-3c
Dish drawer
dishwasher.

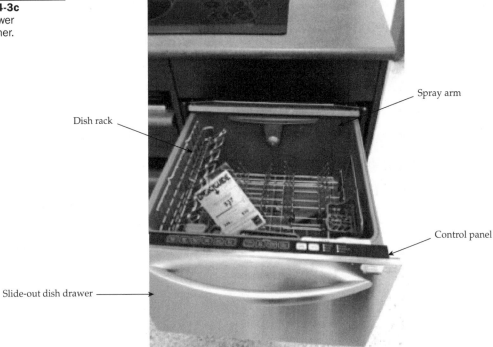

Spray arm

Dish rack

Control panel

Slide-out dish drawer

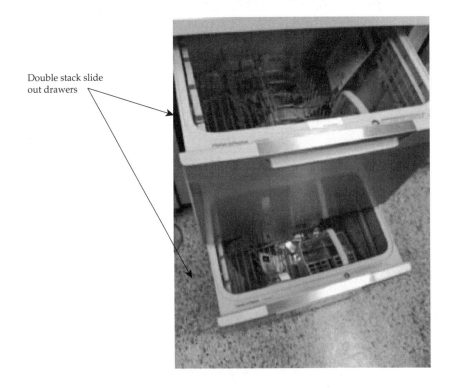

Double stack slide
out drawers

Location and Installation of a Dishwasher

Locate the dishwasher where there is easy access to existing drain, water, and electrical lines (Figure 14-4). Be sure to observe all local codes and ordinances for electrical and plumbing connections. It is strongly recommended that all electrical and plumbing work be done by qualified personnel. The best location for the dishwasher is on either side of the sink.

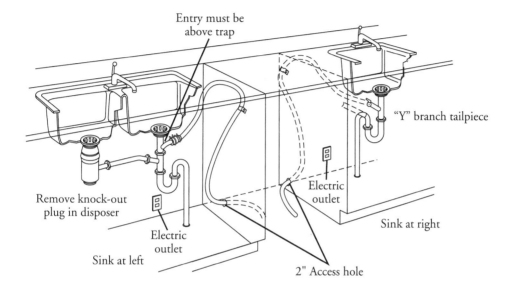

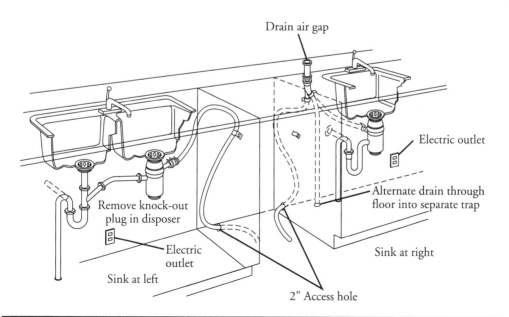

FIGURE 14-4   Typical dishwasher installation for a left or right sink application.

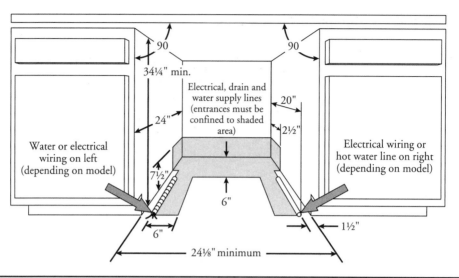

FIGURE 14-5   Undercounter dishwasher cut-out dimensions.

For proper operation and appearance of the dishwasher, the cabinet opening should be square and have the dimensions as shown in Figure 14-5. If the dishwasher is to be installed in a corner, there must be sufficient clearance to open the door (Figure 14-6).

Take the time to read over the installation instructions and the use and care manual that comes with every new dishwasher. These booklets will provide you with very important information, such as:

- Safety
- Tools needed for the installation
- How to remove the panels
- How to change the color of the panels
- Locating drain, water, and electrical supplies

FIGURE 14-6
Corner dishwasher
installation
dimensions.

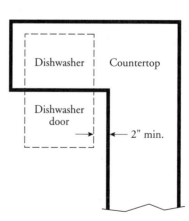

- Positioning, aligning, and leveling the dishwasher
- Drain hose connection
- Water line connection
- Connecting the dishwasher to the electrical supply
- Securing the dishwasher in the opening you have selected
- Proper operation of the dishwasher

Most important, read the warranty information that is supplied with the dishwasher.

Functions and Cycles

Dishwashers are similar to automatic clothes washers. They apply three kinds of energy on the things to be washed. These forces are:

- **Mechanical energy**   Water that is sprayed onto the dishes by the motor and pump assembly to remove the food particles. Detergent is added to aid in cleaning of the dishes.
- **Heat energy**   Using hot water (120 to 140 degrees Fahrenheit) to liquefy the fats and greases on dirty dishes. If the water temperature falls below 120 degrees Fahrenheit, the customer must select the water heating option (if available).
- **Chemical energy**   Detergent dissolves the fat and grease off the dishes and keeps the soils suspended in the water to be removed later in the cycle.

Rinse aids are added to the rinse cycle to prevent water spotting of the dishes and to prevent the water from clinging to the dishes. A rinse aid agent will allow the dishes to dry quicker.

Dishwashers perform four basic functions that are modified and put together in different ways to create the various cycles. The four functions are:

- Fill
- Wash/rinse
- Drain
- Dry

As with clothes washers, the only difference between the wash and rinse cycles is the presence of detergent in the wash water. The mechanical activities that make up a wash and a rinse cycle are basically the same function.

Unlike clothes washers, most dishwashers fill and begin to wash (or rinse) at the same time. The functions are put together in various ways to make up different cycles. For example:

- **Normal wash cycle**   A single or double wash with several rinses and a dry.
- **Heavy wash cycle**   Adds a wash to the normal wash cycle.
- **Pots and pans**   Similar to the heavy wash cycle, but this cycle heats the water in the wash cycles and (on some models) also heats the rinse cycles. On certain models, the timer will not advance until the water temperature is 140 degrees Fahrenheit. This will extend the total time of operation. Depending on the model,

the cycle time can increase from 15 minutes to 2 hours. (Check the use and care manual for exact details.)

- **Light wash cycle**   This cycle is like the normal wash cycle, but minus a wash cycle.
- **Rinse and hold**   Two rinses for holding dishes to wash later.

Water Temperature

The temperature of the incoming water is critical to the operation of a dishwasher. Most dishwashers have heaters, and some have delay periods that extend the time during which water is heated to a specified point, but this does not fully compensate for low temperature of the water supply. You can check the temperature of hot water at the sink nearest to the dishwasher with a thermometer. Open the hot water faucet. Let the water run until it is as hot as possible, and then insert the thermometer into the stream of water. On some models, if the thermometer reading is below 140 degrees Fahrenheit, you will have to raise the water heater thermostat setting. On other models, the dishwasher was designed to operate with water temperatures as low as 120 degrees Fahrenheit. These models have longer detergent wash periods that allow 120-degree Fahrenheit supply water to be heated up to a temperature that gives good washability results. The dishwasher delay periods occur in only one, two, or three of the water fills, and do little for the remaining rinses. Except during delay periods, the water is not in the dishwasher long enough to be heated adequately.

Water Temperature Above 150 Degrees Fahrenheit

It is not recommended to have the water temperature above 150 degrees Fahrenheit in a domestic dishwasher; above this temperature, certain components in the dishwasher might be adversely affected.

Water Temperature of 150 Degrees Fahrenheit

A water temperature of 150 degrees Fahrenheit is the ideal temperature for a mechanical dishwasher. Detergent action and the dissolving of grease are at the maximum at this temperature. Drying of most materials in the dishwasher will be satisfactory.

Water Temperature of 140 Degrees Fahrenheit

Water temperature of 140 degrees Fahrenheit is the minimum temperature recommended by most dishwasher and detergent manufacturers. At this temperature, detergent is still quite active, and most fats are dissolved so that they can be emulsified in the water by the detergent and washed down the drain. Drying will be fair to poor, as water temperature in the last rinse is lowered toward 140 degrees Fahrenheit in some models. Some improvement in drying is possible if a liquid wetting agent is added to the dispenser.

Water Temperature Between 130 and 140 Degrees Fahrenheit

Water temperature between 130 and 140 degrees Fahrenheit is outside the range for most dishwasher operations, and users will have to exert special care if they are to obtain satisfactory results. The cleaning action of detergents and the dissolving of fats are gradually reduced as temperatures drop below 140 degrees Fahrenheit, so the dishes will probably have to be rinsed well before putting them into the dishwasher. Satisfactory drying becomes less likely as water temperature becomes lower. Some dishwashers have an optional feature which will increase the temperature of the water in the tub at different points during the cycle.

Water Temperature Between 120 and 130 Degrees Fahrenheit

Water temperature between 120 and 130 degrees Fahrenheit will aggravate all of the conditions and problems mentioned for the 130- to 140-degree range. Very few fats will dissolve, so the greasy buildup in the lower areas of the tub will be accelerated. Sudsing and foaming are more likely to increase. Detergent action is further reduced, so pre-rinsing of the dishes becomes even more important. The water heat feature will most likely have to be used.

Water Temperature Below 120 Degrees Fahrenheit

Because of poor washing, grease buildup, poor drying, and foaming, it is unlikely that the dishwasher will perform to the user's satisfaction if the water supply is less than 120 degrees Fahrenheit.

The Detergent

The kind and amount of dishwasher detergent that is used is an important part of getting the dishes clean. Different brands of dishwasher detergent contain different amounts of phosphorous, which works to soften water and prevent water spots. If the water is hard, you will have to instruct the customer to use a detergent with a higher phosphorous content—above 12 percent. If the water is soft, the customer can use a low-phosphorous dishwasher detergent. Some areas restrict the phosphate content to 8 percent or less. This means that the customer will have to increase the amount of detergent used in those areas where the water is hard. This is done by adding 1 teaspoon of dishwasher detergent manually in the main wash cycle for each grain of water hardness above 12 grains (general guideline: minimum of three teaspoons) (water hardness is measured in grains):

- 0 to 3 grains for soft water
- 4 to 9 grains for medium-hard water
- 10 to 15 grains for hard water
- Over 15 grains for very hard water

If the hardness of the water supply is unknown, contact the local water department.

If the user is using a concentrated detergent, then advise them to use half the amount recommended earlier.

If the water is above 15 grains, the dishwasher will not perform properly (spots and film will appear on dishes), and a water softener will have to be added to the dishwasher water inlet supply.

Always instruct the user to use automatic dishwasher detergent only. The use of soap, hand dishwashing detergent, or laundry detergent will produce excessive suds and will cause flooding and damage to the dishwasher.

oading the Dishwasher

To obtain the optimum cleaning performance from a dishwasher, the most important factor is loading the dishwasher properly (Figure 14-7). When placed in the dishwasher, the dishes should be positioned in relation to the wash action. The dirtiest side of the dishes should face the source of the water spray. Glasses, cups, and bowls should be positioned slanted and with the bottoms up to prevent any water from collecting on the bottoms. The flatware

FIGURE 14-7
Proper loading of the
dishes in the
dishwasher upper and
lower racks.

Upper dishwasher rack

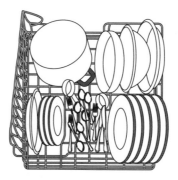

Lower dishwasher rack

should be loaded according to the manufacturer's recommendations as stated in the use and care guide. Any metal items that are loaded into the dishwasher should not touch one another unless they are made from the same materials. Any contact from dissimilar metals could result in permanent damage to the item. Do not place the silverware or dishes crowded together; this will prevent the water from reaching all of the soiled items. Delicate items should be positioned on the racks so they do not move around and cause breakage or chipping. Remember, for optimum cleaning, never overload the dishwasher.

Types of Dishwasher Systems

The basic dishwasher cycles, regardless of manufacturer, will perform the same basic functions when cleaning the dishes. One complete cycle consists of a water fill, water circulation with detergent (wash cycle), then the water is pumped out (drain cycle), again there is a water fill, water circulation without detergent (rinse cycle), and a drying cycle (heated or air only).

Water Fill Cycle

All dishwashers have a timed water fill. The dishwasher timer (mechanical or electronic) controls the amount of water that enters the dishwasher. In most cases, the customer's water pressure entering the home will dictate the quantity of water allowed in the dishwasher. If the home has low water pressure, the dishwasher might encounter problems filling completely, which will lead to not enough water to wash the dishes. Every dishwasher has an overfill protection device called a float and switch to protect the dishwasher from overfilling. If the water pressure is high, the float and switch will turn off the water inlet valve to prevent flooding, regardless of how much fill time is left on the timer. Each dishwasher has a specific time allotment for the amount of water needed for proper washing. The fill system for the dishwasher will consist of the water inlet valve and the overfill protection device (Figure 14-8).

Water Circulation Systems

The wash phase begins after the water that entered the tub has reached the correct level. This is the beginning of the wash cycle with detergent added. Next, the water pump assembly begins to circulate the water and detergent (Figures 14-1 and 14-9). For a specific period of time, the soil is removed from the dishes with water heated to 140 degrees Fahrenheit and mixed with detergent. When the dishwasher stops and goes into the next phase, the dirty dishwater is pumped out through the drain hose. Then the rinse phase begins. This phase is the same as the wash phase but without the detergent. Some models have two or three spray arms to spray water onto the dishes.

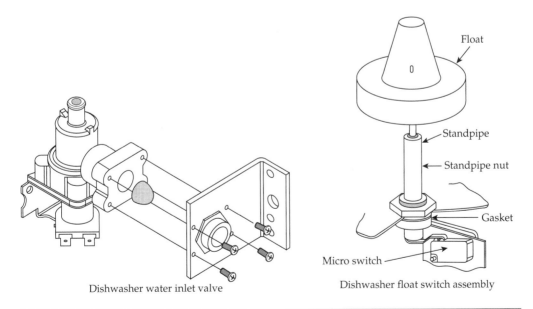

Dishwasher water inlet valve

Dishwasher float switch assembly

FIGURE 14-8   The water inlet valve and the overfill protection device.

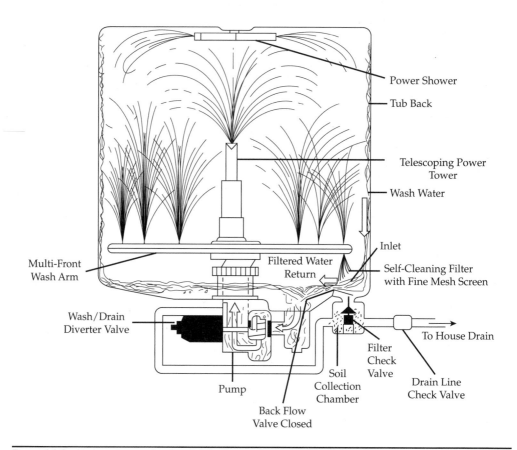

Figure 14-9   Various types of water distribution systems used in dishwashers.

Motor/Pump Assemblies

After the water has stopped entering the dishwasher, the timer contacts close to begin the wash/rinse cycle, and the motor and pump assembly will begin to run. The shaft of the motor is connected directly to the pump assembly. There are many different types of motor and pump assemblies manufactured (Figure 14-10). All of these motor and pump assemblies accomplish the same thing. They circulate water throughout the tub, and most of them drain the water out of the tub, too. Some models add a separate drain pump motor for draining the water from the dishwasher tub.

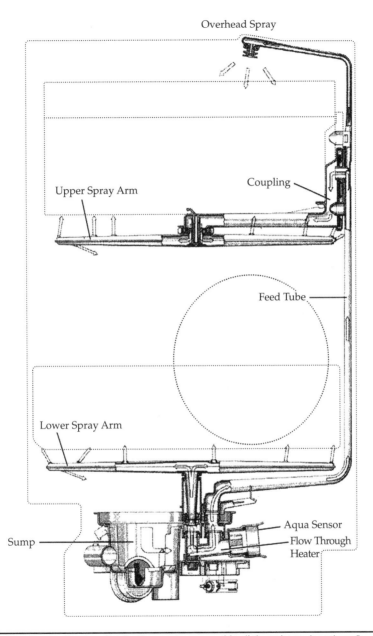

FIGURE 14-9 Various types of water distribution systems used in dishwashers. (*continued*)

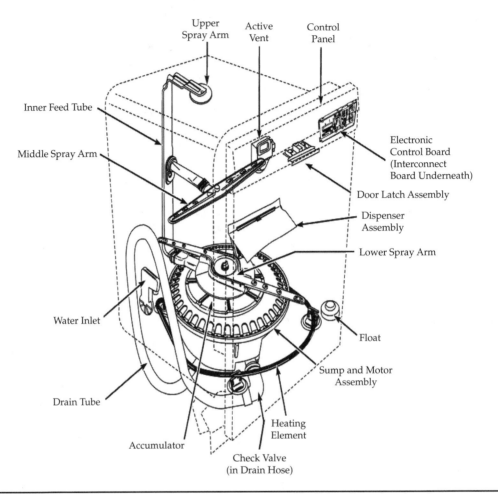

Upper Spray Arm

Active Vent

Control Panel

Inner Feed Tube

Middle Spray Arm

Electronic Control Board (Interconnect Board Underneath)

Door Latch Assembly

Dispenser Assembly

Lower Spray Arm

Water Inlet

Float

Sump and Motor Assembly

Drain Tube

Heating Element

Accumulator

Check Valve (in Drain Hose)

Figure 14-9   Various types of water distribution systems used in dishwashers. (*continued*)

Dishwasher Filtering Systems

Many dishwasher models incorporate some type of wash/rinse water filtering system to separate the food particles from the wash/rinse water (Figure 14-11). This will prevent the food particles from being deposited back on the dishes.

Dispensing Systems

Over the years manufacturers have developed many types of systems used to dispense detergent and rinse agent into the tub (Figure 14-12). A dispensing system must be able to dispense detergent in the wash cycle and also dispense the rinse agent in the rinse cycle. Most dispensing systems used today are electrical or mechanical in nature.

GURE **14-9**
rious types of water
stribution systems
sed in dishwashers.
ontinued)

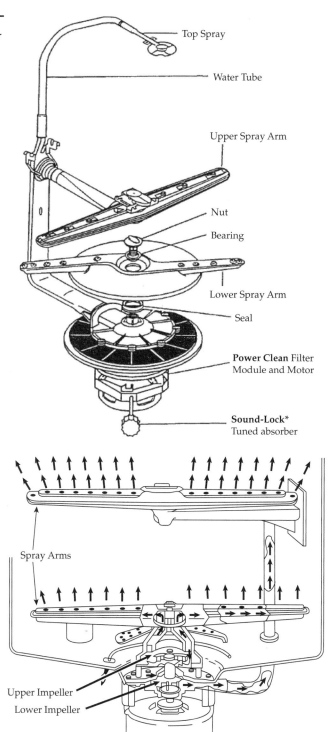

Top Spray

Water Tube

Upper Spray Arm

Nut

Bearing

Lower Spray Arm

Seal

Power Clean Filter
Module and Motor

Sound-Lock*
Tuned absorber

Spray Arms

Upper Impeller

Lower Impeller

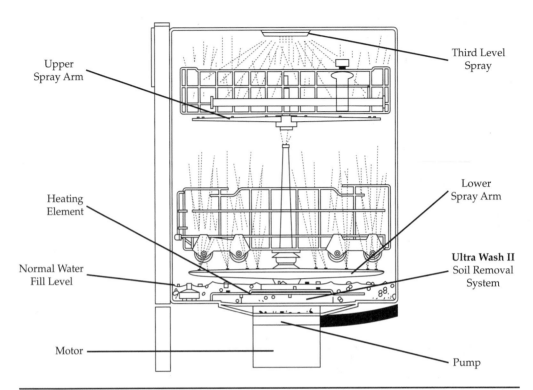

Upper
Spray Arm

Third Level
Spray

Lower
Spray Arm

Heating
Element

Ultra Wash II
Soil Removal
System

Normal Water
Fill Level

Motor

Pump

Figure 14-9   Various types of water distribution systems used in dishwashers.

Drying Systems

Manufacturers also have used various systems for the drying phase. The drying phase begins with the water fill. When a hot water supply is used in the washing phase, that will assist in the drying process. When the dishes are heated by the hot water, they are better able to evaporate the water during the dry cycle. Most dishwashers are equipped with an electric heating element (Figure 14-13). Some models are also equipped with a blower motor assembly to circulate the heat within the dishwasher during the heating cycle. Other models use natural convection to direct the air flow in the drying process. Cool air enters the dishwasher tub and is heated, and then the heated air rises and absorbs the moisture, escaping through the vent in the dishwasher door. Various types of vent systems are used (Figure 14-14).

FIGURE 14-10a

This is an older-model dishwasher motor and pump assembly manufactured by Frigidaire. This system is a direct drive system. In the wash/rinse cycle, this type of motor/pump system turns clockwise. In the drain cycle, the motor/pump system reverses the direction to drain out the water from the tub.

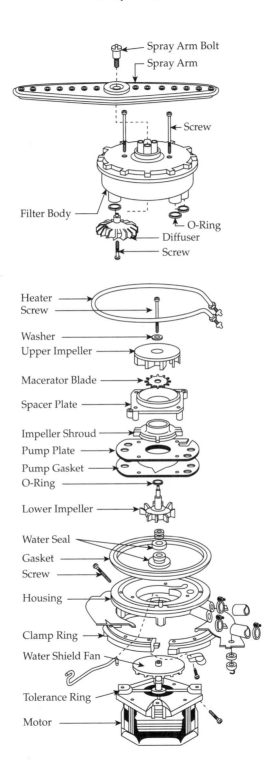

Spray Arm Bolt
Spray Arm
Screw
Filter Body
O-Ring
Diffuser
Screw

Heater
Screw
Washer
Upper Impeller
Macerator Blade
Spacer Plate
Impeller Shroud
Pump Plate
Pump Gasket
O-Ring
Lower Impeller
Water Seal
Gasket
Screw
Housing
Clamp Ring
Water Shield Fan
Tolerance Ring
Motor

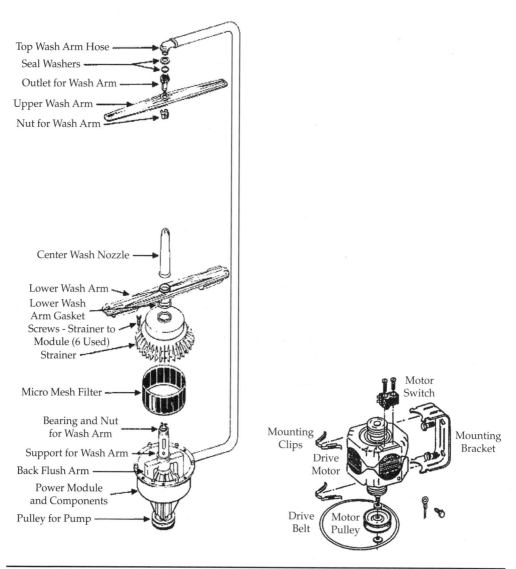

FIGURE 14-10b This is an older model manufactured by Jenn-Air/Maytag. This model utilized a belt drive system, where the motor will operate a separate pump for wash and drain.

FIGURE 14-10c
This dishwasher motor and pump assembly is an older model manufactured by General Electric. This model uses a shaded pole motor and only turns in one direction. To drain the water from the tub, this model uses a solenoid coil to open a valve to drain the water.

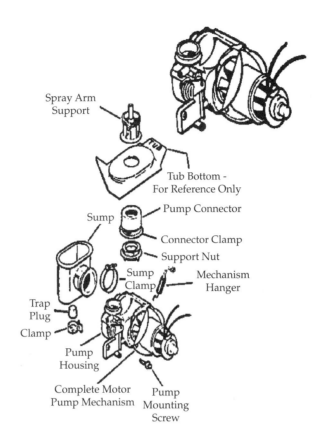

Spray Arm Support

Tub Bottom - For Reference Only

Sump

Pump Connector

Connector Clamp

Support Nut

Sump Clamp

Mechanism Hanger

Trap Plug

Clamp

Pump Housing

Complete Motor Pump Mechanism

Pump Mounting Screw

FIGURE 14-10d This is the newer-style motor and pump assembly manufactured by General Electric. The motor is an induction motor with a capacitor. It is quieter and more energy efficient.

FIGURE 14-10e This is an older model of a dishwasher motor and pump assembly manufactured by Whirlpool. This system is a direct drive system. The motor turns in one direction for wash and turns in the other direction for drain.

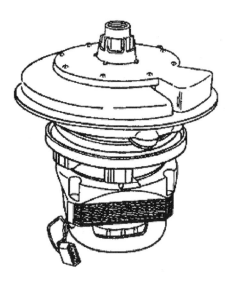

FIGURE 14-10f This motor and pump assembly is a Whirlpool model. This model uses a system for separating the food particles from the recalculating water. It will then flush the food particles down the drain, keeping the water in the tub cleaner during the wash cycle.

FIGURE 14-10g Here is another model manufactured by Whirlpool. This dishwasher motor/pump assembly is a direct drive system without a separate pump-out motor.

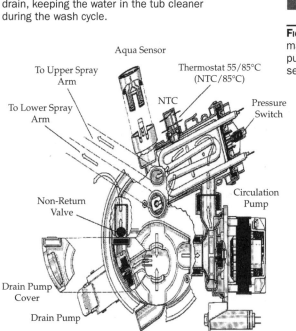

FIGURE 14-10h This model is manufactured by Bosch. It utilizes a direct drive system.

FIGURE 14-10i   This is a Miele model. Due to the closeness of the components, this model will have to be removed from its installation housing to be repaired.

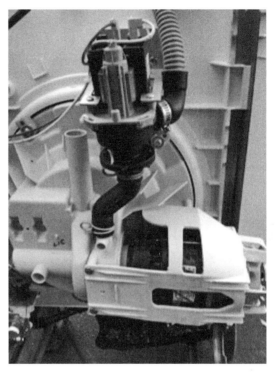

FIGURE 14-10j   This model is manufactured by Frigidaire. It uses a separate drain pump motor to drain the water from the dishwasher tub.

Step-by-Step Troubleshooting by Symptom Diagnosis

In the course of servicing an appliance, you might overlook the simple things that could be causing the problem. Step-by-step troubleshooting by symptom diagnosis is based upon diagnosing the malfunctions, with possible causes arranged into categories relating to the operation of the dishwasher. This section is intended only to serve as a checklist to aid in diagnosing a problem. Look at the symptom that best describes the problem you are experiencing with the dishwasher, and then proceed to correct the problem.

Before testing any electrical component for continuity, disconnect the electrical supply to the appliance.

No Water to Dishwasher

1. Is the water turned on?

2. Is there voltage to the water inlet valve solenoid?

3. Is the water inlet valve solenoid defective? Disconnect the electrical supply, and check the solenoid coil with an ohmmeter.

4. Is the water valve plunger stuck? Disassemble the water valve and check.

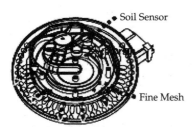

FIGURE 14-11a The LG dishwasher filtering system grinds up the food particles before disposing the food waste down the drain.

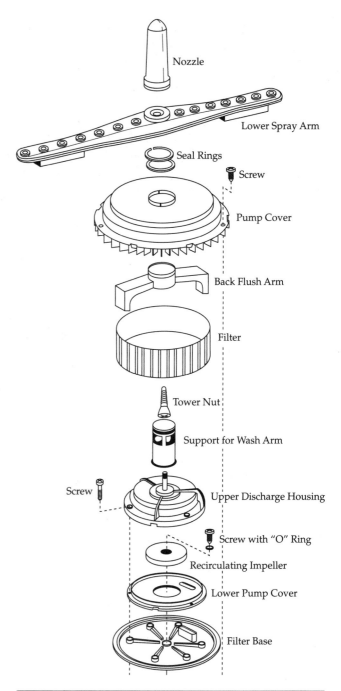

FIGURE 14-11b This type of filtering system used by Maytag allows the food particles to be trapped by the filtering screen before the water enters the recalculating pump.

Wash and Rinse

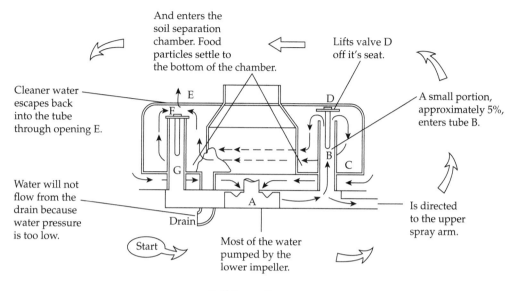

And enters the soil separation chamber. Food particles settle to the bottom of the chamber.

Lifts valve D off it's seat.

Cleaner water escapes back into the tube through opening E.

A small portion, approximately 5%, enters tube B.

Water will not flow from the drain because water pressure is too low.

Is directed to the upper spray arm.

Most of the water pumped by the lower impeller.

Soil Separation

Drain

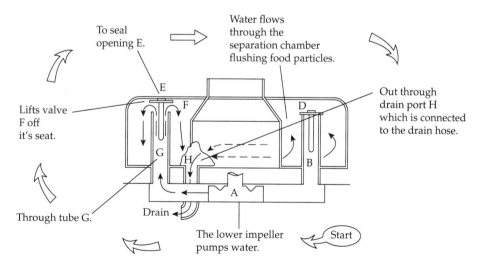

To seal opening E.

Water flows through the separation chamber flushing food particles.

Lifts valve F off it's seat.

Out through drain port H which is connected to the drain hose.

Through tube G.

The lower impeller pumps water.

Note: All drain water must flow through the soil separation chamber.

FIGURE 14-11c The Frigidaire dishwasher filtering device. This is an internal view of the water circulating through the pump as the food particles are separated from the wash/rinse water.

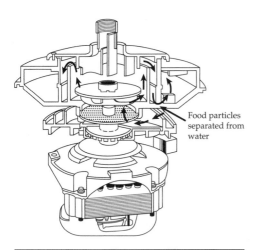

FIGURE 14-11d In this Whirlpool dishwasher model, the food particles are contained in a separate chamber to be disposed of in the drain cycle.

FIGURE 14-11e This type of filtering system manufactured by General Electric will separate the food particles during the wash/rinse cycle and dispose of the food particles in the drain cycle.

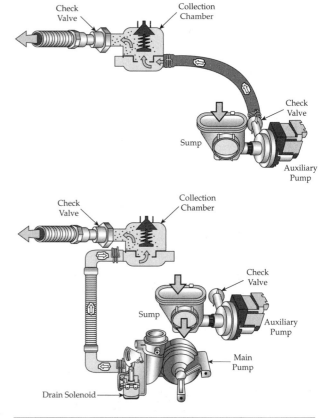

FIGURE 14-11f General Electric models with and without auxiliary pump motor. This motor removes all remaining water in the tub.

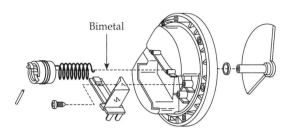

Bimetal

FIGURE 14-12a A bimetal activated dispenser used in a Maytag dishwasher. Current passing through the bimetal will warp it, releasing the detergent.

Rinse agent dispenser

Detergent dispenser

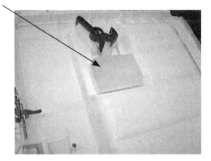

FIGURE 14-12b This mechanical dispenser was manufactured by General Electric. It is the rotation of the timer and cam that moves the lever that releases the detergent cup and rinse agent.

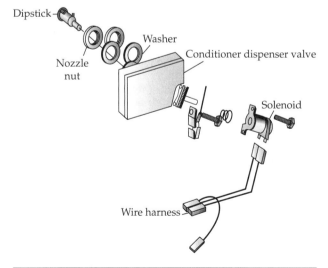

Dipstick

Washer

Conditioner dispenser valve

Nozzle nut

Solenoid

Wire harness

FIGURE 14-12c This Maytag dishwasher uses an electric solenoid to activate the detergent dispenser.

Wax motors are used to activate both the detergent and rinse agent dispensers in a Whirlpool dishwasher.

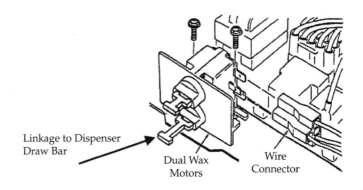

Figure 14-13a The electric heating element is located within the dishwasher tub area on most models.

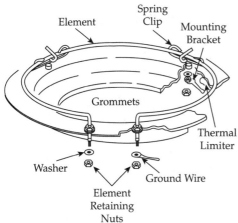

Figure 14-13b Some dishwasher models use a forced air blower to force the moisture out of the dishwasher and through the vent.

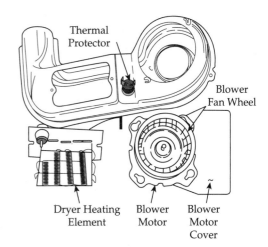

Figure 14-13c
A blower motor assembly used by Jenn-Air to aid in drying the dishes.

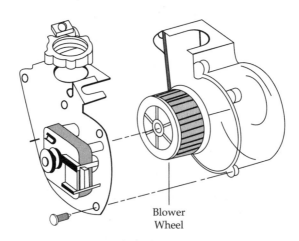

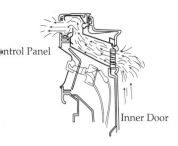

FIGURE 14-14a This vent design by Frigidaire allows heated air to escape but keeps the water in the dishwasher tub during the wash/rinse cycles.

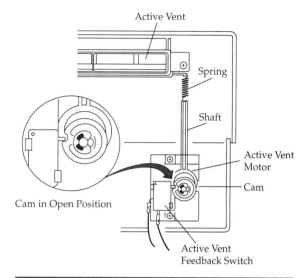

FIGURE 14-14b This active vent designed by General Electric will close during the first fill cycle and remain closed until the drying phase.

FIGURE 14-14c This powered vent by Whirlpool closes the vent during the wash/rinse cycles, and in the drying phase the wax motors will open the vent.

FIGURE 14-14d
The powered vent is located next to the electronic control behind the front panel on the dishwasher.

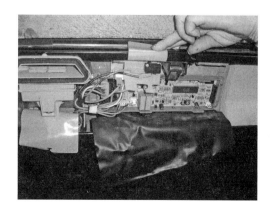

5. Is the water valve inlet screen blocked? Turn off the water supply and remove the water supply line to inspect the screen.

6. Is the water fill line that goes to the water valve kinked? Check visually.

7. Is the water siphoning out of the dishwasher while it is filling? Check the drain hose installation.

8. Are the door latch switches working? Test switch(es) for continuity.

9. Is the float assembly stuck? Check for obstructions. Check assembly and test float switch for continuity.

10. On electronic models, run the diagnostic test mode.

11. Check for loose or burned wires.

12. Check the float assembly and float switch.

Low Water Charge

1. Is there adequate water pressure to the water inlet valve? Pressure should be between 15 and 120 pounds per square inch.

2. Is the water volume adequate? Take a quart container and fill it at the tap. The water must fill the container in seven seconds or less.

3. Is the water valve inlet screen clean? Turn off the water supply and remove the water supply line to inspect the screen.

4. On portable models, check the aerating snap adapter on the faucet.

5. Is the timer defective? Disconnect the electric supply and check the switch contacts for continuity.

6. Is the electronic control functioning properly? Run the test mode.

7. Is the float switch improperly positioned? Is the switch defective? Disconnect the electric supply and check the switch contacts for continuity.

8. Is the water fill line that goes to the water valve kinked? Check visually.

9. Check the service manual for the model you are servicing for the correct water level in the tub when the fill cycle ends.

Poor Washability on the Upper Rack

1. Is the upper spray arm turning?

   a. Are the holes in the spray arm plugged?

   b. Check to see if the spray arm is split.

   c. Is there uneven loading of the dishes?

   d. Check the filter assembly for blockage.

   e. Check the lower impeller to see if it is defective or blocked with debris.

   f. Are any objects protruding down that might prevent the upper spray arm from rotating?

2. Is the water charge okay?

3. What is the temperature of the water entering the tub? Is the temperature at 140 degrees Fahrenheit?

4. Is the consumer using the proper amount of detergent?

5. Is the detergent dispenser functioning properly?

6. Are the dishes loaded properly? Ask the user to load the dishwasher so that you can observe whether he or she is loading the dishwasher properly.

7. On some models you may have to check the filters for blockages.

Poor Washability in the Lower Rack

1. Is the lower spray arm turning?

   a. Are the holes in the spray arm plugged?

   b. Check to see if the spray arm is split.

   c. Is there uneven loading of the dishes?

   d. Is the spray arm binding on the housing?

   e. Are any objects protruding down that might prevent the lower spray arm from rotating?

2. Is the water charge okay?

3. What is the temperature of the water entering the tub? Is the temperature at 140 degrees Fahrenheit?

4. Is the consumer using the proper amount of detergent?

5. Is the detergent dispenser functioning properly?

6. Are the dishes loaded properly? Ask the user to load the dishwasher so that you can observe whether he or she is loading the dishwasher properly.

7. On some models you may have to check the filters for blockages.

Poor Drying

1. Is the water hot enough? Check the water supply temperature. On certain models, check to see if the cycle extender is functioning properly.

2. Is the heater working?

   a. Check the wiring.

   b. Check the timer or the electronic control board.

   c. Check for the correct voltage.

   d. Check the heater itself.

   e. Check the heater fan assembly.

3. Is a wetting agent being used? Check wetting agent level.

4. Check the door baffle.

5. Check the vent system.

6. Suggest that the user open the door a little after the dry cycle has been completed.

Water Leaks at the Front of the Dishwasher

1. Is the spray arm turning? Is it split? Are any of the holes blocked with debris? If so, this can cause the water to spill out of the tub.

2. Is the tub gasket in place? Check the corners especially. Worn tub or corner gaskets will cause water to leak out of the corners.

3. Did you check the vent baffle gasket?

4. Some models have a corner gasket located in the inside front of the tub. Check to see if it is in place.

5. Check to see if the tub is not overcharged with water. Too much water will spill over the front tub flange.

6. Does the door close properly and tightly where the latch fits the strike?

7. On some models, are the gasket clips in place? These clips secure the gasket to the tub.

8. Is the dishwasher loaded properly? Ask the user to load the dishwasher so that you can observe whether he or she is loading the dishwasher properly.

9. Check to see if the dishwasher is draining properly. Inspect the drain hose where it enters the drain.

10. Most important, is the dishwasher level?

11. Check the door liner for any cracks.

12. Check to see if the wetting agent dispenser is working properly. Too much wetting agent will cause sudsing. Also check the gasket between the door and the dispenser.

Water Leaks at Sides, Top, or Bottom of Dishwasher

1. Check the side inlet tube on the side of the dishwasher.

2. Check to see if water is leaking out from the pump assembly.

3. Check the motor seals for water leakage.

4. Does the tub have a hole in it? Is the tub rusted out?

5. Check the nut on the water inlet port. Is it tight?

6. Are the heater element nuts tight?

7. Check to be sure that all screws are tight (motor and pump assembly, upper rack screws, etc.).

The Dishwasher Cycle Will Not Advance (But the Lights Will Come On)

1. Is there voltage to the timer motor or electronic control board?

2. Are the cams in the timer rotating?

3. Is the dishwasher wired correctly?

4. Check to see if the timer is jammed.

5. Check to see if the selector switch is defective.

6. On electronic models, run the test mode.

The Dishwasher Cycle Will Not Advance Past Start

1. Is the thermostat functioning properly? The thermostat must be flush with the underside of the tub.

2. Check for correct timer settings.

The Detergent Cup Won't Open

1. Is there voltage to the solenoid?

2. Is the detergent mechanism properly adjusted?

3. Check to see if there is any binding at any point.

4. Check to see if any dishes are preventing the detergent cup door from opening.

The Detergent Cup Won't Close

1. Check the detergent cup actuator and cams.

2. Check to see if there is any binding at any point.

3. Is the detergent mechanism adjusted properly?

The Main Motor Won't Operate

1. Check for continuity at the start and run windings. Check the motor for a shorted or grounded winding.

2. Is there voltage to the motor?

3. Is the motor jammed? Check for foreign debris in the pump assembly.

4. Check to see if there are any loose wiring terminals or burned wires.

5. Check to see if the motor assembly is wired correctly.

The Motor Runs but Goes into Overload

1. Check the relay.

2. Is there any binding? Check the pump assembly for broken pieces of glass.

3. Be sure you have the correct voltage and the correct polarity.

4. Check to see if the motor windings are shorted.

5. Some models have a run capacitor. Test the run capacitor.

Door Liner Hits Side of Tub in Undercounter Models

1. Check to see if the tub is square.

2. Check the installation, and correct as needed.

Dishwasher Won't Start

1. Is there voltage to the dishwasher? Check the plug, circuit breaker, or fuse box. Also check the wires in the junction box, located behind the lower front panel.

2. Check the door switch for continuity of the switch contacts.

3. Check the selector switch for continuity of the switch contacts.

4. Check to see if the timer and selector switch have been wired correctly.

5. On electronic models, check the electronic control board.

6. Check for any wires that might have come off the timer or switches.

The Dishwasher Repeats the Cycle

1. Check the timer contacts; they should be open in the "off" position.

2. Check to see if the timer motor wires are shorted.

Wetting Agent Assembly Leaks

1. Open the door assembly, and check for leaks in the holding tank.

2. Check the wetting agent assembly for proper operation.

Water Siphons Out Through the Drain Pipe While the Dishwasher Is Trying to Fill

1. Is the drain line properly installed?

2. Check the installation instructions for proper installation.

Common Washability Problems

If there are no mechanical problems with the dishwasher's operation and the complaints are that the dishwasher will not clean the dishes properly or that the glassware is cloudy, etc., the next step will be to look at the best possible cause for the problem that the customer is experiencing with the dishwasher. Then proceed to correct the problem. If necessary, instruct the user on how to get better results from the dishwasher.

Poorly Cleaned Dishes

On occasion, there might be some food particles left on the dishes at the end of the cycle.

- *Possible cause:* Water temperature might be too low. Remember, the water temperature should be 140 to 150 degrees Fahrenheit as it enters the dishwasher tub.

- *Solution:* Check the water temperature at the closest faucet. Let the water from the hot water tap run before starting the dishwasher in order to clear the water line of any cold water.

- *Possible cause:* Not enough detergent for the degree of water hardness or for the amount of dirty dishes to be cleaned.

- *Solution:* Use 1 teaspoon of detergent for each grain of hardness, with 3 teaspoons of detergent, at a minimum, in soft water. The dishwasher will require extra detergent for greasy pans.

- *Possible cause:* The detergent was placed in the wrong side of the dispenser cup.

- *Solution:* Instruct the user on how to fill the dispenser, and have the user reread the use and care manual so that the detergent is placed in the correct dispenser for the cycle that is selected.

- *Possible cause:* Improper loading of the dishes into the dishwasher.

1. Blocking the spray nozzle. If a large bowl or pot is placed over the center of the lower rack, this will block the spray nozzle washing action when the lower rack is pushed in.

2. Larger items that shield smaller items from the washing action.

3. Observe to see if there is a nesting of the bowls or silverware so that the water cannot reach all surfaces.

4. The spray arms are blocked from turning, for example, tall items or an item that fell through the racks.

5. If an item blocks the detergent dispenser from opening, this will not allow the detergent to mix with the water.

- *Solution:* Instruct the user to reread the use and care manual for the proper instructions on how to load the dishes for proper cleaning.

- *Possible cause:* Improper filling of water in the tub. Water pressure must be between 15 and 120 pounds per square inch. After the fill has stopped, check the water level in the tub. On most models, it should be even with the heating element in the bottom of the dishwasher.

- *Solution:* If the water pressure is low, be sure that no other faucets are in use while the dishwasher is operating.

- *Possible cause:* Not enough hot water.

- *Solution:* Instruct the user to use the dishwasher when the hot water is not being used for laundry, baths, or showers.

- *Possible cause:* If the dishwasher detergent is old and caked, it will not dissolve completely.

- *Solution:* Instruct the user to use fresh detergent and always store it in a dry place.

Etching

Etching occurs when the glass is pitted or eroded. It appears as a permanent film on the glass. The beginning stages of etching can be identified by an iridescent look—shades of blue, purple, brown, or pink when the glass is held at an angle to the light. In the advanced stages of etching, the glass surface appears frosted or cloudy.

- *Possible cause:* Certain types of glass will etch in any dishwasher with the combination of soft water, the alkalinity of dishwasher detergents, and heat.

- *Solutions:* There is no way to remove the filmy appearance caused by etching; the damage is permanent. There is no way to predict what glassware might be affected— it is not related to the cost or quality of the glass. To prevent etching from recurring:

  - Adjust the amount of detergent according to the water hardness.

  - Adjust the water temperature so that it enters the dishwasher at approximately 140 degrees Fahrenheit.

  - Recommend to the customer that the Energy Saver dry cycle be used.

  - Instruct the customer not to manually pre-rinse the dishes before loading them into the dishwasher.

PART VI

Discoloration

Discoloration (red, black, or brown) might be present on the dishwasher interior or dishes.

- *Possible cause:* If iron or manganese is in the water, the dishes and/or the interior of the dishwasher might turn red, black, or brown.
- *Solution:* A rust remover can be used to remove the discoloration from the dishwasher interior. With the dishwasher empty, turn to the rinse and hold cycle, and start the dishwasher. During the fill, open the door and add 1/2 cup of rust remover to the water. Allow the dishwasher to complete the cycle. Then start the dishwasher on the normal wash cycle, with detergent but without dishes. To keep this condition from returning, the customer might have to install special filters to remove the iron and manganese. Use a rust remover according to the manufacturer's recommendations to remove the discoloration from the dishes and the glassware.

Lime Deposits on the Dishwasher Interior

- *Possible cause:* If there is a lot of calcium in the water, a lime film or deposit might eventually build up on the interior surfaces of the dishwasher.
- *Solution:* You can try one of the following methods:
  - Use a mild scouring powder and a damp cloth to clean away the lime deposit.
  - With the dishwasher empty, turn the timer to the rinse and hold cycle. During the fill portion, open the door and add 1/2 cup of white vinegar to the water. Let the dishwasher complete the remaining cycle. Do not use detergent. After the cycle is completed, run the dishwasher with a regular load.
  - Use a product that removes lime deposits. Follow the manufacturer's directions.

Suds or Foam in Dishwasher

- *Possible causes:*
  - Sudsing in the dishwasher can be caused by protein foods (milk, eggs, etc.) and an insufficient amount of detergent.
  - The water in the dishwasher is not hot enough to activate the defoaming agents in the dishwasher detergent.
  - The user has used a non-automatic dishwasher detergent.
- *Solutions:*
  - Increase the amount of detergent to reduce sudsing. Dishwasher detergents contain defoaming agents to break down the suds in the dishwasher water.
  - Check that the water temperature is between 140 and 150 degrees Fahrenheit. Only use detergents that are made for automatic dishwashers.

Darkened Aluminum

- *Possible cause:* A combination of water, heat, and alkaline foods will darken or stain aluminum products.

- *Solution:* To remove this discoloration, instruct the customer to use an aluminum cleaner and to clean the item by hand. Never allow undissolved dishwasher detergent to come in direct contact with the metal. Avoid placing aluminum items in the lower rack, directly in front of the detergent dispenser.

Discoloration of Copper

- *Possible cause:* Some copper items will discolor when washed by hand, as well as when washed in the dishwasher, because of the heat and detergent alkalinity.
- *Solution:* Instruct the customer to use a copper cleaner to restore the copper color.

Cracking (Crazing) of China

- *Possible cause: Crazing* is the appearance of tiny cracks that appear over the entire surface of the china. It can occur when porous earthenware, good china that is very old, or lower-quality china is exposed to heat and moisture.
- *Solution:* Once the glaze is cracked or crazed, the damage is permanent. This characteristic is inherent in some clayware; this type of damage can occur during use, handwashing, or automatic dishwashing.

Chipping of China and Crystal

- *Possible cause:* Chipping usually occurs during normal use and handling, and simply might not be noticed until the dishes are removed from the dishwasher. When the dishes are loaded into the dishwasher according to the manufacturer's instructions, there is nothing in the dishwasher that can chip the dishes. The dishes should only come in contact with the cushioned vinyl-coated racks.
- *Solution:* Instruct the user to follow the manufacturer's instructions in the use and care manual for loading the dishwasher properly.

Metal Marks on Dishes and Glassware

- *Possible cause:* If a metal item, especially aluminum, touches a dish in the dishwasher, a metal mark might result. This symptom appears as small black or gray marks or streaks on dishes or glasses. However, most metal marks occur during normal use, when the dishes come in contact with the flatware.
- *Solution:* Instruct the user to load the dishwasher carefully in order to prevent metal items from touching other dishes. There are products on the market that will remove these marks. Have the user read over the use and care manual or check with the manufacturer for this information.

Staining of Melamine Dinnerware

- *Possible cause:* Stains on melamine dinnerware can result from contact with coffee, tea, and some fruit juices. If the surface is worn, it will stain more readily.
- *Solution:* Some specialty products on the market are recommended for removing these stains. Instruct the customer to read over the use and care manual or check with the manufacturer for this information.

Melting or Warping of Plastic Items

- *Possible cause:* Some plastic items cannot be exposed to the temperatures usually found in dishwashers without changing shape.
- *Solution:* Once the plastic item has distorted, it cannot be returned to its original shape. In order to minimize or prevent plastic items from warping or melting, instruct the user to do one of the following:
  - Choose the air-dry cycle to dry the dishes.
  - Place the plastic items on the top rack.
  - Purchase and use plastic items that are labeled "dishwasher-proof."

Discoloration of Silverplate

- *Possible cause:* When silverplate takes on a copper- or bronze-colored appearance, the silverplate has worn thin and the base metal is showing through. The combination of dishwasher detergent and the lack of hand-toweling might result in discoloration of this base metal.
- *Solution:* This discoloration can usually be removed by polishing the item with a silver polish or by soaking the item in vinegar for about 10 minutes. This is only a temporary solution, however. Only a replating with silver by a jeweler will correct the problem.

Tarnishing of Silverware (Sterling or Silverplate)

- *Possible cause:* Sulfur in the water supply might be the cause. This effect might be accelerated by the automatic dishwasher because of the higher water temperature and because the usual hand-drying with a towel has been eliminated.
- *Solution:* Because sulfur cannot be readily removed from the water supply, frequent polishing is the only answer.
- *Possible cause:* Silver will tarnish easily if it is left in contact with foods such as mayonnaise and eggs.
- *Solution:* If silverware has been in contact with such foods, instruct the user to rinse the item thoroughly as soon as possible after its use.

Bluish Discoloration of Stainless Steel

- *Possible cause:* A bluish discoloration of some types of stainless steel is caused by heat and the alkalinity of the automatic dishwasher detergent.
- *Solution:* This discoloration can be removed by using a paste of baking soda and water or a stainless steel cleaner.

Corrosion or Rusting of Stainless Steel

- *Possible cause:* When the protective oxide film on the surface of the steel is removed, corrosion will take place, as with ordinary steels. Certain foods will remove the oxide film, such as table salt, vinegar, salad dressings, milk and milk products, fruits and juices, tomatoes and tomato products, and butter. However, if the stainless steel is washed, rinsed, and dried thoroughly, the oxygen of the air will heal the breaks in the oxide film and return the stainless property to the steel. But if food is not washed off promptly, the air cannot heal the break, and corrosion will occur.

- *Solution:* There is no permanent solution. To minimize rusting, instruct the user to rinse or wash the flatware as soon as possible after use.

Dishwasher Maintenance

The dishwasher's interior is normally self-cleaning. However, there are times when the customer will have to remove food particles or broken glass in the bottom of the tub. Instruct the customer to clean the bottom edge of the dishwasher tub, which is sealed off by the gasket when the door is closed. Food and liquids drip on to this area when the dishwasher is loaded. The control panel should be cleaned with a soft damp cloth. Tell the customer not to use any abrasive powders or cleaning pads. Also, advise the customer to read the use and care manual for proper maintenance procedures for his or her brand of dishwasher. Dishwashers are designed to flush away all normal food soils that have been removed from the dishes. However, on occasion, certain foreign objects, such as fruit pits, bottle caps, etc., might collect in the openings of the pump. These items should be removed periodically to avoid clogging the drain system. Also, on occasion, some of these foreign objects can get caught in the spray arm openings and will have to be cleaned out. Check the racks carefully to see if there are any nicks or cuts in the vinyl. These nicks and cuts can be repaired. A liquid vinyl-repair material is available through the manufacturer or at any appliance supply store.

Repair Procedure

Each of the following repair procedures is a complete inspection and repair process for a single dishwasher component, containing the information you need to test a component that might be faulty and then replace it, if necessary.

Any person who cannot use basic tools should *not* attempt to install, maintain, or repair any dishwashers. Any improper installation, preventative maintenance, or repairs will create a risk of personal injury, as well as property damage. Call the service manager if installation, preventative maintenance, or the repair procedure is not fully understood.

Water Inlet Valve

The water inlet valve controls the flow of water into the dishwasher tub, and is solenoid-operated.

The typical complaints associated with water inlet valve failure (Figure 14-15) are:

- The dishwasher will run, but no water will enter dishwasher.
- The dishwasher will overfill and leak onto the floor.
- When the dishwasher is off, water still enters the tub.
- The dishes are not clean or not enough water enters the tub.

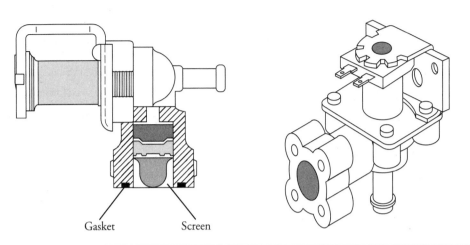

Gasket   Screen

FIGURE 14-15   A typical dishwasher water valve.

When dealing with these complaints, perform the following steps:

1. **Verify the complaint.** Verify the complaint by operating the dishwasher through its cycles. Listen carefully, and you will hear if the water is entering the dishwasher. On electronic models, turn off the electricity to the appliance and wait for two minutes before turning it back on. If a fault code appears, look up the code. If the dishwasher will not power up, locate the technical data sheet behind the control panel or under the tub of the dishwasher for diagnostics information. On some models you will need the actual service manual for the model you are working on to properly diagnose the dishwasher. The service manual will assist you in properly placing the dishwasher in the service test mode for testing the dishwasher functions.

2. **Check for external factors.** You must check for external factors not associated with the appliance. Is the appliance installed properly? Is the voltage supply correct for the dishwasher? Is the water turned on all the way? The voltage at the receptacle is between 108 volts and 132 volts during a load on the circuit. Do you have the correct polarity? (See Chapter 6.)

3. **Disconnect the electricity.** Before working on the dishwasher, disconnect the electricity. This can be done by pulling the plug out of the wall receptacle. Be sure that you only remove the dishwasher plug. Double-check to ensure that the electrical supply has been disconnected before removing any service panels. Another way to disconnect the electricity is at the fuse panel or the circuit breaker panel. Turn off the electricity.

WARNING   *Some diagnostic tests will require you to test the components with the power turned on. When you disassemble the control panel, you can position it in such a way that the wiring will not make contact with metal. This act will allow you to test the components without electrical mishaps.*

4. **Remove the bottom panel.** In order to gain access to the water valve, the bottom panel must be removed (Figure 14-16). The bottom panel is held on with either two or four screws, depending on the model. Remove the screws, and remove the panel.

5. **Remove the wire leads.** In order to check the solenoid coil on the water valve, remove the wire leads that connect to the solenoid coil from the wire harness (Figure 14-17). These are slide-on terminal connectors attached to the ends of the wire—just pull them off.

6. **Test the water valve.** Using the ohmmeter, set the range on R × 1000, and place the probes on the solenoid coil terminals (Figure 14-18). The meter should read between 700- and 900-ohms resistance. If not, replace the water valve.

7. **Inspect the inlet screen.** If you determine that the water valve is good, but there is little water flow through the valve, inspect the inlet screen. If this screen is filled with debris, it must be cleaned out. To accomplish this, use a small flat-blade screwdriver and pry out the screen (Figure 14-19). Then wash out the screen, making sure that all of the debris is removed. Reinstall the screen. Turn on the water supply. Plug in the dishwasher. Allow the water to enter the tub to check the flow rate of the water valve. The tub must be empty, because this will allow you to check the flow rate properly. On a normal fill, the water line (on some models) should be over the heating element within two minutes (Figure 14-20). When you turn the dishwasher on and energize the water valve, if no water enters the dishwasher tub, replace the water valve. If the water valve checks out okay, then check the timer or electronic control board, float switch, and the wiring harness.

8. **Remove the water valve.** Before removing the water valve, turn off the water supply to the dishwasher. Then disconnect the water supply line from the inlet end of the water valve and remove the fill hose from the outlet side (Figure 14-21). Next,

FIGURE 14-16
Removing the bottom panel.

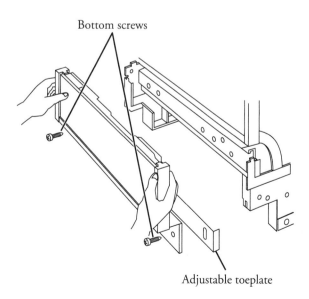

Bottom screws

Adjustable toeplate

Inlet valve solenoid terminals

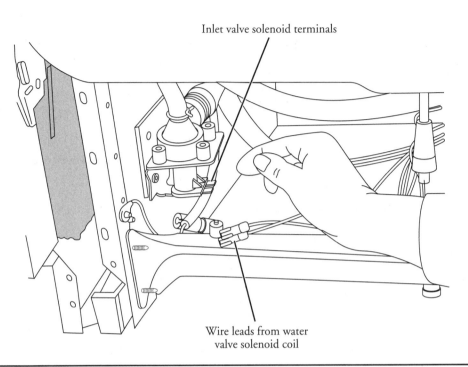

Wire leads from water
valve solenoid coil

Figure 14-17   Removing the wire leads from the solenoid coil.

Figure 14-18
Connect meter probes
to the water valve
solenoid coil.

Inlet valve solenoid terminals

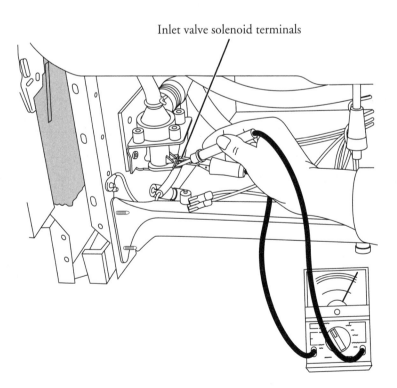

Removing water valve
inlet screen. Be
careful not to distort
the screen.

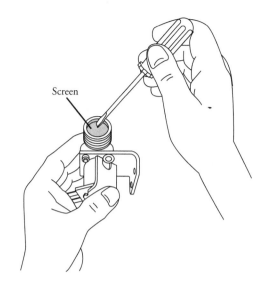

Screen

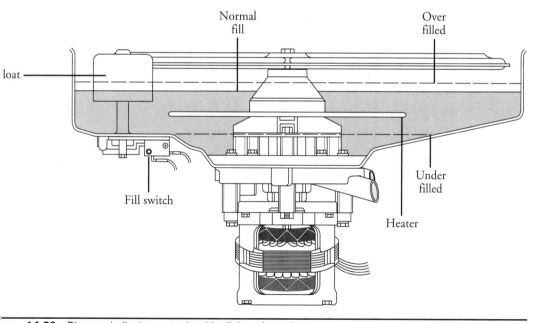

Normal
fill

Over
filled

loat

Fill switch

Under
filled

Heater

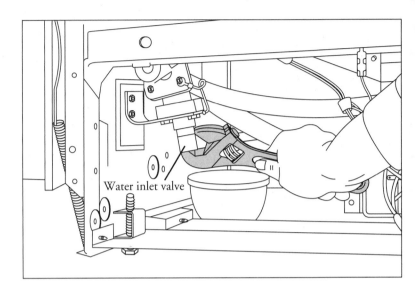

FIGURE 14-21
Disconnect the water
supply line.

Water inlet valve

remove the screws that hold the water valve to the chassis of the dishwasher. To
install the new water valve, just reverse the disassembly procedure and reassemble.
Check for water leaks. If none are found, reinstall the bottom panel and restore the
electricity to the dishwasher.

9. **Operate the dishwasher.** Set the timer and control settings to operate the
dishwasher through its cycle.

Motor and Pump Assembly

The motor and pump assembly today utilizes a direct drive motor/pump assembly. This
means the motor drives the pump, which is attached to the motor. The shaft seal (Figure 14-22)
is a two-part seal needed to keep water from dripping onto the motor. This ceramic and
carbon seal is fragile. The water in the tub acts as a lubricant for the seal.

The typical complaints associated with pump and motor failure are:

- The water will not drain out of the dishwasher.

- Poor washability of the dishes.

- When the motor runs, there are loud noises.

- The fuses or circuit breaker will trip when the dishwasher is started.

- Water is leaking from the bottom of the dishwasher or
leaking on the motor/pump assembly.

FIGURE 14-22
Dishwasher motor and
pump seals. The
seals prevent water
from leaking out of
the dishwasher.

When dealing with these complaints, perform the
following steps:

1. **Verify the complaint.** Verify the complaint by operating
the dishwasher through its cycles. Listen carefully, and

you will hear if there are any unusual noises or if the circuit breaker trips. On electronic models, turn off the electricity to the appliance and wait for two minutes before turning it back on. If a fault code appears, look up the code. If the dishwasher will not power up, locate the technical data sheet behind the control panel or under the tub of the dishwasher for diagnostics information. On some models you will need the actual service manual for the model you are working on to properly diagnose the dishwasher. The service manual will assist you in properly placing the dishwasher in the service test mode for testing the dishwasher functions.

2. **Check for external factors.** You must check for external factors not associated with the appliance. Is the appliance installed properly? Does the appliance have the correct voltage? The voltage at the receptacle is between 108 volts and 132 volts during a load on the circuit. Do you have the correct polarity? (See Chapter 6.)

3. **Disconnect the electricity.** Before working on the dishwasher, disconnect the electricity. This can be done by pulling the plug out of the wall receptacle. Be sure that you only remove the dishwasher plug. Or disconnect the electricity at the fuse panel or at the circuit breaker panel. Turn off the electricity.

WARNING *Some diagnostic tests will require you to test the components with the power turned on. When you disassemble the control panel, you can position it in such a way that the wiring will not make contact with metal. This act will allow you to test the components without electrical mishaps.*

4. **Remove the bottom panel.** In order to gain access to the pump and motor assembly, the bottom panel must be removed (see Figure 14-16). The bottom panel is held on with either two or four screws, depending upon the model. Remove the screws and remove the panel.

5. **Disconnect the motor wire leads.** Disconnect the motor wire leads from the wiring harness. Check the motor windings for continuity (Figure 14-23). Check for resistance from the common wire lead to the run winding (Figure 14-24). Check from the common to the wash winding (Figure 14-25).

    To check for a grounded winding in the motor, take the ohmmeter probes and check from each motor wire lead terminal to the motor housing (Figure 14-26). The ohmmeter will indicate continuity if the windings are grounded (Figure 14-27).

    If the motor/pump assembly shows signs of water leaking, replace the motor/ pump assembly as a complete unit. Most part manufacturers give a one-year warranty on the motor/pump assembly. There is no advantage in tearing down the motor/pump assembly to replace only one or two parts and receiving only a partial warranty. Replacing the motor/pump assembly as a complete assembly will save you time and money in the long run. If the motor/pump assembly checks out okay, then check the timer and the motor relay (if the model you are repairing has one) and check for a kinked or plugged drain line.

6. **Remove motor and pump assembly.** Remove the lower dishwasher rack from the dishwasher. As shown in Figure 14-28, remove the wash tower and spray arm assembly. If there is a filter, remove it also. Remove the motor wiring leads from the wiring harness. Then remove the drain line from the pump assembly.

FIGURE **14-23**
Check the motor
windings for
continuity.

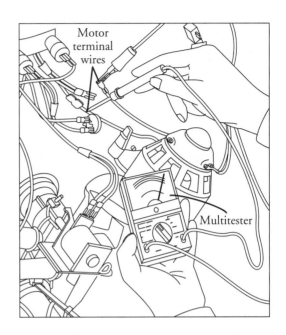

FIGURE **14-23**
Check the motor
windings for
continuity.

Reach underneath the tub and rotate the four pump hold-downs 90 degrees inward
(Figure 14-29). You are now ready to remove the motor/pump assembly. Lift the
motor/pump assembly out from the inside of the tub (Figure 14-30). Keep the work
area dry to help prevent electrical shocks.

FIGURE **14-24**
Check from the
common to the drain
winding.

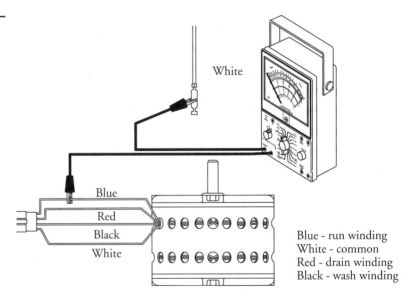

URE **14-25**
eck from the
mmon to the wash
nding.

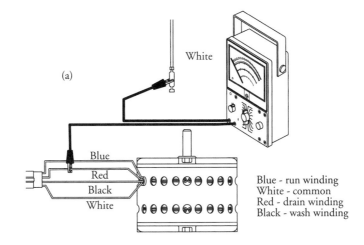

(a)

White

Blue
Red
Black
White

Blue - run winding
White - common
Red - drain winding
Black - wash winding

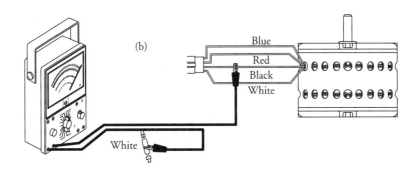

(b)

Blue
Red
Black
White

White

FIGURE 14-26
Check for the
grounded motor.

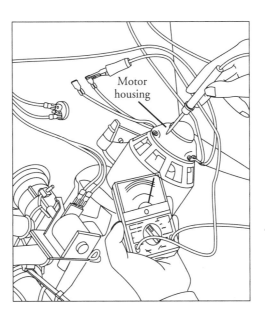

Motor
housing

FIGURE 14-27
The grounded motor.

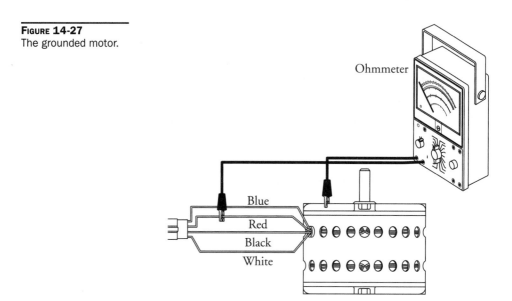

Ohmmeter

Blue
Red
Black
White

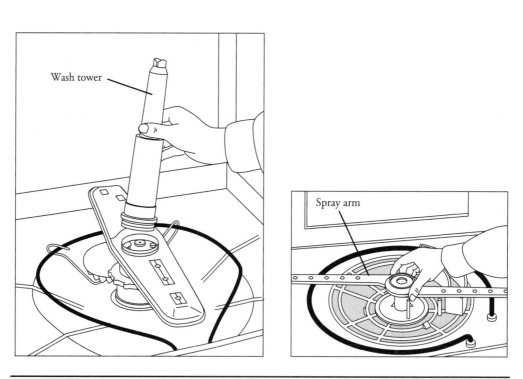

Wash tower

Spray arm

FIGURE 14-28 Remove the wash tower and the spray arm assembly.

FIGURE **14-29**
An exploded view of a
dishwasher motor/
pump assembly.

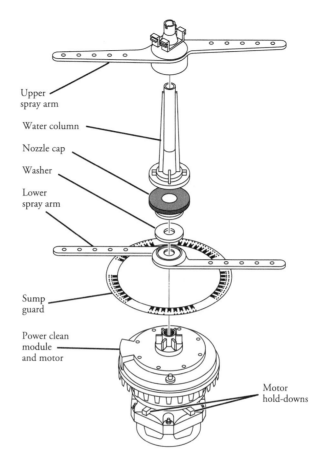

Upper
spray arm

Water column

Nozzle cap

Washer

Lower
spray arm

Sump
guard

Power clean
module
and motor

Motor
hold-downs

FIGURE **14-30**
Remove the motor/
pump assembly. After
removal, inspect the
tub for rust.

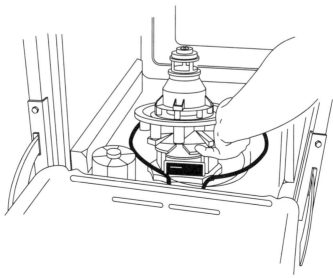

7. **Reinstall motor and pump assembly.** To reinstall the motor/pump assembly, just reverse the order of step 6. Before restoring the electricity to the dishwasher, pour a gallon of water into the tub and check for leaks underneath the tub. Restore the electricity to the dishwasher, and run the dishwasher through a cycle. Check for leaks again. If no leaks are found, reinstall the bottom panel.

Dishwasher Timer

Electromechanical timers are utilized for controlling the different cycles. The timer assembly is driven by a synchronous motor in incremental advances. It controls and sequences the numerous steps and functions involved in each cycle of a dishwasher.

On models with an electronic control board, refer to the service manual for proper removal and testing. On models that use an electronic control along with a sequence switch, refer to the service manual for proper removal and testing.

The typical complaints associated with dishwasher timer failure are:

- The cycle will not advance.
- The dishwasher won't run at all.
- The dishwasher will not fill.
- The dishwasher will not pump the water out.

When dealing with these complaints, perform the following steps:

1. **Verify the complaint.** Verify the complaint by operating the dishwasher through its cycles. Before you change the timer, check the other components controlled by the timer.

2. **Check for other factors.** You must check for other factors that could affect the operation of the appliance. Is the appliance installed properly? Does the appliance have the correct voltage? The voltage at the receptacle is between 108 volts and 132 volts during a load on the circuit. Do you have the correct polarity? (See Chapter 6.)

3. **Disconnect the electricity.** Before working on the dishwasher, disconnect the electricity. This can be done by pulling the plug out of the wall receptacle. Be sure that you only remove the dishwasher plug. Or disconnect the electricity at the fuse panel or circuit breaker panel. Turn off the electricity.

WARNING *Some diagnostic tests will require you to test the components with the power turned on. When you disassemble the control panel, you can position it in such a way that the wiring will not make contact with metal. This act will allow you to test the components without electrical mishaps.*

4. **Remove the console to gain access.** Begin by removing the four screws from the console to access the timer (Figure 14-31). Turn the timer knob counterclockwise to remove it from the timer shaft, and slide the indicator dial off the shaft. Remove the console panel from the dishwasher. On some models, the latch handle knob will also have to be removed.

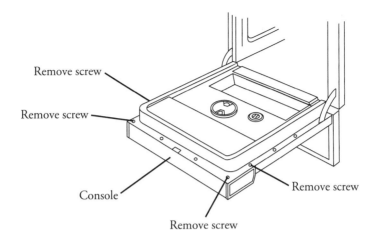

FIGURE 14-31
Remove the screws to access the control panel.

Remove screw

Remove screw

Console

Remove screw

Remove screw

5. **Test the timer.** Disconnect the timer motor wire leads from the timer assembly. Using the ohmmeter, set the range on R × 1000, and place the probes on the timer motor terminals (Figure 14-32). The meter should indicate some resistance. If not, replace the timer. If the motor checks out, then check the door latch switch, cycle extender relay, and the float switch assembly.

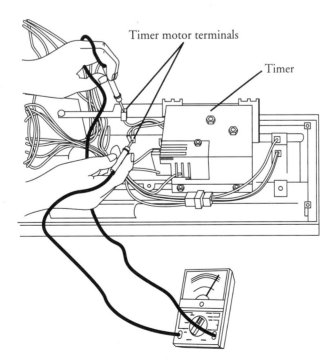

FIGURE 14-32
Attach the meter probes and test for continuity.

Timer motor terminals

Timer

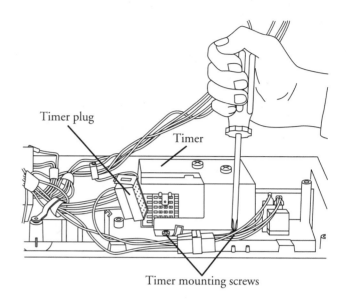

Figure 14-33
Removing the timer.

Timer plug

Timer

Timer mounting screws

6. **Remove the timer.** To remove the timer, remove the timer mounting screws (Figure 14-33). Remove the wire lead terminals from the timer. Mark the wires as to their location on the timer. Some timers have a disconnect terminal block instead of individual wires, which makes it easier to remove the timer wires.

7. **Install the new timer.** To install a new timer, just reverse the disassembly procedure, and reassemble. Reinstall the console panel, and restore the electricity to the dishwasher. Test the dishwasher operation.

The Dishwasher Door Is Hard to Close

Check the latch assembly, door gaskets, tub gaskets, and on some models, the corner gaskets. The typical complaints associated with a dishwasher door being hard to close are:

- The door is hard to close.
- The door won't latch.
- The dishwasher won't run.

When dealing with these complaints, perform the following steps:

1. **Verify the complaint.** Verify the complaint by trying to close and latch the door.

2. **Check for external factors.** You must check for external factors not associated with the appliance. Is the appliance installed properly? Be sure that the dishwasher door is not binding against the side cabinet.

3. **Run the dishwasher.** By pushing on the door, you finally latch the door closed. Next, run the hot water faucet closest to the dishwasher to flush out the supply line. Then turn the timer to the normal wash cycle with heated dry, and let it run through the entire cycle. By doing this procedure, the gasket will soften enough to form itself

FIGURE 14-34
Removing dishwasher
door gasket.

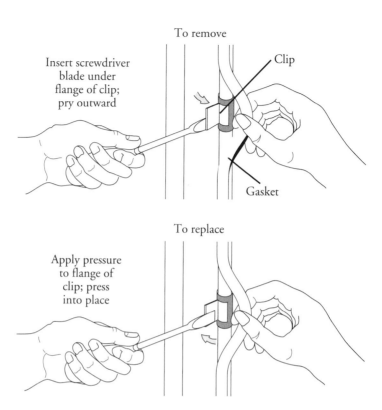

To remove

Insert screwdriver
blade under
flange of clip;
pry outward

Clip

Gasket

To replace

Apply pressure
to flange of
clip; press
into place

to the tub while it is compressed. This should make it easier to latch the door. If not, adjust the latch assembly. If it is still too hard to close, replace the door gasket.

4. **Replace the door gasket.** To remove the old gasket, remove the screws or clips that hold the old gasket in place (Figure 14-34). On some models, the gasket is pressed into the inner door assembly. Remove the inner panel from the dishwasher, and pull off the gasket from the panel (Figure 14-35). Soak the new gasket in warm water to make it more flexible.

5. **Test a new gasket.** After you have installed the new door gasket, close the dishwasher door, and test to see if the door latches without pushing hard against it.

6. **Check for water leaks.** Run the dishwasher through another cycle to check for water leaking out the door. Most models have a baffle, or tub, gasket. If the dishwasher that you're working on has one, check to make sure it is not defective. The tub gasket is located either on the tub or behind the inner door panel.

Dishwasher Heating Element

Most dishwashers use a calrod heating element to heat the water and dry the dishes. The element consists of a tungsten wire packed in magnesium oxide and surrounded by a coated stainless steel sheath. The resistance and voltage can be measured with a multimeter to verify if the element is functioning properly.

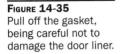

FIGURE 14-35
Pull off the gasket,
being careful not to
damage the door liner.

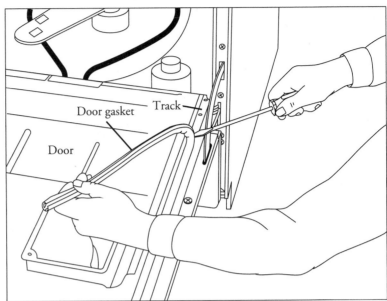

The typical complaints associated with the heating element are:

- Dishes are still wet at the end of the cycle.
- The dishwasher stalls in the middle of the cycle.

When dealing with these complaints, perform the following steps:

1. **Verify the complaint.** Verify the complaint by operating the dishwasher, starting at the dry cycle. On electronic models, turn off the electricity to the appliance and wait for two minutes before turning it back on. If a fault code appears, look up the code. If the dishwasher will not power up, locate the technical data sheet behind the control panel or under the tub of the dishwasher for diagnostics information. On some models you will need the actual service manual for the model you are working on to properly diagnose the dishwasher. The service manual will assist you in properly placing the dishwasher in the service test mode for testing the dishwasher functions.

2. **Check for external factors.** You must check for external factors not associated with the appliance. For example, check the Energy Saver switch. Is it set for heat-dry? Is the appliance installed properly? Does the appliance have the correct voltage? The voltage at the receptacle is between 108 volts and 132 volts during a load on the circuit. Do you have the correct polarity? (See Chapter 6.)

3. **Disconnect the electricity.** Before working on the dishwasher, disconnect the electricity. This can be done by pulling the plug out of the wall receptacle. Be sure that you only remove the dishwasher plug. Or disconnect the electricity at the fuse panel or the circuit breaker panel. Turn off the electricity.

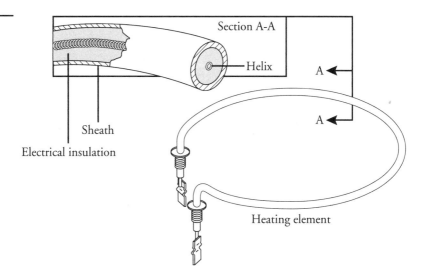

FIGURE 14-36
The dishwasher
heating element.

Section A-A

Helix

A

A

Sheath

Electrical insulation

Heating element

WARNING *Some diagnostic tests will require you to test the components with the power turned on. When you disassemble the control panel, you can position it in such a way that the wiring will not make contact with metal. This act will allow you to test the components without electrical mishaps.*

4. **Remove the bottom panel.** In order to gain access to the heater wire terminals, the bottom panel must be removed (see Figure 14-16). The bottom panel is held on with either two or four screws, depending on the model. Remove the screws and remove the panel.

5. **Test the heating element.** To test the heating element (Figure 14-36), remove the wires from the heating element terminals (Figure 14-37). These are slide-on terminal connectors attached to the ends of the wire. Use the ohmmeter to check for continuity between the two element terminals (Figure 14-38). If the meter indicates no continuity between the terminal ends, replace the heater.

   To check for a shorted-out heating element, take one end of the ohmmeter probe and touch the element terminal; then, with the other probe, touch the sheath (outer cover of element), shown in Figure 14-39. If the meter indicates continuity, the element is shorted out and should be replaced.

6. **Remove the heating element.** To remove the heating element (with the wires already removed from the heater terminals), unscrew the locknuts that hold the element in place. From inside the tub, remove the heating element (Figure 14-40).

7. **Install a new heating element.** To install a new element, just reverse the disassembly procedure and reassemble. Then test the new element by repeating step 1.

Cycle Selector Switch

The dishwasher cycle selector switch allows the consumer to select the various cycles and options for that model dishwasher. When the consumer turns on the timer to start the

PART VI

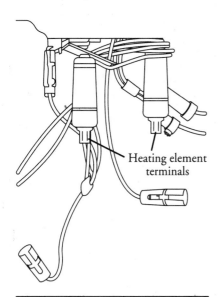

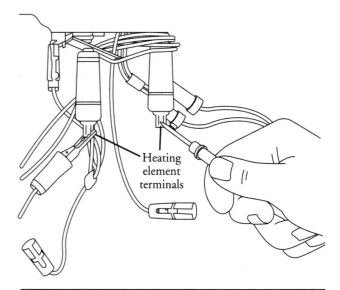

FIGURE 14-37 Remove the wire terminals from the heater element.

FIGURE 14-38 Set the meter on the ohms scale. Connect the probes to the heater terminals.

dishwasher cycle, if the cycle selector switch buttons are not selected correctly, the dishwasher will not perform to the consumer's satisfaction.

The typical complaints associated with the cycle selector switch are:

- Inability to select a different cycle.
- The consumer inadvertently selected the wrong cycle.

FIGURE 14-39
To check for a shorted-out element, attach the meter probe to one terminal and attach the other probe to the sheath.

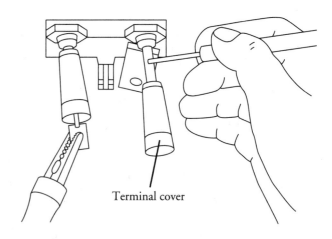

Terminal cover

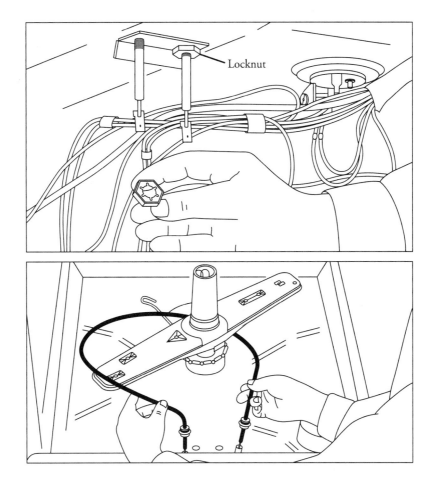

Locknut

When dealing with these complaints, perform the following steps:

1. **Verify the complaint.** As shown in Figure 14-41, verify the complaint by trying to select different cycles. On electronic models, turn off the electricity to the appliance and wait for two minutes before turning it back on. If a fault code appears, look up the code. If the dishwasher will not power up, locate the technical data sheet behind the control panel or under the tub of the dishwasher for diagnostics information. On some models you will need the actual service manual for the model you are working on to properly diagnose the dishwasher. The service manual will assist you in properly placing the dishwasher in the service test mode for testing the dishwasher functions.

2. **Check for external factors.** You must check for external factors not associated with the appliance. For example, is there any physical damage to the component? Did you check for the correct voltage? The voltage at the receptacle is between 108 volts and 132 volts during a load on the circuit. Do you have the correct polarity? (See Chapter 6.)

PART VI

FIGURE 14-41
Selecting a cycle.

3. **Disconnect the electricity.** Before working on the dishwasher, disconnect the electricity. This can be done by pulling the plug out of the wall receptacle. Be sure that you only remove the dishwasher plug. Or disconnect the electricity at the fuse panel or circuit breaker panel. Turn off the electricity.

WARNING *Some diagnostic tests will require you to test the components with the power turned on. When you disassemble the control panel, you can position it in such a way that the wiring will not make contact with metal. This act will allow you to test the components without electrical mishaps.*

4. **Remove the control panel.** To remove the control panel, remove the screws along the top inside edge of the door (see Figure 14-31). On some models, you might have to remove the door latch knob in order to completely remove the control panel.

5. **Test the cycle selector switch.** To test the cycle selector switch, remove all wires from the switch. Just remember—you will have to identify the wires according to the wiring diagram in order to reinstall them back on the cycle selector switch properly. Take your ohmmeter, and check for continuity on the switch contacts; press the switch that coincides with the terminals that are being checked (Figure 14-42). At this point, you have to use the wiring diagram to identify the switch contacts.

6. **Remove the cycle selector switch.** To remove the cycle selector switch, remove the screws that hold the component to the control panel (Figure 14-43).

7. **Reinstall the cycle selector switch.** To reinstall the cycle selector switch, just reverse the disassembly procedure and reassemble. Remember—you will have to identify the wires according to the wiring diagram in order to reinstall them back on the cycle selector switch properly.

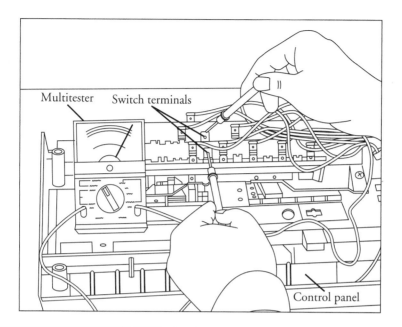

FIGURE 14-42 Testing the selector switch contacts.

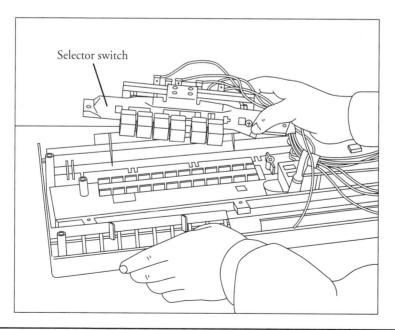

FIGURE 14-43 Removing the selector switch.

Dishwasher Door Switch

Sometimes, the dishwasher door latch assembly and/or switch malfunctions and the door will not close properly, or the door switch will prevent the dishwasher from coming on at all. When dealing with these complaints, perform the following steps:

1. **Verify the complaint.** As shown in Figure 14-44, verify the complaint by closing the dishwasher door and turning the timer dial to start the wash cycle. On electronic models, turn off the electricity to the appliance and wait for two minutes before turning it back on. If a fault code appears, look up the code. If the dishwasher will not power up, locate the technical data sheet behind the control panel or under the tub of the dishwasher for diagnostics information. On some models you will need the actual service manual for the model you are working on to properly diagnose the dishwasher. The service manual will assist you in properly placing the dishwasher in the service test mode for testing the dishwasher functions.

2. **Check for external factors.** You must check for external factors not associated with the appliance. Is the appliance installed properly? Does the appliance have the correct voltage? The voltage at the receptacle is between 108 volts and 132 volts during a load on the circuit. Do you have the correct polarity? (See Chapter 6.) Is there any physical damage to the component?

3. **Disconnect the electricity.** Before working on the dishwasher, disconnect the electricity. This can be done by pulling the plug out of the wall receptacle. Be sure that you only remove the dishwasher plug. Or disconnect the electricity at the fuse panel or circuit breaker panel. Turn off the electricity.

WARNING *Some diagnostic tests will require you to test the components with the power turned on. When you disassemble the control panel, you can position it in such a way that the wiring will not make contact with metal. This act will allow you to test the components without electrical mishaps.*

4. **Remove the control panel.** To remove the control panel, remove the screws along the top inside edge of the door (see Figure 14-31). On some models, you might have to remove the door latch knob in order to completely remove the control panel.

FIGURE 14-44
Closing the door and running the dishwasher.

FIGURE 14-45
The dishwasher latch
assembly in the
closed position.

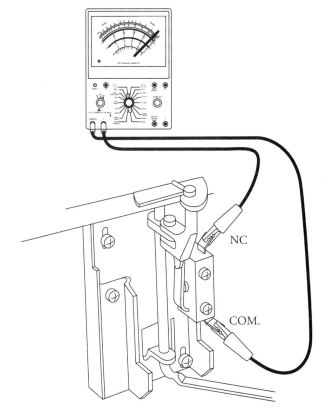

NC

COM.

5. **Test the door switch.** To test the door switch, remove the two wires from the switch (Figure 14-45). Close and latch the door. With your ohmmeter, check for continuity between the two terminals on the switch. Then open the door latch and check for no continuity between the terminals. If the switch fails these tests, replace the switch.

6. **Remove the door switch.** Remove the screws that secure the switch in place. Remove the switch.

7. **Install new door switch.** To install the new door switch, just reverse the disassembly procedure and reassemble. Then reconnect the electricity and test the dishwasher.

Float and Float Switch

Sometimes, the dishwasher float and the float switch malfunctions and water will not enter the tub, or water overfills and spills onto the floor. When dealing with these complaints, perform the following steps:

1. **Verify the complaint.** Verify the complaint by closing the dishwasher door and turning the timer dial to start the wash cycle. On electronic models, turn off the electricity to the appliance and wait for two minutes before turning it back on. If a fault code appears, look up the code. If the dishwasher will not power up, locate the technical data sheet behind the control panel or under the tub of the dishwasher for

diagnostics information. On some models you will need the actual service manual for the model you are working on to properly diagnose the dishwasher. The service manual will assist you in properly placing the dishwasher in the service test mode for testing the dishwasher functions.

2. **Check for external factors.** You must check for external factors not associated with the appliance, as was done in the previous sections. Did you check for the correct voltage? The voltage at the receptacle is between 108 volts and 132 volts during a load on the circuit. Do you have the correct polarity? (See Chapter 6.)

3. **Disconnect the electricity.** Before working on the dishwasher, disconnect the electricity. This can be done by pulling the plug out of the wall receptacle. Be sure that you only remove the dishwasher plug. Or disconnect the electricity at the fuse panel or the circuit breaker panel. Turn off the electricity.

WARNING *Some diagnostic tests will require you to test the components with the power turned on. When you disassemble the control panel, you can position it in such a way that the wiring will not make contact with metal. This act will allow you to test the components without electrical mishaps.*

4. **Check the float.** Check the float from the inside of the tub (Figure 14-46). Be sure that the float moves freely up and down. If there is any soap buildup around the float, clean it off.

FIGURE 14-46
Inspecting the float assembly. Check for soap buildup around the stem and inside of the float.

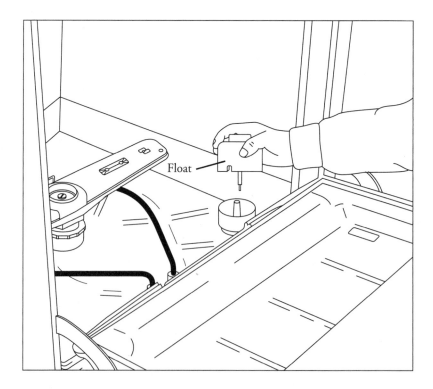

Float

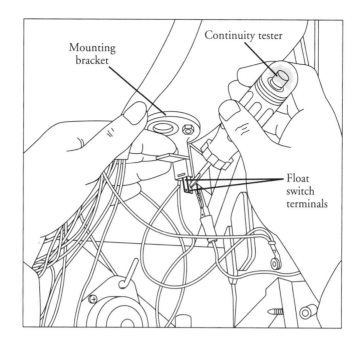

Mounting bracket

Continuity tester

Float switch terminals

5. **Remove the bottom panel.** In order to gain access to the float switch terminals, the bottom panel must be removed (see Figure 14-16). The bottom panel is held on with either two or four screws, depending on the model. Remove the screws, and remove the panel.

6. **Test the float switch.** The float switch is located under the float, underneath the tub (Figure 14-47). Remove the wires from the terminals, and test for continuity. Lift the float—there should be no continuity. Let the float rest, and you should have continuity. If the test fails, replace the float switch.

7. **Remove the float switch.** To remove the float switch, you will have to remove the screws that hold the switch in place (Figure 14-48).

8. **Install a new float switch.** To install the new float switch, just reverse the disassembly procedure and reassemble. Then reconnect the electricity and test a dishwasher cycle that fills the dishwasher.

Wetting Agent and Detergent Dispensers

The dishwasher detergent dispenser will operate by electrical or mechanical means. On some models, the detergent dispenser will operate by a bimetal, motor-driven, or mechanical lever working off the timer assembly. The dishwasher wetting agent dispenser will dispense the wetting agent by an electrical or mechanical lever too. On some models the detergent and wetting agent dispensers are combined into one assembly.

The typical complaints associated with the wetting agent and detergent dispensers are:

- The dishes are dirty.
- The dishes have spots on them.

PART VI

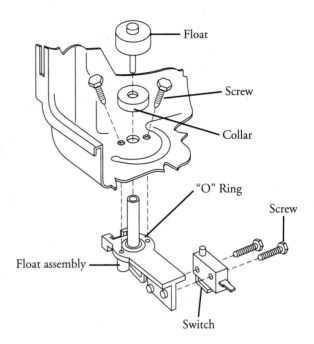

FIGURE 14-48
An exploded view of the float switch assembly.

Float

Screw

Collar

"O" Ring

Screw

Float assembly

Switch

- Detergent remains in the dispenser.
- Wetting agent is leaking onto the floor.

When dealing with these complaints, perform the following steps:

1. **Verify the complaint.** Verify the complaint by operating the dishwasher through its cycles. Make sure that the detergent dispenser door is not blocked by dishes. Also check the wetting agent level in the dispenser. On electronic models, turn off the electricity to the appliance and wait for two minutes before turning it back on. If a fault code appears, look up the code. If the dishwasher will not power up, locate the technical data sheet behind the control panel or under the tub of the dishwasher for diagnostics information. On some models you will need the actual service manual for the model you are working on to properly diagnose the dishwasher. The service manual will assist you in properly placing the dishwasher in the service test mode for testing the dishwasher functions.

2. **Check for external factors.** You must check for external factors not associated with the appliance. For example, is there any physical damage to the component? Is the water in the dishwasher at the correct level? Did the consumer fill the wetting agent dispenser with wetting agent? Did you check the voltage? The voltage at the receptacle is between 108 volts and 132 volts during a load on the circuit. Do you have the correct polarity? (See Chapter 6.)

3. **Disconnect the electricity.** Before working on the dishwasher, disconnect the electricity. This can be done by pulling the plug out of the wall receptacle. Be sure that you only remove the dishwasher plug. Or disconnect the electricity at the fuse panel or circuit breaker panel. Turn off the electricity.

WARNING *Some diagnostic tests will require you to test the components with the power turned on. When you disassemble the control panel, you can position it in such a way that the wiring will not make contact with metal. This act will allow you to test the components without electrical mishaps.*

4. **Remove the control panel.** To remove the control panel, remove the screws along the top inside edge of the door (see Figure 14-31). On some models, you might have to remove the door latch knob in order to completely remove the control panel. Then reconnect the electricity, latch the dishwasher door, and test a dishwasher cycle. A cam follower next to the timer will actuate the detergent dispenser when the wash cycle is in the second wash. The cam follower will engage the draw bar (Figure 14-49) and pull up on the detergent lid latch, releasing the detergent lid and allowing the detergent to enter into the wash cycle. In the final rinse cycle, the same operation occurs to release the wetting agent from the wetting agent dispenser.

Figure 14-49
A side view of the detergent dispenser.

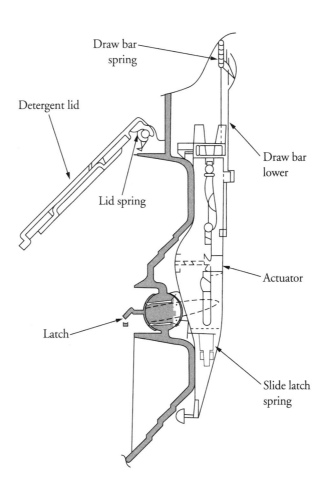

5. **Remove the wetting agent dispenser.** As illustrated in Figure 14-50a, remove the drip cover and drawbar spring. Now begin to remove the upper and lower drawbars by aligning the lower drawbar holes with the locating tabs. In Figure 14-50b, you will note that the locating tabs will align with their respective holes while removing the drawbars. Remember where they reattach when it comes to reassembly later. Begin to slide the drawbars apart, as shown in Figure 14-50c. Next, you will have to remove the wetting agent cap by turning the cap and seal assembly counterclockwise, as in Figure 14-50d. Using a ¾-inch socket wrench, begin to insert it into the wetting agent dispenser locking tabs, shown in Figure 14-50d, to release the wetting agent dispenser from the inner door liner. Now you are ready to remove the wetting agent dispenser actuator. Spread apart the two locking tabs, as shown in Figure 14-50e, and remove the upper and lower slides from the dispenser actuator. Next remove the diaphragm spring and diaphragm (Figure 14-50e). To remove the detergent door and actuator assembly, use a small flat-blade screwdriver to depress and release the lower tabs, noting the position of the door latch level through the hole in the lower slide (Figure 14-50f). Finally, remove the upper slide, slide spring, and the lower slide (Figure 14-50g).

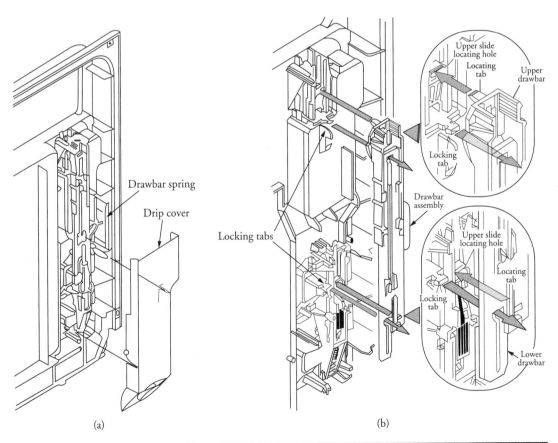

(a)

(b)

Figure 14-50a Remove the drip cover and drawbar spring.

Figure 14-50b Removal of the drawbars. Take note of the alignment of the locating tabs.

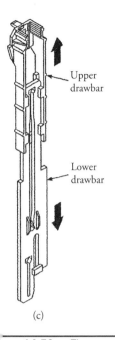

(c)

ᴳᴜʀᴇ 14-50c The upper ᴅd lower drawbars. To ᴍove, follow the ᴇection of the arrows.

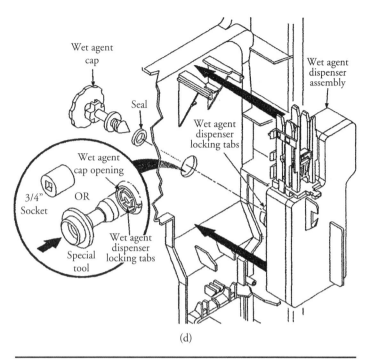

(d)

Fɪɢᴜʀᴇ 14-50d Removal of the wetting agent dispenser.

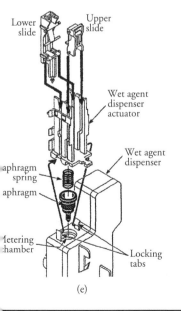

(e)

ᴳᴜʀᴇ 14-50e Removal of the actuator, ᴀphragm spring, and the diaphragm ᴏm the wetting agent dispenser.

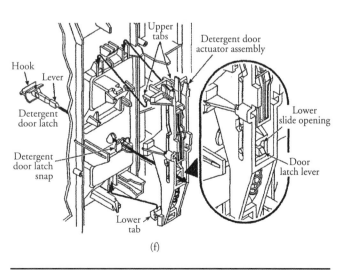

(f)

Fɪɢᴜʀᴇ 14-50f Removal of the detergent door actuator assembly and detergent door latch. Remember the tab alignment when dissembling and reassembling.

Figure 14-50g
The upper slide, slide spring, lower slide, detergent door actuator, and spring will be removed last in disassembly; it will be the first step in reassembly.

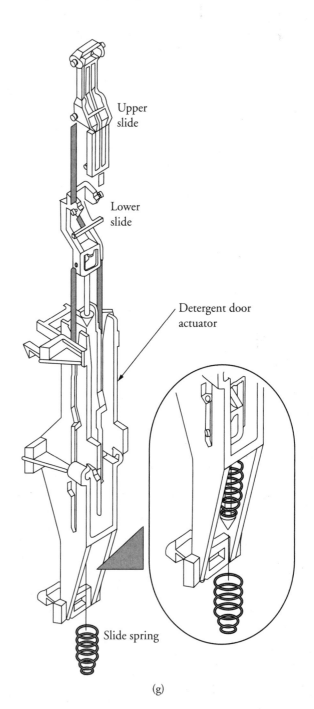

Upper slide

Lower slide

Detergent door actuator

Slide spring

(g)

6. **Install new wetting agent and detergent dispenser.** To install the new dispenser, just reverse the disassembly procedure and reassemble. Then reconnect the electricity and test the dishwasher.

Water Heater Operating Thermostat or Thermistor

The bimetal thermostat is secured and located underneath the tub. The purpose of the thermostat is to either act as a safety net to prevent overheating or is used to raise the water temperature in the tub. The thermistor is used in some models to sense the water temperature on certain cycles. The thermistor sends signals back to the electronic control.

The typical complaints associated with the water heater operating thermostat are:

- Poor washability
- The dishwasher timer will not advance.
- Low inlet water temperature.
- Dishwasher water line hooked up to the cold water line.

When dealing with these complaints, perform the following steps:

1. **Verify the complaint.** Verify the complaint by operating the dishwasher through its cycles. Take the temperature of the water in the dishwasher tub when the dishwasher is in the wash cycle. Also check the hot water temperature at the sink. It should be between 120 to 140 degrees Fahrenheit. On electronic models, turn off the electricity to the appliance and wait for two minutes before turning it back on. If a fault code appears, look up the code. If the dishwasher will not power up, locate the technical data sheet behind the control panel or under the tub of the dishwasher for diagnostics information. On some models you will need the actual service manual for the model you are working on to properly diagnose the dishwasher. The service manual will assist you in properly placing the dishwasher in the service test mode for testing the dishwasher functions.

2. **Check for external factors.** You must check for external factors not associated with the appliance. Test the hot water temperature at the sink. The water temperature should be around 120 to 140 degrees Fahrenheit. The purpose of this thermostat is to delay the wash cycle until the water temperature reaches between 135 and 145 degrees Fahrenheit. The rule of thumb is it takes 20 minutes to raise the water temperature two degrees. The thermostat will then close and complete the timer motor circuit, allowing the dishwasher cycle to continue. Did you check the voltage? The voltage at the receptacle is between 108 volts and 132 volts during a load on the circuit. Do you have the correct polarity? (See Chapter 6.)

3. **Disconnect the electricity.** Before working on the dishwasher, disconnect the electricity. This can be done by pulling the plug out of the wall receptacle. Be sure that you only remove the dishwasher plug. Or disconnect the electricity at the fuse panel or circuit breaker panel. Turn off the electricity.

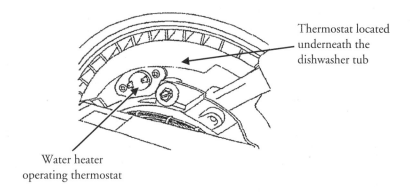

FIGURE 14-51
The heater operating thermostat will allow the water to heat up to 145 degrees Fahrenheit.

Thermostat located underneath the dishwasher tub

Water heater operating thermostat

WARNING *Some diagnostic tests will require you to test the components with the power turned on. When you disassemble the control panel, you can position it in such a way that the wiring will not make contact with metal. This act will allow you to test the components without electrical mishaps.*

4. **Remove the bottom panel.** In order to gain access to the water heater operating thermostat terminals, the bottom panel must be removed (see Figure 14-16). The bottom panel is held on with either two or four screws, depending on the model. Remove the screws, and remove the panel. Inspect the wiring connections first.

5. **Remove and test the water heater operating thermostat.** Locate the thermostat (Figure 14-51) underneath the tub, and remove the wires from the terminals. Remove the thermostat from the dishwasher tub. Take a metal pot of water, and heat it up on the range until the temperature reaches between 150 and 160 degrees Fahrenheit. Next, set your multimeter to the ohm scale. Test the thermostat before placing it against the metal pot—it should read infinite resistance. Now place the thermostat against the metal pot of hot water—the thermostat should close and the multimeter will read continuity. If not, replace the thermostat.

6. **Remove and test the thermistor.** Locate the thermistor underneath the tub, and remove the wires from the terminals. Test the thermistor using the multimeter, and set the meter on the ohm scale. Write down the resistance reading and check it against the technical data sheet for the proper rating. If the resistance does not match the technical data sheet, replace it.

7. **Installing a new water heater operating thermostat or thermistor.** To install the new thermostat or thermistor, just reverse the disassembly procedure and reassemble. Then reconnect the electricity and test the dishwasher.

High-Limit Heater Thermostat

The high-limit heater thermostat is a protective device that will prevent the heater from overheating.

The typical complaints associated with the high-limit heater thermostat are:

- The dishes are extremely hot at the end of the cycle.
- It smells as if something is burning.
- Excessive steam is coming from the vent.

When dealing with these complaints, perform the following steps:

1. **Verify the complaint.** Verify the complaint by operating the dishwasher through its cycles. The purpose of this thermostat is to act as a safety switch to disconnect the heater in the event the temperature rises above 185 degrees Fahrenheit. On electronic models, turn off the electricity to the appliance and wait for two minutes before turning it back on. If a fault code appears, look up the code. If the dishwasher will not power up, locate the technical data sheet behind the control panel or under the tub of the dishwasher for diagnostics information. On some models you will need the actual service manual for the model you are working on to properly diagnose the dishwasher. The service manual will assist you in properly placing the dishwasher in the service test mode for testing the dishwasher functions.

2. **Check for external factors.** You must check for external factors not associated with the appliance. Did you check the voltage? The voltage at the receptacle is between 108 volts and 132 volts during a load on the circuit. Do you have the correct polarity? (See Chapter 6.)

3. **Disconnect the electricity.** Before working on the dishwasher, disconnect the electricity. This can be done by pulling the plug out of the wall receptacle. Be sure that you only remove the dishwasher plug. Or disconnect the electricity at the fuse panel or circuit breaker panel. Turn off the electricity.

WARNING *Some diagnostic tests will require you to test the components with the power turned on. When you disassemble the control panel, you can position it in such a way that the wiring will not make contact with metal. This act will allow you to test the components without electrical mishaps.*

4. **Remove the bottom panel.** In order to gain access to the high-limit heater thermostat terminals, the bottom panel must be removed (see Figure 14-16). The bottom panel is held on with either two or four screws, depending on the model. Remove the screws, and remove the panel. Inspect the wiring connections first.

5. **Remove and test the high-limit heater thermostat.** Locate the high-limit thermostat (Figure 14-52) underneath the tub, and remove the wires from the terminals. Remove the thermostat from the dishwasher tub. Take a metal pot of water, and heat it up on the range until the temperature reaches 185 degrees Fahrenheit. Next, set your multimeter to the ohm scale. Test the thermostat before placing it against the metal pot of hot water—the multimeter will read continuity. Now place the thermostat bimetal section against the metal pot of hot water—the thermostat should open and the multimeter will read infinite resistance. If not, replace the thermostat.

6. **Install a new high-limit heater thermostat.** To install the new thermostat, just reverse the disassembly procedure and reassemble. Then reconnect the electricity and test the dishwasher.

Figure **14-52**
The high-limit heater
thermostat location.

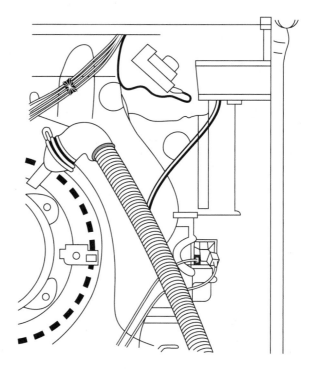

Figure 14-52
The high-limit heater thermostat location.

Dishwasher Motor Start Relay

The dishwasher motor start relay is a solenoid-activated relay used to start the dishwasher motor in some models.

The typical complaints associated with the motor start relay are:

1. The dishwasher motor will not start.

2. The dishwasher motor hums and shuts off on overload.

When dealing with these complaints, perform the following steps:

1. **Verify the complaint.** Verify the complaint by operating the dishwasher. If the motor does not start, turn the dishwasher off immediately. On electronic models, turn off the electricity to the appliance and wait for two minutes before turning it back on. If a fault code appears, look up the code. If the dishwasher will not power up, locate the technical data sheet behind the control panel or under the tub of the dishwasher for diagnostics information. On some models you will need the actual service manual for the model you are working on to properly diagnose the dishwasher. The service manual will assist you in properly placing the dishwasher in the service test mode for testing the dishwasher functions.

2. **Check for external factors.** You must check for external factors not associated with the appliance. Is there electricity to the dishwasher? The voltage at the receptacle is between 108 volts and 132 volts during a load on the circuit. Do you have the

correct polarity? (See Chapter 6.) The purpose of the motor start relay is to protect the dishwasher motor. This current-sensitive relay will close the contacts when the current increases for about one or two seconds when the motor begins to start. When the dishwasher motor starts and runs, the current will decrease in the motor start relay, thus opening the contacts.

3. **Disconnect the electricity.** Before working on the dishwasher, disconnect the electricity. This can be done by pulling the plug out of the wall receptacle. Be sure that you only remove the dishwasher plug. Or disconnect the electricity at the fuse panel or circuit breaker panel. Turn off the electricity.

WARNING *Some diagnostic tests will require you to test the components with the power turned on. When you disassemble the control panel, you can position it in such a way that the wiring will not make contact with metal. This act will allow you to test the components without electrical mishaps.*

4. **Remove the bottom panel.** In order to gain access to the water heater operating thermostat terminals, the bottom panel must be removed (see Figure 14-16). The bottom panel is held on with either two or four screws, depending on the model. Remove the screws, and remove the panel. Inspect the wiring connections first.

5. **Remove and test the motor start relay.** Remove the motor start relay from the dishwasher (Figure 14-53) by removing the screw that holds the relay in place. Disconnect the wires from the relay. Set your ohmmeter to the R x 1 scale, and measure the coil's resistance. The coil will read less than 1 ohm if it is good. Next, attach your ohmmeter probes to the relay contacts; you will check the relay contacts for continuity by turning the relay upside down. If you do not know which way is up or down on the relay, there is an arrow and the word "up" stamped on the relay. The motor start relay contacts should be closed, and the ohmmeter will read zero resistance. With the ohmmeter probes still attached to the relay, turn the relay right-side-up to check the continuity of the contacts—the relay contacts will be open, and the ohmmeter will read infinite resistance.

FIGURE 14-53
The dishwasher motor start relay is located behind the lower front panel.

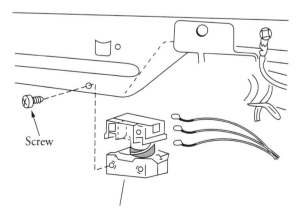

Screw

6. **Install a new dishwasher motor start relay.** To install the new motor start relay, just reverse the disassembly procedure and reassemble. Then reconnect the electricity and test the dishwasher.

Electronic Control Board, Display Board, and Touch Pad Panel

The electronic control board monitors all of the functions and cycles. When servicing this component, check all wiring for loose connections at the wiring harness, connector plugs, and the pins. Loose connections are common and most likely will cause the dishwasher to operate erratically.

The typical complaints associated with the electronic control board, display board, and touch pad panel are:

- Dishwasher won't run or power up.

- Unable to program the dishwasher.

- The display board will not display anything.

- One or more key pads will not accept commands.

- Unusual display readouts and/or error codes.

When dealing with these complaints, perform the following steps:

NOTE *To prevent electrostatic discharge (ESD) from damaging expensive electronic components, follow the steps in Chapters 6, 9, and 11.*

1. **Verify the complaint.** Verify the complaint by operating the dishwasher. Turn off the electricity to the appliance and wait for two minutes before turning it back on. If a fault code appears, look up the code. If the dishwasher will not power up, locate the technical data sheet behind the control panel or under the tub of the dishwasher for diagnostics information. On some models you will need the actual service manual for the model you are working on to properly diagnose the dishwasher. The service manual will assist you in properly placing the dishwasher in the service test mode for testing the dishwasher functions.

2. **Check for external factors.** You must check for external factors not associated with the appliance. Is there electricity to the dishwasher? The voltage at the receptacle is between 108 volts and 132 volts during a load on the circuit. Do you have the correct polarity? (See Chapter 6.) Check for a blown thermal fuse in the circuit.

3. **Disconnect the electricity.** Before working on the dishwasher, disconnect the electricity. This can be done by pulling the plug out of the wall receptacle. Be sure that you only remove the dishwasher plug. Or disconnect the electricity at the fuse panel or circuit breaker panel. Turn off the electricity.

WARNING *Some diagnostic tests will require you to test the components with the power turned on. When you disassemble the control panel, you can position the panel against the door so that the wiring will not make contact with the door. This act will allow you to test the components without electrical mishaps.*

4. **Remove the control panel.** To remove the control panel, remove the screws along the top inside edge of the door (see Figure 14-31). On some models, you might have to remove the door latch knob in order to completely remove the control panel.

5. **Test the electronic control board, display board, or touch pad panel.** If you are able to run the dishwasher diagnostic test mode, check the different functions of the dishwasher. Use the technical data sheet for the model you are servicing to locate the test points from the wiring schematic. Do not forget to check all wiring connections and wiring. Using the technical data sheet, test the key pad matrix, display board LEDs, input voltages, and output voltages on the electronic control board.

6. **Remove the electronic control board, display board, or touch pad panel.** To remove the defective component, remove the covers that shield the control, and then remove the screws that secure the boards to the control panel (Figure 14-54). Disconnect the connectors from the electronic control board, display board, or the touch pad panel. On some models, the touch pad panel might be glued to the control panel. Once you remove the defective part, you will have to replace it with a new one.

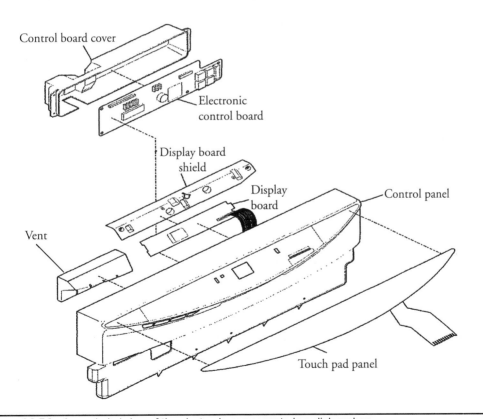

FIGURE 14-54 An exploded view of the electronic components in a dishwasher.

7. **Install the new component.** To install a new electronic control board, display board, or touch pad panel, just reverse the disassembly procedure and reassemble. Reinstall the console panel, and restore the electricity to the dishwasher. Test the dishwasher operation.

Turbidity Sensor

The turbidity sensor (Figure 14-55a and b) is located on the motor and pump assembly. Its main purpose is to measure the amount of suspended food particles in the wash water. If it senses food particles, on some models, an extra wash cycle will be added to the length of the dishwashing cycle.

The typical complaints associated with the turbidity sensor are:

- The dishes are not coming out clean at the end of the cycle.
- Food remains on the dishes.

1. **Verify the complaint.** Verify the complaint by operating the dishwasher through its cycles. On electronic models, turn off the electricity to the appliance and wait for two minutes before turning it back on. If a fault code appears, look up the code. If the dishwasher will not power up, locate the technical data sheet behind the control

Turbidity sensor

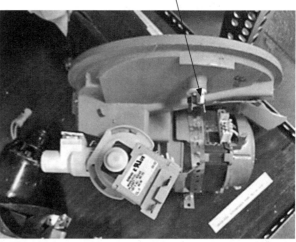

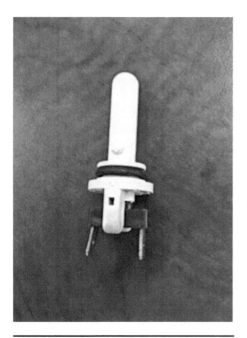

FIGURE 14-55b A turbidity sensor located in the pump assembly.

FIGURE 14-55a A typical turbidity sensor used to detect food particles in the wash water.

panel or under the tub of the dishwasher for diagnostics information. On some models you will need the actual service manual for the model you are working on to properly diagnose the dishwasher. The service manual will assist you in properly placing the dishwasher in the service test mode for testing the dishwasher functions.

2. **Check for external factors.** You must check for external factors not associated with the appliance. Is the appliance installed properly? Does the appliance have the correct voltage? The voltage at the receptacle is between 108 volts and 132 volts during a load on the circuit. Do you have the correct polarity? (See Chapter 6.) Is the water hot enough?

3. **Disconnect the electricity.** Before working on the dishwasher, disconnect the electricity. This can be done by pulling the plug out of the wall receptacle. Be sure that you only remove the dishwasher plug. Or disconnect the electricity at the fuse panel or at the circuit breaker panel. Turn off the electricity.

WARNING *Some diagnostic tests will require you to test the components with the power turned on. When you disassemble the control panel, you can position it in such a way that the wiring will not make contact with metal. This act will allow you to test the components without electrical mishaps.*

4. **Remove the bottom panel.** In order to gain access to the pump and motor assembly, the bottom panel must be removed (see Figure 14-16). The bottom panel is held on with either two or four screws, depending upon the model. Remove the screws and remove the panel.

5. **Test and remove the turbidity sensor.** After locating the turbidity sensor (Figure 14-55b), remove the wires and test the sensor, using the ohms scale on the multimeter. Verify the reading against the technical data sheet or the service manual for proper specifications. Before removing the turbidity sensor, place a pan underneath to collect the water. To remove the sensor on this model, just twist and turn and pull out.

6. **Reinstall the turbidity sensor.** To reinstall the turbidity sensor, just reverse the order of step 5. Before restoring the electricity to the dishwasher, pour a gallon of water into the tub and check for leaks underneath the tub. Restore the electricity to the dishwasher, and run the dishwasher through a cycle. Check for leaks again. If no leaks are found, reinstall the bottom panel.

Diagnostic Charts

The following diagnostic charts will help you to pinpoint the likely causes of the various dishwasher problems (Figures 14-56 through 14-65). Schematics are included in Figures 14-66 through 14-81.

FIGURE 14-56
Dishwasher
diagnostic flowchart:
Dishwasher will not
run at all.

Dishwasher will not run at all

Check for voltage at the outlet. Do you have 115 volts? — No → Check fuse or circuit breakers — No → Replace fuse or reset circuit breakers

Yes ↓ (from fuse/circuit breakers)

Check for loose wire connections or burnt wires — No → Repair or replace loose connections or burnt wires

Yes ↓

Check door switch and latch assembly — No → Replace switch or adjust latch assembly

Yes ↓

Check timer — No → Replace timer

FIGURE 14-57
Dishwasher diagnostic
flowchart: Water not
entering the
dishwasher.

Water not entering dishwasher

Is the water supply turned on to d/w? — No → Turn on water supply

Yes ↓

Turn off power. Check continuity of water valve coil. — No → Replace water valve

Yes ↓

Turn on power. Start dishwasher. Check for voltage at water valve. — No → Check float switch. Replace if necessary. — OK → Check wiring harness. If okay replace the timer.

Yes ↓

Operate the water valve until it is warm. Does the coil open when warm? — Yes → Replace water valve.

No → Clean water inlet screen in valve body.

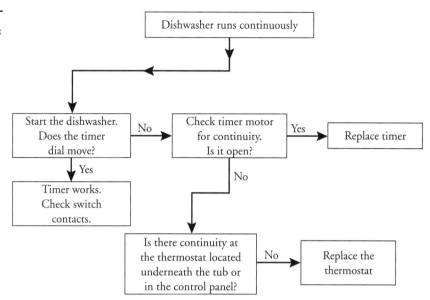

FIGURE 14-58
Dishwasher diagnostic flowchart: Dishwasher runs continuously.

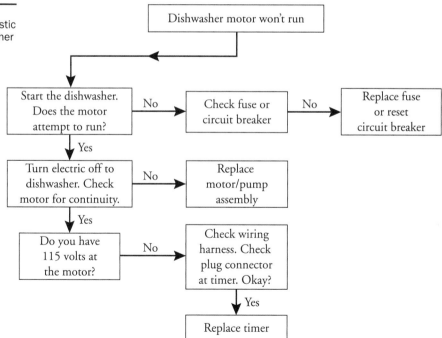

FIGURE 14-59
Dishwasher diagnostic flowchart: Dishwasher motor won't run.

Figure 14-60
Dishwasher diagnostic
flowchart: Dishwasher
won't dry.

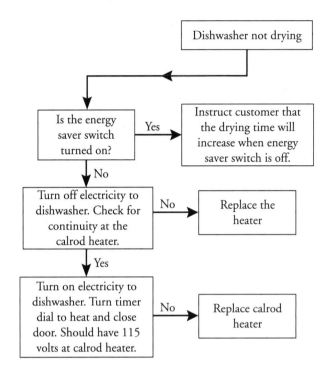

URE 14-61
hwasher diagnostic
wchart: Water in the
at the end of the
le.

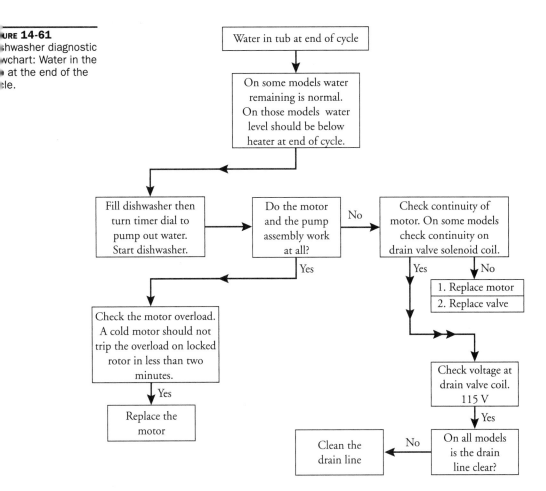

FIGURE 14-62
Dishwasher diagnostic
flowchart: Low water
level.

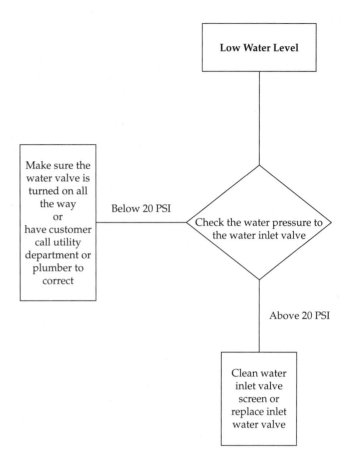

FIGURE **14-63**
Dishwasher diagnostic
flowchart: Low water
temperature.

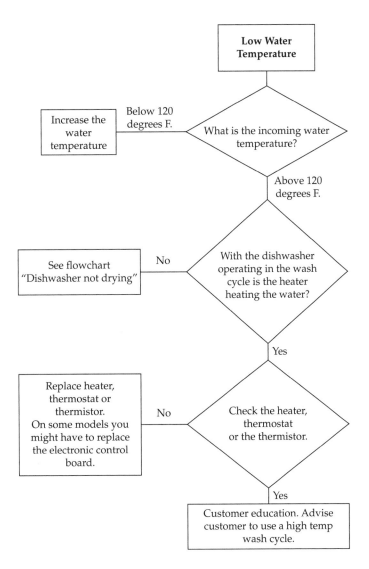

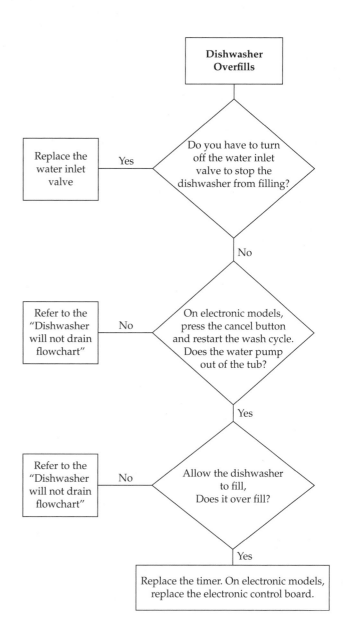

URE **14-65**
shwasher diagnostic
wchart: Dishwasher
not drain.

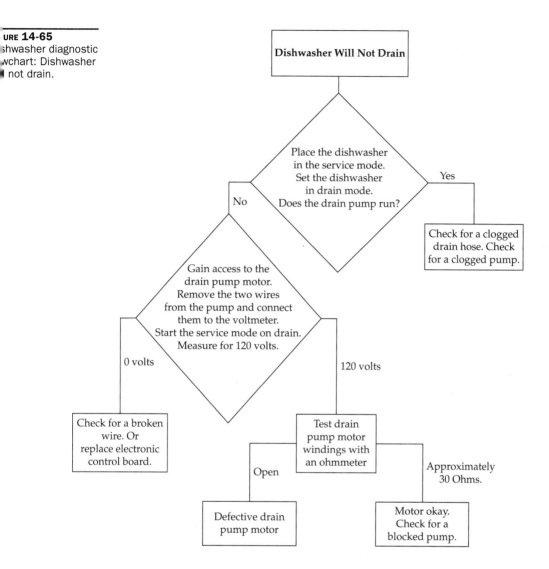

Water valve noise chart

Complaint				
Noise	Description	When noise occurs	Which water valve parts	External factors
Whistle	Makes a "hissing" noise to a shrill whistle.	While the water valve is open.	Outlet assembly and hose to fill funnel, flow-washer.	* Close or open the supply water valve to change the pressure band. * Water heater temperature too high. Set the water heater thermostat to 140 degrees Fahrenheit.
Flutter	Low-frequency "rumble."	When the water valve is about to close.	Flow-washer.	
Chatter	Low-frequency pulsing.	When the water valve closes.	Spring, diaphragm.	* Long runs of unsupported pipes. You must secure the pipes to stop any vibrations. * Water supply line to the water valve is undersized. Must have 3/8" I.P.S. or 1/4" I.D. * Water heater temperature is too high. Set the water heater thermostat at 150 degrees Fahrenheit maximum.
Water hammer	The water valve will make a single "thud."	At closing.	Armature spring, diaphragm.	

FIGURE 14-66　Water valve noise chart.

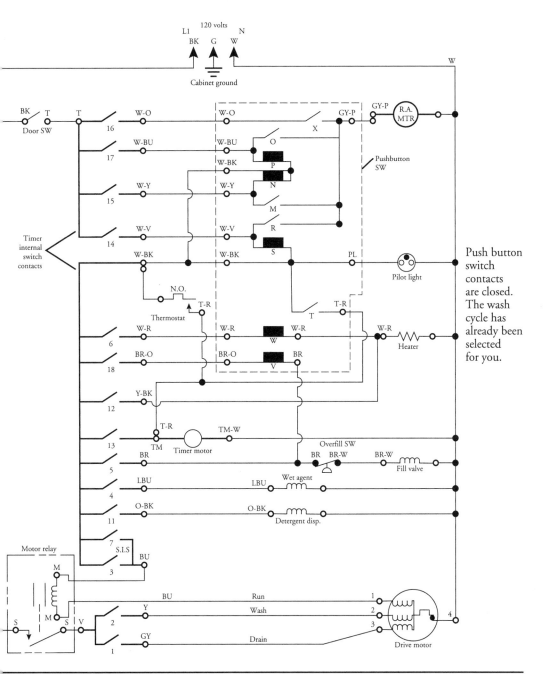

FIGURE 14-67 The typical dishwasher electrical schematic: The wash cycle has already been selected, and the pushbutton switch contacts are closed.

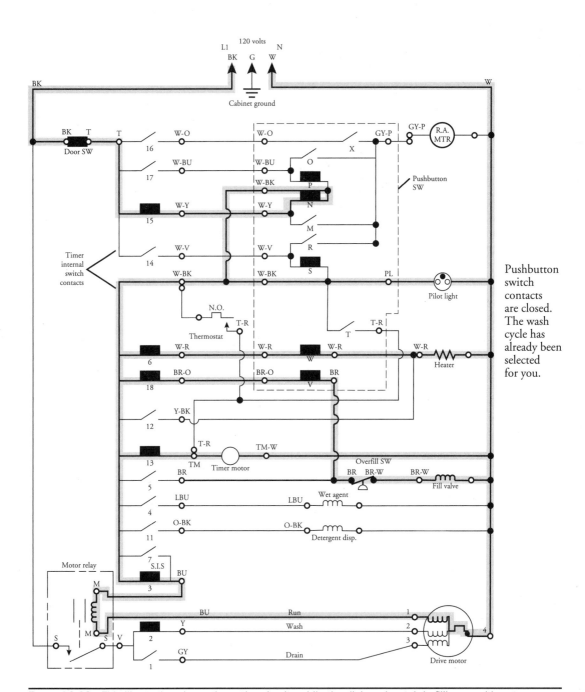

FIGURE 14-68   This illustration shows the active circuits while the dishwasher tub is filling up with water.

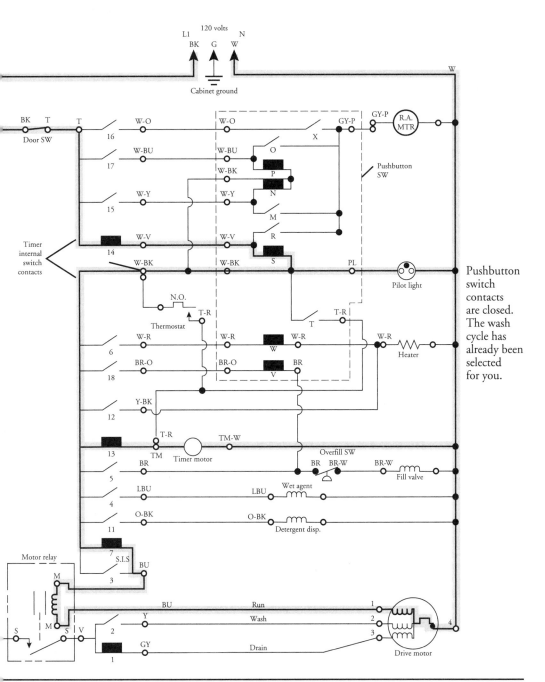

URE 14-69 This illustration shows the active circuits while the dishwasher is draining the water out of the tub.

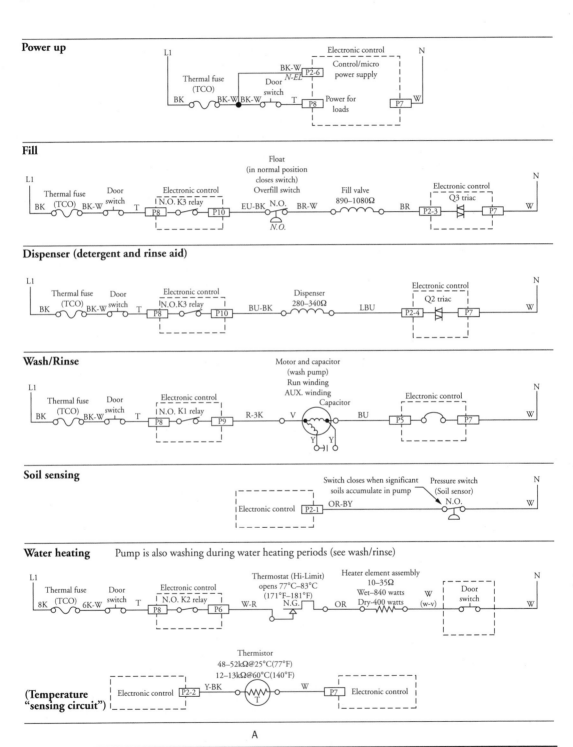

A

FIGURE 14-70 A typical strip circuit wiring diagram for a dishwasher with electronic components. *(continued)*

rbo zone activation   Used only on some models, otherwise open, pump is also washing during sensing periods (see wash/rinse)

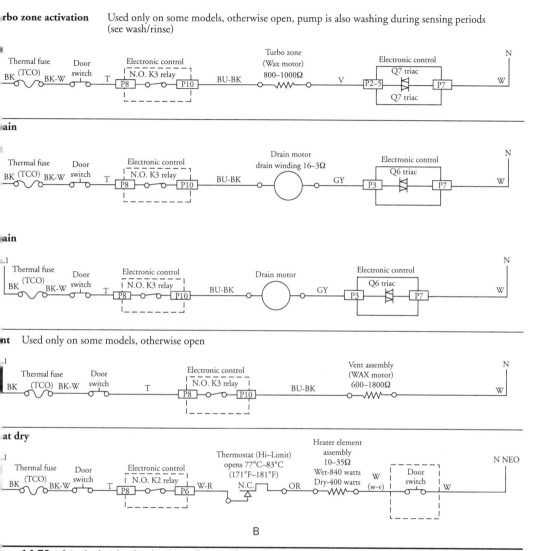

ain

ain

nt   Used only on some models, otherwise open

at dry

B

URE 14-70   A typical strip circuit wiring diagram for a dishwasher with electronic components.

Figure 14-71
A typical wiring diagram for an electronic dishwasher model.

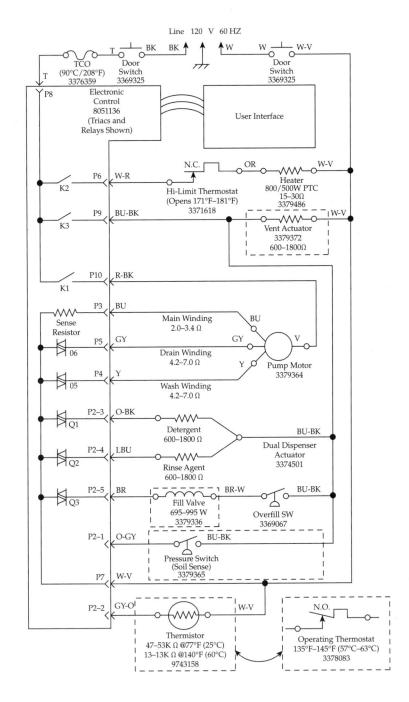

FIGURE 14-72
A typical strip circuit
of the fill cycle for an
electronic dishwasher.

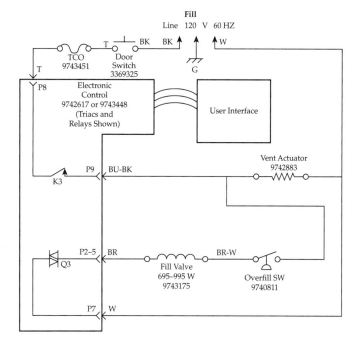

FIGURE 14-73
A typical strip circuit
for the wash circuit in
an electronic
dishwasher.

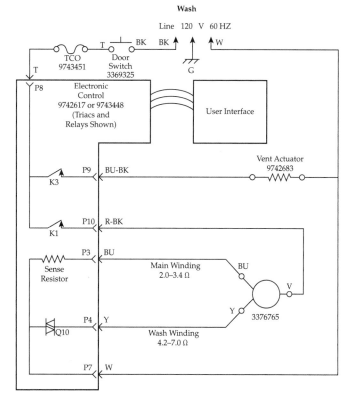

PART VI

Figure 14-74
A typical strip circuit for the water heating circuit for an electronic dishwasher.

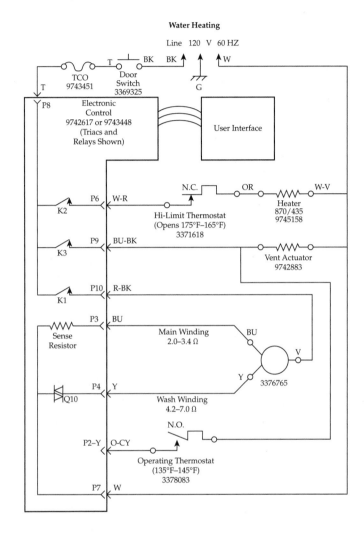

FIGURE 14-75

A typical strip circuit for the drain cycle in an electronic dishwasher.

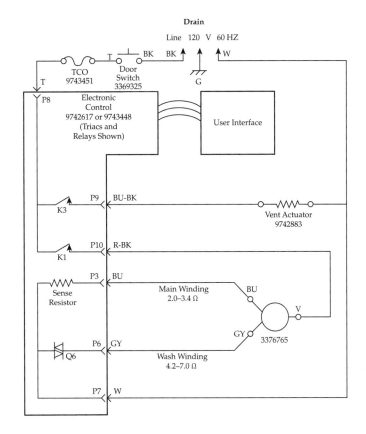

FIGURE 14-76

A typical strip circuit for the air dry in an electronic dishwasher.

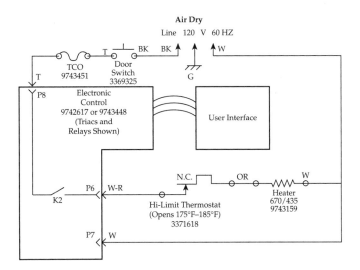

PART VI

FIGURE 14-77

A typical strip circuit
of the dry circuit in an
electronic dishwasher.

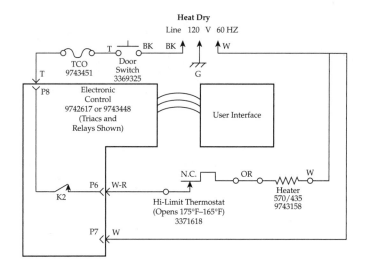

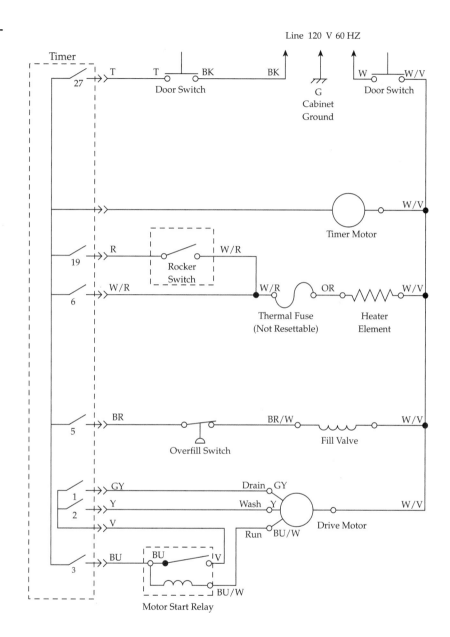

FIGURE 14-78
A typical dishwasher
wiring diagram with
a mechanical timer.

Line 120 V 60 HZ

Timer

27   T   T   BK   BK   Door Switch

G
Cabinet
Ground

W   W/V   Door Switch

Timer Motor   W/V

19   R   Rocker Switch   W/R

6   W/R   W/R   Thermal Fuse (Not Resettable)   OR   Heater Element   W/V

5   BR   Overfill Switch   BR/W   Fill Valve   W/V

1   GY   Drain   GY
2   Y   Wash   Y
     V   Run   BU/W   Drive Motor   W/V

3   BU   BU   V
          BU/W
Motor Start Relay

FIGURE 14-79
A typical timing
sequence chart for a
dishwasher. This chart
and the wiring diagram
in Figure 14-78 are
used together for
diagnosing
dishwasher faults.

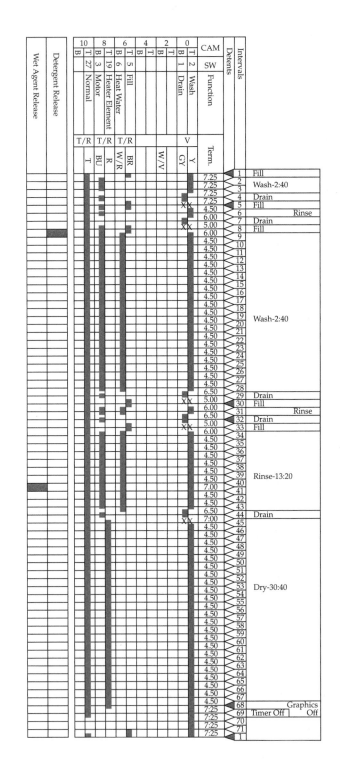

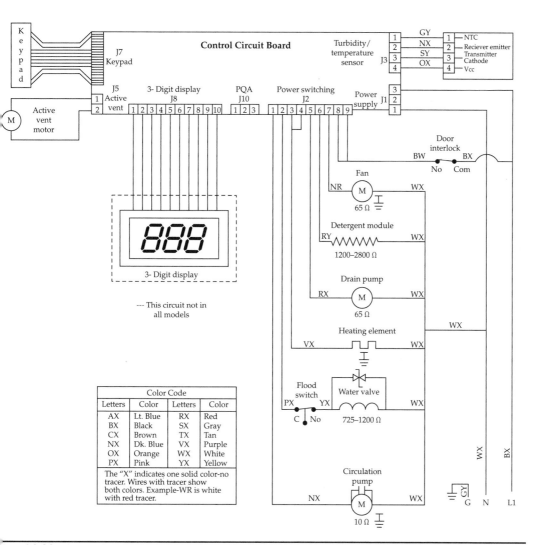

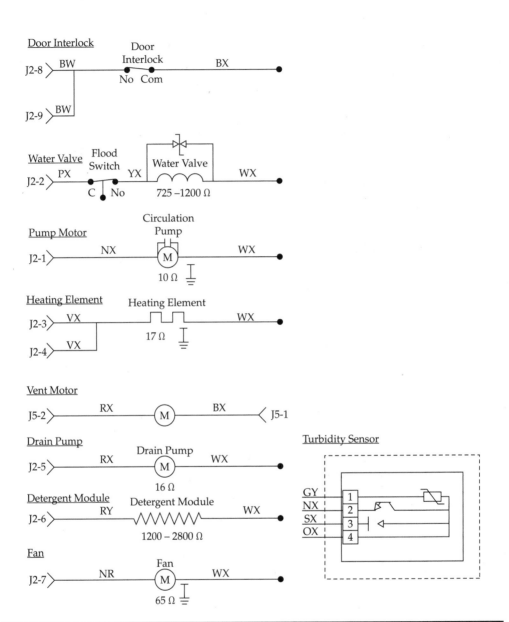

Figure 14-81 A typical strip circuit for the wiring diagram in Figure 14-80.

Garbage Disposers

The garbage disposer provides a convenient and sanitary way to dispose of food waste (Figure 15-1). No sorting or separating of the food waste is necessary. It is designed to handle all types of food waste. If, however, the garbage disposer drains into a septic tank, organic wastes, such as egg shells, lobster, crab, and shrimp shells should be kept at a minimum. Items such as tin cans, glass, china, bottle caps, metal, etc., should never be placed in the garbage disposer because they might damage the appliance and plug the drain.

Types

There are two types of operating methods available in garbage disposers: continuous feed and batch feed.

Continuous Feed Disposers

A continuous feed disposer requires an on-off electrical switch that is remote from the disposer. Before food is placed into the hopper, the user must turn on the cold water faucet and turn on the electrical switch. Then the food waste is placed into the hopper. The food waste will be ground up and expelled into the drain.

Batch Feed Disposers

In a batch feed disposer, the on-off operation of the disposer is controlled by the stopper. The on-off switch is built into the hopper. The stopper is the component that completes the circuit that allows the disposer motor to run. The hopper will hold the food waste until the user is ready to dispose of it. The food waste will be ground up and expelled into the drain.

Principles of Operation

When the food is placed into the hopper and the disposer is running, the centrifugal force throws the food waste outward against the cutting edges of the shredder ring (Figure 15-2). The pivoting impeller arms, attached to the flywheel, push the food waste around and into the teeth of the shredder ring (Figure 15-3). As the food waste is pushed against and cut by the shredder ring, water running into the hopper flushes the ground food waste between the flywheel and the shredder ring, washing it into the drain housing assembly, where it is expelled into the drain.

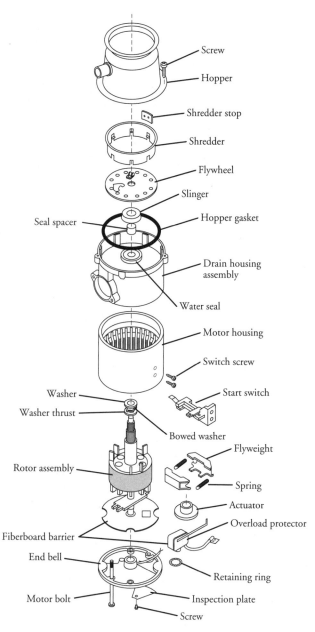

Screw

Hopper

Shredder stop

Shredder

Flywheel

Slinger

Seal spacer

Hopper gasket

Drain housing assembly

Water seal

Motor housing

Switch screw

Washer

Start switch

Washer thrust

Bowed washer

Flyweight

Rotor assembly

Spring

Actuator

Overload protector

Fiberboard barrier

End bell

Motor bolt

Retaining ring

Inspection plate

Screw

FIGURE 15-1   Exploded view of a batch feed garbage disposer.

FIGURE 15-2   Food waste is forced outward against the shredder ring.

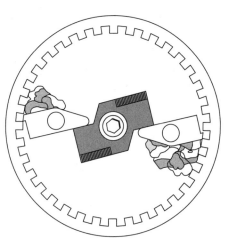

FIGURE 15-3   Impeller directing food waste into the teeth of the shredder ring.

afety First

Any person who cannot use basic tools or follow written instructions should not attempt to install, maintain, or repair any garbage disposers. Any improper installation, preventive maintenance, or repairs could create a risk of personal injury or property damage.

If you do not fully understand the installation, preventive maintenance, or repair procedures in this chapter, or if you doubt your ability to complete the task on the garbage disposer, call your service manager.

These precautions should also be followed:

- Do not put your fingers or hands into the garbage disposer.
- When removing foreign objects, always use a long-handled pair of tongs or pliers.
- To reduce the risk of flying debris that can be expelled by the disposer, always be sure that the splash guard is properly installed.
- When replacing the garbage disposer, always be sure that it is properly grounded.
- Before attempting to free a jam in the garbage disposer, always disconnect the electric supply to the disposer.

arbage Disposers in General

Much of the troubleshooting information in this chapter covers the various types of garbage disposers in general, rather than specific models, in order to present a broad overview of service techniques. The pictures and illustrations that are used in this chapter are for demonstration purposes only, to clarify the description of how to service these units, and in no way reflect on a particular brand's reliability.

To Free Jams from Foreign Objects

When foreign material falls into the disposer and the motor jams, it will trip the motor overload protector—otherwise known as the reset button (see Figure 15-1). When this happens, follow these steps to free the jam:

1. Turn off the electricity to the disposer, and shut off the water.
2. Insert the service wrench into the center hole at the bottom of the disposer (Figure 15-4), and work the wrench back and forth until the motor assembly is turning freely.
3. To free a tight jam, insert a prying tool alongside the grinding protrusion near the outside edge of the flywheel. Be sure that you place the prying tool on the proper side of the protrusion so that when pressure is applied, the flywheel will move (Figure 15-5).
4. Remove the foreign object with tongs.
5. Wait for the motor to cool for a few minutes before you press the reset button.
6. Reconnect the electric supply, turn on the cold water, and test the disposer.

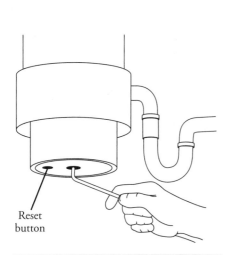

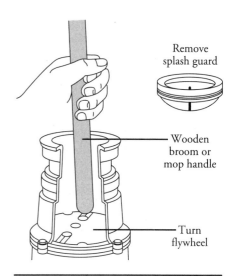

Remove splash guard

Wooden broom or mop handle

Turn flywheel

Reset button

FIGURE 15-4   Insert service wrench, turn wrench, and reset the overload protector button.

FIGURE 15-5   Insert prying tool into the hopper.

Installation of Garbage Disposer

Figure 15-6 shows some typical installations of garbage disposers. Every newly purchased garbage disposer comes with installation instructions, a use and care manual, and a warranty. The steps taken to install a new disposer are as follows:

1. Read the use and care manual.

2. Clean out the sink's drain line.

3. Disconnect the electrical supply:

    a. Continuous feed disposers need a wall switch and a receptacle.

    b. Batch feed disposers need a receptacle or must be wired directly.

4. Be sure to observe all local codes and ordinances for electrical and plumbing connections when installing or repairing disposers.

5. Follow the manufacturer's installation instructions.

6. After completing the installation, check for water leaks.

7. Turn on the electricity, and test the operation of the garbage disposer.

Garbage Disposer Maintenance

The garbage disposer is permanently lubricated; thus, it never has to be oiled. When the disposer is used properly, it cleans itself. If there is an odor coming from the inside of the disposer, you can deodorize it. To do this, take some orange or lemon rinds and grind them up in the disposer. This will dispel unpleasant odors and leave the sink with a sweet smell. Another way to deodorize the disposer is to take about a dozen or so ice cubes sprinkled with a generous amount of household scouring power and grind them up in the disposer

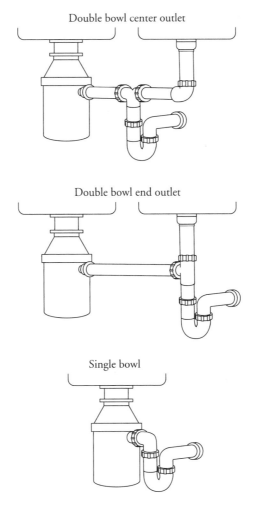

Double bowl center outlet

Double bowl end outlet

Single bowl

without running the water. Flush the disposer for one minute. This will allow any debris to be expelled into the drain.

Drain Blockage

To avoid drain blockage when using the garbage disposer, allow the cold water to flow for a sufficient time after grinding the food waste to be sure that all of the waste is flushed away. The ground waste and water mixture flows at the rate of two seconds per foot in a horizontal drain line. It is recommended that the user allow the water to flow for a minimum of 15 to 30 seconds after grinding the food waste.

The use of cold water in the garbage disposer will congeal and harden the grease, making its disposal easier. Never use chemicals or solvent drain compounds, because they can cause serious damage to the disposer.

The disposer should be used daily to flush the lines. If the dishwasher drain hose is connected to the disposer, it, too, should be used daily to prevent the dishwasher drain hose from becoming clogged.

PART VI

Step-by-Step Troubleshooting by Symptom Diagnosis

In the course of servicing an appliance, don't tend to overlook the simple things that might be causing the problem. Step-by-step troubleshooting by symptom diagnosis is based on diagnosing malfunctions, with their possible causes arranged into categories relating to the operation of the garbage disposer. This section is intended only to serve as a checklist to aid you in diagnosing a problem. Look at the symptom that best describes the problem you are experiencing with the garbage disposer, and then proceed to correct the problem.

Overload Protector Trips

1. Check for voltage at the garbage disposer.

2. Does the motor hum when the disposer is turned on?

3. Was the foreign material removed from the hopper? Make sure that the flywheel turns easily.

4. Does the motor stay in the start winding? Use the ammeter to perform this check. Look at the name plate for the correct amperage rating.

5. Did you check the overload protector?

6. Is there a shorted or open winding in the motor?

7. Is the start relay okay?

Erratic Operation

1. Are there any loose connections? For example: wiring, switch, motor, etc.

2. Is the stopper worn or broken?

3. Is the wall switch defective?

4. Is the garbage disposer wired correctly?

Garbage Disposer Won't Run

1. Did the overload protector trip?

2. Is the fuse or circuit breaker blown or tripped?

3. Is the wall switch okay?

4. Are the motor windings burned out?

5. Check for open or shorted wiring.

6. Is the relay defective?

Garbage Disposer Won't Stop

1. Is the wall switch defective?

2. Check for a short in the wiring.

3. Check for incorrect wiring.

Slow Water Flow from Garbage Disposer

1. Check to see if the drain is partially clogged.
2. Check the shredder teeth, and see if they are clogged with food waste. This is caused by insufficient water flow.

Garbage Disposer Is Grinding Slowly

1. Check for undisposable matter in the hopper.
2. Is the flywheel damaged?
3. Is there sufficient water flow?
4. Is the shredder ring worn or broken?

Abnormal Noise in the Garbage Disposer

1. Check for undisposable matter in the hopper.
2. Are the mounting screws loose?
3. Is the flywheel broken?
4. Check the motor bearings—they might be damaged or worn.

Garbage Disposer Leaks

1. Disconnect the electricity. Then locate the water leak.
2. Check for loose mounting screws.
3. Check the sink flange. There might be insufficient putty around the flange.
4. Check for leaks around the tail pipe gasket.
5. Check for holes in the hopper.
6. Is the dishwasher/disposer connector cracked and leaking?
7. Is water leaking through the motor assembly? The seals might be faulty.

epair Procedure

The repair procedure in this chapter is a complete inspection and repair process for the garbage disposer components. It contains the information you need to test a component that might be faulty and to replace it, if necessary.

Garbage Disposer Assembly

The typical complaints associated with garbage disposer failure are:

- When the motor runs, there are loud noises.
- Water is leaking from the bottom of the disposer.
- The fuses or circuit breaker will trip when the disposer is started.

PART VI

To handle this problem, perform the following steps:

1. **Verify the complaint.** Verify the complaint by operating the disposer. Listen carefully, and you will hear if there are any unusual noises or if the overload protector trips.

2. **Check for external factors.** You must check for external factors not associated with the garbage disposer. Is the disposer installed properly? Does the disposer have the correct voltage?

3. **Disconnect the electricity.** Before working on the garbage disposer, disconnect the electricity. This can be done by pulling the plug out of the wall receptacle. Be sure that you only remove the disposer plug. Or disconnect the electricity at the fuse panel or at the circuit breaker panel. Turn off the electricity. The voltage at the receptacle is between 108 volts and 132 volts during a load on the circuit. Do you have the correct polarity? (See Chapter 6.)

4. **Remove the garbage disposer.** To access the disposer assembly, the disposer must be removed. Start by disconnecting the drain line from the disposer discharge tube. Next, remove the dishwasher drain hose connection, if so equipped (Figure 15-7a). Insert the service wrench into the right side of the flange body.

CAUTION *Place one hand under the disposer to keep it from falling.*

5. Now, turn the flange to the left to free the disposer from its mounting (Figure 15-7b). Turn the disposer upside down, and place it on a protective surface. This will protect the floor in the cabinet from damage. Remove the terminal plate (Figure 15-7c). Remove the ground wire and wire nuts from the service cord (Figures 15-7d and 15-7f). Separate the service cord wires from the motor wires. On models that are wired directly, loosen the cable clamp screws and remove the cable from the disposer (Figure 15-7e).

6. **Disassemble the garbage disposer.** Place the garbage disposer upside down on a protective surface or on a workbench. Some models have an insulated outer shell that must be removed first. Most disposers will separate into four or five sections, depending on the model you are servicing.

- Disposers that separate into four sections consist of:
  - Top container or hopper body
  - Upper-end bell or drain, housing assembly, cutting elements, and rotor and shaft assembly
  - Stator
  - Lower-end bell assembly
- Disposers that separate into five sections consist of:
  - Top container or hopper assembly
  - Stationary shredder
  - Upper-end bell or drain, housing assembly, rotor shredder, and rotor and shaft assembly
  - Stator
  - Lower-end bell assembly

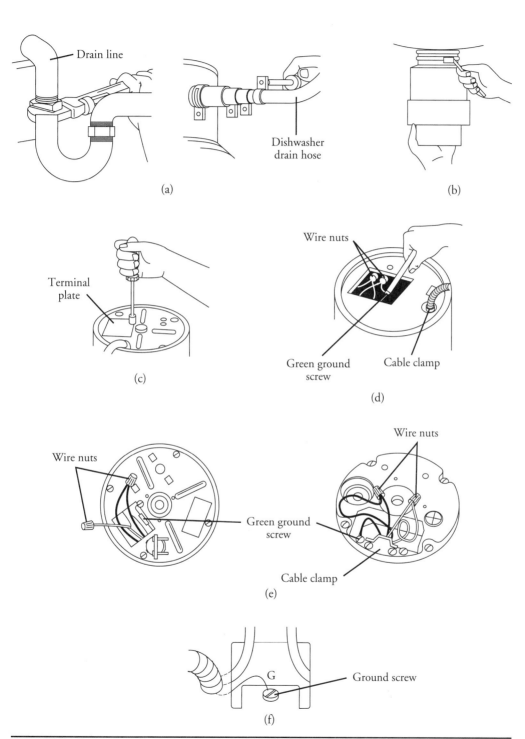

Figure 15-7 The steps taken to remove the garbage disposer.

Before removing any screws from the disposer, you must place scribe marks on the lower-end bell and stator, the upper-end bell and stator, and the upper-end bell and hopper. These scribe marks will be used to align the sections properly when the garbage disposer is reassembled. Remove the four through-bolts from the lower-end bell assembly, and separate the sections of the disposer. Label the wires from the stator to the lower-end bell assembly. Disconnect them. Be cautious if the garbage disposer has a capacitor.

WARNING *Do not touch capacitor terminals. A charged capacitor can cause severe shock when both terminals are touched. A charged capacitor will hold a charge until it is discharged.*

7. **Inspect and test the garbage disposer components.** Inspect the hopper assembly for cracks, holes, and (on some models) corrosion around the dishwasher connection. Also inspect the gasket area for nicks or rough spots. Inspect the gasket or "O" ring for cuts, breaks, or worn areas.

Replace any defective parts. Inspect the upper-end bell or drain housing assembly for defective parts. Test the centrifugal switch for binding by manually operating it. It should slide up and down freely. Inspect the rotor and thrust washer for cracks, breaks, or wear. If the upper drain housing and rotor assembly fail inspection, replace the garbage disposer as a complete unit.

Before testing the capacitor, it must be discharged. Use a screwdriver with an insulated handle. Discharge the capacitor by shorting across both terminals. Remove the wire leads, one at a time, with a pair of needle nose pliers. Set the ohmmeter on the highest scale, place one probe on one terminal, and then attach the other probe to the other terminal (Figure 15-8).

Observe the meter action. While the capacitor is charging, the ohmmeter will dip toward zero ohms for a short period of time. Then the ohmmeter reading will

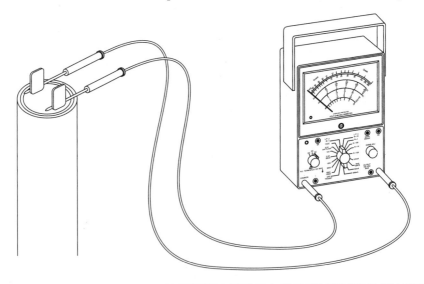

FIGURE 15-8   Place ohmmeter test leads on the capacitor terminals.

return to infinity. If the ohmmeter reading deflects to zero and does not return to infinity, the capacitor is shorted and should be replaced.

If the ohmmeter reading remains at infinity, the capacitor is open and should be replaced. Another way to test the capacitor is to use a capacitor tester.

Inspect the stator windings for nicks, cuts, or burned spots. Set the ohmmeter on R × 1, and test the stator windings for open, grounded, or shorted windings. Test both the start and run windings. If the start or windings fail the test, replace the garbage disposer as a complete unit. Set the ohmmeter on R × 1, and test the start switch. Place the probes on the wires leading to the switch. To ensure an accurate meter reading, isolate the start switch wires from the remainder of the circuitry. The meter reading will indicate an open circuit. If the test fails, replace the start switch. Next, test the overload protector. Isolate the overload protector wires from the remainder of the circuitry. Then place the ohmmeter probes on the overload protector wires. Push the red reset button on the overload protector. The ohmmeter reading will indicate a closed circuit. If this test fails, replace the overload protector.

8. **Reassemble and reinstall the garbage disposer.** To reassemble the disposer, just reverse the disassembly procedure, and reassemble. Check to be sure that all wiring is correct. Remember to use the scribe marks to align the sections properly before inserting the through-bolts. Also, be careful not to damage any of the gaskets or "O" rings when assembling the garbage disposer. Reconnect the service cord wires to the motor wires, and tighten the cable clamp screws (see Figure 15-7e). Next, reconnect the ground wire to the green ground screw (see Figures 15-7d and 15-7f). Reinstall the terminal plate and screw (Figure 15-9a).

Lift the disposer, and position it so that its three mounting ears are lined up under the ends of the sink mounting assembly screws (Figures 15-9b and 15-9c). The disposer will now hang by itself. After the plumbing is reconnected, tighten the ring using the service wrench. Rotate the disposer to align the discharge tube with the drain trap (Figure 15-9d). Reconnect them. Next, reconnect the dishwasher drain hose connection, if so equipped (Figure 15-9e). Finally, place the service wrench into the left side of one of the disposer mounting lugs, located on the top of the disposer. Then turn the wrench to the right until it is firmly secured, engaging the locking notch (Figure 15-9f).

Before you reconnect the electricity, check for leaks by running the water. Inspect all connections. Inspect the disposer for water leaks also. If no leaks are present, turn on the electricity and run the disposer; reinspect for any water leaks.

Before installing a new garbage disposer, the dishwasher connection plug must be removed if the defective disposer was equipped with a dishwasher drain hose connection. The dishwasher connection plug is removed by knocking the plug out with a hammer and some blunt instrument, such as a dowel or a punch (Figure 15-10). Do not use a screwdriver or any sharp tool. This can be driven through the plug and make it difficult to remove. Remove the plug from the hopper before installing or operating the disposer.

agnostic Charts

The following diagnostic flowcharts and wiring diagrams will help you to pinpoint the likely cause of the problem (Figures 15-11, 15-12, and 15-13)

Illustrating the steps taken to reinstall the garbage disposer.

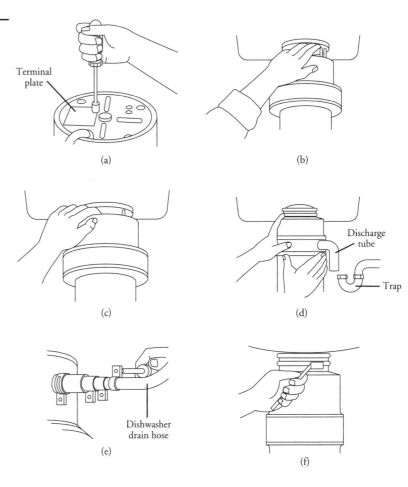

(a)

(b)

(c)

(d)

Terminal plate

Discharge tube

Trap

Dishwasher drain hose

(e)

(f)

Before installing a new garbage disposer, remove the dishwasher connection plug if a dishwasher drain hose will be connected to the disposer.

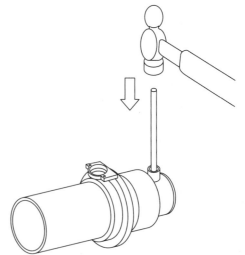

FIGURE 15-11
Garbage disposer
flowchart: overload
protector trips.

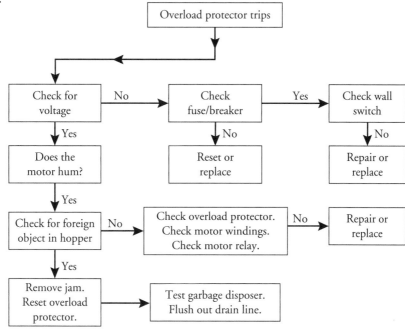

FIGURE 15-12
Garbage disposer
flowchart: Garbage
disposer motor won't
run.

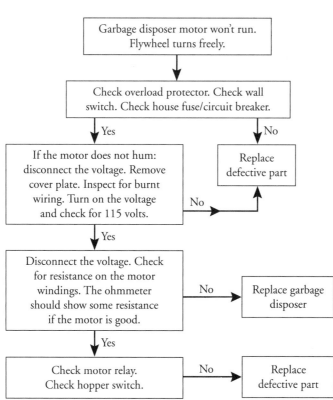

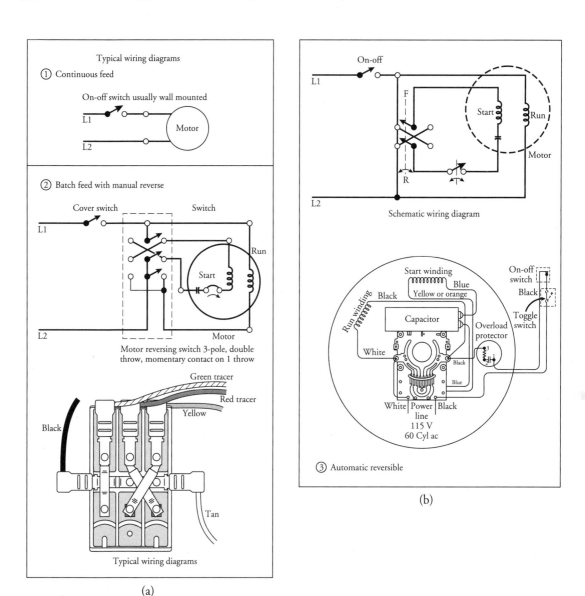

(a)

(b)

Figure 15-13 Typical household garbage disposer wiring diagrams.

Electric Water Heaters

Electric water heaters are heat-producing appliances. They normally have two heating elements (Figure 16-1), with each element controlled by a thermostat. The thermostats are mounted on the outer wall of the water heater tank just above the elements, from where they sense the temperature of the water through the outer wall of the water heater tank. A temperature and pressure-relief valve, which is mounted on the outside of the tank, is a device applied to a water heater, which will open to pass water or steam if excessive pressure or temperature occurs in the water heater tank.

Electric water heaters are available in many different heights, widths, and capacities.

Principles of Operation

The water heater tank is full of water, the electricity is on, and the upper thermostat (see Figure 16-1) senses that the water is cold. This condition energizes the upper element. At the same time, the lower element is not activated. The upper heating element will heat approximately one-quarter of the tank's capacity. When the temperature of the water reaches the thermostat setting, the upper thermostat will shut off. Then the lower thermostat becomes energized and heats the remainder of the water in the tank. When the temperature of the water reaches the thermostat setting, the lower thermostat will deactivate the lower heating element. The tank is now filled with hot water. As the consumer uses the hot water, it is drawn from the top of the tank and is replaced with cold water through a dip tube (see Figure 16-1) located near the bottom of the tank. When the lower thermostat senses the cold water, the thermostat activates the lower heating element, heating the water. Figure 16-2 identifies one of the many types of thermostats. The figure also shows a schematic drawing of the thermostat switch contacts.

Safety First

Any person who cannot use basic tools or follow written instructions should *not* attempt to maintain or repair any electric water heaters. Any improper installation, preventive maintenance, or repairs could create a risk of personal injury or property damage.

If you do not fully understand the preventive maintenance or repair procedures in this chapter, or if you doubt your ability to complete the task on your electric water heater, please call your service manager.

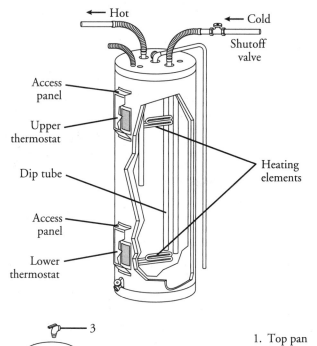

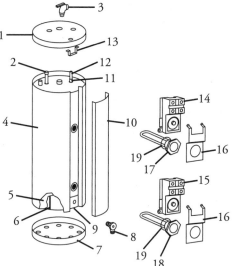

1. Top pan
2. Anode
3. T & P relief valve
4. Jacket
5. Insulation
6. Tank
7. Bottom pan
8. Draincock
9. Draincock panel
10. Front panel
11. Nipple
12. Dip tube
13. Junction bracket
14. Thermostat-upper
15. Thermostat-lower
16. Thermostat-bracket
17. Element-upper
18. Element-lower
19. Element gasket

These precautions should also be followed:

- Before checking the water heater, turn off the electricity.
- Never restore the electricity to the water heater if the tank is empty; do so only after it is full of water.
- Never remove the heating elements with the tank full of water.

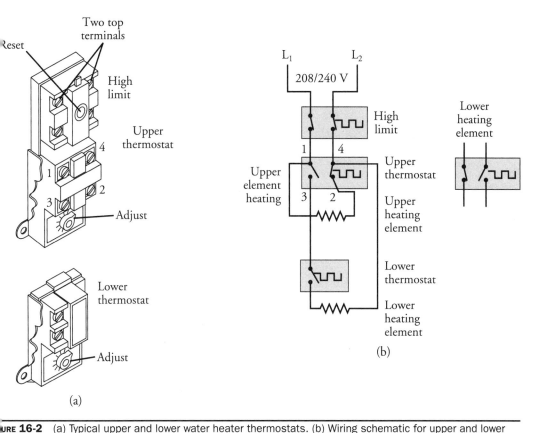

FIGURE 16-2 (a) Typical upper and lower water heater thermostats. (b) Wiring schematic for upper and lower thermostats.

- If a water heater needs to be replaced, it is strongly recommended that all electrical, plumbing, and placement of the tank be done by qualified personnel. Observe all local codes and ordinances for electrical, plumbing, and installation procedures.

Before continuing, take a moment to refresh your memory concerning the safety procedures in Chapter 2.

Electric Water Heaters in General

Much of the troubleshooting information in this chapter covers electric water heaters in general, rather than specific models, in order to present a broad overview of service techniques. The pictures and illustrations that are used in this chapter are for demonstration purposes only, to clarify the description of how to service these heaters. In no way do they reflect on a particular brand's reliability.

Electric Water Heater Maintenance

Every so often, inspect the water heater tank for possible water leaks. Check the following:

- Check all pipe connections to the tank. If corroded, they must be repaired before there is serious water damage to the property.
- Check the temperature/pressure-relief valve. Is it leaking?
- Turn off the electricity; remove the access panels; and inspect the wiring, elements, and insulation for signs of water leakage. If water is leaking, the leak must be repaired or the unit must be replaced immediately.

Once a year, the water heater tank should be drained and flushed out. This will increase the life expectancy of the tank by removing any unwanted sediment. Before you begin draining the tank, turn off the electricity to the water heater. Then turn off the water supply to the tank. Connect a garden hose to the draincock, and open the valve. To increase the water flow draining out of the tank, open the hot water faucet. When you are ready to refill the tank, close the draincock and turn on the water supply to the water heater.

Leave the closest hot water faucet open during refilling; the air that is trapped in the tank will escape through the hot water faucet. When all of the air is out of the tank and water lines, turn off the hot water faucet. Go to all of the other hot water faucets and open them to remove air in the lines. Turn off all water faucets. Now you are ready to turn on the electricity to the water heater.

Step-by-Step Troubleshooting by Symptom Diagnosis

In the course of servicing an appliance, don't tend to overlook the simple things that may be causing the problem. Step-by-step troubleshooting by symptom diagnosis is based upon diagnosing malfunctions, with their possible causes arranged into categories relating to the operation of the water heater. This section is intended only to serve as a checklist to aid you in diagnosing a problem. Look at the symptom that best describes the problem you are experiencing with the water heater, and then proceed to correct the problem.

No Hot Water

1. Do you have the correct voltage at the water heater?
2. Check for loose wiring.
3. Make sure that the reset button did not trip.
4. With the electricity off, check the thermostats for continuity.
5. With the electricity off, check the heating elements for continuity. Also check for a grounded heating element.

Not Enough Hot Water

1. Is the water heater undersized for the usage?
2. Check the lower heating element for continuity. Also check for a grounded heating element.
3. Check for leaking faucets and pipes.

4. Check to see if anyone is wasting water.

5. Drain the tank and remove the lower element. See if any sediment is in the bottom of the tank or around the heating element.

6. Is there any lime formation on the elements?

7. Are the thermostats operating properly? Run a cycle test.

8. Check the thermostat settings. Verify if they are set too low.

Water Too Hot

1. Check to see if the thermostat is snug against the tank.

2. Is the thermostat set too high?

3. Are the thermostats operating properly? Run the cycle test.

4. Check for a grounded element.

Water Heater Element Failure

1. Check for loose or burned wiring connections.

2. Check for the correct voltage to the elements.

3. Is the element shorted or grounded? Test the elements for continuity.

Discolored or Rusty Water

1. Drain the tank and remove the lower element. See if there is any sediment in the bottom of the tank.

2. Is the water supply to the water heater rusty in color?

3. Check the water for softness.

4. Check to see if there are excessive mineral deposits.

5. Inspect the anode rod to note its condition. If it has dissipated, replace the anode.

epair Procedures

Each repair procedure is a complete inspection and repair process for a single water heater component, containing the information you need to test a component that may be faulty and to replace it, if necessary.

Any person who cannot use basic tools should *not* attempt to install, maintain, or repair any electric dryer. Any improper installation, preventative maintenance, or repairs will create a risk of personal injury, as well as property damage. Call the service manager if installation, preventative maintenance, or the repair procedure is not fully understood.

Upper Thermostat

The thermostat is a mechanical device that controls the flow of current to various parts of the water heater. The thermostat senses the heat from the tank wall. When the thermostat senses the heat, it can control the current that is sent to the upper element or to another part of the water heater tank. The upper thermostat is attached to the high-limit thermostat.

The typical complaints associated with the upper thermostat are:

- No hot water
- Burned wires
- Water heater runs continuously
- The fuse blows or the circuit breaker trips

To handle this problem, perform the following steps:

1. **Verify the complaint**. To verify the complaint, turn on the hot water. Does the water get hot?

2. **Check for external factors.** You must check for external factors not associated with the water heater. Is the water heater installed properly? Does the water heater have the correct voltage?

3. **Disconnect the electricity.** Before working on the water heater, disconnect the electricity. This can be done by pulling the plug out of the electrical outlet. Or disconnect the electricity at the fuse panel or at the circuit breaker panel. Many installations have a disconnect switch/breaker box near the water heater. Turn off the electricity. The voltage at the receptacle is between 198 volts and 264 volts during a load on the circuit (see Chapter 6).

4. **Remove the access panel.** To gain access to the upper thermostat, the access panel must be removed (see Figure 16-1). The access panel is held on with two screws on most models. Peel back the insulation and then remove the thermostat's protective cover.

5. **Test the thermostat.** To test the upper thermostat, use your voltmeter. Set the range on the 300-volt scale. Touch the probes to terminals 1 and 3 above the reset button (Figure 16-3). Turn on the electricity. If the meter reads no voltage, have the customer call an electrician to find out where the power failure is. If the meter reads the proper rated voltage, as stated on the name plate, the electricity is okay. Next, touch the probes to terminals 2 and 4 below the reset button (Figure 16-4). If the meter reads the proper rated voltage, the electricity is okay. If the meter reading is "0" volts, press the reset button on the thermostat. If there is still no voltage reading on the meter scale, either the water temperature in the tank is too hot or the thermostat is inoperative. At this stage, turn off the electricity to the tank again. Remove the wires from the upper heating element (Figure 16-5). Be absolutely sure that the electricity is off before any further tests are made using your ohmmeter.

   Use your ohmmeter to test terminals 1 and 2 on the upper thermostat (see Figure 16-5). If the thermostat is calling for heat, the ohmmeter needle will swing to the right, showing continuity. Now touch the probes to terminals 1 and 4 on the thermostat (Figure 16-6). If the water in the tank is cold, you will not get a reading on the ohmmeter scale. If the water in the tank is hot, as called for by the thermostat, the needle will swing to the right, thus showing continuity.

6. **Cycle the upper thermostat.** Reconnect the wires to the heating element. Turn on the electricity to the tank. Take your ammeter and clamp the jaws around the wire that goes from terminal 4 to the upper element (Figure 16-7). With cold water in the tank and the thermostat calling for heat, you should get a reading on the meter in amps.

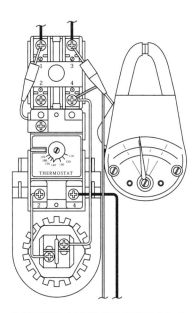

FIGURE 16-3 Place the voltmeter probes on terminals 1 and 3 to check the voltage.

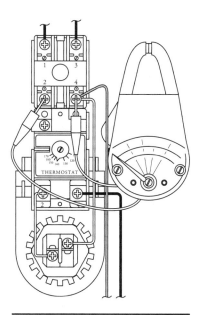

FIGURE 16-4 Place the voltmeter probes on terminals 2 and 4 to check the voltage.

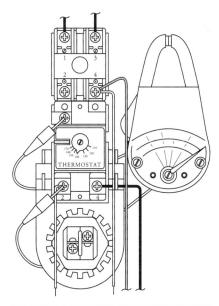

FIGURE 16-5 Remove the wires from the element. Place the ohmmeter probes on terminals 1 and 2. When the thermostat is calling for the element to come on, the ohmmeter scale will read continuity.

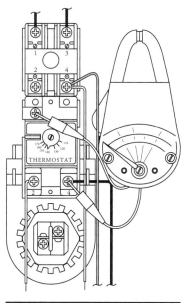

FIGURE 16-6 Place the ohmmeter probes on terminals 1 and 4. If water in the tank is cold, the ohmmeter scale will read no continuity.

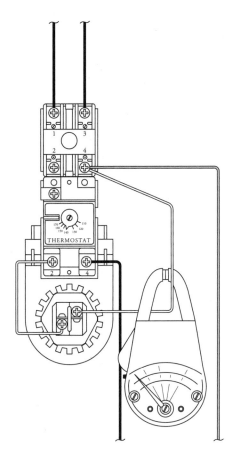

FIGURE 16-7   The ammeter jaws encircling the terminal 4 wire go to the element.

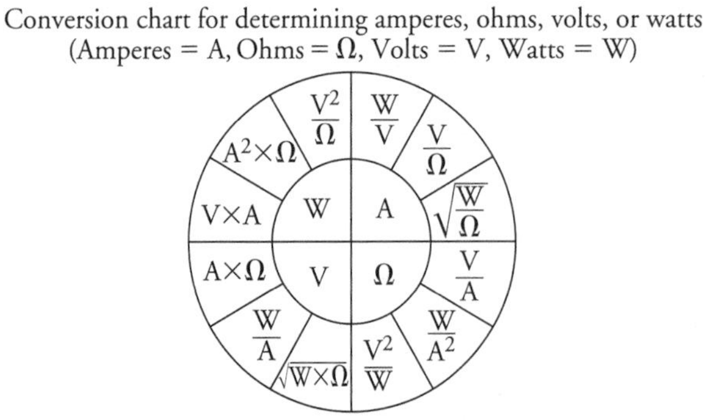

Conversion chart for determining amperes, ohms, volts, or watts
(Amperes = A, Ohms = Ω, Volts = V, Watts = W)

FIGURE 16-8   Ohm's Law Equation Wheel: The conversion chart for determining amperes, ohms, volts, or watts.

The reading you get will depend on the rating (in watts) of the heating element, divided by the voltage supplied (see the formula in Figure 16-8). After the temperature is reached, the needle on the ammeter scale will read "0."

Next, place the ammeter jaws on the terminal 4 wire that goes from the upper thermostat (just below the temperature gauge setting) to the lower element (Figure 16-9). The ammeter should show a reading, indicating that the lower element is working.

7. **Replace the upper thermostat.** Turn off the electricity to the water heater. To replace the upper thermostat, remove all of the wires from the thermostat. Be sure that you mark the wires so that you can replace them exactly the way you took them off. Next, pry back the thermostat retaining bracket with your fingers far enough to slide the thermostat up and out; remove the thermostat.

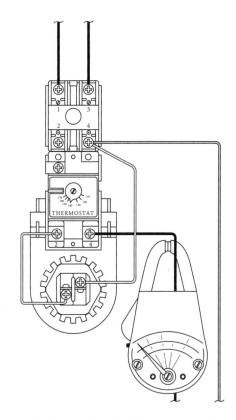

FIGURE 16-9   Place the ammeter jaws around terminal 4. The ammeter reading is 0.

NOTE *Do not pry back too much on the retaining bracket. It might break and the water heater tank will have to be replaced. To install the new thermostat, just reverse the disassembly procedure, and reassemble. Be sure that the thermostat is snugly mounted against the tank so that it can sense the temperature better. Reinstall the protective cover and insulation.*

8. **Test the new upper thermostat.** After you have completed the new installation, turn on the electricity to the water heater. Repeat this testing process as described, and verify that the new unit is working correctly.

Lower Thermostat

The lower thermostat monitors the water temperature in the lower section of the water heater tank. The lower thermostat turns on and off the lower element.

Testing the lower thermostat is similar to testing the upper thermostat, with the following exception. With the wires removed from the element, place the ohmmeter's probes on each terminal. Take a screwdriver, place it on the set screw of the thermostat, and turn it clockwise. The needle on your ohmmeter will swing to the right, indicating that there is continuity. If the needle does not move, the thermostat needs to be replaced. Replace this thermostat as per the upper thermostat replacement procedure previously described. Then test the thermostat by placing the jaws of the ammeter around the wire that connects to the lower thermostat. Check to be sure that the thermostat cycles on and off.

Heating Element

The heating element is made from a nichrome filament, surrounded by magnesium oxide powder for heat transfer, and is encased in an immersion-type casing. This type of heating element cannot be turned on unless there is water present in the water heater tank to submerge the heating element.

The typical complaints associated with the heating element are:

- No hot water
- Burned wires
- The fuse blows or the circuit breaker trips

To handle this problem, perform the following steps:

1. **Verify the complaint.** To verify the complaint, turn on the hot water. Does the water get hot?
2. **Check for external factors.** You must check for external factors not associated with the water heater. Is the water heater installed properly? Does the water heater have the correct voltage?
3. **Disconnect the electricity.** Before working on the water heater, disconnect the electricity. This can be done by pulling the plug out of the electrical outlet. Or disconnect the electricity at the fuse panel, the circuit breaker panel, or the disconnect switch. Turn off the electricity. The voltage at the receptacle is between 198 volts and 264 volts during a load on the circuit (see Chapter 6).
4. **Remove the access panel.** In order to gain access to the heating element, the access panel must be removed (see Figure 16-1). The access panel is held on with two

screws on most models. Peel back the insulation and then remove the wires from the heating element.

5. **Test the heating element.** Use the ohmmeter to test the heating element. Set the range to R × 1. Touch the probes to the element screws (Figure 16-10). If the element is good, the ohmmeter scale will show continuity. If the element is bad, the ohmmeter needle will not move, showing an open element. Now, set the ohmmeter on R × 100. Take one probe and place it on either terminal of the element. Take the other probe and touch the element head (Figure 16-11) or the tank. If you get any reading, the element is grounded and should be replaced. Repeat this procedure for the other terminal on the element. Both terminals must be measured for grounding.

6. **Remove the element.** Turn off the electricity to the water heater. Next, turn off the water supply. Connect a garden hose to the draincock, and open the draincock. At the same time, open a nearby faucet to allow the water to drain faster. After the tank is empty, disconnect the wires from the element and remove it. The screw-in type can be removed by using a 1½-inch socket or a 3/8-inch socket for the four bolt–type elements.

7. **Install the new element.** To install the new element, just reverse the disassembly procedure, and reassemble. When installing the new element, always replace the gasket. Before you insert the new element into the tank, clean the flange of debris.

8. **Test the new element.** When you are done installing the replacement element, close the draincock, and open the water supply. Close the nearby faucet after all of the air in the tank has dissipated. Next, check for water leaks. Turn on the electricity to the water heater. Use the ammeter to check the element. The ammeter should show a reading.

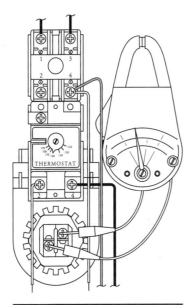

FIGURE 16-10 Testing the heater element. Place the ohmmeter probes on the screw terminals of the element.

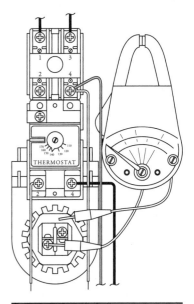

FIGURE 16-11 Testing the heater element for ground.

iagnostic Charts

The following diagnostic flowcharts, wiring diagrams, and tables will help you to pinpoint the likely causes of the problem (Figures 16-12 through 16-26 and Tables 16-1 through 16-12). Figures 16-27 and 16-28 illustrate a typical water heater installation for a series and parallel tanks.

elp on the Internet

To obtain additional technical help or information on selecting an electric water heater, you can go to the following websites:

http://www.aceee.org/consumerguide/topwater.htm
http://www.eere.energy.gov/consumer/your_home/water_heating/index.cfm/
mytopic=12770
http://www.americanwaterheater.com/
http://www.aosmith.com/
http://www.rheem.com/
http://www.hotwater.com/
http://www.tankless-water-heater.com/
http://www.bradfordwhite.com/
http://www.marathonheaters.com/
http://www.statewaterheaters.com/

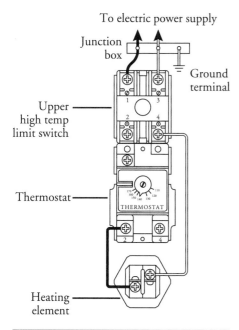

Figure 16-12 Pictorial wiring diagram for a standard two-wire 208/240 volt circuit, single element.

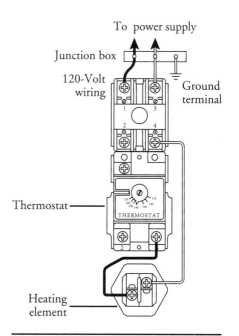

Figure 16-13 Pictorial wiring diagram for a standard two-wire 208/120 volt circuit, single element.

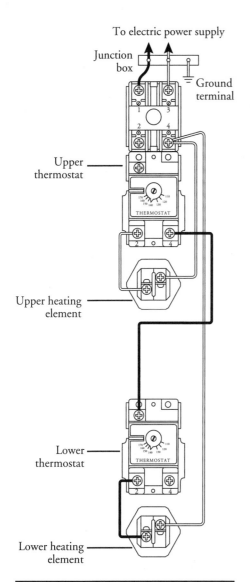

FIGURE 16-14 Pictorial wiring diagram for a standard two-wire interlocking 208/240/277 volt circuit, double element.

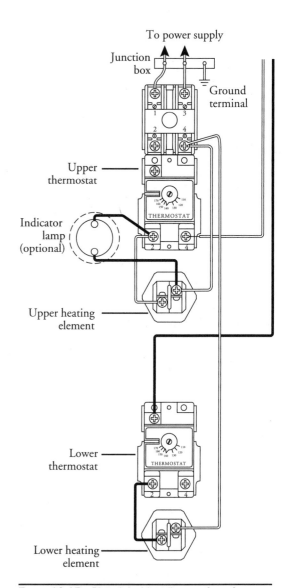

FIGURE 16-15 Pictorial wiring diagram for a four-wire, off-peak double element, nonsimultaneous operation.

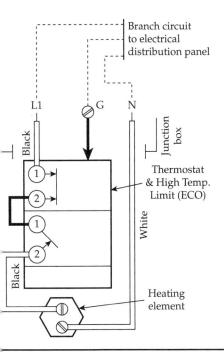

FIGURE 16-16   120-volt single-element water heater wiring diagram.

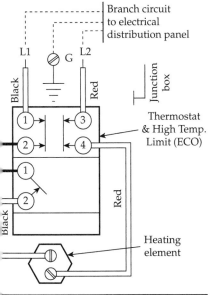

FIGURE 16-17   240-volt single-element water heater wiring diagram.

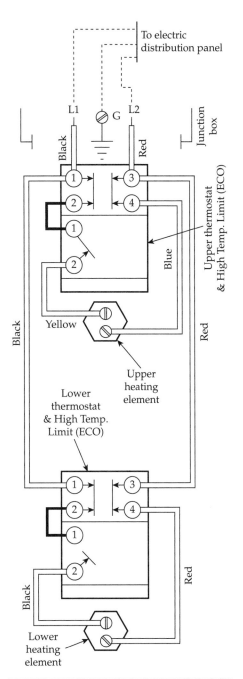

FIGURE 16-18   Double-element simultaneous operation water heater wiring diagram.

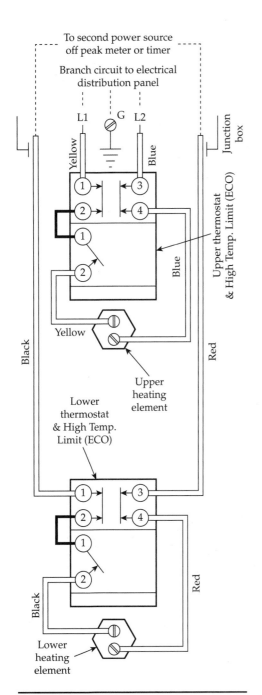

FIGURE 16-19 Double-element simultaneous operation with four-wire outlet water heater wiring diagram.

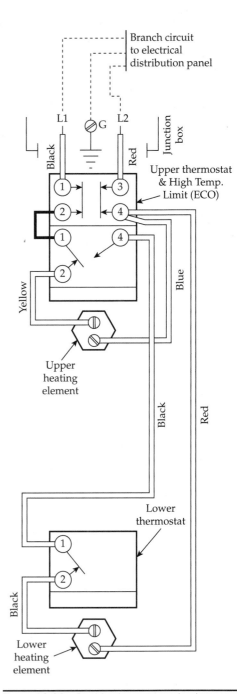

FIGURE 16-20 Double-element nonsimultaneous water heater wiring diagram.

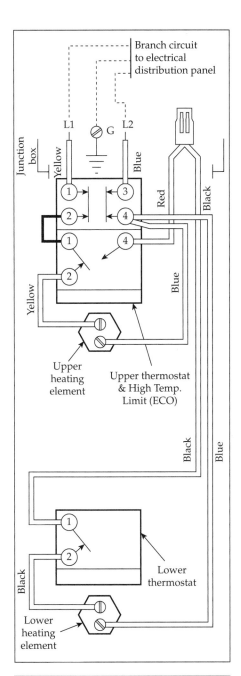

Figure 16-21 Double-element nonsimultaneous with four-wire outlet water heater wiring diagram.

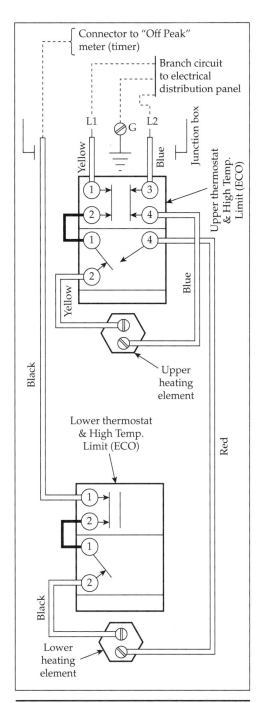

Figure 16-22 Double-element nonsimultaneous with three-wire outlet water heater wiring diagram.

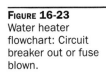

Figure 16-23
Water heater
flowchart: Circuit
breaker out or fuse
blown.

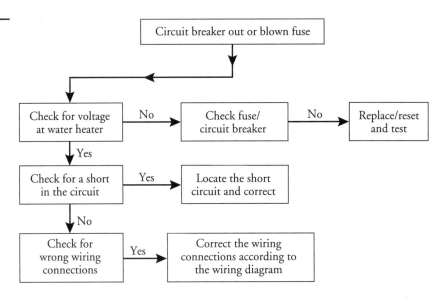

Figure 16-24
Water heater
flowchart: Water
heater runs
continuously.

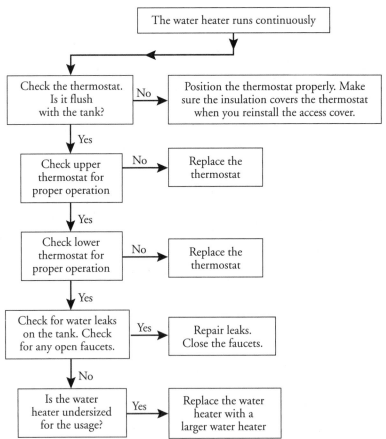

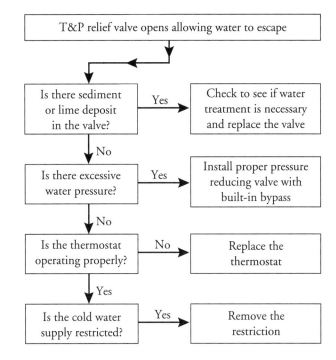

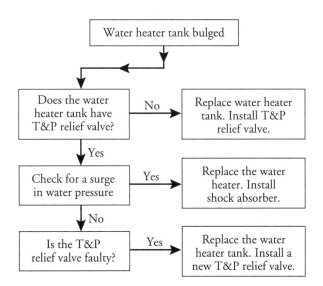

Activity	Gallons
Handwashing	0.9
Shower	5–7
Tub bath	7–12
Shaving	2
One meal per person	2–5
Automatic washer	11–34
Conventional washing	30
Dishwasher	8
Average bathroom use per day	
Adult	12
Child	24

TABLE 16-1   Typical Hot Water Usage Chart

	GPH Recovery at Indicated Temperature Rise				
Heating Element Wattage	**Degrees Fahrenheit**				
	60	70	80	90	100
750	5.1	4.4	3.8	3.4	3.1
1000	6.8	5.8	5.1	4.5	4.1
1250	8.5	7.3	6.4	5.7	5.1
1500	10.2	8.8	7.7	6.8	6.1
2000	13.7	11.7	10.2	9.1	8.2
2500	17.1	14.6	12.8	11.4	10.2
3000	20.5	17.5	15.4	13.6	12.3
3500	23.9	20.5	17.9	15.9	14.3
4000	27.3	23.4	20.5	18.2	16.4
4500	30.7	26.3	23.0	20.5	18.4
5000	34.1	29.2	25.6	22.7	20.5
5500	37.6	32.2	28.2	25.0	22.5
6000	41.0	35.1	30.7	27.3	24.6

TABLE 16-2   Electric Water Heater Recovery Chart

Thermostat Set at 120° F		Water Mixed Temperatures				
		100° F	105° F	110° F	115° F	120° F
Cold water inlet temperature	40° F	75%	81%	88%	94%	100%
	45° F	73%	80%	87%	93%	100%
	50° F	71%	79%	86%	93%	100%
	55° F	69%	77%	85%	92%	100%
	60° F	67%	75%	83%	92%	100%
	65° F	64%	73%	82%	91%	100%
	70° F	60%	70%	80%	90%	100%
	75° F	56%	67%	78%	89%	100%

Example: If the water heater thermostat is set on 120 degrees Fahrenheit and the cold water inlet temperature is 50 degrees Fahrenheit and the bath water is at 110 degrees Fahrenheit, the water mixture would be 86 percent hot water and 14 percent cold water.

TABLE **16-3**   The Percentage of Hot and Cold Water Mixture

Thermostat Set at 125° F		Water Mixed Temperatures				
		100° F	105° F	110° F	115° F	120° F	125° F
Cold water inlet temperature	40° F	71%	76%	82%	88%	94%	100%
	45° F	69%	75%	81%	88%	94%	100%
	50° F	67%	73%	80%	87%	93%	100%
	55° F	64%	71%	79%	86%	93%	100%
	60° F	62%	69%	77%	85%	92%	100%
	65° F	58%	67%	75%	83%	92%	100%
	70° F	55%	64%	73%	82%	91%	100%
	75° F	50%	60%	70%	80%	90%	100%

TABLE **16-4**   The Percentage of Hot and Cold Water Mixture with Water Heater Set on 125 Degrees Fahrenheit

Thermostat Set at 130° F	Water Mixed Temperatures						
		100° F	105° F	110° F	115° F	120° F	125° F	130° F
Cold water inlet temperature	40° F	67%	72%	78%	83%	89%	94%	100%
	45° F	65%	71%	76%	82%	88%	94%	100%
	50° F	63%	69%	75%	81%	88%	94%	100%
	55° F	60%	67%	73%	80%	87%	93%	100%
	60° F	57%	64%	71%	79%	86%	93%	100%
	65° F	54%	62%	69%	77%	85%	92%	100%
	70° F	50%	58%	67%	75%	83%	92%	100%
	75° F	45%	55%	64%	73%	82%	91%	100%

TABLE 16-5   The Percentage of Hot and Cold Water Mixture with Water Heater Set on 130 Degrees Fahrenheit

Thermostat Set at 135° F	Water Mixed Temperatures							
		100° F	105° F	110° F	115° F	120° F	125° F	130° F	135° F
Cold water inlet temperature	40° F	63%	68%	74%	79%	84%	89%	958%	100%
	45° F	61%	67%	72%	78%	83%	89%	94%	100%
	50° F	59%	65%	71%	76%	82%	88%	94%	100%
	55° F	56%	63%	69%	75%	81%	88%	94%	100%
	60° F	53%	60%	67%	73%	80%	87%	93%	100%
	65° F	50%	57%	64%	71%	79%	86%	93%	100%
	70° F	46%	54%	62%	69%	77%	85%	92%	100%
	75° F	42%	50%	58%	67%	75%	83%	92%	100%

TABLE 16-6   The Percentage of Hot and Cold Water Mixture with Water Heater Set on 135 Degrees Fahrenheit

Thermostat set at 140° F		Water Mixed Temperatures								
		100° F	105° F	110° F	115° F	120° F	125° F	130° F	135° F	140°F
Cold water inlet temperature	40° F	60%	65%	70%	75%	80%	85%	90%	95%	100%
	45° F	58%	63%	68%	74%	79%	84%	89%	95%	100%
	50° F	56%	61%	67%	72%	78%	83%	89%	94%	100%
	55° F	53%	59%	65%	71%	76%	82%	88%	94%	100%
	60° F	50%	56%	63%	69%	75%	81%	88%	94%	100%
	65° F	47%	53%	60%	67%	73%	80%	87%	93%	100%
	70° F	43%	50%	57%	64%	71%	79%	86%	93%	100%
	75° F	38%	46%	54%	62%	69%	77%	85%	92%	100%

TABLE 16-7   The Percentage of Hot and Cold Water Mixture with Water Heater Set on 140 Degrees Fahrenheit

TABLE 16-8
Most Water Heater Elements Have a Dual Rating for Single Phase 208 and 240 Volt Operation

Water Heater Element Voltage Ratings		
Voltage	Wattage	Recovery* Gallons Per Hour
208/240	2255/3000	10/14
208/240	2630/3500	12/16
208/240	2855/3800	13/17
208/240	3380/4500	16/21
208/240	3755/5000	17/23
208/240	4130/5500	19/25
208/240	4500/6000	21/27
* The recovery is based on 90° F temperature rise in gallons per hour.		

Water Heater Wattage	Phase	Amperage Rating for Fuse or Circuit Breaker		Copper Wire Size – AWG	
		208 Volt	240 Volt	208 Volt	240 Volt
3000	1	20	20	12	12
3800	1	25	20	10	10
4000	1	25	25	10	10
4500	1	30	25	10	10

TABLE 16-9   Fuse or Circuit Breaker and Wire Size for Electric Water Heater

Heating Element Wattage Rating	120 Volts		208 Volts		240 Volts	
	Amps	Ohms	Amps	Ohms	Amps	Ohms
600	5.0	24.0	2.9	72.1	2.5	96.0
750	6.3	19.2	3.6	57.7	3.1	76.8
1000	8.3	14.4	4.8	43.3	4.2	57.6
1250	10.5	11.5	6.0	34.6	5.2	46.1
1500	12.5	9.6	7.2	28.8	6.3	38.4
2000	16.7	7.2	9.6	21.6	8.3	28.8
2500	20.8	5.8	12.0	17.3	10.4	23.0
3000	25.0	4.8	14.4	14.4	12.5	19.2
3500	–	–	16.8	12.4	14.6	16.5
3800	–	–	18.3	11.4	15.8	15.2
4000	–	–	19.2	10.8	16.7	14.4
4500	–	–	21.6	9.6	18.8	12.8

TABLE 16-10   Voltage, Wattage, Ohms, and Amperage of Heating Elements

TABLE 16-11
Always Mix the Hot and Cold Water when Bathing to Avoid Possible Burns or Scalding

Temperature	Time to Produce Serious Burns
120° F	More than 5 minutes
125° F	1–1½ to 2 minutes
130° F	About 30 seconds
135° F	About 10 seconds
140° F	Less than 5 seconds
145° F	Less than 3 seconds
150° F	About 1½ seconds
155° F	About 1 second
160° F+	Instantaneously

Water temperatures over 110°F–120°F can cause severe burns in seconds.

TABLE 16-12
Water Temperature Guideline for Common Activities Using Hot Water

Hot Water Temperature Guidelines
Handwashing: 110° F to 115° F
Showers: 110° F to 115° F
Dishwasher rinse 130° F to 140° F
Laundry 110° F to 120° F
To check the water heater temperature setting, place a thermometer under a running hot water faucet near the water heater tank.

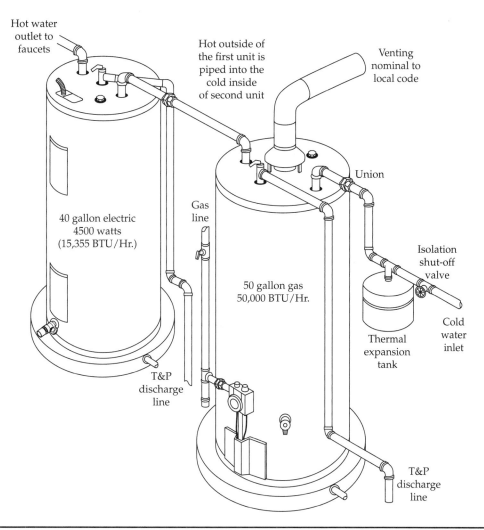

Hot water
outlet to
faucets

Hot outside of
the first unit is
piped into the
cold inside
of second unit

Venting
nominal to
local code

Union

40 gallon electric
4500 watts
(15,355 BTU/Hr.)

Gas
line

50 gallon gas
50,000 BTU/Hr.

Isolation
shut-off
valve

Cold
water
inlet

Thermal
expansion
tank

T&P
discharge
line

T&P
discharge
line

FIGURE 16-27 A typical series water heater installation. This type of installation is used for high-demand households. If the existing single-tank installation cannot handle the demand, then another tank can be added.

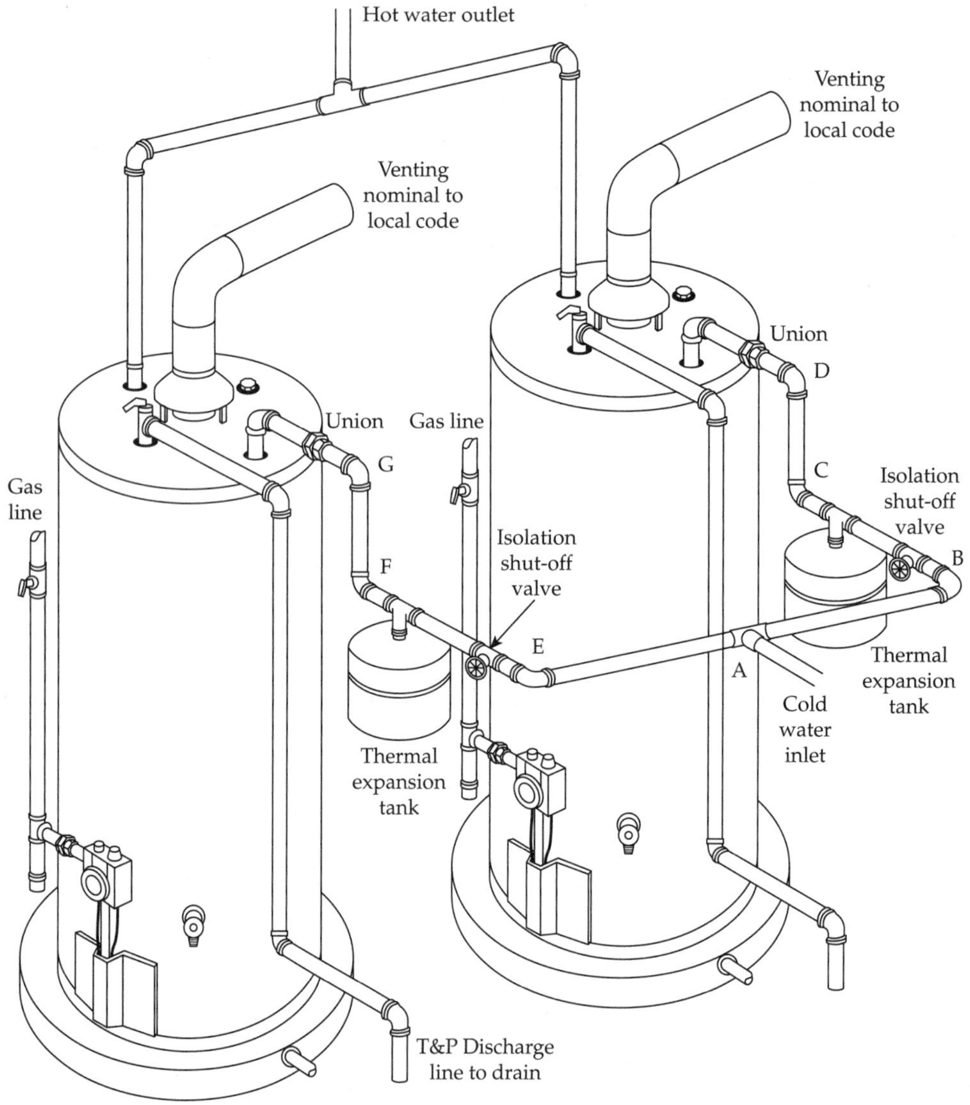

FIGURE 16-28   A typical parallel water heater installation. This type of installation is used for high demand for large quantities of hot water over a short period of time.

Gas Water Heaters

G as water heaters are similar to electric water heaters, with the exception of how the water is heated. They use liquefied petroleum gas (LP or LPG) or natural gas to heat the water. The tank is constructed in such a way that it allows for a vent (chimney) up the middle of the tank, along with a gas burner on the bottom of the tank to heat the water (Figure 17-1). The thermostat that controls the water temperature and gas flow is mounted on the outside of the tank, with the temperature probes inserted into the tank to sense the water temperature. A temperature and pressure-relief valve, which is mounted on the outside of the tank, is a device applied to a water heater that will open to pass water or steam if excessive pressure or temperature occurs in the water heater tank.

Gas water heaters are available in many different sizes. When purchasing a gas water heater, it is important to specify the type of gas fuel that will be used to heat the water.

rinciples of Operation

The gas water heater fills up in the same way as an electric water heater. Water enters the cold-water inlet port through a dip tube, filling the tank from the bottom up. Once the tank is full of water, the gas supply is turned on and the consumer or technician can begin the process of lighting the burner according to the manufacturer's use and care manual. Once the pilot light is lit, the thermocouple, which is near or in the gas flow, will begin to heat up and send a signal to the gas combination valve to open the gas inlet to the pressure regulator section of the combination control valve. At the same time, the thermostat senses the cold water and sends a signal to the combination control to actuate a lever from within the pressure regulator to open the gas valve, allowing the gas to enter the burner for ignition. While the gas flames are heating the water, the thermostat begins monitoring the water temperature. When the water temperature reaches a preset temperature, a signal will be sent to the combination valve to Shutoff the gas flow.

afety First

Any person who cannot use basic tools or follow written instructions should *not* attempt to maintain or repair any gas water heaters. Any improper installation, preventive maintenance, or repairs could create a risk of personal injury, death, or property damage.

FIGURE 17-1
Exploded view of a
gas water heater.

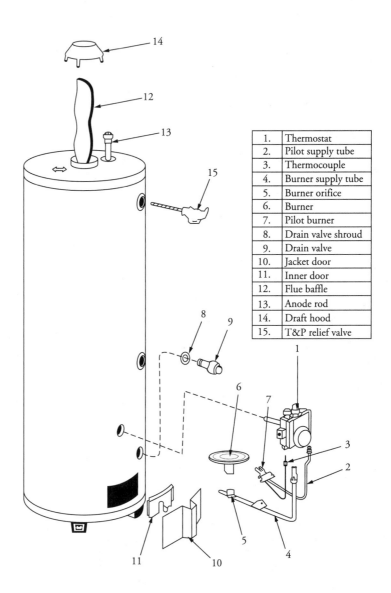

1.	Thermostat
2.	Pilot supply tube
3.	Thermocouple
4.	Burner supply tube
5.	Burner orifice
6.	Burner
7.	Pilot burner
8.	Drain valve shroud
9.	Drain valve
10.	Jacket door
11.	Inner door
12.	Flue baffle
13.	Anode rod
14.	Draft hood
15.	T&P relief valve

If you do not fully understand the preventive maintenance or repair procedures in this chapter, or if you doubt your ability to complete the task on your gas water heater, please call your service manager.

These precautions should also be followed:

- Before servicing the water heater, turn off the gas supply.

- Never turn on the gas to the water heater if the tank is empty; do so only after it is full of water.

- At least once a year perform maintenance on the water heater according to the manufacturer's recommendations.

- If a water heater needs to be replaced, it is strongly recommended that all gas connections, plumbing, ventilation, and placement of the tank be done by qualified personnel. Observe all local codes and ordinances for gas fuels, venting, plumbing, and installation procedures.

Before continuing, take a moment to refresh your memory concerning the safety procedures in Chapter 2. In addition, please follow the manufacturer's recommendations for safety as outlined in the use and care manual.

Do not store any flammable materials near a gas water heater, and avoid ignition sources in the event you smell gas leaking. Make sure you have proper ventilation when servicing the water heater to prevent carbon monoxide (CO) poisoning.

as Water Heaters in General

Much of the troubleshooting information in this chapter covers gas water heaters in general, rather than specific models, in order to present a broad overview of service techniques. The pictures and illustrations that are used in this chapter are for demonstration purposes only, to clarify the description of how to service these heaters. In no way do they reflect on a particular brand's reliability.

as Water Heater Maintenance

Maintenance on gas water heaters *must* be performed annually. The following checklist will assist you in performing the maintenance procedures:

1. A visual inspection must be made to the following components:

   - **The venting system**   A properly working ventilation system will prevent toxic fumes from entering the home. Check the venting system for leakage, corrosion, obstructions, sooting, and rust forming on top of the water heater. If any of these conditions are present, you must take action to correct these problem areas to protect the health and lives of the people living in the home. Most important, check the fitting (vent cap) on the end of the vent pipe for obstructions.

     A properly designed and working vent system will convey the combustion gases from the water heater environment to the atmosphere outside of the home. It will also protect the home from fire hazards due to overheating of walls and other surfaces, prevent condensation of hot combustion gases, and provide clean air circulation and adequate oxygen supply for the water heater.

     Check for adequate air supply to the area in which the water heater is located. A gas water heater will consume about 12.5 cubic feet of air (at sea level) for every 1000 BTUs of heat generated.

   - **The burner components**   Inspect the burner assembly and the pilot assembly. If your inspection reveals that the burner, pilot, or base assemblies need cleaning, turn off the gas supply and remove the burner and pilot assemblies for cleaning.

Some models have a removable screen at the bottom of the tank that keeps dust, lint, and oils from entering the combustion chamber. This filter screen must also be cleaned.

- **Surrounding area**   Clean the area surrounding the water heater of dust, lint, and debris. Make sure that combustible materials are removed completely from the area.

2. Inspect the temperature and pressure-relief valve (T&P) for leaks. If it is leaking water, you must replace it. Make sure that the valve is connected to a drain line that terminates to a drain or to the outside of the residence. Next, you will manually operate the T&P valve and check for proper operation. Be careful—the discharged water is very hot. Lift the T&P lever gently and then release it—the lever should return to its normal closed position. If the water is still leaking out of the valve after you release the lever, replace the valve.

3. Once a year, the water heater tank should be drained and flushed out. This procedure will increase the life expectancy of the tank by removing the unwanted sediment. Before you begin draining the tank, turn off the gas supply to the water heater. To drain the water heater tank, turn off the water supply to the tank. Connect a garden hose to the draincock, and open the valve. To increase the water flow draining out of the tank, open the hot water faucet—this will increase the flow of water draining out of the tank. When you are ready to refill the tank, close the draincock, turn on the water supply to the water heater, and begin to refill the tank.

Leave the closest hot water faucet open during refilling; the air that is trapped in the tank will escape through the hot water faucet. When all of the air is out of the tank and water lines, close the hot water faucet. Go to all of the other hot water faucets, and open them to remove the air in the lines. Close all water faucets. Now you are ready to turn on the gas supply to the water heater. Light the pilot light according to the manufacturer's use and care guide for the model you are servicing.

Thermal Expansion

When water is heated in the water heater tank, it will expand. There is a physical law that states that as the water temperature rises, the water will expand in volume. For example, heating 40 gallons of water to the temperature setting on the thermostat will cause it to expand to 40.5 gallons when the temperature is reached. In the old days, before backflow preventers, water meter check valves, and pressure reducing valves, the excess water flowed back to the city main water supply where it was dissipated. Today, with the added controls and safety valves, the water cannot return to the city water supply. The water in the tank is in a "closed loop system" and has nowhere to go. The thermal expansion of the heated water causes a serious problem to the tank and the residence. The internal pressures in the water heater tank during recovery periods can stress the tank, causing the tank welds and fitting connections to the tank to weaken and fail, in turn causing flooding in the residence and possibly even bodily injury.

WARNING   *Water + heat + pressure + closed loop system = potential explosion and water damage!*

As the pressure builds up in the water heater tank, the temperature pressure relief valve (T&P valve) will open and the excess water will be released from the water heater tank, either on the floor or drained to the outside of the residence. Even though the T&P valve opens during every recovery period, this action will weaken the tank and shorten its life. A good indicator of thermal expansion is when the T&P valve releases about one cup of water for every ten gallons of water in the tank for each heating period.

The best way to control thermal expansion in a water heater tank is to install a thermal expansion tank, shown in Figure 17-2. The thermal expansion tank will accept the excess water and pressure away from the water heater tank, store it, and return the excess water back to the water heater tank in the "closed loop system" as the water is used. This will

FIGURE 17-2
A thermal expansion tank installed to prevent water heater damage and flooding.

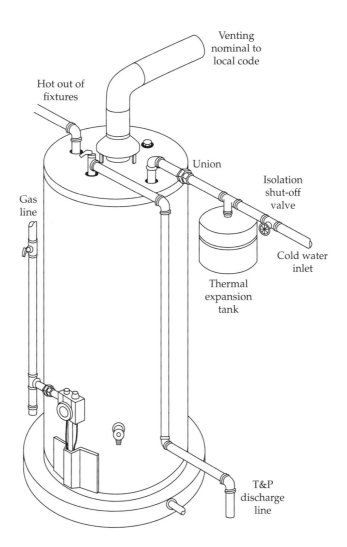

Venting nominal to local code

Hot out of fixtures

Union

Isolation shut-off valve

Gas line

Cold water inlet

Thermal expansion tank

T&P discharge line

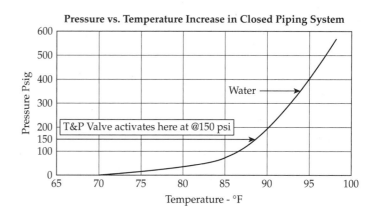

FIGURE **17-3**
This figure points out the danger of what happens to the water pressure when heated in a closed loop system.

prolong the life of the water heater tank. Also, the expansion tank will prolong the life of the temperature pressure relief valve. Figure 17-3 illustrates that water at 70° F in a water heater tank, with no pressure on the tank, will begin to rise in pressure when heated.

Water Hammer

When the hot or cold water valve is turned off quickly, a loud banging noise can be created. This noise is called a water hammer. If you close the water valve slowly, you will not hear a water hammer noise. A water hammer noise occurs when the water in a closed loop system comes to an abrupt stop at the water valve; a high-pressure shockwave reverberates within the piping system until the energy has been spent in frictional losses.

To prevent a water hammer condition, a water hammer arrestor device will have to be installed. These are designed so the water in the closed loop system will not contact the air cushion in the arrestor. This phenomenon is not a water heater issue; it is a plumbing issue. If a water hammer arrestor needs to be installed, contact a certified plumber.

Negative Air Pressure

Air pressure in a residence is affected by different forces, such as blowing wind, forced-air furnaces, clothes dryers, bathroom fans, range hoods, fireplaces, ceiling fans, and even the central vacuum cleaner. When the air pressure inside a residence is greater than the outside pressure, it's called positive pressure. When the pressure in a residence is less than the outside pressure, it's called negative pressure. Any device that can pull air out of a residence can create a negative pressure inside.

Over the past two decades, construction techniques have improved the energy efficiency of buildings, making them tighter and resistant to air leaks. The tighter the building, the greater potential for negative air pressure. This might leave you with a problem: combustion and ventilation air is not able to infiltrate into the building and combustion appliances and heating furnaces will be starved for air. There may be no air left in the building to operate the other devices such as exhaust fans. When this happens, the negative air pressure inside the building draws air in through every crack and crevice it can find to equalize the inside pressure.

Under normal conditions with positive air pressure in the building, the gas water heater will be able to vent the combustible by-products (flue gases) and the heat from the gas burner through the vent system to the outside of the building. When you have negative air pressure in the building, the combustible by-products (flue gases) and the heat from the burner will not vent to the outside. Instead, the vent system will act as a fresh air intake source. The flue gases and heat will stay in the structure, causing poor water heater performance and harmful gases to the occupants. Symptoms of negative air pressure problems are:

- The pilot flame will not stay lit.
- There will be flame rollout at the bottom of the water heater.
- Around the bottom of the heater there will be sooting.
- Slow recovery of the water heater.

The only places to find negative air pressure in a residence are the kitchen and laundry area. In the kitchen there might be an exhaust fan over the range venting air to the outside of the structure. In the laundry area, a gas dryer will vent moist air to the outside of the building.

raft Test

To check and see if the gas water heater is venting properly, perform a draft test (Figure 17-4). This test only applies to a natural vent system, and does not apply to the forced vent system water heater. Symptoms of a blocked vent include:

- There are smells of combustion gases in the room.
- There is melted pipe insulation around the cold and hot nipples on the tank.
- Around the draft hood there are burn or scar marks, as well as paint peeling on the top of the tank and directly under and around the draft hood.

To conduct a draft test, perform the following:

1. Turn the thermostat up high on the combination valve until the burner comes on.
2. Wait 3 to 5 minutes until the venting system gets warm.
3. Place a smoke source halfway between the bottom of the draft hood and the top of the water heater. A smoke source can be a match or an incense stick.
4. Observe how the smoke behaves. If the smoke is drawn into the vent system, then the vent system is okay. If the smoke blows toward you, then there is a venting problem that has to be corrected.
5. When you complete the test, return the thermostat back to its original position (around 120° F).

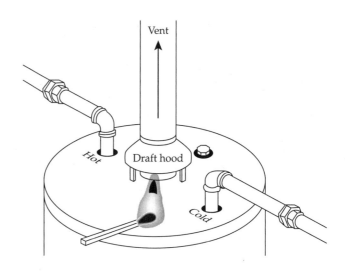

Figure 17-4
To determine if the vent system is working in an atmospherically vented gas water heater, perform a draft test.

Step-by-Step Troubleshooting by Symptom Diagnosis

In the course of servicing an appliance, don't tend to overlook the simple things that may be causing the problem. Step-by-step troubleshooting by symptom diagnosis is based upon diagnosing malfunctions, with their possible causes arranged into categories relating to the operation of the water heater. This section is intended only to serve as a checklist to aid you in diagnosing a problem. Look at the symptom that best describes the problem that you are experiencing with the water heater and then proceed to correct the problem.

Gas Odors

1. Check the pilot light.

2. Check for a gas leak in the supply line.

3. Check the ventilation system.

The Burner Will Not Ignite

1. Make sure that the gas supply to the water heater is turned on.

2. Check the gas line for debris. Also, check the pilot line for debris.

3. Check the line to the burner for a blockage. If blocked, locate the source and correct.

4. Check the thermostat setting. It may be set too low.

5. Test the value of the thermocouple; replace if necessary.

6. Test the thermostat; replace if necessary.

7. Check the installation and fresh air supply.

The Pilot Will Not Light or Stay Lit

1. Make sure that the gas supply to the water heater is turned on.

2. Check the gas pressure to the water heater.

3. Check for air or debris in the lines. Purge the gas line.

4. Check the pilot line and orifice for debris.

5. Test the thermocouple, thermostat, and ignitor. Replace if necessary.

6. Check for drafts that can extinguish the pilot.

7. Check the intake air for obstruction. This is the air that is needed for combustion.

8. Check for flammable vapors.

Not Enough Hot Water

1. Check the thermostat; it might be set too low.

2. Check the installation for correct piping connections. Is the tank adequate for the size of the household and usage?

3. Check the home for hot water leaking from the faucets. Did you notice any wasting of hot water? If so, speak to the consumer.

4. Where are the water pipes located? Are the pipes exposed to the elements? Are they insulated? Check on the length of the piping.

5. Check the tank for sediment or lime. You may have to flush out the tank.

Slow Hot Water Recovery

1. Check for insufficient secondary air supply. Check the venting system and the burner.

2. Check the gas pressure. Is the type of gas correct for the water heater?

3. Check the thermostat setting. Is it out of calibration?

4. If this is a new installation, check the piping connections.

5. Check for wasting of water.

6. Is the water heater the correct size for the household?

The Burner Flame Is Too High

1. Is the orifice too large? You can change the orifice and retest the burner.

2. Check for sufficient secondary air supply. Check the venting system and the burner.

The Flame Burns at the Orifice

1. Check the thermostat. Replace if necessary.

2. Check the gas pressure to the water heater.

The Pilot Flame Is Too Small

1. Check the pilot line and orifice for debris.

2. Check the gas pressure to the water heater.

The Burner Flame Floats and Lifts Off the Ports

1. Is the orifice too large? You can change the orifice and retest the burner.

2. Check for high gas pressure.

3. Check the flue and burner. Locate the source and correct the problem.

4. Check for drafts that can blow the flame off the burner.

A Yellow-Lazy Burner Flame

1. Check for sufficient secondary air supply.

2. Check the gas pressure.

3. Check the venting system. Is the flue clogged?

4. Check the burner line for debris.

5. Check the water heater installation. Is the tank installed in a confined space?

6. Check the burner orifice for obstruction.

The Formation of Soot (Smoke and Carbon)

1. Check for sufficient secondary air supply.

2. Check the flue way, flue baffle, and burner.

3. Test the draft hood for proper ventilation.

4. Check for a yellow-lazy burner flame.

5. Check for low gas pressure to the water heater.

6. Check the thermostat. Replace if necessary.

7. Is the water heater installed in a confined space?

Condensation

Check the thermostat setting. Is it set too low?

The Water Has an Odor

Check for sulfides in the water. If necessary, replace the anode.

The Water Heater Jacket Is Discolored Above the Combustion Area

1. Check that the burner orifice is the proper size.

2. Is the flue clogged?

3. Check for high gas pressure.

Combustion Odors

1. Check for sufficient secondary air. Check the flue way, flue baffle, and the burner.

2. Is the flue clogged?

3. Is the water heater installed in a confined space?

Repair Procedures

Each repair procedure is a complete inspection and repair process for a single water heater component. It contains the information you need to test a component that may be faulty and to replace it, if necessary.

Burner/Manifold Assembly

The typical complaints associated with the burner/manifold assembly are:

- No gas to the burner.
- Yellow-lazy burner flame.
- The flames lift off the burner ports.
- The burner will not ignite.
- High temperatures. Some models have a thermal cutoff device installed to prevent the gas flow if the temperature in the combustion chamber exceeds its limit.
- Poor maintenance. It's time to clean the burner/manifold assembly.

To handle this problem, perform the following steps:

1. **Verify the complaint.** To verify the complaint, run the hot water. Does the burner come on? Check the thermal cutoff device. What color is the burner flame? Does the burner/manifold assembly need cleaning?

2. **Check for external factors.** You must check for external factors not associated with the water heater. Is the gas supply turned on? Is the water heater installed properly? Is there sufficient air supply? Is the surrounding area clear of debris?

3. **Shut off the gas supply.** Before you begin servicing any gas components, shut off the supply of gas to the water heater. Turn the gas control knob on the combination gas control to the off position (Figure 17-5a). See the use and care guide for the model you are servicing.

4. **Remove the burner/manifold.** Remove the screws that secure the manifold door to the skirt (see Figure 17-5a). At the combination gas control, disconnect the following: thermocouple, ignitor wire from the ignitor button, pilot tube, and the manifold tube (Figure 17-5b). Gently push down on the manifold tube to free the manifold, thermocouple, and pilot tube. Next, carefully remove the burner/manifold assembly from the burner compartment. Try not to damage any of the internal parts.

5. **Clean the burner.** Now that the burner/manifold assembly is removed from the water heater, you can inspect the burner and clean it. Disassemble the burner from the manifold assembly by removing the two screws that secure it to the manifold (Figure 17-6). To clean the burner, use soap and hot water, rinse, and let it dry. Clean the remainder of the manifold assembly with a brush, and blow off any debris from the flame trap. Inspect the pilot tube and the manifold tube for a blockage. Before reinstalling the burner/manifold assembly, clean out the combustion chamber (Figure 17-7) with a shop vacuum, making sure that you remove all of the loose debris. Some models are manufactured with a flame arrestor to prevent flames from escaping the combustion chamber. This component must also be cleaned with a vacuum to remove debris.

6. **Reinstall the burner/manifold assembly.** To reinstall the burner/manifold assembly, just reverse the disassembly procedure. Before you get started, inspect the gaskets for wear and tear, and replace them if necessary. Reconnect the pilot tube, manifold tube, thermocouple, and ignitor wire. Next, turn on the gas supply and light

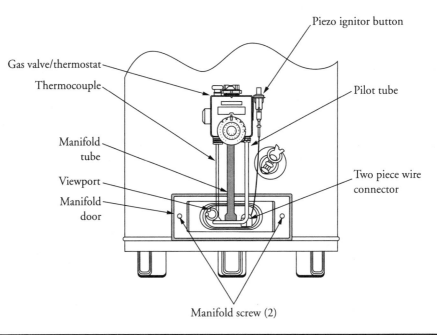

Gas valve/thermostat

Thermocouple

Piezo ignitor button

Pilot tube

Manifold tube

Viewport

Manifold door

Two piece wire connector

Manifold screw (2)

FIGURE 17-5a   Component identification and location on a gas water heater.

FIGURE 17-5b
The combination control valve. The ignitor, thermocouple, manifold tube, pilot tube, and ignitor will have to be disconnected in order to remove the valve assembly.

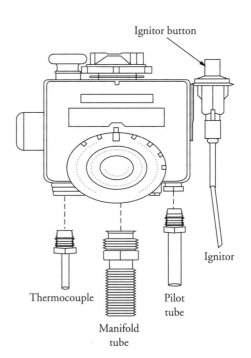

Ignitor button

Ignitor

Thermocouple

Pilot tube

Manifold tube

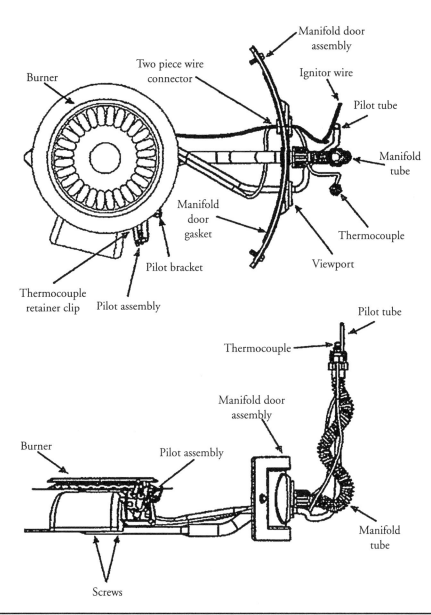

FIGURE 17-6   The top and side view of the burner/manifold assembly.

the burner according to the manufacturer's recommendations for the model you are servicing. Check for gas leaks by using a chloride-free soap solution. *Do not use a flame to check for gas leaks.* The soap solution will begin to bubble if a leak is present. If no leaks are found, you are ready to reinstall the outer door.

Tab

Flame-trap

Bracket

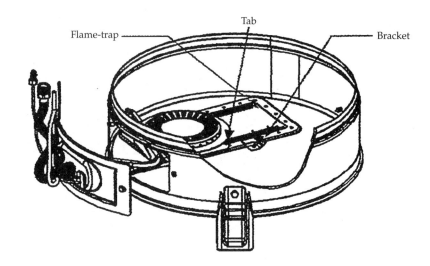

FIGURE 17-7
The burner/manifold assembly located in the combustion chamber.

7. **Test the water heater.** Once assembled, run the water heater for a number of cycles by letting the hot water run for a while and then shutting it off. The water heater will begin cycling. Inspect the flame—it should be a blue flame. Adjust the air shutter as needed. On some models, you will have to remove the burner/manifold assembly to adjust the air shutter.

Combination Control Valve

The combination control valve consists of a gas pressure regulator valve, thermostat, and ignitor. The combination control valve also has a gas cutoff device incorporated into the valve to shut off the gas supply in the event that the thermostat fails. In Figure 17-8, the thermostat does not have numbered temperature markings only notches that indicate 60 degrees and up.

The typical complaints associated with the combination control valve are:

- The burner will not ignite.
- Unable to control the water temperature.
- Slow hot water recovery.
- Smoke and carbon buildup (sooting).
- The combination control will not shut off the gas to the burner.
- Some models have a thermal cutoff device installed to prevent gas flow if the temperature in the combustion chamber exceeds its limit.

To handle this problem, perform the following steps:

1. **Verify the complaint.** To verify the complaint, check the burner operation. Test the water temperature and the thermostat setting. Does the burner cycle at the predetermined temperature? Check the thermal cutoff device. Check for sooting. Can you shut off the burner with the thermostat control knob?

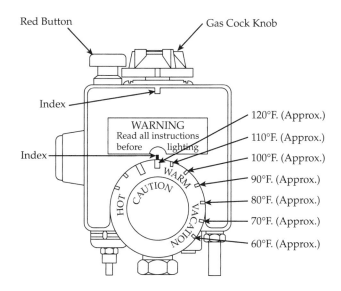

FIGURE 17-8
The notches on the thermostat dial indicate approximately 10-degree increments starting from 60° F to the first index mark as shown.

2. **Check for external factors.** You must check for external factors not associated with the water heater. Is the water heater installed properly? Is there sufficient air supply? Is the surrounding area clear of debris? Is the ventilation system clear of debris?

3. **Shut off the gas supply.** Before you begin servicing any gas components, shut off the supply of gas to the water heater. Turn the gas control knob on the combination gas control to the "off" position (see Figure 17-5a). See the use and care guide for the model you are servicing.

4. **Remove the combination gas control valve.** Turn off the combination gas control valve according to the use and care guide for the model you are servicing. Turn off the main gas supply to the water heater. Before you can remove the control valve, you first have to drain the water heater tank. To drain the water heater tank, turn off the water supply to the tank. Connect a garden hose to the draincock, and open the valve. To increase the water flow draining out of the tank, open the hot water faucet—this will increase the flow of water draining out of the tank. While the tank is draining, disconnect the ignitor wire from the ignitor, and remove the ignitor button and bracket. Remove the manifold tube and pilot tube, and disconnect the thermocouple (see Figures 17-5a and 17-5b). Next, disconnect the ground joint union located between the gas shutoff valve and the combination valve. Disconnect the remaining pipe from the combination gas control valve. When the water heater tank is empty, you are ready to remove the control valve. Turn the valve counterclockwise and remove it from the tank.

5. **Install the combination gas control valve.** To install a combination gas control valve, just reverse the order of disassembly. Make sure that you use approved pipe joint compound or Teflon yellow tape on the pipe connections (according to local codes) and on the threads of the control valve before you install the valve into the water heater tank. This will prevent water from leaking out of the tank. Close the draincock

PART VI

and turn on the water supply to the tank, allowing it to fill up. Keep a faucet open near the tank to allow air to escape from it. As for the gas lines, you will perform a leak check using a chloride-free soap and water solution when the gas supply is turned back on. Finally, check for any water leaks before lighting the burner.

Thermocouple

A thermocouple is a measuring device consisting of two dissimilar metals, which produces a low DC voltage when heated by a gas pilot flame that is measured in millivolts. The average output voltage for a single thermocouple is between 20 and 30 millivolts.

The typical complaints associated with the thermocouple are:

- The burner will not ignite.
- The pilot light will not remain lit.

To handle this problem, perform the following steps:

1. **Verify the complaint.** To verify the complaint, check the burner operation. Does the burner come on? Check the pilot light. Is it lit?

2. **Check for external factors.** You must check for external factors not associated with the water heater. Is the water heater installed properly? Check the gas supply line. Is it turned on? Is there sufficient air supply? Is the surrounding area clear of debris?

3. **Remove the burner/manifold.** Remove the screws that secure the manifold door to the skirt (see Figure 17-5a). At the combination gas control, disconnect the following: thermocouple, ignitor wire from the ignitor button, pilot tube, and the manifold tube (see Figure 17-5b). Gently push down on the manifold tube to free the manifold, thermocouple, and pilot tube. Next, carefully remove the burner/manifold assembly from the burner compartment. Try not to damage any of the internal parts. Remove the thermocouple two-piece connector by removing the retainer clip that secures the thermocouple connector to the manifold door (Figure 17-9). Disconnect the thermocouple from the pilot bracket (Figure 17-10).

4. **Install the new thermocouple.** To install a new thermocouple, just reverse the order of disassembly. Insert the new thermocouple into the bracket holes until it clicks into place. Now you are ready to reinstall the burner/manifold assembly and test.

Ignitor

The typical ignitor that is used on gas water heaters is the piezoelectric ignitor (Figure 17-11). When the button on the ignitor is pushed down, it produces a spark at the burner.

The typical complaint associated with the ignitor is that the burner will not light or remain lit. To handle this problem, perform the following steps:

1. **Verify the complaint.** To verify the complaint, check the burner operation. Does the burner come on? Check the pilot light. Is it lit? Turn off the gas supply and observe the ignitor electrode when activating the ignitor. Do you see a spark at the ignitor?

FIGURE 17-9
The thermocouple assembly. Do not bend the thermocouple as indicated; it has an internal thermal fuse that could be damaged.

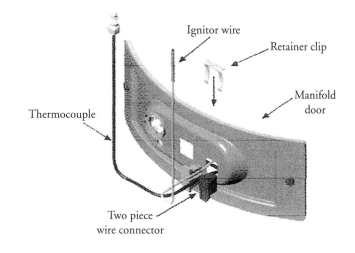

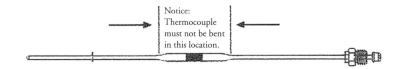

2. **Check for external factors.** You must check for external factors not associated with the water heater. Is the water heater installed properly? Check the wiring connections and wiring. Is the ignitor electrode broken?

3. **Remove the burner/manifold.** Remove the screws that secure the manifold door to the skirt (see Figure 17-5a). At the combination gas control, disconnect the following: thermocouple, ignitor wire from the ignitor button, pilot tube, and the manifold tube (see Figure 17-5b). Gently push down on the manifold tube to free the manifold, thermocouple, and pilot tube. Next, carefully remove the burner/manifold assembly from the burner compartment. Try not to damage any of the internal parts. Remove the ignitor electrode from the pilot bracket (Figure 17-12a and b).

FIGURE 17-10
The thermocouple bracket assembly. The electrode must be positioned 0.125 inches from the pilot.

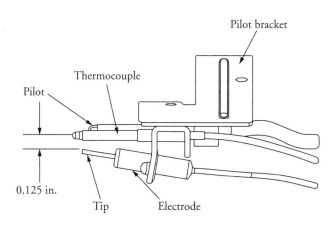

FIGURE 17-11
A piezoelectric ignitor.
This type of ignitor is
used to ignite the gas
burner.

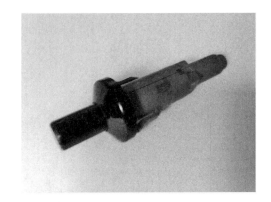

FIGURE 17-11
A piezoelectric ignitor.
This type of ignitor is
used to ignite the gas
burner.

FIGURE 17-12
(a) The natural gas
ignitor assembly. (b)
The LP gas ignitor
assembly.

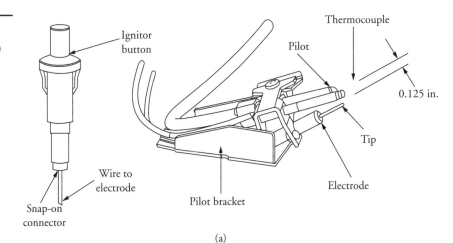

(a)

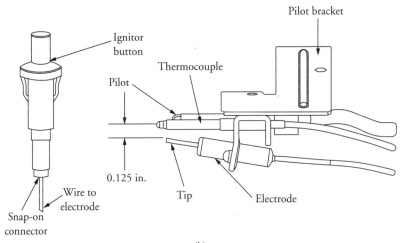

(b)

4. **Install the new ignitor.** To install the new ignitor, just reverse the order of disassembly. When positioning the ignitor electrode in the pilot bracket, the spark gap should be 0.125 inches. Reinstall the burner/manifold assembly and test.

ater Heater Leakage Inspection Points

A technician must check the following areas for leakage when servicing the water heater (Figure 17-13):

A. Check the draft hood for condensation. Water vapor that has formed from combustion products is an indication of a problem in the vent system. If this problem is not corrected, the top of the water heater will begin to rust.

B. On humid days, condensation may appear on the pipes and connections.

C. Check the anode rod fitting for leakage.

D. Check for leakage at the temperature and pressure-relief valve and piping. The leakage could be from thermal expansion or high water pressure to the household. If there is a steady stream of water coming from the valve, the temperature and pressure-relief valve may be defective.

E. Check for leakage from the temperature and pressure-relief valve fitting at the tank.

F. The drain valve may be leaking. Check if the valve is slightly open.

G. Check for leakage from the drain valve fitting at the tank.

H. Check for leakage in the combustion chamber. Combustion products that contain water vapor could form droplets in that area. When cold water enters the tank, condensation may form droplets in that area, too. These droplets can fall on the burner/manifold area, causing damage to the component. In addition, the droplets can run onto the floor. This usually happens on new installation startups.

I. Check the rim around the outside of the water heater tank for leakage, the water heater bottom, and on the floor. Check for leakage at all pipe connections and at the temperature and pressure-relief valve. Check and see if the temperature and pressure-relief valve discharge line is directed over a drain or runs to the outside of the residence. In some cases, I found the discharge line was draining on the floor and was not piped correctly.

Other possibilities for water leakage around the water heater tank can come from other appliances (for example, an automatic washer or air conditioner), water lines, and ground water seepage.

iagnostic Charts

The following diagnostic flowchart (Figure 17-14) will help you to pinpoint the likely causes of the problem. See also Figures 16-25 and 16-26 and Table 16-1.

Table 17-1 illustrates how to figure the hot water recovery rate by hand when you know the BTU input located on the data tag:

Hourly input (BTU's)

11.0 x temperature rise °F

FIGURE 17-13
Leakage inspection
points on a gas water
heater.

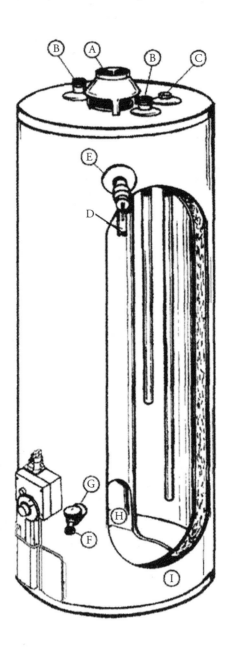

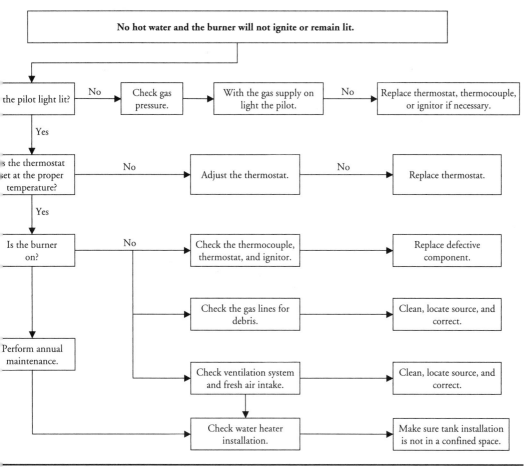

INPUT BTU	40°F	50°F	60°F	70°F	80°F	90°F	100°F	110°F	120°F	130°F	140°F
20,000	45	36	30	26	23	20	18	17	15	14	13
26,000	59	47	39	34	30	26	24	21	20	18	17
28,000	64	51	42	36	32	28	25	23	21	20	18
30,000	68	55	45	39	34	30	27	25	23	21	19
32,000	73	58	48	42	36	32	29	26	24	22	21
34,500	78	63	52	45	39	35	31	29	26	24	22
36,000	82	65	55	47	41	36	33	30	27	25	23
37,000	84	67	56	48	42	37	34	31	28	26	24

BLE 17-1 Recovery Rates in Gallons per Hour for Gas Water Heaters (*continued*)

INPUT BTU	40°F	50°F	60°F	70°F	80°F	90°F	100°F	110°F	120°F	130°F	140°F
40,000	91	73	61	52	45	40	36	33	30	28	26
50,000	114	91	76	65	57	51	45	41	38	35	32
57,000	130	104	86	74	65	58	52	47	43	40	37
60,000	136	109	91	78	68	61	55	50	45	42	39
69,000	157	125	105	90	78	70	63	57	52	48	45
75,000	170	136	114	97	85	76	68	62	57	52	49
98,000	223	178	148	127	111	99	89	81	74	69	64
100,000	227	182	152	130	114	101	91	83	76	70	65
114,000	259	207	173	148	130	115	104	94	86	80	74
156,000	335	284	236	203	177	158	142	129	118	109	101
160,000	364	291	242	208	182	162	145	132	121	112	104
180,000	409	327	273	234	205	182	164	149	136	126	117
199,900	454	363	303	260	227	202	182	165	151	140	130
250,000	568	455	379	325	284	253	227	207	189	175	162
270,000	614	491	409	351	307	273	245	223	205	189	175
300,000	682	545	455	390	341	303	273	248	227	210	195
399,900	909	727	606	519	454	404	364	330	303	280	260
500,000	1136	909	758	649	568	505	455	413	379	350	325

TABLE 17-1   Recovery Rates in Gallons per Hour for Gas Water Heaters

Formulas and Facts

1 gallon of water weighs 8.33 lbs

1 gallon of water has a volume of 231 cubic inches

1 cubic foot of water weighs 62.38 lbs and contains 7.48 gallons of water

100 feet of ¾-inch copper pipe contains 2.5 gallons of water; 1-inch pipe contains 4.3 gallons

8.33 BTU will raise 1 gallon of water 1°F at 100 percent efficiency (electricity)

11 BTUs are required to raise 1 gallon of water 1°F at 75 percent efficiency (gas)

3412 BTU equals 1 kilowatt hour (Kwhr)

1 Kwhr will raise 410 gallons of water 1°F at 100 percent efficiency

1 BTU × 0.293 = watts

1 KW = 1000 watts

2.42 watts are required to raise 1 gallon of water 1°F

1 Kwhr will raise 10.25 gallons of water 40°F at 100 percent efficiency

1 Kwhr will raise 6.8 gallons of water 60°F at 100 percent efficiency

1 Kwhr will raise 5.1 gallons of water 80°F at 100 percent efficiency

1 Kwhr will raise 4.1 gallons of water 100°F at 100 percent efficiency

Electric Water Heaters

Energy Costs:

Kwhr × fuel costs = energy costs

If I use 100 kilowatt hours of electricity, how much will it cost if each kilowatt hour costs $.05?

100 × .05 = $5.00

To obtain gallons per hour (GPH) recovery:

$$\frac{\text{Watts}}{2.42 \times (\text{temp rise } ^\circ F)}$$

I have a 30-gallon electric heater, nonsimultaneous operation, 4500 watt elements. What is the recovery GPH if my cold water is 40° F and my thermostat is set to 120° F?

$$\frac{4500}{2.42 \times 80} = 23 \text{ gallons per hour}$$

Temperature Rise (°F)

$$\frac{\text{Watts}}{2.42 \times \text{GPH}}$$

I have a 30-gallon electric heater, nonsimultaneous operation, 4500 watt elements. What is the maximum temperature rise if the heater can recover 23 gallons per hour?

$$\frac{4500}{2.42 \times 23} = 80 \text{ degrees temperature rise}$$

Gas Water Heaters

Energy Costs:

Cubic feet × fuel costs = energy costs

If I use 100 cubic feet of gas, how much will it cost if each cubic foot costs $.075?

100 × .075 = $7.50

To obtain gallons per hour (GPH) recovery:

$$\frac{\text{Hourly Input (BTUs)}}{11.0 \times (\text{temp rise } ^\circ F)}$$

I have a 30-gallon gas heater rated at 40,000 BTUs. What is the recovery GPH if my cold water is 40° F and my thermostat is set to 120° F?

$$\frac{40,000}{11.0 \times 80} = 45 \text{ gallons per hour}$$

Temperature Rise (°F)

$$\frac{\text{Hourly Input (BTUs)}}{11.0 \times (\text{GPH})}$$

I have a 30-gallon gas heater rated at 40,000 BTUs. What is the maximum temperature rise if the heater can recover 45 gallons per hour?

$$\frac{40,000}{11.0 \times 45} = 80 \text{ degrees temperature rise}$$

Formula for Mixing Hot Water

$$\frac{M - C}{H - C} = \text{Percent of hot water required to produce desired mixed water temperature}$$

where M = mixed water temperature, C = cold water temperature, H = hot water temperature. For example: How much of a shower is hot water and how much is cold water? My shower temperature is 105° F, my water heater thermostat is set on 120° F, and the cold water inlet temperature is 50° F.

$$\frac{105 - 50 = 55}{120 - 50 = 70} = 79\% \text{ of the shower is } 120° \text{ hot water}$$

This formula for mixing hot water is important when explaining to a customer why there might not be enough hot water when the water heater is functioning properly.

Electric Working Formulas That Apply to Water Heaters

To verify circuit breaker or amperage draw on the water heater:

$$\text{Amperage draw} = \frac{\text{Watts}}{\text{Voltage}}$$

$$\text{Amperage draw} = \frac{4500 \text{ watts (heating elements shown on rating plate total)}}{240 \text{ volts (shown on the rating plate)}}$$

Amperage draw = 18.8 amps (circuit breaker should be 20% higher or 25 amp breaker)

Now that you have solved amperage, what is the resistance (Ohms) of the heating element? (If a heating element has no resistance, then it is open, or broken.)

$$\text{Resistance (Ohms)} = \frac{\text{Voltage}}{\text{Amperage}}$$

$$\text{Resistance (Ohms)} = \frac{240 \text{ volts (shown on the data rating plate)}}{18.8 \text{ (answer from previous problem)}}$$

Resistance = 12.8 Ohms

Help on the Internet

To obtain additional technical help or information on selecting a gas water heater, you can go to the following websites:

http://www.aceee.org/consumerguide/topwater.htm
http://www.aosmith.com/

http://www.eere.energy.gov/consumer/your_home/water_heating/index.cfm/
mytopic=12770
http://www.americanwaterheater.com/
http://www.rheem.com/
http://www.hotwater.com/
http://www.tankless-water-heater.com/
http://www.bradfordwhite.com/
http://www.marathonheaters.com/
http://www.statewaterheaters.com/

Top Load Automatic Washers

The automatic washer is a complex electromechanical machine, and Figure 18-1 is used as an example only. The actual construction and features of the washer you are servicing might vary, depending on brand and model. The automatic washer performs various cycles to clean clothes. There are times when a washer fails to operate properly. Don't let its complexity intimidate you. This chapter will provide the basics needed to diagnose and repair the washing machine. On electronic models, you may need to have the service manual and the technical data sheet (located inside the control panel) present to perform the diagnostic test/service modes.

Principles of Operation

The clothes are placed evenly into the washer basket, making sure that the washer is not overloaded and that the proper cycle is selected. The user then adds detergent to the washer (see the use and care instructions), and then the user activates the washer through the timer. The internal switches of the timer distribute the electricity to activate the other components in the washing machine during a given time period, designated by the internal cam of the timer.

The water enters the tub through the water fill hoses, the water inlet valve, and water inlet hose. Hot, warm, or cold water is selected by the user via the water temperature selector switch, located on the console panel. On some models, the water temperature selection is controlled by the timer.

The amount of water that fills the tub is controlled by the water level control (pressure switch). The water level control offers a choice of water levels, depending on the amount of clothing being washed. As the water level rises in the tub, it forces air through the air dome and up the plastic tube to the water level control. The pressure that is exerted on the water level control's diaphragm will trip the water level switch from empty to full, supplying electricity to the washer drive motor, and thus operating the transmission.

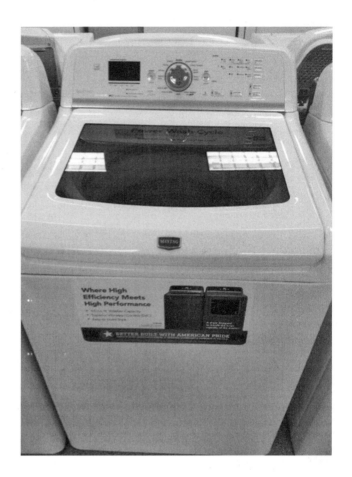

The transmission/gearbox is operated by the drive motor, either by belt drive or by direct drive, depending upon the model. Agitation is accomplished by the agitator, which is located in the center of the basket and which is driven by the transmission/gearbox. The agitator turns clockwise and counterclockwise, creating a water motion that moves the clothes within the basket.

When the washer goes into the drain mode, the agitator will stop agitating, and the water leaves the bottom of the tub through the water pump, to be pumped into the drain. The water pump may operate by belt drive, by direct motor drive, or by an electric motor.

A final deep rinse cycle (the tub fills to the selected water level and begins to agitate) will be introduced to wash off any remaining residue of soap or dirt.

The timer will now select the spin cycle, and the washer will then go into the spin mode. In the spin mode, the washer spins the clothes, removing most of the water from the clothing by centrifugal force. Some models use brief sprays of water to remove any residue of soap or dirt remaining on the clothes in the spin cycle.

Functions and Cycles

The removal of soil from clothing and fabrics is accomplished by a combination of mechanical and chemical processes:

- **Mechanical process**   Soil is removed by agitating and by forcing the detergent through the clothing.
- **Chemical process**   The detergent used will dissolve and loosen the soil in the clothing. As the washing machine operates through its cycles, it is aided by hot, soft water, which increases the chemical processes of the detergent being used.

Top load washing machines perform four basic functions, which are modified and put together in different ways to create the various cycles. The four functions are:

- Fill
- Agitate
- Drain
- Spin

afety First

Any person who cannot use basic tools or follow written instructions should *not* attempt to install, maintain, or repair any top load automatic washers. Any improper installation, preventive maintenance, or repairs could create a risk of personal injury or property damage.

If you do not fully understand the installation, preventive maintenance, or repair procedures in this chapter, or if you doubt your ability to complete the task on the top load automatic washer, please call your service manager.

The following precautions should also be followed:

- Never bypass or disconnect any part or device (originally designed into the washer) as a temporary repair.
- Always reconnect all ground wires, and be sure that they are secure.
- Be careful of moving parts and sharp edges.

Before continuing, take a moment to refresh your memory of the safety procedures in Chapter 2.

utomatic Washers in General

Much of the troubleshooting information in this chapter covers top load automatic washers in general, rather than specific models, in order to present a broad overview of service techniques. The pictures and illustrations that are used in this chapter are for demonstration purposes only, to clarify the description of how to service washing machines, and in no way reflect a particular brand's reliability.

PART VI

Location and Installation of the Top Load Automatic Washer

The following are some general principles that should be followed when performing the installation of a top load washing machine:

- Locate the washing machine where there is easy access to existing drain, water, and electrical lines.
- Be sure you observe all local codes and ordinances for the electrical and plumbing connections.
- The washing machine should be installed and leveled on a firm floor to minimize vibration during operation.
- Do not install the washing machine in an area where the temperature might be below freezing.
- To reduce the risk of a fire, never install a washing machine on any type of carpet.
- Always follow the installation instructions that are provided with every new washing machine model purchased.

Common Installation Problems

Top load automatic washer installations are not complicated. As a service technician, you will come across a top load washer that has not been installed according to the manufacturer's installation instructions. The following sections describe some of the problems that you might run into and how to solve them.

When you arrive at a service call and the consumer tells you that the washer is full of water and it will not drain out, check for the following:

- The drain hose is kinked.
- The drain hose has a blockage.
- The drainpipe might be too small, not allowing for proper venting. The drainpipe must be a minimum of 1½ inches in diameter.
- The drainpipe and the drain hose were installed over six feet above the floor.

If the consumer tells you the washer will not fill up with water or very little water is entering the washer, check the following:

- The water faucets were never turned on or they were turned on only a little bit.
- There is a blockage in the water inlet screen. The consumer did not flush out the water lines before installing the hoses on the washer.
- When the washer is filling with water, the water is siphoning out through the drain line. The drainpipe is too low or the wrong size diameter. You might have to install a siphon break kit at the end of the drain hose.

When you arrive at the service call, you notice water on the floor. Check the following:

- The water hoses to the water inlet valve
- The water faucet might be leaking

- The drain hose is not connected properly to the washer
- The drain hose comes out of the drainpipe
- The drainpipe might have a blockage
- The household drain cannot handle the capacity of the discharge water

When the washer goes into the spin cycle, it will begin to vibrate and walk across the floor. Check the following:

- The feet or leveling legs on the washer are not set properly, causing the washer to teeter-totter.
- Is the washer level?
- Is the floor level?
- Check and see if the packing and shipping straps have been removed. Read the installation instructions on removing the packing straps.
- Are the clothes distributed evenly within the tub?

Water Supply

The water supply for an automatic top load washer should have a hot and cold faucet located within five to seven feet of the washer. The faucets should be a 3/4-inch threaded type to accept the fill hose connection.

The water pressure must be between 25 and 125 pounds per square inch for the washer to operate properly. The water coming out of the fill hoses should be equal in both pressure and in the volume of water to prevent unacceptable water temperature changes when entering and filling the washer.

The hot water supply to the washer should be between 130 and 150 degrees Fahrenheit. If the hot water temperature is below 70 degrees Fahrenheit, the clothes being washed will not clean properly and the detergent will not dissolve properly. You can check the temperature of the hot water by operating the washer in the fill mode with the water temperature setting on hot. Let the water run until it is as hot as possible, and then insert a thermometer into the stream of water. If the thermometer reading is below 130 degrees Fahrenheit, you will have to raise the water heater thermostat setting. The cold water temperature should be between 70 and 100 degrees Fahrenheit. When the washer is in its rinse stage, the cold water will prevent wrinkles from setting into the fabrics. Some fabric manufacturers require that their fabrics be washed in cold water, both to prevent shrinkage and to eliminate the possibility of destroying the fabric. When the user selects the warm fill, the temperature of the water should be 100 degrees Fahrenheit.

It is recommended that the consumer read the use and care manual before performing a wash. Most use and care manuals have a water temperature guide to assist the user in the proper selection of the water temperature.

Drain Requirements

The drain to which the washer's drain hose is connected must be able to accept at least a 20- to 30-gallon-per-minute flow in order to remove the water from the tub. The standpipe should be at least 32 inches high and not exceed 60 inches in height. The internal diameter of the drain pipe should be a minimum of 1½ inches in order to provide an air gap around the drain hose and thus prevent the suction from siphoning the water out of the tub during the wash cycle.

Detergent

The kind and amount of detergent that is used is an important part of getting clothes clean. Some top load automatic washer models are designed to use HE (high efficiency) detergent while other models use regular detergent. Different brands of detergent contain different amounts of phosphorous, which works to soften the water and to boost the cleaning action. If the water is hard, you might need to recommend a detergent with a higher phosphorous content. If the water is soft, the user can use a low-phosphorous detergent.

Some areas restrict the phosphate content to 8 percent or less. This means that the user will have to increase the amount of detergent used in those areas where the water is hard. This is done by adding a certain extra amount of detergent manually to the wash cycle.

It is recommended that the consumer read the use and care manual before performing a wash. Most use and care manuals have a detergent guide to assist the user in the recommended amount of detergent to use.

Water hardness is measured in grains:

- 0 to 3 grains: soft water
- 4 to 9 grains: medium-hard water
- 10 to 15 grains: hard water
- Over 15 grains: very hard water

If you do not know the hardness of the water supply, contact the local water department.

Step-by-Step Troubleshooting by Symptom Diagnosis

In the course of servicing an appliance, don't overlook the simple things that might be causing the problem. Step-by-step troubleshooting by symptom diagnosis is based on diagnosing malfunctions, with their possible causes arranged into categories relating to the operation of the washer. This section is intended only to serve as a checklist to aid you in diagnosing a problem. Look at the symptom that best describes the problem that you are experiencing with the washer, and then proceed to correct the problem.

No Water Entering Washer

1. Is the washer plugged in?
2. Check for proper voltage at the wall receptacle.
3. Check the fuse or reset the circuit breaker.
4. Is the water supply turned on? The fill hoses should feel stiff.
5. Test the water temperature switch contacts for continuity.
6. Check for an open circuit in the timer contacts. On electronic models, run the test mode.
7. Check for loose wires to the water valve solenoid.

8. Check the entire wiring harness for loose connections.

9. Test water valve solenoid coils for continuity.

10. Check the water valve inlet strainer screens. Remove the fill hoses to inspect these filters.

11. Test the water level control switch for continuity.

12. Are the water supply hoses kinked?

13. Check water valves separately for fill.

14. Check for low water pressure at the washer and in the home.

15. Check for frozen pipes and washer fill hoses.

Motor Will Not Run

1. Check for proper voltage at the wall receptacle.

2. Is the washer plugged in? The voltage at the receptacle is between 108 volts and 132 volts during a load on the circuit.

3. Check for a blown fuse or a tripped circuit breaker.

4. Check the line filter.

5. Check for a faulty timer. On electronic models, run the test mode and check the electronic control board.

6. On electronic models, check the motor drive board.

7. Are there any loose wires to the timer, motor, etc.?

8. Test the motor windings for continuity.

9. Test the thermal overload in the motor for continuity.

10. Test the water level control switch contacts for continuity.

11. Test the motor speed selector switch contacts for continuity.

12. Are there any open wires in the washer circuit?

13. Test the capacitor on the motor using a capacitor tester.

14. Check the centrifugal switch in the motor.

15. Check for continuity of the lid switch contacts. On some models, the lid must be closed before the motor will start.

16. On some models, check for a blown motor thermal fuse.

17. Check for obstructions in the drain pump.

18. On some models, turn the transmission pulley by hand in the agitation direction. If it is locked, replace the transmission.

Washer Will Not Agitate

1. Check fuse or circuit breaker. The voltage at the receptacle is between 108 volts and 132 volts during a load on the circuit.

2. On some models you may have to restart the cycle.

3. Check for a broken or worn belt.

4. Check the motor to the transmission drive coupling (direct-drive models).

5. Test the timer contacts for continuity. On electronic models, run the test mode.

6. Are there any loose wires within the wiring harness?

7. Are there any loose wires to the timer, motor, etc.?

8. Test the continuity of the motor windings.

9. Test the agitator solenoid on the transmission.

10. Check for loose pulleys on the transmission and motor.

11. Check the water level control switch.

12. Test for continuity of the start relay.

13. Test the capacitor on the motor using a capacitor tester

14. Check the centrifugal switch in the motor.

15. Test for continuity of the lid switch contacts.

16. Test for continuity of the speed selector switch.

17. Check the clutch assembly.

18. Check the transmission.

Water Will Not Drain

1. Check for a clogged drain connection.

2. In cold climates, check for frozen drain hose.

3. Inspect the pump for obstructions.

4. Check the drain hose and be sure it is not kinked.

5. Check the belt that goes to the pump.

6. Check for suds lock. If this happens, just add cold water and flush the suds out of the pump. (Suds lock is caused by too much soap remaining in the tub, pump, and the connecting hoses. This condition will prevent water from draining effectively.)

7. On direct drive models, check the pump coupling. The pump and motor must be removed for a visual inspection of the coupling.

8. Check for air lock in the pump (air trapped inside the pump caused by debris).

9. Check to be sure that the motor is not running in the agitation direction.

10. Check the height of the drain.

11. Does the pump pulley turn freely?

12. On models with an electric drain pump motor, check for a blockage and also check the pump motor for continuity. Check for voltage at the drain pump motor.

Washer Will Not Spin

1. Check fuse or circuit breaker. The voltage at the receptacle is between 108 volts and 132 volts during a load on the circuit.

2. Check for a loose or broken belt.

3. Check for loose pulleys.

4. Check the clutch assembly.

5. Check for loose or broken wires in the washer circuit.

6. Test the continuity for a faulty lid switch assembly.

7. Test the continuity for a defective spin solenoid.

8. Check for a broken drive coupling (direct-drive models). The pump and motor must be removed for a visual inspection of the coupling.

9. Test the continuity of the water level control switch.

10. Test the continuity of the speed selector switch.

11. Test the continuity in the motor windings and the motor overload protector.

12. Test the timer contacts for continuity.

13. Check for clothing jammed between the inner basket and the outer tub.

14. Check all seals and mechanical linkages.

15. Check the transmission.

16. Check to be sure that the motor is running in the spin direction.

Washer Speed Too Slow

1. Check voltage at washer. The voltage at the receptacle is between 108 volts and 132 volts during a load on the circuit.

2. Check the spin selection. Make sure it was not selected.

3. Check the RPM of the tub in the spin cycle.

4. Check the drive belt.

5. Make sure the drain system in the home is functioning correctly.

6. Check for a kinked drain hose.

7. Check the tub seal. Is it binding?

8. Check the brake assembly for proper operation.

9. Look between the tubs for a foreign object binding the tubs.

10. Check the drain pump for a broken impeller.

Noisy Washer and/or Vibration/Walking Washer

1. Check for proper washer installation.

2. Check the leveling legs on the washer for proper adjustment and that they are locked into position.

3. Check that the rear legs stabilizer is locked into position.

4. Check to make sure that the pads are on the leveling legs.

5. Check flooring. The washer needs to be installed on a solid foundation.

6. On some models, check the belt.

7. Check the transmission, pulleys, and bearings.

8. Check the center tub seal.

9. Check the outer basket suspension.

10. Check for loose screws in the cabinet, front, and rear panels.

11. Check the base of the washer.

Wrong Water Temperature

1. Run the washer and verify the water temperatures.

2. Check that faucets are turned on fully.

3. Check hot water heater for a minimum of 120 degrees Fahrenheit at the tap.

4. Check hot water heater for capacity and recovery rate. May have to purge water prior to starting a cycle.

5. Check the inlet hoses. Are they reversed?

6. Check water valve, timer, or the automatic temperature control.

Washer Leaks Water

1. Check water inlet hoses.

2. Check water inlet hose connections.

3. Check drain hose installation.

4. Check tub seal.

5. Check all water hose connections in the washer.

6. Check for a cracked outer tub.

7. Check the water pump.

Common Washability Problems

If there are no mechanical problems with the washer's operation and the complaints are that the washer does not clean the clothes properly, you have a washability problem. The next step should be to look at the cause that best describes the problem that the customer is experiencing with the washer. Then proceed to correct the problem. If necessary, instruct the user how to get better results from the automatic washer.

Stains on the Clothing

Stains on clothing can be caused by a number of different things. As the servicer, you will have to determine if it is caused by the washer's components or from an external source.

Many stains are blamed on leaking transmissions. This type of problem is related to the increasing use of synthetic fabrics and to the poor washing practices of the user. Many of these stains consist of cooking oil or grease and are not visible when they first occur during cooking or eating. The oil that is embedded in the clothing acts like glue, attracting dirt from the wash water. When the wash cycle is completed, the clothes come out dirty and spotted. If the transmission oil leaked into the wash water, there will be stains on all of the clothes in a random pattern. The color of transmission oil embedded into the clothing is usually a brownish-yellow stain. Transmission oil cannot be washed out of the clothes; a solvent is required to remove the stains.

The following are some stain-removal rules for clothing in general:

- Stains are easier to remove when they first appear on the clothing. If the stains are old, they might never come out of the clothing.
- Before attempting to remove any stain, you must know: what type of stain, what kind of fabric, and how old is the stain?
- Use only cold or warm water to remove stains. Hot water will set the stain permanently into the fabric.
- When bleach is recommended for the removal of the stain, use a bleach that is safe for the fabric. When using a chlorine bleach, always dilute it with water to prevent the bleach from destroying the fibers.
- Always test stain remover products on a hidden corner of the garment to see if the color remains in the fabric.
- When preparing to remove the stain from the fabric, face the stained area down on a paper towel or a white cloth. Then apply the stain remover to the back of the stain so that the stain will be forced off the fabric, instead of through the fabric.
- Some protein stains can be removed with an enzyme pre-soak or with meat tenderizer.
- When using dry-cleaning solvents, always use them in a well-ventilated room, away from flames and sources of ignition to prevent personal injury.
- Alcoholic beverage stains turn brown the longer they stay on the fabric. As soon as the stain appears on the fabric, start treating the stain immediately. Wash or soak the stain in cold water, and then wash the garment.
- To remove blood stains, rinse or soak the garment in cold water with an enzyme pre-soak product. You can use diluted chlorine bleach on white fabrics, if necessary. For colored fabrics, use a powdered oxygen-type bleach. Then wash the garment.
- To remove chewing gum, use ice on the stain to make the chewing gum hard. Then scrape most of it off the fabric. Next, use a nonflammable dry-cleaning solvent with a sponge to remove the excess chewing gum. Wash the garment.
- To remove coffee or chocolate stains, soak the garment in warm or cold water. Next, make a paste of detergent mixed with hot water, and brush it on the stain. Wash the garment.
- To remove a milk product stain, use a nonflammable dry-cleaning solvent with a sponge. Wash the garment.

- To remove antiperspirant and deodorant stains, wash the garment with laundry detergent in the hottest water that is safe for the fabric. If the stain remains on the fabric, place the stain face down on a white towel, and treat the stain with a paste of ammonia and a powdered oxygen-type bleach. Let the paste stay on the stain for 30 minutes, and then wash the garment in the hottest water that is safe for the fabric.

- To remove fruit stains, soak the stain in cool water. Do not use soap—it will set the stain. Wash the garment. If the stain remains, cover the stain with a paste made of a powdered oxygen-type bleach, a few drops of hot water, and a few drops of ammonia. Let the paste stay on the stain for about 15 to 30 minutes. Then wash the garment.

- To remove iron or rust stains, apply some lemon juice mixed with salt. Then place the garment in the sun. Alternatively, a commercial rust-removing solution can be used. Wash the garment.

Yellowing in Fabric

Some causes of yellowing in fabrics are:

- Poor body soil removal
- Clothes washed in water treated with a water softener
- Hard water or minerals in the water, such as iron
- Body oils released into the garment
- The water supply might pick up the color of decaying vegetation

To remove body oils, the user will have to increase the amount of detergent and use 150-degree Fahrenheit wash water. The user must also increase the frequency of using bleach in the wash.

To remove the yellowing from garments that are washed in water treated with a water softener, the user will have to decrease the amount of detergent used, approximately to the point that the decreased amount will not affect the soil removal process. The user must also increase the frequency of using bleach in the wash.

Hard water and minerals in the water can be treated with a water-conditioning apparatus. The user might have to drain the water heater and flush the tank. Never use chlorine bleach to remove hard water stains or iron stains.

To remove body oils from the garment, use a paste made of detergent and water. Let it stay on the fabric for 15 to 30 minutes. Then wash the garment.

To remove the yellowing caused by decaying vegetation, increase the amount of detergent, and bleach more often. White fabrics typically respond quite well to bleaching.

Fabric Softener Stains

Fabric softener stains are becoming more prevalent because it is now being recommended that some fabric softeners be used in the wash cycle, instead of the rinse cycle. These types of stains show up on synthetics as well as cotton fabrics. They can be removed from the fabric by pretreating the stain with liquid detergent and following the washing procedures listed in the use and care manual.

Lint

Lint is cotton fiber that has broken away from the cotton garment. Lint likes to attach itself to synthetic fabrics. When this happens, the user often thinks that the washer is not performing properly. Therefore, to solve the problem of lint on synthetic fabrics, the user must sort the items before washing the clothes. For example:

- The user must separate cottons from permanent press and knits.
- The user must separate light colors from dark colors.

Another cause of lint on clothes is overwashing. This causes the clothes to wear out faster. To correct the overwashing problem, use only one minute of wash time per pound of dry laundry with normal soil. Any more time than this is a waste, and it usually does not get the laundry cleaner.

If the drain cycle is excessive, this, too, will cause lint to remain on the garments. Check for improper drain hose connections. For example:

- Drain hose is too long (over 10 feet)
- Drain hose is too high (over 5 feet)
- Drain hose is kinked

If excessive drain times still exist, check the following:

- Check the filter, located under the wash basket on most newer models.
- Check to be sure that the pump is operating properly.
- Check for any obstructions in the drain system.
- Check for any obstructions within the water circulatory system of the washer.

utomatic Washer Maintenance

The interior is normally self-cleaning. However, there are times when you might have to remove objects from the inner basket. Clean the control panel and outer cabinet with a soft damp cloth. Do not use any abrasive powders or cleaning pads. Clean and inspect the interior underneath the washer. Read the use and care manuals for the proper maintenance of the brands of washers you service.

Cleaning the Interior of the Top Load Automatic Washer

Every one to two months the interior of the washer will need to be cleaned. This periodic cleaning will remove any dirt, soil, mold, mildew, or bacteria residue that may remain in the washer as a result of washing the clothes. To clean and freshen the washer interior:

1. Add a cup of chlorine bleach in the bleach dispenser. On some models, pour the bleach in the wash water as the tub is filling with water.
2. Add ¼ cup of detergent to the tub or detergent dispenser.
3. Run the washer without clothes through a complete cycle using only hot water.
4. Repeat the process if necessary.

Repair Procedures

Each repair procedure is a complete inspection and repair process for a single washer component, containing the information you need to test a component that might be faulty and to replace it, if necessary.

Any person who cannot use basic tools should *not* attempt to install, maintain, or repair any top load washer. Any improper installation, preventative maintenance, or repairs will create a risk of personal injury, as well as property damage. Call the service manager if installation, preventative maintenance, or the repair procedure is not fully understood.

Washer Timer

The washer timer is an electromechanical component controlled by a synchronous motor in incremental advances. It controls and sequences the numerous steps and functions involved in each cycle.

The typical complaints associated with washing machine timer failure are:

- The cycle will not advance.
- The washer won't run at all.
- The washer will not fill.
- The washer will not pump the water out.
- The washer will not shut off.

To handle these problems, perform the following steps:

1. **Verify the complaint.** Verify the complaint by operating the washer through its cycles. Before you change the timer, check the other components controlled by the timer. If the washer will not power up, locate the technical data sheet behind the control panel for diagnostics information. On some models you will need the actual service manual for the model you are working on to properly diagnose the washer.

2. **Check for external factors.** You must check for external factors not associated with the appliance. Is the appliance installed properly? Does the appliance have the correct voltage? The voltage at the receptacle is between 108 volts and 132 volts during a load on the circuit. Do you have the correct polarity? (See Chapter 6.)

3. **Disconnect the electricity.** Before working on the washer, disconnect the electricity. This can be done by pulling the plug out from the electrical outlet. Be sure that you only remove the washer plug. Or disconnect the electricity at the fuse panel or the circuit breaker panel. Turn off the electricity.

WARNING *Some diagnostic tests will require you to test the components with the power turned on. When you disassemble the control panel, you can position it in such a way that the wiring will not make contact with metal. This act will allow you to test the components without electrical mishaps.*

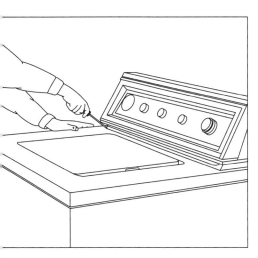
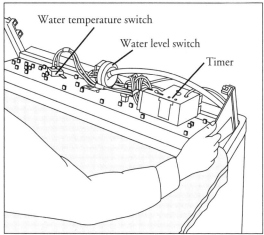

4. **Remove the console panel to gain access.** Begin by removing the screws from the console panel to gain access to the timer (Figure 18-2). Roll the console toward you. On some models, the console will roll upward.

5. **Test the timer.** Remove the timer motor leads from the timer assembly. Test the timer motor by connecting the ohmmeter probes to the timer motor leads (Figure 18-3). Set the range on the ohmmeter to R × 100. The meter should indicate between 2000 and 3000 ohms. Next, test the timer switch contacts using the wiring diagram configuration for the affected cycle. Place the meter probe on each terminal being tested, and turn the timer knob. If the switch contact is good, your meter will read continuity. If the timer motor measures suitably, then connect a 120-volt, fused service cord (Figure 18-4) to the timer motor leads.

FIGURE 18-3
Checking the washer
timer motor.

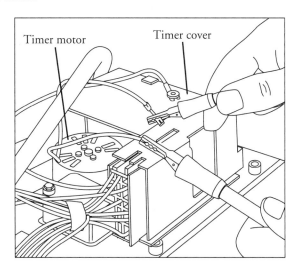

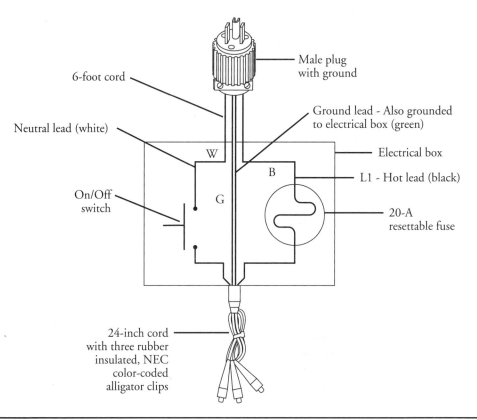

6-foot cord

Male plug
with ground

Ground lead - Also grounded
to electrical box (green)

Neutral lead (white)

Electrical box

W

B

L1 - Hot lead (black)

On/Off
switch

G

20-A
resettable fuse

24-inch cord
with three rubber
insulated, NEC
color-coded
alligator clips

FIGURE 18-4 120-volt fused service test cord.

NOTE *Connect the ground (common) wire test lead to the console ground wire. Be cautious whenever you are working with "live" wires. Avoid any shock hazards.*

If the motor does not operate, replace the timer. If the timer motor runs but does not advance the cams, then the timer has internal defects and should be replaced.

6. **Remove the timer.** To remove the timer, remove the timer mounting screws (Figure 18-5). Remove the wire lead terminals from the timer. Mark the wires as to their location on the timer. Some timers have a disconnect block instead of individual wires, which makes it easier to remove the timer wires.

Turn the timer knob counterclockwise to remove it from the timer shaft, and slide the indicator dial off the shaft.

7. **Install a new timer.** To install a new timer, just reverse the disassembly procedure, and reassemble. Replace the wires on the timer. Reinstall the console panel, and restore the electricity to the washer. Test the washing machine cycles.

RE 18-5
noving the timer.

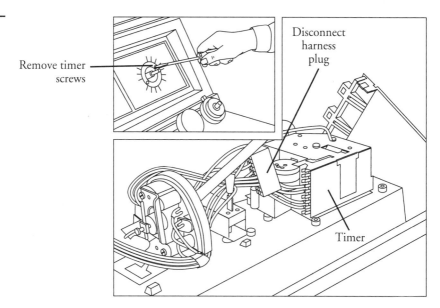

Remove timer screws

Disconnect harness plug

Timer

Electronic Control Board and User Interface Controls

On some models the electronic control board and user interface controls replace the electromechanical timer and rotary selection switches.

The typical complaints associated with the electronic control board or the user interface controls are:

- The washer won't run or power up.
- Unable to program the washer.
- The display board will not display anything.
- One or more key pads will not accept commands.
- Unusual display readouts and/or error codes

To prevent electrostatic discharge (ESD) from damaging expensive electronic components, follow the steps in Chapters 6 and 11.

To handle these problems, perform the following steps:

1. **Verify the complaint.** Verify the complaint by operating the washer. Turn off the electricity to the appliance and wait for two minutes before turning it back on. If a fault code appears, look up the code. If the washer will not power up, locate the technical data sheet behind the control panel for diagnostics information. The service manual will assist you in properly placing the washer in the service test mode for testing the washer functions.

2. **Check for external factors.** You must check for external factors not associated with the appliance. Is there electricity to the washer? The voltage at the receptacle is between 108 volts and 132 volts during a load on the circuit. Do you have the correct polarity? (See Chapter 6.)

3. **Disconnect the electricity.** Before working on the washer, disconnect the electricity. This can be done by pulling the plug out of the wall receptacle. Or disconnect the electricity at the fuse panel or circuit breaker panel. Turn off the electricity.

WARNING *Some diagnostic tests will require you to test the components with the power turned on. When you disassemble the control panel, you can position the panel in such a way so that the wiring will not make contact with metal. This act will allow you to test the components without electrical mishaps.*

4. **Remove the console panel to gain access.** Begin by removing the screws from the washer console to gain access to the electronic control board. On top-loading models, the console will roll upward or toward you after removing the console screws (Figure 18-6).

5. **Test the electronic control board and user interface controls.** If you are able to run the washer diagnostic test mode, check the different functions of the washer. Use the technical data sheet for the model you are servicing to locate the test points from the wiring schematic. Check all wiring connections and wiring. Using the technical data sheet, you can test the electronic control board or user interface controls

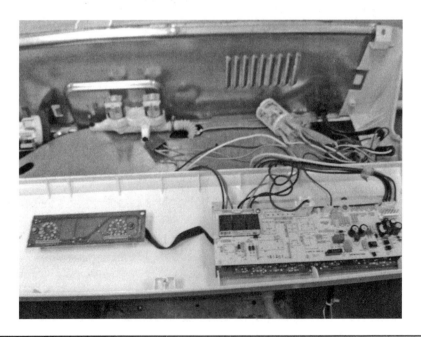

FIGURE 18-6 A view of the control panel parts for a top load automatic washer.

and input and output voltages. On some models, fuses are soldered to the printed circuit board (PCB). These fuses must be tested first before condemning the component.

6. **Remove the electronic control board and user interface controls.** To remove the defective component, remove the screws that secure the board to the control panel or washer frame. On some models you may have to lift a tab and turn the control to remove it. Disconnect the connectors from the electronic control board or user interface control.

7. **Install the new component.** To install a new electronic control board or user interface control, read the data sheet that comes with the part for the proper installation process and just reverse the disassembly procedure and reassemble. Reinstall the console panel, and restore the electricity to the washer. Make sure that the washer is not in the service mode. Test the washer operation.

Water Temperature Selector Switch

The water temperature selector switch will allow the user to choose different water temperatures for the specific wash cycle.

The typical complaints associated with the water temperature selector switch are:

- Inability to select a different water temperature.
- The consumer inadvertently selected the wrong water temperature.

To handle these problems, perform the following steps:

1. **Verify the complaint.** Verify the complaint by trying to select different water temperatures. On electronic models, if a fault code appears, look up the code. If the washer will not power up, locate the technical data sheet behind the control panel for diagnostics information. On some models you will need the actual service manual for the model you are working on to properly diagnose the washer. The service manual will assist you in properly placing the washer in the service test mode for testing the washer functions.

2. **Check for external factors.** You must check for external factors not associated with the appliance. Is the appliance installed properly? Is there any physical damage to the component? Are the fill hoses connected to the hot and cold water supply correctly? Be sure that both the hot and cold water faucets are turned on. The voltage at the receptacle is between 108 volts and 132 volts during a load on the circuit. Do you have the correct polarity? (See Chapter 6.)

3. **Disconnect the electricity.** Before working on the washer, disconnect the electricity. This can be done by pulling the plug out of the electrical outlet. Be sure that you only remove the washer plug. Or disconnect the electricity at the fuse panel or at the circuit breaker panel. Turn off the electricity.

WARNING *Some diagnostic tests will require you to test the components with the power turned on. When you disassemble the control panel, you can position it in such a way that the wiring will not make contact with metal. This act will allow you to test the components without electrical mishaps.*

4. **Remove the console panel to gain access.** Begin by removing the console panel to gain access to the water temperature selector switch (see Figure 18-3).

5. **Test the water temperature selector switch.** To test the water temperature selector switch, remove all wires from the switch. Label the wires.

   Remember, you will have to identify the wires according to the wiring diagram in order to reinstall them onto the water temperature selector switch properly. Take your ohmmeter and check for continuity on the switch contacts. Press or turn on the switch that coincides with the terminals that are being checked (Figure 18-7). At this point, you have to use the wiring diagram to identify the switch contacts.

6. **Remove the water temperature selector switch.** To remove the water temperature selector switch, remove the screws that hold the component to the console panel (Figure 18-8).

7. **Reinstall the water temperature selector switch.** To reinstall the water temperature selector switch, just reverse the disassembly procedure, and reassemble.

NOTE *You will have to identify the wires according to the wiring diagram in order to reinstall them onto the water temperature selector switch properly. Reinstall the console panel, and restore the electricity to the washer. Test the washing machine water temperature cycles.*

Water Valve

The water inlet valve controls the flow of water into the washer, and is solenoid-operated (Figure 18-9). When it is energized, water in the supply line will pass through the valve body and into the washer.

FIGURE 18-7
Checking the water
temperature switch.

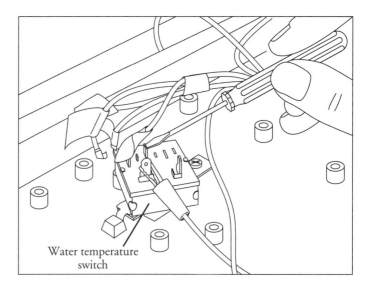

Water temperature
switch

FIGURE 18-8
Removing the water
temperature switch.

The typical complaints associated with water valve failure are:

- The washer will not fill with water.
- The washer overfills and leaks onto the floor.
- When the washer is off, water still enters the tub.

To handle these problems, perform the following steps:

1. **Verify the complaint.** Verify the complaint by operating the washer through its cycles. Listen carefully, and you will hear whether the water is entering the washer. On electronic models, if a fault code appears, look up the code. If the washer will not power up, locate the technical data sheet behind the control panel for diagnostics information. On some models you will need the actual service manual for the model you are working on to properly diagnose the washer. The service manual will assist you in properly placing the washer in the service test mode for testing the washer functions.

FIGURE 18-9
A typical water valve
used in automatic top
load washers.

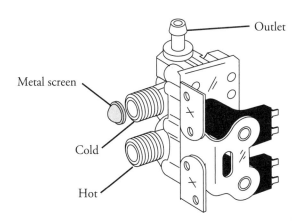

PART VI

2. **Check for external factors.** You must check for external factors not associated with the appliance. Is the appliance installed properly? Does the appliance have the correct voltage? The voltage at the receptacle is between 108 volts and 132 volts during a load on the circuit. Do you have the correct polarity? (See Chapter 6.) Is the water turned on? Both water faucets must be turned all the way counterclockwise.

3. **Disconnect the electricity.** Before working on the washer, disconnect the electricity. This can be done by pulling the plug from the electrical outlet. Be sure that you only remove the washer plug. Or disconnect the electricity at the fuse panel or at the circuit breaker panel. Turn off the electricity.

WARNING *Some diagnostic tests will require you to test the components with the power turned on. When you disassemble the control panel, you can position it in such a way that the wiring will not make contact with metal. This act will allow you to test the components without electrical mishaps.*

4. **Gain access to the water valve.** Turn off the water supply to the washer water valve. To access the water valve, the rear panel must be removed. (On some models, to access the water valve, you will gain access through the control panel.) Disconnect the fill hoses from the inlet end of the water valve (Figure 18-10). Next, remove the screws that hold the water valve to the chassis of the washer (Figure 18-11).

5. **Test the water valve.** In order to check the solenoid coils on the water valve, remove the wire leads (label them) that connect to the coils from the wire harness (see Figure 18-11). These are slide-on terminal connectors attached to the ends of the wire. Just pull them off. Set the ohmmeter on R × 100, and attach the probes to the terminals of one of the solenoid coils (see Figure 18-12). The meter should read between 500 and 2000 ohms. Repeat this test for the second solenoid coil.

FIGURE 18-10
Turn off the water
supply, and remove
the fill hoses.

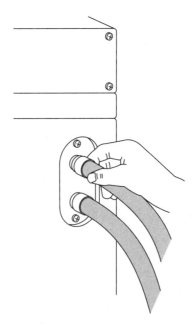

Figure 18-11
Remove the wires
from the solenoid coil.

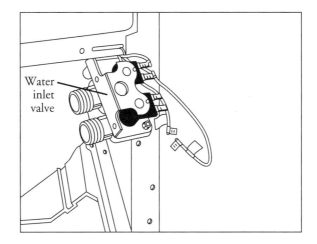

To test the fill rate of the water valve, just reverse the disassembly procedure, and reassemble the water valve. The rear panel does not have to be reinstalled for this test. Attach the 120-volt fused service cord—including the ground wire test lead to the cabinet ground (see Figure 18-4)—to the water valve solenoid coil (see Figure 18-12). Then energize the solenoid coil to allow water to enter the tub and to check the flow rate of the water valve (Table 18-1). This step is repeated for each solenoid coil. If, when you energize the water valve, no water enters the washer tub, replace the water valve. If the water valve checks correctly, check the timer and the wiring harness.

6. **Remove the water valve.** To remove the water valve, follow the instructions in step 4. Remove the water outlet hose from the water valve.

Figure 18-12
Attaching test leads
to the solenoid coil on
the water valve.

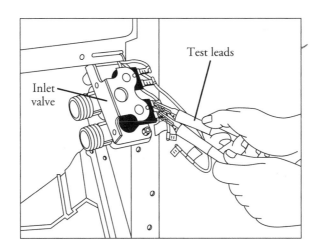

PART VI

Water Input P.S.I.	Gallon per Minute One Side of Valve	Gallon per Minute Both Sides of Valve
20	3.7	4.5
30	4.6	5.5
40	5.3	6.2
50	5.6	6.3
60	5.8	6.7
80	6.6	7.0
100	7.0	7.2
120	7.2	7.4
140	7.0	7.3
160	6.8	7.0

TABLE 18-1   Water Fill Rate for a Typical Water Value

7. **Install a new water valve.** To install the new water valve, just reverse the disassembly procedure, and reassemble. Reconnect the wire leads to the solenoid coils. After the installation of the new valve, turn on the water supply and check for water leaks. If none are found, reinstall the rear panel and restore the electricity to the washer. Set the timer and the water temperature control settings to operate the washer through its cycles.

Washer Motor (Older Models)

The typical complaints associated with motor failure are:

- Fuse is blown or the circuit breaker trips.
- Washer fills up with water but the motor will not run.

To handle these problems, perform the following steps:

1. **Verify the complaint.** Verify the complaint by operating the washer through its cycles. Listen carefully, and you will hear if there are any unusual noises or if the circuit breaker trips. On electronic models, if a fault code appears, look up the code. If the washer will not power up, locate the technical data sheet behind the control panel for diagnostics information. On some models you will need the actual service manual for the model you are working on to properly diagnose the washer. The service manual will assist you in properly placing the washer in the service test mode for testing the washer functions.

2. **Check for external factors.** You must check for external factors not associated with the appliance. Is the appliance installed properly? Does the appliance have the correct voltage? The voltage at the receptacle is between 108 volts and 132 volts during a load on the circuit. Do you have the correct polarity? (See Chapter 6.)

3. **Disconnect the electricity.** Before working on the washer, disconnect the electricity. This can be done by pulling the plug from the electrical outlet. Be sure that you only remove the washer plug. Or disconnect the electricity at the fuse panel or at the circuit breaker panel. Turn off the electricity.

WARNING *Some diagnostic tests will require you to test the components with the power turned on. When you disassemble the control panel, you can position it in such a way that the wiring will not make contact with metal. This act will allow you to test the components without electrical mishaps.*

---

4. **Gain access to the motor.** To access the motor, the back panel must be removed (Figure 18-13). The back panel is held on with screws. Remove the screws and remove the panel.

5. **Disconnect the motor wire leads.** Disconnect the motor wire leads from the wiring harness. Set the ohmmeter on R × 1, and attach the probes to the motor lead wires (Figure 18-14). Refer to the wiring diagram for the common, start, and run motor winding leads identification. Test these for continuity, from the common wire lead to the run winding. Then test for continuity from the common wire lead to the start winding. Next, test for continuity from the start winding to the run winding. To test for a grounded winding in the motor, take the ohmmeter probes and test from each motor wire lead to the motor housing (Figure 18-15). The ohmmeter will indicate continuity if the windings are grounded. If the motor has no continuity between the motor windings, replace the motor. If the motor checks out okay, check the timer and motor relay (if the model that you are repairing has one).

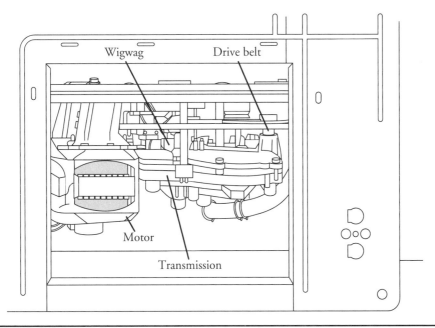

FIGURE 18-13   Removing the screws that hold the back panel.

FIGURE 18-14
Checking the motor
windings for
continuity.

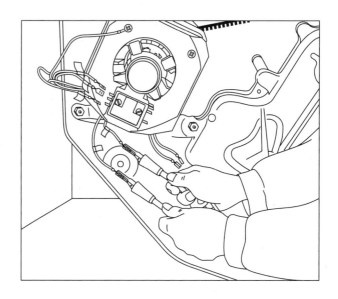

FIGURE 18-14
Checking the motor
windings for
continuity.

6. **Remove the motor.** To remove this type of motor, you must first loosen the two nuts that hold the motor support bracket (Figure 18-16). Then slide the assembly to disengage the belt from the pulley. Next, remove the four nuts that hold the motor to the motor support bracket (Figure 18-17). Remove the motor from the washer. Remember to remove any remaining wires from the motor and label them. Remove the pulley from the motor after loosening the set screw.

7. **Install the new motor.** To install the new motor, just reverse the disassembly procedure, and reassemble. To adjust the belt, refer to the "drive belt" section of this chapter (step 5). Restore the electricity to the washer, and test the motor. If the motor is working, reinstall the back panel.

FIGURE 18-15
Checking the motor
for ground.

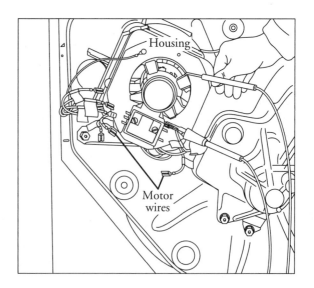

Housing

Motor
wires

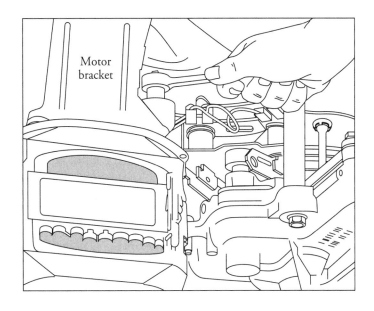

FIGURE 18-16
Removing the bolts
that hold the motor on
the bracket.

Washer Motor (Direct-Drive Models)

The typical complaints associated with motor failure are:

- Fuse is blown or the circuit breaker trips.
- Washer fills up with water but the motor will not run.

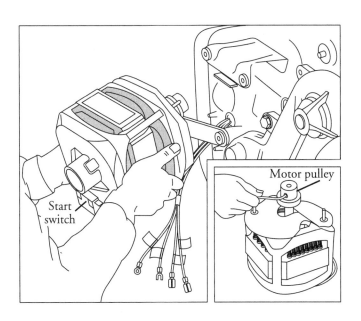

FIGURE 18-17
Removing the motor
and then removing the
pulley.

PART VI

To handle these problems, perform the following steps:

1. **Verify the complaint.** Verify the complaint by operating the washer through its cycles. Listen carefully, and you will hear if there are any unusual noises or if the circuit breaker trips. On electronic models, if a fault code appears, look up the code. If the washer will not power up, locate the technical data sheet behind the control panel for diagnostics information. On some models you will need the actual service manual for the model you are working on to properly diagnose the washer. The service manual will assist you in properly placing the washer in the service test mode for testing the washer functions.

2. **Check for external factors.** You must check for external factors not associated with the appliance. Is the appliance installed properly? Does the appliance have the correct voltage? The voltage at the receptacle is between 108 volts and 132 volts during a load on the circuit. Do you have the correct polarity? (See Chapter 6.)

3. **Disconnect the electricity.** Before working on the washer, disconnect the electricity. This can be done by pulling the plug from the electrical outlet. Be sure that you only remove the washer plug. Or disconnect the electricity at the fuse panel or at the circuit breaker panel. Turn off the electricity.

WARNING *Some diagnostic tests will require you to test the components with the power turned on. When you disassemble the control panel, you can position it in such a way that the wiring will not make contact with metal. This act will allow you to test the components without electrical mishaps.*

4. **Gain access to the motor.** To access the motor, you must first remove the cabinet from the washing machine (Figure 18-18). Remove the two screws that secure the control console to the cabinet. Tilt the control panel upward. Next, insert a flat-blade screwdriver in the cabinet retaining clip and push forward to remove it. Remove the wiring harness connector and ground wire. Finally, remove the cabinet as shown in Figure 18-18.

5. **Test the drive motor.** To test the drive motor, disconnect the wire connector from the motor relay. Refer to the wiring diagram for the motor protector; common; start, low, and high speed; and motor winding leads identification. Set your ohmmeter on the $R \times 1$ scale and place your ohmmeter probes on the start winding. The meter should read around 7 ohms. Next test the high and low speed windings—the resistance should be between 1 and 3 ohms. Now test the motor overload protector—the reading should be zero ohms. After testing the motor, spin the motor shaft; it should spin freely. To test for a grounded winding in the motor, take the ohmmeter probes and test from each motor wire lead to the motor housing. The ohmmeter will indicate continuity if the windings are grounded. If the motor has no continuity between the motor windings, replace the motor. If the motor checks out okay, check the timer and motor relay.

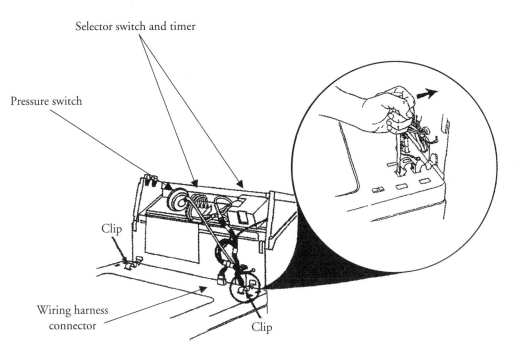

Selector switch and timer

Pressure switch

Clip

Wiring harness connector

Clip

Tilt the cabinet toward you and slide it off the base, and then pull it away from the washer.

Cabinet

FIGURE 18-18 Removing the outer cabinet on a direct-drive automatic washer.

PART VI

6. **Remove the motor.** To remove the motor, you must first remove the water pump and the motor coupler. Before removing the pump, make sure that most of the water is removed from the tub. Use a pinch-off tool to restrict the water lines at the pump hoses. Next, proceed to remove the water pump by removing the clamps that hold the pump in place (Figure 18-19). Remove the three-piece motor coupler (Figure 18-20). Finally, remove the clamps that secure the motor to the gearcase (Figure 18-21). Pull the motor toward you to remove it from the washer.

7. **Install the new motor.** To install the new motor, just reverse the disassembly procedure, and reassemble. Restore the electricity to the washer, and test the motor. If the motor is working, reinstall the cabinet.

Front Serviceable Washer

The typical complaints associated with motor, clutch, and belt failure are:

- Fuse is blown or the circuit breaker trips.
- Washer fills up with water, but the motor will not run.
- Motor runs, but washer will not agitate or spin.

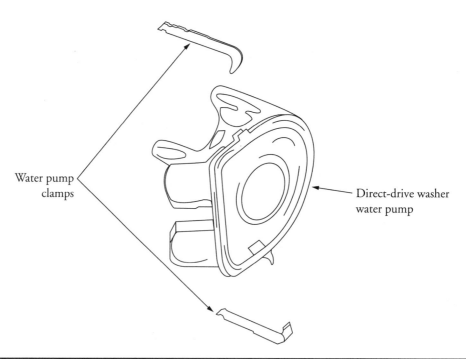

Water pump clamps

Direct-drive washer water pump

FIGURE 18-19   The direct-drive water pump.

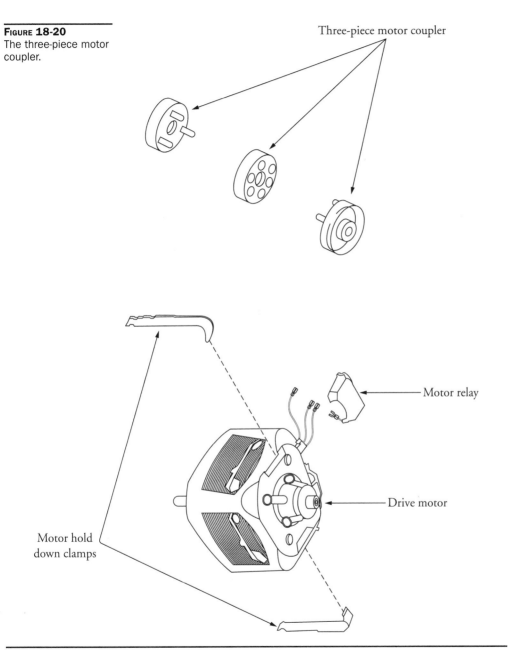

Figure 18-20
The three-piece motor coupler.

Three-piece motor coupler

Motor relay

Drive motor

Motor hold down clamps

Figure 18-21 The drive motor and relay.

To handle these problems, perform the following steps:

1. **Verify the complaint.** Verify the complaint by operating the washer through its cycles. Listen carefully, and you will hear if there are any unusual noises or if the circuit breaker trips. On electronic models, if a fault code appears, look up the code. If the washer will not power up, locate the technical data sheet behind the control panel for diagnostics information. On some models you will need the actual service manual for the model you are working on to properly diagnose the washer. The service manual will assist you in properly placing the washer in the service test mode for testing the washer functions.

2. **Check for external factors.** You must check for external factors not associated with the appliance. Is the appliance installed properly? Does the appliance have the correct voltage? The voltage at the receptacle is between 108 volts and 132 volts during a load on the circuit. Do you have the correct polarity? (See Chapter 6.)

3. **Disconnect the electricity.** Before working on the washer, disconnect the electricity. This can be done by pulling the plug from the electrical outlet. Be sure that you only remove the washer plug. Or disconnect the electricity at the fuse panel or at the circuit breaker panel. Turn off the electricity.

WARNING *Some diagnostic tests will require you to test the components with the power turned on. When you disassemble the control panel, you can position it in such a way that the wiring will not make contact with metal. This act will allow you to test the components without electrical mishaps.*

4. **Gain access to the motor.** To access the motor, you must first remove the front panel. Locate the two spring clips between the top cover and the front cover (Figure 18-22). Insert the putty knife and push in to release the spring clips on both sides. Pull the front panel toward you (Figure 18-23) and remove it from the bottom tabs. Next, remove the two ¼-inch hex screws from the top panel support brackets

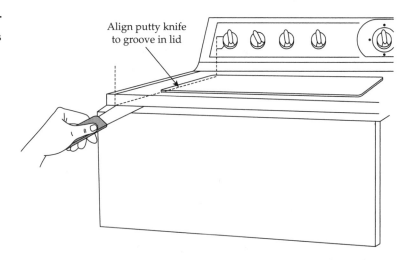

FIGURE 18-22
Locate the spring clips on the left and right side, and press in to release the front panel.

Align putty knife to groove in lid

FIGURE 18-23
To remove the front cover, lift it off the bottom tabs.

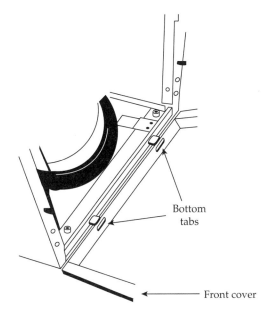

Bottom tabs

Front cover

(left and right sides). Pull the top cover and lid assembly toward you and up enough to clear the locking tabs in the rear (Figure 18-24). Do not pull too much, because you have to remove the bleach hose and the lid switch. Do not cut the lid switch wires; just depress the tab from under the lid switch and remove the lid switch from the top cover. With the top and front panels removed, you now have access to the motor located on the bottom in the front of the cabinet (Figure 18-25).

FIGURE 18-24
To remove the top and lid assembly, (1) lift up about 2 inches and (2) pull toward you and remove from locking tabs.

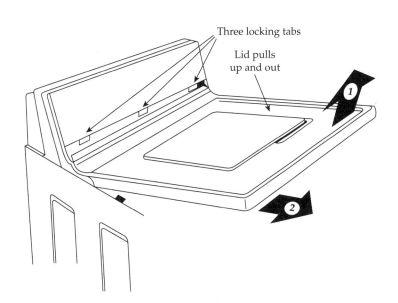

Three locking tabs

Lid pulls up and out

FIGURE 18-25
With the front panel
removed, you now
have access to the
motor and clutch
assembly, drain pump
motor, and belt.

5. **Disconnect the motor wire leads.** Disconnect the motor wire leads from the motor.
   Set the ohmmeter on R × 1, and attach the probes to the motor lead wires. Refer to
   the wiring diagram for the common, start, and run motor winding leads
   identification. Test these for continuity, from the common wire lead to the run
   winding. Then test for continuity from the common wire lead to the start winding.
   Next, test for continuity from the start winding to the run winding. To test for a
   grounded winding in the motor, take the ohmmeter probes and test from each
   motor wire lead to the motor housing. The ohmmeter will indicate continuity if the
   windings are grounded. If the motor has no continuity between the motor
   windings, replace the motor.

6. **Remove the motor.** Loosen the four 3/8-inch mounting nuts that hold the motor in
   place (Figure 18-26). Slide the motor inward and remove the belt. Next, remove the
   four 3/8-inch nuts. Rotate the transmission to a position that allows you to remove
   the motor from the platform.

FIGURE 18-26
To remove the motor, remove the four 3/8-inch nuts from the motor.

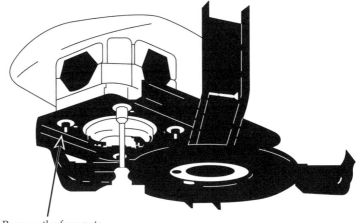

Remove the four nuts

7. **Remove the clutch.** After removing the motor, turn it upright to access the clutch. Remove the one-time-use spring clip from the clutch with pliers (Figures 18-27 and 18-28).

8. **Install the new motor and clutch.** To install the new motor and clutch, just reverse the disassembly procedure, and reassemble (Figure 18-29). Restore the electricity to the washer, and test the motor and clutch assembly. If the motor is working, reinstall the cabinet top and front panels.

Capacitor

A capacitor is a device that stores electricity to provide an electrical boost for motor starting. Most high-torque motors need a capacitor connected in series with the start winding circuit to produce the desired rotation under a heavy starting load.

FIGURE 18-27
The "one-time-use" spring clip is located on the clutch assembly. It must be replaced with a new one.

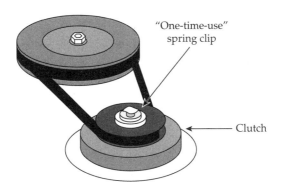

"One-time-use" spring clip

Clutch

FIGURE 18-28
Squeeze and rotate
the pliers to remove
the spring clip.

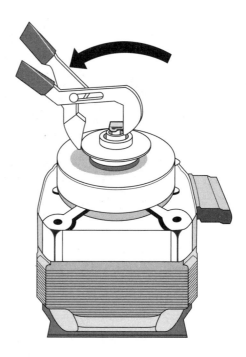

FIGURE 18-29
Squeeze and rotate
the pliers to reinstall
the new spring clip.

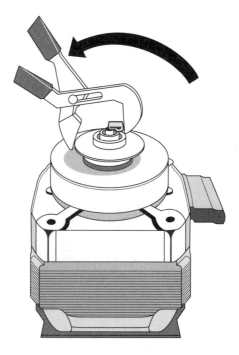

The typical complaints associated with capacitor failure are:

- Fuse is blown or the circuit breaker trips.
- Washer motor will not run.
- Motor has a burning smell.
- Motor will try to start and then shuts off.

To handle these problems, perform the following steps:

1. **Verify the complaint.** Verify the complaint by operating the washer. Listen carefully, and you will hear if there are any unusual noises or if the circuit breaker trips. If you smell something burning, immediately turn off the washer and pull the plug. On electronic models, if a fault code appears, look up the code. If the washer will not power up, locate the technical data sheet behind the control panel for diagnostics information. On some models you will need the actual service manual for the model you are working on to properly diagnose the washer. The service manual will assist you in properly placing the washer in the service test mode for testing the washer functions.

2. **Check for external factors.** You must check for external factors not associated with the appliance. Is the appliance installed properly? Does the appliance have the correct voltage? The voltage at the receptacle is between 108 volts and 132 volts during a load on the circuit. Do you have the correct polarity? (See Chapter 6.)

3. **Disconnect the electricity.** Before working on the washer, disconnect the electricity. This can be done by pulling the plug from the electrical outlet. Be sure that you only remove the washer plug. Or disconnect the electricity at the fuse panel or at the circuit breaker panel. Turn off the electricity.

WARNING *Some diagnostic tests will require you to test the components with the power turned on. When you disassemble the control panel, you can position it in such a way that the wiring will not make contact with metal. This act will allow you to test the components without electrical mishaps.*

4. **Gain access to the capacitor.** Some models have the capacitor mounted on the motor, and some are mounted to the cabinet interior in the rear of the machine. Access might be achieved through the front or rear panel, depending on which model you are working on. Do not touch the capacitor until it's discharged.

WARNING *A capacitor will hold a charge indefinitely, even when it is not currently in use. A charged capacitor is extremely dangerous. Discharge all capacitors immediately any time that work is being conducted in their vicinity. Redischarge after repowering the equipment if further work must be done.*

5. **Test capacitors.** Before testing the capacitor, it must be discharged. Use a screwdriver with an insulated handle to discharge the capacitor by shorting it across both terminals. Remove the wire leads, one at a time, with needle-nose pliers. Set the ohmmeter on the highest scale and then place one probe on one terminal and the other probe on the other terminal (Figure 18-30). Observe the meter action. While the capacitor is charging, the ohmmeter will read nearly zero ohms for a short period of time. Then the ohmmeter reading will slowly begin to return toward infinity. If the ohmmeter reading deflects to zero and does not return to infinity, the capacitor is shorted and should be replaced. If the ohmmeter reading remains at infinity and does not dip toward zero, the capacitor is open and should be replaced.

Another way to test a capacitor is to attach a capacitor tester to it and test the microfarad reading. It should be within +/– 10 percent of the capacitor rating that is stamped on the side of the capacitor. By using a capacitor tester, you will be able to test for a weak capacitor even if it tests out okay with an ohmmeter.

6. **Remove the capacitor.** Remove the capacitor from its mounting bracket.

7. **Install a new capacitor.** To install the new capacitor, just reverse the disassembly procedure, and reassemble. *Note:* A capacitor is rated by its working voltage (WV or WVac) and by its storage capacity in microfarads (μF). Always replace a capacitor with one that has the same voltage rating and the same (or up to 10 percent greater) microfarad rating.

Drive Belt

The typical complaints associated with belt failure are:

- Washer will not agitate.
- Washer will not spin.
- Washer motor spins freely.
- There is a smell of something burning.

FIGURE 18-30
Placing ohmmeter test leads on the capacitor terminals.

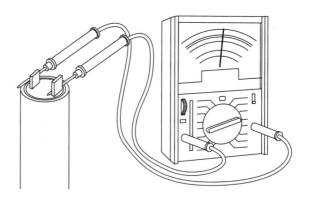

To handle these problems, perform the following steps:

1. **Verify the complaint.** Verify the complaint by operating the washer in the spin cycle. Listen carefully, and you will hear and see if the inner basket is turning or if the circuit breaker trips. On electronic models, if a fault code appears, look up the code. If the washer will not power up, locate the technical data sheet behind the control panel for diagnostics information. On some models you will need the actual service manual for the model you are working on to properly diagnose the washer. The service manual will assist you in properly placing the washer in the service test mode for testing the washer functions.

2. **Check for external factors.** You must check for external factors not associated with the appliance. Is the appliance installed properly? Does the appliance have the correct voltage? The voltage at the receptacle is between 108 volts and 132 volts during a load on the circuit. Do you have the correct polarity? (See Chapter 6.)

3. **Disconnect the electricity.** Before working on the washer, disconnect the electricity. This can be done by pulling the plug out from the electrical outlet. Be sure that you only remove the washer plug. Or disconnect the electricity at the fuse panel or at the circuit breaker panel. Turn off the electricity.

WARNING *Some diagnostic tests will require you to test the components with the power turned on. When you disassemble the control panel, you can position it in such a way that the wiring will not make contact with metal. This act will allow you to test the components without electrical mishaps.*

4. **Gain access to the belt.** You must gain access to the belt—whether by removing the back or the front panel, tilting the washer, or laying it onto its back—depending on which model washer you are working on. The back panel (or the front panel) is usually held on with two screws. Remove the screws, and remove the panel.

5. **Adjust the belt.** Before adjusting the belt, use your finger and press on the belt; it should only deflect about 1/4 inch. To adjust the belt (see Figure 18-16), loosen the motor bracket nut just enough to move the bracket. Take hold of the motor bracket, and pull against the belt just enough to take up the slack in it and to properly re-tension it. If you are unable to adjust the tension, or if the belt is worn, replace the belt. Some models have more than one belt: one is for the drive system that is attached to the motor and the transmission pulleys; the other belt is for the water pump. This belt is attached to the motor pulley and to the water pump pulley (Figure 18-31). To adjust the water pump belt in this type of washer, just loosen the pump mounting screws and adjust to obtain the correct tension, about 1/4-inch deflection. There are even some models that use a direct-drive system (which has no belts) to drive the motor, transmission, and pump (Figure 18-32). This type of washer has the motor and the water pump attached to the transmission with retaining clips.

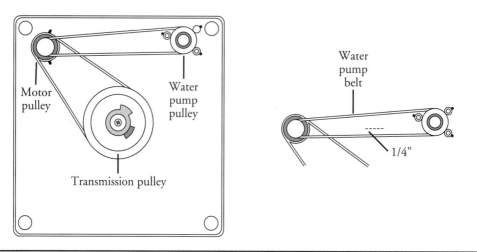

Figure 18-31 View from underneath the washer base. Check belt deflection.

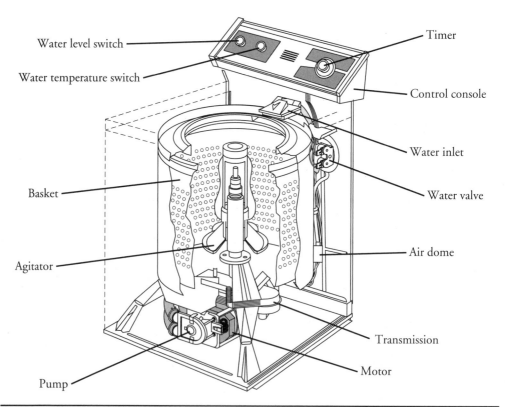

Figure 18-32 The direct-drive washer has no belts.

Between the motor shaft and the transmission, there is a coupling. To replace this coupling, remove the retaining clips that secure the water pump and the motor. Install the new coupling, and reattach the motor and the water pump.

6. **Replace the drive belt.** To replace the drive belt on this type of washer, you must remove the flexible pump coupling (Figure 18-33). Next, loosen the motor bracket nuts, and slide the motor forward to take the tension off the belt. Then remove the belt from the motor pulley and the transmission pulley (Figure 18-34). Install the new belt on the transmission pulley and the clutch pulley. Be sure that the belt is in the pulley grooves. Next, adjust the belt tension, and tighten the motor bracket nuts. Reinstall the flexible pump coupling and clamps, making sure that the coupling is not twisted and that it is seated on the pump and clutch pulleys. The drive belt tension on this type of washer should be approximately 1/2 inch when deflected.

7. **Test the washer.** You are now ready to test the washer. Begin the wash cycle with a full load of laundry in the basket. Check the agitate and the spin cycles. If these check out okay, reinstall the outer panels. If not, readjust the belt tension.

Water Level Control

The water level control starts in the empty position. As the washer fills with water and the water level rises in the tub, it causes the air pressure in the tube and in the air dome to increase (Figure 18-35). The air pressure is transferred from the air hose to the water level control and against the diaphragm, which actuates the water level switch to the full position and agitation begins.

FIGURE 18-33
Loosening the screw on each clamp and removing the flexible coupling.

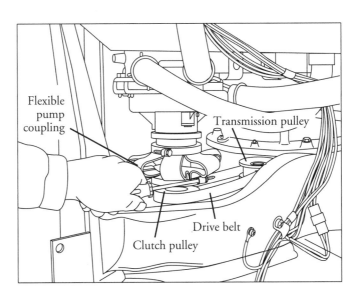

Flexible pump coupling

Transmission pulley

Drive belt

Clutch pulley

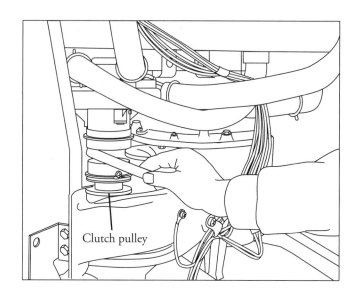

The typical complaints associated with the water level control (pressure switch) are:

- Water is flowing over the top of the tub.
- Tub does not fill to the proper level selected.
- Washer will not agitate.
- Washer will not spin.

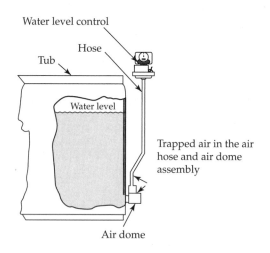

To handle these problems, perform the following steps:

1. **Verify the complaint.** Verify the complaint by trying to select different water levels when operating the washer through its cycles. On electronic models, if a fault code appears, look up the code. If the washer will not power up, locate the technical data sheet behind the control panel for diagnostics information. On some models you will need the actual service manual for the model you are working on to properly diagnose the washer. The service manual will assist you in properly placing the washer in the service test mode for testing the washer functions.

2. **Check for external factors.** You must check for external factors not associated with the appliance. Is the appliance installed properly? Does the appliance have the correct voltage? The voltage at the receptacle is between 108 volts and 132 volts during a load on the circuit. Do you have the correct polarity? (See Chapter 6.) Is there any physical damage to the component? Is the plastic hose connected to the water level control and air dome? Check to be sure that the water is turned on all the way.

3. **Disconnect the electricity.** Before working on the washer, disconnect the electricity. This can be done by pulling the plug from the electrical outlet. Be sure that you only remove the washer plug. Or disconnect the electricity at the fuse panel or at the circuit breaker panel. Turn off the electricity.

WARNING *Some diagnostic tests will require you to test the components with the power turned on. When you disassemble the control panel, you can position it in such a way that the wiring will not make contact with metal. This act will allow you to test the components without electrical mishaps.*

4. **Remove the console panel to gain access.** Begin by removing the console panel to gain access to the water level control (see Figure 18-2).

5. **Test the water level control.** To test the water level control switch, remove and label all of the wires from the switch (Figure 18-36). Remember, you will have to identify the wires according to the wiring diagram in order to reinstall them on the water level control switch properly. With the washer empty, use your ohmmeter (set on R × 1), and test it for continuity on the switch contacts numbered 1 and 2 (Figure 18-37). If you have continuity, this means that the water valve will be energized, allowing the water to enter the tub. Now test for continuity between contacts 1 and 3; the ohmmeter should read no continuity. Reconnect the wires to the water level switch.

   Plug in the washer, and start the wash cycle. Let the console rest on top of the washing machine for this test. Be careful not to touch any live wires or to short them to the washer chassis while the test is being performed. As the water level rises in the tub, it forces air through the air dome and up the plastic tube to the water level control. The pressure that is exerted on the water level control's diaphragm will trip the water level switch from empty to full, which will start the agitation cycle. When the agitation cycle begins, turn off the washer. Pull the power plug from the wall socket to ensure that power has been removed.

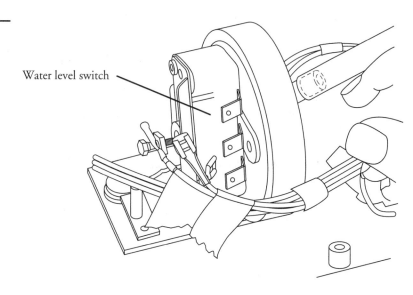

FIGURE 18-36
Removing the
wires from the
water level control
before checking
continuity.

Water level switch

At this point, test for continuity again between the contacts numbered 1 and 2. The ohmmeter should read no continuity. Next, test for continuity between contacts numbered 1 and 3. The ohmmeter will read continuity. If the water level switch checks out, then check the plastic hose that goes from the air dome to the water level control. Ensure that there are no holes or cracks in the line. If the switch does not check out okay, replace it. Remember: Never blow into the water level control switch to activate it. Why? You might activate the switch, but it will not prove that the switch will activate (at lesser pressures) at the proper water level setting selected.

6. **Remove the water level control.** To remove the water level control, remove the wires and then remove the screws that hold the component to the console panel (Figure 18-38). Next, remove the plastic hose.

FIGURE 18-37
Adjustable water
level control terminal
identification for this
model only.

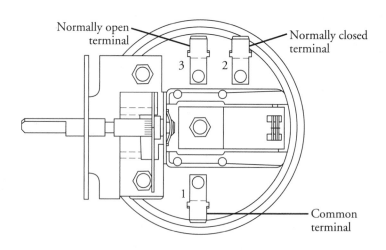

Normally open terminal

Normally closed terminal

Common terminal

FIGURE 18-38
Disconnecting the
wires and removing
the screws that hold
the control to the
console.

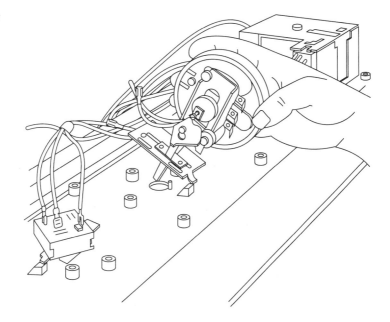

7. **Reinstall the water level control.** To reinstall the water level control, just reverse the disassembly procedure, and reassemble. Remember, you will have to identify the wires according to the wiring diagram in order to properly reinstall them on the water level control.

Lid Switch

On some models, the lid switch will pause the spin operation; on other models, it will pause the wash operations, except water fill. Once the washer cycle is started, the washer lid should remain closed until the end of the wash cycle. The lid switch is a safety device added to protect consumers from getting entangled within the washer. Never bypass this switch.

The typical complaints associated with the lid switch are:

- Washer will not spin.
- Washer will not agitate.

To handle these problems, perform the following steps:

1. **Verify the complaint.** Verify the complaint by closing the washer lid and turning the timer dial to start the spin cycle. On electronic models, if a fault code appears, look up the code. If the washer will not power up, locate the technical data sheet behind the control panel for diagnostics information. On some models you will need the actual service manual for the model you are working on to properly diagnose the washer. The service manual will assist you in properly placing the washer in the service test mode for testing the washer functions.

2. **Check for external factors.** You must check for external factors not associated with the appliance. Is the appliance installed properly? Does the appliance have the correct voltage? The voltage at the receptacle is between 108 volts and 132 volts during a load on the circuit. Do you have the correct polarity? (See Chapter 6.) Is there any physical damage to the component?

3. **Disconnect the electricity.** Before working on the washer, disconnect the electricity. This can be done by pulling the plug from the electrical outlet. Be sure that you only remove the washer plug. Or disconnect the electricity at the fuse panel or at the circuit breaker panel. Turn off the electricity.

WARNING *Some diagnostic tests will require you to test the components with the power turned on. When you disassemble the control panel, you can position it in such a way that the wiring will not make contact with metal. This act will allow you to test the components without electrical mishaps.*

4. **Gain access to the lid switch.** If the washer's top snaps in place, tape the lid shut. Use a putty knife to release the spring clips in each corner, and lift off the top. If the spring clips won't release, open the lid, pull the top toward you, and lift (Figure 18-39). Raise the washer top to gain access to the lid switch (Figure 18-40).

On some models, the top is held down with two screws that are secured from underneath the top. The front of the cabinet is secured in place with two screws; take them out and remove the front panel. Then remove the screws that hold the top in place. On other models, the top is part of the cabinet (Figure 18-41). To gain access to the switch, the cabinet will have to be removed as shown in Figure 18-18.

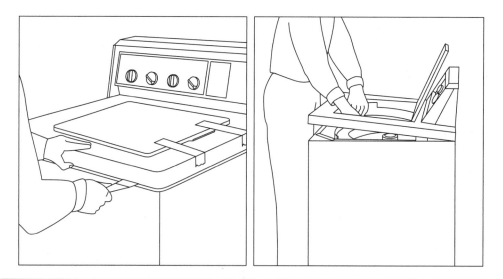

FIGURE 18-39   Tape the lid closed. Then pry the top open and lift the top off.

FIGURE 18-40
Removing the plastic
shield to gain access
to the lid switch.

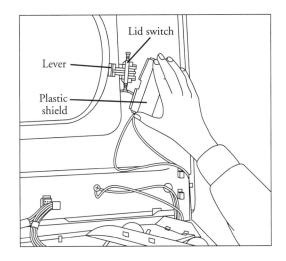

Lid switch

Lever

Plastic
shield

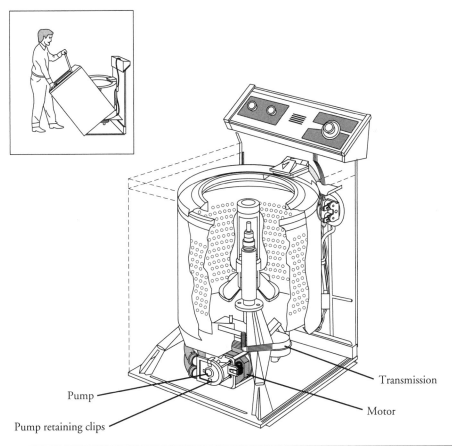

Transmission

Pump

Motor

Pump retaining clips

FIGURE 18-41 Remove the screws that hold the console and then remove the two clips that hold the cabinet in place and remove the cabinet.

FIGURE 18-42
Removing the wires to
check for continuity of
the switch.

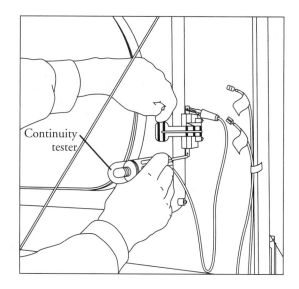

Continuity
tester

5. **Test the lid switch.** To test the lid switch, remove the two wires from the switch (Figure 18-42). With a continuity tester, test for continuity between the two terminals of the switch while moving the lid switch lever. If the switch fails these tests, replace the switch.

6. **Replace the lid switch.** Remove the screws that hold the switch in place (Figure 18-43). Install the new switch, and connect the wires on the switch terminals. Then reconnect the electricity and test the washer.

Water Pump

Two types of water pumps are used on automatic washers: mechanical (direct drive or belt driven; Figure 18-41) and electric (electric motor coupled to a pump; Figure 18-44). The water pump is used for draining the water from the washer.

FIGURE 18-43
Removing the two
screws that hold the
lid switch in place.

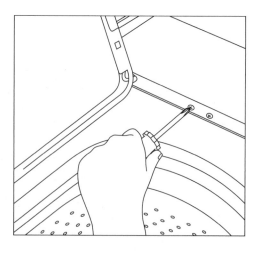

Figure 18-44
Some models use an
electric water pump to
drain the water out of
the tub.

Electric water
pump

The typical complaints associated with the water pump are:

- Washer will not drain the water out.
- It smells like something is burning.
- The water in the washer will not recirculate.

To handle these problems, perform the following steps:

1. **Verify the complaint.** Verify the complaint by operating the washer. On electronic models, if a fault code appears, look up the code. If the washer will not power up, locate the technical data sheet behind the control panel for diagnostics information. On some models you will need the actual service manual for the model you are working on to properly diagnose the washer. The service manual will assist you in properly placing the washer in the service test mode for testing the washer functions.

2. **Check for external factors.** You must check for external factors not associated with the appliance. Is the appliance installed properly? Does the appliance have the correct voltage? The voltage at the receptacle is between 108 volts and 132 volts during a load on the circuit. Do you have the correct polarity? (See Chapter 6.)

3. **Disconnect the electricity.** Before working on the washer, disconnect the electricity to the washer. This can be done by pulling the plug from the electrical outlet. Be sure that you only remove the washer plug. Or disconnect the electricity at the fuse panel or at the circuit breaker panel. Turn off the electricity.

WARNING *Some diagnostic tests will require you to test the components with the power turned on. When you disassemble the control panel, you can position it in such a way that the wiring will not make contact with metal. This act will allow you to test the components without electrical mishaps.*

4. **Testing the electric drain pump motor.** To test the electric drain pump motor (Figure 18-44), remove the wire connector to the pump motor. Next, set your multimeter on the ohms scale R × 1. Place the meter test leads on the motor terminals. You should read between 10 and 13Ω on the meter. Now inspect the impeller inside the pump. If there is any debris, remove it. Make sure that the impeller is not damaged.

5. **Access and remove the direct drive water pump.** In order to gain access to the water pump in this type of washer, the cabinet must be removed (see Figure 18-18). Remove the screws that hold the control panel in place. Next, lift the control panel up. Then remove the clips from each side that hold the cabinet in place. Pull back on the cabinet, and move it out of the way (Figure 18-41). With no water in the tub, remove the hoses from the pump ports (Figure 18-45) (some water will spill out of

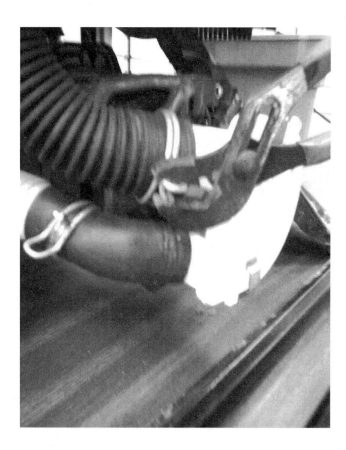

FIGURE 18-45
Loosen the hose clamp on both hoses and remove the hose from the pump.

the pump). Wipe the water up immediately. Check the water pump ports for any obstructions. If no obstructions are found, then disconnect the two clamps that hold the pump in place (Figure 18-46). Remove the pump (Figure 18-47).

To gain access to the water pump in this older type of washer, remove the back panel. With no water in the tub, remove the hoses from the pump ports (some water will spill out of the pump). Wipe the water up immediately. If obstructions are found, remove the obstructions, reconnect the hoses, and test for proper operation. If the unit is still not operating properly, or if the obstruction could not be removed, the pump must be removed. Next, remove the flexible pump coupling (Figure 18-48). Now remove the bolts that hold the pump in place (Figure 18-49). Remove the pump.

6. **Install a new water pump.** To install the new pump, just reverse the disassembly procedure, and reassemble. Test the washer for any water leaks. The direct-drive washer can be operated with the cabinet removed by installing a jumper wire in the lid switch harness connector (Figure 18-50).

FIGURE 18-46
Remove the clamps that secure the water pump to the motor.

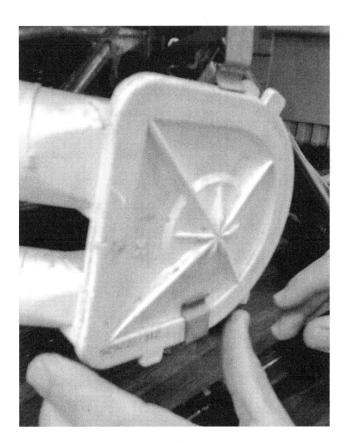

Figure 18-47
Remove the water pump from the motor. Check for obstructions. Inspect the water pump.

Figure 18-48
Loosen the screw on each clamp and remove the flexible coupling.

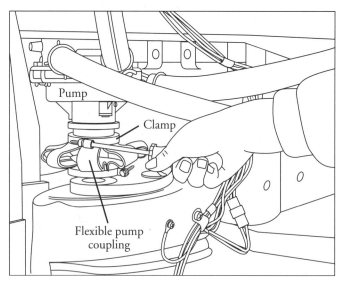

FIGURE 18-49
Removing the pump
hold-down bolts.

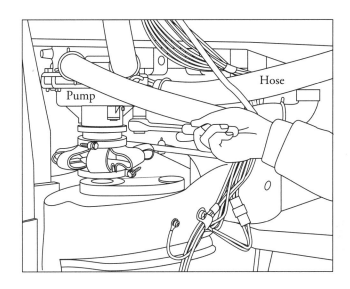

FIGURE 18-49
Removing the pump
hold-down bolts.

Inner Basket

The typical complaints associated with the inner basket are:

- Basket will not spin.
- Water will not drain out of the tub.
- Washer damages the clothing.
- Rust marks on clothing.

FIGURE 18-50
Installing a jumper
wire in the lid switch
harness connector to
test the operation of
a direct-drive washer.

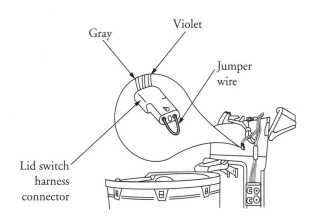

Gray

Violet

Jumper
wire

Lid switch
harness
connector

To handle these problems, perform the following steps:

1. **Verify the complaint.** Verify the complaint by inspecting and operating the washer. Inspect the inner basket.

2. **Check for external factors.** You must check for external factors not associated with the appliance. Is the appliance installed properly? Does the appliance have the correct voltage? The voltage at the receptacle is between 108 volts and 132 volts during a load on the circuit. Do you have the correct polarity? (See Chapter 6.)

3. **Disconnect the electricity.** Before working on the washer, disconnect the electricity. This can be done by pulling the plug from the electrical outlet. Be sure that you only remove the washer plug. Or disconnect the electricity at the fuse panel or at the circuit breaker panel. Turn off the electricity.

WARNING *Some diagnostic tests will require you to test the components with the power turned on. When you disassemble the control panel, you can position it in such a way that the wiring will not make contact with metal. This act will allow you to test the components without electrical mishaps.*

4. **Gain access to inner basket.** To gain access to the inner basket, you must first lift the top of the washer off (see Figure 18-39) and disconnect the fill hose from the water inlet, located on top of the splash guard (Figure 18-51). Remove the splash guard by removing the clips that hold it to the outer tub. Now that you have more room to work with, remove the agitator. On some models, the agitator is held down with a cap and stud assembly, or the agitator might just snap onto the transmission shaft (Figure 18-52). Next, use a spanner wrench to remove the locknut that holds the basket in place (Figure 18-53). On some models, the inner basket is held in place with four bolts. If so, remove the bolts. On some models, the agitator is part of the inner basket (Figure 18-54). Next, lift the inner basket out of the tub. Inspect the outer tub for debris and rust. Be sure that the tub drain opening is clear of debris.

FIGURE 18-51
Removing the clamp
from the inlet hose.

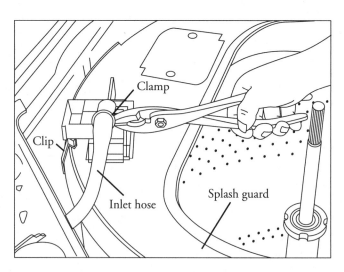

FIGURE 18-52
Removing the agitator.

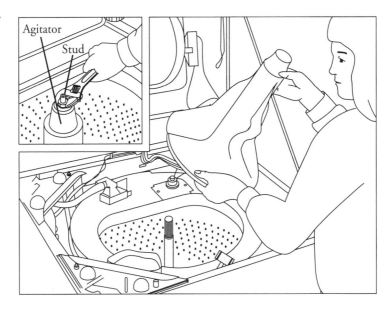

5. **Reinstall the inner basket.** To reinstall the inner basket, just reverse the disassembly procedure, and reassemble. One important note: Do not overtighten the locknut. Just tighten the locknut enough to secure the basket in place. If you overtighten the locknut, the washer will not work properly.

6. **Test the washer.** After you have reassembled the inner basket, splash guard, and agitator, test the washer. Check for leaks around the top rim of the outer tub. Check for agitation and spin with a full load of clothes.

FIGURE 18-53
Always use a spanner wrench to remove the locknut.

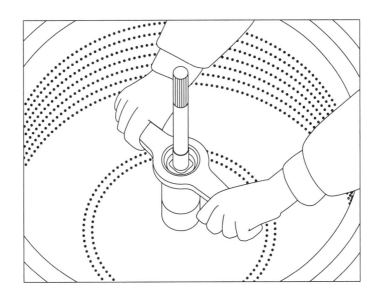

PART VI

FIGURE 18-54
Some models have eliminated the tall type agitator. This type of basket has more room in it for a larger wash load.

Hoses

The typical complaints associated with the hoses are:

- Washer leaks water.
- Water will not pump out.
- Kinked or plugged hoses.

To handle these problems, perform the following steps:

1. **Verify the complaint.** Verify the complaint by inspecting and operating the washer.

2. **Check for external factors.** You must check for external factors not associated with the appliance. Is the appliance installed properly? Does the appliance have the correct voltage? The voltage at the receptacle is between 108 volts and 132 volts during a load on the circuit. Do you have the correct polarity? (See Chapter 6.) Inspect fill and drain hoses.

3. **Disconnect the electricity.** Before working on the washer, disconnect the electricity. This can be done by pulling the plug from the electrical outlet. Be sure that you only remove the washer plug. Or disconnect the electricity at the fuse panel or at the circuit breaker panel. Turn off the electricity.

WARNING *Some diagnostic tests will require you to test the components with the power turned on. When you disassemble the control panel, you can position it in such a way that the wiring will not make contact with metal. This act will allow you to test the components without electrical mishaps.*

---

4. **Remove the defective hose.** To remove the defective hose, the tub must be empty of water. Loosen the clamps that hold the hose in place. On some models, the manufacturer uses a snap ring clamp. To remove this type of clamp, just squeeze the ends together (Figure 18-55).

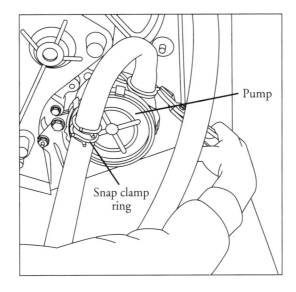

Figure 18-55
Removing the snap ring clamp. If you are just replacing a part, always check the hose for any cracks where the clamp was resting on the hose.

Pump

Snap clamp ring

Clutch

Clutches are used for braking (slowing and stopping the inner basket) and/or allowing the turning of the inner basket to get up to speed.

The typical complaints associated with the clutch are:

- Washer will not agitate.
- Washer will not spin.

To handle these problems, perform the following steps:

1. **Verify the complaint.** Verify the complaint by operating the washer. On electronic models, if a fault code appears, look up the code. If the washer will not power up, locate the technical data sheet behind the control panel for diagnostics information. On some models you will need the actual service manual for the model you are working on to properly diagnose the washer. The service manual will assist you in properly placing the washer in the service test mode for testing the washer functions.

2. **Check for external factors.** You must check for external factors not associated with the appliance. Is the appliance installed properly? Does the appliance have the correct voltage? The voltage at the receptacle is between 108 volts and 132 volts during a load on the circuit. Do you have the correct polarity? (See Chapter 6.)

3. **Disconnect the electricity.** Before working on the washer, disconnect the electricity. This can be done by pulling the plug from the electrical outlet. Be sure that you only remove the washer plug. Or disconnect the electricity at the fuse panel or at the circuit breaker panel. Turn off the electricity.

WARNING *Some diagnostic tests will require you to test the components with the power turned on. When you disassemble the control panel, you can position it in such a way that the wiring will not make contact with metal. This act will allow you to test the components without electrical mishaps.*

4. **Gain access to the motor.** To gain access to the motor, the back panel must be removed. The back panel is held on with screws. Remove the screws, and remove the panel.

5. **Remove the motor and clutch.** In this type of washer, to remove the motor and clutch assembly, you must first disconnect the wires of the motor from the wiring harness. Then remove the flexible pump coupling (see Figure 18-48). Remove the belt from the clutch. Once the motor is isolated, remove the three bolts that hold the motor mounting plate to the suspension. Now remove the motor assembly from the washer. You are now ready to remove the clutch assembly. Remove the clutch drive plate by removing the roll pin. This is accomplished by using a drive pin tool on the roll pin (Figure 18-56) and hitting it with a hammer. On a stubborn clutch drive plate, the use of a wheel puller tool and a horseshoe collar tool will make it easier to remove (Figure 18-57). Disassemble and remove the clutch assembly, and replace with a new assembly.

6. **Install the new clutch assembly.** To install the new clutch assembly, just reverse the disassembly procedure, and reassemble. Adjust the belt tension.

7. **Test the washer.** Test the washer operation by running the washer through a cycle. Be sure that you have some clothes in the washer when testing it.

FIGURE 18-56
When hitting the drive pin tool, be careful not to damage the clutch drive plate.

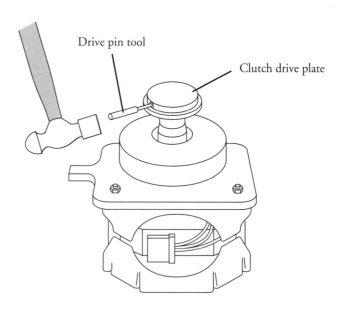

Drive pin tool

Clutch drive plate

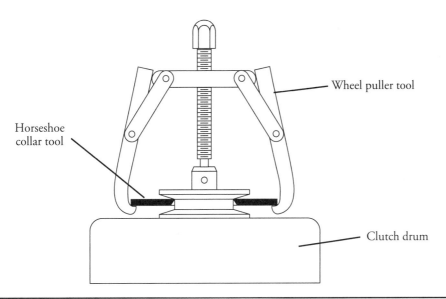

Wheel puller tool

Horseshoe
collar tool

Clutch drum

FIGURE 18-57   Use a horseshoe collar tool to remove a stubborn clutch drum.

Transmission

Transmissions are used for agitation and/or spinning of the inner basket. Some older models have large gear case transmissions while others have solenoid-activated clutch type transmissions. Check with the service manual for the model you are working on for the correct procedures on removing the transmission.

The typical complaints associated with the transmission are:

- Washer will not agitate.
- Washer will not spin.

To handle these problems, perform the following steps:

1. **Verify the complaint.** Verify the complaint by operating the washer. On electronic models, if a fault code appears, look up the code. If the washer will not power up, locate the technical data sheet behind the control panel for diagnostics information. On some models you will need the actual service manual for the model you are working on to properly diagnose the washer. The service manual will assist you in properly placing the washer in the service test mode for testing the washer functions.

2. **Check for external factors.** You must check for external factors not associated with the appliance. Is the appliance installed properly? Does the appliance have the correct voltage? The voltage at the receptacle is between 108 volts and 132 volts during a load on the circuit. Do you have the correct polarity? (See Chapter 6.)

3. **Disconnect the electricity.** Before working on the washer, disconnect the electricity. This can be done by pulling the plug from the electrical outlet. Be sure that you only remove the washer plug. Or disconnect the electricity at the fuse panel or at the circuit breaker panel. Turn off the electricity.

WARNING *Some diagnostic tests will require you to test the components with the power turned on. When you disassemble the control panel, you can position it in such a way that the wiring will not make contact with metal. This act will allow you to test the components without electrical mishaps.*

4. **Remove the inner basket.** To remove the inner basket, see the "Inner Basket" section earlier in this chapter.

5. **Remove the transmission boot.** To remove the transmission boot, loosen the ring clamps and lift the boot off the transmission and the outer tub (Figure 18-58). Examine the boot: if it is damaged, replace it. If the ring clamps are rusted, replace them.

FIGURE 18-58
Loosen the clamps, and lift the boot off. Inspect the boot for holes.

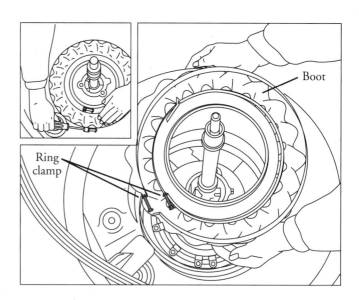

6. **Remove the transmission.** Remove the six bolts that hold the transmission to the washer's suspension. Remember the position that the transmission is in before you remove it. This will help you later for the reinstallation of the new transmission. Next, reach in and remove the drive belt from the transmission pulley. Lift the transmission out of the washer (Figure 18-59).

7. **Install the transmission.** To reinstall the transmission, just reverse the disassembly procedure, and reassemble. Do not forget to reinstall the belt.

8. **Test the washer.** Test the washer operation by running the washer through a cycle. Be sure that you have some clothes in the washer when testing it.

Direct-Drive Washer Transmission and Brake/Drive Assembly

When handling problems related to the direct-drive washer transmission or brake/drive assembly, perform the following steps:

1. **Verify the complaint.** Verify the complaint by operating the washer. On electronic models, if a fault code appears, look up the code. If the washer will not power up, locate the technical data sheet behind the control panel for diagnostics information. On some models you will need the actual service manual for the model you are working on to properly diagnose the washer. The service manual will assist you in properly placing the washer in the service test mode for testing the washer functions.

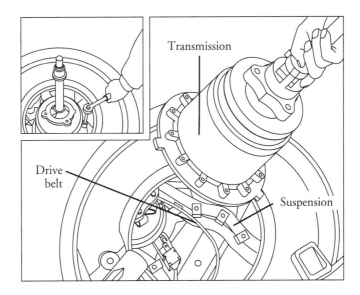

FIGURE 18-59
Removing bolts and lifting the transmission out of the tub.

Transmission

Drive belt

Suspension

2. **Check for external factors.** You must check for external factors not associated with the appliance. Is the appliance installed properly? Does the appliance have the correct voltage? The voltage at the receptacle is between 108 volts and 132 volts during a load on the circuit. Do you have the correct polarity? (See Chapter 6.)

3. **Disconnect the electricity.** Before working on the washer, disconnect the electricity. This can be done by pulling the plug from the electrical outlet. Be sure that you only remove the washer plug. Or disconnect the electricity at the fuse panel or at the circuit breaker panel. Turn off the electricity.

WARNING *Some diagnostic tests will require you to test the components with the power turned on. When you disassemble the control panel, you can position it in such a way that the wiring will not make contact with metal. This act will allow you to test the components without electrical mishaps.*

4. **Remove the outer cabinet.** To remove the cabinet (see Figure 18-18); remove the screws that hold the control panel in place. Next, lift the control panel up. Then remove the clips from each side that hold the cabinet in place. Pull back on the cabinet, and move it out of the way (see Figure 18-41).

5. **Remove the agitator.** To remove the two-piece agitator (Figure 18-60), pull off the agitator cap and the inner cap. Remove the bolt that secures the agitator, pull up on the agitator, and remove it from the inner basket.

6. **Remove the tub ring, inner basket, and transmission.** Before you can remove the inner basket, you must first remove the tub ring. To do this, you must unsnap the tabs around the outer tub (see Figure 18-60). Out of the eight tabs, one of them is a locator tab (the smallest one). The locator tab will help align the tub ring properly on reassembly. Next, remove the spanner nut, expand the drive block, and lift the basket out of the tub. Then remove the motor and pump (Figures 18-19, 18-20, and 18-21) and lay the washer on its back. Locate and remove the three bolts that secure the transmission to the tub support. Pull the transmission out of the basket drive tube, being careful not to damage the bearing.

7. **Remove the brake/drive assembly.** To release the brake, turn the brake cam driver counterclockwise. This will allow you to pull the brake and drive tube out of the base assembly (Figure 18-61). Next, inspect for wear on the drive tube shaft as indicated in Figure 18-62; if it is greater than .005 of an inch, replace the brake/drive assembly.

8. **Install the transmission and the brake/drive assembly.** To reinstall the transmission and the brake/drive assembly, just reverse the disassembly procedure, and reassemble.

9. **Test the washer.** Test the washer operation by running the washer through a cycle. Be sure that you have some clothes in the washer when testing it.

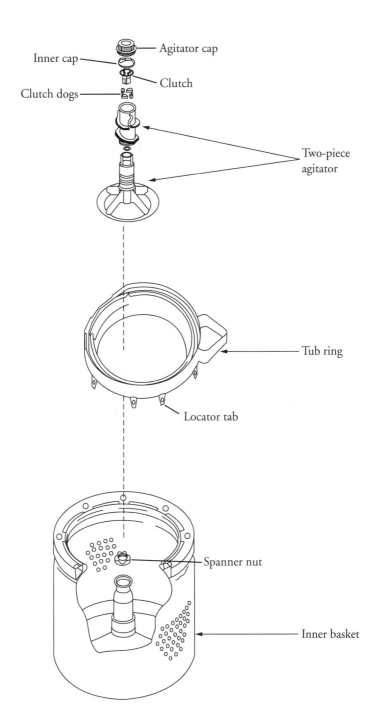

FIGURE 18-60
The agitator, tub ring, and inner basket.

Agitator cap

Inner cap

Clutch

Clutch dogs

Two-piece agitator

Tub ring

Locator tab

Spanner nut

Inner basket

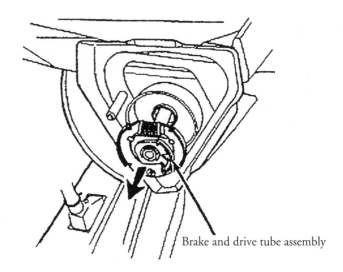

Brake and drive tube assembly

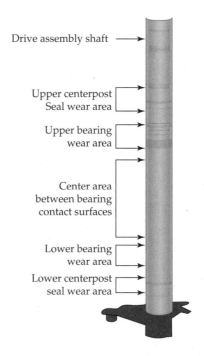

Drive assembly shaft

Upper centerpost
Seal wear area

Upper bearing
wear area

Center area
between bearing
contact surfaces

Lower bearing
wear area

Lower centerpost
seal wear area

gnostic Charts

The following diagnostic flowcharts will help you pinpoint the likely causes of the problem (Figures 18-63, 18-64, 18-65, 18-66, and 18-67).

The wiring diagrams in this chapter are used as examples only. You must refer to the actual wiring diagram on the washer that you are servicing. Figure 18-68 depicts an actual wiring schematic. Figure 18-69 depicts an actual ladder diagram of Figure 18-68. A ladder diagram is generally easier to read. Figures 18-70 through Figure 18-80 depict different types of wiring schematics, timing charts, and strip circuits for automatic washers.

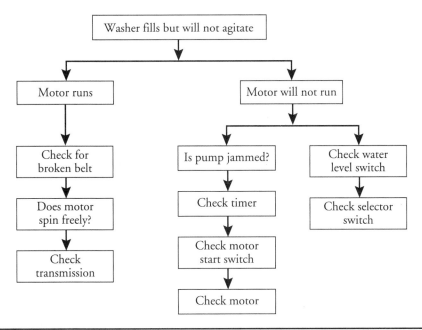

Figure 18-63   Diagnostic flowchart: Washer fills but will not agitate.

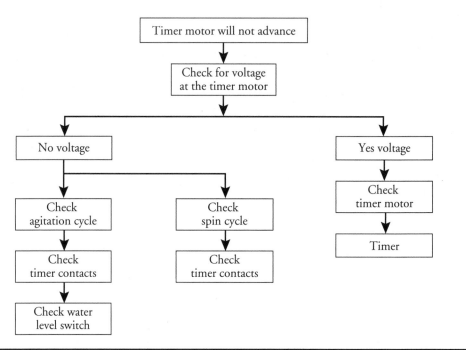

FIGURE 18-64   Diagnostic flowchart: Timer motor will not advance.

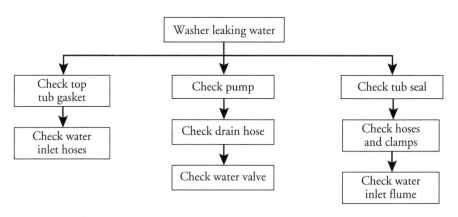

FIGURE 18-65   Diagnostic flowchart: Washer leaking water.

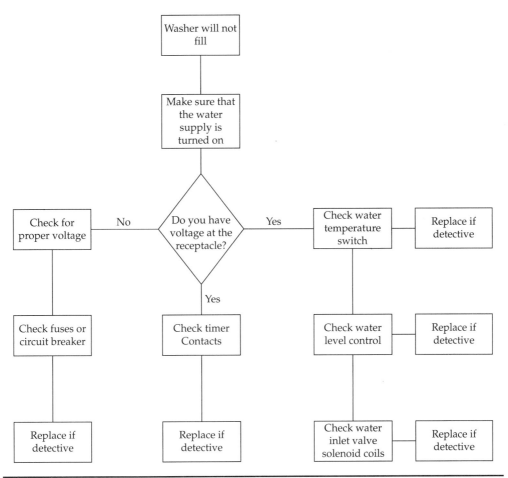

Figure 18-66   Diagnostic flowchart: Washer will not fill.

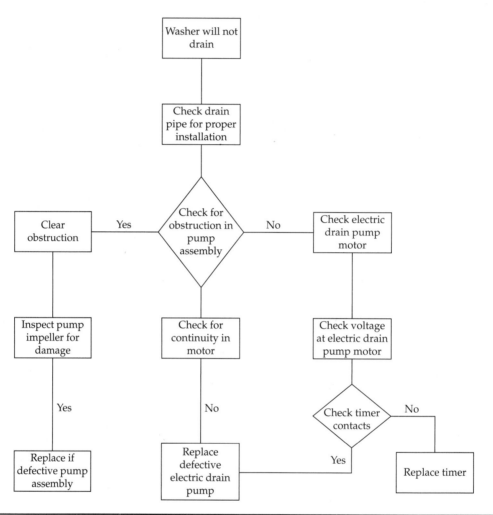

Figure 18-67 Diagnostic flowchart: Washer will not drain.

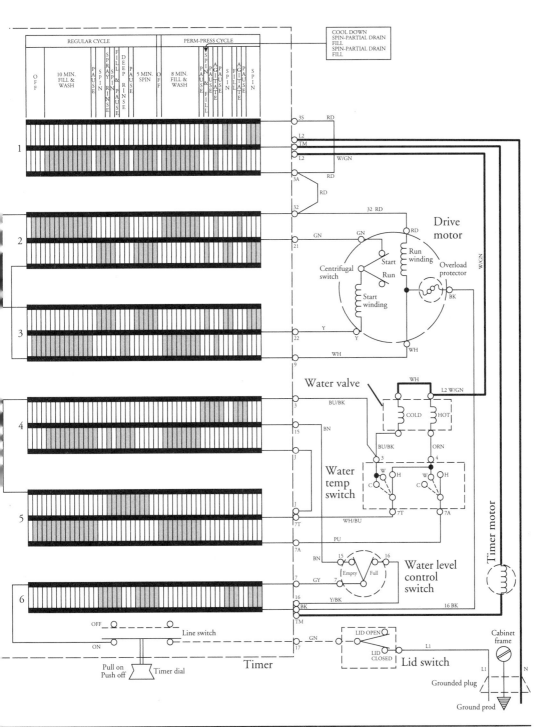

FIGURE 18-68 Automatic washer electrical schematic.

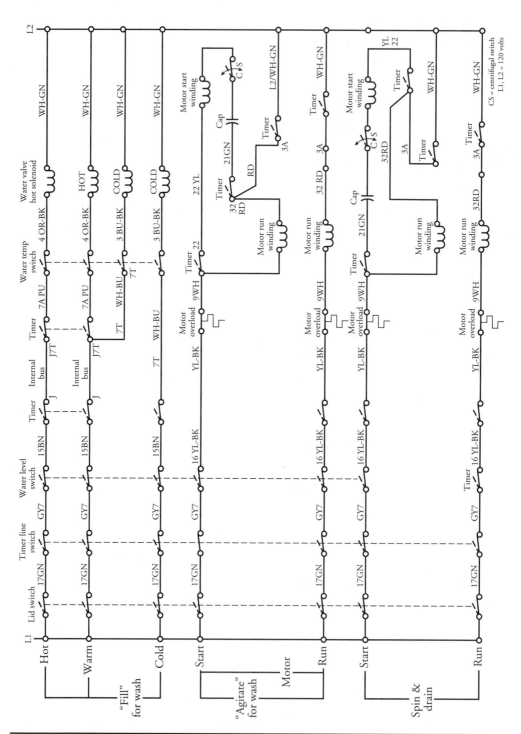

FIGURE 18-69 Automatic washer ladder diagram.

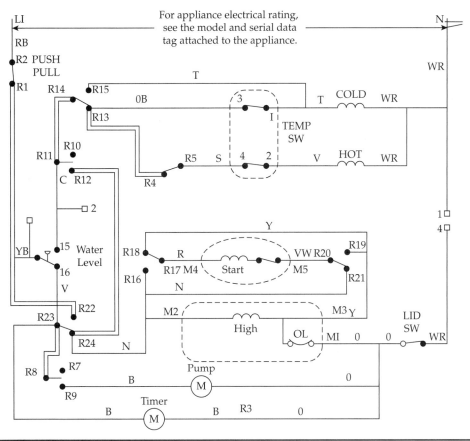

FIGURE 18-70 Automatic washer electrical schematic with a mechanical timer.

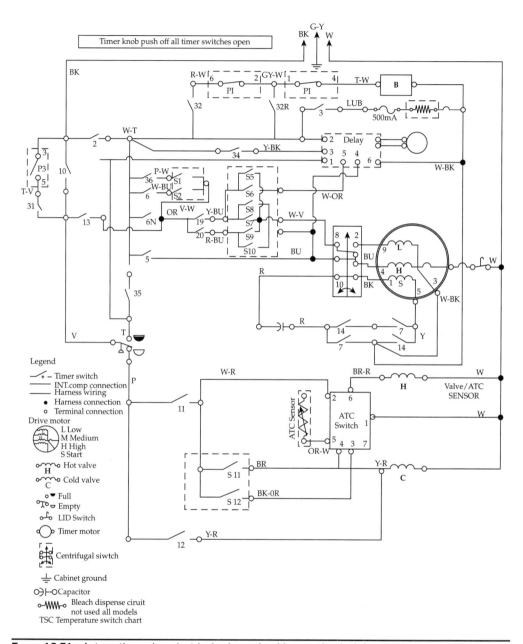

FIGURE 18-71 Automatic washer electrical schematic with a mechanical timer.

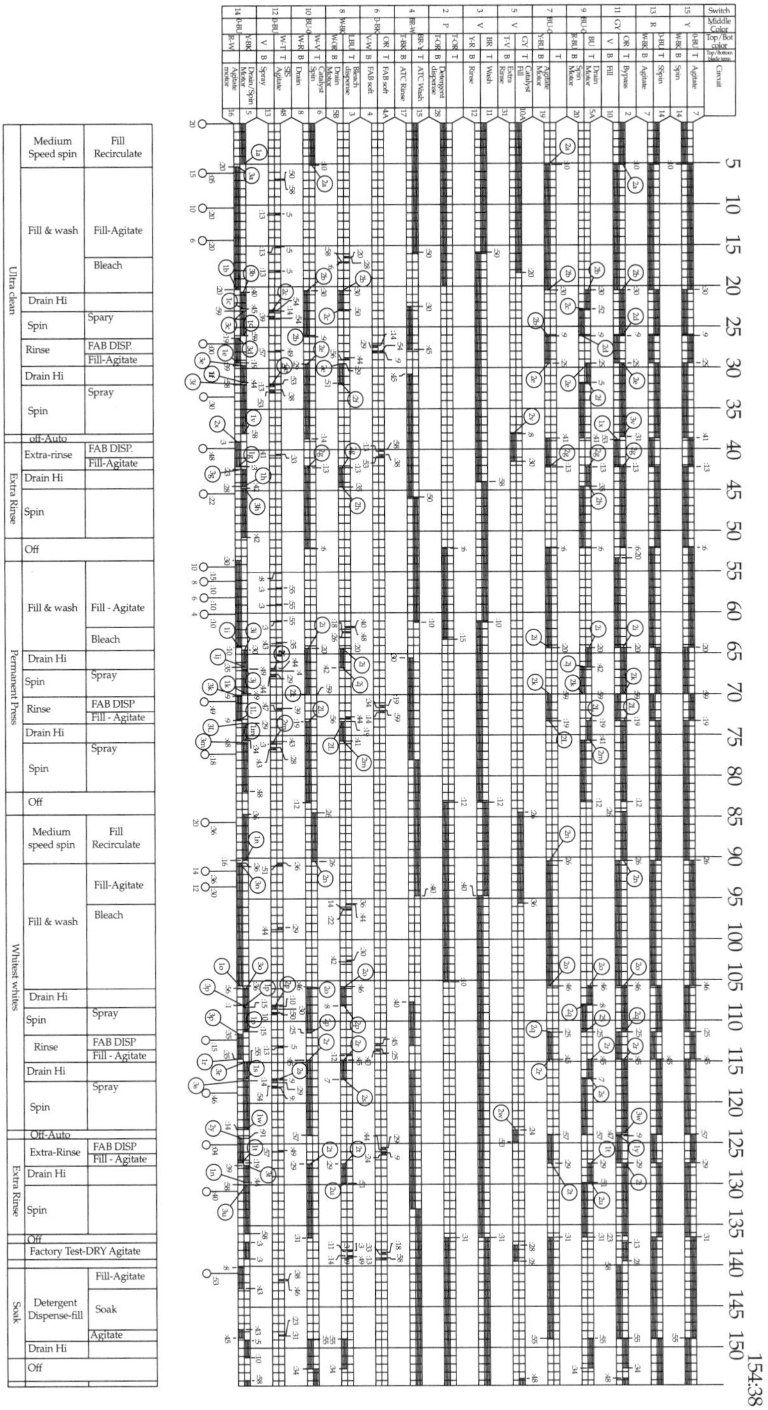

Figure 18-72 Automatic washer timing chart (see wiring schematic in Figure 18-71).

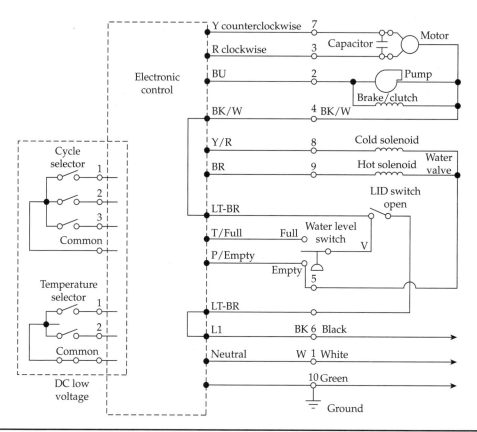

FIGURE 18-73   Automatic washer with an electronic control.

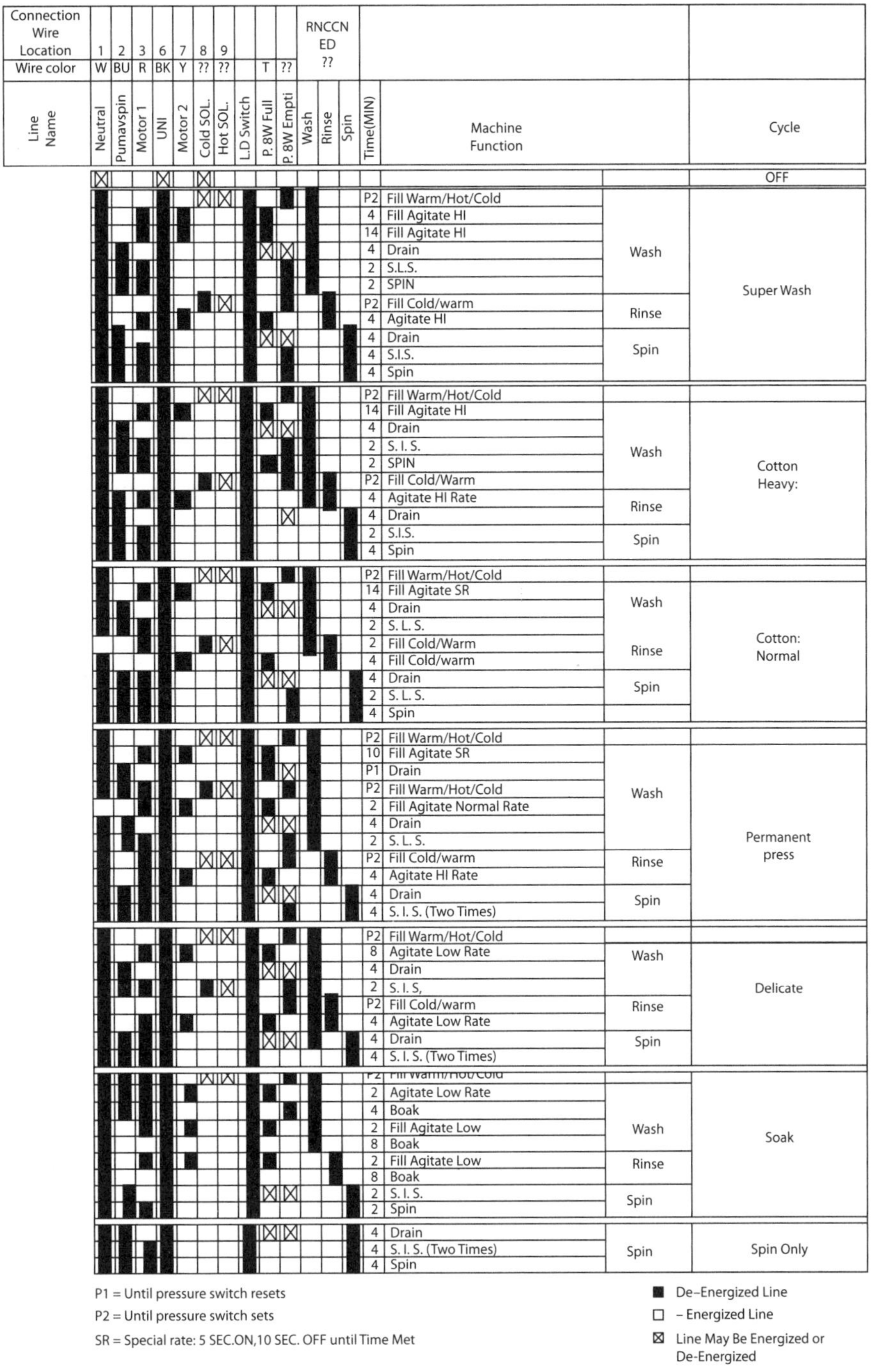

Connection Wire Location	1	2	3	6	7	8	9			RNCCN ED ??		
Wire color	W	BU	R	BK	Y	??	??		T	??		
Line Name	Neutral	Pumavspin	Motor 1	UNI	Motor 2	Cold SOL.	Hot SOL.	L.D Switch	P.8W Full	P.8W Empti	Wash	Rinse	Spin	Time(MIN)	Machine Function	Cycle

Machine Function / Cycle (readable columns):

Time(MIN)	Machine Function		Cycle
⊠			OFF
P2	Fill Warm/Hot/Cold		
4	Fill Agitate HI		
14	Fill Agitate HI		
4	Drain	Wash	
2	S.L.S.		
2	SPIN		Super Wash
P2	Fill Cold/warm	Rinse	
4	Agitate HI		
4	Drain	Spin	
4	S.I.S.		
4	Spin		
P2	Fill Warm/Hot/Cold		
14	Fill Agitate HI		
4	Drain	Wash	
2	S. I. S.		
2	SPIN		Cotton Heavy:
P2	Fill Cold/Warm		
4	Agitate HI Rate	Rinse	
4	Drain		
2	S.I.S.	Spin	
4	Spin		
P2	Fill Warm/Hot/Cold		
14	Fill Agitate SR	Wash	
4	Drain		
2	S. L. S.		
2	Fill Cold/Warm	Rinse	Cotton: Normal
4	Fill Cold/warm		
4	Drain	Spin	
2	S. L. S.		
4	Spin		
P2	Fill Warm/Hot/Cold		
10	Fill Agitate SR		
P1	Drain		
P2	Fill Warm/Hot/Cold	Wash	
2	Fill Agitate Normal Rate		
4	Drain		
2	S. L. S.		Permanent press
P2	Fill Cold/warm	Rinse	
4	Agitate HI Rate		
4	Drain	Spin	
4	S. I. S. (Two Times)		
P2	Fill Warm/Hot/Cold		
8	Agitate Low Rate	Wash	
4	Drain		
2	S.I.S,		Delicate
P2	Fill Cold/warm	Rinse	
4	Agitate Low Rate		
4	Drain	Spin	
4	S. I. S. (Two Times)		
P2	Fill Warm/Hot/Cold		
2	Agitate Low Rate		
4	Boak		
2	Fill Agitate Low	Wash	
8	Boak		Soak
2	Fill Agitate Low	Rinse	
8	Boak		
2	S. I. S.	Spin	
2	Spin		
4	Drain		
4	S. I. S. (Two Times)	Spin	Spin Only
4	Spin		

P1 = Until pressure switch resets

P2 = Until pressure switch sets

SR = Special rate: 5 SEC.ON,10 SEC. OFF until Time Met

■ De–Energized Line

☐ – Energized Line

⊠ Line May Be Energized or De-Energized

FIGURE 18-74   Timing chart for automatic washer with an electronic control (see wiring schematic in Figure 18-73).

Figure 18-75
Strip circuit for
automatic washer
with an electronic
control: Fill cycle.

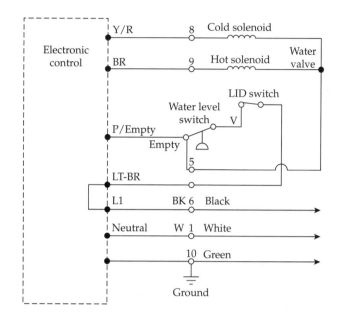

Figure 18-76
Strip circuit for
automatic washer
with an electronic
control: Agitate cycle.

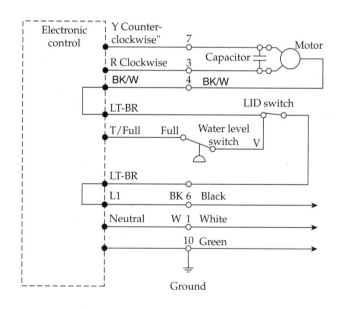

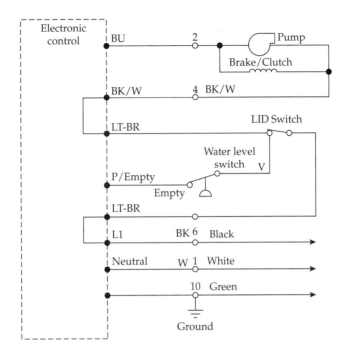

FIGURE 18-77
Strip circuit for
automatic washer
with an electronic
control: Drain cycle.

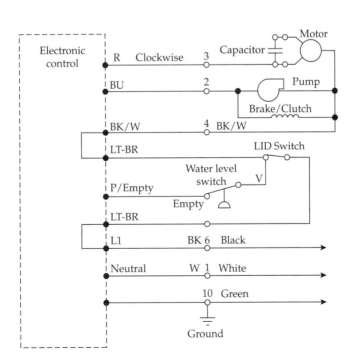

FIGURE 18-78
Strip circuit for
automatic washer
with an electronic
control: Spin cycle.

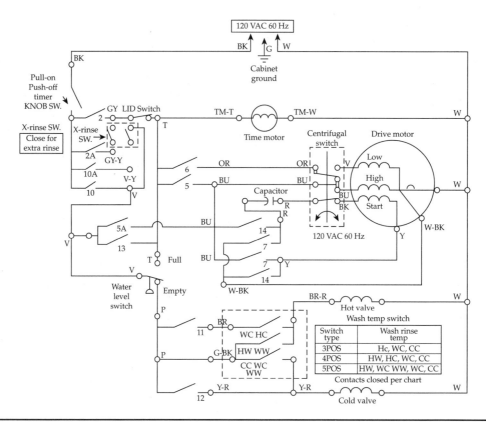

FIGURE 18-79 Typical automatic washer electrical schematic with a mechanical timer.

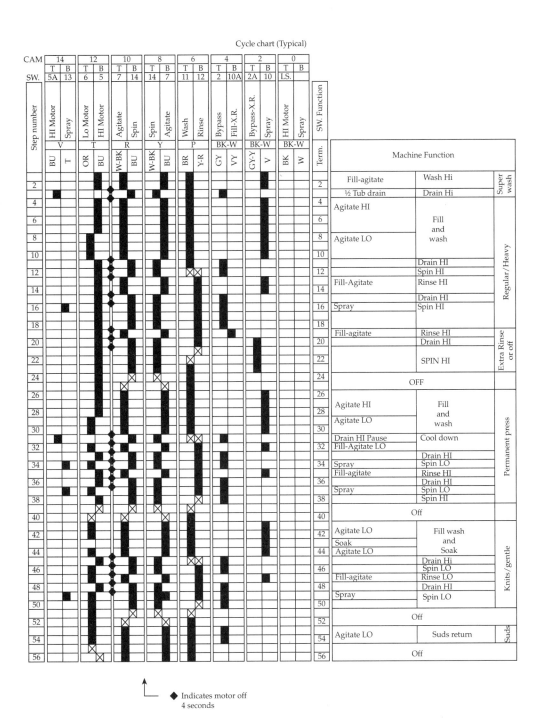

FIGURE 18-80 Typical cycle chart for automatic washer with a mechanical timer (see wiring schematic in Figure 18-79).

19

CHAPTER

Front Load Automatic Washers

The front load automatic washer has been around for decades. The first ones that I ever serviced were a General Electric front load automatic washer and dryer combination and a Westinghouse front load automatic washer back in the late 1970s. Today the front load automatic washer is a complex electromechanical machine, and Figure 19-1 is used as an example only. Today's service technician will need to know how to service electronic controls, thermistors, sensors, inverter boards, and variable speed motors. You must be able to read wiring diagrams, technical data sheets, and service manuals. I also recommend that you attend a training class on front load automatic washers.

Principles of Operation

The front load automatic washer presents a number of new features and operating characteristics quite different from the top load automatic washers (see the use and care guide for the model you are working on). In addition to the front-loading operation, the washer contains a number of unique operating features designed to increase clothes cleaning ability while offering very high water and energy conservation.

The clothes are placed evenly into the washer basket, making sure that the washer is not overloaded and that the proper cycle is selected. The user then adds detergent, bleach, or fabric softener to the dispenser drawer located in the front top section of the washer (read the use and care instructions for the model being serviced). The user will select a cycle and all options for the intended wash load, and then activates the washer through the cycle selector knob and starts the washer. On all front load models, the washer door will lock until the wash cycle ends or the user interrupts the cycle to add clothing to the wash cycle.

All wash and rinse water enters the wash basket through the water fill hoses, the water inlet valves, and the water inlet hose through the dispenser system. The amount of water that enters the wash basket is controlled by a preprogrammed fill through the electronic control board. Some models have a flowmeter sensor that monitors the amount of water entering the wash basket. As the flowmeter tracks the amount of water (about 10.5 gallons maximum), and if the electronic control board has not detected the water level control switch (pressure switch) indicating a full wash basket, an error code will appear and the water will shut off. On some models, the front load washer will maintain the proper fill

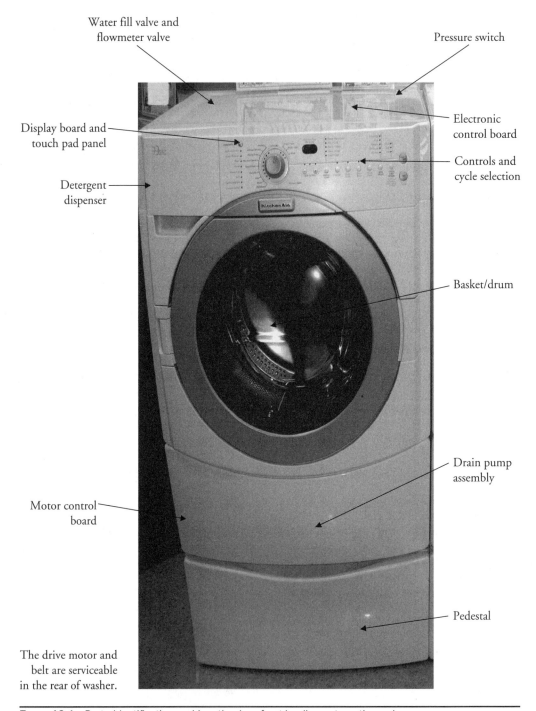

Water fill valve and
flowmeter valve

Pressure switch

Display board and
touch pad panel

Electronic
control board

Controls and
cycle selection

Detergent
dispenser

Basket/drum

Drain pump
assembly

Motor control
board

Pedestal

The drive motor and
belt are serviceable
in the rear of washer.

FIGURE 19-1 Parts identification and location in a front-loading automatic washer.

level using an adaptive fill preprogrammed into the electronic control board. A heating element is used on some models to increase the water temperature on certain wash cycles.

The inner basket will begin to rotate in a direction (clockwise or counterclockwise) at a predetermined speed that is preprogrammed into the electronic control board. Then the inner basket will pause for a predetermined amount of time and then begin to turn in the opposite direction at a predetermined speed. The length of wash tumble time is adaptive to the soil level programmed into the electronic control board at the start of the cycle.

When the preprogrammed tumble wash is completed, the inner basket comes to a halt and the drain pump motor will begin to drain the wash water from the front load washer. The spin cycle is designed to remove as much water and detergent as possible without harming the clothes. Spin speeds can be as low as 90 rpm to as high as 1000 rpm. The spin speeds are controlled by the cycle selection the user chooses at the beginning of the cycle. The spin speed and/or duration of final spin vary from cycle to cycle. Increasing the spin speed will extract more water and detergent, decreasing drying time and conserving energy. Decreasing the spin speed will reduce wrinkling. On some models, if you use no spin at the end of the cycle, the clothes will be very wet and they will have to be line dried. Some models use brief sprays of water to remove any soap or dirt residue remaining on the clothes in the spin cycle.

When the wash cycle ends, the door will unlock and the clothes can be removed from the washer and placed into the dryer for drying.

Functions and Cycles

The removal of soil from clothing and fabrics is accomplished by a combination of mechanical and chemical processes:

- **Mechanical process**   Soil is removed by tumbling the clothes and by forcing the detergent through the clothing.
- **Chemical process**   The HE (high-efficiency) detergent used will dissolve and loosen the soil in the clothing. Also the tumbling action of a front load washer is gentler on the clothing, allowing more air to be added to the water as the clothes are plunged into and lifted out of the water. As the front load washer operates through its cycles, it is aided by hot, soft water, which increases the chemical processes of the detergent being used.

Front load washing machines perform four basic functions, which are modified and put together in different ways to create the various cycles. The four functions are:

- Fill
- Tumble wash (clockwise or counterclockwise)
- Drain
- Spin

Safety First

Any person who cannot use basic tools or follow written instructions should *not* attempt to install, maintain, or repair any automatic front load washers. Any improper installation, preventive maintenance, or repairs could create a risk of personal injury or property damage.

If you do not fully understand the installation, preventive maintenance, or repair procedures in this chapter, or if you doubt your ability to complete the task on the front load automatic washer, please call your service manager.

The following precautions should also be followed:

- Never bypass or disconnect any part or device (originally designed into the washer) as a temporary repair.

- Always reconnect all ground wires, and be sure that they are secure.

- Be careful of moving parts and sharp edges.

Before continuing, take a moment to refresh your memory of the safety procedures in Chapter 2.

Automatic Front Load Washers in General

Much of the troubleshooting information in this chapter covers automatic front load washers in general, rather than specific models, in order to present a broad overview of service techniques. The pictures and illustrations that are used in this chapter are for demonstration purposes only, to clarify the description of how to service front load washing machines and in no way reflect a particular brand's reliability.

Location and Installation of Automatic Front Load Washers

The following are some general principles that should be followed when performing the installation of a washing machine:

- Locate the washing machine where there is easy access to existing drain, water, and electrical lines.

- Be sure you observe all local codes and ordinances for the electrical and plumbing connections.

- The washing machine should be installed and leveled on a firm floor to minimize vibration and possible washer "walk" during operation.

- Do not install the washing machine in an area where the temperature might be below freezing.

- To reduce the risk of a fire, never install a washing machine on any type of carpet.

- Follow the manufacturer's recommendations on washer installation clearances.

- Always follow the installation instructions that are provided with every new front load washing machine model purchased.

Common Installation Problems

Front load automatic washer installations are not complicated. However, as a service technician, you will come across a front load washer that has not been installed according to the manufacturer's installation instructions. The following sections describe some of the problems that you might run into and how to solve them.

When you arrive at a service call and the consumer tells you that the washer is full of water and it will not drain out, check for the following:

- The drain hose is kinked.
- The drain hose has a blockage.
- The drainpipe might be too small, not allowing for proper venting. The drainpipe must be a minimum of 1½ inches in diameter.
- The drainpipe and the drain hose were installed over six feet above the floor.

If the consumer tells you the washer will not fill up with water or it enters slowly, check the following:

- The water faucets were never turned on or they were turned on only a little bit.
- There is a blockage in the water inlet screen. The consumer did not flush out the water lines before installing the hoses on the washer.
- The water might be used somewhere else in the home. Check the water pressure (see use and care guide or installation instructions).
- When the washer is filling with water, the water is siphoning out through the drain line. The drainpipe is too low or the wrong size diameter. You might have to install a siphon break kit at the end of the drain hose.

When you arrive at the service call, you notice water on the floor. Check the following:

- The water hoses to the water inlet valve.
- The water faucet might be leaking.
- The drain hose is not connected properly to the washer.
- The drain hose comes out of the drainpipe.
- The drainpipe might have a blockage.
- The household drain cannot handle the capacity of the discharge water.

When the washer goes into the spin cycle, it will begin to vibrate and walk across the floor. Check the following:

- The feet or leveling legs on the washer are not set properly, causing the washer to teeter-totter.
- Is the washer level?
- Is the floor level?
- Check and see if the packing and shipping straps have been removed. Read the installation instructions on removing the packing straps.

Water Supply

The water supply for an automatic front load washer should have a hot and cold faucet located within four to six feet of the washer. The faucets should be a 3/4-inch threaded type to accept the fill hose connection.

The water pressure must be between 20 and 120 pounds per square inch for the washer to operate properly. The water coming out of the fill hoses should be equal in both pressure and in the volume of water to prevent unacceptable water temperature changes when entering and filling the washer.

The hot water supply to the washer should be between 130 and 150 degrees Fahrenheit. If the hot water temperature is below 70 degrees Fahrenheit, the clothes being washed will not clean properly and the detergent will not dissolve properly. You can check the temperature of the hot water by opening the hot water faucet near the washer. Let the water run until it is as hot as possible, and then insert a thermometer into the stream of water. If the thermometer reading is below 130 degrees Fahrenheit, you will have to raise the water heater thermostat setting. On some front-loading models, the manufacturer has incorporated an electric heater in the washer to heat the water to 153 degrees Fahrenheit for select wash cycles.

The cold water temperature should be between 70 and 100 degrees Fahrenheit. When the washer is in its rinse stage, the cold water will prevent wrinkles from setting into the fabrics. Some fabric manufacturers require that their fabrics be washed in cold water, both to prevent shrinkage and to eliminate the possibility of destroying the fabric. When the user selects the warm fill, the temperature of the water should be 100 degrees Fahrenheit.

It is recommended that the consumer read the use and care manual before performing a wash. Most use and care manuals have a water temperature guide to assist the user in the proper selection of the water temperature.

Drain Requirements

The drain to which the washer's drain hose is connected must be able to accept at least a 17- to 20-gallon-per-minute flow in order to remove the water from the tub. The standpipe should be at least 30 inches in height and no higher than 96 inches. The internal diameter of the drainpipe should be a minimum of 1½ inches in order to provide an air gap around the drain hose and thus to prevent the suction from siphoning the water out of the tub during the wash cycle. To select the proper drain hose installation method for the front load washer model, see the installation instructions that came with the washer.

Electrical Requirements

The front load automatic washer must be plugged into a 120 volt, 60 Hz, alternating current (AC) electrical power supply only. The washer must be connected to a separate 15 or 20 amp time-delay fuse or circuit breaker, and the three-prong receptacle must have the correct polarity (see Chapter 6). The washer must be grounded according to the manufacturer's installation instructions. Two-prong receptacles must not be used. Do not remove the ground prong or plug the washer cord into an adapter. Have an electrician change out the receptacle and replace it with a three-prong receptacle. Also, never use an extension cord on an appliance; this could result in fire, electrical shock, or death.

HE (High-Efficiency) Detergent

The only type of detergent that can be used with a front load automatic washer is HE (high-efficiency) detergent. Soap manufacturers are producing HE detergent for all of the front load automatic washers manufactured today. These detergents are formulated for use in low water volume washers. By using a non-HE detergent, the front load washers wash system, along with less water usage, will create oversudsing, washer errors, longer cycle times, and reduced rinsing performance. In addition, it may cause component failures and mold and mildew problems within the wash system. For more information on detergents go to www.cleaninginstitute.org.

Step-by-Step Troubleshooting by Symptom Diagnosis

Washer Will Not Turn On

1. Is the washer plugged in? The voltage at the receptacle is between 108 volts and 132 volts during a load on the circuit.

2. Check for a blown fuse or a tripped circuit breaker.

3. Check for voltage at the main electronic control board. On some models you will hear a click when the washer is plugged in.

4. Check for continuity in the service cord and the line filter.

5. Check connection and wiring on the electronic control board.

6. Plug in the washer and try starting it again. If necessary, reprogram to another cycle and try to restart.

7. Check touchpad/LED assembly. Try another cycle.

8. If washer has not been turned off, check for an overheated motor. After a cooldown period of 30 minutes, the motor should restart.

Washer Will Not Start the Cycle

1. Open and close the washer door. Between consecutive wash cycles the door has been opened.

2. Place washer in the service mode, and run the diagnostic test to check the door lock assembly for proper operation.

3. If the door is locked, use the diagnostic test to drain the washer.

4. Disconnect the power to the washer.

5. Check the wiring harness and the plug connectors.

6. Reconnect the power to the washer.

7. Check touchpad/LED assembly. Try another cycle.

The Washer Will Not Shut Off

1. Check for an error/fault code on the display console.

2. Cancel the wash cycle.

3. Check touchpad/LED assembly. Try another cycle.

4. Disconnect the power to the washer.

5. Check the drain pump, drain hose, and the drain pump filter for obstructions.

6. Reconnect the power to the washer.

7. Place washer in service mode, and run diagnostic test to verify electronic control board operation.

Electronic Control Will Not Accept Any Selections

1. Check for an error/fault code on the display console.

2. Check the drain pump, drain hose, and the drain pump filter for obstructions.

3. Check touchpad/LED assembly. Try another cycle.

4. Disconnect the power to the washer.

5. Check the wiring harness and the plug connectors.

6. Reconnect the power to the washer.

7. Place washer in service mode, and run diagnostic test to verify electronic control board operation.

The Washer Will Not Dispense Chemicals

1. Is the washer level? Check washer installation.

2. Check the dispenser drawer for chemicals that might be clogged in the dispenser.

3. Check all water connections to the washer and throughout the washer for obstructions. Also, check for a clogged water valve inlet screen.

4. Check dispenser motor assembly.

5. Disconnect the power to the washer.

6. Check the wiring harness and the plug connectors.

7. Reconnect the power to the washer.

8. Place washer in service mode, and run diagnostic test to verify electronic control board operation.

Washer Will Not Fill or Enters Slowly

1. Check washer installation. Check that both water faucets are open all the way. Check for low water pressure.

2. Disconnect the power to the washer.

3. Check water inlet valves.

4. Check all water connections to the washer and throughout the washer for obstructions or kinked hoses. Also, check for a clogged water valve inlet screen.

5. Reconnect the power to the washer.

6. Check the pressure switch for proper operation.

7. Check the drain pump motor.

8. Place washer in service mode, and run diagnostic test to verify electronic control board operation.

9. Check under the problem "The Washer Will Not Dispense Chemicals."

The Washer Overfills

1. Check washer installation.

2. Place washer in service mode, and run diagnostic test to verify drain pump system.

3. Check the pressure switch for proper operation.

4. Check the pressure switch hose from the outer drum to the switch.

5. Check the flowmeter for proper operation.

6. Place washer in service mode, and run diagnostic test to verify electronic control board operation.

Washer Drum Will Not Rotate

1. Check drive belt.

2. Check the drive motor for proper operation.

3. Disconnect the power to the washer.

4. Check the wiring harness and the plug connectors.

5. Reconnect the power to the washer.

6. Place washer in service mode, and run diagnostic test to verify electronic control board operations for the drive motor.

Washer Drive Motor Overheats

1. Check the drive motor.

2. Disconnect the power to the washer.

3. Check the wiring harness and the plug connectors.

4. Check the drive motor belt.

5. Reconnect the power to the washer.

6. Place washer in service mode, and run diagnostic test to verify electronic control board operations for the drive motor.

7. If washer has not been turned off, check for an overheated motor. After a cooldown period of 30 minutes, the motor should restart.

Washer Will Not Drain or Drains Slowly

1. Disconnect the power to the washer.

2. Check the wiring harness and the plug connectors.

3. Check the drain pump, drain hose, and the drain pump filter for obstructions.

4. Reconnect the power to the washer.

5. Place washer in service mode, and run diagnostic test to verify electronic control board operations.

6. Check drain hose installation. The maximum standpipe height is 96 inches. See washer installation instructions.

Washer Will Not Spin

1. Check dispenser drawer. On some models the drawer must be completely closed.

2. Check washer door. It must be closed.

3. On some models the wash load may be too small. Add more clothes.

4. Disconnect the power to the washer.

5. Check belt.

6. Check drive motor.

7. Reconnect the power to the washer.

8. Check for error/fault code.

9. Place washer in service mode, and run diagnostic test to verify electronic control board operations.

Front Load Washer Vibrates and Walks

1. Check for removal of shipping bolts and packing materials.

2. Check washer installation.

3. Make sure washer is level. Check the leveling feet.

4. Wash load may not be evenly distributed in drum. Stop washer and rearrange clothing.

Washer Has Incorrect Water Temperature

1. Check that the water inlet hoses are connected properly.

2. Disconnect the power to the washer.

3. Check the water heater element and the wire connections.

4. Check the water temperature sensor.

5. Reconnect the power to the washer.

6. Place washer in service mode, and run diagnostic test to verify electronic control board operations.

Flashing Display

1. Check for an error/fault code.

High-Pitched Noise

1. When the motor goes into spin, a certain high-pitched whining noise is normal.

Clanking and Rattling Noises

1. Stop the washer and check for foreign objects in the drum.

2. Restart the washer; if noises continue, stop the washer again.

3. Disconnect the power to the washer.

4. Check pump assembly for foreign objects.

5. Reconnect the power to the washer. Run washer.

6. Sometimes belt buckles or metal fasteners are hitting the drum. Check use and care instructions for proper handling of items.

Squealing Noises or Burning Rubber Odor

1. Check and see if washer is overloaded.
2. Reduce wash load.
3. Disconnect the power to the washer.
4. Inspect drive motor belt.
5. Reconnect the power to the washer.
6. Test washer operation.

Thumping Sounds While Washer Is Running

1. Heavy wash loads may produce thumping sounds. This is normal.
2. Check washer installation.
3. Stop washer and redistribute wash load.

Water Leaks

1. Check for loose fill hoses.
2. Disconnect the power to the washer.
3. Check the drain pump, drain hose, and the drain pump filter.
4. Check the tub gasket (boot) for holes.
5. Reconnect the power to the washer.
6. Test washer operation and inspect for other water leaks. If still leaking water, immediately turn off the power and repair washer.

Oversudsing

1. Check for wrong detergent used. Only use HE detergent.
2. Check for too much detergent used.
3. Run washer through several rinse cycles to remove excess suds.
4. Advise customer to use less detergent for future wash loads.

Incorrect Wash and Rinse Water Temperatures

1. Check fill hoses are connected correctly.
2. Check water pressures.

Water Is Entering Tub But Tub Will Not Fill with Water

1. Drain hose was improperly installed. The standpipe must be a minimum of 24 inches high to prevent siphoning.
2. See installation instructions for proper installation.

Common Washability Problems

If there are no mechanical problems with the washer's operation and the complaints are that the washer does not clean the clothes properly, you have a washability problem. The next step should be to look at the cause that best describes the problem that the customer is experiencing with the washer. Then proceed to correct the problem. If necessary, instruct the user on how to get better results from the automatic front load washer.

Stains on the Clothing

Stains on clothing can be caused by a number of different things. As the servicer, you will have to determine if it is caused by the washer's components or an external source. This type of problem is related to the increasing use of synthetic fabrics and to the poor washing practices of the user. Many of these stains consist of cooking oil or grease and are not visible when they first occur during cooking or eating. The oil that is embedded in the clothing acts like glue, attracting dirt from the wash water. When the wash cycle is completed, the clothes come out dirty and spotted.

Chemical Safety

There are some safety concerns that should be followed to reduce the risk of fire or serious injury to people or property:

- When using stain removal products always read and comply with the instructions listed on the container.
- Keep stain removal products in their original container and out of reach of children.
- Utensils used to assist in removing stains should be washed thoroughly.
- Never combine stain removal products such as chlorine bleach and ammonia. The fumes from both of these chemicals can make you very sick or kill you.
- Never wash clothing or items that have been soaked in, washed in, or spotted in gasoline, dry-cleaning solvents, or other flammable or explosive substances. These substances give off vapors that will ignite and explode causing property damage and/or personal injury or death.
- Never use flammable solvents or chemicals inside the home. Vapors can explode on contact with flames or sparks. Be careful and safe.

Stain Removal

The following are some stain-removal rules for clothing in general:

- Stains are easier to remove when they first appear on the clothing. If the stains are old, they might never come out of the clothing.
- Before attempting to remove any stain, you must know what type of stain, what kind of fabric, and how old is the stain?
- Use only cold or warm water to remove stains. Hot water will set the stain permanently into the fabric.

- When bleach is recommended for the removal of the stain, use a bleach that is safe for the fabric. When using a chlorine bleach, always dilute it with water to prevent the bleach from destroying the fibers.

- Always test stain remover products on a hidden corner of the garment to see if the color remains in the fabric.

- When preparing to remove the stain from the fabric, face the stained area down on a paper towel or a white cloth. Then apply the stain remover to the back of the stain so that the stain will be forced off the fabric, instead of through the fabric.

- Some protein stains can be removed with an enzyme presoak or with meat tenderizer.

- When using dry-cleaning solvents, always use them in a well-ventilated room, away from flames and sources of ignition to prevent personal injury or death.

- Alcoholic beverage stains turn brown the longer they stay on the fabric. As soon as the stain appears on the fabric, start treating it immediately. Wash or soak the stain in cold water, and then wash the garment.

- To remove blood stains, rinse or soak the garment in cold water with an enzyme presoak product. You can use diluted chlorine bleach on white fabrics, if necessary. For colored fabrics, use a powdered oxygen-type bleach. Then wash the garment.

- To remove chewing gum, use ice on the stain to make the chewing gum hard. Then scrape most of it off the fabric. Next, use a nonflammable dry-cleaning solvent with a sponge to remove the excess chewing gum. Wash the garment.

- To remove coffee or chocolate stains, soak the garment in warm or cold water. Next, make a paste of detergent mixed with hot water, and brush it on the stain. Wash the garment.

- To remove a milk product stain, use a nonflammable dry-cleaning solvent with a sponge. Wash the garment.

- To remove antiperspirant and deodorant stains, wash the garment with laundry detergent in the hottest water that is safe for the fabric. If the stain remains on the fabric, place the stain face down on a white towel, and treat the stain with a paste of ammonia and a powdered oxygen-type bleach. Let the paste stay on the stain for 30 minutes, and then wash the garment in the hottest water that is safe for the fabric.

- To remove fruit stains, soak the stain in cool water. Do not use soap—it will set the stain. Wash the garment. If the stain remains, cover the stain with a paste made of a powdered oxygen-type bleach, a few drops of hot water, and a few drops of ammonia. Let the paste stay on the stain for about 15 to 30 minutes. Then wash the garment.

- To remove iron or rust stains, apply some lemon juice mixed with salt. Then place the garment in the sun. Alternatively, a commercial rust-removing solution can be used. Wash the garment.

- To remove the stains caused by dairy products other than milk or baby formula, use a product containing enzymes to pretreat or soak the stains for 30 minutes or more and then place in the wash.

PART VI

- To remove crayons or candle wax, scrape off excess wax from fabric. Then place the stain face down between paper towels. Use a warm iron on fabric until the wax is absorbed into the paper towel. Replace the paper towels frequently. Next, pretreat the fabric with a prewash stain remover or a nonflammable dry-cleaning fluid. Hand-wash the fabric to remove the solvent and then place in the washer.
- To remove chocolate or grass stains, pretreat or soak the fabric in warm water using a product containing enzymes. Next, place fabric in the washer and use a bleach safe for the fabric.

Yellowing in Fabric

Some causes of yellowing in fabrics are:

- Poor body soil removal
- Clothes washed in water treated with a water softener
- Hard water or minerals in the water, such as iron
- Body oils released into the garment
- The water supply might pick up the color of decaying vegetation

To remove body oils, the user will have to increase the amount of detergent and use 150-degree Fahrenheit wash water. The user must also increase the frequency of using bleach in the wash.

To remove the yellowing from garments that are washed in water treated with a water softener, the user will have to decrease the amount of detergent used, approximately to the point that the decreased amount will not affect the soil removal process. The user must also increase the frequency of using bleach in the wash.

Hard water and minerals in the water can be treated with a water-conditioning apparatus. The user might have to drain the water heater and flush the tank. Never use chlorine bleach to remove hard water stains or iron stains.

To remove body oils from the garment, use a paste made of detergent and water. Let it stay on the fabric for 15 to 30 minutes. Then wash the garment.

To remove the yellowing caused by decaying vegetation, increase the amount of detergent, and bleach more often. White fabrics typically respond quite well to bleaching.

Fabric Softener Stains

Fabric softener stains are becoming more prevalent because it is now being recommended that some fabric softeners be used in the wash cycle, instead of the rinse cycle. These types of stains show up on synthetics as well as cotton fabrics. They can be removed from the fabric by pretreating the stain with liquid detergent and following the washing procedures listed in the use and care manual.

Lint

Lint is cotton fiber that has broken away from the cotton garment. Lint likes to attach itself to synthetic fabrics. When this happens, the user often thinks that the washer is not performing properly. Therefore, to solve the problem of lint on synthetic fabrics, the user must sort the items before washing the clothes. For example:

- The user must separate cottons from permanent press and knits.
- The user must separate light colors from dark colors.

Another cause of lint on clothes is overwashing. This causes the clothes to wear out faster. To correct the overwashing problem, use only one minute of wash time per pound of dry laundry with normal soil. Any more time than this is a waste, and it usually does not get the laundry cleaner.

If the drain cycle is excessive, this, too, will cause lint to remain on the garments. Check for improper drain hose connections. For example:

- Drain hose is too long (over 10 feet).
- Drain hose is too high (over 8 feet).
- Drain hose is kinked.

If excessive drain times still exist, check the following:

- Check the drain pump filter, located at the bottom front of the washer.
- Check that the drain pump is operating properly.
- Check for any obstructions in the drain system.
- Check for any obstructions within the water circulatory system of the washer.

Automatic Washer Maintenance

The interior is normally self-cleaning. However, there are times when you might have to remove objects from the inner basket. Clean the control panel and outer cabinet with a soft damp cloth. Do not use any abrasive powders or cleaning pads. Clean and inspect the interior underneath the washer. Read the use and care manuals for the proper maintenance of the brands of front load washers you service.

Cleaning the Door Seal Gasket (Boot)

Before cleaning the door seal gasket (Figure 19-2), remove all clothing from the washer. Now, inspect the door seal gasket for any rips or damage. If stains are found on the door seal gasket use the following procedure to clean the door seal gasket:

1. Take a gallon of warm tap water and mix it with ¾ of a cup of chlorine bleach; mix it well.
2. Use a cloth with the diluted solution and wipe the door seal gasket.
3. Let the diluted solution stand on the door seal gasket for 5 to 8 minutes.
4. After the time is up, wipe down the door seal gasket area thoroughly with a dry cloth, leave the door open, and let the interior of the washer air dry.

FIGURE 19-2
Front load washer
door seal gasket
(boot).

Door seal
gasket (boot)

SAFETY NOTE *Wear rubber gloves when using bleach for cleaning. Refer to the bleach manufacturer's instructions for proper handling and use.*

Cleaning the Dispenser Drawer

Fabric softener and detergent have a tendency to accumulate and build up in the dispenser drawer. To clean the dispenser drawer, you will have to remove it from the washer console. Open the dispenser drawer by first sliding the safety latch release lever to the left, and pull out the drawer until it stops. (On some models just pull out the drawer.) Next, press down on the locking tab located in the left rear of the dispenser drawer; now pull it out of the console (Figure 19-3). Take the dispenser drawer over to the sink and rinse it off using hot tap water. The fabric softener and detergent will dissolve away from the dispenser. Tell the customer that if the fabric softener and detergent continue to build up, a more frequent cleaning will be required. After you have cleaned the dispenser drawer, you will begin to

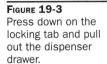

Figure 19-3
Press down on the
locking tab and pull
out the dispenser
drawer.

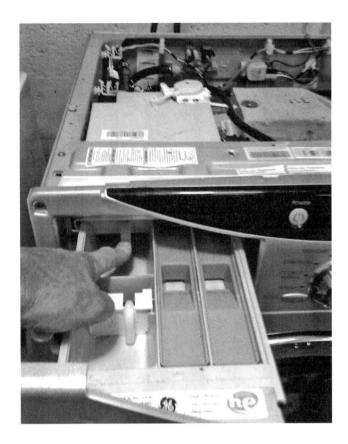

clean the drawer opening (Figure 19-4). Use a toothbrush to clean the upper and lower
recess in the drawer opening. Reinstall the dispenser drawer and test the dispenser drawer
operation.

Maintain Washer Freshness

To maintain the freshness of the washer always use HE detergent and at the end of the wash
cycle, leave the washer door open slightly. On some models the washer manufacturer has
added a special cycle in the washer's programming to raise the water level, along with chlorine
bleach added to begin the process of thoroughly cleaning the interior of the washer. To access
this special cycle you must read the use and care manual before beginning the cleaning process.

On other models, you can add 1/3 of a cup of chlorine bleach to the dispenser drawer and
run a short wash cycle. Do not add detergent during this process. At the end of the cycle
always leave the door open slightly to have better ventilation and drying of the washer's

FIGURE 19-4
Use a toothbrush to
clean the excess
fabric softener and
detergent from the
recess in the
dispenser drawer.

interior. Always read the use and care guide that comes with the washer for the proper
maintenance procedures. If this cleaning procedure does not sufficiently improve the washer's
freshness, then you will have to evaluate the washer for other causes.

Winterize Washer

If the washer is going to be stored or moved in the winter months, you will have to prepare
the washer in the following way:

1. Add about one quart to one gallon of nontoxic recreational vehicle (RV) antifreeze
   to the empty washer drum.

2. Select the drain/spin cycle and press the start button. Allow the washer to drain for
   about 1 minute. This procedure will allow the water to drain out of the washer. This
   process will not drain out all of the antifreeze.

3. Open the washer door and dry the interior of the drum.

4. Remove the dispenser drawer and pour out any water remaining in the dispenser
   compartments and dry thoroughly.

5. Unplug the washer.

6. Turn off the water inlet and remove the hoses and drain them.

7. Store or move the washer in an upright position. Remember the transport bolts have been removed when the washer was originally installed. If you kept them, reinstall the bolts to protect the washer from damage.

8. To reinstall the washer always follow the installation and the use and care instructions provided by the manufacturer. Don't forget to remove the transport bolts.

9. To remove the antifreeze from the washer after storage, run an empty washer through a complete cycle, add HE detergent, and do not add clothing to the wash cycle.

By winterizing the washer with antifreeze, any liquid remaining in the drain pump and hoses will not freeze up, causing damage to the washer.

Repair Procedures

Each repair procedure is a complete inspection and repair process for a single washer component, containing the information you need to test a component that might be faulty and to replace it, if necessary.

Any person who cannot use basic tools should *not* attempt to install, maintain, or repair any front load washer. Any improper installation, preventative maintenance, or repairs will create a risk of personal injury, as well as property damage. Call the service manager if installation, preventative maintenance, or the repair procedure is not fully understood.

Washer Timer

The washer timer is an electromechanical component controlled by a synchronous motor in incremental advances. It controls and sequences the numerous steps and functions involved in each cycle.

The typical complaints associated with washing machine timer failure are:

- The cycle will not advance.
- The washer won't run at all.
- The washer will not fill.
- The washer will not pump the water out.
- The washer will not shut off.

To handle these problems, perform the following steps:

1. **Verify the complaint.** Verify the complaint by operating the washer through its cycles. Before you change the timer, check the other components controlled by the timer. If the washer will not power up, locate the technical data sheet behind the control panel for the diagnostics information. On some models you will need the actual service manual for the model you are working on to properly diagnose the washer.

2. **Check for external factors.** You must check for external factors not associated with the appliance. Is the appliance installed properly? Does the appliance have the correct voltage? The voltage at the receptacle is between 108 volts and 132 volts during a load on the circuit.

3. **Disconnect the electricity.** Before working on the washer, disconnect the electricity. This can be done by pulling the plug out from the electrical outlet. Be sure that you only remove the washer plug. Or disconnect the electricity at the fuse panel or the circuit breaker panel. Turn off the electricity.

WARNING *Some diagnostic tests will require you to test the components with the power turned on. When you disassemble the control panel, you can position it in such a way that the wiring will not make contact with metal. This act will allow you to test the components without electrical mishaps.*

4. **Remove the top panel to gain access.** Begin by removing the screws from the top panel to gain access to the timer (Figure 19-5). Roll the console toward you. On some models, the front console can be removed.

5. **Test the timer.** Remove the timer motor leads from the timer assembly. Test the timer motor by connecting the ohmmeter probes to the timer motor leads (Figure 19-6). Set the range on the ohmmeter to R × 100. The meter should indicate between 2000 and 3000 ohms. Next, test the timer switch contacts using the wiring diagram configuration

FIGURE 19-5
Remove the screws that hold the top panel. Remove the top cover to gain access.

FIGURE **19-6**
Checking the washer
timer motor.

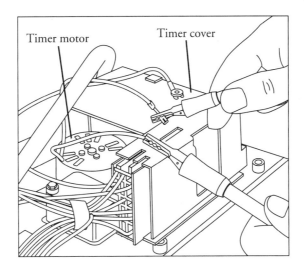

for the affected cycle. Place the meter probe on each terminal being tested, and turn the
timer knob. If the switch contact is good, your meter will read continuity. If the timer
motor measures suitably, then connect a 120-volt, fused service cord (Figure 18-4) to
the timer motor leads.

NOTE *Connect the ground (common) wire test lead to the console ground wire. Be cautious
whenever you are working with "live" wires. Avoid any shock hazards.*

If the motor does not operate, replace the timer. If the timer motor runs but does not
advance the cams, then the timer has internal defects and should be replaced.

6. **Remove the timer.** To remove the timer, remove the timer mounting screws. On
   some front load models, you might have to remove the control panel to gain access
   to the timer. Remove the wire lead terminals from the timer. Mark the wires as to
   their location on the timer. Some timers have a disconnect block instead of individual
   wires, which makes it easier to remove the timer wires.

   Turn the timer knob counterclockwise or pull the knob toward you to remove it
   from the timer shaft, and slide the indicator dial off the shaft.

7. **Install a new timer.** To install a new timer, just reverse the disassembly procedure,
   and reassemble. Replace the wires on the timer. Reinstall the console panel, and
   restore the electricity to the washer. Test the washing machine cycles. Make sure to
   take the washer out of the service test mode when the repair is completed.

Electronic Control Board and User Interface Controls

On some models the electronic control board and user interface controls replace the
electromechanical timer and rotary selection switches.

The typical complaints associated with the electronic control board or the user interface controls are:

- The washer won't run or power up.
- Unable to program the washer.
- The display board will not display anything.
- One or more keypads will not accept commands.
- Unusual display readouts and/or error codes.

To prevent electrostatic discharge (ESD) from damaging expensive electronic components, follow the steps in Chapters 6 and 11.

To handle these problems, perform the following steps:

1. **Verify the complaint.** Verify the complaint by operating the washer. Turn off the electricity to the appliance and wait for two minutes before turning it back on. If a fault code appears, look up the code. If the washer will not power up, locate the technical data sheet behind the control panel or in the rear of the washer for diagnostics information. On some models you will need the actual service manual for the model you are working on to properly diagnose the washer. The service manual will assist you in properly placing the washer in the service test mode for testing the washer functions.

2. **Check for external factors.** You must check for external factors not associated with the appliance. Is there electricity to the washer? The voltage at the receptacle is between 108 volts and 132 volts during a load on the circuit. Do you have the correct polarity? (See Chapter 6.)

3. **Disconnect the electricity.** Before working on the washer, disconnect the electricity. This can be done by pulling the plug out of the wall receptacle. Or disconnect the electricity at the fuse panel or circuit breaker panel. Turn off the electricity.

WARNING *Some diagnostic tests will require you to test the components with the power turned on. When you disassemble the control panel, you can position it in such a way that the wiring will not make contact with metal. This act will allow you to test the components without electrical mishaps.*

4. **Remove the console panel to gain access**. Begin by removing the screws from the washer top to gain access to the electronic control board. Next remove the top (Figure 19-7). Now remove the console panel screws to gain access to the user interface controls (Figure 19-8).

5. **Test the electronic control board and user interface controls.** If you are able to run the washer diagnostic test mode, check the different functions of the washer. Use the technical data sheet for the model you are servicing to locate the test points from the wiring schematic. Check all wiring connections and wiring. Using the technical data sheet, you can test the electronic control board or user interface controls and input and output voltages. On some models, fuses are soldered to the printed circuit board (PCB). These fuses must be tested first before condemning the component.

FIGURE 19-7
Removing the top from
the front load washer.

6. **Remove the electronic control board and user interface controls.** To remove the
   defective component, remove the screws that secure the board to the control panel
   or washer frame. On some models you may have to lift a tab and turn the control to
   remove it. Disconnect the connectors from the electronic control board or user
   interface control (Figure 19-9).

7. **Install the new component.** To install a new electronic control board or user
   interface control, read the parts data sheet that comes with the part for the proper
   installation process and just reverse the disassembly procedure and reassemble.
   Reinstall the console panel, and restore the electricity to the washer. Test the washer
   operation. Make sure to take the washer out of the service test mode when the
   repair is completed.

Door Lock

The door lock is mounted behind the front panel of the washer. It contains the door switch
and the solenoid operating mechanism that opens and closes the washer door. A release
ring located at the bottom of the lock mechanism can be pulled down to release the door
lock in case of failure. The door lock on some models will not open until the cycle is

PART VI

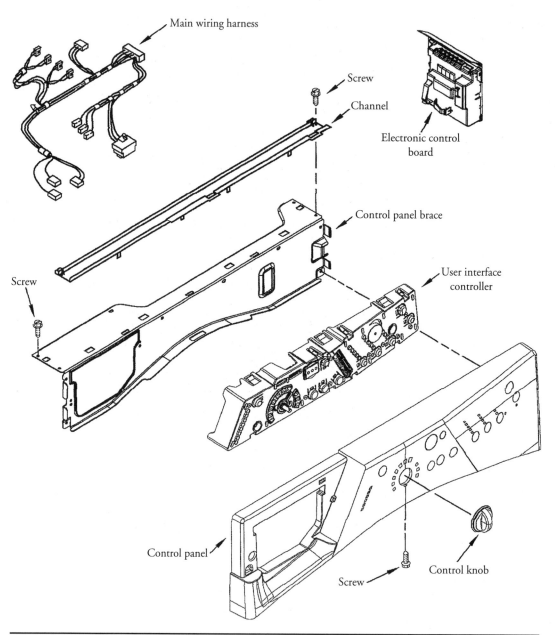

FIGURE 19-8   An exploded view of the control panel parts for a front load automatic washer.

FIGURE 19-9
When removing the wire connectors from the board, label them so you will remember where to plug them in. In today's age, you could use your smartphone to take a picture before you start removing wires.

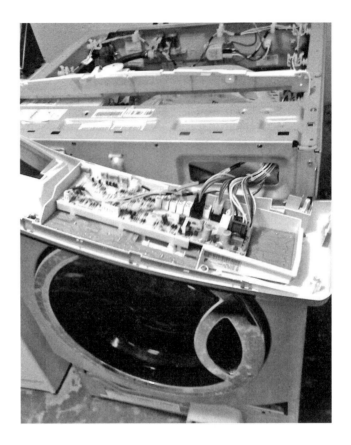

completed and during a high suds condition. Also, the door lock will not open when the basket is rotating or heating the water.

The typical complaints associated with the door lock are:

- The door will not lock.
- The door will not unlock.
- The door lock can initiate error codes.

To prevent electrostatic discharge (ESD) from damaging expensive electronic components, follow the steps in Chapters 6 and 11.

PART VI

To handle these problems, perform the following steps:

1. **Verify the complaint.** Verify the complaint by operating the washer. On electronic models, turn off the electricity to the appliance and wait for two minutes before turning it back on. If a fault code appears, look up the code. If the washer will not power up, locate the technical data sheet behind the control panel or in the rear of the washer for diagnostics information. On some models you will need the actual service manual for the model you are working on to properly diagnose the washer. The service manual will assist you in properly placing the washer in the service test mode for testing the washer functions.

2. **Check for external factors.** You must check for external factors not associated with the appliance. Is there electricity to the washer? The voltage at the receptacle is between 108 volts and 132 volts during a load on the circuit. Do you have the correct polarity? (See Chapter 6.)

3. **Disconnect the electricity.** Before working on the washer, disconnect the electricity. This can be done by pulling the plug out of the wall receptacle. Or disconnect the electricity at the fuse panel or circuit breaker panel. Turn off the electricity.

WARNING *Some diagnostic tests will require you to test the components with the power turned on. When you disassemble the control panel, you can position it in such a way that the wiring will not make contact with metal. This act will allow you to test the components without electrical mishaps.*

4. **Test the door lock switches.** Disconnect the wire connectors from the door lock assembly. Set your multimeter on the ohms scale to R × 1. Next touch the meter test leads to the two terminals for each switch. You will need the wiring diagram for this test. The wiring diagram will identify the switch number contacts and the electronic control board connector contacts. The switch contacts will read 0Ω when closed. When the switch is open you will read infinite (open circuit). Next, test the door lock and unlock solenoids. Touch the meter test leads to the indicated connector terminals. Each solenoid should read about 60Ω.

5. **Remove the door lock to gain access.** First, remove the bottom panel and reach in and locate the ring at the bottom of the lock mechanism and pull it down to release the door lock. This will open the washer door. The door lock assembly is attached to the front panel by three Phillips-head screws (Figure 19-10). Next, peel back the door seal gasket and locate the wire loop and spring. Using long-nose pliers, remove the wire loop and spring from around the gasket (Figure 19-11a). Now, pull the door seal gasket away from the front panel (Figure 19-11b). Remove the three Phillips-head screws that secure the door lock assembly to the front panel. Reach in behind the front panel and pull out the door lock assembly and remove the three wire harnesses (Figures 19-12 and 19-13). Once this lock assembly has failed it should be replaced with a duplicate of the original.

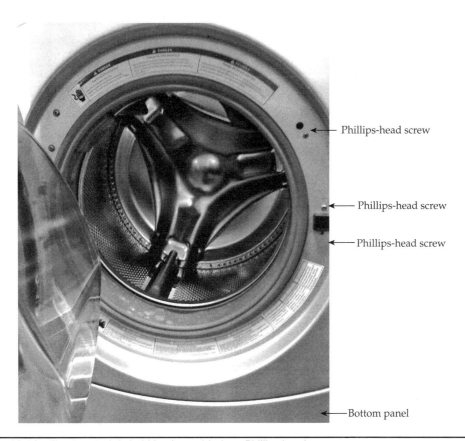

Phillips-head screw

Phillips-head screw

Phillips-head screw

Bottom panel

FIGURE 19-10   The door lock is held in place with three Phillips-head screws.

FIGURE 19-11a
Using a long-nose pliers to remove the wire loop and spring from the door seal gasket.

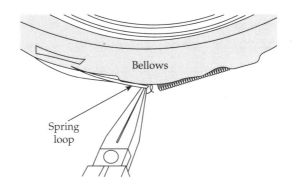

Bellows

Spring loop

Figure 19-11b
Peel back the gasket to gain access to the door lock located on the right side of the front panel.

6. **Install a new door lock.** To install a new door lock, just reverse the disassembly procedure, and reassemble. Replace the wires on the door lock assembly. Reinstall the door seal gasket, and restore the electricity to the washer. Inspect and test the washing machine for water leaks and proper operation. Make sure to take the washer out of the service test mode when the repair is completed.

Line Filter

The line filter is located on the interior of the rear panel next to the water valve. The purpose of a line filter in a front load washer is to smooth out any electrical fluctuations in the electrical supply voltage to the washer. The line filter also protects the electronic components and provides for a more reliable operation of the washer cycles.

To prevent electrostatic discharge (ESD) from damaging expensive electronic components, follow the steps in Chapters 6 and 11.

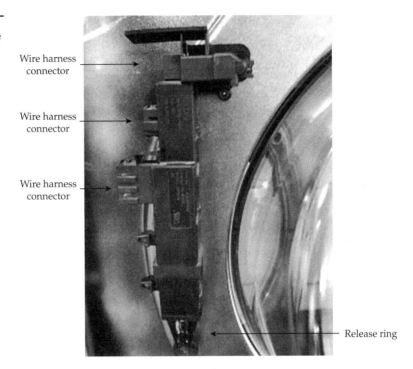

Figure 19-12
A view from the inside of the washer looking out toward the front panel. The door lock assembly is attached to the front panel.

Wire harness connector

Wire harness connector

Wire harness connector

Release ring

The typical complaints associated with the line filter are:

- The washer has error codes.
- Electronic control board malfunctions.

To handle these problems, perform the following steps:

1. **Verify the complaint.** Verify the complaint by operating the washer. Turn off the electricity to the appliance and wait for two minutes before turning it back on. If a fault code appears, look up the code. If the washer will not power up, locate the technical data sheet behind the control panel or in the rear of the washer for diagnostics information. On some models you will need the actual service manual for the model you are working on to properly diagnose the washer. The service manual will assist you in properly placing the washer in the service test mode for testing the washer functions.

FIGURE 19-13
Top view of the door
lock assembly.

FIGURE 19-13
Top view of the door
lock assembly.

2. **Check for external factors.** You must check for external factors not associated with the appliance. Is there electricity to the washer? The voltage at the receptacle is between 108 volts and 132 volts during a load on the circuit. Do you have the correct polarity? (See Chapter 6.)

3. **Disconnect the electricity.** Before working on the washer, disconnect the electricity. This can be done by pulling the plug out of the wall receptacle. Or disconnect the electricity at the fuse panel or circuit breaker panel. Turn off the electricity.

WARNING *Some diagnostic tests will require you to test the components with the power turned on. When you disassemble the control panel, you can position it in such a way that the wiring will not make contact with metal. This act will allow you to test the components without electrical mishaps.*

4. **Remove the top panel to gain access to the line filter.** Begin by removing the screws from the top panel to gain access to the line filter (Figure 19-7). Locate the line filter in the rear next to the water valve (Figure 19-14).

FIGURE 19-14
The line filter is located in the rear, near the top of the washer. If you hold an AM radio next to the filter and hear static when the washer is running, the line filter is defective.

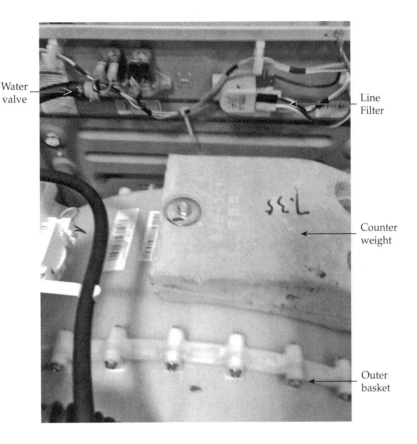

Water valve

Line Filter

Counter weight

Outer basket

5. **Test the line filter.** Inspect the line filter body for burn marks caused by heat or a power surge. Next, set your multimeter on the ohms scale, remove the wires from the terminals on the line filter, and check the resistance between the black wire terminals and the white wire terminals. The resistance should read 0Ω between the terminals.

6. **Remove the line filter.** Remove the wires from the terminals on the line filter. Then remove the screws that secure the line filter to the back panel. Slide the line filter to the right to remove it from the back panel.

7. **Install a new line filter.** To install a new line filter, just reverse the disassembly procedure, and reassemble. Replace the wires on the line filter. Restore the electricity to the washer. Test the washing machine for proper operation. Next reinstall the top panel and trim (Figure 19-7). Make sure to take the washer out of the service test mode when the repair is completed.

Dispenser Assembly

Front load washers have an automatic dispenser distribution system that allows for the automatic dispensing of detergent, bleach, and fabric softener. The dispenser assembly is located in the upper left side of the washer (Figure 19-3). A water diverter sprays the correct amount of water into the proper compartment at the correct time, which will dilute the chemicals before they are dispensed into the wash.

The typical complaints associated with the dispenser assembly are:

- The washer has error codes.
- Detergent will not dispense into wash.
- Bleach will not dispense into wash.
- Fabric softener will not dispense into wash.

To prevent electrostatic discharge (ESD) from damaging expensive electronic components, follow the steps in Chapters 6 and 11.

1. **Verify the complaint.** Verify the complaint by operating the washer. On electronic models, turn off the electricity to the appliance and wait for two minutes before turning it back on. If a fault code appears, look up the code. If the washer will not power up, locate the technical data sheet behind the control panel or in the rear of the washer for diagnostics information. On some models you will need the actual service manual for the model you are working on to properly diagnose the washer. The service manual will assist you in properly placing the washer in the service test mode for testing the washer functions.

2. **Check for external factors.** You must check for external factors not associated with the appliance. Is there electricity to the washer? The voltage at the receptacle is between 108 volts and 132 volts during a load on the circuit. Do you have the correct polarity? (See Chapter 6.)

3. **Disconnect the electricity.** Before working on the washer, disconnect the electricity. This can be done by pulling the plug out of the wall receptacle. Or disconnect the electricity at the fuse panel or circuit breaker panel. Turn off the electricity.

WARNING *Some diagnostic tests will require you to test the components with the power turned on. When you disassemble the control panel, you can position it in such a way that the wiring will not make contact with metal. This act will allow you to test the components without electrical mishaps.*

4. **Gain access to the dispenser assembly.** Begin by removing the screws from the top panel to gain access to the dispenser assembly (Figures 19-3 and 19-7).

5. **Test the dispenser assembly.** On the dispenser tank (Figure 19-16), the water diverter movement is controlled by a motor-driven cam. By placing the washer into the service test mode, you can observe the water diverter movement. The dispenser motor operates on 120 VAC, 60 Hz. The dispenser motor receives its commands from the electronic control board. You can plug in the washer and check for 120 VAC at the dispenser motor terminals when the washer is placed in the service test mode. Use the wiring diagram to find the correct wire connections to place your multimeter test leads on. Set your meter on AC voltage. If you read 120 volts on your meter and the dispenser motor does not advance the water diverter, replace the dispenser motor and/or dispenser assembly. When the washer is unplugged, the dispenser motor should read approximately 1500Ω resistance.

6. **Remove the dispenser assembly.** Remove the top and the control panel (Figures 19-4 and 19-15). Lay the control panel forward (Figure 19-9). Be careful not to damage the wiring harness. Next remove the gasket from the front panel (Figures 19-10 and 19-11a, b). Now remove all hoses from the dispenser (Figure 19-17). Disconnect the wiring harness and remove any screws holding the dispenser in place along with removing the dispenser motor from the tank.

7. **Install the new dispenser assembly.** To install a new dispenser assembly, just reverse the disassembly procedure, and reassemble. Replace the wires on the dispenser motor. Restore the electricity to the washer. Test the washing machine for proper operation. Next reinstall the top panel and trim (Figure 19-7). Make sure to take the washer out of the service test mode when the repair is completed.

Water Level Control

The water level control (Figure 19-18) is located on the upper-right brace and it starts in the empty position. As the washer fills with water and the water level rises in the tub, it causes the air in the tube to increase the air pressure in the air dome. The air pressure is transferred

Figure 19-15
Remove the eight Phillips-head screws that hold the front bracket and the control panel rear cover.

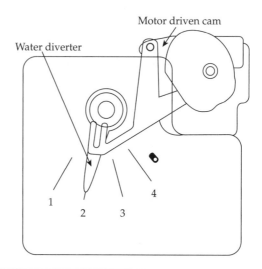

Water diverter arrow position	Dispenser distribution function
1	Pre wash
2	Main wash
3	Bleach
4	Fabric softener

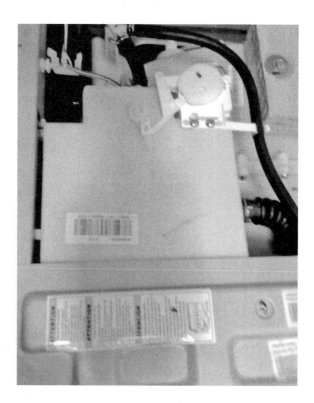

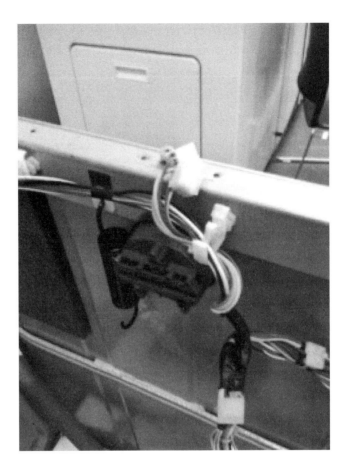

Figure 19-18
Water level control
(pressure switch)
located under the
top cover.

from the air hose to the water level control and against the diaphragm, which actuates three switches: foam (suds), main, and overflow (Figure 19-19). The electronic control board monitors four water level conditions: empty, main, foam (suds), and overflow. Most front load washer models have overflow protection, which will turn on the drain pump motor and drain the water from the washer if an overfill condition exists, whether the washer is running or not. On other washer models, the water level control is electronic (Figure 19-20). Both controls operate similarly. The electronic board is mounted inside the water level control unit. The electronic board reads the pressure in the diaphragm bellows and converts it into an electronic signal. This signal is monitored by the electronic control board, which turns on or off the water valve solenoids.

The typical complaints associated with the water level control (pressure switch) are:

- Tub does not fill to the proper level selected.
- Washer basket will not rotate.
- Washer will not spin.
- Control panel indicates an error/fault code.

PART VI

FIGURE 19-19
The air pressure in the air hose operates three internal switches, which monitor the four water level conditions listed.

Level	Switch position		
	Foam	Main	Overflow
Empty	Closed	Open	Open
Foam	Open	Open	Open
Main	Open	Closed	Open
Overflow	Open	Closed	Closed

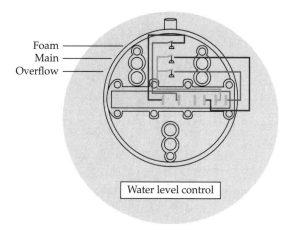

FIGURE 19-20
Electronic water level control (sensor). The air hose is attached to the sensor and to the drain boot in the washer.

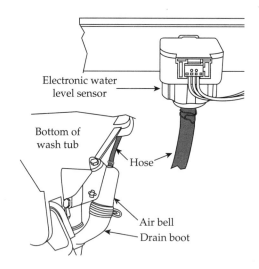

To handle these problems, perform the following steps:

1. **Verify the complaint.** Verify the complaint by trying to select different water levels when operating the washer through its cycles. On electronic models, turn off the electricity to the appliance and wait for two minutes before turning it back on. If a fault code appears, look up the code. If the washer will not power up, locate the technical data sheet behind the control panel or in the rear of the washer for diagnostics information. On some models you will need the actual service manual for the model you are working on to properly diagnose the washer. The service manual will assist you in properly placing the washer in the service test mode for testing the washer functions.

2. **Check for external factors.** You must check for external factors not associated with the appliance. Is the appliance installed properly? Does the appliance have the correct voltage? The voltage at the receptacle is between 108 volts and 132 volts during a load on the circuit. Do you have the correct polarity? (See Chapter 6.) Is there any physical damage to the component? Is the plastic hose connected to the water level control and air dome? Check to be sure that the water is turned on all the way.

3. **Disconnect the electricity.** Before working on the washer, disconnect the electricity. This can be done by pulling the plug from the electrical outlet. Be sure that you only remove the washer plug. Or disconnect the electricity at the fuse panel or at the circuit breaker panel. Turn off the electricity.

WARNING *Some diagnostic tests will require you to test the components with the power turned on. When you disassemble the control panel, you can position it in such a way that the wiring will not make contact with metal. This act will allow you to test the components without electrical mishaps.*

4. **Remove the top panel to gain access.** Begin by removing the top panel to gain access to the water level control (see Figure 19-7).

5. **Test the water level control.** Disconnect the air hose from the air dome and the wire terminal connectors from the water level control (pressure switch). Set the multimeter to the ohms scale R × 1. Locate on the wiring diagram the correct switch terminals to test. Touch the test leads to the correct switch terminals. Gently blow in the air hose until you hear the diaphragm activate. The switches will activate also. The meter should indicate 0Ω for each measurement while the diaphragm is activated.

WARNING *Do not blow too hard into the air hose; you might damage the diaphragm in the water level control.*

6. **Remove the water level control.** To remove the water level control (Figures 19-18 and 19-19), remove the wires and then remove the screws that hold the component to the washer frame. On some models you may have to twist the component and remove it. Next, remove the plastic hose.

7. **Reinstall the water level control.** To install a water level control (pressure switch), just reverse the disassembly procedure, and reassemble. Replace the wires on the water level control. Restore the electricity to the washer. Test the washing machine for proper operation. Next reinstall the top panel and trim (Figure 19-7). Make sure to take the washer out of the service test mode when the repair is completed.

Water Valve

The water inlet valve controls the flow of water into the washer, and is solenoid-operated (Figure 19-21). When it is energized, water in the supply line will pass through the valve body and into the washer.

The typical complaints associated with water valve failure are:

- The washer will not fill with water.
- The washer overfills and leaks onto the floor.
- When the washer is off, water still enters the tub.
- The washer has error codes.

To handle these problems, perform the following steps:

1. **Verify the complaint.** Verify the complaint by operating the washer through its cycles. Listen carefully, and you will hear whether the water is entering the washer. On electronic models, turn off the electricity to the appliance and wait for two minutes

FIGURE 19-21
Front load washer
water valve located
on the rear cabinet.

before turning it back on. If a fault code appears, look up the code. If the washer will not power up, locate the technical data sheet behind the control panel for the diagnostics information. On some models you will need the actual service manual for the model you are working on to properly diagnose the washer. The service manual will assist you in properly placing the washer in the service test mode for testing the washer functions.

2. **Check for external factors.** You must check for external factors not associated with the appliance. Is the appliance installed properly? Does the appliance have the correct voltage? The voltage at the receptacle is between 108 volts and 132 volts during a load on the circuit. Do you have the correct polarity? (See Chapter 6.) Is the water turned on? Both water faucets must be turned all the way counterclockwise.

3. **Disconnect the electricity.** Before working on the washer, disconnect the electricity. This can be done by pulling the plug from the electrical outlet. Be sure that you only remove the washer plug. Or disconnect the electricity at the fuse panel or at the circuit breaker panel. Turn off the electricity.

WARNING *Some diagnostic tests will require you to test the components with the power turned on. When you disassemble the control panel, you can position it in such a way that the wiring will not make contact with metal. This act will allow you to test the components without electrical mishaps.*

4. **Gain access to the water valve.** Turn off the water supply to the washer water valve. To access the water valve, the top panel must be removed (Figure 19-7).

5. **Test the water valve.** In order to check the solenoid coils on the water valve, remove the wire leads (label them) that connect to the coils from the wire harness (see Figure 19-21). These are slide-on terminal connectors attached to the ends of the wire. Just pull them off. Set the ohmmeter on R × 1000, and attach the probes to the terminals of one of the solenoid coils. The meter should read between 750 and 1100 ohms. Repeat this test for the second solenoid coil.

To test the fill rate of the water valve, the top panel does not have to be reinstalled. The washer can be placed into the service mode and the water valve activated. Another way to test the water valve is to attach the 120-volt fused service cord (Figure 18-4)—including the ground wire test lead to the cabinet ground—to the water valve solenoid coil. Then energize the solenoid coil to allow water to enter the tub and to check the flow rate of the water valve. The water valve flow rate is 2.1 gallons per minute. This step is repeated for each solenoid coil. If, when you energize the water valve, no water enters the washer tub, replace the water valve. If the water valve checks correctly, check the timer or the electronic control board and the wiring harness.

6. **Remove the water valve.** Remove the wires from both solenoid coils, and disconnect the outlet hose. Next, remove the water inlet hoses from the water valve. Remove the screws that secure the water valve to the washer. On some models there is only one screw that holds the water valve in place, and then you must slide the water valve horizontally to the right to remove it.

7. **Install a new water valve.** To install the new water valve, just reverse the disassembly procedure, and reassemble. Reconnect the wire leads to the solenoid coils. After the installation of the new valve, turn on the water supply and check for water leaks. If none are found, reinstall the top panel and trim (Figure 19-7) and restore the electricity to the washer. Test the washing machine for proper operation. Make sure to take the washer out of the service test mode when the repair is completed.

Drive Belt

The drive belt is a six- or seven-rib flat Poly-V design belt. It is used to transmit power from the motor pulley to the tub. The belt is constructed of a material that stretches, which makes belt tension adjustments unnecessary. It also makes it easier to remove and reinstall a belt.

The typical complaints associated with belt failure are:

- Washer basket will not rotate.
- Washer basket will not spin.
- Washer motor spins freely.
- There is a smell of something burning.
- The washer has error codes.

To handle these problems, perform the following steps:

1. **Verify the complaint.** Verify the complaint by operating the washer in the spin cycle. Listen carefully, and you will hear and see if the inner basket is turning or if the circuit breaker trips. On electronic models, turn off the electricity to the appliance and wait for two minutes before turning it back on. If a fault code appears, look up the code. If the washer will not power up, locate the technical data sheet behind the control panel or in the rear of the washer for diagnostics information. On some models you will need the actual service manual for the model you are working on to properly diagnose the washer. The service manual will assist you in properly placing the washer in the service test mode for testing the washer functions.

2. **Check for external factors.** You must check for external factors not associated with the appliance. Is the appliance installed properly? Does the appliance have the correct voltage? The voltage at the receptacle is between 108 volts and 132 volts during a load on the circuit. Do you have the correct polarity? (See Chapter 6.)

3. **Disconnect the electricity.** Before working on the washer, disconnect the electricity. This can be done by pulling the plug out from the electrical outlet. Be sure that you only remove the washer plug. Or disconnect the electricity at the fuse panel or at the circuit breaker panel. Turn off the electricity.

WARNING *Some diagnostic tests will require you to test the components with the power turned on. When you disassemble the control panel, you can position it in such a way that the wiring will not make contact with metal. This act will allow you to test the components without electrical mishaps.*

4. **Gain access to the belt.** Pull the washer away from the wall. You must gain access to the belt by removing the back panel. The back panel is usually held on with two to four screws. Remove the screws, and remove the panel (Figure 19-22).

5. **Replace the drive belt.** To replace the drive belt on this type of washer, you must remove the belt. Pull out on the belt and turn the basket drive pulley until the belt comes off the pulley. To reinstall the belt, just reverse the disassembly procedure, and reassemble (Figure 19-23). The drive belt tension on this type of washer does not need adjusting. Also, remember to place the rib side of the belt facing into the pulleys.

6. **Test the washer.** You are now ready to test the washer. Begin the wash cycle with a full load of laundry in the basket. Check the wash and the spin cycles. If these check out okay, reinstall the outer panel. Make sure to take the washer out of the service test mode when the repair is completed.

FIGURE 19-22
Remove the screws that secure the back panel to the rear of the washer.

Basket drive pulley

Outer tub

Drive motor and belt

FIGURE 19-23 A properly installed drive belt.

Drain Pump Motor

The drain pump motor consists of a 120 volt 60 Hz motor, impeller, and impeller housing. It also contains a filter strainer that prevents objects from entering the impeller and the drain outlet hose (Figure 19-24 and Figure 19-25). The drain pump motor can drain between 8 and 17 gallons per minute with a standpipe height between 2 and 8 feet. The drain pump motor will operate in the spin cycle and provides overflow protection when signaled from the water level control (pressure switch) and the electronic control board. The washer has to be plugged in for the overflow protection to work.

The typical complaints associated with the water pump are:

- Washer will not drain the water out.
- It smells like something is burning.
- The water in the washer will not recirculate.
- The washer has an error code.

FIGURE 19-24
The filter strainer is located in the front of the washer at the bottom behind the lower panel.

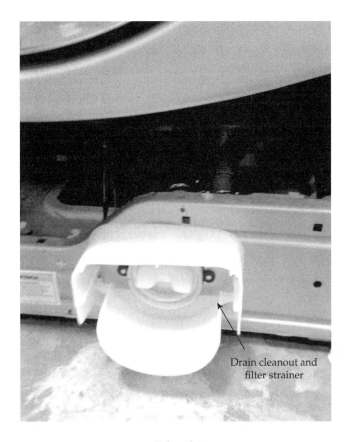

Drain cleanout and filter strainer

FIGURE 19-25
A top view of the drain pump motor and filter strainer. The drain pump motor is accessible through the lower panel in the front of the washer.

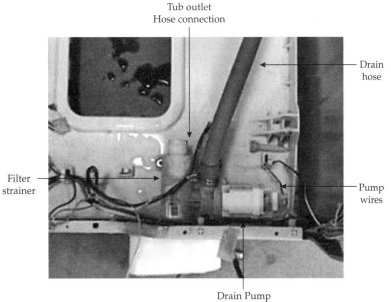

Tub outlet
Hose connection

Drain hose

Filter strainer

Pump wires

Drain Pump motor

To handle these problems, perform the following steps:

1. **Verify the complaint.** Verify the complaint by operating the washer. On electronic models, turn off the electricity to the appliance and wait for two minutes before turning it back on. If a fault code appears, look up the code. If the washer will not power up, locate the technical data sheet behind the control panel or in the rear of the washer for diagnostics information. On some models you will need the actual service manual for the model you are working on to properly diagnose the washer. The service manual will assist you in properly placing the washer in the service test mode for testing the washer functions.

2. **Check for external factors.** You must check for external factors not associated with the appliance. Is the appliance installed properly? Does the appliance have the correct voltage? The voltage at the receptacle is between 108 volts and 132 volts during a load on the circuit. Do you have the correct polarity? (See Chapter 6.)

3. **Disconnect the electricity.** Before working on the washer, disconnect the electricity to the washer. This can be done by pulling the plug from the electrical outlet. Be sure that you only remove the washer plug. Or disconnect the electricity at the fuse panel or at the circuit breaker panel. Turn off the electricity.

WARNING *Some diagnostic tests will require you to test the components with the power turned on. When you disassemble the control panel, you can position it in such a way that the wiring will not make contact with metal. This act will allow you to test the components without electrical mishaps.*

4. **Access and remove the drain pump motor.** In order to gain access to the drain pump motor assembly in this type of washer, remove the lower panel (Figure 19-26). Place a flat pan under the drain cleanout cap (Figure 19-24). About one quart of water will drain out when you remove the drain cleanout cap. Next, turn the drain cleanout cap counterclockwise about two to three turns and pull toward you. The water will drain in the flat pan. Check for debris or foreign objects in the filter strainer. If you found debris in the filter strainer, clean it out and retest the drain pump motor. It could have been blocked only.

5. **Test the drain pump motor.** To test the drain pump motor, remove the wire connector to the pump motor. Next, set your multimeter on the ohms scale R × 1. Place the meter test leads on the motor terminals. You should read between 10 and 13Ω on the meter. Now inspect the impeller inside the pump. If there is any debris, remove it. Make sure that the impeller is not damaged.

6. **Remove the drain pump motor.** With the pump wires disconnected and the water drained from the pump, remove the tub outlet hose and drain hose from the pump (Figure 19-25). Next, remove the screws that secure the drain pump motor to the washer. You may have to move the drain pump motor rearward to clear the pins and grommets in the front frame. Remove the drain pump motor from the washer in the opening in the front frame.

Figure 19-26
This illustration shows the bottom access panel has already been removed.

7. **Install the drain pump motor.** To install the drain pump motor, just reverse the disassembly procedure, and reassemble. Reconnect the wire leads to the motor. Reconnect the hoses to the pump. Restore the electricity to the washer. Check for water leaks. Test the washing machine for proper operation. Make sure to take the washer out of the service test mode when the repair is completed.

Heater Assembly

The heater assembly is located above the drain pump motor attached to the bottom of the outer tub (Figure 19-27). The heater assembly consists of a 900 to 1100 watt, 120 VAC heater and water temperature thermistor. The heater will operate on certain cycles, including the sanitize cycle, when selected by the customer.

The typical complaints associated with the heater assembly are:

- The washer has error codes.
- The laundry at the end of the cycle did not meet the customer's expectations.
- Water temperature is not correct.

To handle these problems, perform the following steps:

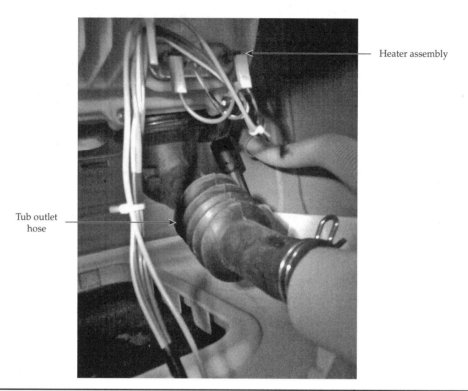

Tub outlet
hose

Heater assembly

FIGURE 19-27   The heater is held in place and sealed to the tub by compressing a gasket between two metal brackets to the outer tub with a 10 mm nut.

1. **Verify the complaint.** Verify the complaint by operating the washer through its cycles. On electronic models, turn off the electricity to the appliance and wait for two minutes before turning it back on. If a fault code appears, look up the code. If the washer will not power up, locate the technical data sheet behind the control panel or in the rear of the washer for diagnostics information. On some models you will need the actual service manual for the model you are working on to properly diagnose the washer. The service manual will assist you in properly placing the washer in the service test mode for testing the washer functions.

2. **Check for external factors.** You must check for external factors not associated with the appliance. Is the appliance installed properly? Does the appliance have the correct voltage? The voltage at the receptacle is between 108 volts and 132 volts during a load on the circuit. Do you have the correct polarity? (See Chapter 6.) Is the water turned on? Both water faucets must be turned all the way counterclockwise. Check the hot and cold water temperature at the nearest water faucet.

3. **Disconnect the electricity.** Before working on the washer, disconnect the electricity. This can be done by pulling the plug from the electrical outlet. Be sure that you only remove the washer plug. Or disconnect the electricity at the fuse panel or at the circuit breaker panel. Turn off the electricity.

WARNING *Some diagnostic tests will require you to test the components with the power turned on. When you disassemble the control panel, you can position it in such a way that the wiring will not make contact with metal. This act will allow you to test the components without electrical mishaps.*

4. **Gain access to the heater assembly.** In order to gain access to the heater assembly in this type of washer, remove the lower panel (Figure 19-26). Look underneath at the bottom of the tub to the right; you will see the heater assembly (Figure 19-27).

5. **Test the heater assembly.** To test the heater, remove the wires from the heating element. Set your multimeter on ohms R × 1 and place the test leads on the heater terminals. The meter should read between 13 and 16Ω resistance. To test the heater element with the washer running, turn on the washer in the sanitize cycle with no load in the drum. Let the washer fill and run for about 5 minutes; then take a wattage measurement or amperage measurement. The wattage reading should be between 900 and 1100 watts. The amperage reading should be between 7 and 9 amps. If there is less than 1½ inches of water in the basket or if the water level is below the vanes, the heater will not come on. To measure the height of the water in the basket when the heater turns on, connect an ammeter or wattmeter to the heater element wire. Next, place the washer into the service test mode and allow the water to enter the tub. Once the ammeter or wattmeter indicates a reading (900 to 1100 watts or 7 to 9 amps), stop the washer from filling. Now open the door and measure the water level. To test the thermistor on the heater assembly, just remove the wires from the terminals and take a resistance reading. For the correct reading, see the technical data sheet. See sample charts in Figure 19-28 and Figure 19-29. Remember that resistance goes down when the temperature increases.

6. **Remove the heater assembly.** First, place the washer in the drain mode to drain out the water. Remove all the wires from the heater assembly. Take a 10-mm socket and loosen the nut on the heater assembly until it is flush with the end of the stud. To relax the gasket, push in on the nut. Now, grab hold of the heater assembly and pull toward you to remove it from the tub.

FIGURE 19-28
A component resistance chart found on a wiring diagram. Locate the correct resistance chart for the model you are servicing.

Component Resistance Chart

Electrical component	Resistance @ 77°F (25°C)
Dispenser valve soleniods	800 ± 7%
Door lock solenoid	46.8 ± 10%
Pump motor	12 ± 7%
NTC thermistor dispenser	3K
NTC thermistor heater	4.8K
Water heater	14 ± 10%
Motor– M1 to M2	5.3 ± 7%
Motor– M2 to M3	5.3 ± 7%
Motor– M1 to M3	5.3 ± 7%
Motor– M4 to M5	118 ± 7%

FIGURE 19-29
A temperature/
resistance chart
found on a technical
data sheet. Locate
the correct chart for
the model you are
servicing.

Temperature	Results
32°F (0°C)	35.9K Ω
86°F (30°C)	9.7K Ω
104°F (40°C)	6.6K Ω
122°F (50°C)	4.6K Ω
140°F (60°C)	3.2K Ω
158°F (70°C)	2.3K Ω
203°F (95°C)	1K Ω

7. **Install a heater assembly.** To install the heater assembly, just reverse the disassembly procedure, and reassemble. Reconnect the wire leads to the heater and the thermistor. When tightening the 10-mm nut, use a torque wrench to secure the nut to 31-in. lbs. of torque. Undertorquing the nut will cause the tub to leak water, and if it is overtorqued, the tub will crack and leak water. Restore the electricity to the washer. Check for water leaks. Test the washing machine for proper operation. Make sure to take the washer out of the service test mode when the repair is completed.

Inverter (Motor Speed Control Board)

The inverter board, also known as the motor speed control unit, is located in the base of the washer in the rear (Figure 19-30). It receives its commands from the electronic control board (Figure 19-9) and controls the speed and direction of the drive motor. For the inverter to operate the motor correctly, it needs 120 VAC supply voltage, a DC input signal from the electronic control board, and a good working three-phase motor. To operate a preprogrammed speed and direction of a motor, you will need to vary the amount, frequency, and polarity of the voltage and compare the input from the tachometer (sensor) on the drive motor.

The typical complaints associated with the inverter (motor speed control board) are:

- The motor will not run.
- The washer has an error code.
- Fuse is blown or circuit breaker trips.
- The motor runs at the wrong speed.

To handle these problems, perform the following steps:

1. **Verify the complaint.** Verify the complaint by operating the washer through its cycles. Turn off the electricity to the appliance and wait for two minutes before turning it back on. If a fault code appears, look up the code. If the washer will not power up, locate the technical data sheet behind the control panel or in the rear of the washer for diagnostics information. On some models you will need the actual service manual for the model you are working on to properly diagnose the washer. The service manual will assist you in properly placing the washer in the service test mode for testing the washer functions.

2. **Check for external factors.** You must check for external factors not associated with the appliance. Is the appliance installed properly? Does the appliance have the correct voltage? The voltage at the receptacle is between 108 volts and 132 volts during a load on the circuit. Do you have the correct polarity? (See Chapter 6.)

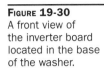

A front view of
the inverter board
located in the base
of the washer.

3. **Disconnect the electricity.** Before working on the washer, disconnect the electricity. This can be done by pulling the plug from the electrical outlet. Be sure that you only remove the washer plug. Or disconnect the electricity at the fuse panel or at the circuit breaker panel. Turn off the electricity.

WARNING *Some diagnostic tests will require you to test the components with the power turned on. When you disassemble the control panel, you can position it in such a way that the wiring will not make contact with metal. This act will allow you to test the components without electrical mishaps.*

4. **Gain access to the inverter.** In order to gain access to the inverter in this type of washer, remove the lower panel (Figure 19-26). Locate the inverter on the left or right side.

5. **Test the inverter board.** Locate the junction box near the inverter board (Figure 19-31). Remove the junction box cover by pressing in on the tabs. At this point you will need to enter into the service test mode to check for 120 VAC at the AC input wires. The service manual will assist you in properly placing the washer in the service test mode for testing this function. With the motor running in the test mode, test for 120 VAC at the AC input wires. Next, unplug the washer and test the motor resistance. The resistance value should be between 3 and 8Ω between any two of the three windings. If it is correct, replace the inverter.

FIGURE 19-31
A top view showing
the inverter board
and junction box.

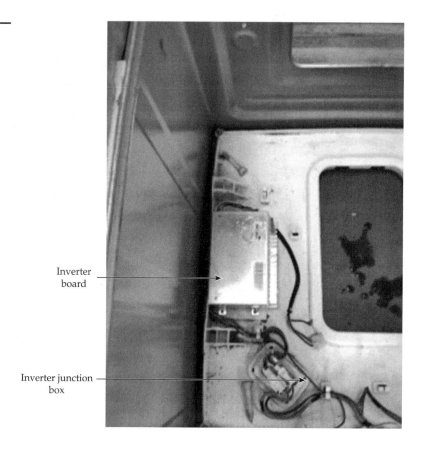

FIGURE 19-31
A top view showing the inverter board and junction box.

Inverter board

Inverter junction box

6. **Remove the inverter board.** With the junction box cover off, disconnect the wire connections (Figure 19-31) in the junction box. Next, remove the wire connector from the inverter to the motor and remove the ground wire. Remove the screws that secure the inverter to the base and slide the inverter toward you.

7. **Install an inverter board.** To install the inverter board, just reverse the disassembly procedure, and reassemble. Reconnect the wire connectors to the inverter board assembly. Test the washing machine for proper operation. Make sure to take the washer out of the service test mode when the repair is completed.

Drive Motor

The drive motor (Figure 19-32) is a three-phase induction type motor operating at variable speeds and direction, depending on the input voltages and frequencies from the inverter board. The tachometer (sensor) attached to the motor monitors the motor speed and direction and sends that information back to the inverter board. The drive motor drives the tub pulley with a belt.

Figure 19-32
Typical drive motor
used in a front load
washer.

The typical complaints associated with the drive motor are:

- The motor will not run.
- The washer has an error code.
- Fuse is blown or circuit breaker trips.

To handle these problems, perform the following steps:

1. **Verify the complaint.** Verify the complaint by operating the washer through its cycles. On electronic models, turn off the electricity to the appliance and wait for two minutes before turning it back on. If a fault code appears, look up the code. If the washer will not power up, locate the technical data sheet behind the control panel or in the rear of the washer for diagnostics information. On some models you will need the actual service manual for the model you are working on to properly diagnose the washer. The service manual will assist you in properly placing the washer in the service test mode for testing the washer functions.

2. **Check for external factors.** You must check for external factors not associated with the appliance. Is the appliance installed properly? Does the appliance have the correct voltage? The voltage at the receptacle is between 108 volts and 132 volts during a load on the circuit. Do you have the correct polarity? (See Chapter 6.)

3. **Disconnect the electricity.** Before working on the washer, disconnect the electricity. This can be done by pulling the plug from the electrical outlet. Be sure that you only remove the washer plug. Or disconnect the electricity at the fuse panel or at the circuit breaker panel. Turn off the electricity.

WARNING *Some diagnostic tests will require you to test the components with the power turned on. When you disassemble the control panel, you can position it in such a way that the wiring will not make contact with metal. This act will allow you to test the components without electrical mishaps.*

4. **Gain access to the drive motor.** In order to gain access to the drive motor, you must gain access to the front lower panel and pull out the washer to gain access to the rear panel (Figures 19-22 and 19-26).

5. **Test the drive motor.** There are two methods for testing the drive motor: at the motor harness connector or at the inverter board. The operation of the drive motor can also be checked by placing the washer in the service test mode. The service manual will assist you in properly placing the washer in the service test mode for testing this function. To test at the drive motor, separate the motor harness connector and place your multimeter on the ohms scale R × 1. Using the wiring diagram, locate the three windings of the drive motor and the tachometer (sensor) as it relates to the wiring connector. The resistance value should be between 3 and 8Ω between any two of the three windings. Next, test the tachometer (sensor); the meter should read approximately between 115 and 120Ω. If any of the motor windings vary in resistance from ½ to 1Ω the motor might be bad, depending on the model of washer serviced. Check the technical data sheet for the exact readings.

6. **Remove the drive motor.** Remove the rear access panel to gain access to the drive motor. Remove the drive belt by turning it off the drive pulley (Figure 19-33). The belt will stretch; this is normal. Next, disconnect the ground wire and the motor electrical connector (Figure 19-34). Now, remove the four bolts that secure the drive motor to the outer tub (Figure 19-35). Support the motor when you pull the drive motor forward to remove it from the washer. On other models, you may have to remove only one bolt and slide the drive motor rearward to remove it from its motor mounts.

7. **Install the drive motor.** To install the drive motor, just reverse the disassembly procedure, and reassemble. Reconnect the wire connector to the drive motor and reinstall the ground wire. Test the washing machine for proper operation. Make sure to take the washer out of the service test mode when the repair is completed.

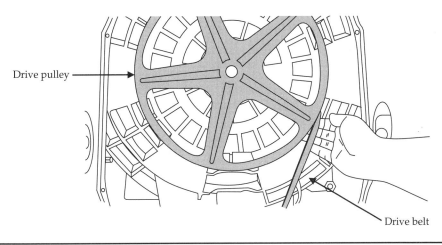

FIGURE 19-33   Removing the drive belt from the tub drive pulley.

FIGURE 19-34
Before removing the drive motor, remove the ground wire and the electrical connector from the drive motor.

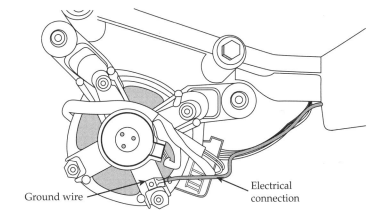

FIGURE 19-35
Remove the bolts that secure the motor to the frame and slide the motor out of the washer.

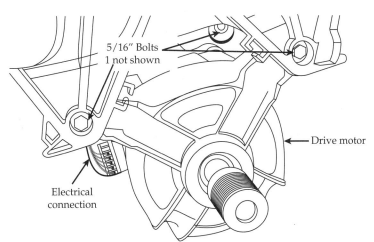

Wiring Schematic Diagrams

The following wiring diagrams (Figures 19-36 through 19-40) will be helpful in pinpointing the location of the components in relationship to the electronic control board.

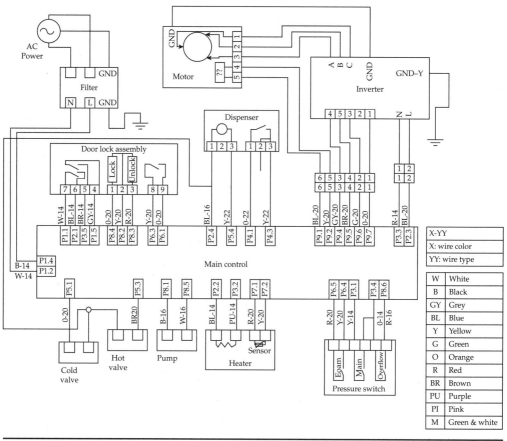

FIGURE 19-36   Front load washer schematic diagram #1.

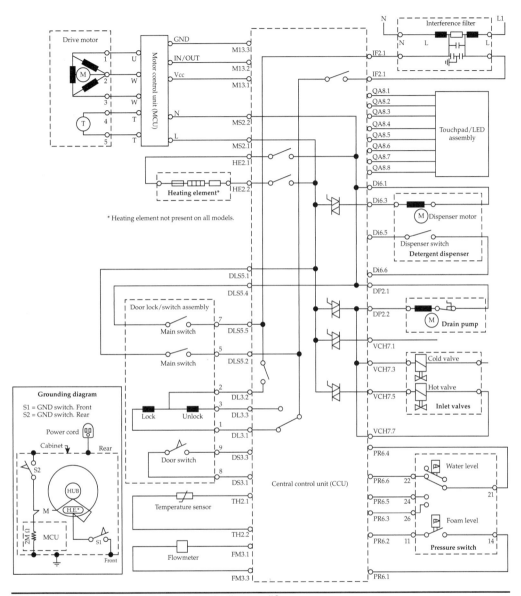

FIGURE 19-37   Front load washer schematic diagram #2.

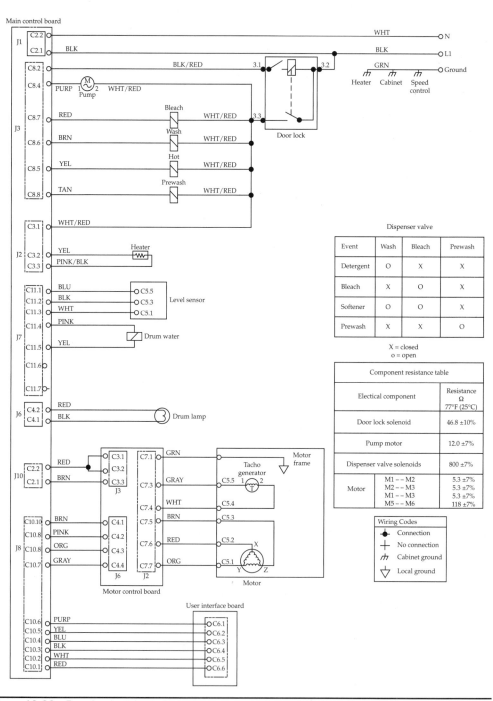

Figure 19-38   Front load washer schematic diagram #3.

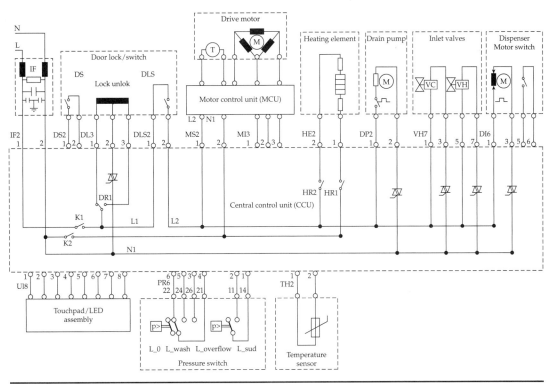

FIGURE 19-39   Front load washer schematic diagram #4.

Figure 19-40
Front load washer
typical grounding
schematic diagram #5.

Grounding System

Grounding system without heater

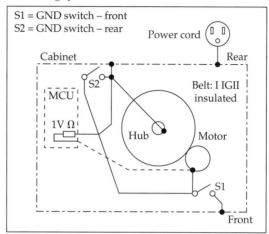

Grounding system with heater

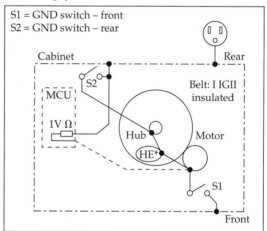

* Heating element not present on all models

Automatic Electric Dryers

Automatic electric dryers are not complicated to repair. The more you know about the electrical, mechanical, and operational basics of the dryer, the easier it is to solve the problem. Some models are designed to operate on 240 volts; other models operate on 120 volts. To dry clothing, an automatic dryer must have drum rotation (clockwise rotation at 40 to 60 RPM), electric heat (approximately 5600 watts), and air circulation. This chapter will provide the basic skills needed to diagnose and repair automatic electric dryers. Figures 20-1 and 20-2 identify where some of the components are located within the automatic electric dryer. However, this illustration is used as an example only. The actual construction and features might vary, depending on what brand and model you are servicing.

Principles of Operation

The clothes are placed into the dryer according to the manufacturer's recommendation for the proper loading of the dryer. Next, the proper cycle is selected, and the dryer start button is pressed. The combination of the timer, the switches, thermistors, and/or the thermostats regulates the air temperature within the drum and the duration of the drying cycle. During the drying cycle, room air is pulled into the dryer drum from the lower rear (or sometimes the front) of the cabinet, depending on which model the consumer owns (Figure 20-3). The air is pulled through the heater, drum, lint screen, down through the lint chute, and through the fan housing. It is then pushed out of the exhaust duct. The drive motor, blower wheel, belt, and pulleys cause the drum to turn in a clockwise rotation at 40 to 60 RPM and the air to move through the dryer. The belt wraps around the drum, motor pulley, and idler pulley. The blower wheel is secured to one end of the drive motor shaft, or there may be a separate motor to turn the blower wheel. As the drive motor turns, the drum rotates, moving the clothes. At the same time, the blower wheel turns, moving the air. The heating element will cycle on and off, according to the temperature selected.

Functions and Cycles

Electric automatic dryers use three basic functions to operate:

- Heat is supplied by a resistance-type heating element.

FIGURE 20-1
The location of
components in a
typical automatic
electric dryer. The
consumer has a
choice between a
drop-down or side-
open door.

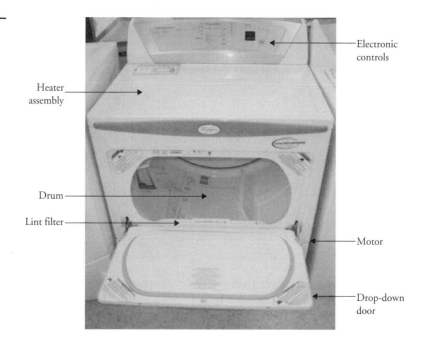

Electronic
controls

Heater
assembly

Drum

Lint filter

Motor

Drop-down
door

FIGURE 20-2
Typical component
location in an electric
dryer.

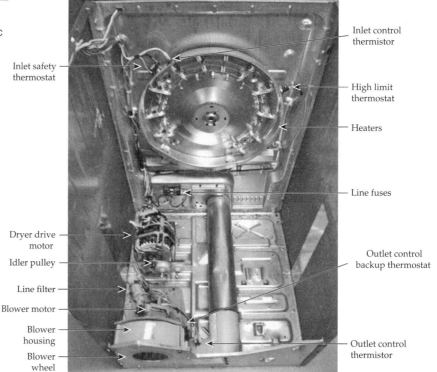

Inlet control
thermistor

Inlet safety
thermostat

High limit
thermostat

Heaters

Line fuses

Dryer drive
motor

Idler pulley

Line filter

Blower motor

Blower
housing

Blower
wheel

Outlet control
backup thermostat

Outlet control
thermistor

FIGURE **20-3**
Typical air flow pattern in an automatic electric dryer.

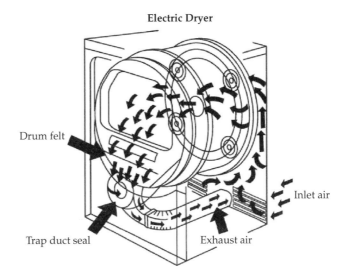

Electric Dryer

Drum felt

Inlet air

Trap duct seal

Exhaust air

- Air is drawn into the dryer. It is heated and circulated through the tumbling clothes. Then the warm, moisture-laden air is drawn through the lint screen and is vented through the duct system to the outside.

- Tumbling of the clothes is accomplished with a motor that drives a belt, which rotates the drum.

The cycles of an electric automatic dryer are as follows:

- **Timed dry cycle** The timed dry cycle is controlled by the amount of time selected on the timer. The temperature in the dryer is controlled by a thermostat or a thermistor, which turns the heating element on and off throughout the timed cycle.

- **Automatic dry cycle** The automatic dry cycle is not controlled by the timer. This cycle is controlled by the cycling thermostats or thermistors. Heat is supplied to dry the clothes, and it will continue until the temperature in the drum reaches the selected cutout setting of the thermostat or thermistor. When the thermostat or thermistor is satisfied, the heater shuts off and the timer motor is activated. However, certain variables can control the cycle, which will cause the cycling of the thermostat or thermistor before the end of the cycle.

- **Permanent press and knit cycles** The permanent press and knit cycles are controlled by the amount of time selected on the timer. The temperatures of these cycles are controlled by the temperature rating of the cycling thermostats or by the thermistors that are located within the cabinet, on the exhaust duct, and in the air supply. On some models, the user can select the desired type of heat setting with the temperature selector switch, located on the control panel.

- **Air dry** The air dry cycle is controlled by the amount of time selected on the timer. This cycle uses the air to dry the clothes. The heating element is not used at all during this cycle.

Safety First

Any person who cannot use basic tools or follow written instructions should *not* attempt to install, maintain, or repair any automatic dryers. Any improper installation, preventive maintenance, or repairs could create a risk of personal injury or property damage.

If you do not fully understand the installation, preventive maintenance, or repair procedures in this chapter, or if you doubt your ability to complete the task on the automatic dryer, please call your service manager.

The following precautions should also be followed:

- Never bypass (or interfere with) the operation of any switch, component, or feature.
- The dryer exhaust should be vented properly. Never exhaust the dryer into a chimney, a common duct, an attic, or a crawl space.
- Be careful of sharp edges when working on the dryer.
- The dryer produces combustible lint, and the area should be kept clean.
- Never remove any ground wires from the dryer or the third (grounding) prong from the service cord.
- Never use an extension cord to operate a dryer.
- The wiring used in dryers is made with a special heat-resistant insulation. Never substitute it with ordinary wire.

Before continuing, take a moment to refresh your memory of the safety procedures in Chapter 2.

Automatic Electric Dryers in General

Much of the troubleshooting information in this chapter covers automatic electric dryers in general, rather than specific models, in order to present a broad overview of service techniques. The illustrations that are used in this chapter are for demonstration purposes only to clarify the description of how to service dryers. They in no way reflect upon a particular brand's reliability.

Electrical Requirements

Electrical dryers can be connected to an electrical power source in three ways:

- Electrical cord and plug, which plugs into a 240-volt wall receptacle.
- Directly wired to a fused electrical disconnect box with a built-in shut-off switch.
- Directly wired to a circuit breaker or to a fuse panel. All dryers must be properly wired, grounded, and polarized, according to the manufacturer's installation instructions and per all local codes and ordinances.

Electrical Plug Connection – Three-Wire Cord

Always install a UL-approved, 30-amp. power cord, NEMA 10-30–type SRDT with a strain relief on the dryer electrical terminal block located in the rear of the dryer (Figures 20-4 and 20-5). You must follow the manufacturer's installation instructions for installing a power cord on the dryer properly.

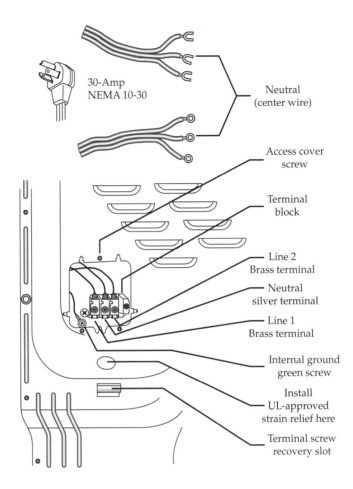

FIGURE 20-4
Attach the three-wire power cord to the two brass and silver terminals as shown in the figure.

Electrical Plug Connection – Four-Wire Cord

The new power cord is a UL-approved, 30-amp, power cord, NEMA 14-30–type ST or SRDT (Figures 20-6 and 20-7). The plug end is a four-prong plug. This new type of power cord will better protect the dryer and the consumer from electrical mishaps. You must follow the manufacturer's installation instructions for installing a power cord on the dryer properly. You will need an electrician to replace the old three-prong receptacle with a new four-prong receptacle.

Electrical Tests for Electric Dryers

While performing a service call on an electric dryer, the following electrical tests should be performed:

- **Continuity**   Use your multimeter and set it on the ohms scale. When measuring resistance between the terminal block ground and any exposed unpainted metal cabinet component, including control shafts and switches, there should be no more than .1 ohms resistance reading.

Figure 20-5
A properly attached
three-wire power cord.

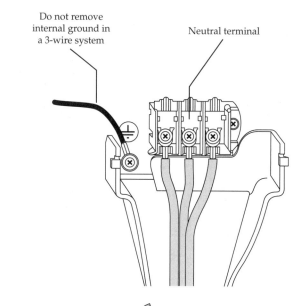

Do not remove
internal ground in
a 3-wire system

Neutral terminal

Figure 20-6
Attach the four-wire
power cord to the two
brass and silver
terminals as shown in
the figure. Attach the
green wire to the
cabinet.

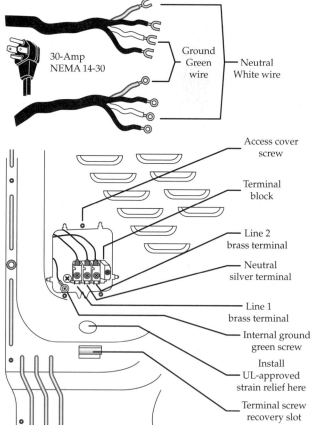

30-Amp
NEMA 14-30

Ground
Green
wire

Neutral
White wire

Access cover
screw

Terminal
block

Line 2
brass terminal

Neutral
silver terminal

Line 1
brass terminal

Internal ground
green screw

Install
UL-approved
strain relief here

Terminal screw
recovery slot

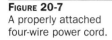

FIGURE 20-7
A properly attached
four-wire power cord.

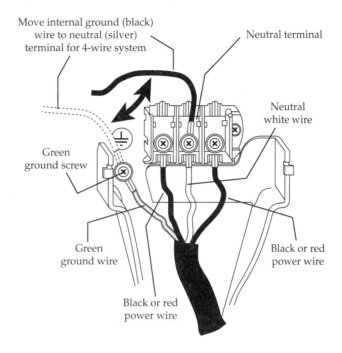

Move internal ground (black)
wire to neutral (silver)
terminal for 4-wire system

Neutral terminal

Neutral
white wire

Green
ground screw

Green
ground wire

Black or red
power wire

Black or red
power wire

- **Motor power**   Perform an audit on the motor (power required after 20 seconds with no load, and the motor is turning in the clockwise direction, and the electric heater is turned off). The reading should be between 200 and 280 watts. Next, perform a line test (power required after 1 to 5 seconds with no clothes in the dryer, the motor is turning in a clockwise direction, and the electric heater is turned off). The reading should be between 210 and 290 watts. If you do not have a wattmeter, use the Ohm's Law formula to convert to watts.

- **Low voltage start**   The minimum voltage needed to start the dryer motor is 100 VAC, 60 Hz (200 volts for 230 VAC, 50 Hz dryers). Place about 25 pounds of clothing with 100 percent moisture into the dryer and set the dryer at its maximum heat setting. Turn on the dryer and measure the voltage at the start of the dryer cycle. The voltage should not be below the listed voltage for any part of the cycle.

- **Drum temperatures**   The following test will be run with the exhaust duct vent disconnected from the dryer. Take your multimeter temperature thermal couple and place it in the lint grill about 1 to 2 inches. You will need to set the dryer timer as needed for this test. The recordable temperature is to be the maximum temperature reading after the third cycle when the heater cycles off. There should not be any clothes in the dryer for this test. Each time the thermostat or thermistor cycles open, there should be equal to or higher than opening temperature for the next lower temperature setting.

  - Maximum heating temperature when the heater cycles off is 120 to 160 degrees Fahrenheit.

- Medium heating temperature when the heater cycles off is 110 to 140 degrees Fahrenheit.
- Low heating temperature when the heater cycles off is 95 to 130 degrees Fahrenheit.

Location and Installation of Dryer

The following are some general principles that should be followed when installing a dryer:

- Locate the dryer where there is easy access to existing electrical lines.
- Be sure that you observe all local codes and ordinances for the electrical, plumbing, and venting connections.
- The dryer should be installed and leveled on a firm floor to minimize vibration during operation.
- To reduce the risk of fire, never install a dryer on any type of carpeting.
- For proper operation of the dryer, be sure that there is adequate make-up (new, replacement) air in the room where it is installed.
- Do not install a dryer in an area where the make-up air is below 45 degrees Fahrenheit. This will greatly reduce the drying efficiency and increase the cost of operating the dryer.
- Always follow the installation instructions that are provided with every automatic dryer purchased. If the installation instructions are not available, order a copy from the manufacturer.
- When installing a dryer in a mobile home, the dryer must have an outside exhaust. If the dryer exhaust goes through the floor, and the area under the mobile home is enclosed, the exhaust must terminate outside of the enclosed area.
- If the dryer is to be installed in a recessed area or closet area, follow the manufacturer's recommendation for its installation.
- When relocating the dryer to a new location, test the voltage at the new location, and be sure that it matches the dryer voltage specifications as listed on the nameplate (serial plate). Also inspect the entire vent system for any obstructions.

Proper Exhausting of the Dryer

Proper exhausting instructions for the model being installed are available through the manufacturer. Each manufacturer has its own specifications for the size and the length of the ductwork needed to run its dryer properly. The maximum length of the exhaust system depends upon the type of duct, the number of elbows, and the type of exhaust hood used. Figure 20-8 illustrates a typical dryer exhaust installation.

The following guide is recommended for the exhausting of a dryer:

- Keep the duct length as short as possible.
- Keep the number of elbows to a minimum to minimize the air resistance.

FIGURE 20-8
Typical dryer exhaust
installation.

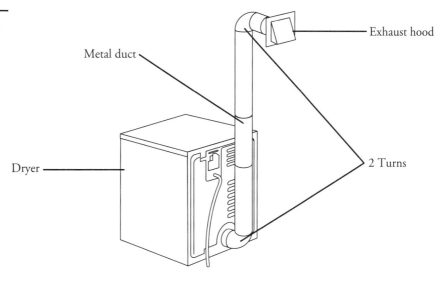

- Never reduce the diameter of round ductwork below 4 inches.

- Install all exhaust hoods at least 12 inches above ground level.

- The exhaust duct and exhaust hood should be inspected periodically and cleaned, if necessary.

- All duct joints should be taped. Never use screws to join the duct joints together. Screws protruding into the duct will cause lint buildup, and this will eventually clog the duct.

- Never exhaust the dryer into any wall, ceiling, attic, or under a building.

- Accumulated lint could become a fire hazard, and moisture could cause damage.

- If the exhaust duct is adjacent to an air conditioning duct, the exhaust duct must be insulated to prevent moisture buildup.

Exhaust Ducting

The more that the dryer is used, the more frequently the exhaust ducts and vent hood must be inspected to prevent duct blockages and lint fires. The exhaust ducting and vent hood must be cleaned at least once a year for peak performance in drying.

When installing ductwork, separate all turns by at least 4 feet of straight duct, including the distance between the last turn and the dampened wall hood. If a turn is 45° or less, it can be ignored in figuring out duct distance. If you have two 45° turns, count it as one 90° turn. Also, each turn over 45° should be counted as one 90° turn (Figures 20-9, 20-10, 20-11, 20-12, and 20-13).

Figure 20-9
Recommended maximum length exhaust duct and hood for a direct drive blower in a dryer (one drive motor).

Recommended maximum length				
Exhaust hood types				
Recommended			Use only for short run installations	
No. of 90° Elbows	Rigid metal	Flexible metal	Rigid metal	Flexible metal
0	45 Feet	30 Feet	30 Feet	15 Feet
1	35 Feet	20 Feet	20 Feet	10 Feet
2	25 Feet	10 Feet	10 Feet	
3	15 Feet			

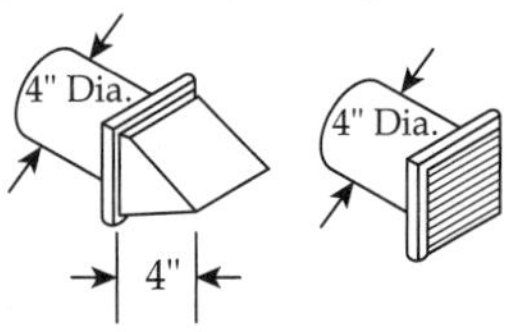

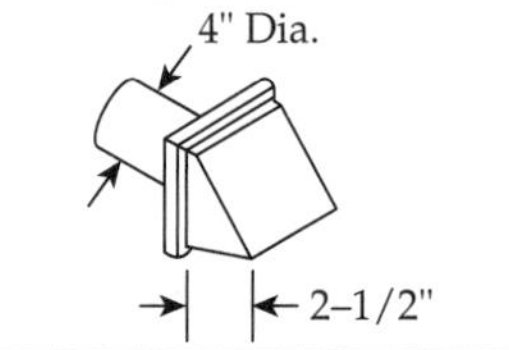

Figure 20-10
Recommended maximum length rigid exhaust duct and hood for a dryer with a separate drive motor for the blower assembly.

Maximum Length of 4" (10.2 cm) Dia. Rigid Metal Duct		
Preferred vent hood type		
Number of 90° turns	4" (10.2cm) 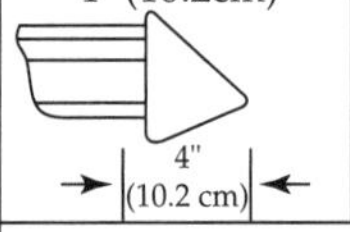 / Louvered 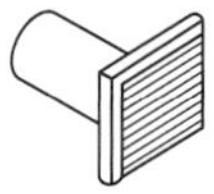	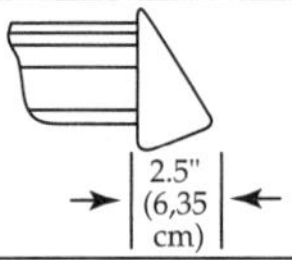
0	60' (18.28 m)	48' (14.63 m)
1	52' (15.84 m)	40' (12.19 m)
2	44' (13.41 m)	32' (9.75 m)
3	32' (9.75 m)	24' (7.31 m)
4	28' (8.53 m)	16' (4.87 m)

Figure 20-11
Recommended maximum length for flexible metal exhaust duct for a dryer with a separate drive motor for the blower assembly.

Maximum Length of 4" (10.2 cm) Dia. Flexible Metal Duct		
Preferred vent hood type		
Number of 90° turns	4" (10.2 cm) / Louvered	
0	30' (9.14 m)	18' (5.49 m)
1	22' (6.71 m)	14' (4.27 m)
2	14' (4.27 m)	10' (3.05 m)
3	not recommended	

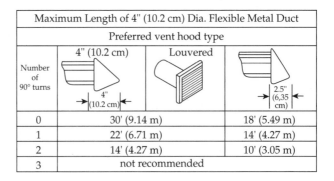

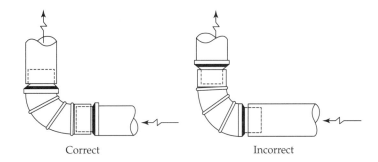

FIGURE 20-12
Make sure that the male fittings are installed in the correct direction.

Correct   Incorrect

When sealing the joints of the exhaust ducts with duct tape, the male end of each section of duct must point away from the dryer to prevent lint buildup. Also, never use screws to connect the ducts together; this will cause lint buildup. Any ductwork that runs through an unheated area or is near air conditioning must be insulated to prevent condensation and lint buildup.

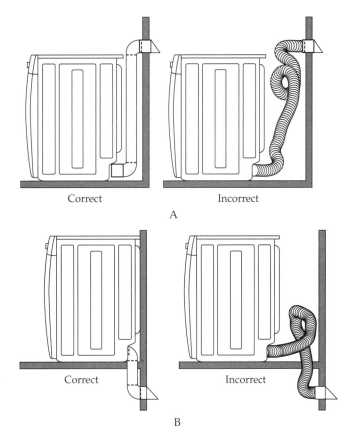

FIGURE 20-13
Some examples of correct and incorrect exhaust duct installations.

Correct   Incorrect

A

Correct   Incorrect

B

SAFETY NOTE   All exhaust ductwork must exhaust to the outside of a building to prevent lint buildup and fire. Do not terminate exhaust ductwork in a chimney, any gas vent, under enclosed floor (crawl space), common duct with a kitchen exhaust, or into an attic because the lint accumulation will cause a fire. Finally, never block incoming or exhausted air supplies; this will cause the dryer to run inefficiently and may cause a fire.

Step-by-Step Troubleshooting by Symptom Diagnosis

In the course of servicing an appliance, don't overlook the simple things that might be causing the problem. Step-by-step troubleshooting by symptom diagnosis is based on diagnosing malfunctions, with possible causes arranged into categories relating to the operation of the dryer. This section is intended only to serve as a checklist to aid you in diagnosing a problem. Look at the symptom that best describes the problem that you are experiencing with the dryer, and then correct the problem.

Dryer Will Not Run

- Do you have the correct voltage at the dryer?
- Test the door switch for continuity of the switch contacts.
- Test for continuity of the motor windings. Also check for a grounded motor.
- Test the timer for continuity of the switch contacts.
- Test the "push to start" switch for continuity of the switch contacts.
- Check for broken or loose wiring. Also check the wire terminal connections that connect to the different components.
- For electronic control models, check for an error code. Check the electronic control board.

No Heat

- Do you have the correct voltage? The motor in a dryer runs on 120 volts. The heating element works on 240 volts; compact models run on 120/240 volts. Check voltage at the wall receptacle. Also check the nameplate voltage rating for the model that you are working on.
- Check for any in-line fuses that might have blown. Refer to the wiring diagram, located on the rear of the dryer or behind the console panel.
- Are the thermostats functioning properly?
- Test the timer for continuity of the switch contacts.
- Test for continuity of the motor windings. Also test for a grounded motor.
- Test the heating element for continuity.
- Test the temperature selector switch for continuity.
- Check for broken or loose wiring.
- Check the exhaust vent duct system.
- For electronic control models, check for an error code.

Drum Will Not Rotate

- Check the drive belt. Is it broken or worn?
- Check the idler pulley. Is the pulley seized up?
- Check the drum support bearings. Be sure that they rotate freely.
- Is the dryer overloaded with clothing?
- Check for foreign objects that might be lodged between the drum and the bulkheads.
- For electronic control models, check for an error code.

Dryer Is Noisy

- Check for loose components. Is everything secure and in its proper place?
- Check the idler pulley. Lint buildup can cause the idler pulley shaft to squeak.
- Check the drive belt. Is the belt partially torn?
- Check the drive motor. Is the pulley secured to the shaft? Also check the fan assembly. Is it loose?
- Check for lint or foreign objects lodged between the drum and the bulkheads.
- Check the front and rear bearings.
- Check the blower assembly. Be sure that the blower wheel is tight on the motor shaft end.
- Is the dryer level?

Dryer Runs and Heats, but the Clothes Won't Dry

- Check for a defective thermostat. Use a thermometer to check the duct temperatures during the cycling of the dryer.
- Check for a loose pulley or blower wheel. Inspect and tighten the blower wheel.
- Check for restricted air flow in the dryer.
- Check the exhaust vent duct system.
- For electronic control models, check for an error code.

Clothes Too Hot or Fabric Damage

- Check the exhaust system for restrictions.
- Check for air leaks in the ductwork and the front and rear seals on the drum.
- Check the thermostats.
- Check the temperature setting with the customer.
- Check the door latch and strike.
- Inspect the interior of the drum for any foreign objects protruding into it.

Common Drying Problems

The drying of clothes can be affected by several factors:

- Clogged lint screen.
- Exhaust duct too long, collapsed, or crimped.
- Dryer air is exhausted into the area where the dryer is located. For example, the dryer is located inside a closet and operating with the closet door closed.
- Overloading of clothing in the dryer.
- Dryer is located in an area where the temperature is below 45 degrees Fahrenheit.
- Laundry washed in cold water will take slightly longer to dry.
- The moisture content of certain clothes. Towels and denim retain more moisture, for example.
- Wrong spin speed selected for the type of laundry being washed. For example, towels, denim, blankets, small rugs, etc. use the normal wash setting to extract most of the water from the laundry. If there is an extra option for high-speed spin it should also be selected to extract as much water as possible from the laundry. The drying cycle will be increased greatly if the permanent press cycle or the gentle cycle is selected for these items.
- Electric supply is less than what is needed to operate the dryer efficiently.
- Load type (towels, bedspreads, jeans, etc.)

All of these must be taken into account when diagnosing a complaint of clothes not drying properly.

Lint

Lint consists of fibers that have broken away from the fabric. It can collect inside the dryer cabinet and base and create a fire hazard. This lint should be removed every one to two years by cleaning it out. Some lint can also collect in the door opening, the drum, the heater assembly, the blower assembly, and the duct assembly. This accumulation of lint should be removed periodically when performing maintenance on the dryer; otherwise, it could create problems with the future use of the dryer.

The lint screen should be cleaned every time the dryer is used. The duct system that exhausts the air to the outside of the dwelling should also be cleaned out every one to two years.

Shrinkage

Shrinkage is caused by overdrying and by the type of fabric being dried. To reduce the shrinking of cotton and rayon knits, it is recommended that the user remove them from the dryer while the clothing is still damp and lay them on a flat surface to air-dry.

Also, you might suggest that the user set the heat setting on a lower setting. Before drying any synthetic clothing in a dryer, the user should read the label on the garment for the proper drying instructions and for the proper heat selection. The user should also read the use and care manual for the proper instructions when operating the dryer.

Stains

Greasy-looking stains are often caused by:

- Fabric softener that is designed for use in the dryer.
- Fabric softener that was undiluted when the clothes were washed.
- Not enough detergent being used in the wash. This will cause the soil in the water to stick to the outer tub and to return to the next load being washed.
- Drying of clothing that already has a stain on it.

Brown stains on the clothing are often caused by:

- Fabric softener that is designed for use in the dryer.
- Leaving the wash in the washer tub after the cycle is completed.
- Leaving the wet clothing in the dryer for an extended amount of time.
- A leaking transmission seal.

Static Electricity

Static electricity in the clothes is caused by overdrying them. To reduce this condition, you must instruct the user to reduce the amount of time the clothes are to be dried and to use a lower temperature setting. Static electricity in synthetics is normal. To reduce this condition, use a liquid fabric softener in the wash cycle or use a fabric softener that is designed for use in the dryer.

Wrinkling

Wrinkling of the clothes is often caused by:

- Not using the permanent press cycle, which has a cooling-down period during which the heating element is not on.
- Clothing not removed from the dryer after it has completed the cycle.
- Overloading the dryer. The clothes need room to tumble freely without getting tangled.
- The quality of the permanent press material in the garment might be poor.
- On some models that have steam, check the water supply, water valve, and steam port.

Electric Dryer Maintenance

Automatic dryers must be cleaned periodically. The excess lint must be removed to prevent the possibility of a fire and the possibility of the dryer not functioning properly. Most models have a service panel on the front of the dryer for accessibility. The outside of the cabinet should be wiped with a damp cloth. It is recommended that the components be inspected for wear and tear; if repairs are needed, they should be made as soon as possible.

PART VI

Repair Procedures

Each repair procedure is a complete inspection and repair process for a single dryer component, containing the information you need to test a component that might be faulty and to replace it, if necessary.

Any person who cannot use basic tools should *not* attempt to install, maintain, or repair any electric dryer. Any improper installation, preventative maintenance, or repairs will create a risk of personal injury, as well as property damage. Call the service manager if installation, preventative maintenance, or the repair procedure is not fully understood.

Dryer Timer

The dryer timer is an electromechanical component controlled by a synchronous motor in incremental advances. It controls and sequences the numerous steps and functions involved in each cycle.

The typical complaints associated with the dryer timer are:

- The dryer will not run at all.
- The clothes are not drying.
- The dryer will not stop at the end of the cycle.
- The timer will not advance through the cycle.

To handle these problems, perform the following steps:

1. **Verify the complaint.** Verify the complaint by operating the dryer through its cycles. Before you change the timer, check the other components controlled by the timer. If the dryer will not power up, locate the technical data sheet behind the control panel or for diagnostics information. On some models you will need the actual service manual for the model you are working on to properly diagnose the dryer.

2. **Check for external factors.** You must check for external factors not associated with the appliance. Is the appliance installed properly? Does the appliance have the correct voltage? The voltage at the receptacle is between 198 volts and 264 volts during a load on the circuit (see Chapter 6).

3. **Disconnect the electricity.** Before working on the dryer, disconnect the electricity. This can be done by pulling the plug from the electrical outlet. Or disconnect the electricity at the fuse panel or at the circuit breaker panel. Turn off the electricity.

WARNING *Some diagnostic tests will require you to test the components with the power turned on. When you disassemble the control panel, you can position it in such a way that the wiring will not make contact with metal. This act will allow you to test the components without electrical mishaps.*

4. **Remove the console panel to gain access.** To gain access to the timer, remove the screws that secure the console to the top of the dryer (Figure 20-14). On other models, to gain access, remove the back panel behind the console.

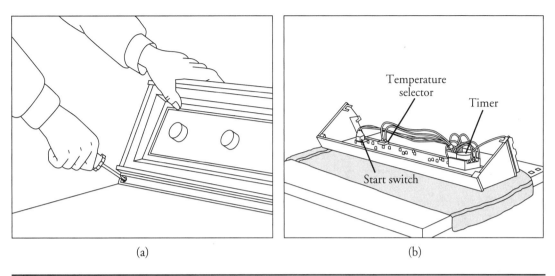

(a)   (b)

FIGURE 20-14   To gain access to the component, remove the screws. Rest the console panel on its face to gain access to the component.

5. **Test the timer.** There are two ways to check the timer:

   - Connect a 120-volt fused service cord to the timer motor to see if the timer motor is operating (Figure 20-15). Connect the ground wire of the 120-volt fused service cord to the console ground. Be cautious when working with live wires. Avoid getting shocked. The timer motor operates on 120 volts. If the motor does not operate, replace the timer. If the timer motor runs but does not advance the cams, the timer has internal defects and should be replaced.

   - Set the ohmmeter range to R × 100, disconnect the timer motor leads, and check for resistance (Figure 20-16). The meter should read between 2000 and 3000 ohms. Next, test the timer switch contacts with the wiring diagram's configuration for the affected cycle. Place the meter probe on each terminal being tested, and turn the timer knob. If the switch contact is good, your meter will show continuity.

6. **Remove the timer.** To remove the timer, pull the timer knob from the timer stem; then remove the timer mounting screws. Remove the wire lead terminals from the timer. Mark the wires as to their location on the timer. Some timers have a quick disconnect, instead of individual wires, which makes it easier to remove the timer wires (Figure 20-17).

7. **Install a new timer.** To install a new timer, just reverse the disassembly procedure, and reassemble. Replace the wires on the timer. Reinstall the console panel, and restore the electricity to the dryer. Test the dryer for proper operation. Make sure to take the dryer out of the service test mode when the repair is completed.

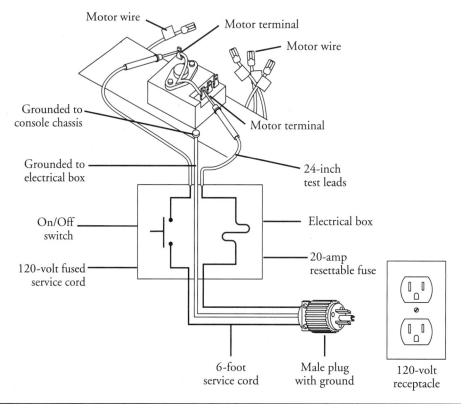

FIGURE 20-15 The test cord attached to the timer motor leads.

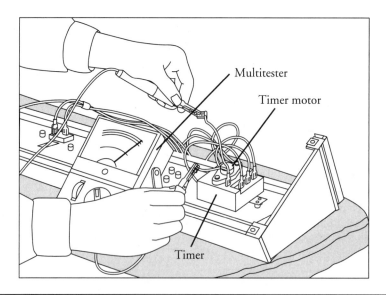

FIGURE 20-16 Checking for continuity between the timer motor and the switch contacts.

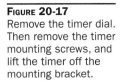

FIGURE 20-17
Remove the timer dial.
Then remove the timer
mounting screws, and
lift the timer off the
mounting bracket.

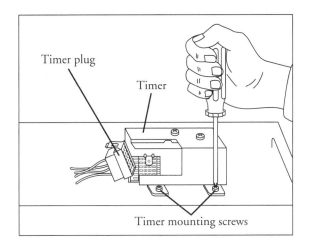

Electronic Control Board and User Interface Controls

On some models the electronic control board and user interface controls replace the electromechanical timer and rotary selection switches.

The typical complaints associated with the electronic control board or the user interface controls are:

- The dryer won't run or power up.
- Unable to program the dryer.
- The display board will not display anything.
- One or more key pads will not accept commands.
- Unusual display readouts.
- Unusual display readouts and/or error codes.

To prevent electrostatic discharge (ESD) from damaging expensive electronic components, follow the steps in Chapters 6 and 11.

To handle these problems, perform the following steps:

1. **Verify the complaint.** Verify the complaint by operating the dryer. Turn off the electricity to the appliance and wait for two minutes before turning it back on. If a fault code appears, look up the code. If the dryer will not power up, locate the technical data sheet behind the control panel or for diagnostics information. On some models you will need the actual service manual for the model you are working on to properly diagnose the dryer. The service manual will assist you in properly placing the dryer in the service test mode for testing the dryer functions.

2. **Check for external factors.** You must check for external factors not associated with the appliance. Does the appliance have the correct voltage? The voltage at the receptacle is between 198 volts and 264 volts during a load on the circuit (see Chapter 6).

3. **Disconnect the electricity.** Before working on the dryer, disconnect the electricity. This can be done by pulling the plug out of the wall receptacle. Or disconnect the electricity at the fuse panel or circuit breaker panel. Turn off the electricity.

WARNING *Some diagnostic tests will require you to test the components with the power turned on. When you disassemble the control panel, you can position the panel in such a way so that the wiring will not make contact with metal. This will allow you to test the components without electrical mishaps.*

4. **Remove the console panel to gain access.** Begin by removing the screws from the dryer top (for front-loading models only) to gain access to the electronic control board. Remove the console panel screws to gain access to the user interface controls (Figure 20-18). On top-loading models, the console will roll upwards or toward you after you remove the console screws.

5. **Test the electronic control board and user interface controls.** If you are able to run the dryer diagnostic test mode, check the different functions of the dryer. Use the technical data sheet for the model you are servicing to locate the test points from the wiring schematic. Check all wiring connections and wiring. Using the technical data sheet, test the electronic control or user interface controls, input voltages, and output voltages. On some models, fuses are soldered to the printed circuit board (PCB). These fuses must be tested first before condemning the component.

6. **Remove the electronic control board and user interface controls.** To remove the defective component, remove the screws that secure the board to the control panel or dryer frame. Disconnect the connectors from the electronic control board or user interface control.

7. **Install the new component.** To install a new electronic control board or user interface control, read the parts data sheet that comes with the part for the proper installation process, and just reverse the disassembly procedure and reassemble. Reinstall the console panel, and restore the electricity to the dryer. Make sure that the dryer in not in the service mode. Test the dryer operation. Make sure to take the dryer out of the service test mode when the repair is completed.

Dryer Motor

The dryer motor on most models is a dual-shaft, single-speed, ¼-horsepower, 1725-rpm motor with an automatic thermal reset protector. One side of the motor shaft is threaded to hold the blower wheel, and the opposite end of the shaft holds the belt pulley, which is pressed onto the shaft. Inside the motor is a centrifugal switch, which serves three purposes. It disengages the motor start winding when the motor reaches 75 percent of its rated speed, engages the run winding, and closes the circuit for the heater element. On other models, there are dual motors (Figure 20-2): one is dedicated to the drum and the other is dedicated to the blower assembly. The purpose of the dedicated motor for the blower assembly is to be able to alter the blower wheel speed whenever necessary to optimize air flow within the dryer.

FIGURE 20-18
An exploded view of the electronic controls in a front-loading dryer.

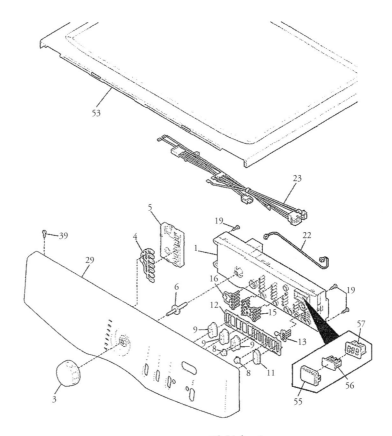

1. Electronic control board assembly
3. Knob
4. Program light pipe
5. Auxiliary interface board
6. Pin
8. Push button
9. Push button
11. Push button
12. Push button spring
13. Start light pipe
15. Light pipe
16. Light pipe
19. Screw
22. Interface board wiring harness
23. Wiring harness
29. Control panel
39. Screw
53. Top panel
55. Digital display lens
56. Digital display light pipe
57. Digital display (LED)

The typical complaints associated with motor failure are:

- Fuse is blown or the circuit breaker trips.
- Motor will not start; it only hums.
- The dryer will not run.

To handle these problems, perform the following steps:

1. **Verify the complaint.** Verify the complaint by operating the dryer through its cycles. Listen carefully, and you will hear if there are any unusual noises or if the circuit breaker trips. On electronic models, turn off the electricity to the appliance and wait for two minutes before turning it back on. If a fault code appears, look up the code. If the dryer will not power up, locate the technical data sheet behind the control panel or for diagnostics information. On some models you will need the actual service manual for the model you are working on to properly diagnose the dryer. The service manual will assist you in properly placing the dryer in the service test mode for testing the dryer functions.

2. **Check for external factors.** You must check for external factors not associated with the appliance. Is the appliance installed properly? Does the appliance have the correct voltage? The voltage at the receptacle is between 198 volts and 264 volts during a load on the circuit (see Chapter 6).

3. **Disconnect the electricity.** Before working on the dryer, disconnect the electricity. This can be done by pulling the plug from the electrical outlet. Or disconnect the electricity at the fuse panel or at the circuit breaker panel. Turn off the electricity.

WARNING *Some diagnostic tests will require you to test the components with the power turned on. When you disassemble the control panel, you can position it in such a way that the wiring will not make contact with metal. This act will allow you to test the components without electrical mishaps.*

4. **Gain access to the motor.** In order to gain access to the motor, the top must be raised (Figure 20-19) by removing the screws from the lint screen slot. Insert a putty knife about two inches from each corner; then disengage the retaining clips and lift the top. On some models, the top is held down with screws (see inset in Figure 20-19). Remove the screws and lift the top. Now the front panel of the dryer must be removed (Figure 20-20). To remove the top panel, insert a putty knife and disengage the retaining clip (Figure 20-20a).

   On some models, the top panel is held in place with two screws. As Figure 20-20b shows, remove the screws that hold the lower part of the front panel in place. Next, remove the screws that hold the upper part of the front panel in place (Figure 20-20c), and then disconnect the door switch wires (label them). With the front panel out of the way, you can now disconnect the drive belt (Figure 20-21). Push on the idler pulley to release the tension from the drive belt, and remove the belt from the motor pulley. Grab the drum and remove it from the cabinet (Figure 20-22). On some models, you will have to remove the back panel because the drum comes out through the rear of the cabinet.

5. **Disconnect the motor wire leads.** Disconnect the motor wire leads from the wiring harness. Set the ohmmeter on R × 1. Figure 20-23 illustrates testing the motor windings and the centrifugal switch for continuity. When testing for resistance on the motor, test from the common wire lead to the run winding. Then test for resistance from the common wire lead to the start winding. Next, test for resistance from the

FIGURE 20-19
Insert a putty knife to disengage the retaining clips that hold the top down. Don't forget to remove the two screws under the lint screen cover. On some models, remove the screws from underneath the top (see insert).

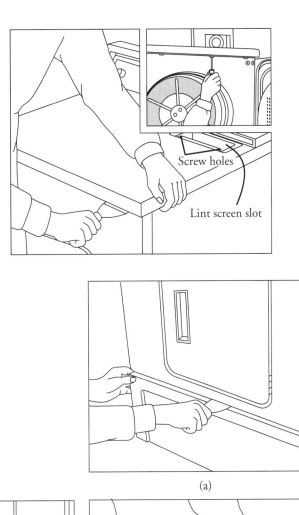

Screw holes

Lint screen slot

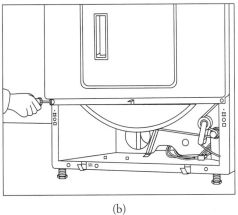

(a)

(b)

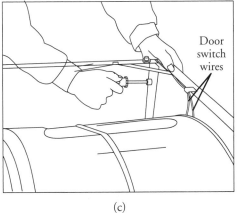

Door switch wires

(c)

FIGURE 20-20 Remove the top panel. Remove the screws from the lower front panel. Remove the screws from the upper front panel. Remember to remove the door switch wires.

FIGURE 20-21
Push the idler pulley
assembly toward the
drive motor pulley to
release the belt
tension. Then
disconnect the drive
belt from the idler
pulley and drive motor
pulley.

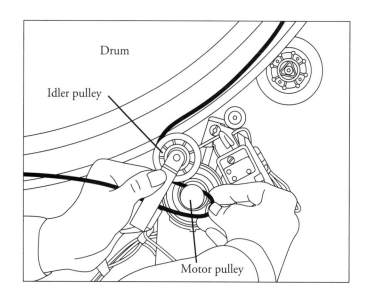

FIGURE 20-22
Remove the drum
from the dryer cavity.

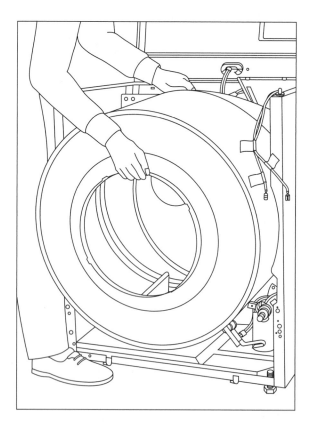

Figure **20-23**
Test the centrifugal
switch and drive motor
windings.

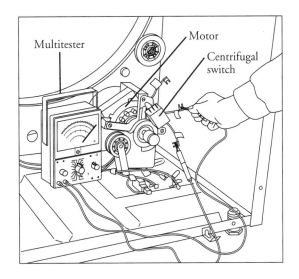

start winding to the run winding. To test for a grounded winding in the motor, take
the ohmmeter probes and test from each motor wire lead terminal to the motor
housing. The ohmmeter will indicate continuity if the windings are grounded. If the
ohmmeter reading shows no resistance between the motor windings, then replace the
motor. If the motor checks out okay, then check the timer.

6. **Remove the motor.** To remove this type of motor, you must first disconnect the
blower assembly by holding the motor shaft stationary and then turning the
blower wheel to remove it from the rear of the motor shaft (Figure 20-24). Then
remove the spring clamps that hold the motor in the motor bracket (Figure 20-25).
On some models, the motor pulley must be removed (new motors come without the
pulley attached) by loosening the allen set screw.

Figure **20-24**
Disconnect the blower
assembly from the
drive motor shaft.

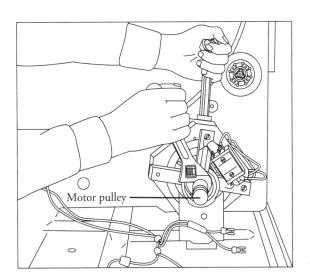

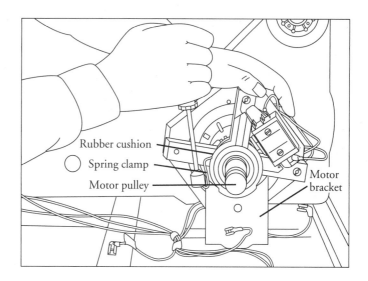

FIGURE 20-25
Hold the motor and
remove the spring
clamps.

Rubber cushion

Spring clamp

Motor pulley

Motor
bracket

7. **Install the new motor.** To install the new motor (Figure 20-26), just reverse the
disassembly procedure, and reassemble. Then reassemble the dryer in the reverse
order of its disassembly. Restore the electricity to the dryer and test the motor. Make
sure to take the dryer out of the service test mode when the repair is completed.

Drive Belt

The drive belt extends from the motor pulley, past the idler pulley, and around the perimeter
of the dryer drum. The tension on the belt is maintained by the idler pulley and driven by a
pulley attached to the one end of the motor shaft.

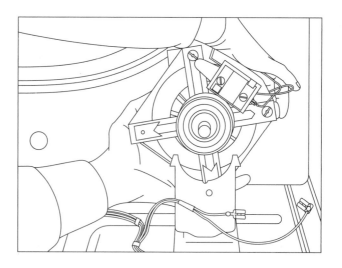

FIGURE 20-26
Place the motor in the
cradle and position it
to fit into the slots
properly.

The typical complaints associated with belt failure are:

- The drum will not turn.
- Dryer motor spins freely.
- It smells like something is burning.

To handle these problems, perform the following steps:

1. **Verify the complaint.** Verify the complaint by operating the dryer through its cycles. Listen carefully, and you will hear if there are any unusual noises. Then, with the door open, press the door switch and start the dryer. The drum should rotate. On electronic models, turn off the electricity to the appliance and wait for two minutes before turning it back on. If a fault code appears, look up the code. If the dryer will not power up, locate the technical data sheet behind the control panel or for diagnostics information. On some models you will need the actual service manual for the model you are working on to properly diagnose the dryer. The service manual will assist you in properly placing the dryer in the service test mode for testing the dryer functions.

2. **Check for external factors.** You must check for external factors not associated with the appliance. Is the appliance installed properly? Does the appliance have the correct voltage? The voltage at the receptacle is between 198 volts and 264 volts during a load on the circuit (see Chapter 6). Check for foreign objects lodged between the drum and bulkhead, etc.

3. **Disconnect the electricity.** Before working on the dryer, disconnect the electricity. This can be done by pulling the plug from the electrical outlet. Or disconnect the electricity at the fuse panel or at the circuit breaker panel. Turn off the electricity.

WARNING *Some diagnostic tests will require you to test the components with the power turned on. When you disassemble the control panel, you can position it in such a way that the wiring will not make contact with metal. This act will allow you to test the components without electrical mishaps.*

4. **Gain access to the drive belt.** To gain access to the drive belt, the top must be raised (see Figure 20-19) by removing the screws from the lint screen slot. Then insert a putty knife about two inches from each corner, disengage the retaining clips, and lift the top. On some models, the top is held down with screws (see the inset in Figure 20-19). Remove the screws and lift the top. Now the front panel of the dryer must be removed (see Figure 20-20). To remove the top panel, insert a putty knife and disengage the retaining clip (see Figure 20-20a). As shown in Figure 20-20b, remove the screws that hold the lower part of the front panel in place. Next, remove the screws that hold the upper part of the front panel in place (see Figure 20-20c), and then disconnect the door switch wires.

5. **Remove the drive belt.** To remove the drive belt on this type of dryer, you can disconnect the belt (see Figure 20-21). Push on the idler pulley to release the tension from the drive belt. Now remove the belt from the motor pulley and from around the drum (Figure 20-27). If the drive belt is broken, just remove the belt.

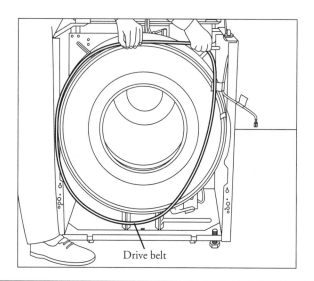

Figure 20-27   Removing the drive belt from around the drum.

Figure 20-28   Dual-heater element mounted on the rear of the dryer.

6. **Install a new drive belt.** To install the new drive belt, just reverse the disassembly procedure, and reassemble. Then reassemble the dryer in the reverse order of its disassembly. Restore the electricity to the dryer and test. Make sure to take the dryer out of the service test mode when the repair is completed.

Heater Element

The heater assembly on some models is located behind the dryer drum (Figure 20-28), while on other models the heater assembly is located in the back of the dryer, and on other models the heater assembly is located on the bottom of the dryer (Figure 20-29).

The typical complaints associated with heater failure are:

- The dryer will not dry the clothes properly.
- There is no heat at all when a heat cycle is selected.
- On electronic models, an error code appears.

FIGURE 20-29
Heating element located on the bottom of the dryer.

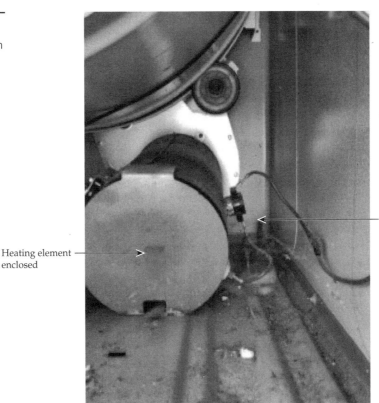

Thermostat

Heating element enclosed

To handle these problems, perform the following steps:

1. **Verify the complaint.** Verify the complaint by operating the dryer through its cycles. Then open the dryer door, and place your hand inside to see if it is warm in the drum. On electronic models, turn off the electricity to the appliance and wait for two minutes before turning it back on. If a fault code appears, look up the code. If the dryer will not power up, locate the technical data sheet behind the control panel or for diagnostics information. On some models you will need the actual service manual for the model you are working on to properly diagnose the dryer. The service manual will assist you in properly placing the dryer in the service test mode for testing the dryer functions.

2. **Check for external factors.** You must check for external factors not associated with the appliance. Is the appliance installed properly? Is the exhaust vent clogged? Does the appliance have the correct voltage? The voltage at the receptacle is between 198 volts and 264 volts during a load on the circuit (see Chapter 6).

NOTE *The dryer motor runs on 120 volts, but the heater works on 240 volts. Check the fuses or the circuit breakers in the home.*

3. **Disconnect the electricity.** Before working on the dryer, disconnect the electricity. This can be done by pulling the plug from the electrical outlet. Or disconnect the electricity at the fuse panel or at the circuit breaker panel. Turn off the electricity.

WARNING *Some diagnostic tests will require you to test the components with the power turned on. When you disassemble the control panel, you can position it in such a way that the wiring will not make contact with metal. This act will allow you to test the components without electrical mishaps.*

4. **Gain access to the heater assembly.** You must gain access to the heater assembly through the back. Remove the exhaust duct from the dryer. Then remove the screws from the back panel (Figure 20-30).

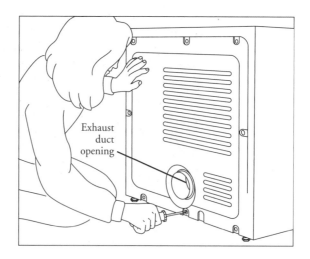

FIGURE 20-30
Remove the screws that hold the back panel in place.

Exhaust duct opening

FIGURE 20-31
Test the heater
element for continuity.

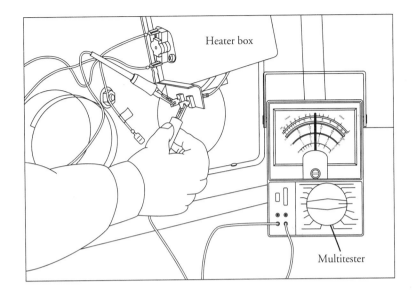

Heater box

Multitester

5. **Test the heater element.** To test the heater element, set the ohmmeter on R × 1. Remove the wires from the heater (Figure 20-31), and test for continuity. If it checks out okay, check the thermostats.

6. **Remove the heater element.** To remove the heater in this type of dryer, you must first remove the screw and the bracket that holds the heater box in place. It is located on top of the heater box. You can gain access through the back of the dryer or lift the top and reach in near the back bulkhead. Then slide the heater box up, and pull it away at the bottom (Figure 20-32). With the heater box out of the dryer, remove the screw that holds the heater element in place (Figure 20-33). Pull the heater element out of the heater box.

7. **Install new heater element.** To install the new heater element, just reverse the disassembly procedure, and reassemble. Then reinstall the back panel. Restore the electricity to the dryer and test it. Make sure to take the dryer out of the service test mode when the repair is completed.

Door Switch

The door switch is a normally open switch wired in series with the dryer motor. When the dryer door is closed, the door switch contacts will close completing the circuit through the run winding of the drive motor.

The typical complaints associated with switch failure are:

- Dryer will not operate at all.
- Dryer light does not work when the door is open.

PART VI

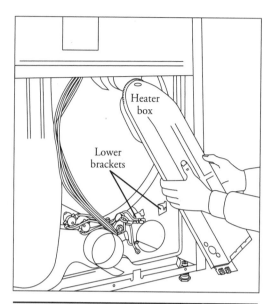

FIGURE 20-32 Remove the heater box.

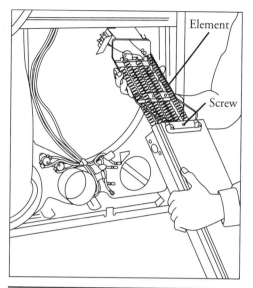

FIGURE 20-33 Remove the heater coil from the heater box.

To handle these problems, perform the following steps:

1. **Verify the complaint.** Verify the complaint by operating the dryer through its cycles. Open the door to see if the light is working. On electronic models, turn off the electricity to the appliance and wait for two minutes before turning it back on. If a fault code appears, look up the code. If the dryer will not power up, locate the technical data sheet behind the control panel or for diagnostics information. On some models you will need the actual service manual for the model you are working on to properly diagnose the dryer. The service manual will assist you in properly placing the dryer in the service test mode for testing the dryer functions.

2. **Check for external factors.** You must check for external factors not associated with the appliance. Is the appliance installed properly? Does the dryer have the correct voltage supply? The voltage at the receptacle is between 198 volts and 264 volts during a load on the circuit (see Chapter 6).

3. **Disconnect the electricity.** Before working on the dryer, disconnect the electricity. This can be done by pulling the plug from the electrical outlet. Or disconnect the electricity at the fuse panel or at the circuit breaker panel. Turn off the electricity.

WARNING *Some diagnostic tests will require you to test the components with the power turned on. When you disassemble the control panel, you can position it in such a way that the wiring will not make contact with metal. This act will allow you to test the components without electrical mishaps.*

4. **Gain access to the door switch.** You must gain access to the door switch. On this model, the top must be raised (see Figure 20-19) by removing the screws from the lint

screen slot. Then insert a putty knife about two inches from each corner, disengage the retaining clips, and lift the top. On some models, the top is held down with screws (see the inset in Figure 20-19). Remove the screws and lift the top.

5. **Test the door switch.** The door switch is located in one of the upper corners of the inside of the front panel (Figure 20-34a). Once you have located the door switch, disconnect the wires from the terminals. Set the ohmmeter on R x 1, attach the probes of the ohmmeter to the terminals of the door switch, and then close the door. With the door closed, there should be continuity. With the door open, the ohmmeter should not show continuity. Some dryer models have a light inside the drum. This light circuit is also part of the door switch circuitry. Three terminals are located on the door switch. Take your ohmmeter probe and place it on the common terminal of the door switch. To locate the common terminal, read the wiring diagram. It will indicate which terminal is the common, which is the light, and which is for the motor circuit. Take your probe and place it on the switch terminal (the light circuit); then close the dryer door. The reading should indicate no continuity. Now open the dryer door—you should have a continuity reading on the meter.

6. **Remove the door switch.** To remove the lever-type door switch (Figure 20-34b); remove the screws that hold the switch in place. Now lift the switch out from behind the front panel. To remove the cylindrical type of door switch (Figure 20-34c), squeeze the retaining clips on the back side of the front panel, and pull out the switch. To remove the hinge type of door switch (Figure 20-34d), the front panel must be removed. Next, remove the screws that hold the switch assembly in place.

FIGURE 20-34
(a) Test the door switch for continuity.
(b) Remove the screws that secure the door switch.
(c) Removing a cylindrical door switch.
(d) Removing a hinge-mounted door switch.

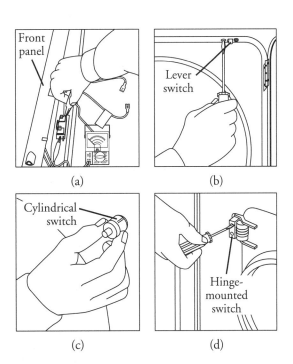

(a)   (b)

(c)   (d)

7. **Install the new door switch.** To install the new door switch, reverse the disassembly procedure, and reassemble. Reassemble the dryer, and install the wires on the new switch terminals. Plug in the dryer and test it. Make sure to take the dryer out of the service test mode when the repair is completed.

Thermostat

Thermostats can be mounted on the heater assembly (Figure 20-2 and Figure 20-29) and/or on the blower assembly (Figures 20-2 and 20-35), or in the inlet filter screen (Figure 20-37). The thermostats monitor the air temperatures within the dryer drum, exhaust vent, and the incoming air supply. The thermostats are bimetal switches that are automatic reset, one-time use, or manual reset.

The typical complaints associated with thermostat failure are:

- The clothes are not drying.
- The dryer will not shut off.
- The dryer will not heat at all.
- The drying temperature is too high.
- Moisture retention (of fabrics) is unsatisfactory.
- On electronic models, an error code appears.

FIGURE 20-35
The thermostat is mounted on the exhaust vent pipe along with the thermal fuse.

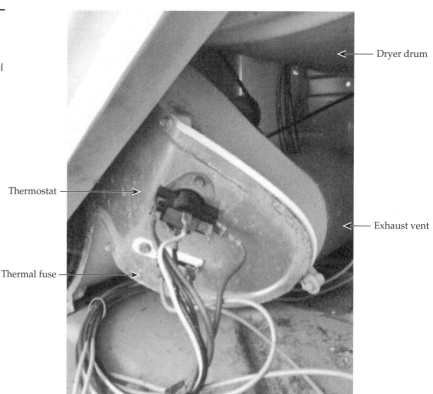

To handle these problems, perform the following steps:

1. **Verify the complaint.** Verify the complaint by operating the dryer through its cycles. On electronic models, turn off the electricity to the appliance and wait for two minutes before turning it back on. If a fault code appears, look up the code. If the dryer will not power up, locate the technical data sheet behind the control panel or for diagnostics information. On some models you will need the actual service manual for the model you are working on to properly diagnose the dryer. The service manual will assist you in properly placing the dryer in the service test mode for testing the dryer functions.

2. **Check for external factors.** You must check for external factors not associated with the appliance. Is the appliance installed properly? Does the dryer have the correct voltage supply? The voltage at the receptacle is between 198 volts and 264 volts during a load on the circuit (see Chapter 6). Is the exhaust vent blocked?

3. **Disconnect the electricity.** Before working on the dryer, disconnect the electricity. This can be done by pulling the plug from the electrical outlet. Or disconnect the electricity at the fuse panel or at the circuit breaker panel. Turn off the electricity.

WARNING *Some diagnostic tests will require you to test the components with the power turned on. When you disassemble the control panel, you can position it in such a way that the wiring will not make contact with metal. This act will allow you to test the components without electrical mishaps.*

4. **Gain access to the thermostat.** You must gain access to the thermostat through the back on this model. Remove the screws from the back panel (see Figure 20-30).

5. **Test the thermostat.** To test the thermostat for continuity, remove the wires from the thermostat terminals. With a continuity tester or an ohmmeter, test the thermostat for continuity (Figure 20-36). Do this to all of the thermostats in the dryer. On some models, the thermostats might be located on the heater housing, behind the drum, or in the lint screen opening (Figure 20-37). If they all check out okay, then reassemble the dryer and test for temperature operation.

NOTE *Do not reinstall the back panel at this point. Take a piece of paper and write down the temperature ratings of the thermostats. These ratings are printed on the thermostats (L140, L290, etc.) or on the wiring schematic. To test for temperature operation, you will need a voltmeter and a temperature tester.*

First, set up the instruments for testing the thermostatic operation. Take the temperature tester thermocouple lead and insert it between the thermostat mounting ear (Figure 20-38) and the plate against which it mounts. Then connect the voltmeter probes across the thermostat terminals. Use alligator clips attached to the probe tips. This will allow you freedom of movement. Do not disconnect the wires from the thermostat. Set the voltmeter range to AC voltage and the selector switch on 300 volts. Remember: If there is more than one thermostat (in series with other thermostats), the thermostats not under test must be electrically isolated and jumped out with an insulated jumper wire with alligator clips attached. Review the wiring schematic to determine which wires to remove and which ones to isolate

FIGURE 20-36
Use a continuity tester
to check the
thermostats.

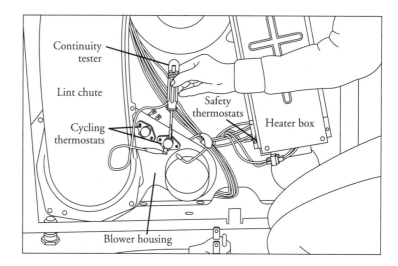

and jump out. With the test meters in place, you are now ready to remove the
exhaust vent duct from the dryer. Seal off 75 percent of the exhaust opening from
the dryer (Figure 20-39). This will simulate a load of clothing in the dryer. On some
models, you do not have to block off the exhaust opening. Check the technical data
sheet or the service manual for the model you are servicing.

Restore the electricity to the dryer. This test requires that the electricity be turned on
for its duration. Always be cautious when working with live wires. Avoid getting
shocked. Set the controls on the dryer to operate at a high heat. Turn on the dryer
and let it cycle. When the voltmeter is reading voltage, the thermostat has opened.
When there is no voltage reading on the voltmeter, the thermostat is closed. As the
dryer is cycling via the thermostat, record the temperature.

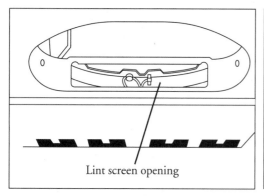

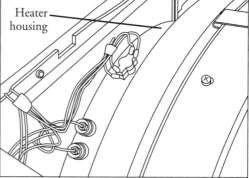

FIGURE 20-37   The location of thermostats on some models.

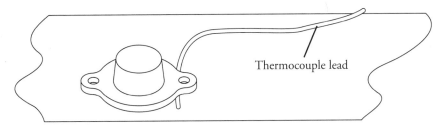

FIGURE 20-38   Insert the thermocouple lead under the thermostat ear.

The thermostat should open at the preset temperature listed on the thermostat. Table 20-1 illustrates the types of thermostats, along with their opening and closing temperature settings. The temperature range of the thermostat should be within ±10 percent of the printed setting. If not, replace the thermostat.

CAUTION   *Turn off the electricity before replacing any parts of the dryer.*

On some models, the thermostat will have four wires attached to it. Two wires are for the bimetal switch contacts and the other two wires are for the bias heater. The purpose of the bias heater is to apply a small amount of heat to the bimetal switch contacts when the timer or fabric temperature switch is set to a lower temperature setting. This will allow the bimetal switch contacts to open up sooner to maintain a lower drum temperature. To test the bias heater, remove the wires from the thermostat, set your multimeter to the ohms scale, and test across the bias heater terminals. The resistance of the bias heater should be between 7000 and 28,000 ohms.

6. **Replace the thermostat.** Remove the screws that hold the thermostat in place. Replace the thermostat with an exact replacement with the same temperature rating. Reconnect the wires to their correct terminal positions. Then reverse the disassembly procedure to reassemble the dryer, and test the thermostat.

7. **Test the new thermostats.** To test the new thermostats, repeat step 5. Make sure to take the dryer out of the service test mode when the repair is completed.

FIGURE 20-39
Block (or tape) 75 percent of the dryer exhaust to simulate a load.

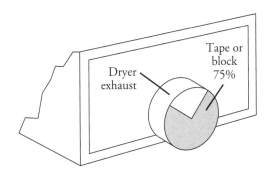

Open	Close	Type	Open	Close	Type
260°	210°	L260	120°	110°	L120
270°	230°	L270	125°	115°	L125
290°	250°	L290	130°	115°	L130
300°	250°	L300	135°	120°	L135
310°	270°	L320	140°	130°	L140
340°	300°	L340	145°	125°	L145
120°	105°	LD120	150°	130°	L150
130°	115°	LD130	155°	135°	L155
140°	120°	LD140	160°	120°	L160
145°	125°	LD145	170°	130°	L170
155°	135°	LD155	180°	170°	L180
170°	150°	LD170	190°	150°	L190
200°	160°	LD200	200°	160°	L200
225°	185°	LD225	205°	165°	L205
240°	195°	LD240	225°	185°	L225
270°	225°	LD270	240°	200°	L240
290°	250°	LD290	250°	210°	L250

TABLE 20-1   Thermostat Opening and Closing Temperatures

Thermistor

The thermistor (Figure 20-2) monitors the incoming and outgoing air temperatures and relays that information back to the electronic control board.

The typical complaints associated with thermistor failure are:

- The clothes are not drying.
- The dryer will not shut off.
- The dryer will not heat at all.
- The drying temperature is too high.
- Moisture retention (of fabrics) is unsatisfactory.
- On electronic models, an error code appears.

To handle these problems, perform the following steps:

1. **Verify the complaint.** Verify the complaint by operating the dryer through its cycles. On electronic models, turn off the electricity to the appliance and wait for two minutes before turning it back on. If a fault code appears, look up the code. If the dryer will not power up, locate the technical data sheet behind the control panel or for diagnostics information. On some models you will need the actual service

manual for the model you are working on to properly diagnose the dryer. The service manual will assist you in properly placing the dryer in the service test mode for testing the dryer functions.

2. **Check for external factors.** You must check for external factors not associated with the appliance. Is the appliance installed properly? Does the dryer have the correct voltage supply? The voltage at the receptacle is between 198 volts and 264 volts during a load on the circuit (see Chapter 6). Is the exhaust vent blocked?

3. **Disconnect the electricity.** Before working on the dryer, disconnect the electricity. This can be done by pulling the plug from the electrical outlet. Or disconnect the electricity at the fuse panel or at the circuit breaker panel. Turn off the electricity.

WARNING *Some diagnostic tests will require you to test the components with the power turned on. When you disassemble the control panel, you can position it in such a way that the wiring will not make contact with metal. This act will allow you to test the components without electrical mishaps.*

4. **Gain access to the thermistor.** To gain access to the thermistor on this model you will have to remove the top panel, control panel (Figure 20-40), front panel (Figure 20-41), and the dryer drum (Figure 20-2).

5. **Test the thermistor.** To test the thermistor for resistance, remove the wires from the thermistor terminals. Set your multimeter on the ohms scale and place the test leads on the thermistor terminals. Match the reading to the technical data sheet or the service manual values. If you are reading an open or infinity, replace the thermistor. Another way to test the thermistor is by placing the dryer into the service mode. On some models, the thermistor might be located on the heater housing, behind the

FIGURE 20-40
After removing the control panel, place it on a table so as not to damage the electronics inside the panel.

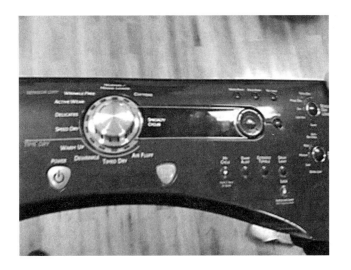

FIGURE 20-41
A top view of the dryer
with the front panel
being removed.

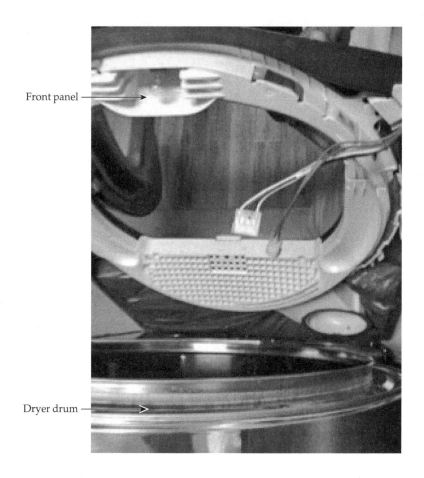

FIGURE 20-41
A top view of the dryer
with the front panel
being removed.

drum, or in the lint screen opening (Figures 20-2 and 20-37). If they all check out
okay, then reassemble the dryer and test for temperature operation.

6. **Replace the thermistor.** Remove the screws that hold the thermistor in place.
   Replace the thermistor with an exact replacement with the same temperature
   rating. Reconnect the wires to their correct terminal positions. Then reverse the
   disassembly procedure to reassemble the dryer, and test the thermistor.

7. **Test the new thermistor.** To test the new thermistor, repeat step 5. Make sure to take
   the dryer out of the service test mode when the repair is completed.

Start Switch

The start switch is a rotary or push button momentary contact switch used to start the dryer
and to energize the start winding in the drive motor. This switch is located in the control
panel on the dryer. The dryer door must be closed for the dryer to operate. If and when the
dryer door is opened, the start switch must be turned on to restart the dryer operation.

The typical complaint associated with the start switch is that the dryer will not start. To handle this problem, perform the following steps:

1. **Verify the complaint.** Verify the complaint by operating the dryer. Before you change the start switch, check the other components that are in the start circuit. On electronic models, turn off the electricity to the appliance and wait for two minutes before turning it back on. If a fault code appears, look up the code. If the dryer will not power up, locate the technical data sheet behind the control panel for the diagnostics information. On some models you will need the actual service manual for the model you are working on to properly diagnose the dryer. The service manual will assist you in properly placing the dryer in the service test mode for testing the dryer functions.

2. **Check for external factors.** You must check for external factors not associated with the appliance. Is the appliance installed properly? Does it have the correct voltage? The voltage at the receptacle is between 198 volts and 264 volts during a load on the circuit (see Chapter 6). Check the fuses or circuit breaker.

3. **Disconnect the electricity.** Before working on the dryer, disconnect the electricity. This can be done by pulling the plug from the electrical outlet. Or disconnect the electricity at the fuse panel or at the circuit breaker panel. Turn off the electricity.

WARNING *Some diagnostic tests will require you to test the components with the power turned on. When you disassemble the control panel, you can position it in such a way that the wiring will not make contact with metal. This act will allow you to test the components without electrical mishaps.*

4. **Remove the console panel to gain access.** Begin by removing the console panel to gain access to the start switch. With this type of dryer, remove the screws from the console (see Figure 20-14a). On some models, the console will be able to lie flat (see Figure 20-14b).

5. **Test the start switch.** Remove the wires from the start switch terminals. Set the ohmmeter on R × 1. Begin testing the start switch for continuity by placing one probe of the ohmmeter on the common terminal (C) of the switch; then connect the other probe to the normally open (NO) terminal. There should be no continuity. With the ohmmeter probes attached to the switch, press the start switch button. You should have continuity (Figure 20-42). If the switch fails the test, replace it.

6. **Remove the start switch.** To remove the start switch, pull the knob off the switch stem, remove the start switch mounting screws, and remove the switch.

7. **Install a new start switch.** To install a new start switch, just reverse the disassembly procedure, and reassemble. Replace the wires on the switch. Reinstall the console panel, and restore the electricity to the dryer. Test the dryer operation.

Drum Roller

The drum roller is located on the rear bulkhead of the dryer. Its main purpose is to support the dryer drum and assist with easy drum rotation.

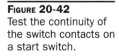

FIGURE 20-42
Test the continuity of
the switch contacts on
a start switch.

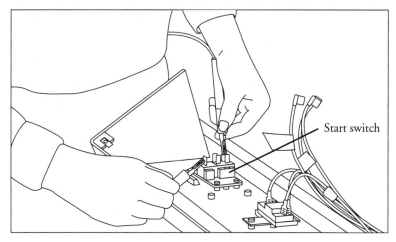

Start switch

The typical complaints associated with the drum roller are:

- Dryer is very noisy when operating.
- Dryer has a burning smell.
- The dryer drum is hard to turn manually.

To handle these problems, perform the following steps:

1. **Verify the complaint.** Verify the complaint by operating the dryer through its cycles. On electronic models, turn off the electricity to the appliance and wait for two minutes before turning it back on. If a fault code appears, look up the code. If the dryer will not power up, locate the technical data sheet behind the control panel or for diagnostics information. On some models you will need the actual service manual for the model you are working on to properly diagnose the dryer. The service manual will assist you in properly placing the dryer in the service test mode for testing the dryer functions.

2. **Check for external factors.** You must check for external factors not associated with the appliance. Is the appliance installed properly? Does the dryer have the correct voltage supply? The voltage at the receptacle is between 198 volts and 264 volts during a load on the circuit (see Chapter 6).

3. **Disconnect the electricity.** Before working on the dryer, disconnect the electricity. This can be done by pulling the plug from the electrical outlet. Or disconnect the electricity at the fuse panel or at the circuit breaker panel. Turn off the electricity.

WARNING *Some diagnostic tests will require you to test the components with the power turned on. When you disassemble the control panel, you can position it in such a way that the wiring will not make contact with metal. This act will allow you to test the components without electrical mishaps.*

4. **Gain access to the drum roller.** To gain access to the drum roller, the top must be raised (see Figure 20-19) by removing the screws from the lint screen slot. Then insert a putty knife about two inches from each corner to disengage the retaining clips, and lift the top. On some models, the top is held down with screws (see the inset in Figure 20-19). Remove the screws and lift the top. Now the front panel of the dryer must be removed (see Figure 20-20). To remove the top panel, insert a putty knife and disengage the retaining clip (see Figure 20-20a). As in Figure 20-20b, remove the screws that hold the lower part of the front panel in place; then remove the screws that hold the upper part of the front panel in place (see Figure 20-20c). Disconnect the door switch wires.

5. **Remove the drive belt and drum.** To remove the drive belt and the drum on this type of dryer, you must first disconnect the belt (see Figure 20-21). Push on the idler pulley to release the tension from the drive belt. Remove the belt from the motor pulley and from around the drum (see Figure 20-27). Grab hold of the drum and remove it from the cabinet (see Figure 20-22).

6. **Remove the drum roller.** Once the drum has been removed, the drum rollers, which are located on the rear bulkhead of this model (Figure 20-43), can then be removed. Take a pair of needle nose pliers and remove the tri-ring from the shaft. Then slide the roller off the shaft.

7. **Install a new drum roller.** To install the new roller, clean the shaft and lubricate it with a small amount of oil. Slide the new drum roller onto the shaft, and reinstall the tri-ring. To reassemble the dryer, just reverse the disassembly procedure, and reassemble. Test the dryer for proper operation. Make sure to take the dryer out of the service test mode when the repair is completed.

FIGURE 20-43
You must remove the tri-ring first in order to remove the drum roller.

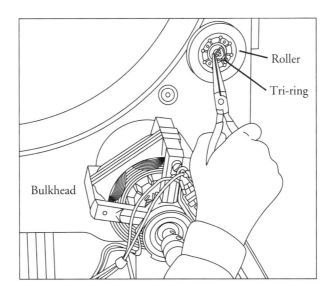

Roller

Tri-ring

Bulkhead

Idler Pulley

The idler pulley assembly is positioned on the dryer chassis and inserted into a slot in the base located next to the drive motor; it maintains the proper tension on the drive belt and minimizes belt slippage.

The typical complaints associated with the idler pulley are:

- When dryer is running, you will hear a squealing noise.
- The dryer belt burns and might snap.
- The dryer drum will not turn.

To handle these problems, perform the following steps:

1. **Verify the complaint.** Verify the complaint by operating the dryer through its cycles. On electronic models, turn off the electricity to the appliance and wait for two minutes before turning it back on. If a fault code appears, look up the code. If the dryer will not power up, locate the technical data sheet behind the control panel or for diagnostics information. On some models you will need the actual service manual for the model you are working on to properly diagnose the dryer. The service manual will assist you in properly placing the dryer in the service test mode for testing the dryer functions.

2. **Check for external factors.** You must check for external factors not associated with the appliance. Is the appliance installed properly? Does the dryer have the correct voltage supply? The voltage at the receptacle is between 198 volts and 264 volts during a load on the circuit (see Chapter 6).

3. **Disconnect the electricity.** Before working on the dryer, disconnect the electricity. This can be done by pulling the plug from the electrical outlet. Or disconnect the electricity at the fuse panel or at the circuit breaker panel. Turn off the electricity.

WARNING *Some diagnostic tests will require you to test the components with the power turned on. When you disassemble the control panel, you can position it in such a way that the wiring will not make contact with metal. This act will allow you to test the components without electrical mishaps.*

4. **Gain access to the idler pulley.** To gain access to the idler pulley, in this type of dryer, remove the top panel, insert a putty knife, and disengage the retaining clip (see Figure 20-20a).

5. **Remove the idler pulley.** To remove the idler pulley, you must first remove the drive belt. To disconnect the belt on this model (see Figure 20-21), push on the idler pulley to release the tension from the drive belt. Now remove the belt from the motor pulley and from around the idler pulley. If the drive belt is broken, just remove the belt. Next, lift the idler pulley up and out of the dryer (Figure 20-44). This type of idler pulley is replaced as a complete assembly. Another type of idler pulley (Figure 20-45a) has a tension spring. To remove this type of idler, you must first remove the tension spring (Figure 20-45b), and then lift the idler assembly out of the dryer. Inspect the pulley for wear. If it is worn, remove the screw from the axle, and pull the axle out (Figure 20-45c). On this type of idler pulley, only replace the worn-out part.

FIGURE 20-44
After removing the drive belt, lift the idler pulley off the base.

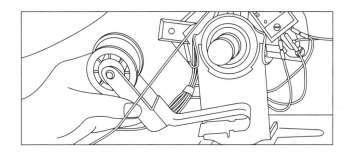

6. **Install the new idler pulley.** To install the repaired idler pulley, just reverse the disassembly procedure, and reassemble it. Reassemble the dryer in the reverse order of its disassembly, and test it for proper operation. Make sure to take the dryer out of the service test mode when the repair is completed.

FIGURE 20-45
(a) This type of idler pulley uses a spring to hold tension against the drive belt.
(b) Remove the spring and lift the idler assembly out of the dryer for inspection.
(c) Remove the axle to replace the pulley.

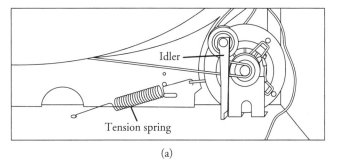

(a)

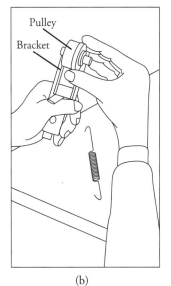

(b)

(c)

PART VI

Temperature Selector Switch

The temperature selector switch is located in the console panel. This switch allows the consumer to select different temperature settings for drying.

The typical complaints associated with the temperature selector switch are:

- Inability to select a certain temperature setting.
- No heat.
- The switch is stuck.

To handle these problems, perform the following steps:

1. **Verify the complaint.** Verify the complaint by operating the dryer through its cycles. Before you change the temperature selector switch, check the other components controlled by this switch. On electronic models, turn off the electricity to the appliance and wait for two minutes before turning it back on. If a fault code appears, look up the code. If the dryer will not power up, locate the technical data sheet behind the control panel or for diagnostics information. On some models you will need the actual service manual for the model you are working on to properly diagnose the dryer. The service manual will assist you in properly placing the dryer in the service test mode for testing the dryer functions.

2. **Check for external factors.** You must check for external factors not associated with the appliance. Is the appliance installed properly? Does it have the correct voltage? The voltage at the receptacle is between 198 volts and 264 volts during a load on the circuit (see Chapter 6).

3. **Disconnect the electricity.** Before working on the dryer, disconnect the electricity to it. This can be done by pulling the plug from the electrical outlet. Or disconnect the electricity at the fuse panel or at the circuit breaker panel. Turn off the electricity.

WARNING *Some diagnostic tests will require you to test the components with the power turned on. When you disassemble the control panel, you can position it in such a way that the wiring will not make contact with metal. This act will allow you to test the components without electrical mishaps.*

4. **Gain access to the temperature selector switch.** To access the temperature selector switch, remove the console panel. On this type of dryer, remove the screws in the console (see Figure 20-14a). On some models, the console will be able to lie flat (see Figure 20-14b). On other models, your access is through the rear panel on the console.

5. **Test the temperature selector switch.** To test the temperature selector switch, locate the selector switch circuit on the wiring diagram. Identify the terminals that are regulating the temperature setting to be tested. Set the ohmmeter on the R × 1 scale. Next, place the ohmmeter probes on those terminals. Then select that temperature setting by either rotating the dial or by depressing the proper button on the switch (Figure 20-46). If the switch contacts are good, your meter will show continuity. Test all of the remaining temperature settings on the temperature selector switch. Remember to check the wiring diagram for the correct switch contact terminals (those that correspond to the setting that you are testing).

FIGURE 20-46
Test the temperature
selector switch for
continuity.

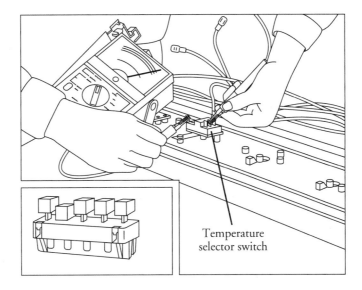

Temperature
selector switch

6. **Remove the temperature selector switch.** To remove the temperature selector switch, remove the screws that hold the switch to the console base, and remove the switch.

7. **Install the new temperature selector switch.** To install the new temperature selector switch, just reverse the disassembly procedure, and reassemble it. Then reattach the wires to the switch terminals according to the wiring diagram. Reassemble the console panel. Be sure when reassembling the console panel that the wires do not become pinched between the console panel and the top of the dryer. Make sure to take the dryer out of the service test mode when the repair is completed.

Belt Switch

The belt switch is located on the drive motor baseplate (Figure 20-47) and held in place with two Phillips-head screws. The belt switch is activated by the movement of the idler pulley assembly. If the belt breaks or comes off the pulley, the belt switch disconnects the power to the motor, shutting down the dryer.

The typical complaints associated with the idler pulley are:

- The dryer will not run.
- The clothes are wet.

1. **Verify the complaint.** Verify the complaint by operating the dryer through its cycles. On electronic models, turn off the electricity to the appliance and wait for two minutes before turning it back on. If a fault code appears, look up the code. If the dryer will not power up, locate the technical data sheet behind the control panel or for diagnostics information. On some models you will need the actual service manual for the model you are working on to properly diagnose the dryer. The service manual will assist you in properly placing the dryer in the service test mode for testing the dryer functions.

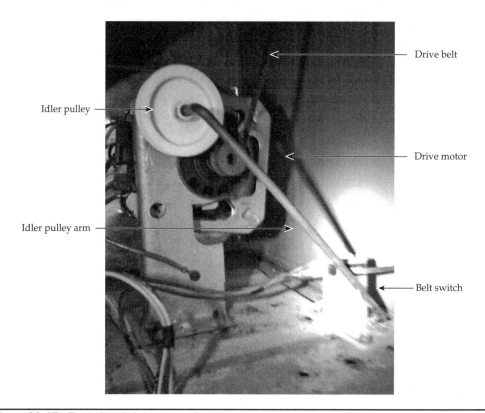

Drive belt

Idler pulley

Drive motor

Idler pulley arm

Belt switch

FIGURE 20-47   The belt switch is a normally closed switch. When the belt breaks, the belt switch opens the circuit, shutting off the dryer motor.

2. **Check for external factors.** You must check for external factors not associated with the appliance. Is the appliance installed properly? Does the dryer have the correct voltage supply? The voltage at the receptacle is between 198 volts and 264 volts during a load on the circuit (see Chapter 6).

3. **Disconnect the electricity.** Before working on the dryer, disconnect the electricity. This can be done by pulling the plug from the electrical outlet. Or disconnect the electricity at the fuse panel or at the circuit breaker panel. Turn off the electricity.

WARNING *Some diagnostic tests will require you to test the components with the power turned on. When you disassemble the control panel, you can position it in such a way that the wiring will not make contact with metal. This act will allow you to test the components without electrical mishaps.*

4. **Gain access to the belt switch.** To gain access to the belt switch, in this type of dryer, remove the top panel, insert a putty knife, and disengage the retaining clip (see Figure 20-20a).

5. **Remove the idler pulley.** Before removing the belt switch you must first remove the idler pulley and drive belt. To disconnect the belt on this model (see Figure 20-21), push on the idler pulley to release the tension from the drive belt. Now remove the belt from the motor pulley and from around the idler pulley. If the drive belt is broken, just remove the belt. Next remove the two Phillips-head screws from the belt switch (Figure 20-47).

6. **Test the belt switch.** Remove the two wires from the belt switch. Place your multimeter test leads on the terminals and turn your meter to the ohms scale. When activating the belt switch, the meter reading will indicate continuity. When deactivating the belt switch, the meter reading will indicate an infinity reading. If the meter indicates infinity when the belt switch is activated and deactivated, replace the switch.

7. **Install the new belt switch.** To install the belt switch, just reverse the disassembly procedure, and reassemble it. Reassemble the dryer in the reverse order of its disassembly, and test it for proper operation. Make sure to take the dryer out of the service test mode when the repair is completed.

Diagnostic Charts

The following diagnostic flowcharts will help you to pinpoint the likely causes of dryer problems (Figures 20-48, 20-49, 20-50, 20-51, and 20-52).

The wiring diagram in this chapter is only an example. You must refer to the wiring diagrams on the dryer that you are servicing. Figure 20-53 depicts an actual wiring schematic diagram and an actual pictorial electrical wiring diagram. Figure 20-54 depicts an actual ladder diagram. A ladder diagram is simpler and is usually easier to read. Figure 20-55 depicts a typical wiring diagram for a new-style front load dryer with electronic controls and two electric heating elements. Figure 20-56 depicts a typical wiring diagram for a standard electric dryer with a mechanical timer. Figure 20-57 depicts a typical wiring diagram for a new-style front load dryer with electronic controls and three electric heating elements.

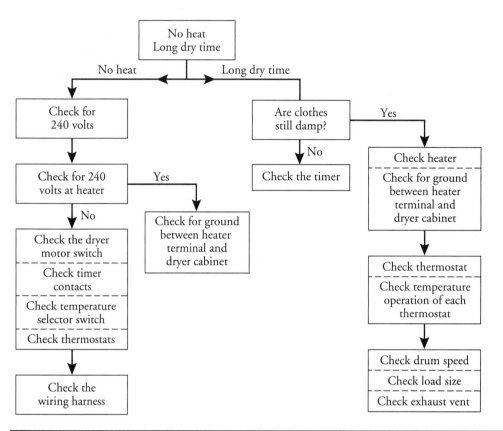

Figure 20-48 The diagnostic flowchart: No heat, long dry time.

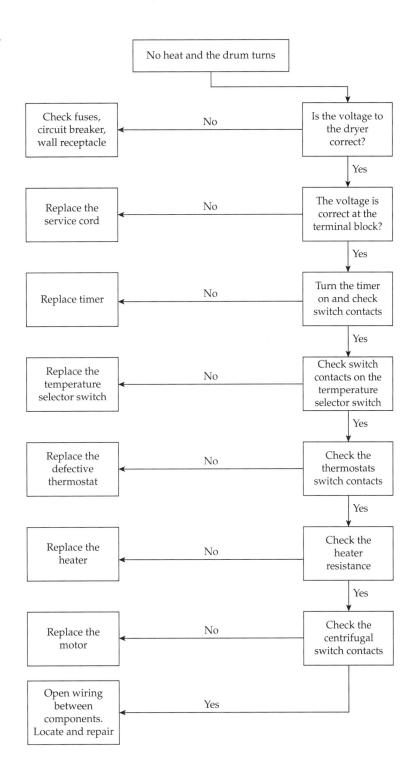

PART VI

FIGURE **20-49**
The diagnostic
flowchart: No heat
and the drum turns.

Figure 20-50
The diagnostic
flowchart: The timer
will not advance on
auto-cycle.

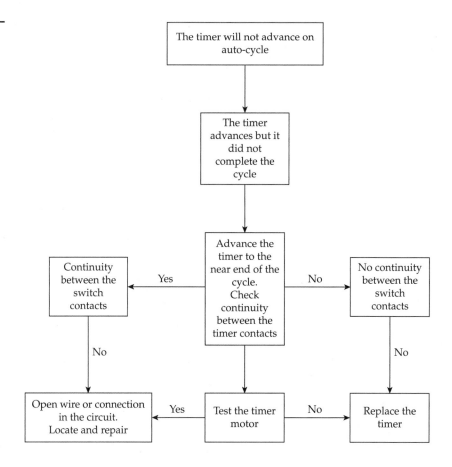

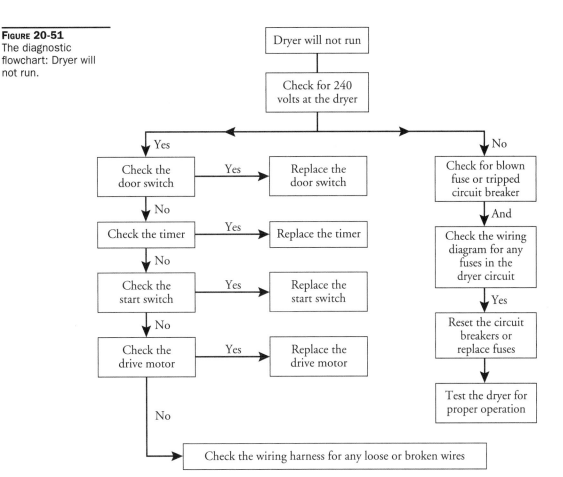

FIGURE 20-51
The diagnostic flowchart: Dryer will not run.

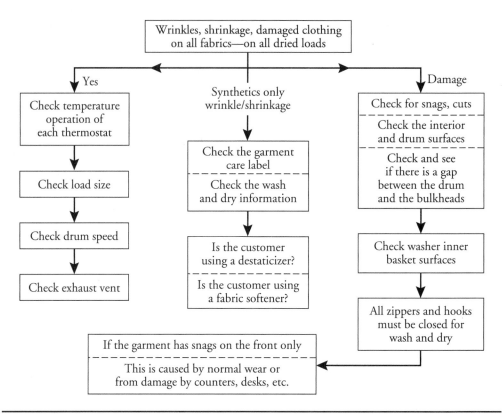

FIGURE 20-52 The diagnostic flowchart: Wrinkles, shrinkage, and damaged clothing with all fabrics on all dried loads.

FIGURE 20-53
The automatic dryer
wiring diagram.

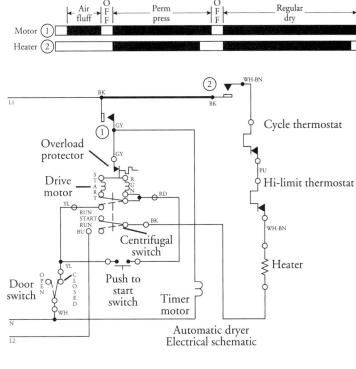

Automatic dryer
Electrical schematic

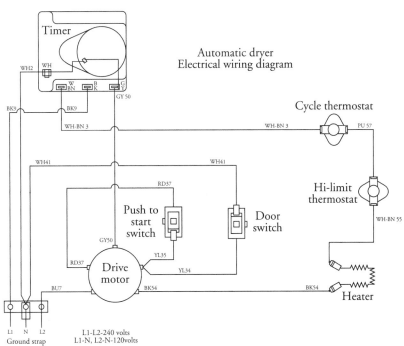

Automatic dryer
Electrical wiring diagram

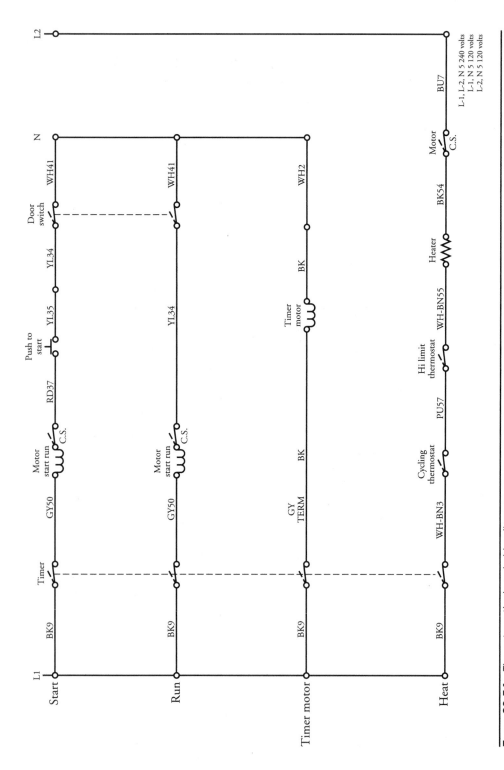

FIGURE 20-54 The automatic dryer ladder diagram.

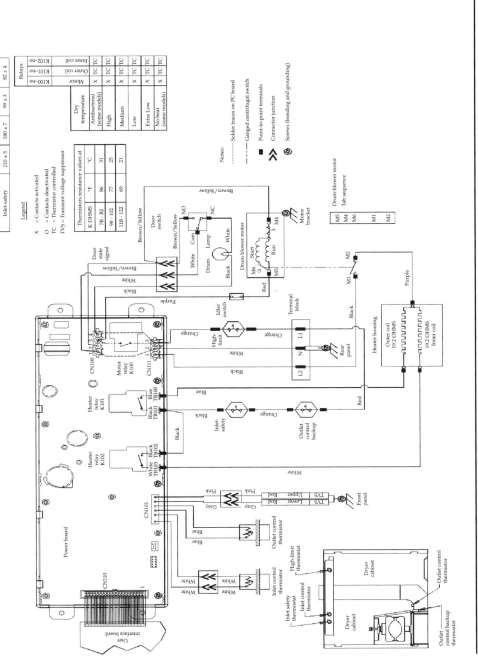

Figure 20-55 A typical wiring diagram for a new-style front load dryer with electronic controls and two electric heating elements.

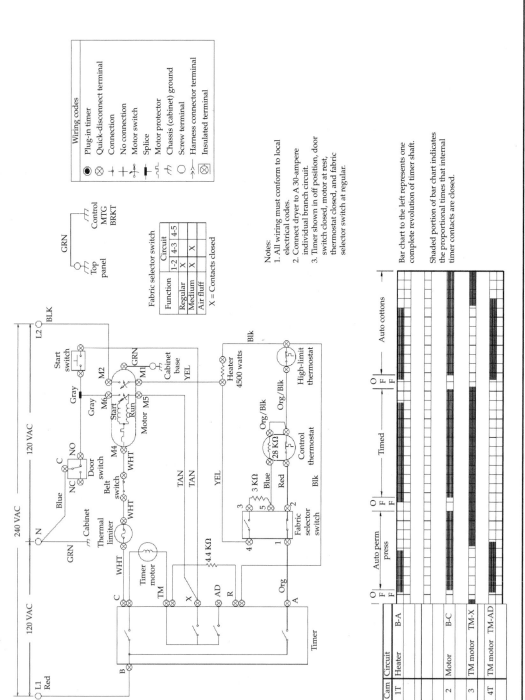

Figure 20-56 A typical wiring diagram for a standard electric dryer with a mechanical timer.

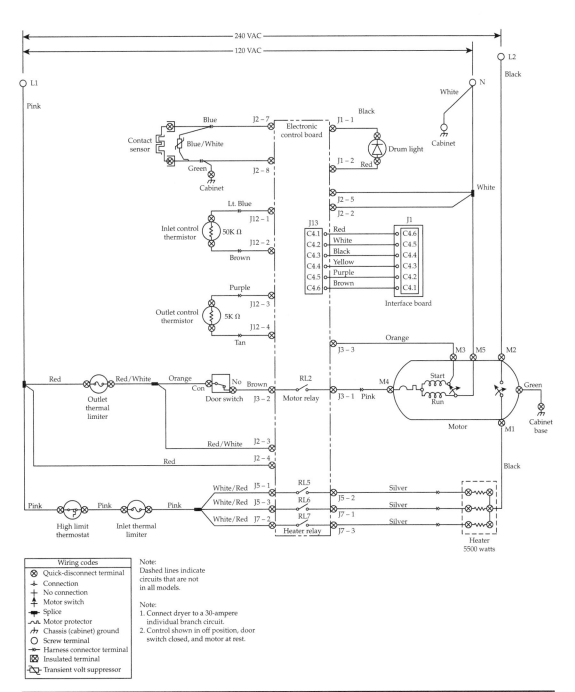

FIGURE 20-57 A typical wiring diagram for a new-style front load dryer with electronic controls and three electric heating elements.

Automatic Gas Dryers

Gas and electric dryers are similar in operation, with the exception of how the air is heated for drying the clothes. The electric components in a gas dryer will operate on 120 volts only. They use liquefied petroleum gas (LP or LPG) or natural gas as a fuel, air (oxygen), and an ignition source to produce a flame. The gas burner can produce between 18,000 and 30,000 BTU/hr of heat, whereas the electric dryer produces between 17,000 and 19,000 BTU/hr of heat. All of the mechanics of the electric dryer, with the exception of the electric heater, are incorporated into the gas dryer.

The dryer will circulate the air, apply heat, and tumble the clothes to increase the surface exposure to remove water from the clothes by evaporation, and then vent the combustion by-products to the outside of the residence. The gas burner assembly includes a gas shutoff valve, burner, automatic solenoid gas control valve, pressure regulator, an ignition system, temperature controls, and safety limit controls.

The operating controls will prevent the burner from coming on if certain conditions are not met. Before the dryer can start, the door must be closed (closing the door switch), and then the consumer will set the timer and press the start button, starting the cycle. If for any reason the motor does not start, the burner operation will not commence. On some models, a belt switch is incorporated into the circuitry to prevent the burner from operating if the dryer belt breaks.

The safety switches will prevent burner operation if the temperature exceeds their pre-set limit. These safety switches are located on the burner chamber, bulkhead, and exhaust passages.

Top-of-the-line models offer electronic controls, moisture sensors to sense how much water is remaining in the clothes, and automatic cycles designed for custom drying a variety of fabrics and load sizes.

This chapter will provide the basic skills needed to diagnose and repair automatic gas dryers. Figure 21-1 identifies where the components are located within the automatic gas dryer. However, this illustration is used as an example only. The actual construction and features might vary, depending on what brand and model you are servicing.

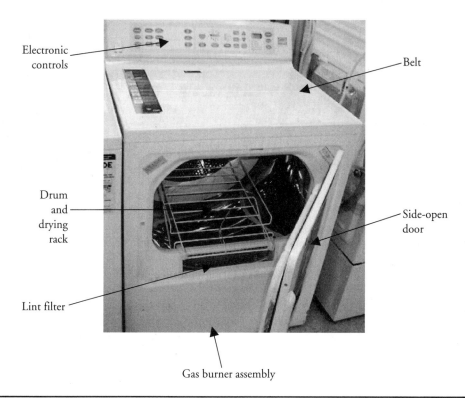

Electronic controls

Belt

Drum and drying rack

Side-open door

Lint filter

Gas burner assembly

Figure 21-1 The location of components in a typical automatic gas dryer. The consumer has a choice between a drop-down or side-open door.

Safety First

Any person who cannot use basic tools or follow written instructions should *not* attempt to install, maintain, or repair any automatic dryers. Any improper installation, preventive maintenance, or repairs could create a risk of personal injury or property damage.

If you do not fully understand the installation, preventive maintenance, or repair procedures in this chapter, or if you doubt your ability to complete the task on the automatic dryer, please call your service manager.

Do not store any flammable materials near a gas dryer, and avoid ignition sources in the event you smell gas leaking. Make sure you have proper ventilation when servicing the gas dryer to prevent carbon monoxide (CO) poisoning.

The following additional precautions should also be followed:

- Never bypass (or interfere with) the operation of any switch, component, or feature.

- Be careful of sharp edges when working on the dryer.

- The dryer produces combustible lint, and the area should be kept clean.

- Never remove any ground wires from the dryer or the third (grounding) prong from the service cord.

- Never use an extension cord to operate a dryer.
- The wiring used in dryers is made with a special heat-resistant insulation. Never substitute it with ordinary wire.
- Always follow the use and care guide instructions from the dryer manufacturer.
- Always keep combustible products away from the gas dryer.
- Keep the dryer clean of soot, grease, and food spillages.
- Teach your children not to play near or with the dryer.
- Always have a fire extinguisher nearby just in case of mishaps that might lead to a fire.
- Have a smoke detector and a carbon monoxide detector installed in the home, and check the batteries yearly.
- Make sure the gas dryer is properly vented according to the manufacturer's recommendations. Never exhaust the dryer into a chimney, a common duct, an attic, or a crawl space.

Before continuing, take a moment to refresh your memory of the safety procedures in Chapter 2.

Automatic Gas Dryers in General

Much of the troubleshooting information in this chapter covers automatic gas dryers in general, rather than specific models, in order to present a broad overview of service techniques. The illustrations that are used in this chapter are for demonstration purposes only to clarify the description of how to service dryers. They in no way reflect upon a particular brand's reliability.

Gas Supply Requirements

The gas dryer is manufactured and shipped for natural gas installation. If the gas supply to the home is liquefied petroleum gas (LP or LPG), the dryer will then have to be converted. When installing the gas supply line to the dryer, the shutoff valve must be within six feet of the product. This will make it easier to locate and shut off the supply of gas to the product for servicing. The recommended gas supply line size should be ½-inch IPS pipe; it is acceptable to use ⅜-inch tubing for less than 20 feet to connect to the shutoff valve, only if it is approved by the local codes in your area. The gas supply line must include a ⅛-inch NPT-plugged connection between the dryer and shutoff valve for testing the gas pressure (Figure 21-2).

For more information regarding gas appliance installations and gas supply line installations, refer to the following sources:

- National Fuel Gas Code
- International Fuel Gas Code
- The local gas codes in your area
- The manufacturer's recommendations
- The AGA (American Gas Association)

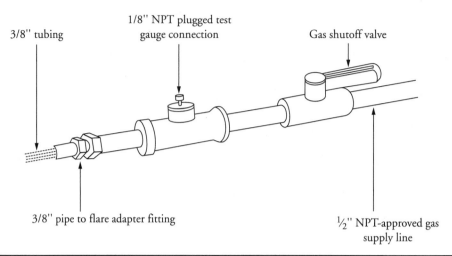

FIGURE 21-2   A typical gas supply line to the dryer. The plugged test gauge connection will be used for testing the gas pressure.

Gas Tests for Gas Dryers

The following testing procedures should be carried out on every service call after repairing the gas dryer:

- To test the manifold pressure on the dryer, connect a manometer to the pressure tap on the gas valve. During the burner operation, the manometer reading should be between 2.5 and 4.0 inches water column.

- Inspect the flames entering the drying chamber, block off the exhaust vent, and inspect the flame length. The flames should not be entering the drying chamber.

- Inspect for flashback into the burner, block off the exhaust vent, and inspect for flashback into the burner during ignition.

- Checking for power interruption, interrupt the power momentarily for two seconds. The motor should not continue to run when the power is restored. The gas valve should not open when the power is restored until the ignitor is re-energized.

- Check for carbon deposit, and inspect the heater housing and burner assembly for carbon deposits.

- Inspect the ignition time of the ignitor (glow-bar); the time required for the glow-bar to light the gas burner should be between 50 and 90 seconds depending on the model serviced. Check with the technical data sheet.

- Check the ignition power; this is the power required for the glow-bar to light the gas (300 to 850 watts depending on which glow-bar ignitor is used in the dryer).

- Check the static pressure of the air flow in the dryer. With the dryer exhaust 100 percent blocked, measure the static pressure with a manometer. The minimum reading should be 1.6-inch water column.

Principles of Operation

The clothes are placed into the dryer according to the manufacturer's recommendation for proper loading. Next, the proper cycle is selected, and the dryer start button is pressed. The combination of the timer, the switches, thermistors, and/or the thermostats regulates the air temperature within the drum and the duration of the drying cycle. During the drying cycle, room air is pulled into the dryer drum from the lower rear (or sometimes the front) of the cabinet, depending on which model the consumer owns (Figures 21-3 and 21-4). The air is pulled through the gas burner assembly (combustion chamber), the drum, the lint screen, down through the lint chute, and through the fan housing. It is then pushed out of the exhaust duct. The drive motor, blower wheel, belt, and pulleys cause the drum to turn in a clockwise rotation at 40 to 60 RPM, and the air to move through the dryer. The belt wraps around the drum, motor pulley, and idler pulley. The blower wheel is secured to one end of the drive motor shaft, or there may be a separate motor to turn the blower wheel. As the drive motor turns, the drum rotates, moving the clothes. At the same time, the blower wheel turns, moving the air. The gas burner assembly will cycle on and off, according to the temperature selected.

Functions and Cycles

Gas automatic dryers use three basic functions to operate:

- Heat is supplied by a gas burner.
- Air is drawn into the dryer. It is heated and circulated through the tumbling clothes. Then the warm, moisture-laden air is drawn through the lint screen and is vented through the duct system to the outside.
- Tumbling of the clothes is accomplished with a motor that drives a belt, which rotates the drum.

FIGURE 21-3

Typical air flow pattern in an automatic gas dryer.

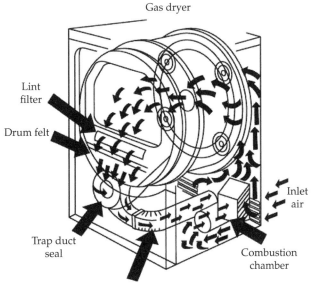

Gas dryer

Lint filter

Drum felt

Trap duct seal

Exhaust air

Inlet air

Combustion chamber

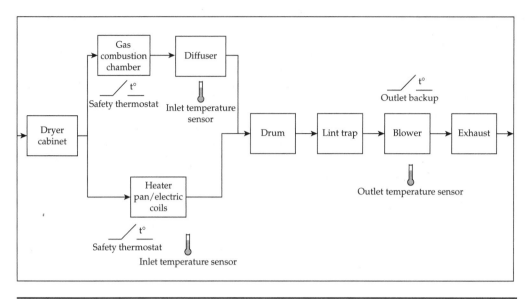

Figure 21-4   Gas dryer air flow system.

The cycles of a gas automatic dryer are as follows:

- **Timed dry cycle**   The timed dry cycle is controlled by the amount of time selected on the timer. The temperature in the dryer is controlled by a thermostat or a thermistor, which turns the gas burner on and off throughout the timed cycle.

- **Automatic dry cycle**   The automatic dry cycle is not controlled by the timer. This cycle is controlled by the cycling thermostats or thermistors. Heat is supplied to dry the clothes, and it will continue until the temperature in the drum reaches the selected cutout setting of the thermostat or thermistor. When the thermostat or thermistor is satisfied, the gas burner shuts off and the timer motor is activated. However, certain variables can control the cycle, which will cause the cycling of the thermostat or thermistor before the end of the cycle. *(You may want to check the service manual for the model you are servicing.)*

- **Permanent press and knit cycles**   The permanent press and knit cycles are controlled by the amount of time selected on the timer. The temperatures of these cycles are controlled by the temperature rating of the cycling thermostats or by the thermistors that are located within the cabinet, on the exhaust duct, and in the air supply. On some models, the user can select the desired type of heat setting with the temperature selector switch, located on the control panel.

- **Air dry**   The air-dry cycle is controlled by the amount of time selected on the timer. This cycle uses the air to dry the clothes. The gas burner is not used at all during this cycle.

Electrical Requirements and Tests for Gas Dryers

The dryer must be grounded. In the event of a malfunction, grounding will reduce the risk of electrical shock by a path of least resistance for electrical current. The dryer is equipped with a 120-volt power supply cord having an equipment-grounding conductor and a grounding three-prong plug. The plug must be plugged into a three-prong receptacle (120 volts) that is properly grounded in accordance with all local codes and ordinances (Figure 21-5). Never remove the grounding prong from the power cord.

While performing a service call on a gas dryer the following electrical tests should be performed:

- **Continuity**   Use your multimeter and set it on the ohms scale. When measuring resistance between the terminal block ground and any exposed unpainted metal cabinet component, including control shafts and switches, there should be no more than 0.1 ohms resistance reading.

- **Motor power**   Perform an audit on the motor (power required after 20 seconds with no load, and the motor is turning in the clockwise direction, and the electric heater is turned off). The reading should be between 200 and 280 watts. Next, perform a line test (power required after 1 to 5 seconds with no clothes in the dryer, the motor is turning in a clockwise direction, and the electric heater is turned off). The reading should be between 210 and 290 watts. If you do not have a wattmeter, use the Ohm's Law formula to convert to watts.

- **Low voltage start**   The minimum voltage needed to start the dryer motor is 100 VAC, 60 Hz (200 volts for 230 VAC, 50 Hz dryers). Place about 25 pounds of clothing with 100 percent moisture into the dryer and set the dryer at its maximum heat setting. Turn on the dryer and measure the voltage at the start of the dryer cycle. The voltage should not be below the listed voltage for any part of the cycle.

- **Drum temperatures**   The following test will be run with the exhaust duct vent disconnected from the dryer. Take your multimeter temperature thermal couple and place it in the lint grill about 1 to 2 inches. You will need to set the dryer timer as

FIGURE 21-5
This figure illustrates the proper way to plug in a gas dryer power cord.

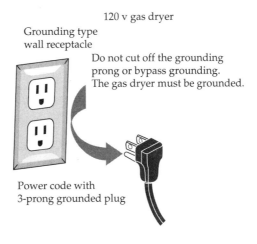

120 v gas dryer

Grounding type
wall receptacle

Do not cut off the grounding
prong or bypass grounding.
The gas dryer must be grounded.

Power code with
3-prong grounded plug

needed for this test. The recordable temperature is to be the maximum temperature reading after the third cycle when the heater cycles off. There should not be any clothes in the dryer for this test. Each time the thermostat or thermistor cycles open, there should be equal to or higher than opening temperature for the next lower temperature setting.

- Maximum heating temperature when the heater cycles off is 120 to 180 degrees Fahrenheit.

- Medium heating temperature when the heater cycles off is 105 to 145 degrees Fahrenheit.

- Low heating temperature when the heater cycles off is 95 to 130 degrees Fahrenheit.

Gas Dryer Maintenance

Automatic dryers must be cleaned periodically. The excess lint must be removed to prevent the possibility of a fire and the possibility of the dryer not functioning properly. Most models have a service panel on the front of the dryer for accessibility. On other models, the front panel will have to be removed. The outside of the cabinet should be wiped with a damp cloth. It is recommended that the components be inspected for wear and tear; if repairs are needed, they should be made as soon as possible. The gas burner assembly must be inspected and cleaned yearly. The dryer gas line must also be inspected and checked for gas leaks yearly—sooner if you suspect a gas leak.

Vent System

A clogged or restricted vent system can cause the dryer not to function properly. The following symptoms can occur:

- The clothes take too long to dry.
- The clothes are too hot at the end of the cycle.
- Unusual noises coming from the dryer.
- At the end of the cycle, the clothes are damp and hot.
- The sensors and thermostats fail to operate properly.
- Complications in the burner assembly.

Proper Exhausting of the Dryer

Proper exhausting instructions for the model being installed are available through the manufacturer. Each manufacturer has its own specifications for the size and the length of the ductwork needed to run its dryer properly. The maximum length of the exhaust system depends upon the type of duct, the number of elbows, and the type of exhaust hood used. Figure 20-8 illustrates a typical dryer exhaust installation.

The following guide is recommended for the exhausting of a dryer:

- Keep the duct length as short as possible.
- Keep the number of elbows to a minimum to minimize the air resistance.
- Never reduce the diameter of round ductwork below 4 inches.

- Install all exhaust hoods at least 12 inches above ground level.
- The exhaust duct and exhaust hood should be inspected periodically and cleaned, if necessary.
- All duct joints should be taped. Never use screws to join the duct joints together. Screws protruding into the duct will cause lint buildup, and this will eventually clog the duct.
- Never exhaust the dryer into any wall, ceiling, attic, or under a building.
- Accumulated lint could become a fire hazard, and moisture could cause damage.
- If the exhaust duct is adjacent to an air conditioning duct, the exhaust duct must be insulated to prevent moisture buildup.

Exhaust Ducting

The more that the dryer is used, the more the exhaust ducts and vent hood must be inspected to prevent duct blockages and lint fires. The exhaust ducting and vent hood must be cleaned at least once a year for peak performance in drying.

When installing ductwork, separate all turns by at least 4 feet of straight duct, including the distance between the last turn and the dampened wall hood. If a turn is 45° or less, it can be ignored in figuring out duct distance. If you have two 45° turns, count it as one 90° turn. Also, each turn over 45° should be counted as one 90° turn (see Figures 20-9, 20-10, 20-11, 20-12, and 20-13).

When sealing the joints of the exhaust ducts with duct tape, the male end of each section of duct must point away from the dryer to prevent lint buildup. Also, never use screws to connect the ducts together; this will cause lint buildup. Any ductwork that runs through an unheated area or is near air conditioning must be insulated to prevent condensation and lint buildup (check with your local codes and ordinances).

Safety note: All exhaust ductwork must exhaust to the outside of a building to prevent lint buildup and fire. Do not terminate exhaust ductwork in a chimney, any gas vent, under enclosed floor (crawl space), common duct with a kitchen exhaust, or into an attic because the lint accumulation will cause a fire. Finally, never block incoming or exhausted air supplies; this will cause the dryer to run inefficiently and may cause a fire.

Step-by-Step Troubleshooting by Symptom Diagnosis

In the course of servicing an appliance, don't overlook the simple things that might be causing the problem. Step-by-step troubleshooting by symptom diagnosis is based on diagnosing malfunctions, with possible causes arranged into categories relating to the operation of the dryer. This section is intended only to serve as a checklist to aid you in diagnosing a problem. Look at the symptom that best describes the problem that you are experiencing with the dryer, and then correct the problem.

Gas Odors

- Check the pilot light and gas valve.
- Check for a gas leak in the supply line.
- Check the ventilation system.

Dryer Will Not Run

- Do you have 120 volts at the dryer?
- Test the door switch for continuity of the switch contacts. In addition, make sure the dryer door closes properly.
- Test for continuity of the motor windings. Also check for a grounded motor.
- Test the timer for continuity of the switch contacts.
- Test the "push to start" switch for continuity of the switch contacts.
- Check for broken or loose wiring. Also check the wire terminal connections that connect to the different components.
- Check the thermal fuse.
- On electronic models, check for unusual display readouts and/or error codes.

No Heat

- Check the controls setting.
- Do you have 120 volts at the dryer?
- Check the gas valve. Is the gas turned on?
- Check the thermal fuse.
- Check the thermostats.
- Check the ignitor. Does it light up?
- Check the flame sensor.
- Check for proper ventilation.
- On electronic models, check for unusual display readouts and/or error codes.

Dryer Will Not Turn Off

- Check the control settings. Does the timer advance?
- Check the electronics. Run the self-diagnostic test mode.
- Check the thermistor. Is it open or shorted?
- Check the thermostats. Check continuity.
- On electronic models, check for unusual display readouts and/or error codes.

Dryer Runs and Heats, but the Clothes Won't Dry

- Check for a defective thermostat or thermistor. Use a thermometer to check the duct temperatures during the cycling of the dryer.
- Check for a loose pulley or blower wheel. Inspect and tighten the blower wheel.
- Check for restricted air flow in the dryer.
- Check the vent system. Is the vent hose behind the dryer kinked?
- Check the thermostats and thermistors.

- Check the lint filter. Does the lint filter need cleaning?
- On electronic models, check for unusual display readouts and/or error codes.

The Burner Will Not Ignite

- Make sure that the gas supply to the dryer is turned on.
- Check the gas line for debris.
- Check the line to the burner for a blockage. If blocked, locate the source and correct.
- Test the value of the thermocouple; replace if necessary.
- Test the thermostat; replace if necessary.
- Check the installation and fresh air supply.
- Check the ignition components.
- On electronic models, check for unusual display readouts and/or error codes.

The Burner Flame Is Too High

- Check the dryer installation. Is the orifice too large? You can change the orifice and retest the burner.
- Check for sufficient secondary air supply.
- Check the venting system.
- Check the burner air shutter. Is it adjusted correctly?

The Pilot Flame Is Too Small

- Check the pilot line and orifice for debris.
- Check the gas pressure to the dryer.
- Test the gas pressure at the regulator.

A Yellow Burner Flame

- Check for sufficient secondary air supply.
- Check the gas pressure.
- Check the venting system.
- Check the burner line for debris.
- Check the dryer installation.
- Check the burner orifice for obstruction.

The Burner Flame Floats and Lifts Off the Ports

- Check the dryer installation. Is the orifice too large? You can change the orifice and retest the burner.
- Check for high gas pressure.
- Check the vent system and burner. Locate the source and correct the problem.
- Check for drafts that can blow the flame off the burner.

The Formation of Soot (Smoke and Carbon)

- Check for sufficient secondary air supply.
- Check the ventilation system.
- Check for a yellow burner flame.
- Check for low gas pressure to the dryer.
- Is the dryer installed in a confined space?

Combustion Odors

- Check for sufficient secondary air.
- Check the ventilation system and the burner.
- Is the dryer installed in a confined space?

For additional symptoms, see Chapter 20.

Repair Procedures

Each repair procedure is a complete inspection and repair process for a single dryer component, containing the information you need to test a component that might be faulty and to replace it, if necessary.

Any person who cannot use basic tools should *not* attempt to install, maintain, or repair any electric dryer. Any improper installation, preventative maintenance, or repairs will create a risk of personal injury, as well as property damage. Call the service manager if installation, preventative maintenance, or the repair procedure is not fully understood.

Electronic Control Board and User Interface Controls

On some models the electronic control board and user interface controls replace the electromechanical timer and rotary selection switches (see Figures 21-1 and 20-18). The electronic control board is located behind the control panel.

The typical complaints associated with the electronic control board or the user interface controls are:

- The dryer won't run or power up.
- Unable to program the dryer.
- The display board will not display anything.
- One or more key pads will not accept commands.
- Unusual display readouts and/or error codes.

To prevent electrostatic discharge (ESD) from damaging expensive electronic components, follow the steps in Chapters 6 and 11.

To handle these problems, perform the following steps:

1. **Verify the complaint.** Verify the complaint by operating the dryer. Turn off the electricity to the appliance and wait for two minutes before turning it back on. If a

fault code appears, look up the code. If the dryer will not power up, locate the technical data sheet behind the control panel or for diagnostics information. On some models you will need the actual service manual for the model you are working on to properly diagnose the dryer. The service manual will assist you in properly placing the dryer in the service test mode for testing the dryer functions.

2. **Check for external factors.** You must check for external factors not associated with the appliance. Are there 120 volts of electricity to the dryer? The voltage at the receptacle is between 108 volts and 132 volts during a load on the circuit. Do you have the correct polarity? (See Chapter 6.)

3. **Disconnect the electricity.** Before working on the dryer, disconnect the electricity. This can be done by pulling the plug out of the wall receptacle. Or disconnect the electricity at the fuse panel or circuit breaker panel. Turn off the electricity.

WARNING *Some diagnostic tests will require you to test the components with the power turned on. When you disassemble the control panel, you can position it in such a way that the wiring will not make contact with metal. This act will allow you to test the components without electrical mishaps.*

4. **Shut off the gas supply.** Before you begin servicing any gas components, shut off the supply of gas to the appliance. The shutoff valve should be within 6 feet of the gas appliance.

5. **Remove the console panel to gain access.** Begin by removing the screws from the dryer top (for front-loading models only) to gain access to the electronic control board. Remove the console panel screws to gain access to the user interface controls (see Figure 20-18). On top-loading models, the console will roll upwards or toward you after you remove the console screws.

6. **Test the electronic control board and user interface controls.** If you are able to run the dryer diagnostic test mode, check the different functions of the dryer. Use the technical data sheet for the model you are servicing to locate the test points from the wiring schematic. Check all wiring connections and wiring. Using the technical data sheet, test the electronic control or user interface controls, input voltages, and output voltages. On some models, fuses are soldered to the printed circuit board (PCB). These fuses must be tested first before condemning the component.

7. **Remove the electronic control board and user interface control.** To remove the defective component, remove the screws that secure the board to the control panel or dryer frame. Disconnect the connectors from the electronic control board or user interface control.

8. **Install the new component.** To install a new electronic control board or user interface control, read the parts data sheet that comes with the part for the proper installation process, and just reverse the disassembly procedure and reassemble. Reinstall the console panel, and restore the electricity and gas supply to the dryer. Make sure that the dryer is not in the service mode. Test the dryer operation. Make sure to take the dryer out of the service test mode when the repair is completed.

Dryer Motor

The dryer motor on most models is a dual-shaft, single-speed, ¼-horsepower, 1725 RPM motor with an automatic thermal reset protector. One side of the motor shaft is threaded to hold the blower wheel, and the opposite end of the shaft holds the belt pulley, which is pressed onto the shaft. Inside the motor is a centrifugal switch, which serves three purposes. It disengages the motor start winding when the motor reaches 75 percent of its rated speed, it engages the run winding, and it also closes the circuit for the heater element. On other models, there are dual motors (see Figure 20-2): one is dedicated to the drum and the other one is dedicated to the blower assembly. The purpose of the dedicated motor for the blower assembly is to be able to alter the blower wheel speed whenever necessary to optimize air flow within the dryer.

The typical complaints associated with motor failure are:

- Fuse is blown or the circuit breaker trips.
- Motor will not start; it only hums.
- The dryer will not run.
- On electronic models, check for unusual display readouts and/or error codes.

To handle these problems, perform the following steps:

1. **Verify the complaint.** Verify the complaint by operating the dryer through its cycles. Listen carefully, and you will hear if there are any unusual noises or if the circuit breaker trips. On electronic models, turn off the electricity to the appliance and wait for two minutes before turning it back on. If a fault code appears, look up the code. If the dryer will not power up, locate the technical data sheet behind the control panel or for diagnostics information. On some models you will need the actual service manual for the model you are working on to properly diagnose the dryer. The service manual will assist you in properly placing the dryer in the service test mode for testing the dryer functions.

2. **Check for external factors.** You must check for external factors not associated with the appliance. Is the appliance installed properly? Does the appliance have the correct voltage? The voltage at the receptacle is between 108 volts and 132 volts during a load on the circuit. Do you have the correct polarity? (See Chapter 6.)

3. **Disconnect the electricity.** Before working on the dryer, disconnect the electricity. This can be done by pulling the plug from the electrical outlet. Or disconnect the electricity at the fuse panel or at the circuit breaker panel. Turn off the electricity.

WARNING *Some diagnostic tests will require you to test the components with the power turned on. When you disassemble the control panel, you can position it in such a way that the wiring will not make contact with metal. This act will allow you to test the components without electrical mishaps.*

4. **Think safety. Shut off the gas supply.** As a precaution before you begin servicing the dryer, shut off the supply of gas to the dryer. The shutoff valve should be within 6 feet of the gas appliance.

5. **Gain access to the motor.** In order to gain access to the motor, the top must be raised (see Figure 20-19) by removing the screws from the lint screen slot. Insert a putty knife about 2 inches from each corner; then disengage the retaining clips and lift the top. On some models, the top is held down with screws (see inset in Figure 20-19). Remove the screws and lift the top. Now the front panel of the dryer must be removed (see Figure 20-20). To remove the top panel, insert a putty knife and disengage the retaining clip (see Figure 20-20a).

   On some models, the top panel is held in place with two screws. Figure 20-20b shows the removal of the screws that hold the lower part of the front panel in place. Next, remove the screws that hold the upper part of the front panel in place (see Figure 20-20c), and then disconnect the door switch wires (label them). With the front panel out of the way, you can now disconnect the drive belt (see Figure 20-21). Push on the idler pulley to release the tension from the drive belt, and remove the belt from the motor pulley. Grab the drum and remove it from the cabinet (see Figure 20-22). On some models, you will have to remove the back panel because the drum comes out through the rear of the cabinet.

6. **Disconnect the motor wire leads.** Disconnect the motor wire leads from the wiring harness. Set the ohmmeter on R × 1. Figure 20-23 illustrates testing the motor windings and the centrifugal switch for continuity. When testing for resistance on the motor, test from the common wire lead to the run winding. Then test for resistance from the common wire lead to the start winding. Next, test for resistance from the start winding to the run winding. To test for a grounded winding in the motor, take the ohmmeter probes and test from each motor wire lead terminal to the motor housing. The ohmmeter will indicate continuity if the windings are grounded. If the ohmmeter reading shows no resistance between the motor windings, then replace the motor. If the motor checks out okay, then check the timer.

7. **Remove the motor.** To remove this type of motor, you must first disconnect the blower assembly by holding the motor shaft stationary and then turning the blower wheel to remove it from the rear of the motor shaft (see Figure 20-24). Then remove the spring clamps that hold the motor in the motor bracket (see Figure 20-25). On some models, the motor pulley must be removed (new motors come without the pulley attached) by loosening the allen set screw.

8. **Install the new motor.** To install the new motor (see Figure 20-26), just reverse the disassembly procedure and reassemble. Then reassemble the dryer in the reverse order of its disassembly. Restore the electricity and gas supply to the dryer and test the motor. Make sure to take the dryer out of the service test mode when the repair is completed.

Drive Belt

The drive belt extends from the motor pulley, past the idler pulley, and around the perimeter of the dryer drum. The tension on the belt is maintained by the idler pulley and driven by a pulley attached to one end of the motor shaft.

The typical complaints associated with belt failure are:

- The drum will not turn.
- Dryer motor spins freely.

- It smells like something is burning.
- On electronic models, check for unusual display readouts and/or error codes.

To handle these problems, perform the following steps:

1. **Verify the complaint.** Verify the complaint by operating the dryer through its cycles. Listen carefully, and you will hear if there are any unusual noises. Then, with the door open, press the door switch and start the dryer. The drum should rotate. On electronic models, turn off the electricity to the appliance and wait for two minutes before turning it back on. If a fault code appears, look up the code. If the dryer will not power up, locate the technical data sheet behind the control panel or for diagnostics information. On some models you will need the actual service manual for the model you are working on to properly diagnose the dryer. The service manual will assist you in properly placing the dryer in the service test mode for testing the dryer functions.

2. **Check for external factors.** You must check for external factors not associated with the appliance. Is the appliance installed properly? Does the appliance have the correct voltage? The voltage at the receptacle is between 108 volts and 132 volts during a load on the circuit. Do you have the correct polarity? (See Chapter 6.) Check for foreign objects lodged between the drum and bulkhead, etc.

3. **Disconnect the electricity.** Before working on the dryer, disconnect the electricity. This can be done by pulling the plug from the electrical outlet. Or disconnect the electricity at the fuse panel or at the circuit breaker panel. Turn off the electricity.

4. **Think safety. Shut off the gas supply.** As a precaution before you begin servicing the dryer, shut off the supply of gas to the dryer. The shutoff valve should be within 6 feet of the gas appliance.

WARNING *Some diagnostic tests will require you to test the components with the power turned on. When you disassemble the control panel, you can position it in such a way that the wiring will not make contact with metal. This act will allow you to test the components without electrical mishaps.*

5. **Gain access to the drive belt.** To gain access to the drive belt, the top must be raised (see Figure 20-19) by removing the screws from the lint screen slot. Then insert a putty knife about 2 inches from each corner, disengage the retaining clips, and lift the top. On some models, the top is held down with screws (see the inset in Figure 20-19). Remove the screws and lift the top. Now the front panel of the dryer must be removed (see Figure 20-20). To remove the top panel, insert a putty knife and disengage the retaining clip (see Figure 20-20a). As shown in Figure 20-20b, remove the screws that hold the lower part of the front panel in place. Next, remove the screws that hold the upper part of the front panel in place (see Figure 20-20c), and then disconnect the door switch wires.

6. **Remove the drive belt.** To remove the drive belt on this type of dryer, you can disconnect the belt (see Figure 20-21). Push on the idler pulley to release the tension from the drive belt. Now remove the belt from the motor pulley and from around the drum (see Figure 20-27). If the drive belt is broken, just remove the belt.

7. **Install a new drive belt.** To install the new drive belt, just reverse the disassembly procedure, and reassemble. Then reassemble the dryer in the reverse order of its disassembly. Restore the electricity and gas supply to the dryer and test. Make sure to take the dryer out of the service test mode when the repair is completed.

Door Switch

The door switch is a normally open switch wired in series with the dryer motor. When the dryer door is closed, the door switch contacts will close, completing the circuit through the run winding of the drive motor. The door switch is located behind the front panel.

The typical complaints associated with switch failure are:

- Dryer will not operate at all.
- Dryer light does not work when the door is open.
- On electronic models, check for unusual display readouts and/or error codes.

To handle these problems, perform the following steps:

1. **Verify the complaint.** Verify the complaint by operating the dryer through its cycles. Open the door to see if the light is working. On electronic models, turn off the electricity to the appliance and wait for two minutes before turning it back on. If a fault code appears, look up the code. If the dryer will not power up, locate the technical data sheet behind the control panel or for diagnostics information. On some models you will need the actual service manual for the model you are working on to properly diagnose the dryer. The service manual will assist you in properly placing the dryer in the service test mode for testing the dryer functions.

2. **Check for external factors.** You must check for external factors not associated with the appliance. Is the appliance installed properly? Does the dryer have the correct voltage supply? The voltage at the receptacle is between 108 volts and 132 volts during a load on the circuit. Do you have the correct polarity? (See Chapter 6.)

3. **Disconnect the electricity.** Before working on the dryer, disconnect the electricity. This can be done by pulling the plug from the electrical outlet. Or disconnect the electricity at the fuse panel or at the circuit breaker panel. Turn off the electricity.

4. **Think safety. Shut off the gas supply.** As a precaution before you begin servicing the dryer, shut off the supply of gas to the dryer. The shutoff valve should be within 6 feet of the gas appliance.

***

WARNING *Some diagnostic tests will require you to test the components with the power turned on. When you disassemble the control panel, you can position it in such a way that the wiring will not make contact with metal. This act will allow you to test the components without electrical mishaps.*

5. **Gain access to the door switch.** You must gain access to the door switch. On this model, the top must be raised (see Figure 20-19) by removing the screws from the lint screen slot. Then insert a putty knife about 2 inches from each corner, disengage the retaining clips, and lift the top. On some models, the top is held down with screws (see the inset in Figure 20-19). Remove the screws and lift the top.

6. **Test the door switch.** The door switch is located in one of the upper corners of the inside of the front panel (see Figure 20-34a). Once you have located the door switch, disconnect the wires from the terminals. Set the ohmmeter on R × 1, attach the probes of the ohmmeter to the terminals of the door switch, and then close the door. With the door closed, there should be continuity. With the door open, the ohmmeter should not show continuity. Some dryer models have a light inside the drum. This light circuit is also part of the door switch circuitry. Three terminals are located on the door switch. Take your ohmmeter probe and place it on the common terminal of the door switch. To locate the common terminal, read the wiring diagram. It will indicate which terminal is the common, which is the light, and which is for the motor circuit. Take your probe and place it on the switch terminal (the light circuit); then close the dryer door. The reading should indicate no continuity. Now open the dryer door—you should have a continuity reading on the meter.

7. **Remove the door switch.** To remove the lever-type door switch (see Figure 20-34b), remove the screws that hold the switch in place. Now lift the switch out from behind the front panel. To remove the cylindrical type of door switch (see Figure 20-34c), squeeze the retaining clips on the back side of the front panel, and pull out the switch. To remove the hinge type of door switch (see Figure 20-34d), the front panel must be removed. Next, remove the screws that hold the switch assembly in place.

8. **Install the new door switch.** To install the new door switch, reverse the disassembly procedure, and reassemble. Reassemble the dryer, and install the wires on the new switch terminals. Restore the electricity and gas supply to the dryer and test. Make sure to take the dryer out of the service test mode when the repair is completed.

Thermostat

Thermostats can be mounted on the heater assembly (see Figure 20-2 and Figure 20-29) and/or on the blower assembly (see Figures 20-2 and 20-35), or in the inlet filter screen (see Figure 20-37). The thermostats monitor the air temperatures within the dryer drum, exhaust vent, and the incoming air supply. The thermostats are bimetal switches that are automatic reset, one-time use, or manual reset.

The typical complaints associated with thermostat failure are:

- The clothes are not drying.
- The dryer will not shut off.
- The dryer will not heat at all.
- The drying temperature is too high.
- Moisture retention (of fabrics) is unsatisfactory.
- On electronic models, an error code appears.

To handle these problems, perform the following steps:

1. **Verify the complaint.** Verify the complaint by operating the dryer through its cycles. On electronic models, turn off the electricity to the appliance and wait for two minutes before turning it back on. If a fault code appears, look up the code. If the dryer will not power up, locate the technical data sheet behind the control panel

or for diagnostics information. On some models you will need the actual service manual for the model you are working on to properly diagnose the dryer. The service manual will assist you in properly placing the dryer in the service test mode for testing the dryer functions.

2. **Check for external factors.** You must check for external factors not associated with the appliance. Is the appliance installed properly? Does the dryer have the correct voltage supply? The voltage at the receptacle is between 108 volts and 132 volts during a load on the circuit. Do you have the correct polarity? (See Chapter 6.) Is the exhaust vent blocked?

3. **Disconnect the electricity.** Before working on the dryer, disconnect the electricity. This can be done by pulling the plug from the electrical outlet. Or disconnect the electricity at the fuse panel or at the circuit breaker panel. Turn off the electricity.

4. **Think safety. Shut off the gas supply.** As a precaution before you begin servicing the dryer, shut off the supply of gas to the dryer. The shutoff valve should be within 6 feet of the gas appliance.

WARNING *Some diagnostic tests will require you to test the components with the power turned on. When you disassemble the control panel, you can position it in such a way that the wiring will not make contact with metal. This act will allow you to test the components without electrical mishaps.*

5. **Gain access to the thermostat.** You must gain access to the thermostat through the back on this model. Remove the screws from the back panel (see Figure 20-30). On other models you will have to remove the drum to gain access to the thermostats.

6. **Test the thermostat.** To test the thermostat for continuity, remove the wires from the thermostat terminals. With a continuity tester or an ohmmeter, test the thermostat for continuity (see Figure 20-36). Do this to all of the thermostats in the dryer. On some models, the thermostats might be located on the heater housing, behind the drum, or in the lint screen opening (see Figure 20-37). If they all check out okay, then reassemble the dryer and test for temperature operation. *Note:* Do not reinstall the back panel at this point. Take a piece of paper and write down the temperature ratings of the thermostats. These ratings are printed on the thermostats (L140, L290, etc.) or on the wiring schematic. To test for temperature operation, you will need a voltmeter and a temperature tester.

First, set up the instruments for testing the thermostatic operation. Take the temperature tester thermocouple lead and insert it between the thermostat mounting ear (see Figure 20-38) and the plate against which it mounts. Then connect the voltmeter probes across the thermostat terminals. Use alligator clips attached to the probe tips. This will allow you freedom of movement. Do not disconnect the wires from the thermostat. Set the voltmeter range to AC voltage and the selector switch on 300 volts. Remember: If there is more than one thermostat (in series with other thermostats), the thermostats not under test must be electrically isolated and jumped out with an insulated jumper wire with alligator clips attached. Review the wiring schematic to determine which wires to remove and which ones to isolate and jump out.

With the test meters in place, you are now ready to remove the exhaust vent duct from the dryer. Seal off 75 percent of the exhaust opening from the dryer (see Figure 20-39). This will simulate a load of clothing in the dryer. On some models, you do not have to block off the exhaust opening. Check the technical data sheet or the service manual for the model you are servicing.

Restore the electricity and gas supply to the dryer. This test requires that the electricity be turned on for its duration. Always be cautious when working with live wires. Avoid getting shocked. Set the controls on the dryer to operate at a high heat. Turn on the dryer and let it cycle. When the voltmeter is reading voltage, the thermostat has opened. When there is no voltage reading on the voltmeter, the thermostat is closed. As the dryer is cycling via the thermostat, record the temperature.

The thermostat should open at the preset temperature listed on the thermostat. Table 20-1 illustrates the types of thermostats, along with their opening and closing temperature settings. The temperature range of the thermostat should be within ±10 percent of the printed setting. If not, replace the thermostat.

Caution *Turn off the electricity and gas supply before replacing any parts of the dryer.*

On some models, the thermostat will have four wires attached to it. Two wires are for the bimetal switch contacts and the other two wires are for the bias heater. The purpose of the bias heater is to apply a small amount of heat to the bimetal switch contacts when the timer or fabric temperature switch is set to a lower temperature setting. This will allow the bimetal switch contacts to open up sooner to maintain a lower drum temperature. To test the bias heater, remove the wires from the thermostat, set your multimeter to the ohms scale, and test across the bias heater terminals. The resistance of the bias heater should be between 7000 and 28,000 ohms.

7. **Replace the thermostat.** Remove the screws that hold the thermostat in place. Replace the thermostat with an exact replacement with the same temperature rating. Reconnect the wires to their correct terminal positions. Then reverse the disassembly procedure to reassemble the dryer, and test the thermostat.

8. **Test the new thermostats.** To test the new thermostats, repeat step 6. Restore the electricity and gas supply to the dryer and test. Make sure to take the dryer out of the service test mode when the repair is completed.

Thermistor

The thermistor (see Figure 20-2) monitors the incoming and outgoing air temperatures and relays that information back to the electronic control board. Check the technical data sheet for the location of the thermistors.

The typical complaints associated with thermistor failure are:

• The clothes are not drying.

• The dryer will not shut off.

• The dryer will not heat at all.

- The drying temperature is too high.
- Moisture retention (of fabrics) is unsatisfactory.
- On electronic models, an error code appears.

To handle these problems, perform the following steps:

1. **Verify the complaint.** Verify the complaint by operating the dryer through its cycles. On electronic models, turn off the electricity to the appliance and wait for two minutes before turning it back on. If a fault code appears, look up the code. If the dryer will not power up, locate the technical data sheet behind the control panel or for diagnostics information. On some models you will need the actual service manual for the model you are working on to properly diagnose the dryer. The service manual will assist you in properly placing the dryer in the service test mode for testing the dryer functions.

2. **Check for external factors.** You must check for external factors not associated with the appliance. Is the appliance installed properly? Does the dryer have the correct voltage supply? The voltage at the receptacle is between 108 volts and 132 volts during a load on the circuit. Do you have the correct polarity? (See Chapter 6.) Is the exhaust vent blocked?

3. **Disconnect the electricity.** Before working on the dryer, disconnect the electricity. This can be done by pulling the plug from the electrical outlet. Or disconnect the electricity at the fuse panel or at the circuit breaker panel. Turn off the electricity.

4. **Think safety. Shut off the gas supply.** As a precaution before you begin servicing the dryer, shut off the supply of gas to the dryer. The shutoff valve should be within 6 feet of the gas appliance.

WARNING *Some diagnostic tests will require you to test the components with the power turned on. When you disassemble the control panel, you can position it in such a way that the wiring will not make contact with metal. This act will allow you to test the components without electrical mishaps.*

5. **Gain access to the thermistor.** To gain access to the thermistor on this model, you will have to remove the top panel, control panel (see Figure 20-40), front panel (see Figure 20-41), and dryer drum (see Figure 20-2).

6. **Test the thermistor.** To test the thermistor for resistance, remove the wires from the thermistor terminals. Set your multimeter on the ohms scale and place the test leads on the thermistor's terminals. Match the reading to the technical data sheet or the service manual values. If you are reading an open or infinity, replace the thermistor. Another way to test the thermistor is by placing the dryer into the service mode. On some models, the thermistor might be located on the heater housing, behind the drum, or in the lint screen opening (see Figures 20-2 and 20-37). If they all check out okay, then reassemble the dryer and test for temperature operation.

7. **Replace the thermistor.** Remove the screws that hold the thermistor in place. Replace the thermistor with an exact replacement with the same temperature rating.

Reconnect the wires to their correct terminal positions. Then reverse the disassembly procedure to reassemble the dryer, and test the thermistor.

8. **Test the new thermistor.** To test the new thermostats, repeat step 6. Restore the electricity and gas supply to the dryer and test. Make sure to take the dryer out of the service test mode when the repair is completed.

Start Switch

The start switch is a rotary or push button momentary contact switch used to start the dryer and to energize the start winding in the drive motor. This switch is located in the control panel on the dryer. The dryer door must be closed for the dryer to operate. If and when the dryer door is opened, the start switch must be turned on to restart the dryer operation.

The typical complaints associated with the start switch are:

- The dryer will not start.

- On electronic models, check for unusual display readouts and/or error codes.

To handle these problems, perform the following steps:

1. **Verify the complaint.** Verify the complaint by operating the dryer. Before you change the start switch, check the other components that are in the start circuit. On electronic models, turn off the electricity to the appliance and wait for two minutes before turning it back on. If a fault code appears, look up the code. If the dryer will not power up, locate the technical data sheet behind the control panel for the diagnostics information. On some models you will need the actual service manual for the model you are working on to properly diagnose the dryer. The service manual will assist you in properly placing the dryer in the service test mode for testing the dryer functions.

2. **Check for external factors.** You must check for external factors not associated with the appliance. Is the appliance installed properly? Does it have the correct voltage? The voltage at the receptacle is between 108 volts and 132 volts during a load on the circuit. Do you have the correct polarity? (See Chapter 6.) Check the fuses or circuit breaker.

3. **Disconnect the electricity.** Before working on the dryer, disconnect the electricity. This can be done by pulling the plug from the electrical outlet. Or disconnect the electricity at the fuse panel or at the circuit breaker panel. Turn off the electricity.

4. **Think safety. Shut off the gas supply.** As a precaution before you begin servicing the dryer, shut off the supply of gas to the dryer. The shutoff valve should be within 6 feet of the gas appliance.

WARNING *Some diagnostic tests will require you to test the components with the power turned on. When you disassemble the control panel, you can position it in such a way that the wiring will not make contact with metal. This act will allow you to test the components without electrical mishaps.*

5. **Remove the console panel to gain access.** Begin by removing the console panel to gain access to the start switch. With this type of dryer, remove the screws from the console (see Figure 20-14a). On some models, the console will be able to lie flat (see Figure 20-14b).

6. **Test the start switch.** Remove the wires from the start switch terminals. Set the ohmmeter on R × 1. Begin testing the start switch for continuity by placing one probe of the ohmmeter on the common terminal (C) of the switch; then connect the other probe to the normally open (NO) terminal. There should be no continuity. With the ohmmeter probes attached to the switch, press the start switch button. You should have continuity (see Figure 20-42). If the switch fails the test, replace it.

7. **Remove the start switch.** To remove the start switch, pull the knob off the switch stem, remove the start switch mounting screws, and remove the switch.

8. **Install a new start switch.** To install a new start switch, just reverse the disassembly procedure, and reassemble. Replace the wires on the switch. Reinstall the console panel, and restore the electricity and gas supply to the dryer and test. Make sure to take the dryer out of the service test mode when the repair is completed.

Temperature Selector Switch

The temperature selector switch is located in the console panel. This switch allows the consumer to select different temperature settings for drying.

The typical complaints associated with the temperature selector switch are:

- Inability to select a certain temperature setting.
- No heat.
- The switch is stuck.
- On electronic models, check for unusual display readouts and/or error codes.

To handle these problems, perform the following steps:

1. **Verify the complaint.** Verify the complaint by operating the dryer through its cycles. Before you change the temperature selector switch, check the other components controlled by this switch. On electronic models, turn off the electricity to the appliance and wait for two minutes before turning it back on. If a fault code appears, look up the code. If the dryer will not power up, locate the technical data sheet behind the control panel or for diagnostics information. On some models you will need the actual service manual for the model you are working on to properly diagnose the dryer. The service manual will assist you in properly placing the dryer in the service test mode for testing the dryer functions.

2. **Check for external factors.** You must check for external factors not associated with the appliance. Is the appliance installed properly? Does it have the correct voltage? The voltage at the receptacle is between 108 volts and 132 volts during a load on the circuit (see Chapter 6).

3. **Disconnect the electricity.** Before working on the dryer, disconnect the electricity to it. This can be done by pulling the plug from the electrical outlet. Or disconnect the electricity at the fuse panel or at the circuit breaker panel. Turn off the electricity.

4. **Think safety. Shut off the gas supply.** As a precaution before you begin servicing the dryer, shut off the supply of gas to the dryer. The shutoff valve should be within 6 feet of the gas appliance.

WARNING *Some diagnostic tests will require you to test the components with the power turned on. When you disassemble the control panel, you can position it in such a way that the wiring will not make contact with metal. This act will allow you to test the components without electrical mishaps.*

5. **Gain access to the temperature selector switch.** To access the temperature selector switch, remove the console panel. On this type of dryer, remove the screws in the console (see Figure 20-14a). On some models, the console will be able to lie flat (see Figure 20-14b). On other models, your access is through the rear panel on the console.

6. **Test the temperature selector switch.** To test the temperature selector switch, locate the selector switch circuit on the wiring diagram. Identify the terminals that are regulating the temperature setting to be tested. Set the ohmmeter on the R × 1 scale. Next, place the ohmmeter probes on those terminals. Then select that temperature setting by either rotating the dial or by depressing the proper button on the switch (see Figure 20-46). If the switch contacts are good, your meter will show continuity. Test all of the remaining temperature settings on the temperature selector switch. Remember to check the wiring diagram for the correct switch contact terminals (those that correspond to the setting that you are testing).

7. **Remove the temperature selector switch.** To remove the temperature selector switch, remove the screws that hold the switch to the console base, and remove the switch.

8. **Install the new temperature selector switch.** To install the new temperature selector switch, just reverse the disassembly procedure and reassemble it. Then reattach the wires to the switch terminals according to the wiring diagram. Reassemble the console panel. Be sure when reassembling the console panel that the wires do not become pinched between the console panel and the top of the dryer. Restore the electricity and gas supply to the dryer and test. Make sure to take the dryer out of the service test mode when the repair is completed.

Belt Switch

The belt switch is located on the drive motor baseplate (see Figure 20-47) and held in place with two Phillips-head screws. The belt switch is activated by the movement of the idler pulley assembly. If the belt breaks or comes off the pulley, the belt switch disconnects the power to the motor, shutting down the dryer.

The typical complaints associated with the idler pulley are:

- The dryer will not run.
- The clothes are wet.
- On electronic models, check for unusual display readouts and/or error codes.

To handle these problems, perform the following steps:

1. **Verify the complaint.** Verify the complaint by operating the dryer through its cycles. On electronic models, turn off the electricity to the appliance and wait for two minutes before turning it back on. If a fault code appears, look up the code. If the dryer will not power up, locate the technical data sheet behind the control panel or for diagnostics information. On some models you will need the actual service manual for the model

you are working on to properly diagnose the dryer. The service manual will assist you in properly placing the dryer in the service test mode for testing the dryer functions.

2. **Check for external factors.** You must check for external factors not associated with the appliance. Is the appliance installed properly? Does the dryer have the correct voltage supply? The voltage at the receptacle is between 108 volts and 132 volts during a load on the circuit (see Chapter 6).

3. **Disconnect the electricity.** Before working on the dryer, disconnect the electricity. This can be done by pulling the plug from the electrical outlet. Or disconnect the electricity at the fuse panel or at the circuit breaker panel. Turn off the electricity.

4. **Think safety. Shut off the gas supply.** As a precaution before you begin servicing the dryer, shut off the supply of gas to the dryer. The shutoff valve should be within 6 feet of the gas appliance.

WARNING   *Some diagnostic tests will require you to test the components with the power turned on. When you disassemble the control panel, you can position it in such a way that the wiring will not make contact with metal. This act will allow you to test the components without electrical mishaps.*

5. **Gain access to the belt switch.** To gain access to the belt switch, in this type of dryer, remove the top panel, insert a putty knife, and disengage the retaining clip (see Figure 20-20a).

6. **Remove the idler pulley.** Before removing the belt switch, you must first remove the idler pulley and drive belt. To disconnect the belt on this model (see Figure 20-21), push on the idler pulley to release the tension from the drive belt. Now remove the belt from the motor pulley and from around the idler pulley. If the drive belt is broken, just remove the belt. Next remove the two Phillips-head screws from the belt switch (see Figure 20-47).

7. **Testing the belt switch.** Remove the two wires from the belt switch. Place your multimeter test leads on the terminals and turn your meter to the ohms scale. When activating the belt switch, the meter reading will indicate continuity. When deactivating the belt switch, the meter reading will indicate an infinity reading. If the meter indicates infinity when the belt switch is activated and deactivated, replace the switch.

8. **Install the new belt switch.** To install the belt switch, just reverse the disassembly procedure, and reassemble it. Reassemble the dryer in the reverse order of its disassembly, and test it for proper operation. Restore the electricity and gas supply to the dryer and test. Make sure to take the dryer out of the service test mode when the repair is completed.

Burner Assembly

The heating system for a gas dryer is composed of four main components: the burner, the ignitor, the flame sensor, and the gas valve. The burner assembly is located on the baseplate of the dryer.

The typical complaints associated with the burner assembly are:

- Burner will not light.
- The burner goes on and off.

- Soot buildup.
- The ignitor will not glow.
- The burner trips on the high-limit thermostat.
- On electronic models, check for unusual display readouts and/or error codes.

To handle these problems, perform the following steps:

1. **Verify the complaint.** Verify the complaint by operating the dryer through its cycles. Then open the dryer door, and place your hand inside to see if it is warm in the drum. On electronic models, turn off the electricity to the appliance and wait for two minutes before turning it back on. If a fault code appears, look up the code. If the dryer will not power up, locate the technical data sheet behind the control panel or for diagnostics information. On some models you will need the actual service manual for the model you are working on to properly diagnose the dryer. The service manual will assist you in properly placing the dryer in the service test mode for testing the dryer functions.

2. **Check for external factors.** You must check for external factors not associated with the appliance. Is the appliance installed properly? Is the exhaust vent clogged? Check the voltage and gas supply to the dryer. Does the dryer have the correct voltage supply? The voltage at the receptacle is between 108 volts and 132 volts during a load on the circuit (see Chapter 6).

3. **Disconnect the electricity.** Before working on the dryer, disconnect the electricity. This can be done by pulling the plug from the electrical outlet. Or disconnect the electricity at the fuse panel or at the circuit breaker panel. Turn off the electricity.

4. **Shut off the gas supply.** Before you begin servicing any gas components, shut off the supply of gas to the dryer. The shutoff valve should be within 6 feet of the gas appliance.

5. **Gain access to the burner assembly.** To gain access to the burner assembly, you must remove the top panel (or lower panel or access panel). The panel is held in place with two retaining clips or screws, depending on the model. Using a flat-blade screwdriver, insert it about 4 inches from each side (Figure 21-6). Depress the clips, pull the panel toward you, and remove the panel.

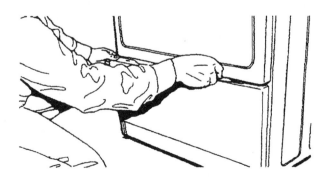

FIGURE 21-6
To open the top panel, insert a flat-blade screwdriver, depress the retaining clips, and remove the panel.

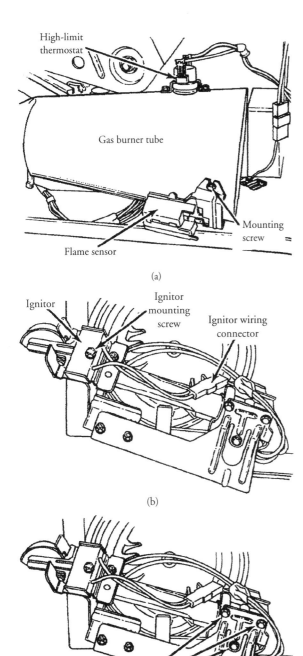

High-limit thermostat

Gas burner tube

Mounting screw

Flame sensor

(a)

Ignitor

Ignitor mounting screw

Ignitor wiring connector

(b)

Gas valve mounting screws

(c)

Figure 21-7 Removing a typical gas burner assembly.

6. **Remove the burner assembly.** The burner assembly is located on the lower-front side of the cabinet on the right. Some models place the burner assembly on the lower-front side on the left. Remove the wires from the flame sensor (Figure 21-7a), and then disconnect the wiring harness connector from the burner assembly (Figure 21-7b). Disconnect the gas line from the gas valve. Locate the two screws that secure the burner assembly to the dryer frame, and remove the screws. Remove the burner assembly by pulling it toward you carefully (Figure 21-7c). With the burner assembly removed from the cabinet, you can now inspect and access the burner, high-limit thermostat, ignitor, gas valve, pilot assembly (older models), and flame sensor. Before you reinstall the burner assembly, do the maintenance on the burner assembly and the lower cabinet, removing lint and debris.

7. **Test the gas valve solenoid coils (Type 1 gas valve).** This gas valve solenoid assembly is located on top of the gas valve. This part is a separate replaceable component. The solenoid coils consist of a double coil (safety and booster coils combined) and a single main solenoid coil. In order to check the solenoid coils on the gas valve, remove the wire leads (and label them) that connect to the coils from the wiring harness. These are slide-on terminal connectors attached to the ends of the wire. Set your ohmmeter on R × 100, and attach the meter probes as follows (Figure 21-8):

   • Terminals 1 and 2 (safety coil): The resistance will be between 1340 and 1390 ohms.

   • Terminals 1 and 3 (booster coil): The resistance will be between 535 and 585 ohms.

   • Terminals 4 and 5 (main coil): The resistance will be between 1170 and 1270 ohms.

PART VI

FIGURE 21-8
Type 1 gas valve. The test points on the solenoid coils, which are mounted on the gas valve. Terminals 1 and 2 are the safety coil, terminals 1 and 3 are the booster coil, and terminals 4 and 5 are the main coil.

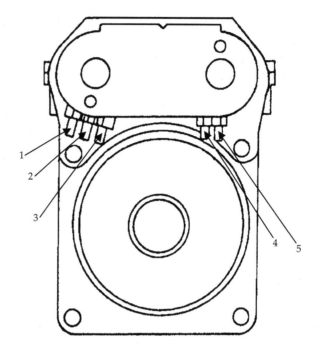

If the meter readings differ, replace the solenoid coils.

8. **Test the gas valve solenoid coils (Type 2 gas valve) (Figure 21-9).** This type of gas valve consists of three valves that are operated by 12 VDC solenoid coils. Two of the gas valves are either in an open or closed state, depending on whether the solenoid coils are energized by the electronic control board. The third gas valve is a linear valve.

FIGURE 21-9
Type 2 gas valve. This type of gas valve has three separate solenoid coils and operates on 12 volts DC.

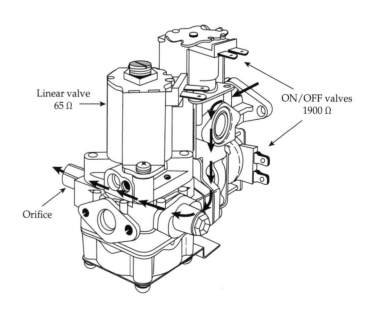

Linear valve
65 Ω

ON/OFF valves
1900 Ω

Orifice

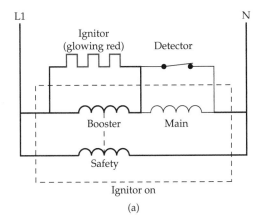

(a)

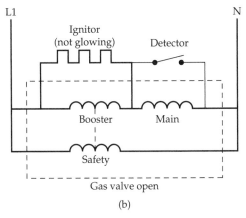

(b)

FIGURE 21-10 A wiring schematic of a dual solenoid gas valve with an ignitor. (a) Indicates the ignitor is on. (b) Indicates the gas valve is open.

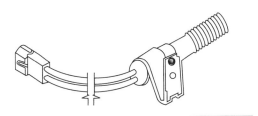

FIGURE 21-11 The gas flame ignitor.

To open this gas valve, the voltage will vary depending on the amount of voltage it receives from the electronic control board. The voltages are based on clothes load and temperature settings. This is why it is difficult to determine specific voltages at the linear solenoid coil. The voltages are derived from the software algorithms built into the electronic control board program chip.

9. **Test the ignitor.** The ignitor is made of silicon carbide. The ignitor is very fragile and it is susceptible to damage by skin oils. Hold the ignitor by the insulated support when handling to prevent damage. Electrically the ignitor is in parallel with the booster solenoid coil on the gas valve (see Figure 21-8). The parallel circuit of the ignitor and booster solenoid coil on the gas valve is in series with the main solenoid coil on the gas valve. The series-parallel circuit of the ignitor, booster solenoid coil, and main solenoid coil is, in turn, in parallel with the safety solenoid coil on the gas valve (Figure 21-10). Both the booster solenoid coil and the main solenoid coils are needed to open the gas valve. Once energized, the safety solenoid coil alone will hold the gas valve open allowing the gas to flow to the burner. Disconnect the ignitor wiring connection, and measure the resistance between the terminals on the ignitor. Set your ohmmeter on R × 100— the resistance, depending on room temperatures, will be between 50 and 400 ohms. If the meter readings differ, replace the ignitor (Figure 21-11). Another way to test the ignitor is by turning on the dryer and measuring the amperage of the ignitor. The amperage should be between 2½ and 4 amps to open the gas valves. If the amperage is less, the ignitor is defective and must be replaced.

10. **Test the flame sensor.** A flame sensor (Figure 21-12) is a thermostatically controlled single-pole, single-throw, normally closed switch. This device is mounted over a window cut out in the burner tube. The switch will open within 15 to 90 seconds, when the switch senses the radiant heat from the ignitor, and once open, the flame sensor will be held open by the radiant heat produced by the gas flame. The temperature of the ignitor will be at 1800 to 2500 degrees Fahrenheit. To test the flame sensor, set the ohmmeter on R × 1, remove the wire terminals from the sensor, and measure the

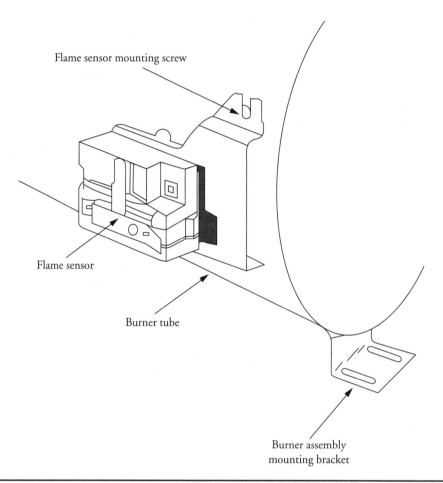

FIGURE 21-12   The flame sensor is mounted over a slot opening in the side of the burner tube.

resistance between the terminals. The flame sensor should measure zero ohms, indicating that the sensor is good. If the meter readings differ, replace the flame sensor.

11. **Test the high-limit thermostat.** The high-limit thermostat is located on the burner tube (see Figure 21-7a) or in the rear of the dryer (Figure 21-13); or there may be two high-limit thermostats mounted in both places, depending on the model. If the temperature exceeds the pre-set limit, the high-limit thermostat will open to shut down the gas burner. To test the thermostat, remove the wire terminals from the high-limit thermostat. Set the ohmmeter on R x 1, and measure the resistance between the thermostat terminals. The ohmmeter will read continuity, indicating that the high-limit thermostat is good. If the meter readings differ, replace the thermostat.

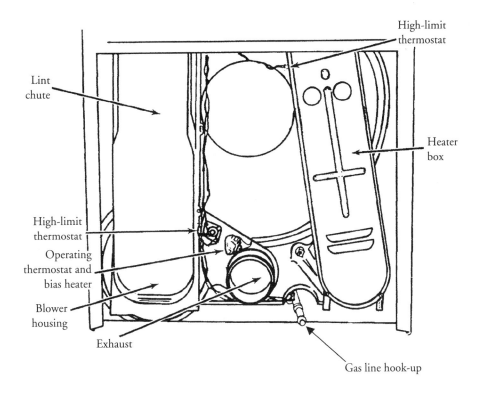

High-limit
thermostat

Lint
chute

Heater
box

High-limit
thermostat

Operating
thermostat and
bias heater

Blower
housing

Exhaust

Gas line hook-up

FIGURE 21-13   The rear view of a dryer, illustrating the location of the high-limit thermostat and the location of the other components.

12. **Install the burner assembly.** To install the burner assembly (see Figure 21-7), just reverse the disassembly procedure, and reassemble. If you are replacing the gas valve, make sure that you use approved pipe joint compound or yellow Teflon tape on the pipe connections (according to local codes) and on the threads of the gas valve before you install the valve. Turn on the gas supply. Check for gas leaks by using a chloride-free soap solution. *Do not use a flame to check for gas leaks.* The soap solution will begin to bubble if a leak is present. Restore the electricity and the gas supply to the dryer, and test the burner assembly. Check the burner flame for proper operation. Then reinstall the top panel (see Figure 21-6) in the reverse order of its disassembly. Make sure to take the dryer out of the service test mode when the repair is completed.

Diagnostic Charts

The following diagnostic flowcharts will help you to pinpoint the likely causes of various dryer problems (Figures 21-14 and 21-15; see also Figure 20-52).

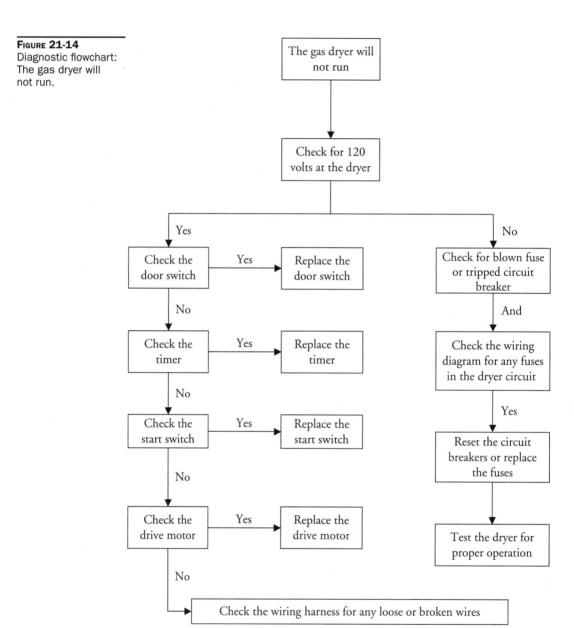

Figure 21-14
Diagnostic flowchart:
The gas dryer will
not run.

The wiring diagram in this chapter is only an example. You must refer to the actual wiring diagrams on the dryer that you are servicing. Figures 21-16, 21-17, and 21-18 depict actual strip circuit diagrams. The strip circuit diagram is simpler, and it is usually easier to read. Figures 21-19, 21-20, 21-21, 21-22, and 21-23 depict actual wiring schematic diagrams.

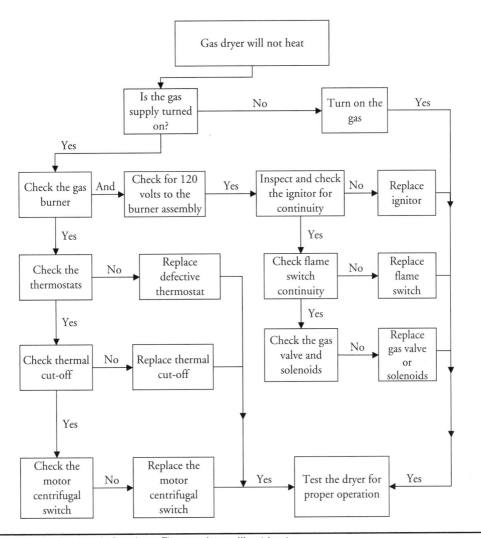

FIGURE 21-15 Diagnostic flowchart: The gas dryer will not heat.

PART VI

Main motor circuit—at start-up

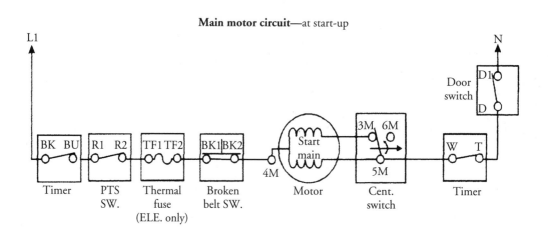

Main motor circuit—at running speed

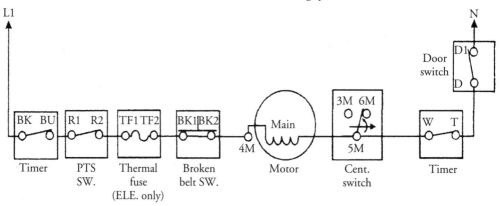

Figure 21-16   Typical strip circuits illustrating the drive motor circuit.

Temperature control circuit—the thermostat heater energized

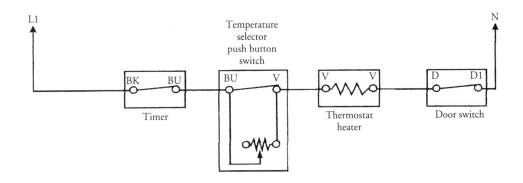

Gas burner circuit—the burner assembly energized

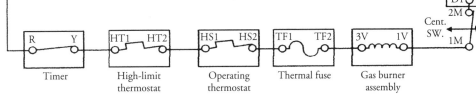

Push-to-start—with the relay coil energized

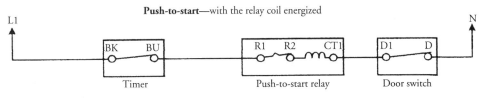

FIGURE 21-17 Typical strip circuits illustrating the temperature control circuit, heating circuit, and the push-to-start circuit.

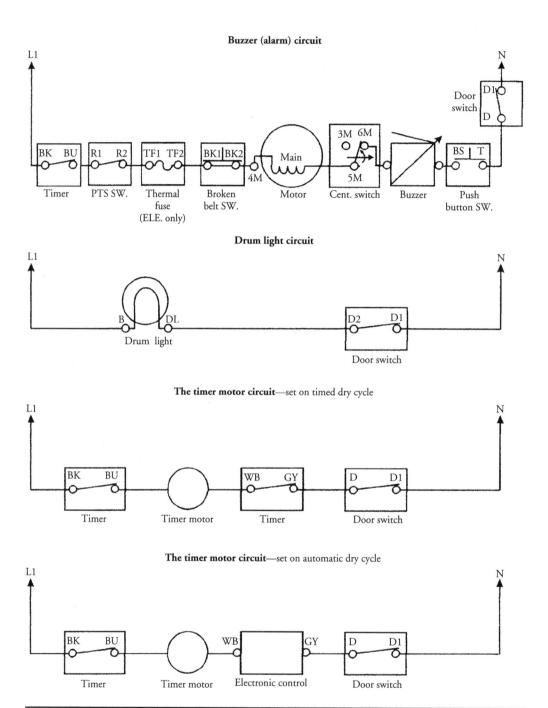

FIGURE 21-18 Typical strip circuits illustrating the buzzer circuit, drum lamp circuit, and the timer circuit.

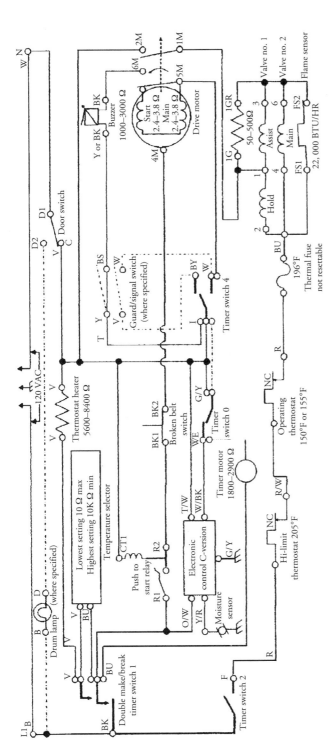

Figure 21-19 A typical gas dryer wiring schematic with electronic components.

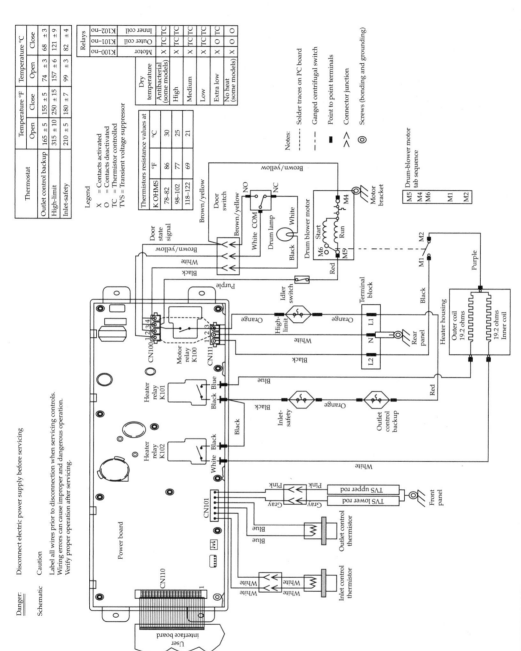

FIGURE 21-20 A typical gas dryer wiring schematic for the new-style front load dryer.

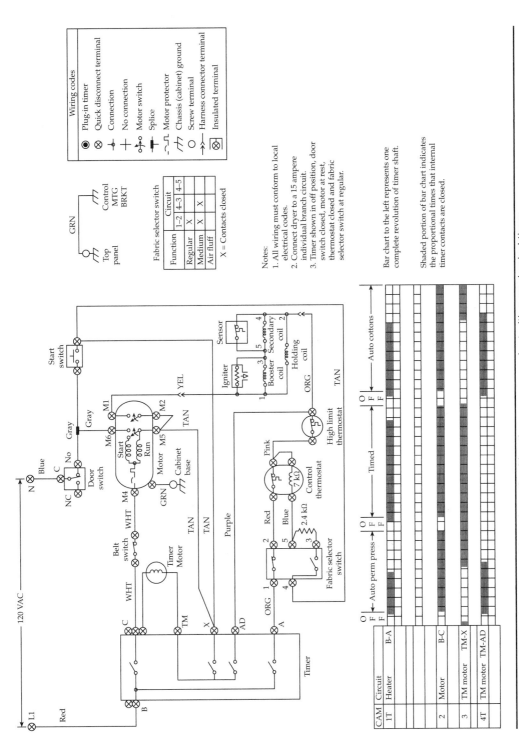

Figure 21-21 A typical gas dryer wiring schematic with a timer sequence chart with a mechanical timer.

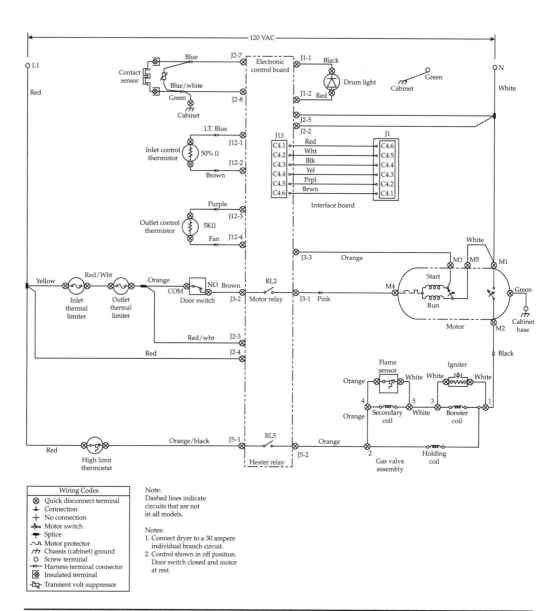

FIGURE 21-22 A typical gas wiring schematic with an electronic control board.

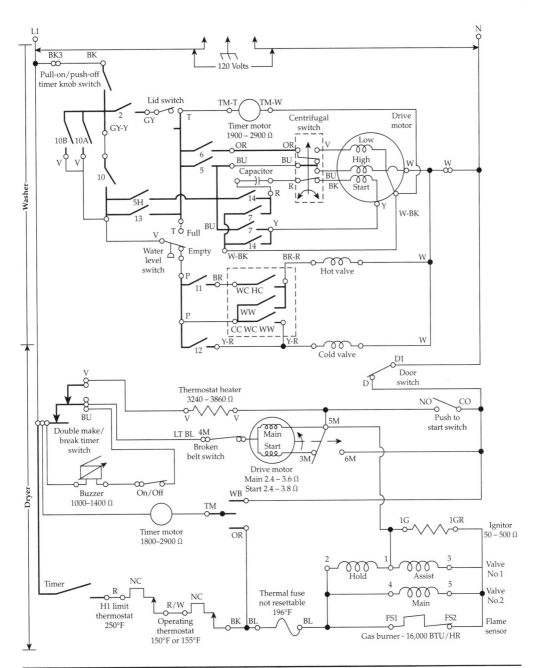

Figure 21-23   A typical washer and gas dryer wiring schematic for a stack washer and dryer combination.

Electric Ranges, Cooktops, and Ovens

Electric ranges (Figures 1-2 and 22-1), cooktops (Figure 1-4), and ovens (Figure 1-6) are available in a variety of styles. Electric ranges and ovens might seem to be complicated, but they are not. The more you know about the electrical and mechanical operation of the product, the easier it will be to solve its problems. An electric range, cooktop, or oven operates on 240 volts for the heating elements and 120 volts for the accessories (clock, lights, etc.). Most repairs are electrical in nature. This chapter will provide the basics needed to diagnose and repair these appliances. Figure 22-2 identifies where components are located within the range—this illustration is used as an example only. The actual construction and features might vary, depending on what brand and model you are servicing.

Principles of Operation

The surface units are simple to operate. After placing a pot on the surface element, you turn the surface unit switch to the desired setting. When the surface unit switch is on, current flows from the wall receptacle, through the wiring and the surface unit switch, and then through the heating element. To properly cook on the surface unit, the pot or pan must lie flat across the heating element (Figure 22-3), making contact with the entire cooking surface. If not, the food will not cook evenly and there will be greater heat losses (resulting in wasted electricity).

To bake in the oven, place the food on the oven rack, and close the oven door. Set the oven selector switch to bake and set the oven temperature control on the desired temperature. Current flows from the wall receptacle, through the wiring and the selector switch, through the thermostat, and to the bake element. When the selected temperature in the oven is reached, the bake element will go off. When the temperature in the oven decreases, the thermostat then reactivates the bake element. On some models, both the bake and broil elements operate at the same time to first preheat the oven cavity.

The broiling operation in the range is accomplished by placing the food on the top oven rack and partially closing the oven door. When closing the door, there is usually a stop in the hinge that allows the door to remain open a certain amount; this is done so that the

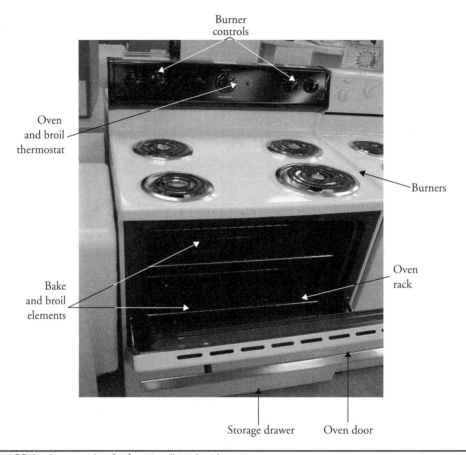

FIGURE 22-1   An example of a freestanding electric range.

broiler element will not cycle on temperature. The broiler element stays on for continuous operation until the user turns off the controls. When broiling, current will flow from the wall receptacle, through the wiring and the selector switch, through the thermostat, and then to the broil element.

Self-cleaning ranges and ovens differ—both from manufacturer to manufacturer and from model to model—in how the heating element operates to clean the oven. Pyrolytic cleaning is the true self-cleaning system. It uses high heat during a special one- to three-hour cycle to decompose food soil and grease. During the cycle, which is clock-controlled, the oven door is latched and locked. It cannot be opened until the oven cools down (below 550 degrees F). All of the oven walls, racks, and the door (except for a small area outside the door gasket) are completely cleaned. After cleaning, you might find a small bit of white ash, which can be easily wiped out.

Catalytic, or continuous, cleaning uses a special porous coating on the oven walls that partially absorbs and disperses the soil. This process occurs during normal baking and keeps

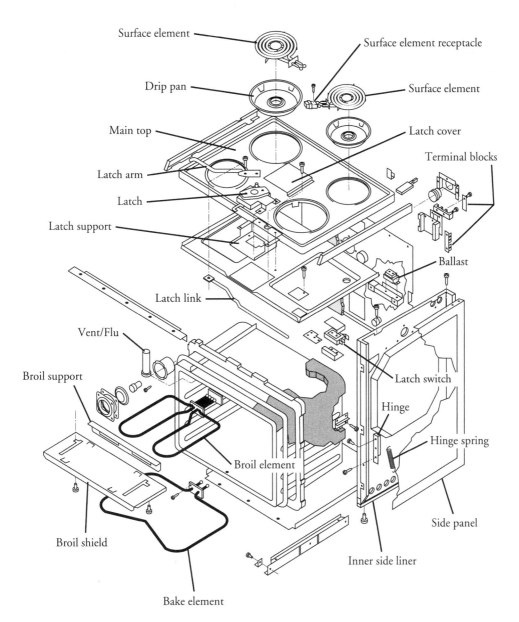

Surface element

Surface element receptacle

Drip pan

Surface element

Main top

Latch cover

Terminal blocks

Latch arm

Latch

Latch support

Ballast

Latch link

Vent/Flu

Broil support

Latch switch

Hinge

Hinge spring

Broil element

Broil shield

Side panel

Inner side liner

Bake element

Example only: Construction and features may vary depending on brand and model

FIGURE 22-2 An example of a range chassis.

the oven presentably clean; however, the racks and door parts must be cleaned by hand. Some manufacturers recommend occasionally operating an empty oven at 500 degrees Fahrenheit to remove any buildup of soil. This special oven coating cannot be cleaned with soap, detergent, or commercial oven cleaners without causing permanent damage.

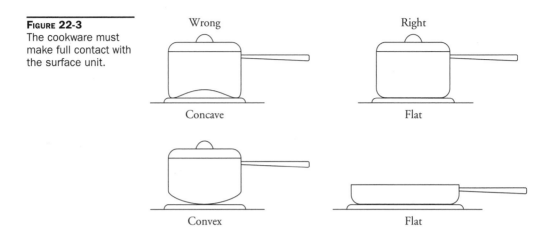

Figure 22-3
The cookware must make full contact with the surface unit.

On pyrolytic clean models, you simply close and latch the oven door and then set the controls to clean. Before using the self-cleaning cycle, the consumer must read the use and care manual in order to set the controls properly. If the controls are not set according to the manufacturer's recommendations, the oven will not be properly cleaned.

Next, you set the timer clock for the proper time when you want the clean cycle to begin. The clock in a self-cleaning oven has two functions: it controls when the oven will operate in the timed-bake cycle and it controls the clean cycle. The bake element will come on, and the temperature will begin to rise. This will take approximately 60 minutes. The reason for the slow temperature rise is to prevent damage to the oven cavity and the door.

When the temperature is above 550 degrees Fahrenheit, the oven lock light will come on, indicating that the latch on the oven door cannot be opened until the oven cavity cools below that temperature. When the oven cavity temperature stabilizes (between 840 and 920 degrees Fahrenheit), the cleaning process begins. This process requires approximately two to three hours. Some models have a cooling fan in the circuit to aid in keeping the exterior temperature low. Self-cleaning ovens have a nonelectric, "catalytic" smoke eliminator in the vent to consume the smoke from the soil load. The catalytic smoke eliminator will begin to operate when the oven cavity temperature is between 300 and 400 degrees Fahrenheit. The types of oven door lock mechanisms used on self-cleaning models are:

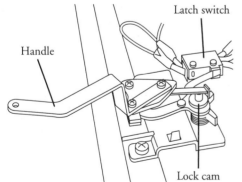

Figure 22-4   The manual latch system for the self-cleaning range is used on some models.

- **Manual door lock**   With this type of system, the user manually moves the oven door latch assembly to lock the door (see Figure 22-4).

- **Electromechanical door lock**   With this type of system, the user will either press a lock switch button (located on the control panel) in order to move the latch handle or, alternatively, move the latch handle to lock or unlock the oven door. This action will activate a solenoid coil, allowing the door latch mechanism and linkage to operate (Figure 22-5).

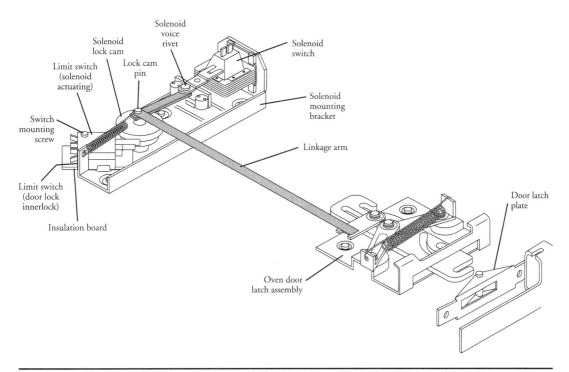

FIGURE 22-5   The latch solenoid system for the self-cleaning range is used on some models.

- **Electric door lock**   With this type of system, the user sets the cleaning controls, and the door lock assembly will operate through the control of an electric motor to activate the locking mechanism. This system is similar to the electromechanical door lock; however, instead of a latch solenoid, an electric motor is used.

Safety First

Any person who cannot use basic tools or follow written instructions should *not* attempt to install, maintain, or repair any electric range, oven, or cooktop. Any improper installation, preventive maintenance, or repairs could create a risk of personal injury or property damage.

If you do not fully understand the installation, preventive maintenance, or repair procedures in this chapter, or if you doubt your ability to complete the task on your appliance, please call your service manager.

The following precautions should also be followed:

- Never use a range to heat the home; it simply wasn't designed for that purpose.
- Keep the cooking area clean of spills and grease.
- Do not use flammable liquids near a cooking appliance.
- When repairing the range, always use the proper tools.

- Always reconnect the ground wire to the range after repairs have been made.
- Never use aluminum foil to line drip bowls; it could cause an electrical shock or become a fire hazard.

Before continuing, take a moment to refresh your memory of the safety procedures in Chapter 2.

Electric Ranges, Ovens, and Cooktops in General

Much of the troubleshooting information in this chapter covers electric ranges, ovens, and cooktops in general, rather than specific models, in order to present a broad overview of service techniques. The illustrations that are used in this chapter are for demonstration purposes only to clarify the description of how to service these appliances. They in no way reflect on a particular brand's reliability.

Location and Installation of Electric Range, Oven, or Cooktop

Locate the range, oven, or cooktop where it will be well lighted and have access to proper ventilation. The range must be level for proper baking and cooking results. The range might be installed adjacent to the left and/or right base cabinets and against a rear vertical wall (for the "antitip" cleat). A wall oven must be installed on a supporting surface that is strong enough to support the weight of the oven and its contents while remaining level from side to side and from front to rear. A cooktop must be installed on a flat surface, supported by the countertop, and it should be level.

Contact your flooring company to check if the flooring can withstand a minimum of 200 pounds. It is also recommended that you contact your builder to determine if the cabinets and walls can withstand the heat produced by the appliance. Some kitchen cabinets and building materials were not designed to withstand the heat that is given off by the product.

The proper installation instructions for your model are included in the use and care manual. These instructions will assist you with the installation requirements (dimensions, electrical requirements, cutout dimensions, venting, etc.) needed to complete the installation according to the manufacturer's specifications.

Step-by-Step Troubleshooting by Symptom Diagnosis

When servicing an appliance, don't overlook the simple things that might be causing the problem. Step-by-step troubleshooting by symptom diagnosis is based on diagnosing malfunctions, with their possible causes arranged into categories relating to the operation of the range. This section is intended only to serve as a checklist to aid you in diagnosing a problem. Look at the symptom that best describes the problem that you are experiencing with the range, oven, or cooktop, and then proceed to correct the problem.

Oven Will Not Heat

- Are the oven controls set properly?
- If the entire range is inoperative, check for voltage and check the circuit breakers or fuses.
- Test oven thermostat switch contacts for continuity.

- Test the heating elements for continuity and for a short.
- Check for loose or broken wiring connections.
- Test the other components in the circuits that operate the heating elements.
- Test the oven selector switch.

Oven Temperature Is Not Accurate

- Test the oven thermostat for accuracy by using an oven temperature tester.
- Check to be sure that the oven door is closed for baking.
- Is the oven vent blocked with aluminum foil?

Oven Will Not Cycle Off

- Test the oven thermostat switch contacts for continuity.
- Test the oven selector switch contacts for continuity.
- Test the time clock switch contacts for continuity.
- Test the oven cycling relay.

Broil Element Is Not Heating

- Are the broil controls set properly?
- If the entire range is inoperative, check for voltage and check circuit breakers or fuses.
- Test the oven thermostat switch contacts for continuity.
- Test the heating element for continuity and for a short.
- Check for loose or broken connections.
- Test other components in the circuits that operate the heating elements.
- Does the user have the door closed completely?
- Check the oven selector switch.
- Check to be sure that the oven door is open for broiling.

Surface Unit Will Not Operate

- Is the right surface unit switch turned on?
- Check for loose or broken wiring connections.
- Test the surface element for continuity and for a short.
- If the entire range is inoperative, check for voltage and check circuit breakers or fuses.
- Test the surface unit switch for continuity.

Heat Is Escaping from Oven Door

- Is the oven door aligned properly with the range body?
- Check the oven door gasket.

- Check the oven door hinge for damage.
- Check the oven door spring. Is it broken?
- Is the consumer closing the door completely for baking?

Cooking Performance

If there are no electrical or mechanical problems with the oven's operation but it will not bake the food properly or if the food is partially cooked, etc., your next step will be to look at the symptom that best describes the problem that you are experiencing with the oven. Then correct the problem. If necessary, instruct the user on how to get better results from the oven. This information is covered in the use and care manual.

Satisfactory Baking Results

To have satisfactory baking results, the following conditions must exist:

- Proper oven door seal.
- The oven vent not blocked off.
- The oven thermostat must be calibrated properly.
- Food preparation must be done correctly.
- The proper type of cookware must be used.
- Follow the manufacturer's recommended cooking instructions.

Satisfactory Broiling Results

To have satisfactory broiling results, the following conditions must exist:

- On most models, the oven door must be open partially to ensure that the broiling element will not cycle on temperature.
- Food preparation must be done correctly.
- The proper type of cookware must be used.
- Follow the manufacturer's recommended broiling instructions.

Satisfactory Surface Cooking Results

To have satisfactory surface cooking results, the following conditions must exist:

- The reflector pans (bowls) must be used under all of the surface units.
- Use cookware with flat bottoms that make correct contact with the heating element.
- Use cookware large enough to entirely cover the surface units.
- Food preparation must be done correctly.
- Follow the manufacturer's recommended cooking instructions.

Surface Cookware

One of the keys to successful cooking is the use of proper cookware. For efficiency and best results, use pans that have tight-fitting lids and lightweight handles that do not tilt the pans.

The following is a brief review of the various types of cookware that are available. The different characteristics of the cookware will aid you when diagnosing cooking complaints.

- **Aluminum** is a metal that spreads heat quickly and evenly, and responds to temperature changes. This cookware is best for frying, braising, and pot roasts. The inside of an aluminum pan might be a natural finish or a nonstick coating.
- **Cast iron** is slow to change temperature and holds heat. This makes good cookware for browning, frying, stewing, and other cooktop cooking. Cast iron cookware is also available with an enamel finish.
- **Copper** is excellent for gourmet cooking, wine sauces, and egg cookery, and it is quick to change temperature.
- **Glass cookware** is slow to change temperature. This type of cookware works best for a long time on a low heat when cooking with a liquid.
- **Porcelain enamel** is long-lasting and is used for cooking soups and other liquids.
- **Stainless steel** is exceptionally strong. It is used for frying, sauces, soups, vegetables, and egg cooking.

Oven Cookware

Sometimes, a technician must educate the consumer on how to be a successful cook. The use of the correct type of cookware is very important. The following are some guidelines for choosing cookware:

- Always use the correct size pan as given in the recipe.
- When baking foods, use flat-bottomed pans to keep the food level.
- Aluminum oven cookware that is shiny (not cast) produces delicate browning, tender crusts, and reduces the spattering of roasts. This type of cookware is best for cakes, muffins, some quick breads, cookies, and roasting.
- Pottery, ceramic, cast metal, and ovenproof glass cookware give food a deep, crusty brown surface. When this type of cookware is used, the oven temperature should be reduced by 25 degrees Fahrenheit.
- The use of dull or darkened cookware is suitable for pies and other foods baked in pastry shells.
- Shallow-sided pans and flat baking sheets are best for cookies and biscuits, where top and side browning is wanted.

Food Preparation

Proper food preparation is the other key to successful cooking. If you suspect a baking problem, ask the cook to see the cooked food. If needed, ask the cook to prepare some food for baking, and watch how the food is prepared. The following descriptions will assist you in determining the problem.

Flat Cake

The cake comes out of the oven, and you notice that the cake is flat or that it has no volume to it. The reasons are:

- There might be too much liquid or not enough liquid.
- During preparation, the mix was overbeaten or underbeaten.

- The pan used to make the cake in was too large.
- The cook did not adjust the oven racks properly or placed the pan on the wrong rack.
- The oven temperature is too high or too low.
- The mix was stored improperly and/or was exposed to high humidity.
- The cook might have forgotten to use eggs.

Cake Has Fallen

The cake comes out of the oven, and you notice that the cake has fallen or the center has a dip in it. The reasons are:

- The cake is underbaked.
- There might be too much liquid or not enough liquid.
- The cook tested the cake too soon.
- The cake pan was too small.
- The cook moved the cake before it was completely baked.

Sticky Crust

The cake comes out of the oven, and you notice that it has a sticky top crust. The reasons are:

- The cake was underbaked.
- The cook stored the cake in a sealed container when the cake was still warm.
- The humidity in the air is high.
- Too much liquid in the mix.
- The cook might have substituted sweet fruit juices for other liquids.

Holes in Cake

The cake comes out of the oven, and you notice that it has holes or tunnels in it. The reasons are:

- Baking temperature selected was too high.
- The cook did not adjust the oven racks properly or placed the pan on the wrong rack.
- During preparation, the mix was overbeaten or underbeaten.
- Very large air bubbles became trapped in the batter.
- The cook might have used the wrong type of pan, which might have caused uneven temperature conductance.

Cake Shrinkage

The cake comes out of the oven, and you notice that it shrank or pulled away from the sides of the pan. The reasons are:

- The cake was overbaked.
- The cake pan was too close to the oven wall or too close to other pans.
- Extreme overbeating of the mixture.

Cake Peaked in the Center

The cake comes out of the oven, and you notice that the center of the cake has a peak in it. The reasons are:

- The oven temperature selected was too high.
- The pan was too small.
- The cook might have used the wrong type of pan, which might have caused uneven temperature conductance.
- There might not have been enough liquid used in the mixture.
- The cake pan was too close to the oven wall or too close to other pans.
- Extreme overbeating of the mixture.

Crust Not Brown

The cake comes out of the oven, and you notice that the crust on top of the cake is not brown. The reasons are:

- Oven door opened too many times during baking.
- Too much liquid in the mix.
- Oven temperature too low when baking.
- The cake pan was too deep.
- Extreme overbeating of the mixture.

Cake Too Dry

The cake comes out of the oven, and you notice that it is too dry and that it falls apart. The reasons are:

- The cake was overbaked.
- Not enough liquid in the mix.
- The cook might have forgotten to use eggs.

Cookies Brown Rapidly

The cookies come out of the oven, and you notice that they are browned more at the sides and/or the end of the cookie sheet. The reasons are:

- The cook might have used the wrong size cookie sheet, which might have caused uneven temperature conductance.
- Cookie sheets with sides will cause rapid browning at the edges. Are the cookies placed at least one inch from the sides?

Cookies Brown Slowly

The cookies come out of the oven, and you notice that they browned slowly. The reasons are:

- The heat might be leaking out of the oven. Check the oven door for heat leakage.
- The racks are uneven. Check the oven racks. Are they level when cold and when hot?

Cookies Are Dark on the Bottom

The cookies come out of the oven, and you notice that they are dark on the bottom. The reasons are:

- The oven racks were not adjusted properly, or the cookie sheet was placed on the wrong rack.
- More than one cookie sheet was placed in the oven.

Cooking Meats

The cook complains that the meats are not cooking properly; ask to see some of the food. This section is intended only to serve as a checklist to aid you in diagnosing a problem. Look at the symptom that best describes the problem that the cook is experiencing with the meats, and then correct the problem.

The Meat Burns on the Bottom

The possible reasons are:

- Check the size of the pan used to cook the meat. There should be at least one or two inches of oven rack visible around the pan when it is placed on the oven rack so that the oven cavity is evenly heated.
- Was the oven preheated?
- Was the door opened and closed frequently?
- Was the meat elevated off the bottom of the pan?

Meats Are Undercooked

The possible reasons are:

- Check the oven temperature. Is it calibrated correctly?
- Check the size of the pan used to cook the meat. There should be at least one or two inches of oven rack visible around the pan when it is placed on the oven rack so that the oven cavity is evenly heated.
- Was the type of pan used too deep? Was the pan covered?

Roasting of Meats Takes Too Long

The possible reasons are:

- Check the oven temperature. Is it calibrated correctly?
- Check for inadequate ventilation.
- Check for the improper use of aluminum foil.

Range, Oven, and Cooktop Maintenance

The range, oven, or cooktop can be cleaned with warm water, mild detergent, and a soft cloth on all cleanable parts, as recommended in the use and care manual. Never use scouring pads on these surfaces, except where recommended in the use and care manual. Also, never use abrasive cleaners that are not specifically recommended by the manufacturer. Do not allow grease spillovers to accumulate after cooking on top of the range; this will become a fire hazard.

On glass surfaces, you can use a glass cleaner to clean any soil stains. Stubborn soil stains on glass surfaces can be removed with a paste of baking soda and water. Never use abrasive cleaners on glass surfaces. Never clean the heating elements. When you turn on the elements, the soil will burn off.

On self-cleaning models, never use oven cleaners to clean the oven cavity. This will cause hazardous fumes when the oven is in the cleaning cycle. After the oven cavity has cooled, use only soap and water to clean small spills on the inside of the cavity.

Repair Procedures

Each repair procedure is a complete inspection and repair process for a single range, oven, or cooktop component. It contains the information you need to test and replace components.

Any person who cannot use basic tools should *not* attempt to install, maintain, or repair any electric dryer. Any improper installation, preventative maintenance, or repairs will create a risk of personal injury, as well as property damage. Call the service manager if installation, preventative maintenance, or the repair procedure is not fully understood.

Surface Element (Burner) Control Systems

There are three types of surface (burner) control systems used on ranges and cooktops:

1. Standard infinite (burner) switch

2. Dual infinite (burner) switch

3. Electronic element (burner) control system

The standard infinite (burner) switch has been around for many years. This type of switch has a built-in safety; you must push in and turn the knob to turn the switch on. Voltage is supplied to the surface elements through the infinite (burner) switch contacts L1-H1 and L2-H2 (Figure 22-6).

The dual infinite switch is used to control the expandable and bridge surface elements on glass-top models (Figure 22-7). This type of switch can provide an infinite choice of heating selections for cooking and provide a choice of two surface element sizes.

The electronic element (burner) control system on some models is more accurate and allows for a lower simmer temperature selection than the conventional infinite switches. This type of system uses an electronic motherboard, user interface board, and potentiometers (Figure 22-8) to turn on and control the surface elements. This type of system communicates with the electronic control board to turn on a double line break relay as a safety to prevent the surface (burner) elements and the oven from coming on when activated by other means than the consumer.

Standard Infinite (Burner) Switch

All infinite (burner) switches are mounted in the control panel and are marked by their respective surface element. During the actual surface element operation, if the infinite

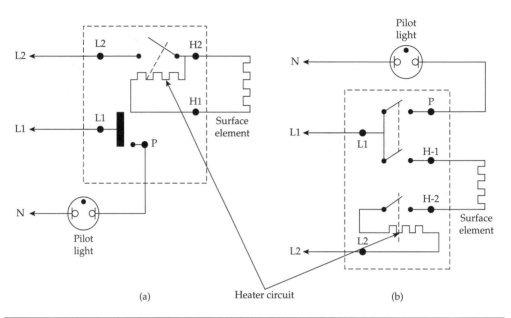

FIGURE 22-6 The difference between these two types of infinite switches is the position of the small heater in the circuit, located inside the switch. In figure (a), the heater is in parallel with the surface element and is voltage sensitive. In figure (b), the heater is in series with the surface element and is current sensitive.

switch is set to the high position, contacts L2-H2 are closed providing continuous voltage to the surface (burner) element. If any other position on the infinite switch is selected, contacts L2-H2 will cycle to maintain the correct heat setting. Contacts L1-P will provide voltage to the indicator light when the infinite switch is turned on. See Figure 22-6.

On some models, the infinite switch terminal numbers may look a little different from what's displayed in this text. Table 22-1 will help you in identifying the switch terminals.

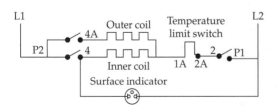

FIGURE 22-7 The dual infinite (burner) switch is used on glass-top models.

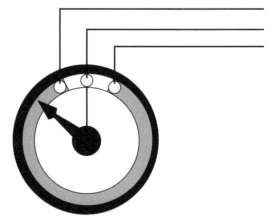

FIGURE 22-8 Four potentiometers (variable resistors) are used to turn on the surface elements. The consumer turns the knob on the potentiometer to adjust the surface elements.

TABLE 22-1 Infinite
(Burner) Switch
Terminal Identification
Chart

Infinite (Burner) Switch Terminal Identification	
H1	1
H2	4
L1	3
L2	5
P	2

The typical complaints associated with failure of the infinite (burner) switch are:

- The surface element will not heat at all.
- The surface element will stay on high in any position.
- Intermittent switch operation.
- The infinite switch does not cycle the surface element off and on when set to a position other than high.
- Unusual display readouts and/or error codes.
- The infinite switch operates correctly, but the indicator light does not glow.
- The indicator light will glow very dimly with all of the infinite switches in the off position.

To handle these problems, perform the following steps:

1. **Verify the complaint.** Verify the complaint by turning the infinite switch on. Is the surface element heating? On electronic models, turn off the electricity to the appliance and wait for two minutes before turning it back on. If a fault code appears, look up the code. If the range/oven will not power up, locate the technical data sheet behind the control panel for the diagnostics information. On some models you will need the actual service manual for the model you are working on to properly diagnose the range/oven. The service manual will assist you in properly placing the range/oven in the service test mode for testing the range/oven functions.

2. **Check for external factors.** You must check for external factors not associated with the appliance. Is the appliance installed properly? Does it have the correct voltage? The voltage at the receptacle is between 198 volts and 264 volts during a load on the circuit (see Chapter 6).

3. **Disconnect the electricity.** Before working on the range or cooktop, disconnect the electricity to the appliance. This can be done by pulling the plug from the receptacle. Or disconnect the electricity at the fuse panel or at the circuit breaker panel. Turn off the electricity.

WARNING *Some diagnostic tests will require you to test the components with the power turned on. When you disassemble the control panel, you can position it in such a way that the wiring will not make contact with metal. This act will allow you to test the components without electrical mishaps.*

PART VI

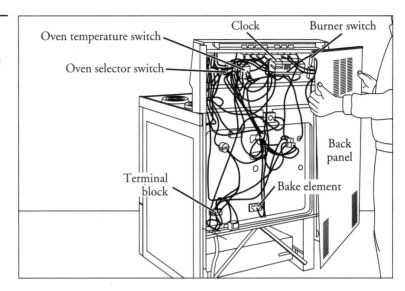

FIGURE 22-9
Remove the screws
from the back panel
to gain access to the
components.

4. **Gain access to the infinite switch.** Depending on the model you are repairing, you can access the component by removing the back panel (Figure 22-9). On models with front-mounted controls (Figure 22-10), the panel is attached with screws on both ends. Remove the screws, and tilt the control panel. Be very careful not to let the wires disengage from the components.

   Some built-in models have a removable backsplash (Figure 22-11); just lift the backsplash, and rest it on the cooktop. It would be a good idea to place something on the cooktop to protect it from getting damaged. Next, remove the screws from

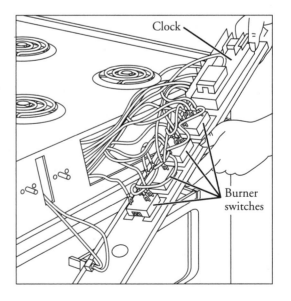

FIGURE 22-10
Remove the screws
from both ends to
remove the control
panel and gain access
to the components.

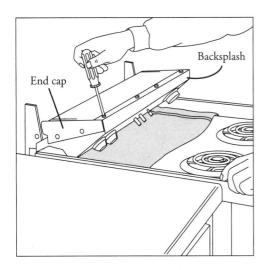

Figure 22-11 Remove the screws from the back panel to gain access to the components.

Figure 22-12 Remove the screws that secure the control panel to the oven to gain access to the components.

the backsplash, which hold the rear panel, to access the components. If you are repairing a wall oven or an eye-level range, the control panel can be removed (Figure 22-12) by opening the door and removing the screws that secure the panel. These screws might be underneath the front of the exhaust hood or just below the control panel. Some control panels are hinged—just tilt the control panel toward you for servicing. To access the components on other models, both the rear panel and the front control panel (usually glass) will have to be removed. To pull out the glass, remove the screws that secure the trim piece that holds the glass in place.

5. **Test the infinite switch (continuity).** To test the infinite switch for continuity between the switch contacts, remove all of the wires from the switch terminals (label them). Set the infinite switch in the "high" position. Using your ohmmeter, place the probes on the L1 and H1 (Figure 22-13a) terminals; there should be continuity. Next, place the probes on the L2 and H2 (Figure 22-13b) terminals; there should be continuity. Now place the probes on the L1 and P (Figure 22-13c) terminals; there should be continuity. The infinite switch is defective if there is no continuity between L1 and H1 or between L2 and H2. If there is no continuity between P and H1, the indicator light circuit is defective.

6. **Test the infinite switch (voltage).** Plug in the range and test the supply voltage. Set your multimeter on the voltage scale, and place the test leads on L1 and L2 terminals on the infinite switch. If the meter reads zero volts, the wiring between the terminal block (rear/bottom of range) and the infinite switch is open. If the meter reads line voltage at the infinite switch terminals, then place the test leads on terminals H1 and H2. Set the infinite switch to the high position and measure the voltage drop across H1 and H2. If your multimeter reads zero, the infinite switch is defective. If the multimeter reads line-to-line voltage, then the infinite switch

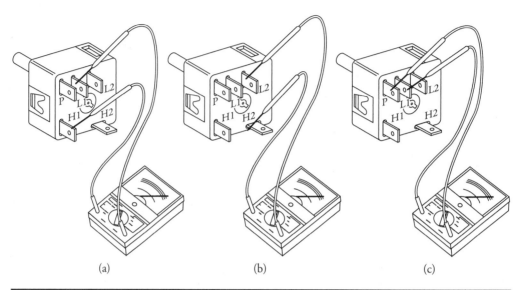

(a)   (b)   (c)

Figure 22-13   Testing an infinite switch for continuity.

is good. Next, remove the surface element and measure the voltage drop between the terminals by placing your multimeter leads on the terminal block terminals. If the multimeter reading reads zero, the terminal block or the wiring between the infinite switch and the terminal block is defective. If the multimeter reads line-to-line voltage, then the surface element is defective. On glass-top models, raise the top to access the elements. Locate the correct wires from the infinite switch to the element, and place your test leads on the element terminals. Measure the voltage drop between the two terminals. If the multimeter reads zero, the wires between the infinite switch and the element are open. If the multimeter reads line-to-line voltage, the element is defective.

7. **Remove the infinite switch.** To remove the infinite switch on most models, you must remove the two screws that secure the switch to the control panel. On some models, the infinite switch is secured with a nut to the control panel. Unscrew the nut to remove this type of infinite switch.

8. **Install a new infinite switch.** To install the new infinite switch, just reverse the disassembly procedure, and reassemble. Be sure that you install the wires onto the correct terminals according to the wiring diagram. On electronic models, make sure to take the range/oven out of the service test mode when the repair is completed.

Dual Infinite (Burner) Switch

On some glass-top models, the dual infinite (burner) switch controls the expandable and bridge surface elements. When the burner knob is turned clockwise, less than 180 degrees, switch contacts P2 to 4, P2 to 4A, and P1 to 2 close, turning on both surface elements.

When the burner knob is turned counterclockwise, less than 180 degrees, switch contacts P2 to 4 and P1 to 2 close, providing voltage to the inner surface element. See Figures 22-7 and 22-14. When you set the dual infinite switch to high, switch contacts P1 and 2 are locked closed, providing continuous voltage to the surface element. Switch contacts 4 to L2 provide voltage to the indicator lamp.

The typical complaints associated with failure of the infinite (burner) switch are:

- Both surface elements will not heat at all.
- The surface element will stay on high in any position.
- Intermittent switch operation.
- The dual infinite switch does not cycle the surface element off and on when set to a position other than high.
- Unusual display readouts and/or error codes.
- The dual infinite switch operates correctly, but the indicator light does not glow.
- The indicator light will glow very dimly with all of the infinite switches in the off position.
- The outer surface element will not heat.
- The inner surface element will not heat.

To handle these problems, perform the following steps:

1. **Verify the complaint.** Verify the complaint by turning the infinite switch on. Is the surface element heating? On electronic models, turn off the electricity to the appliance and wait for two minutes before turning it back on. If a fault code appears, look up the code. If the range/oven will not power up, locate the technical data sheet behind the control panel for the diagnostics information. On some models you will need the actual service manual for the model you are working on to properly diagnose the range/oven. The service manual will assist you in properly placing the range/oven in the service test mode for testing the range/oven functions.

2. **Check for external factors.** You must check for external factors not associated with the appliance. Is the appliance installed properly? Does it have the correct voltage? The voltage at the receptacle is between 198 volts and 264 volts during a load on the circuit (see Chapter 6).

FIGURE 22-14
(a) Rear view of the back of a dual infinite switch. (b) Rear view of the back of a standard infinite switch.

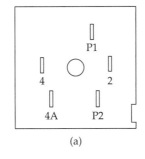

(a)

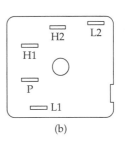

(b)

3. **Disconnect the electricity.** Before working on the range or cooktop, disconnect the electricity to the appliance. This can be done by pulling the plug from the receptacle. Or disconnect the electricity at the fuse panel or at the circuit breaker panel. Turn off the electricity.

WARNING *Some diagnostic tests will require you to test the components with the power turned on. When you disassemble the control panel, you can position it in such a way that the wiring will not make contact with metal. This act will allow you to test the components without electrical mishaps.*

4. **Gain access to the infinite switch.** Depending on the model you are repairing, you can access the component by removing the back panel (Figure 22-9). On models with front-mounted controls (Figure 22-10), the panel is attached with screws on both ends. Remove the screws, and tilt the control panel. Be very careful not to let the wires disengage from the components.

5. **Test the dual infinite switch (continuity).** First remove all wires from the switch terminals. Next, set the multimeter to the ohms scale R × 1. Then place the multimeter test leads on the following terminals on the dual infinite switch to test the switch contacts. See Table 22-2.

6. **Test the dual infinite switch (voltage).** Plug in the range and test the supply voltage. Set your multimeter on the voltage scale, place the test leads on P1 and P2 terminals, and measure the voltage drop (see Figures 22-7 and 22-14). If the multimeter reads zero volts, the wiring between the main terminal block in the rear of the range and the dual infinite switch is open. If the multimeter reads line-to-line voltage, turn the dual infinite switch clockwise to high position. Next, measure the voltage drop between switch terminals 4 and 2. If the multimeter reads zero, then the dual infinite switch is defective. If the multimeter reads line-to-line voltage around 240 volts, the dual infinite switch is good. Finally, raise the glass top and locate the two terminals on the surface element with the wires from the switch terminals 4 and 2 that are connected on the surface element. Now measure the voltage drop between these two terminals on the surface element. If the multimeter reads zero, the wires between the dual infinite switch and the surface element are open. If the multimeter reads line-to-line voltage, then the surface element is bad.

Dual Infinite Switch Position	Switch Contacts Closed	Condition
Clockwise less than 180°	P2 to 4, P2 to 4A, P1 to 2	Both elements are on
Counterclockwise, less than 180°	P2 to 4, P1 to 2	Inner element on
Switch set on "High" position	P1 to 2 locked closed	Element on high
Switch on in either position	4 to L2	Indicator lamp on

TABLE 22-2 Switch Contact Position for a Dual Infinite Switch

7. **Remove the dual infinite switch.** To remove the dual infinite switch on most models, you must remove the two screws that secure the switch to the control panel. On some models, the dual infinite switch is secured with a nut to the control panel. Unscrew the nut to remove this type of infinite switch.

8. **Install a new dual infinite switch.** To install the new dual infinite switch, just reverse the disassembly procedure and reassemble. Be sure that you install the wires onto the correct terminals according to the wiring diagram. On electronic models, make sure to take the range/oven out of the service test mode when the repair is completed.

Electronic Element (Burner) Control System

The electronic element (burner) control system is made up of five components: electronic control board, a main circuit board, a user interface board, potentiometers, and the surface elements.

The typical complaints associated with failure of the electronic element (burner) control system are:

- Both surface elements will not heat at all.
- The surface element will stay on high in any position.
- Intermittent burner operation.
- The electronic element control system does not cycle the surface element.
- Unusual display readouts and/or error codes.
- The outer surface element will not heat.
- The inner surface element will not heat.

To handle these problems, perform the following steps:

1. **Verify the complaint.** Verify the complaint by turning the burner switch on. Is the surface element heating? On electronic models, turn off the electricity to the appliance and wait for two minutes before turning it back on. If a fault code appears, look up the code. If the range/oven will not power up, locate the technical data sheet behind the control panel for the diagnostics information. On this model you will need the actual service manual for the model you are working on to properly diagnose the range/oven/cooktop. The service manual will assist you in properly placing the range/oven/cooktop in the service test mode for testing the range/oven/cooktop functions.

2. **Check for external factors.** You must check for external factors not associated with the appliance. Is the appliance installed properly? Does it have the correct voltage? The voltage at the receptacle is between 198 volts and 264 volts during a load on the circuit (see Chapter 6).

3. **Disconnect the electricity.** Before working on the range or cooktop, disconnect the electricity to the appliance. This can be done by pulling the plug from the receptacle. Or disconnect the electricity at the fuse panel or at the circuit breaker panel. Turn off the electricity.

WARNING *Some diagnostic tests will require you to test the components with the power turned on. When you disassemble the control panel, you can position it in such a way that the wiring will not make contact with metal. This act will allow you to test the components without electrical mishaps.*

4. **Gain access to the infinite switch.** Depending on the model you are repairing, you can access the component by removing the back panel (Figure 22-9). On models with front-mounted controls (Figure 22-10), the panel is attached with screws on both ends. Remove the screws, and tilt the control panel. Be very careful not to let the wires disengage from the components. On cooktop models, you will have to dissemble the cooktop to gain access to the components.

5. **Test the electronic element (burner) control system.** To test this type of system, you will have to refer to the technical data sheet and/or the service manual. This will provide you with the necessary information to correctly diagnose this type of system.

6. **Remove the defective part.** To remove the defective part on most models, you must remove the screws that secure the component to the control panel or back panel.

7. **Install a new component.** To install a new component, just reverse the disassembly procedure and reassemble. Be sure that you install the wires onto the correct terminals according to the wiring diagram. On electronic models, make sure to take the range out of the service test mode when the repair is completed.

Surface Calrod (Burner) Element

Surface calrod (burner) elements are available in 4-, 6-, and 8-inch diameter, single or dual element design, hard wired or a plug-in style depending on manufacturer specifications.

The typical complaints associated with surface element (burner) failure are:

- The surface element will not heat at all.
- When the surface element is turned on, it trips the circuit breaker or blows the fuses.
- On dual-surface elements, only part of the element heats up.
- Unusual display readouts and/or error codes.

To handle these problems, perform the following steps:

1. **Verify the complaint.** Verify the complaint by turning on the infinite switch. Is the surface element heating? On electronic models, turn off the electricity to the appliance and wait for two minutes before turning it back on. If a fault code appears, look up the code. If the range/oven will not power up, locate the technical data sheet behind the control panel for the diagnostics information. On some models you will need the actual service manual for the model you are working on to properly diagnose the range/oven. The service manual will assist you in properly placing the range/oven in the service test mode for testing the range/oven functions.

2. **Check for external factors.** You must check for external factors not associated with the appliance. Is the appliance installed properly? Does it have the correct voltage? The voltage at the receptacle is between 198 volts and 264 volts during a load on the circuit (see Chapter 6).

3. **Disconnect the electricity.** Before working on the range or cooktop, disconnect the electricity to the appliance. This can be done by pulling the plug from the receptacle. Or disconnect the electricity at the fuse panel or at the circuit breaker panel. Turn off the electricity.

WARNING *Some diagnostic tests will require you to test the components with the power turned on. When you disassemble the control panel, you can position it in such a way that the wiring will not make contact with metal. This act will allow you to test the components without electrical mishaps.*

4. **Remove the surface element.** On most models, the surface element can be removed by simply pulling the element out of its receptacle (Figure 22-15). Other models have the surface elements connected to the cooktop and directly wired to their infinite switch wires (Figure 22-16). To remove this type of element, remove the screw that secures the element to the cooktop. Then remove the clips that hold the insulators to the element terminals (see inset in Figure 22-16). Next, unscrew the wires from the terminals without bending the terminals.

5. **Test the surface element.** Using the ohmmeter, set the range scale on R × 1. Place the probes on the element terminals (Figure 22-17). The meter reading should show continuity. The actual readings will vary from manufacturer to manufacturer, according to the size of the unit and the wattage used, but the readings should generally be between 15 and 115 ohms, approximately ± 20 percent. To test for a grounded surface element, place one probe on the sheath and the other probe on each element terminal, in turn (Figure 22-18). If continuity exists at either terminal, the element is shorted and should be replaced.

FIGURE 22-15
To remove a plug in a surface element, just lift and slide the element out of the receptacle.

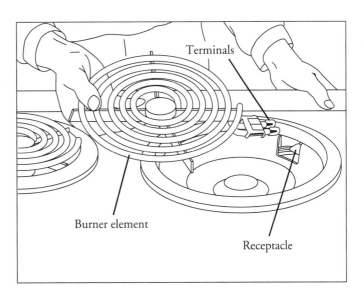

Terminals

Burner element

Receptacle

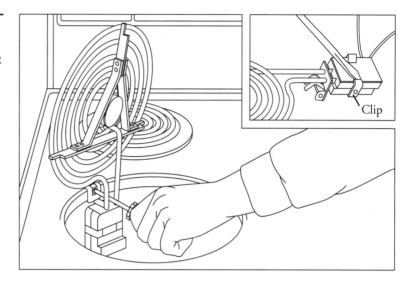

Clip

6. **Install a new surface element.** To install the new surface element, just reverse the disassembly procedure, and reassemble. Be sure that you install the wires on their correct terminals, according to the wiring diagram. On electronic models, make sure to take the range out of the service test mode when the repair is completed.

Solid-Disc Elements

Solid-disc elements are a solid closed burner with the heater encased in the burner. They are available in 6- and 8-inch diameter.

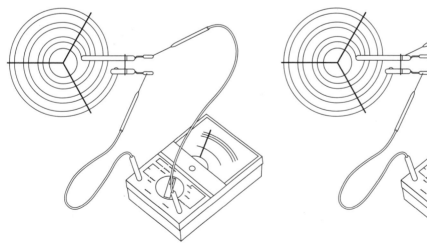

FIGURE 22-17   Testing the surface element for resistance.

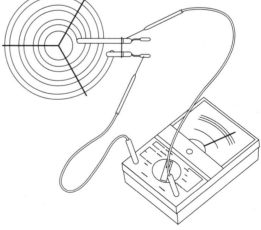

FIGURE 22-18   Testing the surface element for ground.

The typical complaints associated with solid-disc elements are the same as those for the traditional surface elements:

- Unusual display readouts and/or error codes.
- The surface element will not heat at all.
- When the surface element is turned on, it trips the circuit breaker or blows the fuses.

To handle problems pertaining to them, perform the following steps:

1. **Verify the complaint.** Verify the complaint by turning the infinite switch on. Is the solid-disc element heating? On electronic models, turn off the electricity to the appliance and wait for two minutes before turning it back on. If a fault code appears, look up the code. If the range/oven will not power up, locate the technical data sheet behind the control panel for the diagnostics information. On some models you will need the actual service manual for the model you are working on to properly diagnose the range/oven. The service manual will assist you in properly placing the range/oven in the service test mode for testing the range/oven functions.

2. **Check for external factors.** You must check for external factors not associated with the appliance. Is the appliance installed properly? Does it have the correct voltage? The voltage at the receptacle is between 198 volts and 264 volts during a load on the circuit (see Chapter 6).

3. **Disconnect the electricity.** Before working on the range or cooktop, disconnect the electricity from the appliance. This can be done by pulling the plug from the receptacle. Or disconnect the electricity at the fuse panel or the circuit breaker panel. Turn off the electricity.

WARNING *Some diagnostic tests will require you to test the components with the power turned on. When you disassemble the control panel, you can position it in such a way that the wiring will not make contact with metal. This act will allow you to test the components without electrical mishaps.*

4. **Gain access to the solid-disc element.** To access the element on this model, you will have to raise the cooktop. To do this, remove the screws from under the front edge of the cooktop (Figure 22-19); then lift the cooktop and prop it up. On some models, the trim might have to be removed first.

5. **Test the solid-disc element.** Remove the wires from the element. Using the ohmmeter, set the range scale on R × 1. Place the probes on the element terminals; there should be continuity. The actual readings will vary from manufacturer to manufacturer and with the size of the unit and the wattage used, but the readings should generally be between 15 and 115 ohms, approximately ± 20 percent. To test for a grounded element, place one probe on the sheath, and attach the other probe to each element terminal. If continuity exists at either terminal, the element is shorted and should be replaced.

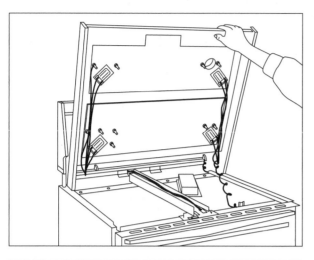

FIGURE 22-19 Remove the screws from under the front edge of the cooktop to gain access to the solid-disc elements.

6. **Remove the solid-disc element.** To remove the solid-disc element, remove the wires from the element (Figure 22-20). Remove any brackets that secure the element to the cooktop. The element is removed from the top of the cooktop (see inset in Figure 22-20).

7. **Install a new solid-disc element.** To install the new solid-disc element, just reverse the disassembly procedure, and reassemble. Be sure that you install the wires on their correct terminals according to the wiring diagram. On electronic models, make sure to take the range out of the service test mode when the repair is completed.

Radiant Heating Element

Radiant heating elements are installed in glass-top ranges and cooktops. The element consists of a ribbon-type resistance wire embedded in an insulation media housed in a metal dish. They are mounted under the glass with special mounting brackets. They are available in 4-, 6-, 7-, 8-, and 9-inch diameter and they can be manufactured in any combination of multiple sizes in one metal dish. To access this type of burner, the technician will have to disassemble the top of the range or cooktop. This style of burner will have a temperature limiter attached to the burner. This limiter will tell the customer when it is safe to touch the glass. Also, it protects the glass from breakage and overheating if there is any other component failure associated with the burner operation.

The typical complaints associated with the radiant heating element are the same as those for surface elements.

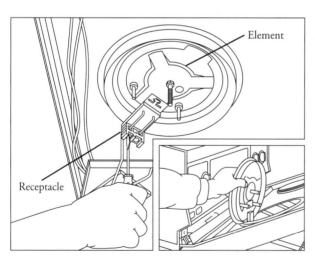

FIGURE 22-20 Remove the wires from the solid-disc element before you test for continuity. Remove the screws that secure the element; then remove the element from the top of the cooktop.

- The surface element will not heat at all.

- When the surface element is turned on, it trips the circuit breaker or blows the fuses.

- On dual-surface elements, only part of the element heats up.

- Unusual display readouts and/or error codes.

To handle problems pertaining to it, perform the following steps:

1. **Verify the complaint.** Verify the complaint by turning the infinite switch on. Is the radiant element heating? On electronic models, turn off the electricity to the appliance and wait for two minutes before turning it back on. If a fault code appears, look up the code. If the range/oven will not power up, locate the technical data sheet behind the control panel for the diagnostics information. On some models you will need the actual service manual for the model you are working on to properly diagnose the range/oven. The service manual will assist you in properly placing the range/oven in the service test mode for testing the range/oven functions.

2. **Check for external factors.** You must check for external factors not associated with the appliance. Is the appliance installed properly? Does it have the correct voltage? The voltage at the receptacle is between 198 volts and 264 volts during a load on the circuit (see Chapter 6).

3. **Disconnect the electricity.** Before working on the range or cooktop, disconnect the electricity to the appliance. This can be done by pulling the plug from the receptacle. Or disconnect the electricity at the fuse panel or at the circuit breaker panel. Turn off the electricity.

WARNING *Some diagnostic tests will require you to test the components with the power turned on. When you disassemble the control panel, you can position it in such a way that the wiring will not make contact with metal. This act will allow you to test the components without electrical mishaps.*

4. **Gain access to the radiant heating element.** To access the element, you will have to raise the cooktop (see Figure 22-19). On some models, you will have to remove the trim first. Remove the screws from under the front edge of the cooktop; then lift the cooktop and disconnect the element's wiring harness plug. Remove the cooktop, and place it upside-down on a table. Don't forget to place a blanket under the cooktop to prevent damage.

5. **Test the radiant heating element.** In order to test the element, remove the element mounting bracket from the cooktop. This is held on with two screws on either end of the bracket. Then carefully turn over the heating element assembly. This will give you access to the limiter and the element (Figure 22-21). Be careful—the element is embedded in an insulated casting, which is used to prevent the cooktop glass from overheating. Remove the wires from the element. On the ohmmeter, set the range scale on R × 1. Place the probes on the element terminals; there should be continuity. The actual readings will vary from manufacturer to manufacturer and according to the size of the unit and the wattage used, but the readings should generally be between 15 and 115 ohms, approximately ± 20 percent. If the element fails the test, replace it along with the limiter.

6. **Remove the radiant heating element.** To remove the element, remove the screws that secure the element to the bracket. Lift the element out. The new element comes as a complete assembly.

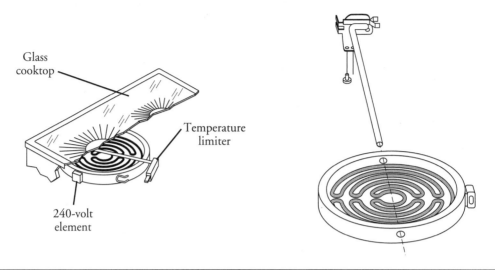

Glass cooktop

Temperature limiter

240-volt element

FIGURE 22-21 The radiant heating element must be flat against the glass.

7. **Install a new radiant heating element.** To install the new radiant heating element, just reverse the disassembly procedure, and reassemble. The element must be flat against the glass and located under the heater panel (see Figure 22-21). Be sure that you install the wires on their correct terminals according to the wiring diagram. On electronic models, make sure to take the range/cooktop out of the service test mode when the repair is completed.

Bake Element

The bake element is a calrod heating element located in the oven cavity on the bottom. It may have plug-on terminals or screw-type connections to hold the wires in place.

The typical complaints associated with bake element failure are:

- The bake element will not heat at all.
- When the bake element is turned on, it trips the circuit breaker or it blows the fuses.
- Unusual display readouts and/or error codes.

To handle these problems, perform the following steps:

1. **Verify the complaint.** Verify the complaint by turning the selector switch to "bake" and setting the thermostat. Is the bake element heating? On some models, the clock must be set to "manual." Check the use and care manual for the model you are servicing. On electronic models, turn off the electricity to the appliance and wait for two minutes before turning it back on. If a fault code appears, look up the code. If the range/oven will not power up, locate the technical data sheet behind the control panel for the diagnostics information. On some models, you will need the actual service manual for the model you are working on to properly diagnose the range/oven. The service manual will assist you in properly placing the range/oven in the service test mode for testing the range/oven functions.

2. **Check for external factors.** You must check for external factors not associated with the appliance. Is the appliance installed properly? Does it have the correct voltage? The voltage at the receptacle is between 198 volts and 264 volts during a load on the circuit (see Chapter 6).

3. **Disconnect the electricity.** Before working on the range, disconnect the electricity to it. This can be done by pulling the plug from the receptacle. Or disconnect the electricity at the fuse panel or at the circuit breaker panel. Turn off the electricity.

WARNING *Some diagnostic tests will require you to test the components with the power turned on. When you disassemble the control panel, you can position it in such a way that the wiring will not make contact with metal. This act will allow you to test the components without electrical mishaps.*

4. **Gain access to the bake element.** To access the bake element, open the oven door and remove the oven racks. Begin by removing the two screws that secure the element to the oven cavity (Figure 22-22).

5. **Remove and test the bake element.** Slide the element forward, and remove the wires from the bake element terminals (label them), either by removing the screws from the terminals or by pulling the wires off the bake element terminals (Figure 22-23). Using the ohmmeter, set the range scale on R × 1. Place the probes on the element terminals (Figure 22-24); there should be continuity. The actual readings will vary from manufacturer to manufacturer and according to the size of the unit and the wattage used, but the readings should generally be between 15 and 115 ohms, approximately ± 20 percent. To test for a grounded bake element, place one probe on the sheath and the other probe on an element terminal (Figure 22-25). If continuity exists, the element is shorted and should be replaced. Test both terminals.

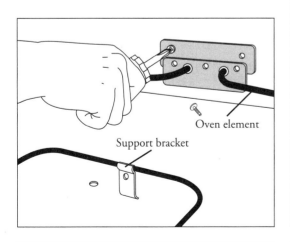

FIGURE 22-22 Remove the two screws and slide the element forward to get at the element terminals.

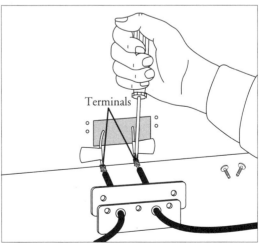

FIGURE 22-23 Remove the screws and wires from the element before testing.

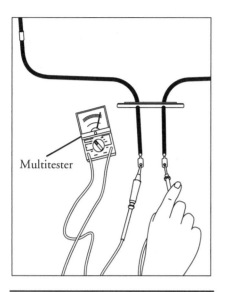

Multitester

Figure 22-24   Testing the bake element for resistance.

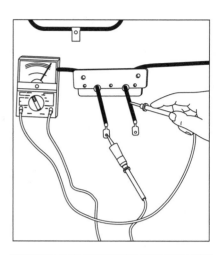

Figure 22-25   Testing the element for ground.

6. **Install a new bake element.** To install the new bake element, just reverse the disassembly procedure, and reassemble. Be sure that you install the wires on their correct terminals according to the wiring diagram. On electronic models, make sure to take the range out of the service test mode when the repair is completed

Broil Element

The broil element is a calrod heating element located in the oven cavity at the top. It may have plug-on terminals or screw-type connections to hold the wires in place.

The typical complaints associated with failure of the broil element are:

- The broil element will not heat at all.
- When the broil element is turned on, it trips the circuit breaker or it blows the fuses.
- On dual broil elements, only one element heats.
- Unusual display readouts and/or error codes.

To handle these problems, perform the following steps:

1. **Verify the complaint.** Verify the complaint by turning the selector switch to "broil" and/or by setting the thermostat to "broil." Is the broil element heating? On some models, the clock must be set to "manual." Check the use and care manual for the model you are servicing. On electronic models, turn off the electricity to the appliance and wait for two minutes before turning it back on. If a fault code appears, look up the code. If the range/oven will not power up, locate the technical data sheet behind the control panel for the diagnostics information. On some models you will need the actual service manual for the model you are working on to properly diagnose the

range/oven. The service manual will assist you in properly placing the range/oven in the service test mode for testing the range/oven functions.

2. **Check for external factors.** You must check for external factors not associated with the appliance. Is the appliance installed properly? Does it have the correct voltage? The voltage at the receptacle is between 198 volts and 264 volts during a load on the circuit (see Chapter 6).

3. **Disconnect the electricity.** Before working on the range, disconnect the electricity. This can be done by pulling the plug from the receptacle. Or disconnect the electricity at the fuse panel or at the circuit breaker panel. Turn off the electricity.

WARNING *Some diagnostic tests will require you to test the components with the power turned on. When you disassemble the control panel, you can position it in such a way that the wiring will not make contact with metal. This act will allow you to test the components without electrical mishaps.*

4. **Gain access to the broil element.** To access the broil element, open the oven door and remove the oven racks. The broil element is located at the top of the oven cavity. Begin by removing the two screws that secure the element to the oven cavity. Then remove the holding brackets from the element and slide the element forward.

5. **Remove and test the broil element.** Remove the wires from the broil element terminals, either by removing the screws from the terminals or by pulling the wires off the broil element terminals (label them). Using the ohmmeter, set the range scale on R × 1. Place the probes on the element terminals (Figure 22-26); there should be continuity. The actual readings will vary from manufacturer to manufacturer and according to the size of the unit and the wattage used. The readings should generally be between 15 and 115 ohms, approximately ± 20 percent. To test for a grounded broil element, place one probe on the sheath and the other probe on an element terminal (see Figure 22-25). If continuity exists, the element is shorted and should be replaced. Test both terminals.

To test a dual broil element, remove the wires from their terminals. Using the ohmmeter, set the range scale on R × 1. Place the probes on terminals A and C (Figure 22-27); there should be continuity. Then place the probes on terminals

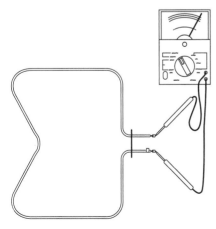

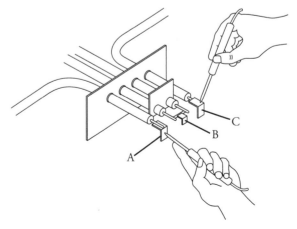

FIGURE 22-26   Testing the broil element for resistance.

FIGURE 22-27   Testing a dual broil element for resistance between the A and C terminals.

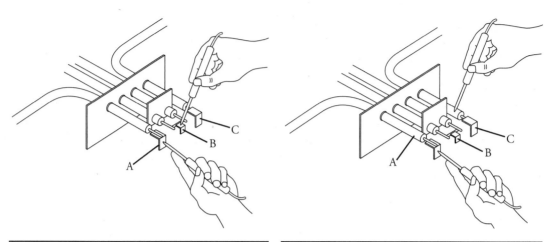

FIGURE 22-28 Testing a dual broil element for resistance between the A and B terminals.

FIGURE 22-29 Testing the dual broil element for ground.

A and B (Figure 22-28); there should be continuity. Finally, place the probes on terminals B and C; there should be continuity. To test for a grounded element, place one probe on the sheath. With the other probe, touch terminal A, then B, then C (Figure 22-29). If continuity exists, the element is shorted and should be replaced.

6. **Install a new broil element.** To install the new broil element, just reverse the disassembly procedure, and reassemble. Be sure that you install the wires on their correct terminals according to the wiring diagram. On electronic models, make sure to take the range out of the service test mode when the repair is completed.

Oven Selector Switch

The oven selector switch is located in the control panel. Its main purpose is to turn on bake, broil, time bake, or clean.

The typical complaints associated with failure of the oven selector switch are:

- Bake element will not heat.
- Broil element will not heat.
- Self-cleaning feature will not work.
- Timed baking will not work.
- Unusual display readouts and/or error codes.

To handle these problems, perform the following steps:

1. **Verify the complaint.** Verify the complaint by turning the oven selector switch to the desired setting. Set the thermostat to the desired temperature setting. Does that part of the range/oven work? Test all of the cycles on the selector switch. Turn off the electricity to the appliance and wait for two minutes before turning it back on. If a fault code appears, look up the code. If the range/oven will not power up, locate

the technical data sheet behind the control panel for the diagnostics information. On some models you will need the actual service manual for the model you are working on to properly diagnose the range/oven. The service manual will assist you in properly placing the range/oven in the service test mode for testing the range/oven functions.

2. **Check for external factors.** You must check for external factors not associated with the appliance. Is the appliance installed properly? Does it have the correct voltage? The voltage at the receptacle is between 198 volts and 264 volts during a load on the circuit (see Chapter 6).

3. **Disconnect the electricity.** Before working on the range or oven, disconnect the electricity to the appliance. This can be done by pulling the plug from the receptacle. Or disconnect the electricity at the fuse panel or at the circuit breaker panel. Turn off the electricity.

WARNING *Some diagnostic tests will require you to test the components with the power turned on. When you disassemble the control panel, you can position it in such a way that the wiring will not make contact with metal. This act will allow you to test the components without electrical mishaps.*

4. **Gain access to the oven selector switch.** Depending on which model you are repairing, you can access the component by removing the back panel (see Figure 22-9) on a freestanding range. On models with front-mounted controls (see Figure 22-10), the panel is attached with screws on both ends. Remove the screws, and tilt the control panel. Be careful not to let the wires come off their components. Some built-in models have a removable backsplash (see Figure 22-11). Lift the backsplash and rest it on the cooktop. It would be a good idea to place something on the cooktop first to protect it from damage. Next, remove the screws from the backsplash that hold the rear panel to gain access to the components. If you are repairing a wall oven or an eye-level range, the control panel can be removed (see Figure 22-12) by opening the door and removing the screws that secure the panel. These might be underneath the front of the exhaust hood or just below the control panel. Some control panels are hinged; just tilt the control panel toward you for servicing. On other models, both the rear panel and the front control panel (usually glass) will have to be removed to gain access to the components. To remove the glass, remove the screws that secure the trim piece that holds the glass in place, and then remove the trim ring. On some models, the trim might have to be removed first.

5. **Test the oven selector switch.** To test the oven selector switch for continuity between certain switch contacts, remove only those wires that are being tested and label them (Figure 22-30). To test each switch contact, refer to the wiring diagram for the proper switch contact terminals. Only remove one pair of wires at a time to test for continuity, as the oven selector switch is checked in each position.

6. **Remove the oven selector switch.** To remove the oven selector switch on most models, you must remove the two screws that secure the switch to the control panel. On some models, the oven selector switch is secured to the control panel with a nut. Unscrew this nut to remove the oven selector switch (Figure 22-31). Do not remove the

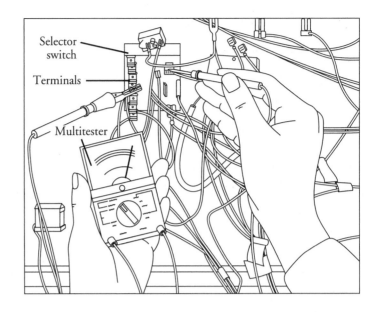

FIGURE 22-30
When testing the oven selector switch, only remove one wire from each pair of terminals being tested.

wires until the new switch is ready for installation. Then transfer the wires to the correct terminals.

7. **Install a new oven selector switch.** To install the new oven selector switch, just reverse the disassembly procedure, and reassemble. Be sure that you install the wires on the correct terminals according to the wiring diagram. On electronic models, make sure to take the range out of the service test mode when the repair is completed.

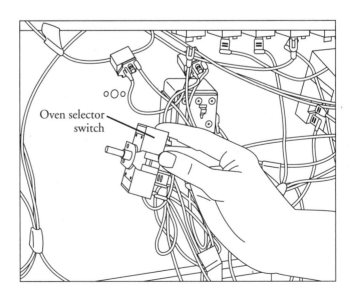

FIGURE 22-31
Removing the two screws that secure the oven selector switch from the control panel.

Oven Thermostat

The oven thermostat is located in the control panel. The sensing bulb from the thermostat is located in the upper-left or upper-right side of the oven cavity, and it is attached to a capillary tube that leads back to the thermostat. As heat rises in the oven, the gas within the thermostat bulb will expand and travel up to the thermostat. The increased pressure will push against a diaphragm, which will activate or deactivate switch contacts, turning on or off the heating elements in the oven cavity at a predetermined temperature.

The typical complaints associated with failure of the oven thermostat are:

- Bake element will not heat.
- Broil element will not heat.
- Timed baking will not work.
- Oven temperature is not accurate.
- Unusual display readouts and/or error codes.

To handle these problems, perform the following steps:

1. **Verify the complaint.** Verify the complaint by turning the oven selector switch to the desired setting and setting the oven thermostat to the desired temperature. Do those functions of the range/oven work? On some models, the clock must be set to "manual." Check the use and care manual for the model you are servicing. Turn off the electricity to the appliance and wait for two minutes before turning it back on. If a fault code appears, look up the code. If the range/oven will not power up, locate the technical data sheet behind the control panel for the diagnostics information. On some models you will need the actual service manual for the model you are working on to properly diagnose the range/oven. The service manual will assist you in properly placing the range/oven in the service test mode for testing the range/oven functions.

2. **Check for external factors.** You must check for external factors not associated with the appliance. Is the appliance installed properly? Does it have the correct voltage? The voltage at the receptacle is between 198 volts and 264 volts during a load on the circuit (see Chapter 6).

3. **Disconnect the electricity.** Before working on the range or oven, disconnect the electricity to the appliance. This can be done by pulling the plug from the receptacle. Or disconnect the electricity at the fuse panel or at the circuit breaker panel. Turn off the electricity.

WARNING   *Some diagnostic tests will require you to test the components with the power turned on. When you disassemble the control panel, you can position it in such a way that the wiring will not make contact with metal. This act will allow you to test the components without electrical mishaps.*

4. **Gain access to the oven thermostat.** You can gain access to the component by removing the back panel (see Figure 22-9) on a freestanding range. On models with front-mounted controls (see Figure 22-10), the panel is attached with screws on both

ends. Remove the screws, and tilt the control panel. Be careful not to let the wires come off their components. Some built-in models have a removable backsplash (see Figure 22-11); lift the backsplash and rest it on the cooktop. It would be a good idea to place something on the cooktop first to protect it from damage. Next, remove the screws from the backsplash, which holds the rear panel, to access the components. If you are repairing a wall oven or an eye-level range, the control panel can be removed (see Figure 22-12) by opening the door and removing the screws that secure the panel. These might be underneath the front of the exhaust hood or just below the control panel. Some control panels are hinged—just tilt the control panel toward you for servicing. On other models, both the rear panel and the front control panel (usually glass) will have to be removed to gain access to the components. To remove the glass, remove the screws that secure the trim that holds the glass in place. Then remove the trim piece. On some models, the trim might have to be removed first.

5. **Test the oven thermostat switch contacts.** To test the oven thermostat switch for continuity between certain switch contacts, remove only those wires that are being tested from their terminals (Figure 22-32). To test each switch contact, refer to the wiring diagram for the proper switch contact terminals. Only remove one pair of wires at a time (label them) to test for continuity.

6. **Calibrate the oven thermostat.** Before making any adjustments to the thermostat, test the oven temperature. With an oven temperature tester, place the thermocouple tip in the center of the oven cavity. Be sure that it does not touch any metal. Close the oven door, set the oven to bake, and adjust the thermostat setting to the 350-degree mark. Let the oven cycle for 20 to 30 minutes. Then record the minimum and maximum temperatures of three cycles. Next, add these temperatures and divide by 6. This will give you the average temperature of the oven.

Figure 22-32
Testing the oven thermostat switch contacts for continuity.

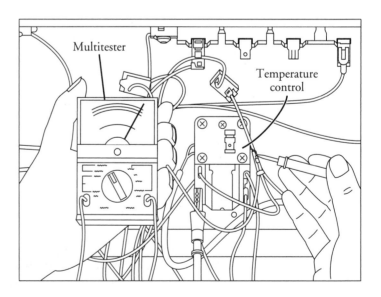

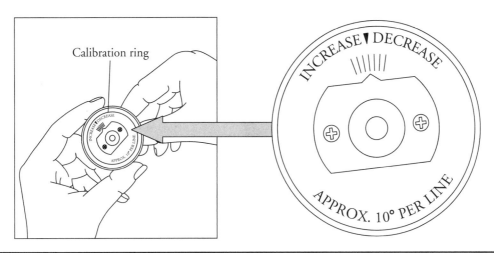

FIGURE 22-33   Turn the oven thermostat knob over. On the back of the knob are the words "increase" and "decrease." By moving the pointer, you can increase or decrease the temperature setting of the thermostat. Each line indicates 10-degree increments.

The average temperature calculated should be within ±25 degrees of the temperature setting selected (rotary dial type). If not, try calibrating the thermostat. To calibrate the thermostat, pull the oven thermostat knob off. Turn it over (Figure 22-33). The calibration ring is on the back of the knob. Loosen the two screws, and move the pointer in the direction needed. Each line that the pointer is moved indicates 10 degrees of change in the calibration. Tighten the screws and place the knob back on the thermostat stem. Retest the oven temperature. On models that do not have the calibration ring on the back of the dial, replace the thermostat if the temperature is more than 25 degrees out of calibration.

7. **Remove the oven thermostat.** To remove the oven thermostat on most models, you must disconnect the thermostat capillary tube from its supports (Figure 22-34) and push it through the back wall of the oven cavity. Be careful. Do not break the capillary tube wire because the contents inside are flammable. Next, remove the two screws that secure the thermostat to the control panel. Remove the thermostat (Figure 22-35). Leave the wires on the thermostat for now.

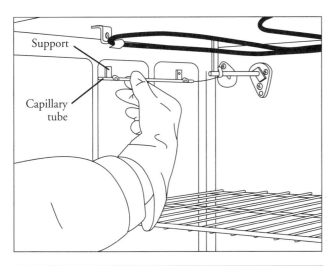

FIGURE 22-34   Lift the thermostat capillary tube off the supports, being careful not to break it.

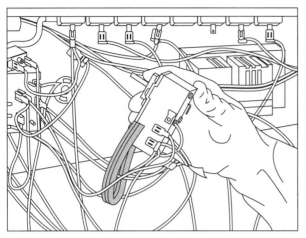

FIGURE 22-35   Remove the oven thermostat from the range. Leave the wires on the thermostat until the new thermostat is ready for installation. Then transfer the wires to the correct terminals.

8. **Install a new oven thermostat.** To install the new oven thermostat, just reverse the disassembly procedure, and reassemble. Transfer the wires, one at a time, from the old thermostat to the new one. Be sure that you install the wires on their correct terminals according to the wiring diagram. Reassemble the control panel in the reverse order of disassembly, and test. On electronic models, make sure to take the range out of the service test mode when the repair is completed.

Oven Temperature Sensor

The oven temperature sensor (Figure 22-36) is a thermistor. The thermistor is a thermally sensitive resistor that exhibits a change in electrical resistance with a change in its temperature. The oven temperature sensor is located in the upper left or upper right, attached to the rear wall of the oven cavity. The oven temperature sensor sends the resistance readings to the electronic oven control board. The electronic control board monitors the resistance readings of the oven temperature sensor and turns the bake or broil on or off at the desired temperature.

The typical complaints associated with failure of the oven temperature sensor are:

- Bake element will not heat.
- Broil element will not heat.
- Timed baking will not work.
- Oven temperature is not accurate.
- Unusual display readouts and/or error codes.

FIGURE 22-36
A typical oven temperature sensor used in electronic ranges and ovens.

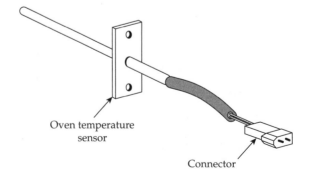

Oven temperature
sensor

Connector

To handle these problems, perform the following steps:

1. **Verify the complaint.** Verify the complaint by turning the oven selector switch to the desired setting and setting the oven thermostat knob to the desired temperature. Do those functions of the range/oven work? Check the use and care manual for the model you are servicing to make sure that you have set the controls properly. Turn off the electricity to the appliance and wait for two minutes before turning it back on. If a fault code appears, look up the code. If the range/oven will not power up, locate the technical data sheet behind the control panel for the diagnostics information. On some models you will need the actual service manual for the model you are working on to properly diagnose the range/oven. The service manual will assist you in properly placing the range/oven in the service test mode for testing the range/oven functions.

2. **Check for external factors.** You must check for external factors not associated with the appliance. Is the appliance installed properly? Does it have the correct voltage? The voltage at the receptacle is between 198 volts and 264 volts during a load on the circuit (see Chapter 6).

3. **Disconnect the electricity.** Before working on the range or oven, disconnect the electricity to the appliance. This can be done by pulling the plug from the receptacle. Or disconnect the electricity at the fuse panel or at the circuit breaker panel. Turn off the electricity.

WARNING *Some diagnostic tests will require you to test the components with the power turned on. When you disassemble the control panel, you can position it in such a way so that the wiring will not make contact with metal. This act will allow you to test the components without electrical mishaps.*

4. **Gain access to the oven temperature sensor.** You can gain access to the component by removing the back panel (see Figure 22-9) on a freestanding range. To locate the oven temperature sensor, look in the upper-right or upper-left side of the oven cavity from the back of the range (Figure 22-37).

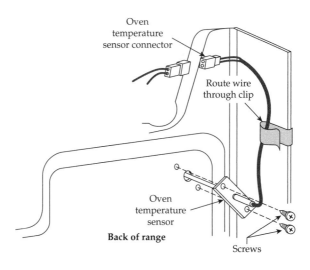

FIGURE 22-37
The technician can service the oven temperature sensor from the back of the range or oven.

Oven temperature sensor connector

Route wire through clip

Oven temperature sensor

Back of range

Screws

Temperature (°F)	Resistance (Ω)
32	1000
75	1100
250	1450
350	1650
450	1850
550	2050
650	2230
900	2700

Figure 22-38 Sample temperature resistance chart. For the correct meter reading, you must refer to the technical data sheet or the service manual for the model you are servicing.

5. **Test the oven temperature sensor.** To test the temperature sensor, unplug the connector (see Figure 22-37). Set the multimeter to the ohms scale, and set the range to R × 1. Next, touch the multimeter test leads to the connector plug pins on the oven temperature sensor. The meter will indicate the resistance at the temperature tested (see Figure 22-38).

6. **Remove the oven temperature sensor.** Unplug the connector (Figure 22-37) from the oven temperature sensor. Now, remove the two screws from the oven temperature sensor, pull the sensor toward you, and remove it from the back of the oven.

7. **Install a new oven temperature sensor.** To install the new oven temperature sensor, just reverse the disassembly procedure and reassemble. Do not forget to plug in the oven temperature sensor connector. Reassemble the back panel in the reverse order of disassembly, and test. Make sure to take the range out of the service test mode when the repair is completed.

Oven Cycling Relay

The oven cycling relay is located in the back of the range (older models). This relay is used for turning on and off the bake and broil elements in the oven.

The typical complaints associated with oven cycling relay failure are:

- Bake element will not heat.
- Broil element will not heat.
- Bake element stays on all the time.
- Oven temperature is not accurate.
- Unusual display readouts and/or error codes.

To handle these problems, perform the following steps:

1. **Verify the complaint.** Verify the complaint by turning the oven selector switch to the desired setting and setting the oven thermostat to the desired temperature. Do these functions of the oven work? Test all of the cycles. Turn off the electricity to the appliance and wait for two minutes before turning it back on. If a fault code appears, look up the code. If the range/oven will not power up, locate the technical data sheet behind the control panel for the diagnostics information. On some models you will need the actual service manual for the model you are working on to properly diagnose the range/oven. The service manual will assist you in properly placing the range/oven in the service test mode for testing the range/oven functions.

2. **Check for external factors.** You must check for external factors not associated with the appliance. Is the appliance installed properly? Does it have the correct voltage? The voltage at the receptacle is between 198 volts and 264 volts during a load on the circuit (see Chapter 6).

3. **Disconnect the electricity.** Before working on the range or oven, disconnect the electricity to the appliance. This can be done by pulling the plug from the receptacle. Or disconnect the electricity at the fuse panel or at the circuit breaker panel. Turn off the electricity.

WARNING *Some diagnostic tests will require you to test the components with the power turned on. When you disassemble the control panel, you can position it in such a way that the wiring will not make contact with metal. This act will allow you to test the components without electrical mishaps.*

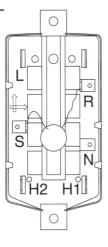

FIGURE 22-39
The oven cycling relay.

4. **Gain access to the oven cycling relay.** To access the oven cycling relay (Figure 22-39), remove the back panel (see Figure 22-9). On some older models, this relay is located under the oven cavity, where the utility drawer is usually found.

5. **Test the oven cycling relay.** To test the oven cycling relay, remove all of the wires from the relay (label them). Using the ohmmeter, set the range scale on R × 1. Place the probes on the L and H1 terminals; there should not be continuity. Next, place the probes on the L and H2 terminals; there should not be continuity. Then place the probes on the H1 and H2 terminals. There should not be continuity. Now place the probes on the S and R terminals. There should be continuity. If your meter indicates no reading at all between the S and R terminals, this indicates that the heater wire inside the relay is defective and that the relay should be replaced.

6. **Remove the oven cycling relay.** To remove the oven cycling relay, remove the two screws that secure the relay to the cabinet.

7. **Install a new oven cycling relay.** To install the new oven cycling relay, just reverse the order of step 6. Transfer the wires from the old relay to the new relay. Be sure that you install the wires on their correct terminals according to the wiring diagram. Reassemble the remainder of the range/oven in the reverse order of its disassembly, and test. On electronic models, make sure to take the range out of the service test mode when the repair is completed.

Range/Oven Time Clock

The range/oven time clock is located in the control panel. The clock runs on 120 volts and controls the timer, time bake, clean cycle, and the time of day. On electronic models the clock will display error/fault codes also.

The typical complaints associated with failure of the range/oven time clock are:

- Timed bake cycle will not operate.
- Unable to use the self-cleaning cycle.
- Clock loses time.
- Clock is not functioning.
- The minute reminder is not functioning.
- Unusual display readouts and/or error codes.

To handle these problems, perform the following steps:

1. **Verify the complaint.** Verify the complaint by turning the clock controls to the correct time. Does the clock time advance? Some models have the minute reminder located on the same clock. Set this timer to test it. Does it advance? Turn off the electricity to the appliance and wait for two minutes before turning it back on. If a fault code appears, look up the code. If the range/oven will not power up, locate the technical data sheet behind the control panel for the diagnostics information. On some models you will need the actual service manual for the model you are working on to properly diagnose the range/oven. The service manual will assist you in properly placing the range/oven in the service test mode for testing the range/oven functions.

2. **Check for external factors.** You must check for external factors not associated with the appliance. Is the fuse blown? Does the appliance have the correct voltage? The voltage at the receptacle is between 198 volts and 264 volts during a load on the circuit (see Chapter 6).

3. **Disconnect the electricity.** Before working on the range or oven, disconnect the electricity to the appliance. This can be done by pulling the plug from the receptacle. Or disconnect the electricity at the fuse panel or at the circuit breaker panel. Turn off the electricity.

WARNING *Some diagnostic tests will require you to test the components with the power turned on. When you disassemble the control panel, you can position it in such a way that the wiring will not make contact with metal. This act will allow you to test the components without electrical mishaps.*

4. **Gain access to the range/oven clock.** You can access the clock by removing the back panel (see Figure 22-9) on a freestanding range. On models with front-mounted controls (see Figure 22-10), the panel is attached with screws on both ends. Remove the screws, and tilt the control panel. Be careful not to let the wires come off their components. Some built-in models have a removable backsplash (see Figure 22-11); just lift the backsplash and rest it on the cooktop. It would be a good idea to place something on the cooktop first to protect it from damage. Next, remove the screws from the backsplash that hold the rear panel to gain access to the clock. If you are repairing a wall oven or an eye-level range, the control panel can be removed (see Figure 22-12) by opening the door and removing the screws that secure the panel.

These might be underneath the front of the exhaust hood or just below the control panel. Some control panels are hinged; just tilt the control panel toward you for servicing. On other models, to gain access to the clock, both the rear panel and the front control panel (usually glass) will have to be removed. Remove the screws that secure the trim piece that holds the glass in place. Then remove the trim piece. On some models, the trim might have to be removed first. Figure 22-40 illustrates the different types of clock faces available on ranges/ovens.

FIGURE 22-40
(a) Standard time clock. (b) Analog clock with minute reminder, timed baking, and self-cleaning. (c) and (d) Digital clock with timed baking, self-cleaning, and minute reminder. (e) Electronic clock, with or without additional features.

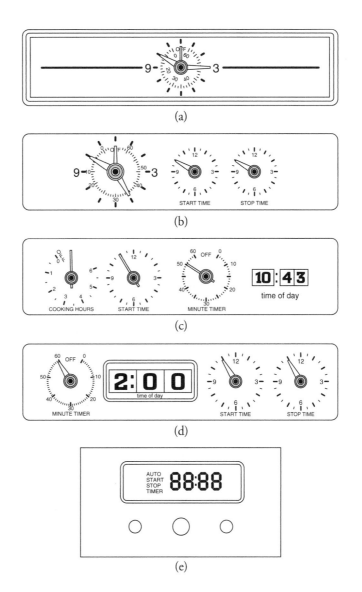

5. **Test the range/oven clock.** On some models, if the clock does not run, check for a fuse in the circuit. Locate the clock motor wire leads, and isolate them from the circuit (label them). Using the ohmmeter, set the range scale on R × 1. Place the probes on the clock motor leads; there should be continuity. If not, replace the clock. On some models, the clock is used to control the timed baking and the self-cleaning cycles (see Figures 22-40b, 22-40c, and 22-40d). To check the switch mechanism of the clock, remove the wires from the switch terminals (label them). Using the ohmmeter, set the range scale on R × 1. Place the probes on the terminals. Look at the wiring diagram for the correct terminals to test. Some models have one switch; other models have two sets of switches. Test for continuity of the switch contacts when you push in and turn the start and stop knobs on the clock and when the knobs pop out. On electronic control models, refer to the technical data sheet located in the range for the testing procedure.

6. **Remove the range/oven time clock.** First remove the clock knobs from the clock stems. To remove the range/oven time clock in this model (Figure 22-41), use a screwdriver and depress the clips that hold the clock to the control panel. On other models, the clock is secured to the control panel by screws or nuts. Pull the clock toward the front of the appliance (Figure 22-42).

7. **Install a new range/oven clock.** Transfer the wires from the old clock to the new clock. Be sure that you install the wires on their correct terminals according to the wiring diagram. To install the new range/oven clock, just reverse the disassembly procedure, and reassemble. On electronic models, make sure to take the range out of the service test mode when the repair is completed.

Electronic Control Board and User Interface Controls

The electronic control board and the user interface controls operate the bake, broil, time bake, and the self-clean functions of the range or oven. The electronic control board also displays the time of day, timer, and any error or fault codes. The electronic control board and the user interface controls are located in the control panel.

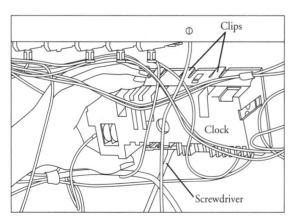

The typical complaints associated with the electronic control board or the user interface controls are:

- The range or oven won't run or power up.
- Unable to program the range or oven.
- The display board will not display anything.
- One or more key pads will not accept commands.
- Unusual display readouts.
- The control displays error codes.

To prevent electrostatic discharge (ESD) from damaging expensive electronic components, follow the steps in Chapter 11.

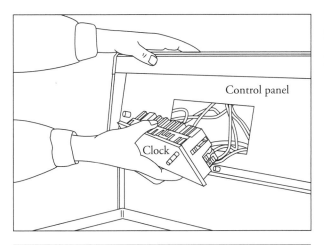

FIGURE 22-42 Removing the clock from the front of the control panel.

To handle these problems, perform the following steps:

1. **Verify the complaint.** Verify the complaint by operating the range or oven. Turn off the electricity to the appliance and wait for two minutes before turning it back on. If a fault code appears, look up the code. If the range/oven will not power up, locate the technical data sheet behind the control panel for the diagnostics information. On some models you will need the actual service manual for the model you are working on to properly diagnose the range/oven. The service manual will assist you in properly placing the range/oven in the service test mode for testing the range/oven functions.

2. **Check for external factors.** You must check for external factors not associated with the appliance. The voltage at the receptacle is between 198 volts and 264 volts during a load on the circuit (see Chapter 6).

3. **Disconnect the electricity.** Before working on the range or oven, disconnect the electricity. This can be done by pulling the plug out of the wall receptacle. Or disconnect the electricity at the fuse panel or circuit breaker panel. Turn off the electricity.

WARNING *Some diagnostic tests will require you to test the components with the power turned on. When you disassemble the control panel, you can position the panel in such a way that the wiring will not make contact with metal. This will allow you to test the components without electrical mishaps.*

4. **Remove the console panel to gain access.** You can access the electronic control (some models have multiple circuit boards) by removing the back panel (see Figure 22-9) on a freestanding range. On models with front-mounted controls (see Figure 22-10), the panel is attached with screws on both ends. Remove the screws, and tilt the control panel. Be careful not to let the wires come off their components. Some built-in models have a removable backsplash (see Figure 22-11); just lift the backsplash and rest it on the cooktop. It would be a good idea to place something on the cooktop first to protect it from damage. Next, remove the screws from the backsplash that hold the rear panel to gain access to the electronic control. If you are repairing a wall oven or an eye-level range, the control panel can be removed (see Figure 22-12) by opening the door and removing the screws that secure the panel. These might be underneath the front of the exhaust hood or just below the control panel. Some control panels are hinged; just tilt the control panel toward you for servicing. On other models, to gain access to the electronic control, both the rear panel and the front control panel (usually glass) will

have to be removed. Remove the screws that secure the trim piece that holds the glass in place. Then remove the trim piece. On some models, the trim might have to be removed first.

5. **Test the electronic control board and user interface controls.** If you are able to run the range/oven diagnostic test mode, check the different functions of the range/oven. Use the technical data sheet for the model you are servicing to locate the test points from the wiring schematic. Check all wiring connections and wiring. Using the technical data sheet, test the electronic control or user interface controls, input voltages, and output voltages. On some models, fuses are soldered to the printed circuit board (PCB). These fuses must be tested first before condemning the component.

6. **Remove the electronic control board or user interface control.** To remove the defective component, remove the screws that secure the board to the control panel or range/oven frame. Disconnect the connectors from the electronic control board or user interface control.

7. **Install the new component.** To install a new electronic control board or user interface control, just reverse the disassembly procedure, and reassemble. You must also follow the installation instructions that are packed with every component. The manufacturers like to use a universal board that fits most of their models. The technician will have to program the new board to the model being serviced. Reinstall the console panel, and restore the electricity to the range/oven. Test the range operation. Make sure to take the range out of the service test mode when the repair is completed

Diagnostic Charts and Wiring Diagrams

The following diagnostic charts will help you to pinpoint the likely causes of the various problems associated with electric ranges and ovens (Figures 22-43, 22-44, 22-45, and 22-46).

The wiring diagrams in this chapter are examples only. You must refer to the actual wiring diagram on the range, oven, or cooktop that you are servicing. Figures 22-47 and 22-48 are for an identical appliance. Figure 22-47 depicts an actual wiring schematic. Figure 22-48 depicts a cross-section of a ladder wiring diagram (strip circuit). This type of diagram is easier to read and understand. Figure 22-49 depicts a typical wiring schematic for an electronic range. Figures 22-50, 22-51, 22-52, 22-53, and 22-54 depict the strip circuits for the various functions of the electronic range.

Table 22-3 illustrates a self-cleaning cycle chart indicating a light soil cycle. On some models, the cleaning cycle will take two hours to complete. The heavy soil cycle will take two to three hours to complete. The average stabilized temperature in a self-cleaning cycle should be between 840 and 920 degrees Fahrenheit. The door can only be opened when the oven temperature has dropped below 520 degrees Fahrenheit.

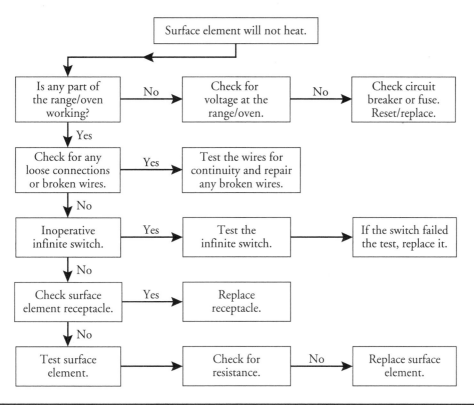

Figure 22-43   Diagnostic flowchart: Surface element will not heat.

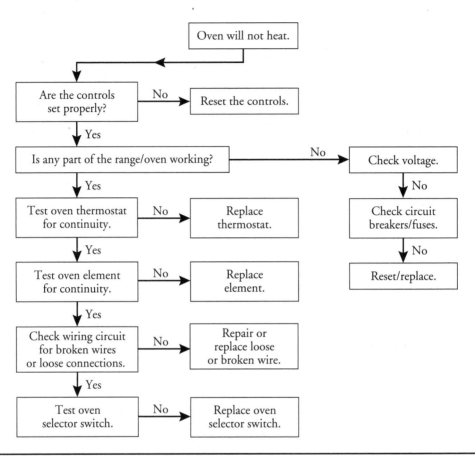

FIGURE 22-44 Diagnostic flowchart: Oven will not heat.

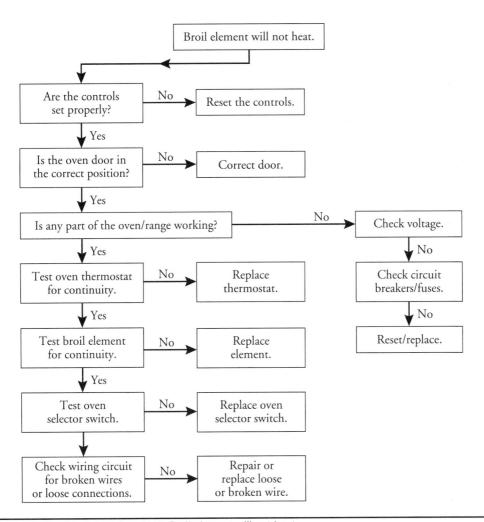

FIGURE 22-45   Diagnostic flowchart: Broil element will not heat.

PART VI

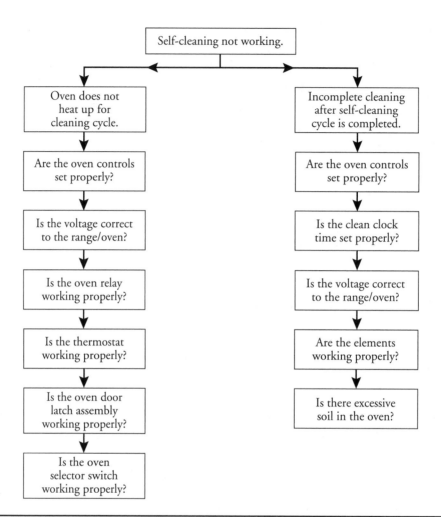

FIGURE 22-46 Diagnostic flowchart: Self-cleaning function not working.

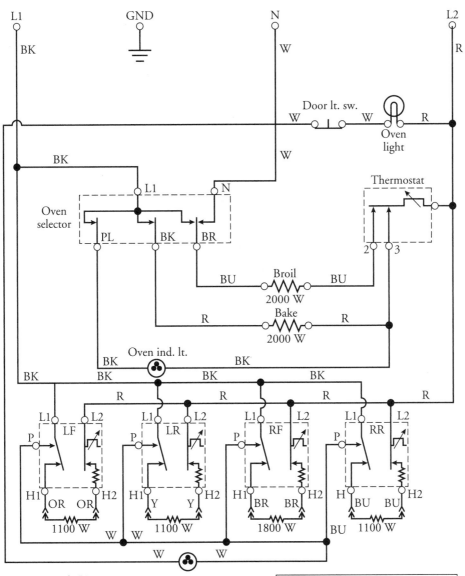

L-1, L-2 = 240 V
L-1, N = 120 V
L-2, N = 120 V

Selector switch operation	
Position	Connection
Off	No connection
Bake	L1-BK, L1-PL, N-BR
Broil	L1-BR, L1-PL

FIGURE 22-47 Wiring schematic of a typical electric range.

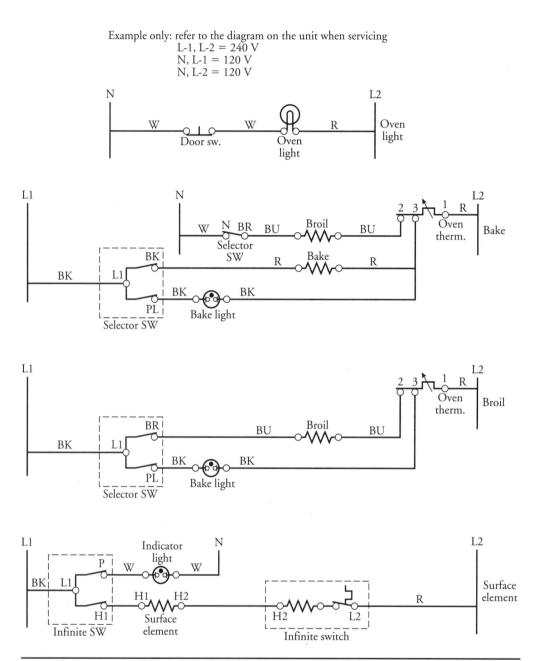

FIGURE 22-48   The oven light, bake, broil, and surface element circuits.

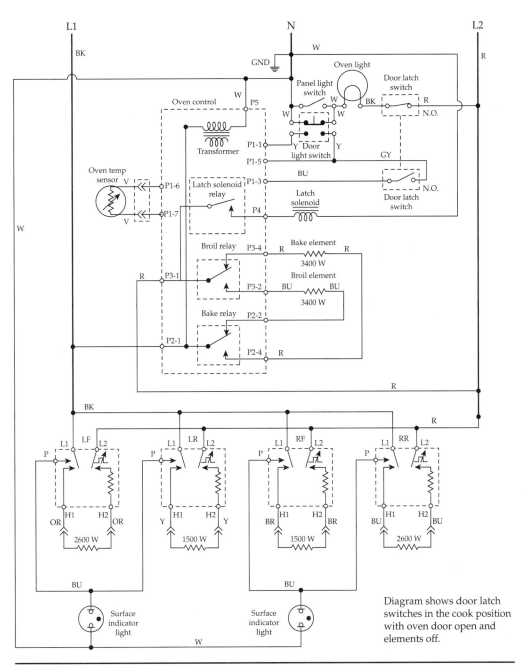

FIGURE 22-49 A typical wiring schematic depicting an electronic oven control board.

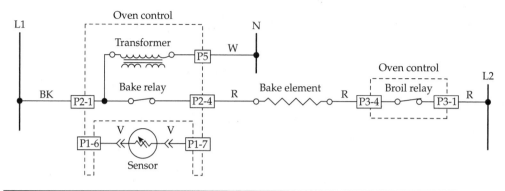

FIGURE 22-50   A typical strip circuit for the bake function.

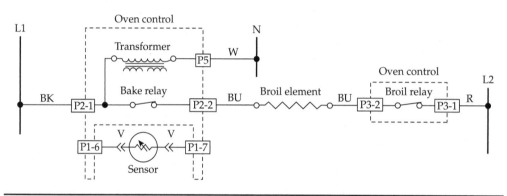

FIGURE 22-51   A typical strip circuit for the broil function.

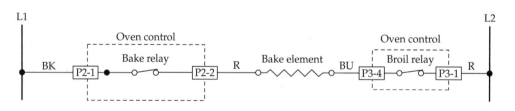

FIGURE 22-52   A typical strip circuit for the clean function.

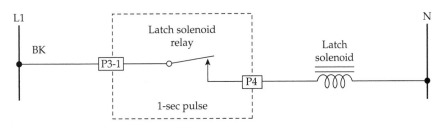

Figure 22-53 A typical strip circuit for the latch solenoid circuit.

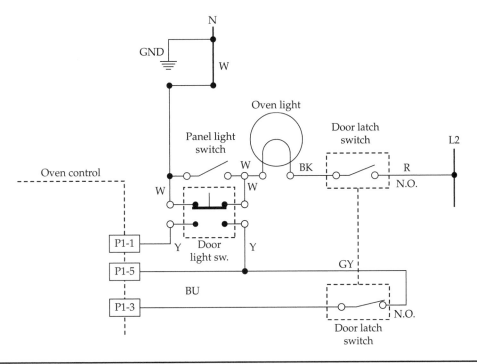

Figure 22-54 A typical strip circuit for the latch circuit.

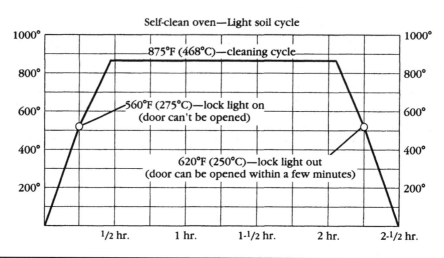

TABLE 22-3 Self-Cleaning Cycle Chart

Gas Ranges and Ovens

Gas ranges, ovens, and cooktops are designed on the basis of using the heat generated by combustion to cook the food. The types of fuels that feed gas appliances are:

- **Natural gas**   Natural gas is lighter than air and has a heating value between 900 and 1200 BTUs.

- **Liquefied petroleum (LP or LPG)**   LP gas is heavier than air and has a heating value between 2500 and 3200 BTUs.

- **Mixed gas**   Mixed gas is lighter than air and has a heating value between 700 and 900 BTUs.

- **Manufactured gas**   Manufactured gas is lighter than air and has a heating value between 500 and 700 BTUs.

To sustain combustion in a range, oven, or cooktop, an ignition source, such as a flame or by electrical means, is used to ignite the gas vapors. To ignite these gases, the temperature will have to be between 900 and 1200 degrees Fahrenheit.

The pressure of natural gas that is supplied to a residence will vary between a five- and nine-inch water column. LP or LPG gas pressure for residential appliances, as established by the gas industry, will be between a nine- and eleven-inch water column. To determine the correct pressure rating, the technician must refer to the manufacturer's specifications or the installation instructions for that product. The two most common types of gas used in homes are LP and natural gas.

On some models, electricity (120 volts) is supplied to the gas appliance for ignition, temperature control, safety valve, electronic controls, and accessories (clock, lights, etc.). Gas cooking appliances are available in a wide variety of styles also. Figures 1-1, 1-4, and 23-1 are just a small sampling.

This chapter will provide the basics needed to diagnose and repair gas cooking appliances. Figure 23-2 identifies where components are located within the gas range. This illustration is used as an example only. The actual construction and features might vary, depending on the brand and model you are servicing.

FIGURE 23-1 A five-burner gas cooktop with sealed burners.

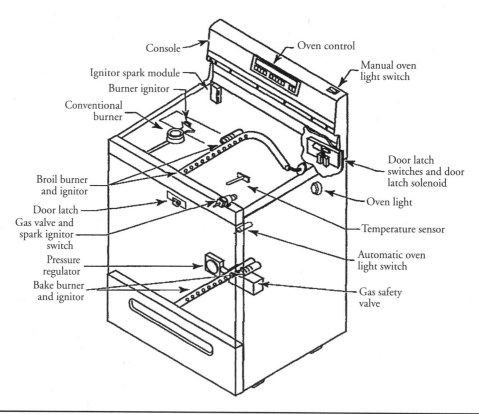

FIGURE 23-2 The locations of components in a gas freestanding range.

Principles of Operation

When the gas is turned on (standing pilot ignition system), whether by a knob or automatically by a control device, gas will begin to flow from the supply line through a gas pressure regulator into the gas manifold for distribution to the burners or to the oven safety valve (Figure 23-3). The gas travels from the manifold through the controls, passing through the orifice into the venturi throat and mixing tube. The air-gas mixture then enters the burner head, exits through the burner ports, and travels through the flash tube to the pilot flame (Figure 23-4). Once ignited, the flame begins to travel back through the flash tube to the charge ports and climber ports and burns evenly around the burner head.

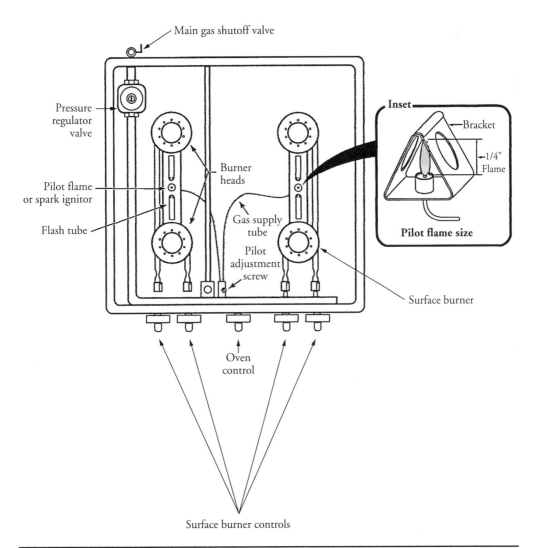

FIGURE 23-3 A top view of a standing pilot ignition system, illustrating the component locations in a freestanding gas range. See inset for pilot flame height.

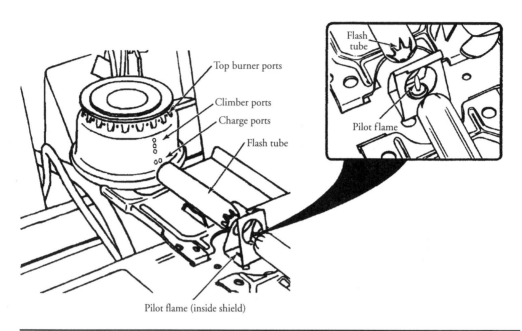

Top burner ports

Climber ports

Charge ports

Flash tube

Pilot flame (inside shield)

Flash tube

Pilot flame

FIGURE 23-4 The air-gas mixture is traveling through the burner ports and flash tube, encountering the pilot flame and igniting the burner.

When the oven thermostat is turned on, gas flows from the manifold (Figure 23-5) through the thermostat and to the safety valve. The thermostat also feeds gas through the pilot gas line to the pilot assembly. The pilot flame will increase in size, heating the safety valve-sensing bulb (Figure 23-6), opening up the safety valve and allowing gas to flow to the burner. When the air-gas mixture flows out of the oven burner, it encounters the pilot flame and the burner ignites. When the oven cavity temperature is satisfied, the gas flow will stop to the oven burner and the gas flow to the pilot assembly will decrease to the standing pilot light.

On models with electronic ignition, the standing pilot is replaced with ignitor switches, surface burner electrodes, and a spark module (Figures 23-7, 23-8, 23-9, and 23-10). When the consumer turns the burner control knob to the "lite" position, a switch within the ignitor switch will close, supplying 120 volts through the ignitor switch to the spark module. Some spark modules will produce a high-voltage pulse to the ignitor every two seconds at very low amperage (4 milliamps). Some models use a spark module that will produce two to three sparks per second. This pulse will produce a spark between the ignitor electrode and the ignitor-grounding strap (see Figure 23-8). At the same time, the gas will begin to flow from the supply line through a gas pressure regulator into the gas manifold for distribution to the burners. When the burner is lit, the control knob will then be turned to the "on" position, where the height of the flame can be adjusted. When this occurs, the switch within the ignitor switch will open, turning off the voltage to the spark module (see Figure 23-7).

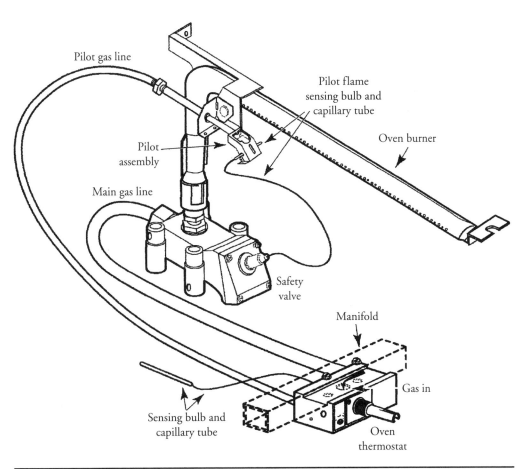

Figure 23-5 The components that make up the oven burner system with a standing pilot ignition system.

When the oven thermostat is turned on and set to the desired temperature, a switch contact from within the thermostat will close. This action will complete the circuit, supplying 120 volts to the glow-bar ignition system (Figure 23-11). The glow-bar ignitor (Figure 23-12) must heat up to 2000 degrees Fahrenheit, allowing 2.5 to 4.0 amps to flow to the safety valve. As the bimetal warps, the safety valve will open, allowing the gas to flow through the orifice into the venturi throat and mixing tube. The air-gas mixture then enters the oven burner head, exits through the burner ports, and travels to the glow-bar ignitor, igniting the oven burner. Once ignited, the flame begins to burn evenly around the burner head. The entire process takes about 60 to 90 seconds to complete.

The self-cleaning operation in a gas range is similar to that of the electric range (see Chapter 22). The only difference is that some models use two burners (Figure 23-13). The broil burner allows the consumer to broil food in the same oven cavity. In the self-cleaning mode, both burners do not come on at the same time. As Figure 23-13 shows, the burners are controlled by the electronic control to operate the dual safety valve and the glow-bar

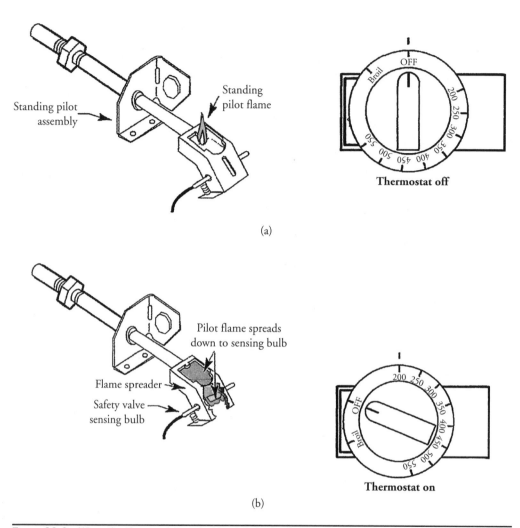

(a)

(b)

Figure 23-6   When the thermostat control knob is in the "off" position, the pilot flame stands up. With the thermostat control knob turned on, the pilot flame will increase in size, heating the safety valve-sensing bulb.

ignitors. When the oven cavity temperature reaches 600 degrees Fahrenheit, the door latch lock light will come on, locking the oven door and preventing the consumer from opening it. When the oven cavity temperature stabilizes (between 840 and 920 degrees Fahrenheit), the cleaning process begins. This cleaning process requires approximately two to three hours. When the self-cleaning cycle is complete, the burners turn off and the oven begins to cool down. When the oven cavity temperature drops below 600 degrees Fahrenheit, the oven latch lock light will go off, allowing the consumer to open the oven door.

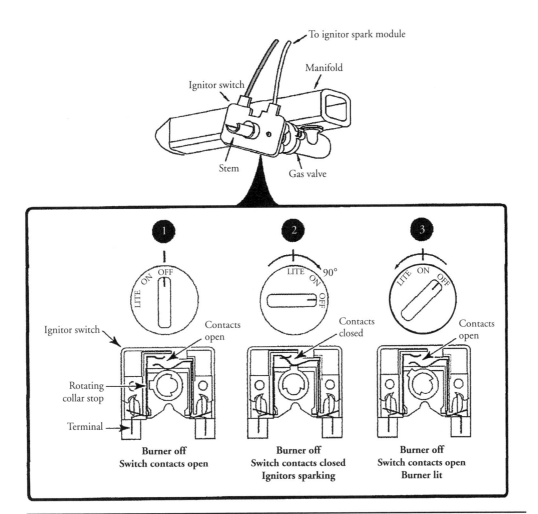

FIGURE 23-7   The electronic ignition switch is mounted over the surface burner valve stem. 1. When the knob is in the "off" position, the contact switch is open and there is no gas flow. 2. The knob is set in the "lite" position, the contact switch is closed, and gas flows to the burner. 3. When the knob is turned to the "on" position, the ignitor switch will open, breaking the circuit, and gas flows to the burner.

Safety First

Any person who cannot use basic tools or follow written instructions should *not* attempt to install, maintain, or repair any gas range, oven, or cooktop. Any improper installation, preventive maintenance, or repairs could create a risk of personal injury or property damage.

If you do not fully understand the installation, preventive maintenance, or repair procedures in this chapter, or if you doubt your ability to complete the task on your appliance, please call your service manager.

PART VI

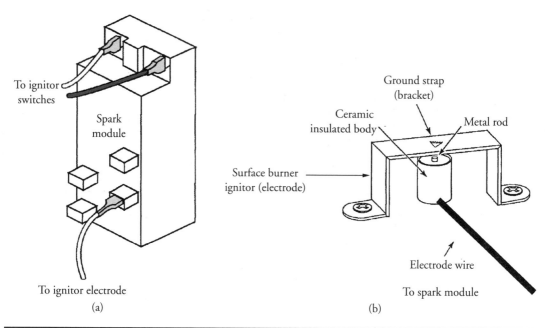

FIGURE 23-8 (a) A typical spark module. (b) The surface burner ignitor and grounding bracket.

The following precautions should also be followed:

- Always follow the use and care guide instructions from the manufacturer.
- Never use a range to heat the home; it simply was not designed for that purpose.
- Keep the cooking area clean of soot, spills, and grease.
- Do not use flammable liquids near a cooking appliance.
- When repairing the range, always use the proper tools.
- Always reconnect the wires to the range after repairs have been made.
- Before servicing any gas components in the range, turn off the gas supply.
- Always have a fire extinguisher nearby in case of mishaps that might lead to a fire.
- Make sure that gas appliances have proper venting according to the manufacturer's recommendations.
- To eliminate personal injury or fire when operating the appliance, avoid using the storage cabinets above the product.

Before continuing, take a moment to refresh your memory of the safety procedures in Chapter 2.

Do not store any flammable materials near a gas appliance, and avoid ignition sources in the event you smell gas leaking. Make sure you have proper ventilation when servicing the gas appliance to prevent carbon monoxide (CO) poisoning.

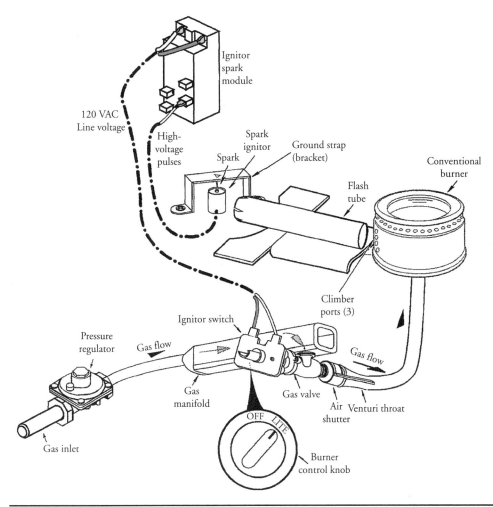

FIGURE 23-9   The components that make up the standard burner system using an electronic ignition system to light the burner.

Gas Ranges, Ovens, and Cooktops in General

Much of the troubleshooting information in this chapter covers gas ranges, ovens, and cooktops in general, rather than specific models, in order to present a broad overview of service techniques. The illustrations that are used in this chapter are for demonstration purposes only to clarify the description of how to service these appliances. They in no way reflect on a particular brand's reliability.

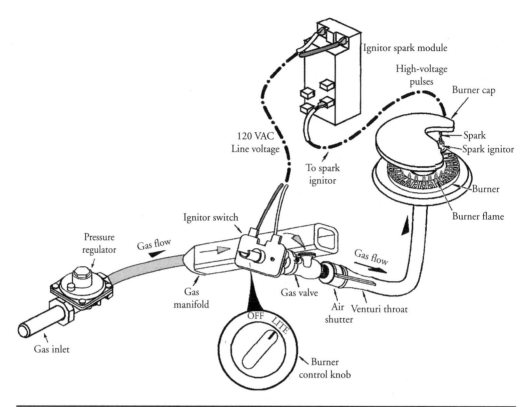

Figure 23-10   The components that make up the sealed burner system using an electronic ignition system to light the burner.

Location and Installation of Gas Range, Oven, and Cooktop

Locate the range, oven, or cooktop where it will be well lighted and have access to proper ventilation. The range must be level for proper baking and cooking results. The range might be installed adjacent to the left and/or right base cabinets and against a rear vertical wall (for the "anti-tip" cleat). A wall oven must be installed on a supporting surface that is strong enough to support the weight of the oven and its contents while remaining level from side to side and from front to rear. A cooktop must be installed on a flat surface, supported by the countertop, and it should be level. Contact your flooring company to check if the flooring can withstand a minimum of 200 pounds. It is also recommended that you contact your builder to determine if the cabinets and walls can withstand the heat produced by the gas appliance. Some kitchen cabinets and building materials were not designed to withstand the heat that is given off by the product.

The proper installation instructions for your model are included with the appliance. These instructions will assist you with the installation requirements (dimensions, electrical requirements, cutout dimensions, venting, etc.) needed to complete the installation according to the manufacturer's specifications.

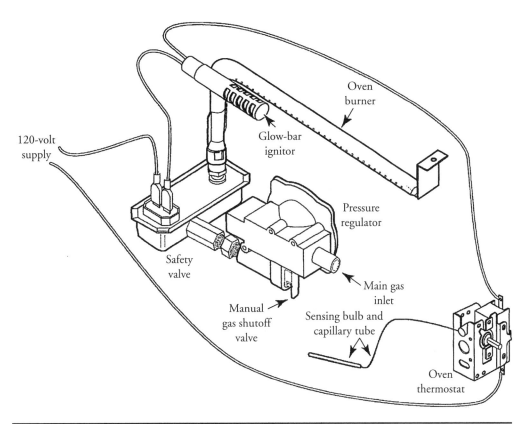

FIGURE 23-11   The components that make up the oven burner system using a glow-bar ignition system to light the burner.

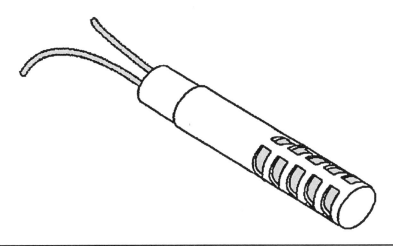

FIGURE 23-12   A typical glow-bar ignitor.

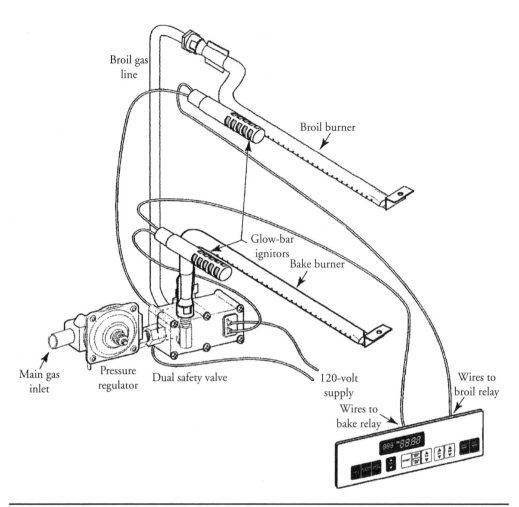

FIGURE 23-13   The components that make up the self-cleaning oven system using two glow-bar ignitors to light the burners. The consumer can bake or broil in the same oven cavity.

Some gas appliance models are equipped with electronic controls and electronic ignition. The electrical plug that is considered for use must be properly polarized and grounded in order for the electronic controls to function properly. The appliance chassis must also be grounded so that the electronics in the product will function properly.

Step-by-Step Troubleshooting by Symptom Diagnosis

When servicing an appliance, don't overlook the simple things that might be causing the problem. Step-by-step troubleshooting by symptom diagnosis is based on diagnosing malfunctions, with their possible causes arranged into categories relating to the operation of the gas cooking appliance. This section is intended only to serve as a checklist to aid you in diagnosing a problem. Look at the symptom that best describes the problem that you are experiencing with the range, oven, or cooktop, and then proceed to correct the problem.

Gas Odors

- Check the pilot light.
- Check for a gas leak in the supply line.
- Check the ventilation system.
- Check the burner controls. When not in use, the controls must be completely turned off.
- Check the regulator.

Surface Burner Will Not Light

- Check the pilot light.
- Check the pilot burner port holes for a blockage.
- Check for 120 volts to the appliance. Check fuses or circuit breakers.
- On electronic models, check the ignition system.
- Check the burner openings for a blockage.
- Check the air-gas mixture.
- Check for a blockage in the gas supply to the appliance and burner.
- Make sure that the gas supply is turned on.

The Pilot for the Surface Burner Will Not Stay Lit

- Check the burner for blockages.
- Check the setting and adjust the pilot light.
- Check the air-gas mixture.
- Check for a blockage in the gas supply to the appliance and burner.
- Check for a wind source that may affect the pilot light.

The Surface Burner Flame Is Too Low

- Check the burner assembly for blockages.
- Check the air-gas mixture.
- Check for a blockage in the gas supply to the appliance and burner.
- Test the gas pressure to the appliance and make corrections, if necessary.
- Check and make sure that the range/oven is converted to the correct gas type (LP or natural).

The Surface Burner Flame Is Too High

- Check the air-gas mixture.
- Test the gas pressure to the appliance and make corrections, if necessary.
- Check and make sure that the range/oven is converted to the correct gas type (LP or natural).

The Oven Burner Will Not Light or Stay Lit

- Check the pilot light. If the pilot is too small, adjust the height.
- Check the control settings on the clock or electronic controls.
- Check for 120 volts to the appliance.
- Check the ignition system.
- Check the fuses in the appliance.
- Check the glow-bar ignitor.
- Check the selector switch, thermostat, and the safety valve for proper operation.
- Check for a blockage in the pilot light.
- Test the gas pressure to the appliance and make corrections, if necessary.
- Check and make sure that the range/oven is converted to the correct gas type (LP or natural).
- On electronic models, check for unusual display readouts and/or error codes.

Oven Temperature Will Not Hold

- Check the oven temperature and thermostat control.
- Check for a blockage in the oven burner.
- Check the oven door and gasket.
- On electronic models, check for unusual display readouts and/or error codes.

Self-Cleaning Function Is Not Working

- Check the controls for the proper settings.
- Check the door latch assembly. Make sure the latch engages.
- Check the thermostat and the selector switch for proper operation.
- Check for 120 volts to the appliance.
- Check the door alignment and gasket.
- On electronic models, check for unusual display readouts and/or error codes.

The Burner Orifice Squeals

- Check the orifice opening for debris.
- Make sure that the correct orifice size is in use.
- Test the gas pressure to the appliance and make corrections, if necessary.

Supply Air Is Noisy

- Check the venturi throat and mixing tube for debris and rough edges.
- Check the burner positioning over the orifice.

Erratic Burner Flame

- Check the gas pressure at the gas regulator.
- Check the burner controls for proper operation.

Yellow or Sooty Burner Flame

- Make sure that the burner is in the correct position.
- Check the air-gas mixture.
- Test the gas pressure to the appliance and make corrections, if necessary.

Gas Appliance Maintenance

To maintain gas appliances, always follow the manufacturer's recommendations for periodic maintenance as stated in the use and care manual. The range, oven, or cooktop can be cleaned with warm water, mild detergent, and a soft cloth on all cleanable parts, as recommended in the use and care manual. Also, never use abrasive cleaners that are not recommended by the manufacturer.

Do not allow grease spillovers to accumulate on top of the range after cooking; they will become a fire hazard. When cleaning the burners, always make sure that all of the portholes are free of debris. If, for any reason, the burner portholes are blocked, the flame appearance will be different. Blocked portholes will reduce gas flow and the heating valve of the burner will be reduced.

Repair Procedures

Each repair procedure is a complete inspection and repair process for a single range, oven, or cooktop component, containing the information you need to test a component that might be faulty and replace it, if necessary.

Any person who cannot use basic tools should *not* attempt to install, maintain, or repair any electric dryer. Any improper installation, preventative maintenance, or repairs will create a risk of personal injury, as well as property damage. Call the service manager if installation, preventative maintenance, or the repair procedure is not fully understood.

Gas Burner Valve

The gas burner valve is mounted on the manifold and it supplies gas to the burners.

The typical complaints associated with the burner valve are:

- It is hard to turn on.
- When the gas burner valve is on, little or no gas comes out.
- There is a gas smell around the burner valve.
- Unusual display readouts and/or error codes.

PART VI

To handle these problems, perform the following steps:

1. **Verify the complaint.** Verify the complaint by testing for gas leaks. Next, turn on the burners. Is the pilot light on? Are the electronic switches working?

2. **Check for external factors.** You must check for external factors not associated with the appliance. Is the appliance installed properly? Are there 120 volts to the appliance? The voltage at the receptacle is between 108 volts and 132 volts during a load on the circuit. Do you have the correct polarity? (See Chapter 6.) Is the gas turned on?

3. **Disconnect the electricity.** Before working on the appliance, disconnect the electricity. This can be done by pulling the plug from the electrical outlet. Or disconnect the electricity at the fuse panel or at the circuit breaker panel. Turn off the electricity.

WARNING *Some diagnostic tests will require you to test the components with the power turned on. When you disassemble the control panel, you can position it in such a way that the wiring will not make contact with metal. This act will allow you to test the components without electrical mishaps.*

4. **Shut off the gas supply.** Before you begin servicing any gas components, shut off the supply of gas to the appliance. The shutoff valve should be within 6 feet of the gas appliance.

5. **Gain access to the gas burner valve.** To gain access to the gas burner valve, lift the cooktop, pull off the control knobs, and then remove the screws from the front panel (Figure 23-14). Remove the control panel. This will expose the gas burner valve and manifold (Figure 23-15).

6. **Remove the gas burner valve.** To remove the gas burner valve, remove the screw that secures the burner to the burner bar, and then lift the burner head up and out of the range (Figure 23-16). With the gas burner valve still attached to the gas manifold, remove the orifice with a wrench, turning it counterclockwise. Now remove the gas burner valve by removing the mounting screw that is secured to the manifold right above the valve (see Figure 23-15). Remove the valve. On some models you might have to completely remove the manifold from the range in order to remove the screw that secures the valve to the manifold.

7. **Install a new gas burner valve.** To install a new gas burner valve, just reverse the disassembly procedure, and reassemble. Be careful not to damage the washers when reinstalling the gas burner valve (see Figure 23-15). Most gas ranges, ovens, and cooktops have universal orifices. When you remove the orifice, you must reinstall that orifice for the type of gas supplied to the range. For natural gas, the orifice must be tightened down clockwise about two to three turns, making sure that the gas LP insert rests against the orifice cap, then unscrew counterclockwise the orifice cap about two and a half turns (Figure 23-17a). For LP gas, the orifice will tighten down only two and a half times, resting against the gas LP insert (Figure 23-17b). If the orifice is tightened down too much the gas flow will be restricted. Reinstall the front

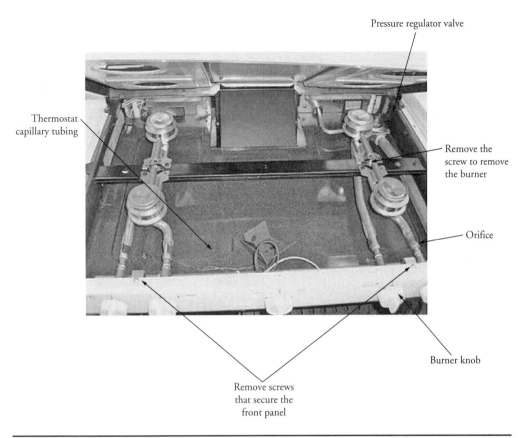

Pressure regulator valve

Thermostat
capillary tubing

Remove the
screw to remove
the burner

Orifice

Burner knob

Remove screws
that secure the
front panel

FIGURE 23-14 A top view of the components in an electronic ignition system.

panel, and turn on the gas supply to the appliance. Test for any gas leaks. Then turn on the electricity, test the operation, and adjust the burner flame if needed. For additional information, see Chapter 6. On electronic models, make sure to take the range/oven out of the service test mode when the repair is completed.

Ignitor Switch

The ignitor switches (four) are wired in parallel and mounted on each burner valve stem. These switches are rotary-actuated (Figure 23-7). The ignitor switches control the 120 volt supply voltage to the spark module by opening and closing the circuit.

The typical complaints associated with the ignitor switch are:

- The ignitor switch is hard to turn on.
- The surface burner will not light.
- Unusual display readouts and/or error codes.

PART VI

FIGURE 23-15
An exploded view of the surface burner valve.

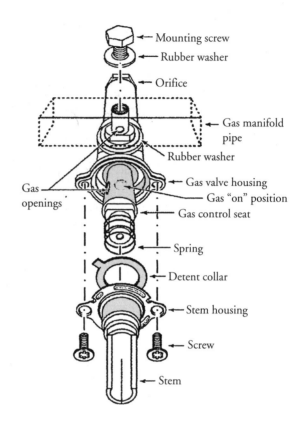

- Mounting screw
- Rubber washer
- Orifice
- Gas manifold pipe
- Rubber washer
- Gas valve housing
- Gas "on" position
- Gas control seat
- Spring
- Detent collar
- Stem housing
- Screw
- Stem

Gas openings

FIGURE 23-16
To remove the surface burner, remove the screw that secures the burner to the burner bar, and then lift the burner head up and out of the range.

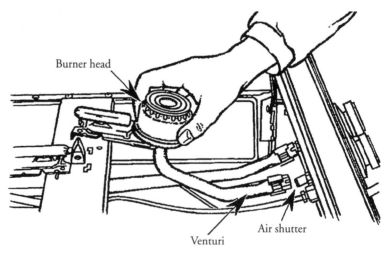

Burner head

Venturi

Air shutter

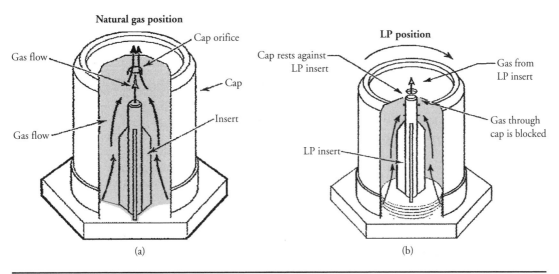

FIGURE 23-17   (a) The universal orifice in the natural gas position. (b) The universal orifice in the LP position.

To handle these problems, perform the following steps:

1. **Verify the complaint.** Verify the complaint by turning on the surface burners. Do any of the burners come on? Do you hear or see the ignitors sparking? Are the ignitor electrodes clear of debris? Is there good contact between the ignitor bracket and the burner bar? On electronic models, turn off the electricity to the appliance and wait for two minutes before turning it back on. If a fault code appears, look up the code. If the range/oven will not power up, locate the technical data sheet behind the control panel or for diagnostics information. On some models you will need the actual service manual for the model you are working on to properly diagnose the range/oven. The service manual will assist you in properly placing the range/oven in the service test mode for testing the range/oven functions.

2. **Check for external factors.** You must check for external factors not associated with the appliance. Is the appliance installed properly? Are there 120 volts to the appliance? The voltage at the receptacle is between 108 volts and 132 volts during a load on the circuit. Do you have the correct polarity? (See Chapter 6.)

3. **Disconnect the electricity.** Before working on the range or cooktop, disconnect the electricity to the appliance. This can be done by pulling the plug from the receptacle. Or disconnect the electricity at the fuse panel or at the circuit breaker panel. Turn off the electricity.

WARNING *Some diagnostic tests will require you to test the components with the power turned on. When you disassemble the control panel, you can position it in such a way that the wiring will not make contact with metal. This act will allow you to test the components without electrical mishaps.*

4. **Shut off the gas supply.** Before you begin servicing any gas components, shut off the supply of gas to the appliance. The shutoff valve should be within six feet of the gas appliance.

5. **Gain access to the ignitor switch.** To gain access to the ignitor switch, lift the cooktop, pull off the control knobs, and then remove the screws from the front panel (see Figures 23-7 and 23-14). Remove the control panel to expose the ignitor switches.

6. **Test the ignitor switch.** Remove the wires from the ignitor switch terminals. Set your ohmmeter to the R × 1 scale; reinstall the surface burner knob on the gas valve. Place the ohmmeter probes on the ignitor switch terminals. With the burner knob in the "off" position, the ohmmeter should read no continuity. Turn the burner knob to the "lite" position; the ohmmeter should read continuity. If your readings differ, replace the ignitor switch.

7. **Install a new ignitor switch.** To install a new ignitor switch, remove the screws that secure the ignitor switch, reverse the disassembly procedure, and reassemble. Turn on the electricity and the gas supply, and test the surface burner. On some models, the ignitor will produce between two and three sparks per second. On electronic models, make sure to take the range/oven out of the service test mode when the repair is completed.

Surface Burner Ignitor

A surface burner ignitor is a metal rod with a ceramic insulating body that is wired to a spark module. Electrical pulses from the spark module cause the surface burner ignitor to arc to the ground strap, which is mounted above the ignitor. These sparks will ignite the surface burner. On a gas range there are two ignitors to light the four surface burners.

The typical complaints associated with the surface burner ignitor are:

- The surface burner will not light.
- Only one surface burner ignitor is working.
- Intermittent operation of the surface burner ignitor.
- Unusual display readouts and/or error codes.

To handle these problems, perform the following steps:

1. **Verify the complaint.** Verify the complaint by turning on the surface burners. Do the burners come on? Do you hear or see the ignitors sparking? Are the ignitor electrodes clear of debris? Is there a good ground contact between the ignitor bracket and the burner bar? If all of the ignitors are sparking, check for a shorted ignitor switch. On electronic models, turn off the electricity to the appliance and wait for two minutes before turning it back on. If a fault code appears, look up the code. If the range/oven will not power up, locate the technical data sheet behind the control panel or for diagnostics information. On some models you will need the actual service manual for the model you are working on to properly diagnose the range/oven. The service manual will assist you in properly placing the range/oven in the service test mode for testing the range/oven functions.

2. **Check for external factors.** You must check for external factors not associated with the appliance. Is the appliance installed properly? Are there 120 volts to the appliance? The voltage at the receptacle is between 108 volts and 132 volts during a load on the circuit. Do you have the correct polarity? (See Chapter 6.)

3. **Disconnect the electricity.** Before working on the range or cooktop, disconnect the electricity to the appliance. This can be done by pulling the plug from the receptacle. Or disconnect the electricity at the fuse panel or the circuit breaker panel. Turn off the electricity.

WARNING   *Some diagnostic tests will require you to test the components with the power turned on. When you disassemble the control panel, you can position it in such a way that the wiring will not make contact with metal. This act will allow you to test the components without electrical mishaps.*

4. **Shut off the gas supply.** Before you begin servicing any gas components, shut off the supply of gas to the appliance. The shutoff valve should be within six feet of the gas appliance.

5. **Gain access to the ignitor.** To gain access to the ignitor (see Figure 23-8), lift up the cooktop (see Figure 23-14). The ignitor is located on the burner bar, between the two burners' flash tubes.

6. **Test the ignitor.** Set your ohmmeter on R × 100. Test the ignitor bracket to ground; the meter should indicate continuity. Disconnect the ignitor electrode wire and test for continuity.

7. **Install a new ignitor.** Remove the screws that secure the ignitor bracket to the burner bar, and disconnect the ignitor electrode wire. To reinstall the ignitor, just reverse the disassembly procedure, and reassemble. Test for a good ground between the ignitor bracket and the burner bar. Turn on the electricity and the gas supply, and test the surface burners. The ignitor should produce two sparks per second or, on some models, two to three sparks per second. On electronic models, make sure to take the range/oven out of the service test mode when the repair is completed.

Spark Module

The spark module is located in the rear of the range or underneath the cooktop on a cooktop model. Whenever the spark module is energized by the ignitor switches through an ignitor cable, a solid-state circuit and pulse transformer within the module housing will send electronic pulses to both surface ignitors at the same time. The spark module operates at very low amperage (4 milliamps). The high-voltage pulses or sparks present no severe shock hazard to the consumer or to the service technician.

The typical complaints associated with the spark module are:

- You hear a "clicking" noise but the surface burner will not light.
- You hear an erratic "clicking" noise from the surface burner ignitor.
- You do not hear the "clicking" noise and the surface burner will not light.
- Unusual display readouts and/or error codes.

To handle these problems, perform the following steps:

1. **Verify the complaint.** Verify the complaint by turning on the surface burners. Do the burners come on? Do you hear or see the ignitors sparking? Are the ignitor electrodes clear of debris? Is there good contact between the ignitor bracket and the burner bar?. On electronic models, turn off the electricity to the appliance and wait for two minutes before turning it back on. If a fault code appears, look up the code. If the range/oven will not power up, locate the technical data sheet behind the control panel or for diagnostics information. On some models you will need the actual service manual for the model you are working on to properly diagnose the range/oven. The service manual will assist you in properly placing the range/oven in the service test mode for testing the range/oven functions.

2. **Check for external factors.** You must check for external factors not associated with the appliance. Is the appliance installed properly? Are there 120 volts to the appliance? Does the electrical outlet have the correct polarity? The voltage at the receptacle is between 108 volts and 132 volts during a load on the circuit. Do you have the correct polarity? (See Chapter 6.) Is the gas to the appliance turned on?

3. **Disconnect the electricity.** Before working on the range/oven or cooktop, disconnect the electricity to the appliance. This can be done by pulling the plug from the receptacle. Or disconnect the electricity at the fuse panel or at the circuit breaker panel. Turn off the electricity.

WARNING *Some diagnostic tests will require you to test the components with the power turned on. When you disassemble the control panel, you can position it in such a way that the wiring will not make contact with metal. This act will allow you to test the components without electrical mishaps.*

4. **Shut off the gas supply.** Before you begin servicing any gas components, shut off the supply of gas to the appliance. The shutoff valve should be within six feet of the gas appliance.

5. **Gain access to the spark module.** You can access the component by removing the back panel (see Figure 22-9). Locate the spark module (see Figure 23-8).

6. **Test the spark module.** Before you test the spark module, test the surface burner igniter electrodes and the surface burner igniter switches first. Next, check the grounding strap on the ignitor. Now, check the ignition system wiring for a short or loose wiring. If they test okay, proceed to test the spark module (Figures 23-8 and 23-18). Set your voltmeter on the 120-volt scale. Remove the L1 wire from the spark module, and attach one of the meter probes to that wire. Attach the other meter probe to the neutral terminal on the spark module. Next, turn on the electricity to the appliance, turn on the surface burner control knob to the "lite" position, and test for 120 volts. The voltmeter should read 120 volts. If it does not, check the wiring connections and wiring. After checking the wiring and connections, if you still do not have the ignitors sparking, then replace the spark module.

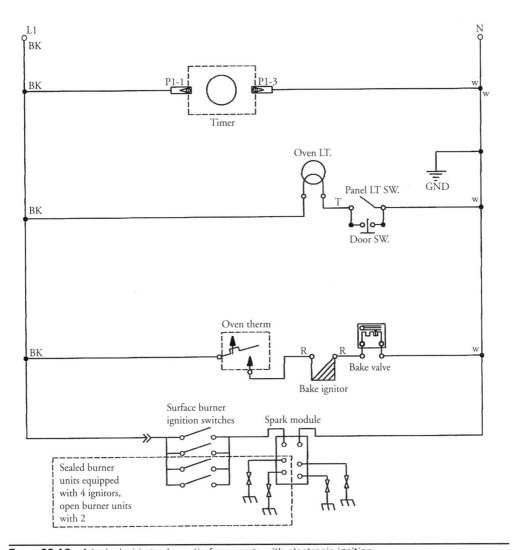

FIGURE 23-18 A typical wiring schematic for a range with electronic ignition.

7. **Remove the spark module.** With the electricity turned off, disconnect the wiring terminals from the spark module one at a time and label them. Remove the screws that secure the spark module to the appliance.

8. **Install a new spark module.** To install the new spark module, just reverse the disassembly procedure, and reassemble. Be sure that you install the wires onto the correct terminals according to the wiring diagram for the product you are servicing. Turn on the electricity and gas supply, and test the surface burners. Make sure that the spark at the ignitor has a sharp blue appearance. This indicates a good working ignition system. On electronic models, make sure to take the range/oven out of the service test mode when the repair is completed.

Glow-Bar Ignitor

The glow-bar ignitor is constructed of a silicon-carbide material. When 120 volts is applied to the glow-bar, it heats up to about 2000 degrees F. As the temperature increases, its resistance decreases, allowing 2.5 to 4.0 amps to flow to the safety valve.

The typical complaints associated with the glow-bar ignitor are:

- The oven/broiler will not light.
- The glow-bar ignitor is on but, the oven/broiler burner will not ignite.
- Unusual display readouts and/or error codes.

To handle these problems, perform the following steps:

1. **Verify the complaint.** Verify the complaint by turning on the oven. Does the glow-bar ignitor turn an orange-red color? On electronic models, turn off the electricity to the appliance and wait for two minutes before turning it back on. If a fault code appears, look up the code. If the range/oven will not power up, locate the technical data sheet behind the control panel or for diagnostics information. On some models you will need the actual service manual for the model you are working on to properly diagnose the range/oven. The service manual will assist you in properly placing the range/oven in the service test mode for testing the range/oven functions.

2. **Check for external factors.** You must check for external factors not associated with the appliance. Is the appliance installed properly? Are there 120 volts to the appliance? The voltage at the receptacle is between 108 volts and 132 volts during a load on the circuit. Do you have the correct polarity? (See Chapter 6.) Is the gas to the appliance turned on?

3. **Disconnect the electricity.** Before working on the range, oven, or cooktop, disconnect the electricity to the appliance. This can be done by pulling the plug from the receptacle. Or disconnect the electricity at the fuse panel or at the circuit breaker panel. Turn off the electricity.

WARNING *Some diagnostic tests will require you to test the components with the power turned on. When you disassemble the control panel, you can position it in such a way that the wiring will not make contact with metal. This act will allow you to test the components without electrical mishaps.*

4. **Shut off the gas supply.** Before you begin servicing any gas components, shut off the supply of gas to the appliance. The shutoff valve should be within six feet of the gas appliance.

5. **Gain access to the glow-bar ignitor.** To gain access to the glow-bar ignitor (see Figures 23-11 and 23-13), open the oven door, remove the oven racks, and remove the oven floor panel. Next, remove the nut (Figure 23-19) from the oven burner baffle, and remove the baffle. You now have access to all of the components shown in Figure 23-19.

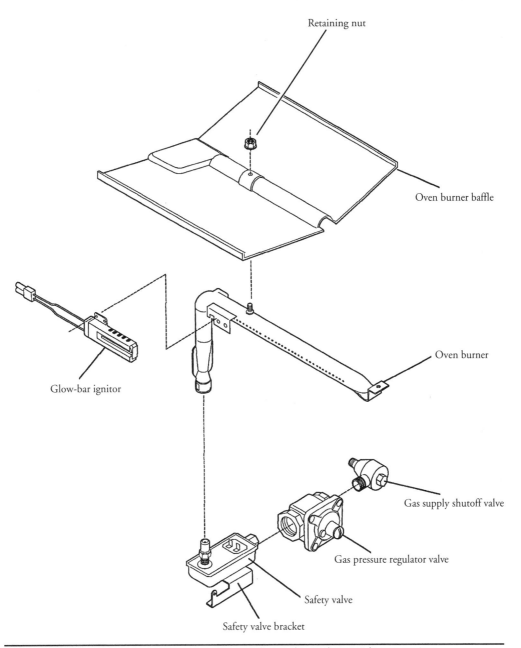

FIGURE 23-19   An exploded view of the components that make up the oven burner system.

6. **Test the glow-bar ignitor.** Disconnect the glow-bar ignitor wires from the wiring harness. Set your ohmmeter on the R × 100 scale, and check for continuity at the glow-bar ignitor wire terminals. Read the meter scale—at room temperature of 75 degrees F, the resistance of the glow-bar ignitor should between 50 and 1100 ohms. If it is under 50 or over 1200 ohms, replace the ignitor. If you are testing the glow-bar ignitor for amperage and it's below 2.5 amps, replace it.

7. **Remove the glow-bar ignitor.** To remove the ignitor, remove the screws that secure the glow-bar ignitor to the oven burner (see Figures 23-11, 23-13, and 23-19).

8. **Install the new glow-bar ignitor.** To install the new glow-bar ignitor, just reverse the disassembly procedure, and reassemble. Be careful—do not touch the silicon carbide bar within the metal enclosure. If you do, it will create a hot spot on the silicon carbide bar (from body oils), and the ignitor might burn out. Turn on the electric and gas supply, and test the glow-bar ignitor. On electronic models, make sure to take the range/oven out of the service test mode when the repair is completed.

Pressure Regulator Valve

This mechanical device will perform two functions: it will reduce the higher incoming gas pressure to a desired lower pressure, and it will maintain a steady and an even flow of gas through the regulator valve. This valve is located in the bottom rear of the range or it can be located under the cooktop in the left or right rear.

The typical complaints associated with the pressure regulator valve are:

- The oven/broiler burner will not ignite. No gas is supplied to the burner.
- Erratic oven/broiler burner flame. The flame will fluctuate between high and low.
- Unusual display readouts and/or error codes.

To handle these problems, perform the following steps:

1. **Verify the complaint.** Verify the complaint by turning on the appliance. Do the surface burners come on? Are the burner flames small and blue in color? Did you check the gas pressure at the gas supply line, surface burners, and the oven burner? Do the burner flames fluctuate in size? Is the pilot light staying lit? Is there a gas smell near the pressure regulator valve? On electronic models, turn off the electricity to the appliance and wait for two minutes before turning it back on. If a fault code appears, look up the code. If the range/oven will not power up, locate the technical data sheet behind the control panel or for diagnostics information. On some models you will need the actual service manual for the model you are working on to properly diagnose the range/oven. The service manual will assist you in properly placing the range/oven in the service test mode for testing the range/oven functions.

2. **Check for external factors.** You must check for external factors not associated with the appliance. Is the appliance installed properly? Are there 120 volts to the appliance? The voltage at the receptacle is between 108 volts and 132 volts during a load on the circuit. Do you have the correct polarity? (See Chapter 6.) Is the gas to the appliance turned on?

3. **Disconnect the electricity.** Before working on the range or cooktop, disconnect the electricity to the appliance. This can be done by pulling the plug from the receptacle. Or disconnect the electricity at the fuse panel or at the circuit breaker panel. Turn off the electricity.

WARNING *Some diagnostic tests will require you to test the components with the power turned on. When you disassemble the control panel, you can position it in such a way that the wiring will not make contact with metal. This act will allow you to test the components without electrical mishaps.*

4. **Shut off the gas supply.** Before you begin servicing any gas components, shut off the supply of gas to the appliance. The shutoff valve should be within six feet of the gas appliance.

5. **Gain access to the pressure regulator valve.** Locate the pressure regulator valve. On some models, lift the cooktop or gain access through the oven cavity (see Figures 23-3, 23-11, 23-14, and 23-19).

6. **Test the pressure regulator.** The pressure regulator valve is a non-serviceable component. If you suspect a problem with the regulator, test the gas pressures with a manometer or a magnehelic gauge (see Figures 8-22 and 8-24). Test the gas pressure at the supply line before the pressure regulator valve and at the surface burner orifices. On some models using natural gas, the reading will be between a four- and six-inch water column. For LP models, the pressure will be between a nine- and eleven-inch water column. Next, test the gas pressure from the burner orifice and turn on all the burners; the pressure should remain constant. If the readings are not within the ranges mentioned, replace the pressure regulator.

7. **Remove the gas pressure regulator valve.** To remove the gas pressure valve (see Figure 23-14), pull out the range from the wall and lift up the cooktop. Disconnect the gas supply line from the range. Remove the gas line from the intake side of the regulator. Remove the surface burners, pull off the control knobs, and then remove the screws from the front panel (see Figure 23-14). This will expose the gas burner valves, thermostat, and manifold. Next, remove the thermostat from the manifold. If the model you are servicing has electronic ignition, remove the wires from the ignitor switches and label them. Finally, remove the screws that secure the manifold to the range. Figure 8-20 shows an exploded view of the gas components. Remove the manifold with the pressure regulator. Be careful—do not damage the capillary tube in the thermostat. To remove the gas pressure regulator valve (see Figures 23-11, 23-13, and 23-19), disconnect the gas line to the regulator. Remove the oven burner and ignitor. Next, remove the screws that secure the safety valve bracket to the range, and remove the safety valve with the regulator attached.

8. **Install the new gas pressure regulator.** To install the gas pressure regulator, just reverse the disassembly procedure, and reassemble. Make sure that you use approved pipe joint compound or yellow Teflon tape on the pipe connections (according to local codes) and on the threads of the pressure regulator valve before you install the valve in the range. Turn on the gas supply to the range. Check for gas leaks by using a chloride-free soap solution. *Do not use a flame to check for gas leaks.* The soap solution will begin to bubble if a leak is present. Turn on the electric and test the range operation. On electronic models , make sure to take the range/oven out of the service test mode when the repair is completed.

Safety Valve

The safety valve is located in the bottom at the rear of the range or oven (Figures 23-5 and 23-11). The electric safety valve is wired in series with the ignitor or glow-bar. The safety valve will only open to allow gas to flow to the burner when there is enough current (2.5 to 4 amps) to cause the bimetal-controlled diaphragm within the valve body to warp open and allow the gas to flow through the safety valve to the burner.

The typical complaints associated with the safety valve are:

- The oven/broiler burner will not ignite. No gas is supplied to the burner.
- Broken capillary tube.
- Unusual display readouts and/or error codes.

To handle these problems, perform the following steps:

1. **Verify the complaint.** Verify the complaint by turning on the appliance. Does the oven or broil burner come on? On electronic models, turn off the electricity to the appliance and wait for two minutes before turning it back on. If a fault code appears, look up the code. If the range/oven will not power up, locate the technical data sheet behind the control panel or for diagnostics information. On some models you will need the actual service manual for the model you are working on to properly diagnose the range/oven. The service manual will assist you in properly placing the range/oven in the service test mode for testing the range/oven functions.

2. **Check for external factors.** You must check for external factors not associated with the appliance. Is the appliance installed properly? Are there 120 volts to the appliance? The voltage at the receptacle is between 108 volts and 132 volts during a load on the circuit. Do you have the correct polarity? (See Chapter 6.) Is the gas to the appliance turned on?

3. **Disconnect the electricity.** Before working on the range or cooktop, disconnect the electricity to the appliance. This can be done by pulling the plug from the receptacle. Or disconnect the electricity at the fuse panel or at the circuit breaker panel. Turn off the electricity.

WARNING *Some diagnostic tests will require you to test the components with the power turned on. When you disassemble the control panel, you can position it in such a way that the wiring will not make contact with metal. This act will allow you to test the components without electrical mishaps.*

4. **Shut off the gas supply.** Before you begin servicing any gas components, shut off the supply of gas to the appliance. The shutoff valve should be within six feet of the gas appliance.

5. **Gain access to the safety valve.** To gain access to the safety valve (see Figure 23-20), open the oven door, remove the oven racks, and remove the oven floor panel. Next, remove the nut (see Figure 23-19) from the oven burner baffle, and remove the baffle. You now have access to all of the components shown in Figure 23-19.

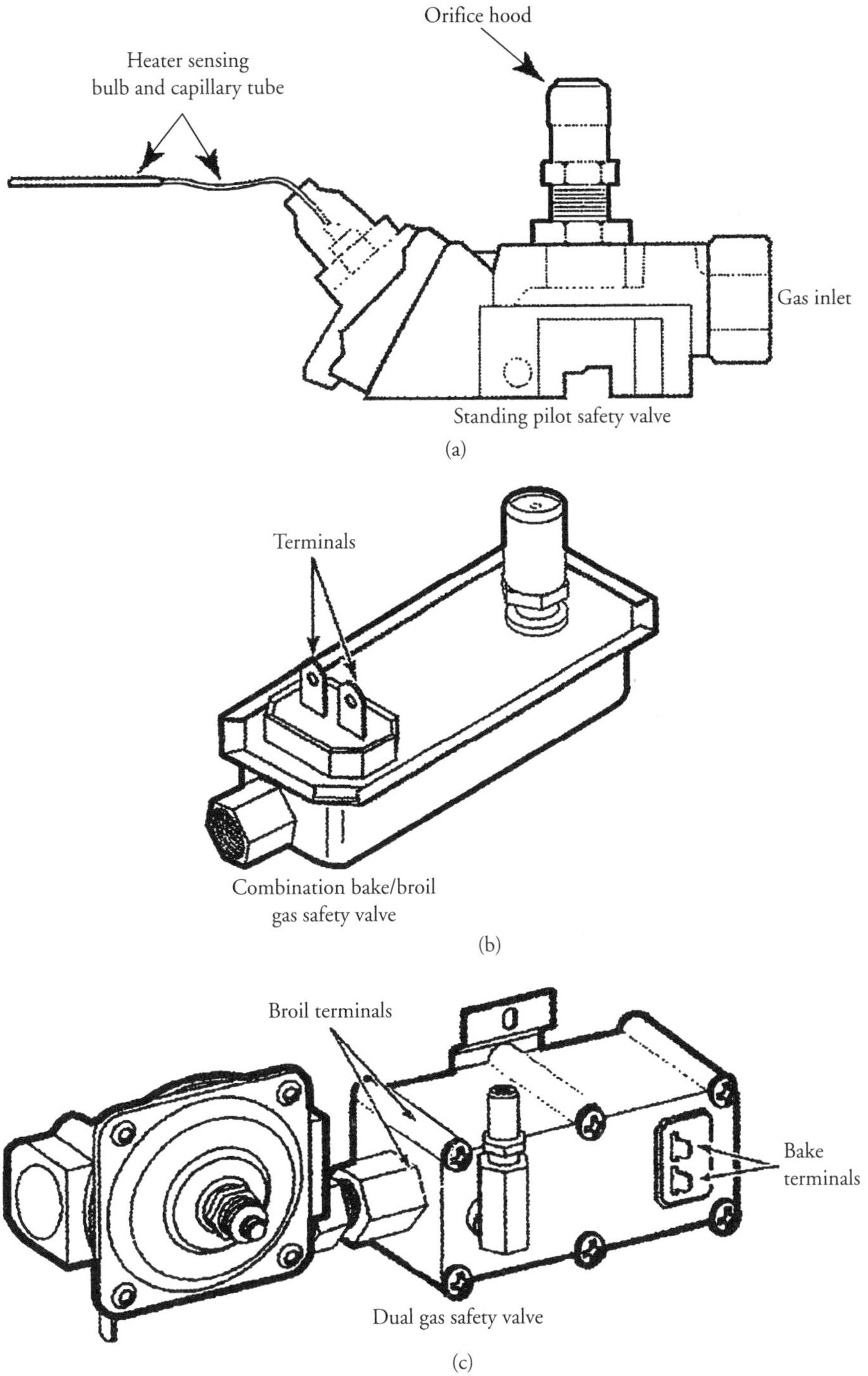

Heater sensing
bulb and capillary tube

Orifice hood

Gas inlet

Standing pilot safety valve

(a)

Terminals

Combination bake/broil
gas safety valve

(b)

Broil terminals

Bake
terminals

Dual gas safety valve

(c)

FIGURE 23-20   (a) A standing pilot safety valve. (b) A combination bake/broil safety valve. (c) A dual gas safety valve with the pressure regulator attached.

6. **Test the safety valve.** The standing pilot safety valve cannot be tested: it is a temperature-controlled device. If you notice that the capillary tube is broken, replace the safety valve. Other models use an electrically controlled safety valve (bimetal operated). To test this type of valve (see Figure 23-20), set your ohmmeter to the R × 1 scale. Disconnect the wires from the valve terminals, and touch the ohmmeter probes to the terminals. The reading should be between 1 and 5 ohms. If your meter reads an open circuit or high resistance, the safety valve will have to be replaced.

7. **Remove the safety valve.** To remove the gas safety valve (see Figures 23-11, 23-13, and 23-19), disconnect the gas line to the regulator. Remove the oven burner and ignitor. Next, remove the screws that secure the safety valve bracket to the range, and remove the safety valve with the regulator attached.

8. **Install the new safety valve.** To install the safety valve, just reverse the disassembly procedure, and reassemble. Make sure that you use approved pipe joint compound or yellow Teflon tape on the pipe connections (according to local codes) and on the threads of the pressure regulator valve and the safety valve before you install the valve(s) in the range. Turn on the gas supply to the range. Check for gas leaks by using a chloride-free soap solution. *Do not use a flame to check for gas leaks.* The soap solution will begin to bubble if a leak is present. Turn on the electric and test the operation of the range/oven. On electronic models, make sure to take the range/oven out of the service test mode when the repair is completed.

Electronic Control Board and User Interface Controls

The electronic control board and the user interface controls operate the bake, broil, time bake, and the self-clean functions of the range or oven. The electronic control board also displays the time of day, timer, and any error or fault codes. The electronic control board and the user interface controls are located in the control panel.

The typical complaints associated with the electronic control board or the user interface controls are:

- The range or oven won't run or power up.
- Unable to program the range or oven.
- The display board will not display anything.
- One or more key pads will not accept commands.
- Unusual display readouts and/or error codes.

To prevent electrostatic discharge (ESD) from damaging expensive electronic components, follow the steps in Chapter 11.

To handle these problems, perform the following steps:

1. **Verify the complaint.** Verify the complaint by operating the range or oven. Turn off the electricity to the appliance and wait for two minutes before turning it back on. On electronic models, turn off the electricity to the appliance and wait for two minutes before turning it back on. If a fault code appears, look up the code. If the range/oven will not power up, locate the technical data sheet behind the control panel or for diagnostics information. On some models you will need the actual service manual for the model you are working on to properly diagnose the range/oven. The service

manual will assist you in properly placing the range/oven in the service test mode for testing the range/oven functions.

2. **Check for external factors.** You must check for external factors not associated with the appliance. Are there 120 volts of electricity to the range or oven? The voltage at the receptacle is between 108 volts and 132 volts during a load on the circuit. Do you have the correct polarity? (See Chapter 6.)

3. **Disconnect the electricity.** Before working on the range or oven, disconnect the electricity. This can be done by pulling the plug out of the wall receptacle. Or disconnect the electricity at the fuse panel or circuit breaker panel. Turn off the electricity.

WARNING *Some diagnostic tests will require you to test the components with the power turned on. When you disassemble the control panel, you can position the panel in such a way that the wiring will not make contact with metal. This will allow you to test the components without electrical mishaps.*

4. **Shut off the gas supply.** Before you begin servicing any gas components, shut off the supply of gas to the appliance. The shutoff valve should be within six feet of the gas appliance.

5. **Remove the console panel to gain access.** You can access the electronic control board (some models have multiple circuit boards) by removing the back panel (see Figure 22-9) on a freestanding range. On models with front-mounted controls (see Figure 22-10), the panel is attached with screws on both ends. Remove the screws, and tilt the control panel. Be careful not to let the wires come off their components. Some built-in models have a removable backsplash (see Figure 22-11); just lift the backsplash and rest it on the cooktop. It would be a good idea to place something on the cooktop first to protect it from damage. Next, remove the screws from the backsplash that hold the rear panel to gain access to the electronic control board. If you are repairing a wall oven or an eye-level range, the control panel can be removed (see Figure 22-12) by opening the door and removing the screws that secure the panel. These might be underneath the front of the exhaust hood or just below the control panel. Some control panels are hinged; just tilt the control panel toward you for servicing. On other models, to gain access to the electronic control board, both the rear panel and the front control panel (usually glass) will have to be removed. Remove the screws that secure the trim piece that holds the glass in place.

6. **Test the electronic control board and user interface control.** If you are able to run the range/oven diagnostic test mode, check the different functions. Use the technical data sheet for the model you are servicing to locate the test points on the wiring schematic. Check all wiring connections and wiring. Using the technical data sheet, test the electronic control or user interface controls, input voltages, and output voltages. On some models, fuses are soldered to the printed circuit board (PCB). These fuses must be tested first before condemning the component.

7. **Remove the electronic control board or user interface control.** To remove the defective component, remove the screws that secure the boards to the control panel or range/oven frame. Disconnect the connectors from the electronic control board or user interface control.

8. **Install the new component.** To install a new electronic control board or user interface control, read the part installation literature that came with the part and just reverse the disassembly procedure, and reassemble. Reinstall the console panel, and restore the electricity and gas supply to the range/oven. Test the range/oven operation. On electronic models, make sure to take the range/oven out of the service test mode when the repair is completed.

Oven Temperature Sensor

The oven temperature sensor (Figure 22-36) is a thermistor. The thermistor is a thermally sensitive resistor that exhibits a change in electrical resistance with a change in its temperature. The oven temperature sensor is located in the upper left or right attached to the rear wall of the oven cavity. The oven temperature sensor sends the resistance readings to the electronic oven control board. The electronic control board monitors the resistance readings of the oven temperature sensor and turns the bake or broil on or off at the desired temperature.

The typical complaints associated with the oven temperature sensor are:

- Erratic temperature in the oven cavity.
- The oven/broiler burner will not ignite.
- Unusual display readouts and/or error codes.

To handle these problems, perform the following steps:

1. **Verify the complaint.** Verify the complaint by turning on the appliance. Does the oven operate? Test the oven temperature. On electronic models, turn off the electricity to the appliance and wait for two minutes before turning it back on. If a fault code appears, look up the code. If the range/oven will not power up, locate the technical data sheet behind the control panel or for diagnostics information. On some models you will need the actual service manual for the model you are working on to properly diagnose the range/oven. The service manual will assist you in properly placing the range/oven in the service test mode for testing the range/oven functions.

2. **Check for external factors.** You must check for external factors not associated with the appliance. Is the appliance installed properly? Are there 120 volts to the appliance? The voltage at the receptacle is between 108 volts and 132 volts during a load on the circuit. Do you have the correct polarity? (See Chapter 6.) Is the gas to the appliance turned on?

3. **Disconnect the electricity.** Before working on the range or cooktop, disconnect the electricity to the appliance. This can be done by pulling the plug from the receptacle. Or disconnect the electricity at the fuse panel or at the circuit breaker panel. Turn off the electricity.

WARNING *Some diagnostic tests will require you to test the components with the power turned on. When you disassemble the control panel, you can position it in such a way that the wiring will not make contact with metal. This act will allow you to test the components without electrical mishaps.*

4. **Shut off the gas supply.** Before you begin servicing any gas components, shut off the supply of gas to the appliance. The shutoff valve should be within six feet of the gas appliance.

5. **Gain access to the oven temperature sensor.** To gain access, open the oven door. The temperature sensor is located in the upper corners of the oven cavity.

6. **Remove the oven temperature sensor.** Remove the oven racks. Remove the two screws that secure the sensor to the oven cavity wall. Pull the sensor toward you gently, disconnect the connector, and remove the temperature sensor from the oven. There may be times when the connector will not pass through the hole in the oven cavity. If this is the case, then you will have to access the connector from the rear of the range/oven. On some models, to access the temperature sensor, you will have to remove the temperature sensor from the rear of the range/oven by removing the screws that secure the sensor to the oven cavity in the rear of the range/oven.

7. **Test the oven temperature sensor.** Set your ohmmeter on the R × 10 scale, and measure the resistance between the two pin connectors on the sensor wire connector. For most sensors, the meter reading at room temperature should match the chart in Figure 23-21. For example: If the room temperature is approximately 75 degrees Fahrenheit, the resistance in the sensor will be 1100 ohms. If the reading is less than 1000 ohms, replace the sensor. If the reading is above 1100 ohms at 75 degrees, replace the sensor.

8. **Install the new oven temperature sensor.** To install the oven temperature sensor, just reverse the disassembly procedure, and reassemble. Turn on the electric and the gas supply and test the oven temperature. On electronic models, make sure to take the range/oven out of the service test mode when the repair is completed. With an oven temperature tester, place the thermocouple tip in the center of the oven cavity. Be sure that the thermocouple tip does not touch any metal. Close the oven door, set the oven to bake, and adjust the thermostat setting to the 350-degree mark. Let the oven cycle for 20 to 30 minutes. Then record the minimum and maximum

Figure 23-21
A typical oven temperature sensor. This type of sensor is only used in ranges and ovens with electronic controls.

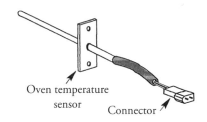

Oven temperature sensor   Connector

Temperature (°F)	Resistance (Ω)
32	1000
75	1100
250	1450
350	1650
450	1850
550	2050
650	2230
900	2700

PART VI

temperatures of three cycles. Next, add these temperatures, and divide by 6. This will give you the average temperature of the oven:

$$\frac{370 + 335 + 350 + 340 + 360 + 335}{6} = 348.3 \text{ degrees Fahrenheit}$$

The average temperature calculated should be within ±5 degrees of the temperature setting selected. Oven temperature sensors (RTDs) are used with electronically controlled ovens. To make any temperature adjustments in the electronic control, the technician needs to locate the technical data sheet. This sheet will show you how to make the temperature adjustments for the model you are servicing.

Thermostat (Standing Pilot)

The oven thermostat is located in the control panel and it is mounted on the manifold pipe. The thermostat is a hydraulic valve that has two separate gas lines and a sensing bulb (Figure 23-5). The pilot gas line maintains a pilot flame in the oven, and the main gas line provides gas to the safety gas valve. The sensing bulb is a mercury-filled tube attached to a capillary tube that controls the thermostat to open or close the gas supply to the oven burner. The thermostat operates on temperature change it senses from within the oven cavity.

The typical complaints associated with the thermostat are:

- The oven/broiler burner will not come on.
- Erratic oven cavity temperatures.

To handle these problems, perform the following steps:

1. **Verify the complaint.** Verify the complaint by turning on the appliance. Does the oven operate? Test the oven temperature.

2. **Check for external factors.** You must check for external factors not associated with the appliance. Is the appliance installed properly? Are there 120 volts to the appliance? The voltage at the receptacle is between 108 volts and 132 volts during a load on the circuit. Do you have the correct polarity? (See Chapter 6.) Is the gas to the appliance turned on?

3. **Disconnect the electricity.** Before working on the range or cooktop, disconnect the electricity to the appliance. This can be done by pulling the plug from the receptacle. Or disconnect the electricity at the fuse panel or at the circuit breaker panel. Turn off the electricity.

WARNING *Some diagnostic tests will require you to test the components with the power turned on. When you disassemble the control panel, you can position it in such a way that the wiring will not make contact with metal. This act will allow you to test the components without electrical mishaps.*

4. **Shut off the gas supply.** Before you begin servicing any gas components, shut off the supply of gas to the appliance. The shutoff valve should be within six feet of the gas appliance.

5. **Gain access to the oven thermostat.** To gain access to the oven thermostat, lift up the cooktop, pull off the control knobs, and then remove the screws from the front panel (see Figure 23-14). Remove the control panel. This will expose the gas burner valves, thermostat, and manifold (see Figures 8-20 and 23-5).

6. **Remove the thermostat.** Before removing the thermostat, remove the capillary tube and sensor from the oven cavity. The sensing bulb will be attached by two retaining clips. Squeeze the clips, pull away from the oven cavity, and remove the sensing bulb and capillary tubing. Feed the capillary tubing through the oven cavity to the top of the range. Disconnect the main gas line and the pilot gas line from the thermostat. Then remove the screws that secure the thermostat to the manifold (see Figures 8-20 and 23-5).

7. **Install the new thermostat.** To install the thermostat, just reverse the disassembly procedure, and reassemble. Make sure that you use approved pipe joint compound or yellow Teflon tape on the pipe connections (according to local codes) and on the threads of the gas lines (if needed) before you install the thermostat on the manifold. Turn on the gas supply to the range. Check for gas leaks by using a chloride-free soap solution. *Do not use a flame to check for gas leaks.* The soap solution will begin to bubble if a leak is present.

8. **Test the oven temperature.** Turn on the electric and gas supply to the range/oven. With an oven temperature tester, place the thermocouple tip in the center of the oven cavity. Be sure that the thermocouple tip does not touch any metal. Close the oven door, set the oven to bake, and adjust the thermostat setting to the 350-degree mark. Let the oven cycle for 20 to 30 minutes. Then record the minimum and maximum temperatures of three cycles. Next, add these temperatures, and divide by 6. This will give you the average temperature of the oven:

$$\frac{370 + 335 + 350 + 340 + 360 + 335}{6} = 348.3 \text{ degrees Fahrenheit}$$

The average temperature calculated should be within ±25 degrees of the temperature setting selected. If not, try calibrating the thermostat. To calibrate the thermostat, pull the oven thermostat knob off, and look at the thermostat (see inset in Figure 23-22). The slotted screw on the thermostat is for converting the thermostat for the type of gas supplied to the range. Set it to the proper setting. If you look into the thermostat stem, you will see a small adjustment screw. This screw is used for calibrating the thermostat. Turning the screw slightly to the right will increase the temperature. To decrease the temperature setting, turn the screw to the left slightly. For every adjustment made, you must test the oven temperature as stated above.

Range/Oven Time Clock

The range/oven time clock is located in the control panel. The clock runs on 120 volts and controls the timer, time bake, clean cycle, and shows the time of day. On electronic models the clock will display error/ fault codes also.

The typical complaints associated with failure of the range/oven time clock are:

- Timed bake cycle will not operate.
- Unable to use the self-cleaning cycle.

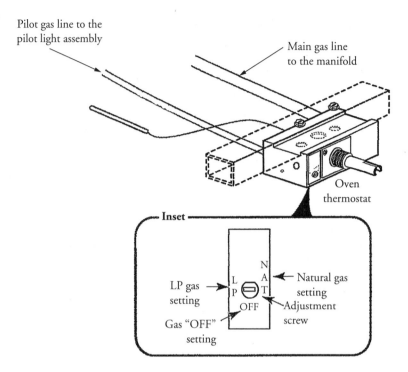

FIGURE 23-22
When converting the range to the type of gas that is supplied to the home, you must convert the thermostat as well.

- Clock loses time.

- Clock is not functioning.

- The minute reminder is not functioning.

- Unusual display readouts and/or error codes.

To handle these problems, perform the following steps:

1. **Verify the complaint.** Verify the complaint by turning the clock controls to the correct time. Does the clock time advance? Some models have the minute reminder located on the same clock. Set this timer to test it. Does it advance? Turn off the electricity to the appliance and wait for two minutes before turning it back on. If a fault code appears, look up the code. If the range/oven will not power up, locate the technical data sheet behind the control panel for the diagnostics information. On some models you will need the actual service manual for the model you are working on to properly diagnose the range/oven. The service manual will assist you in properly placing the range/oven in the service test mode for testing the range/oven functions.

2. **Check for external factors.** You must check for external factors not associated with the appliance. Is the fuse blown? Does the appliance have the correct voltage? The voltage at the receptacle is between 198 volts and 264 volts during a load on the circuit (see Chapter 6).

3. **Disconnect the electricity.** Before working on the range or oven, disconnect the electricity to the appliance. This can be done by pulling the plug from the receptacle. Or disconnect the electricity at the fuse panel or at the circuit breaker panel. Turn off the electricity.

WARNING *Some diagnostic tests will require you to test the components with the power turned on. When you disassemble the control panel, you can position it in such a way that the wiring will not make contact with metal. This act will allow you to test the components without electrical mishaps.*

4. **Gain access to the range/oven clock.** You can access the clock by removing the back panel (see Figure 22-9) on a freestanding range. On models with front-mounted controls (see Figure 22-10), the panel is attached with screws on both ends. Remove the screws, and tilt the control panel. Be careful not to let the wires come off their components. Some built-in models have a removable backsplash (see Figure 22-11); just lift the backsplash and rest it on the cooktop. It would be a good idea to place something on the cooktop first to protect it from damage. Next, remove the screws from the backsplash that hold the rear panel to gain access to the clock. If you are repairing a wall oven or an eye-level range, the control panel can be removed (see Figure 22-12) by opening the door and removing the screws that secure the panel. These might be underneath the front of the exhaust hood or just below the control panel. Some control panels are hinged; just tilt the control panel toward you for servicing. On other models, to gain access to the clock, both the rear panel and the front control panel (usually glass) will have to be removed. Remove the screws that secure the trim piece that holds the glass in place. Then remove the trim piece. On some models, the trim might have to be removed first. Figure 22-40 illustrates the different types of clock faces available on ranges/ovens.

5. **Test the range/oven clock.** On some models, if the clock does not run, check for a fuse in the circuit. Locate the clock motor wire leads, and isolate them from the circuit (label them). Using the ohmmeter, set the range scale on R × 1. Place the probes on the clock motor leads; there should be continuity. If not, replace the clock. On some models, the clock is used to control the timed baking and the self-cleaning cycles (see Figures 22-40b, 22-40c, and 22-40d). To check the switch mechanism of the clock, remove the wires from the switch terminals (label them). Using the ohmmeter, set the range scale on R × 1. Place the probes on the terminals. Look at the wiring diagram for the correct terminals to test. Some models have one switch; other models have two sets of switches. Test for continuity of the switch contacts when you push in and turn the start and stop knobs on the clock and when the knobs pop out. On electronic control models, refer to the technical data sheet located in the range for the testing procedure.

6. **Remove the range/oven time clock.** First remove the clock knobs from the stems. To remove the range/oven time clock in this model (Figure 22-41), use a screwdriver and depress the clips that hold the clock to the control panel. On other models, the clock is secured to the control panel by screws or nuts. Pull the clock toward the front of the appliance (Figure 22-42).

7. **Install a new range/oven clock.** Transfer the wires from the old clock to the new clock. Be sure that you install the wires on their correct terminals according to the wiring diagram. To install the new range/oven clock, just reverse the disassembly procedure and reassemble. Turn on the electric and gas supply and test the operation of the product.

Flame Adjustments

There may be times when the surface, oven, or broil burner flame needs to be adjusted. The proper surface burner flame should be about 5/8 inch high with a well-defined blue flame. If the gas pressure to the burner is correct, all you have to do is adjust the air shutter on the surface burner (see Figure 23-16) to obtain the correct size and color. The oven and broil burners' flame should be about 3/4 inch high with a well-defined blue flame. With the correct gas pressure to the burners, all that is needed is to adjust the air shutters on the burners (Figures 23-13 and 23-23) to obtain the correct size and color. For sealed burner cooktops, you cannot make any air adjustments.

Diagnostic Charts and Wiring Diagrams

The following diagnostic flowcharts will help you to pinpoint the likely causes of the various problems associated with gas ranges and ovens (Figures 23-24 and 23-25). The wiring diagrams in this chapter are examples only. You must refer to the actual wiring diagram for the range, oven, or cooktop that you are servicing. Figure 23-26 depicts samples of actual strip circuit diagrams. The strip circuit diagram is simpler and is usually easier to read. Figures 23-18 and 23-27 depict actual wiring schematic diagrams.

FIGURE 23-23
Adjusting the air shutter on the oven burner.

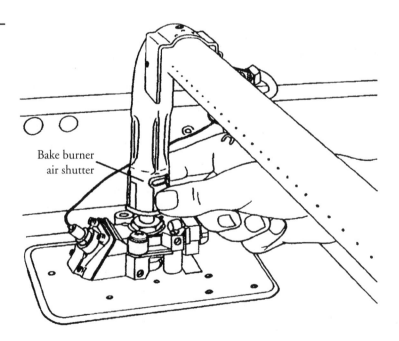

Bake burner
air shutter

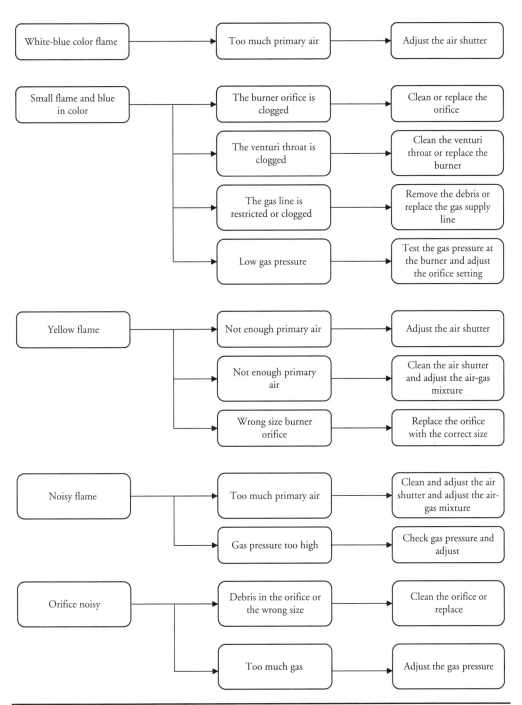

FIGURE 23-24 Flowchart identifying the causes of various flame characteristics. (*continued*)

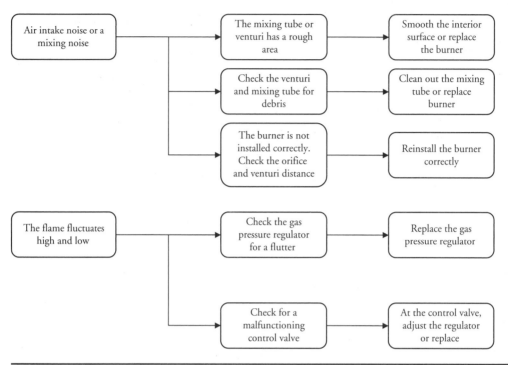

FIGURE 23-24   Flowchart identifying the causes of various flame characteristics.

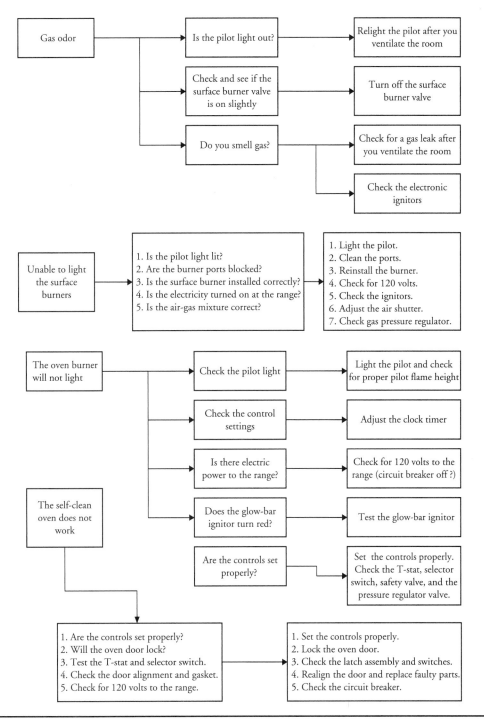

FIGURE 23-25   Diagnostic flowchart: gas odor; unable to light the surface burners; the oven burner will not light; the self-cleaning feature does not work.

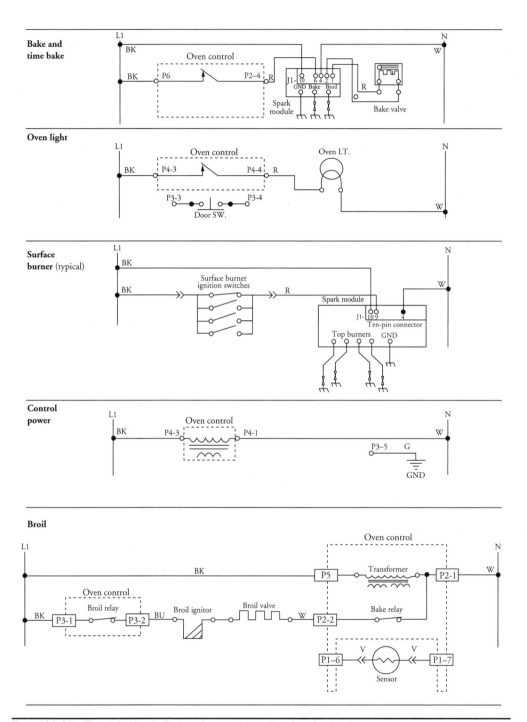

Figure 23-26   The strip circuit diagram is easy to read and understand.

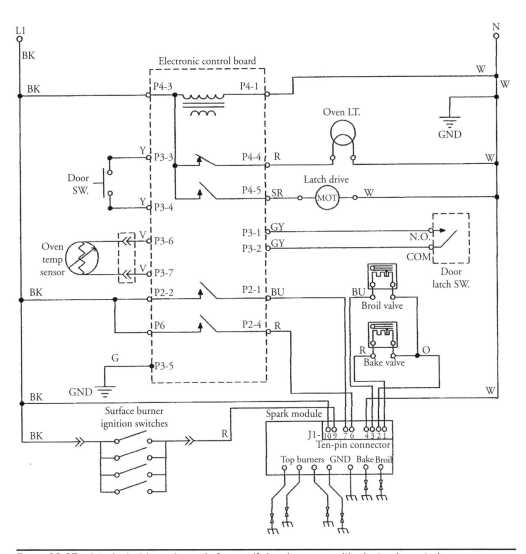

FIGURE 23-27 A typical wiring schematic for a self-cleaning range with electronic controls.

Microwave Ovens

Microwave ovens have been around since the late 1940s, and their popularity increased in the mid-1970s. Over the years, the cost to purchase a microwave oven has decreased. Today, you can purchase a microwave oven for under $50, as compared to the 1940s price of around $3000.

Microwave ovens produce electromagnetic non-ionizing radiation to cook the food, operating at approximately 3000 volts. They operate in the microwave region in the electromagnetic spectrum at a frequency of 2450 MHz. For domestic use, the microwave oven (see Figures 1-6, 24-1, 24-2, 24-3, and 24-4) is manufactured in different sizes and styles, countertop, hanging, and built-in models with electromechanical or electronic controls.

Microwave energy can travel at the speed of light in straight lines; it is produced in the oven by a device called a magnetron tube and reflected by metal channels to enter the oven cavity. Microwaves can pass through non-metallic objects and be absorbed into the food.

This chapter will provide the basics needed to diagnose and repair the components in a microwave oven. Figure 24-5 identifies where various components are typically located within the microwave oven. However, this illustration is used as an example only. The actual construction and features might vary, depending on the brand and model you are servicing.

Safety First

Any person who cannot use basic tools or follow written instructions should *not* attempt to install, maintain, or repair any microwave oven. Any improper installation, preventive maintenance, or repairs could create a risk of death, personal injury, death to others, or property damage.

If you do not fully understand the installation, preventive maintenance, or repair procedures in this chapter, or if you doubt your ability to complete the task on the microwave, do not attempt to make the repairs. Call your service manager instead.

- The following precautions should also be followed: Do not operate the microwave with the door open.
- Do not operate the microwave if the glass in the door is broken.
- Do not allow children under the age of 10 to operate a microwave oven.
- Do not jump out or bypass any components to operate the microwave oven.

FIGURE 24-1
A countertop microwave oven with electronic controls.

- Do not operate the microwave oven if the door hinges and door assembly are loose.

- Always perform a microwave leakage test before servicing and when repairs are completed.

- Before you begin to service any high-voltage components within the microwave oven, you must discharge the high-voltage capacitors and inverter first.

- Never touch any wires or components with your hands or tools when the microwave oven is operating.

- Never take voltage readings of the high-voltage circuit when the microwave oven is running. It is unnecessary and unsafe to do so.

FIGURE 24-2
A countertop microwave oven with electromechanical controls.

FIGURE 24-3
Over-the-range
microwave ovens are
available in various
colors and with
various features.

- Never run the microwave oven with the oven cavity empty. The microwave energy needs to be absorbed into food or a liquid; otherwise, the energy will bounce around, making its way to the high-voltage system and causing damage.

Before continuing, take a moment to refresh your memory of the safety procedures in Chapter 2.

Principles of Operation

The microwave oven circuitry consists of two primary circuits divided by the high-voltage transformer. The high-voltage circuit consists of the secondary winding of the high-voltage

FIGURE 24-4
An under-the-counter
microwave oven.

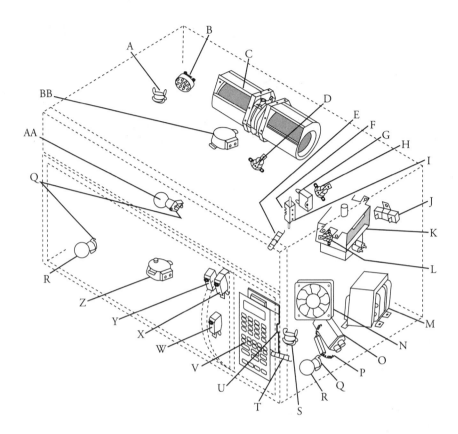

A. Oven cavity thermostat	H. Cavity thermostat	O. High-voltage capacitor	V. Touch panel
B. Humidity sensor	I. Fuse holder	P. High-voltage diode	W. Interlock switch (secondary)
C. Exhaust hood fan motor assembly	J. Resistor		X. Interlock switch (monitor)
	K. Magnetron	Q. Lamp holder	
D. Cavity thermostat	L. Magnetron thermostat	R. Cooktop lamp	Y. Interlock switch (primary)
E. Fuse		S. Exhaust fan thermostat	
F. Fuse	M. High-voltage transformer		Z. Turntable motor
G. Motor capacitor	N. Cooling fan motor	T. Fuse	AA. Cavity lamp
		U. Electronic control	BB. Stirrer motor

Figure 24-5 An exploded view of an over-the-range microwave oven, illustrating the various component locations.

transformer (or inverter on some models), magnetron, high-voltage capacitor, and the high-voltage diode. The low-voltage circuit consists of the primary side of the high-voltage transformer, fans, lights, switches, timer, motors, fuses, thermostats, and a low-voltage transformer that controls the electronic components in some models.

When the microwave oven is turned on and running, the low-voltage alternating current (120 VAC) is stepped up to a high-voltage alternating current (1800 to 2000 VAC) and converted into an even higher negative direct current (–3000 to –4500 VDC, depending on the model).

WARNING *Extreme care must be taken when handling high-voltage components.*

The direct current travels to the magnetron. The output of the magnetron produces microwave energy out of the antenna (2450 MHz) into the waveguide. From the waveguide, the non-ionizing energy will then be channeled to the stirrer blade (on some models). The stirrer blade will begin to spin around, dispersing the microwaves into the oven cavity evenly. Then the microwaves will reflect off the interior walls, ceiling, and floor of the oven cavity, penetrating the food. On some models, a glass tray is used that will rotate within the oven cavity so that the food will cook evenly.

Cooking the food is achieved by three methods: conduction, convection, and radiation. The microwave energy will penetrate the food approximately ¾ inch to two inches, depending on the density of the food and how much water it contains. Microwaves cook the food from the outside toward the center. The molecules within the food will begin to resonate violently, producing heat to cook the food. When the cooking cycle is completed, the oven will turn off and the magnetron will stop producing non-ionizing energy.

Microwave High-Voltage Circuit

Figures 24-6, 24-7, 24-8, 24-9, and 24-10 depict how the high-voltage circuitry operates in a microwave oven. Figures 24-11 and 24-12 are the inverter board that replaces the high-voltage transformer, high-voltage capacitor, and high-voltage diode in some models.

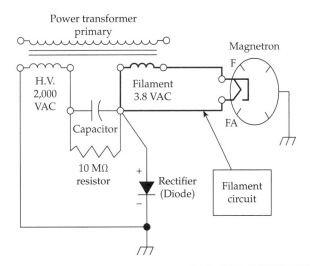

FIGURE 24-6 The high-voltage transformer's secondary winding is divided into two sections. The first section is a stepped-up winding of 2000 VAC used in the voltage doubler circuit; the second section is a stepped-down winding to 3.8 VAC. The stepped-down winding is used for the filament circuit for the magnetron filament. As the magnetron filament heats up, it causes an increase in molecular activity within the magnetron tube.

FIGURE 24-7
The voltage doubler circuit consists of a high-voltage capacitor and diode. As shown in the wiring diagram, when the high-voltage secondary winding is on, the voltage is being stored in the capacitor. The diode will only allow voltage to pass through it in one direction as indicated in the diagram.

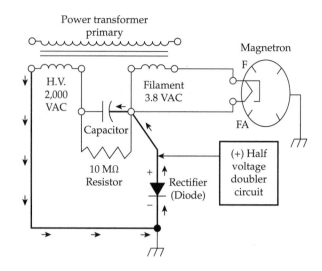

FIGURE 24-8
AC current sine wave for the high-voltage transformer. The positive 2000 volts will be stored in the capacitor in the voltage doubler circuit. The negative 2000 volts will be added to the positive 2000 volts to start the magnetron tube to produce microwave energy.

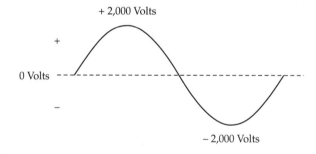

FIGURE 24-9
When the AC current changes to the negative voltage, the negative 2000 volts will be added to the already stored positive 2000 volts in the capacitor.

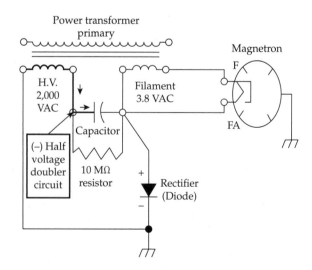

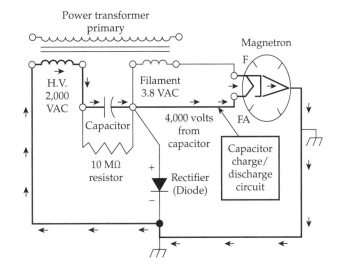

FIGURE 24-10
When the voltage is at negative 4000 volts, the magnetron will start producing microwave energy. This wiring diagram illustrates that the high-voltage circuit is operating and producing microwave energy.

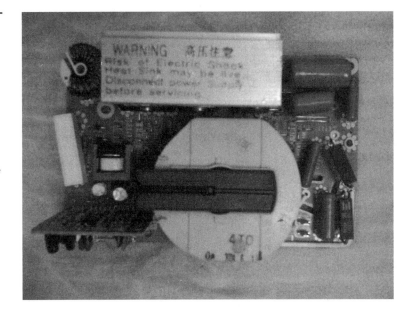

FIGURE 24-11
Inverter board (front view) used in some microwave models. The inverter board replaces the high-voltage transformer, high-voltage capacitor, and high-voltage diode. Consult the technical data sheet for the model you are servicing to locate the high-voltage capacitor discharge test points.

Location and Installation of Microwave Ovens

Locate the microwave oven where it will be well lighted and have access to proper ventilation. The oven must be level for proper baking and cooking results. The microwave oven must also have the proper electrical and grounding requirements, as described in the manufacturer's installation instructions.

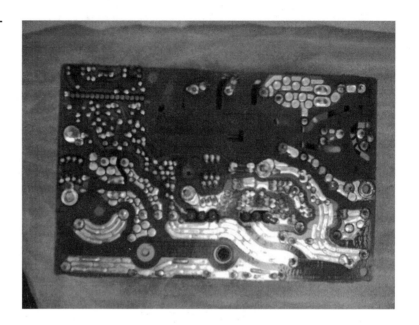

The proper installation instructions for your model are included with your purchase. These instructions will assist you with the installation requirements (dimensions, cutout dimensions, venting, etc.) needed to complete the installation according to the manufacturer's specifications.

Microwave Oven Cookware, Utensils, and Coverings

The following "do not use" list for microwave cooking should be followed for safety reasons:

- Do not use metal pots, pans, and bakeware.
- Do not use dishes with metallic trim.
- Do not use non-heat-resistant glass.
- Do not use non-microwave-safe plastics.
- Do not use recycled paper products.
- Do not use brown paper bags.
- Do not use food storage bags.
- Do not use metal twist-ties.

Any of these items can cause arcing, fire, and damage to the microwave oven. If you are uncertain if a dish is safe for microwaving, place the dish in the microwave on high for 30 seconds. If the dish is hot, it is unsafe for microwaving and should not be used.

The following items are safe to use in microwave ovens:

- Glass ceramic manufactured out of PyroCeram (for example, CorningWare products)
- Pyrex heat-resistant glass products
- Any plastic product with "microwave-safe" stamped on it

- Paper plates, paper towels, wax paper
- Any pottery, porcelain, or stoneware that has "microwave-safe" stamped on it
- A browning dish (follow the manufacturer's directions)
- Wood, straw, and wicker for short-term reheating purposes only
- Plastic wrap that is approved for microwave use only
- Oven cooking bags approved for microwave cooking

If you are not sure if the item can be used in the microwave oven, check your use and care instructions or go to the manufacturer's website.

Can aluminum foil be used in the microwave? Yes, it can. Small, flat pieces of aluminum foil can be placed on the food in areas that cook too quickly to prevent the food from overcooking. When using aluminum foil, you must make sure to keep it from contacting any part of the oven's interior surfaces (about one inch clearance to prevent arcing). For best cooking results, always follow the manufacturer's recommendations in the use and care guide.

Microwave Ovens in General

Much of the troubleshooting information in this chapter covers microwave ovens in general, rather than specific models, in order to present a broad overview of service techniques. The illustrations that are used in this chapter are for demonstration purposes only to clarify the description of how to service these appliances. They in no way reflect on a particular brand's reliability.

Step-by-Step Troubleshooting by Symptom Diagnosis

When servicing an appliance, don't overlook the simple things that might be causing the problem. Step-by-step troubleshooting by symptom diagnosis is based on diagnosing malfunctions, with their possible causes arranged into categories relating to the operation of the microwave oven. This section is intended only to serve as a checklist to aid you in diagnosing a problem. Look at the symptom that best describes the problem that you are experiencing with the oven, and then proceed to correct the problem.

Microwave Oven Dead in the "Off" Position

- Check for the proper voltage and grounding to the oven.
- Check the fuses in the microwave oven.
- Test the thermal fuses within the oven.
- Test the electronic control board or electromechanical timer.
- Check the power supply.
- Check the door alignment and the door interlock switches.
- Check the service cord, wiring, and wiring connectors.
- Test all of the high-voltage components.

Microwave Oven Will Not Operate After the Controls Are Set

- Check for proper voltage and grounding to the oven.
- Check the door interlock switches.
- Make sure that the door closes properly, with the latch assembly engaging the interlock switches.
- Test the electronic control board or electromechanical timer.
- Check the power supply.
- Check all the fuses.
- Check the start button.
- Try reprogramming the control panel. The programming may have timed out.
- Check the wiring and wiring connectors.
- Test the service cord for an open or short circuit.
- On electronic models, run the diagnostic test mode. Check for fault codes.

Unable to Program the Touchpad Controls

- Does the oven door close properly?
- Disconnect the voltage and wait two minutes. Can you reprogram the touchpad control?
- Check the oven door interlock switches.
- Test the electronic control board and the touchpad panel.

Microwave Oven Works, but There Is No Display

- Check the electronic control board.
- Check the display board and power supply.
- Check the wiring and wiring connectors.
- Disconnect the electricity and wait two minutes.
- Try resetting the microwave controls.

Microwave Oven Still Runs with the Door Open

- Check the oven door interlock switches.
- Check the electromechanical timer.
- Check the electronic controls for a stuck relay contact.
- Check the electromechanical timer.
- Check for a bad sensor.

Microwave Oven Starts Running as Soon as the Door Is Closed

- Check the microwave relay or triac.
- Check the electronic control board or electromechanical timer.

- Test the power supply.
- Test the touchpad panel.

Microwave Oven Is Running but Will Not Cook

- Check the magnetron.
- Test the diode.
- Check the triac or relay.
- Check the thermal fuse.
- Check the high-voltage transformer fuse.
- Check the door interlock switches.
- Check the high-voltage capacitor.
- Check the electronic control board or electromechanical timer.
- On electronic models, run the diagnostic test mode. Check for fault codes.

Microwave Oven Is Running, but It Takes Too Long to Cook the Food

- Check for proper voltage to the oven.
- Check the thermostats.
- Check the high-voltage transformer or inverter.
- Check the magnetron.
- Check the diode.
- Check the triac or relay.
- Check the electronic control board.
- Check the fans.
- Check the stirrer blade. Is it turning?
- Check the turntable. Does it rotate?
- Check the vents for blockage.
- On electronic models, run the diagnostic test mode. Check for fault codes.

Microwave Oven Is Running, but the Operation Is Erratic

- Test the voltage to the microwave oven.
- Check the magnetron performance.
- Check the electronic control board or electromechanical timer.
- Is the stirrer blade turning?
- Check the high-voltage transformer or inverter.
- Check the wiring and wiring connections.
- Check the diode.
- Check the thermostats.
- Check the turntable operation.

PART VI

- Is the oven cavity clean?
- Check the door alignment and check for microwave leakage.
- On electronic models, run the diagnostic test mode. Check for fault codes.

Microwave Oven Keeps Blowing Fuses

- Check the door interlock switches.
- Check the oven door and the door latch assembly for proper operation.
- Check the diode.
- Check for a shorted magnetron.
- Check the high-voltage transformer.
- Check the high-voltage capacitor.
- Check the triac or relay.
- Check the incoming voltage supply for power surges.

Microwave Oven Is Running, but Will Not Cook and Makes a Loud Humming Noise

- Check the triac or relay.
- Check the diode.
- Check the magnetron.
- Check the high-voltage transformer or inverter.
- Check the wiring or wiring connections.

Microwave Oven Has a Burning Smell

- Ask the customer what was cooking in the oven.
- Inspect in and above the oven cavity for carbonized food or grease.
- Inspect the wiring and wiring connections.
- Inspect the waveguide and cover for grease.
- Check the high-voltage system.

Microwave Oven Is Arcing

- Ask the customer if aluminum foil or any other metal product was used for cooking.
- Inspect the oven cavity.
- Check for arcing in the waveguide.
- Check the magnetron.
- Check for any loose screws in the door.
- Inspect for sharp exposed metal edges.

Microwave Oven Is Very Noisy

- Check the fan blade(s) and assembly.
- Check the stirrer blade. Is it turning?

- Check for unusual noises coming from the high-voltage transformer.
- Check the turntable motor.
- Check the electromechanical timer motor.

Microwave Oven Door Will Not Close Properly

- Check the door latch assembly and catch.
- Check the door hinges.
- Check the door alignment.
- Check for loose or missing screws in the door.

Electromechanical Timer Will Not Advance in the Cooking Cycle

- Check for voltage to the timer motor.
- Check wiring and wiring connections.
- Test the motor on the timer for an open winding.
- Does the timer knob turn easily or is it jammed?

Microwave Oven Performs the Wrong Functions

- Ask the customer if the cycle was interrupted and then someone else tried to run the oven.
- Disconnect the electricity to the oven; then turn on the electricity and try to reprogram the controls. Does it operate okay?
- Check the polarity of the 120 volt receptacle.
- Check the electronic control board.
- Check the power supply to the electronic control board.
- Test the touchpad control panel and display board for proper operation.
- Check all sensors.
- On electronic models, run the diagnostic test mode. Check for fault codes.

Touchpad Does Not Function or the Wrong Commands Appear

- Turn off the electricity to the oven, wait two minutes, turn the electricity back on, and try to reprogram the electronic controls.
- Check the wiring and wiring connections from the touchpad membrane to the electronic control board.
- Test the touchpad membrane circuitry.
- Check the electronic control board.
- On electronic models, run the diagnostic test mode. Check for fault codes.

Microwave Oven Turntable Is Not Working

- Check the wiring and wiring connections.
- Is the glass tray installed properly?

PART VI

- Check the turntable motor.
- Check the thermostat.
- Check the electronic control board.
- On older models, check the turntable bearings.

Microwave Oven Runs Okay, but Randomly Shuts Off During the Cooking Cycle

- Check the cooling fan motor.
- Check the stirrer blade. Is it turning?
- Check the stirrer blade motor.
- Check the air vents for a blockage.
- Check the thermostats and check the sensors.
- Check the door alignment and the door interlock switches.
- Check for proper voltage to the oven.
- Check the electronic control board.
- Check the wiring and wiring connections.
- Check the magnetron.
- On electronic models, run the diagnostic test mode. Check for fault codes.

Stirrer Blade Will Not Rotate

- Check for obstructions.
- Check the stirrer blade. Is it bent or broken?
- Check the stirrer motor.
- Check for a loose or broken belt.
- Check the wiring and wiring connections.

Microwave Oven Is in Cook Mode and Running, but the Light Is Out in the Oven

- Check the light bulb.
- Check the light socket.
- Check the wiring and wiring connections.

Microwave Oven Only Heats on High Power

- Check the control settings.
- Check the electronic control board.
- Check the electromechanical timer.
- On older models, check for a stuck switch.
- Check the triac or relay.
- On electronic models, run the diagnostic test mode. Check for fault codes.

Microwave Oven Maintenance

The maintenance on a microwave oven is simple: keep the interior and exterior clean. Make sure that the air vents are clear of debris for proper ventilation. The interior can be cleaned with a damp cloth and a mild detergent. If food particles are allowed to collect in the oven cavity, on the stirrer blade or cover, or waveguide, the food will become carbonized, causing arcing, improper heating, and possible damage to the components. When cleaning the control panel, use a damp cloth only, and do not use a spray cleaner or harsh chemical on the touch panel. This will prevent the touch panel from shorting out or preventing some of the keypad functions from working properly. Inspect the service cord and outlet; check the prongs of the plug for a tight fit. A loose connection can cause the service cord to heat up and short out. This is especially important on built-in models, given the amount of heat that some of these ovens can generate. It is also recommended that the consumer keep the area around the microwave oven clean.

After cooking in the microwave oven, the smell of food lingers for quite a while. There have been cases where cockroaches have penetrated into the microwave oven, traveling to the electronics area, and eating the wires and the metal traces off the printed circuit boards, causing expensive repairs. This can be prevented by placing a fine mesh over the intake and discharge vent areas. You might have to clean the mesh more often, however, because dust and dirt will build up on the mesh much faster.

Recipe for Microwave Oven Odor Removal

To remove cooking odors from the microwave, combine one cup of water, grated lemon peels, lemon juice, and several slices of lemon. Stir and place in a microwave-approved glass container. Cook on high power, and let the mixture boil for several minutes. When the oven shuts off, let the mixture cool, remove the container from the oven, and wipe the interior with a damp cloth.

Microwave Leakage Testing

On every microwave oven service call, you must perform a radio frequency (RF) leakage test (Figure 24-13), before and after the repairs, to determine that the energy emission level is below 4 mw/cm^2 at a distance of five centimeters, with a scan rate not to exceed 2.54 cm/second (approx. 1 in/sec). Make sure that you use an approved microwave leak detector for testing.

Take a glass and fill it with water, about 275 ml (±15—just under 9.5 ounces). Place it in the center of the floor in the oven cavity. Set the microwave power level to high, set the timer for three minutes, and press the start button. Take the RF leakage detector probe, and scan the following areas at a rate of one inch per second:

- Scan around the front of the cabinet
- Scan around the oven door
- Scan horizontally across the oven door
- Scan vertically across the oven door
- Scan across the control panel

PART VI

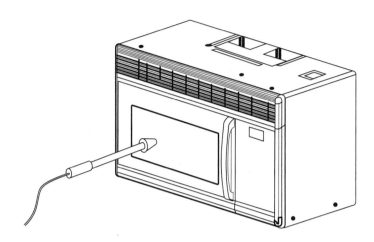

FIGURE 24-13
Performing a microwave leakage test. Move the wand at 1 inch/second around the door, front and rear of the cabinet, intake and discharge vents, oven feet (countertop models), and the bottom of the microwave oven (over-the-range models).

- Scan diagonally across the oven door
- Scan across the oven intake and discharge vents
- Scan all seams and welds
- Scan the bottom plate (for hanging microwave ovens)
- Scan the microwave oven feet

WARNING *Never let a consumer operate a microwave oven if it fails the RF leakage test.*

Microwave Oven Magnetron Power Output Test

The magnetron's performance can be tested for proper operation by performing a power output test in the following manner:

1. The oven cavity must be cool and clean. Inspect the waveguide and stirrer blade.
2. The voltage to the oven should be around 120 volts (±10 percent).
3. Place an 8-ounce glass of water heated to about 75 degrees Fahrenheit in the middle of the cavity floor (remove the metal rack).
4. Take a thermometer and stir the water for 30 seconds. Record the water temperature.
5. Set the oven controls to 100 percent power and the timer to one minute, and press the start button.
6. When the oven turns off, remove the glass, and stir the water with the thermometer for 30 seconds. Record the water temperature.
7. Subtract the two recorded water temperatures to determine the temperature rise.

The temperature rise should be between 19 and 26 degrees Fahrenheit at 120 volts (±10 percent), indicating that the magnetron output power is operating properly.

Repair Procedures

Each of the following repair procedures is a complete inspection and repair process for a single microwave oven component, containing the information you need to test a component that might be faulty and to replace it, if necessary.

Any person who cannot use basic tools should *not* attempt to install, maintain, or repair any microwave ovens. Any improper installation, preventative maintenance, or repairs will create a risk of personal injury, as well as property damage. Call the service manager if installation, preventative maintenance, or the repair procedure is not fully understood.

Electronic Control Board, Touchpad Panel, or Display Board

The electronic control board, touchpad, or display board operate the microwave oven functions. The electronic control board also displays the cooking or defrost settings, the time of day, timer, preprogrammed functions, and any error or fault codes. The electronic control board, touchpad, or display board are located in the control panel.

The typical complaints associated with the electronic control board, touchpad panel, or display board are:

- The microwave oven will not run or power up.
- Unable to program the touchpad panel functions.
- The display board will not display anything.
- One or more key pads will not accept commands. Unusual display readouts and/or error codes.

To prevent electrostatic discharge (ESD) from damaging expensive electronic components, follow the steps in Chapter 11.

To handle these problems, perform the following steps:

1. **Verify the complaint.** Verify the complaint by operating the microwave oven. On electronic models, turn off the electricity to the appliance and wait for two minutes before turning it back on. If a fault code appears, look up the code. If the microwave oven will not power up, locate the technical data sheet behind the control panel for diagnostics information. On some models you will need the actual service manual for the model you are working on to properly diagnose the microwave oven. The service manual will assist you in properly placing the microwave oven in the service test mode for testing the microwave oven functions.

2. **Check for external factors.** You must check for external factors not associated with the appliance. Is there electricity to the microwave oven? The voltage at the receptacle is between 108 volts and 132 volts during a load on the circuit. Do you have the correct polarity? (See Chapter 6.)

3. **Disconnect the electricity.** Before working on the microwave oven, disconnect the electricity. This can be done by pulling the plug out of the wall receptacle. Or disconnect the electricity at the fuse panel or circuit breaker panel. Turn off the electricity.

WARNING *Before you begin to service any high-voltage components within the microwave oven, you must discharge the high-voltage capacitors or inverter board first. Failure to do so will cause injury or death.*

WARNING *Some diagnostic tests will require you to test the components with the power turned on. When you disassemble the control panel or remove the outer cabinet, you can position the panel in such a way that the wiring will not make contact with metal. This will allow you to test the components without electrical mishaps.*

4. **Remove the outer cabinet or control panel to gain access.** You can gain access to the electrical components, touchpad panel, or electronic control board (some models have multiple circuit boards) by removing the access panel or the outer cabinet on a microwave oven or combination oven. If you are repairing a wall oven or an eye-level combination range, the control panel can be removed (see Figures 11-1 and 22-12) by opening the door and removing the screws that secure the panel. These might be underneath the front of the exhaust hood or just below the control panel. Some control panels are hinged; just tilt the control panel toward you for servicing.

5. **Test the electronic control board, touchpad panel, or display board.** If you are able to run the microwave oven's diagnostic test mode, check the different functions of the oven. Use the technical data sheet for the model you are servicing to locate the test points on the wiring schematic. Check all wiring connections and wiring. Using the technical data sheet, test the electronic control and display board, input voltages, and output voltages. On some models, fuses are soldered to the printed circuit board (PCB). These fuses must be tested before condemning the component. The touchpad panel can also be tested for proper operation by checking the continuity for each keypad function.

6. **Remove the defective component.** To remove the defective component, remove the screws that secure the boards to the control panel or oven frame. Disconnect the connectors from the electronic control board, touchpad membrane, or display board.

7. **Install the new component.** To install a new component, just reverse the disassembly procedure, and reassemble. Reinstall all panels or the console panel, and restore the electricity to the oven. Test the microwave oven operation. Perform the microwave performance and leakage test. Make sure to take the microwave oven out of the service test mode when the repair is completed.

Magnetron

The magnetron produces the microwave energy that will cook the food. The magnetron is attached to the outer oven cavity (Figure 24-14).

The typical complaints associated with the magnetron are:

- Unable to cook the food.
- Failed magnetron power output test.

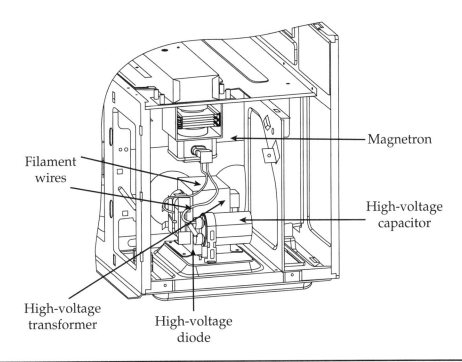

Magnetron

Filament
wires

High-voltage
capacitor

High-voltage
transformer

High-voltage
diode

Figure 24-14   Location of the high-voltage components.

- Noisy operation.
- Unusual display readouts and/or error codes.

To handle these problems, perform the following steps:

1. **Verify the complaint.** Verify the complaint by operating the microwave oven with a glass of water and the microwave test light bar centered on the oven floor (Figures 24-15 and 24-16). The lights in the light bar will light up if the magnetron and high-voltage system are working correctly. At the same time, place an ammeter on one wire of the service cord—the meter should indicate approximately 15 amps at 100 percent power. It may be necessary to turn off the electricity to the appliance and wait for two minutes before turning it back on. If a fault code appears, look up the code. If the microwave oven will not power up, locate the technical data sheet behind the panel or cabinet for diagnostics information. On some models you will need the actual service manual for the model you are working on to properly diagnose the microwave oven. The service manual will assist you in properly placing the microwave oven in the service test mode for testing the microwave oven functions.

2. **Check for external factors.** You must check for external factors not associated with the appliance. Is there electricity to the microwave oven? With over-the-range models, is the vent system blocked? The voltage at the receptacle is between 108 volts and 132 volts during a load on the circuit. Do you have the correct polarity? (See Chapter 6.)

FIGURE 24-15
Place a glass of water in the microwave cavity along with the microwave test light bar.

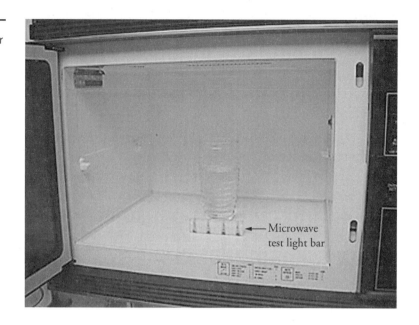

Microwave test light bar

3. **Disconnect the electricity.** Before working on the microwave oven, disconnect the electricity. This can be done by pulling the plug out of the wall receptacle. Or disconnect the electricity at the fuse panel or circuit breaker panel. Turn off the electricity.

WARNING *Before you begin to service any high-voltage components within the microwave oven, you must discharge the high-voltage capacitors and/or inverter board first. Failure to do so will cause injury or death.*

FIGURE 24-16
The microwave oven door is closed, the controls are set, and the oven is on. This illustration shows that the microwave test lights are on, indicating that the magnetron and the high-voltage system are functioning correctly.

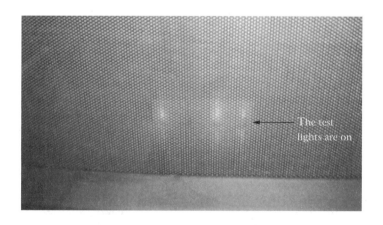

The test lights are on

WARNING *Some diagnostic tests will require you to test the components with the power turned on. When you disassemble the control panel or remove the outer cabinet, you can position the panel in such a way that the wiring will not make contact with metal. This will allow you to test the components without electrical mishaps.*

4. **Remove the outer cabinet or control panel to gain access.** You can gain access to the magnetron by removing the access panel or the outer cabinet on a microwave oven or combination oven. If you are repairing a wall oven or an eye-level combination range, the control panel can be removed (see Figures 11-1 and 22-12) by opening the door and removing the screws that secure the panel. These might be underneath the front of the exhaust hood or just below the control panel. Some control panels are hinged; just tilt the control panel toward you for servicing. On some combination models, you might have to uninstall the appliance to gain access.

5. **Test the magnetron.** Locate the magnetron tube within the high-voltage section of the microwave oven (see Figure 24-5 and 24-14). Set your ohmmeter on the R × 1 scale. Carefully remove the wires from the magnetron filament terminals (Figure 24-17). Inspect the terminal seal for any damage. Take your ohmmeter probes, and place one probe on each filament terminal. The ohmmeter should read less than 1 ohm. Next, leave one probe on the filament terminal, and place the other probe on the chassis. The meter should read infinite ohms. Do the same procedure for the other filament terminal. The meter should read infinite ohms also. If the testing results differ, replace the magnetron.

FIGURE 24-17
The magnetron.

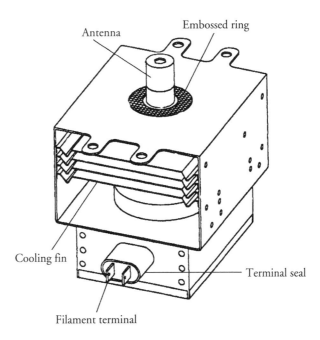

Antenna

Embossed ring

Cooling fin

Terminal seal

Filament terminal

6. **Remove the defective magnetron.** To remove the defective magnetron, remove the screws that secure the magnetron to the oven frame. Carefully remove the magnetron from the oven so as not to damage any other components. Inspect the magnetron tube antenna and embossed ring; replace the magnetron if there are signs of arcing, burn marks, or a broken antenna. Also, inspect the wave guide for burn marks. If there are burn marks, replace the microwave oven.

7. **Install the new magnetron.** To install a new magnetron, just reverse the disassembly procedure, and reassemble. Reinstall all panels or the console panel, and restore the electricity to the oven. Test the microwave oven operation. Perform the microwave performance and leakage test. Make sure to take the microwave oven out of the service test mode when the repair is completed.

High-Voltage Transformer

The high-voltage transformer will step up the voltage to 2000 VAC on one part of the secondary coil. On the other secondary coil, the voltage will be stepped down to 3 to 4 VAC. The primary winding is connected to 120 VAC. The high-voltage transformer is located behind the control panel, mounted to the frame of the microwave oven (Figure 24-14).

The typical complaints associated with the high-voltage transformer are:

- The microwave oven operates but is unable to cook the food.
- Erratic cooking.
- The oven turns off the electricity or blows a fuse in the circuitry when the oven door is closed.
- You hear a constant buzzing noise coming from the high-voltage section.
- Unusual display readouts and/or error codes.

To handle these problems, perform the following steps:

1. **Verify the complaint.** Verify the complaint by operating the microwave oven with a glass of water and the microwave test light bar centered on the oven floor (see Figures 24-15 and 24-16). The lights in the light bar will light up if the magnetron and high-voltage system are working correctly. At the same time, place an ammeter on one wire of the service cord; the meter should indicate approximately 15 amps at 100 percent power. It may be necessary to turn off the electricity to the appliance and wait for two minutes before turning it back on. If a fault code appears, look up the code. If the microwave oven will not power up, locate the technical data sheet behind the panel or cabinet for diagnostics information. On some models you will need the actual service manual for the model you are working on to properly diagnose the microwave oven. The service manual will assist you in properly placing the microwave oven in the service test mode for testing the microwave oven functions.

2. **Check for external factors.** You must check for external factors not associated with the appliance. Is there electricity to the microwave oven? The voltage at the receptacle is between 108 volts and 132 volts during a load on the circuit. Do you have the correct polarity? (See Chapter 6.)

3. **Disconnect the electricity.** Before working on the microwave oven, disconnect the electricity. This can be done by pulling the plug out of the wall receptacle. Or disconnect the electricity at the fuse panel or circuit breaker panel. Turn off the electricity.

WARNING Before you begin to service any high-voltage components within the microwave oven, you must discharge the high-voltage capacitors first and/or inverter board first. Failure to do so will cause injury or death.

WARNING Some diagnostic tests will require you to test the components with the power turned on. When you disassemble the control panel or remove the outer cabinet, you can position the panel in such a way that the wiring will not make contact with metal. This will allow you to test the components without electrical mishaps.

4. **Remove the outer cabinet or control panel to gain access.** You can gain access to the high-voltage transformer by removing the access panel or the outer cabinet on a microwave oven or combination oven. If you are repairing a wall oven or an eye-level combination range, the control panel can be removed (see Figures 11-1 and 22-12) by opening the door and removing the screws that secure the panel. These might be underneath the front of the exhaust hood or just below the control panel. Some control panels are hinged; just tilt the control panel toward you for servicing. On some combination models, you might have to uninstall the appliance to gain access.

5. **Test the high-voltage transformer.** To test the high-voltage transformer (Figure 24-18), disconnect the wires from the primary winding terminals on the transformer (label them). Set your ohmmeter on the R × 1 scale, and measure the resistance at the primary winding terminals on the transformer. The ohmmeter scale should indicate resistance in the winding of less than 1 ohm. Now remove the wires from the secondary winding terminals on the transformer (label them as well), and measure the resistance of the winding. The ohmmeter scale should indicate resistance in the secondary winding from 50 to 130 ohms, depending on the type of transformer used. Next, test the resistance of the filament winding—the ohmmeter should indicate zero to 1 ohm. Finally, take the ohmmeter probes and test each winding to ground: you should have an infinite reading for each winding. If the readings differ, replace the transformer.

6. **Remove the defective high-voltage transformer.** To remove the defective transformer, remove the screws that secure it to the oven frame. Be careful: the high-voltage transformer is heavy.

7. **Install the new high-voltage transformer.** To install a new high-voltage transformer, just reverse the disassembly procedure, and reassemble. Reinstall all panels or the console panel, and restore the electricity to the oven. Test the microwave oven operation. Perform the microwave performance and leakage test. On electronic models, make sure to take the microwave oven out of the service test mode when the repair is completed.

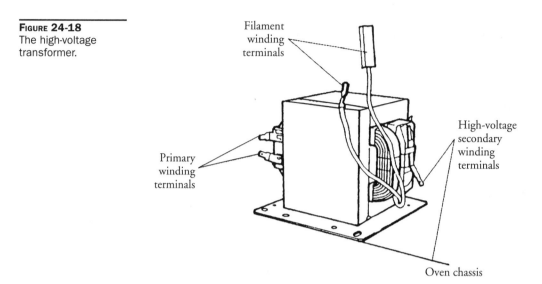

FIGURE 24-18
The high-voltage
transformer.

Filament
winding
terminals

High-voltage
secondary
winding
terminals

Primary
winding
terminals

Oven chassis

Inverter Board

On some models the inverter board replaces the high-voltage transformer, high-voltage capacitor, and high-voltage diode (Figures 24-11, 24-12, and 24-19). The inverter board is located behind the control panel, mounted to the microwave oven frame.

The typical complaints associated with the inverter are:

- The microwave oven operates, but is unable to cook the food.
- Erratic cooking.
- The oven turns off the electricity or blows a fuse in the circuitry when the oven door is closed.
- You hear a constant buzzing noise coming from the high-voltage section.
- Unusual display readouts and/or error codes.

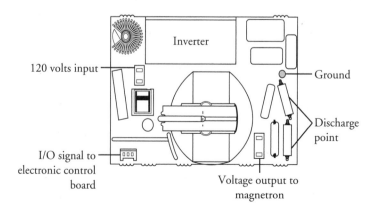

FIGURE 24-19
The high-voltage
inverter board.

Inverter

120 volts input

Ground

Discharge
point

I/O signal to
electronic control
board

Voltage output to
magnetron

To handle these problems, perform the following steps:

1. **Verify the complaint.** Verify the complaint by operating the microwave oven with a glass of water and the microwave test light bar centered on the oven floor (see Figures 24-15 and 24-16). The lights in the light bar will light up if the magnetron and high-voltage system are working correctly. At the same time, place an ammeter on one wire of the service cord; the meter should indicate approximately 15 amps at 100 percent power. It may be necessary to turn off the electricity to the appliance and wait for two minutes before turning it back on. If a fault code appears, look up the code. If the microwave oven will not power up, locate the technical data sheet behind the panel or cabinet for diagnostics information. On some models you will need the actual service manual for the model you are working on to properly diagnose the microwave oven. The service manual will assist you in properly placing the microwave oven in the service test mode for testing the microwave oven functions.

2. **Check for external factors.** You must check for external factors not associated with the appliance. Is there electricity to the microwave oven? The voltage at the receptacle is between 108 volts and 132 volts during a load on the circuit. Do you have the correct polarity? (See Chapter 6.)

3. **Disconnect the electricity.** Before working on the microwave oven, disconnect the electricity. This can be done by pulling the plug out of the wall receptacle. Or disconnect the electricity at the fuse panel or circuit breaker panel. Turn off the electricity.

WARNING *Before you begin to service any high-voltage components within the microwave oven, you must discharge the high-voltage capacitors first and/or inverter board first. Failure to do so will cause injury or death.*

WARNING *Some diagnostic tests will require you to test the components with the power turned on. When you disassemble the control panel or remove the outer cabinet, you can position the panel in such a way that the wiring will not make contact with metal. This will allow you to test the components without electrical mishaps.*

4. **Remove the outer cabinet or control panel to gain access.** You can gain access to the inverter by removing the access panel or the outer cabinet on a microwave oven or combination oven. If you are repairing a wall oven or an eye-level combination range, the control panel can be removed (see Figures 11-1 and 22-12) by opening the door and removing the screws that secure the panel. These might be underneath the front of the exhaust hood or just below the control panel. Some control panels are hinged; just tilt the control panel toward you for servicing. On some combination models, you might have to uninstall the appliance to gain access.

5. **Test the inverter.** The inverter replaces the high-voltage transformer, high-voltage capacitor, and high-voltage diode. To test the inverter (Figure 24-19), locate the service cord wires, and connect an ammeter to the white wire. Then place a glass of water in the oven cavity, turn on the electricity, and run the microwave oven at 100 percent

power. If the reading on the ammeter is above 0.5 amps, check the magnetron and the wiring. If the reading is below 0.5 amps, there is no 120-volt input voltage to the inverter. In this case, you will have to check the cooking relay, electronic control board, or power relay board. Under normal operating conditions the high-voltage circuit can be measured with an ammeter and it will read 12–15 amps.

Some models use a smaller inverter (about 40 watts) to provide 12 volts DC to the hood lamp and the cooling fan motor. To test this type of inverter, set your voltmeter to the AC voltage scale, and test for 120 volts AC to the input connector on the inverter. Then set your voltmeter to the DC voltage scale, and check for 12 volts DC from the hood lamp and cooling fan motor connectors on the inverter.

WARNING *Do not measure the high-voltage side of the circuit with your multimeter. Only test the 120-volt input voltage.*

6. **Remove the defective inverter.** Before you begin to remove the inverter, you must use a 20,000-ohm 2-watt resistor to discharge the inverter. Hold the resistor body with a pair of insulated pliers (see Figure 3-8f), and discharge the inverter at the discharge points, as shown in Figure 24-19. To remove the inverter, remove the inverter cover, wire connectors, grounding wire, and screws; then slide the inverter out of its holding container. On some models, you might have to remove other components to gain access to the inverter.

7. **Install the new inverter.** To install a new inverter, just reverse the disassembly procedure, and reassemble. Reinstall all panels or the console panel, and restore the electricity to the oven. Test the microwave oven operation. Perform the microwave performance and leakage test. On electronic models, make sure to take the microwave oven out of the service test mode when the repair is completed.

High-Voltage Diode

The high-voltage diode is an electrical device that allows current to flow in one direction only, and it is part of the voltage doubler circuit. The high-voltage diode is located behind the control panel attached to the high-voltage capacitor and to a ground screw (Figures 24-14 and 24-20).

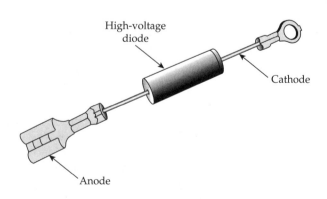

FIGURE 24-20
The high-voltage diode is attached to the high-voltage capacitor and to ground.

High-voltage diode

Cathode

Anode

The typical complaints associated with the high-voltage diode are:

- The microwave oven operates but is unable to cook the food.
- Unusual display readouts and/or error codes.

To handle this problem, perform the following steps:

1. **Verify the complaint.** Verify the complaint by operating the microwave oven with a glass of water and the microwave test light bar centered on the oven floor (see Figures 24-15 and 24-16). The lights in the light bar will light up if the magnetron and high-voltage system are working correctly. At the same time, place an ammeter on one wire of the service cord; the meter should indicate approximately 15 amps at 100 percent power. It may be necessary to turn off the electricity to the appliance and wait for two minutes before turning it back on. If a fault code appears, look up the code. If the microwave oven will not power up, locate the technical data sheet behind the panel or cabinet for diagnostics information. On some models you will need the actual service manual for the model you are working on to properly diagnose the microwave oven. The service manual will assist you in properly placing the microwave oven in the service test mode for testing the microwave oven functions.

2. **Check for external factors.** You must check for external factors not associated with the appliance. Is there electricity to the microwave oven? The voltage at the receptacle is between 108 volts and 132 volts during a load on the circuit. Do you have the correct polarity? (See Chapter 6.)

3. **Disconnect the electricity.** Before working on the microwave oven, disconnect the electricity. This can be done by pulling the plug out of the wall receptacle. Or disconnect the electricity at the fuse panel or circuit breaker panel. Turn off the electricity.

WARNING *Before you begin to service any high-voltage components within the microwave oven, you must discharge the high-voltage capacitors first and/or inverter board first. Failure to do so will cause injury or death.*

WARNING *Some diagnostic tests will require you to test the components with the power turned on. When you disassemble the control panel or remove the outer cabinet, you can position the panel in such a way that the wiring will not make contact with metal. This will allow you to test the components without electrical mishaps.*

4. **Remove the outer cabinet or control panel to gain access.** You can gain access to the high-voltage diode by removing the access panel or the outer cabinet on a microwave oven or combination oven. If you are repairing a wall oven or an eye-level combination range, the control panel can be removed (see Figures 11-1 and 22-12) by opening the door and removing the screws that secure the panel. These might be underneath the front of the exhaust hood or just below the control panel. Some

control panels are hinged; just tilt the control panel toward you for servicing. On some combination models, you might have to uninstall the appliance to gain access.

5. **Test the diode.** To test a diode, you must first disconnect one side of the diode to isolate it from the remainder of the circuit. Set your ohmmeter to the R × 1K scale, and connect the leads to the diode. The meter may or may not read continuity. Then reverse the leads on the diode and check for continuity. You should read continuity in one direction only. If continuity is not indicated in either direction, the diode is open (see Figure 7-20a). If continuity is indicated in both directions, the diode is shorted (see Figure 7-20b). Some high-voltage diodes may have to be tested with a 9-volt battery. Place the battery in series with the diode and measure the DC voltage. Then, reverse the diode on the battery and take another meter reading. You should read voltage in one direction only. If not, the diode is defective.

6. **Remove the high-voltage diode.** To remove the diode from the oven, disconnect the wires and remove.

7. **Install the new high-voltage diode.** To install a new high-voltage diode, just reverse the disassembly procedure, and reassemble. Reinstall all panels or the console panel, and restore the electricity to the oven. Test the microwave oven operation. Perform the microwave performance and leakage test. Make sure to take the microwave oven out of the service test mode when the repair is completed.

High-Voltage Capacitor

The high-voltage capacitor stores electricity, up to 2100 VAC, and it is part of the voltage doubler circuit. The high-voltage capacitor is located behind the control panel attached to the oven frame (Figure 24-14).

The typical complaints associated with the high-voltage capacitor are:

- The microwave oven operates but is unable to cook the food.
- Unusual display readouts and/or error codes.

To handle this problem, perform the following steps:

1. **Verify the complaint.** Verify the complaint by operating the microwave oven with a glass of water and the microwave test light bar centered on the oven floor (see Figures 24-15 and 24-16). The lights in the light bar will light up if the magnetron and high-voltage system are working correctly. At the same time, place an ammeter on one wire of the service cord; the meter should indicate approximately 15 amps at 100 percent power. It may be necessary to turn off the electricity to the appliance and wait for two minutes before turning it back on. If a fault code appears, look up the code. If the microwave oven will not power up, locate the technical data sheet behind the panel or cabinet for diagnostics information. On some models you will need the actual service manual for the model you are working on to properly diagnose the microwave oven. The service manual will assist you in properly placing the microwave oven in the service test mode for testing the microwave oven functions.

2. **Check for external factors.** You must check for external factors not associated with the appliance. Is there electricity to the microwave oven? The voltage at the receptacle is between 108 volts and 132 volts during a load on the circuit. Do you have the correct polarity? (See Chapter 6.)

3. **Disconnect the electricity.** Before working on the microwave oven, disconnect the electricity. This can be done by pulling the plug out of the wall receptacle. Or disconnect the electricity at the fuse panel or circuit breaker panel. Turn off the electricity.

WARNING *Before you begin to service any high-voltage components within the microwave oven, you must discharge the high-voltage capacitor first and/or inverter board first. Failure to do so will cause injury or death.*

WARNING *Some diagnostic tests will require you to test the components with the power turned on. When you disassemble the control panel or remove the outer cabinet, you can position the panel in such a way that the wiring will not make contact with metal. This will allow you to test the components without electrical mishaps.*

4. **Remove the outer cabinet or control panel to gain access.** You can gain access to the high-voltage capacitor by removing the access panel or the outer cabinet on a microwave oven or combination oven. If you are repairing a wall oven or an eye-level combination range, the control panel can be removed (see Figures 11-1 and 22-12) by opening the door and removing the screws that secure the panel. These might be underneath the front of the exhaust hood or just below the control panel. Some control panels are hinged; just tilt the control panel toward you for servicing. On some combination models, you might have to uninstall the appliance to gain access.

5. **Test the high-voltage capacitor.** Do not touch the high-voltage capacitor until it's discharged.

WARNING *A capacitor will hold a charge indefinitely, even when it is not currently in use. A charged capacitor is extremely dangerous. Discharge the capacitor immediately any time that work is being conducted in its vicinity. Redischarge after repowering the microwave oven if further work must be done.*

Do not remove the wires from the capacitor yet. Use two screwdrivers with insulated handles to discharge the capacitor by placing one screwdriver blade on one capacitor terminal and the other screwdriver blade on the other capacitor terminal. Now cross both screwdrivers and touch them together to discharge the capacitor. Next, place one screwdriver on one capacitor terminal and the other screwdriver to chassis ground. Now cross the screwdrivers together to short out the capacitor. Next, repeat this procedure for the other capacitor terminal. If you are not sure that you discharged the capacitor, repeat the procedures until you feel comfortable with removing the wires from the capacitor. Remove the wire leads (label them), one at a time, with needle nose pliers. Set the ohmmeter on the highest scale, and place one probe on one terminal and the other probe on the other terminal (see Figure 24-21). Observe the meter action. While the capacitor is charging, the ohmmeter will read nearly zero ohms for a short period of time. Then the ohmmeter reading will slowly return toward infinity. If the ohmmeter reading deflects to zero and does not return to infinity, the capacitor is shorted and should be replaced with a duplicate of the original. If the ohmmeter reading remains at infinity and does not dip toward zero, the capacitor is open and should be replaced with a duplicate of the original.

6. **Remove the high-voltage capacitor.** To remove the capacitor, remove the wires from the capacitor terminals. Then remove the screws from the capacitor-mounting bracket.

7. **Install the new high-voltage capacitor.** To install a new high-voltage capacitor, just reverse the disassembly procedure, and reassemble. Reinstall all panels or the console panel, and restore the electricity to the oven. Test the microwave oven operation. Perform the microwave performance and leakage test. On electronic models, make sure to take the microwave oven out of the service test mode when the repair is completed.

Microwave Oven Thermostats

Microwave oven thermostats operate a switch within the body of the thermostat. The switch is actuated by a change in temperature. There are between three and five thermostats (each model may vary) located within the microwave oven (oven cavity thermostats, magnetron thermostat, and exhaust fan thermostat) (see Figure 24-5 part # A, D, H, L, and S).

The typical complaints associated with the thermostats are:

- The microwave oven operates but turns off after a while.
- The microwave oven will not cook.
- The microwave oven takes too long to cook the food.

FIGURE 24-21
Placing ohmmeter test leads on the capacitor terminals.

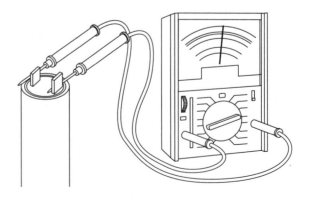

- Erratic operation.
- Unusual display readouts and/or error codes.

To handle these problems, perform the following steps:

1. **Verify the complaint.** Verify the complaint by operating the microwave oven with a glass of water and the microwave test light bar centered on the oven floor (see Figures 24-15 and 24-16). The lights in the light bar will light up if the magnetron and high-voltage system are working correctly. At the same time, place an ammeter on one wire of the service cord; the meter should indicate approximately 15 amps at 100 percent power. Check the operation of the cooling fans and stirrer blade. It may be necessary to turn off the electricity to the appliance and wait for two minutes before turning it back on. If a fault code appears, look up the code. If the microwave oven will not power up, locate the technical data sheet behind the panel or cabinet for diagnostics information. On some models you will need the actual service manual for the model you are working on to properly diagnose the microwave oven. The service manual will assist you in properly placing the microwave oven in the service test mode for testing the microwave oven functions.

2. **Check for external factors.** You must check for external factors not associated with the appliance. Is there electricity to the microwave oven? Is the vent system blocked? The voltage at the receptacle is between 108 volts and 132 volts during a load on the circuit. Do you have the correct polarity? (See Chapter 6.)

3. **Disconnect the electricity.** Before working on the microwave oven, disconnect the electricity. This can be done by pulling the plug out of the wall receptacle. Or disconnect the electricity at the fuse panel or circuit breaker panel. Turn off the electricity.

WARNING *Before you begin to service any high-voltage components within the microwave oven, you must discharge the high-voltage capacitor or inverter board first. Failure to do so will cause injury or death.*

WARNING *Some diagnostic tests will require you to test the components with the power turned on. When you disassemble the control panel or remove the outer cabinet, you can position the panel in such a way that the wiring will not make contact with metal. This will allow you to test the components without electrical mishaps.*

4. **Remove the outer cabinet or control panel to gain access.** You can gain access to the thermostats by removing the access panel or the outer cabinet on a microwave oven or combination oven. If you are repairing a wall oven or an eye-level combination range, the control panel can be removed (see Figures 11-1 and 24-12) by opening the door and removing the screws that secure the panel. These might be underneath the front of the exhaust hood or just below the control panel. Some control panels are hinged; just tilt the control panel toward you for servicing. On some combination models, you might have to uninstall the appliance to gain access.

5. **Test the thermostats.** The microwave oven can have up to five thermostats (see Figure 24-5) in the oven that operate as a safety switch to turn the oven off if the oven temperature exceeds its limits. The cavity thermostats monitor the airflow temperatures in the oven. The magnetron thermostat monitors the operating temperature of the magnetron. Some models have an exhaust fan thermostat that will monitor the temperature from beneath the oven and the airflow temperature (this is common in over-the-range models).

   The cavity and magnetron thermostats are normally closed—when the temperature exceeds its rated limits, it will open the circuit. When the temperature drops below its rating, it will reset and close the circuit. The exhaust fan thermostat is normally open; it closes on temperature rise, turning on the exhaust fan. When the temperature drops below the rated temperature, the thermostat will reset to open the circuit.

   To test the thermostat, set the ohmmeter on R x 1, remove the wires from the thermostat terminals, and test for continuity. Most wire connectors are a locking-type connector. See Figure 24-22 for instructions on how to remove them without breaking the connectors or the components. The lock connector will prevent the connector and wire from coming off the component terminal connector, avoiding circuit faults.

6. **Remove the thermostat.** Locate the thermostat (see Figure 24-5), remove the wires from the terminals, and remove the screws that secure the thermostat to the cavity or magnetron.

7. **Install the new thermostat.** To install a new thermostat, just reverse the disassembly procedure, and reassemble. Reinstall all panels or the console panel, and restore the electricity to the oven. Test the microwave oven operation. Perform the microwave performance and leakage test. On electronic models, make sure to take the microwave oven out of the service test mode when the repair is completed.

FIGURE 24-22
To remove the lock connector: (1) push on the lever and (2) pull down to remove the lock connector from the terminal. When installing the lock connector, always have the lever face you for easy removal the next time you service the microwave.

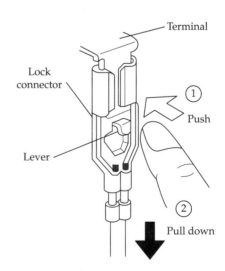

Microwave Oven Door Switches

The oven door switches are electro mechanical devices used for directing and controlling the flow of current in a circuit. Simply put, the door switches are used for turning a component on or off (Figures 24-23 and 24-24). The monitor switch is intended to render the microwave oven inoperative by means of blowing the fuse when the contacts of the primary switch or secondary switch fail to open when the door is opened.

The typical complaints associated with the door switches are:

- The microwave oven will not operate.

- The oven goes off when the oven door is closed.

- Unusual display readouts and/or error codes.

To handle these problems, perform the following steps:

1. **Verify the complaint.** Verify the complaint by operating the microwave oven with a glass of water and the microwave test light bar centered on the oven floor (see Figures 24-15 and 24-16). The lights in the light bar will light up if the magnetron and high-voltage system are working correctly. At the same time, place an ammeter on one wire of the service cord; the meter should indicate approximately 15 amps at 100 percent power. It may be necessary to turn off the electricity to the appliance and wait for two minutes before turning it back on. If a fault code appears, look up the code. If the microwave oven will not power up, locate the technical data sheet behind the panel or cabinet for diagnostics information. On some models you will need the actual service manual for the model you are working on to properly diagnose the microwave oven. The service manual will assist you in properly placing the microwave oven in the service test mode for testing the microwave oven functions.

FIGURE 24-23
An exploded view of a microwave oven door switch.

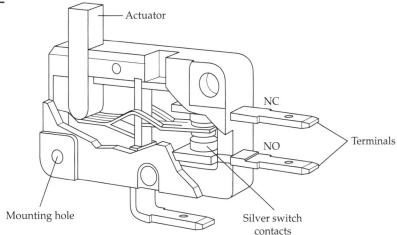

Actuator

NC

NO

Terminals

Mounting hole

Silver switch contacts

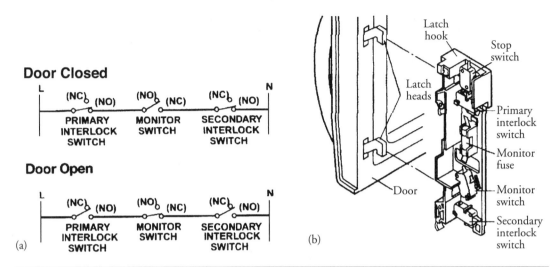

FIGURE 24-24 (a) The strip circuits for the microwave oven door interlock switches with the oven door closed and open. (b) A typical view of one oven model with the door interlock switches. The latch heads and the door switch assembly can be adjusted.

2. **Check for external factors.** You must check for external factors not associated with the appliance. Is there electricity to the microwave oven? The voltage at the receptacle is between 108 volts and 132 volts during a load on the circuit. Do you have the correct polarity? (See Chapter 6.)

3. **Disconnect the electricity.** Before working on the microwave oven, disconnect the electricity. This can be done by pulling the plug out of the wall receptacle. Or disconnect the electricity at the fuse panel or circuit breaker panel. Turn off the electricity.

WARNING *Before you begin to service any high-voltage components within the microwave oven, you must discharge the high-voltage capacitor or inverter board first. Failure to do so will cause injury or death.*

WARNING *Some diagnostic tests will require you to test the components with the power turned on. When you disassemble the control panel or remove the outer cabinet, you can position the panel in such a way that the wiring will not make contact with metal. This will allow you to test the components without electrical mishaps.*

4. **Remove the outer cabinet or control panel to gain access.** You can gain access to the door switches by removing the access panel or the outer cabinet on a microwave oven or combination oven. If you are repairing a wall oven or an eye-level combination range, the control panel can be removed (see Figures 11-1 and 22-12) by opening the door and removing the screws that secure the panel. These might be underneath the front of the exhaust hood or just below the control panel. Some control panels are hinged; just tilt the control panel toward you for servicing. On some combination models, you might have to uninstall the appliance to gain access.

5. **Test the microwave door switches.** Locate the primary interlock, monitor, and secondary interlock switches (see Figures 24-5 and 24-24). Identify each switch using the technical data sheet for the model you are servicing. The switch housing (Figure 24-23) is usually marked with the terminal identification numbers that correspond to the wiring diagram. It identifies the contacts by number: normally open (NO) contacts, normally closed (NC) contacts, or common (COM). Internally, the switch can house many contact points for controlling more than one circuit.

   In Figure 24-24a, the wiring diagram illustrates the door switches in the closed and open positions. Figure 24-24b illustrates the location of the door switches in a countertop model. Set your ohmmeter to the R × 1 scale, remove the wires from the primary interlock switch terminals, and place your meter probes on the terminals. With the door closed, the ohmmeter will indicate continuity. When the door is opened, the ohmmeter reads no continuity. Next, test the monitor switch. Remove the wires from the switch terminals (Figure 24-22), and place your meter probes on the terminals. With the door closed, the ohmmeter will indicate no continuity. When the door is opened, the ohmmeter reads continuity. Finally, test the secondary interlock switch. Remove the wires from the switch, and place the meter probes on the terminals. With the door closed, the ohmmeter will indicate continuity. When the door is opened, the ohmmeter reads no continuity. If any of the readings differ from what is stated here, replace all of the switches in the microwave.

6. **Remove the door switches.** Remove the wires from the door switches and label them. Depending on the model you are servicing, remove the door switch assembly as a complete unit or remove each switch separately.

7. **Install the new door switch.** To install a new door switch, just reverse the disassembly procedure, and reassemble. Check the in-and-out play in the door: it should be less than 0.5 mm when the door is closed. Reinstall all panels or the console panel, and restore the electricity to the oven. Test the microwave oven operation. Perform the microwave performance and leakage test. On electronic models, make sure to take the microwave oven out of the service test mode when the repair is completed.

Temperature Probe

The temperature probe is a thermistor device that is used to penetrate the food to monitor the cooking temperatures. One end of the probe is plugged into the oven receptacle which sends signals to the electronic control board.

The typical complaints associated with the temperature probe are:

- Erratic cooking of the food.
- The food will not cook.
- The food burns.
- Unusual display readouts and/or error codes.

The temperature probe is a separate component that plugs into a receptacle located in the oven cavity (on some models). The consumer inserts the probe into the food product being cooked, programs the controls, and starts the oven. The temperature probe will sense

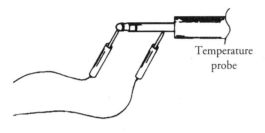

Temperature probe

FIGURE 24-25 Testing the resistance in the temperature probe.

the temperature of the food cooking and sends a signal back to the electronic control to turn off the microwave oven when the food is fully cooked. To test the temperature probe, set your ohmmeter on R × 1K, and place the meter probes on the male end that goes into the receptacle (Figure 24-25). With some models, at room temperature the ohmmeter scale will read between 19,000 and 75,000 ohms. Next, with the ohmmeter probes still attached to the male jack end, insert the probe tip in a glass of hot water, and measure the resistance. The resistance reading should decrease. If the reading does not decrease, replace the temperature probe.

Humidity Sensor

The humidity sensor (Figure 24-26) is attached to the outer microwave oven cavity in the airflow to detect how much humidity (vapor) comes off the food being cooked. The humidity sensor will then send a signal back to the electronic control board that the food is cooked properly and to stop the sensor cooking process.

The typical complaints associated with the humidity sensor are:

- The microwave oven will not cook the food properly.
- The microwave oven takes too long to cook the food.
- Erratic operation.
- Unusual display readouts and/or error codes.

To handle these problems, perform the following steps:

1. **Verify the complaint.** Verify the complaint by operating the microwave oven with a glass of water and the microwave test light bar centered on the oven floor (see Figures 24-11 and 22-12). The lights in the light bar will light up if the magnetron and high-voltage system are working correctly. At the same time, place an ammeter

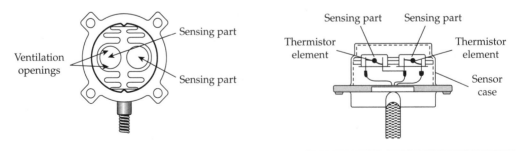

FIGURE 24-26 A humidity sensor. This sensor is mounted in the air stream of the microwave oven on the outside of the oven cavity.

on one wire of the service cord; the meter should indicate approximately 15 amps at 100 percent power. Check the operation of the cooling fans and stirrer blade. It may be necessary to turn off the electricity to the appliance and wait for two minutes before turning it back on. If a fault code appears, look up the code. If the microwave oven will not power up, locate the technical data sheet behind the panel or cabinet for diagnostics information. On some models you will need the actual service manual for the model you are working on to properly diagnose the microwave oven. The service manual will assist you in properly placing the microwave oven in the service test mode for testing the microwave oven functions.

2. **Check for external factors.** You must check for external factors not associated with the appliance. Is there electricity to the microwave oven? Is the vent system blocked? On countertop models, is there enough open space surrounding the microwave oven for proper ventilation? The voltage at the receptacle is between 108 volts and 132 volts during a load on the circuit. Do you have the correct polarity? (See Chapter 6.)

3. **Disconnect the electricity.** Before working on the microwave oven, disconnect the electricity. This can be done by pulling the plug out of the wall receptacle. Or disconnect the electricity at the fuse panel or circuit breaker panel. Turn off the electricity.

WARNING *Before you begin to service any high-voltage components within the microwave oven, you must discharge the high-voltage capacitor or inverter board first. Failure to do so will cause injury or death.*

WARNING *Some diagnostic tests will require you to test the components with the power turned on. When you disassemble the control panel or remove the outer cabinet, you can position the panel in such a way that the wiring will not make contact with metal. This will allow you to test the components without electrical mishaps.*

4. **Remove the outer cabinet or control panel to gain access.** You can gain access to the humidity sensor (see Figure 24-5) by removing the access panel or the outer cabinet on a microwave oven or combination oven. If you are repairing a wall oven or an eye-level combination range, the control panel can be removed (see Figures 11-1 and 22-12) by opening the door and removing the screws that secure the panel. These might be underneath the front of the exhaust hood or just below the control panel. Some control panels are hinged; just tilt the control panel toward you for servicing. On some combination models, you might have to uninstall the appliance to gain access.

5. **Test the humidity sensor.** The humidity sensor can be tested at the electronic control board behind the control panel. Locate the sensor wires, trace them to the electronic control board, and disconnect the three-wire connector from the board. Set your ohmmeter on the R × 1K scale. Place your meter probes on pins 1 and 3 (check the technical data sheet for the pin designations), located on the humidity sensor wire connector. The ohmmeter scale should read approximately 2800 ohms at 77 degrees Fahrenheit (±2 percent). Next, place the meter probes on the connector

pins 2 and 3—the ohmmeter scale should read approximately 2800 ohms at 77 degrees Fahrenheit (±2 percent). If the ohmmeter readings differ, replace the humidity sensor.

6. **Remove the humidity sensor.** Disconnect the sensor wire connector from the electronic control board and trace it back to the sensor. Remove the screws that secure the sensor to the oven cavity.

7. **Install the new humidity sensor.** To install a new humidity sensor, just reverse the disassembly procedure, and reassemble. Make sure that you place the wires in the same route that you removed them and do not pinch the wires. Reinstall all panels or the console panel, and restore the electricity to the oven. Test the microwave oven operation. Perform the microwave performance and leakage test. On electron models, make sure to take the microwave oven out of the service test mode when the repair is completed.

Stirrer Motor

The stirrer motor is located under the access panel (Figure 24-5 part # BB). The stirrer motor turns the stirrer blade in the microwave oven.

The typical complaints associated with the stirrer motor are:

- The microwave oven will not cook the food properly.
- The microwave oven takes too long to cook the food.
- Erratic operation.
- Unusual display readouts and/or error codes.

To handle these problems, perform the following steps:

1. **Verify the complaint.** Verify the complaint by operating the microwave oven with a glass of water and the microwave test light bar centered on the oven floor (see Figures 24-15 and 24-16). The lights in the light bar will light up if the magnetron and high-voltage system are working correctly. At the same time, place an ammeter on one wire of the service cord; the meter should indicate approximately 15 amps at 100 percent power. Check the operation of the cooling fans and stirrer blade. It may be necessary to turn off the electricity to the appliance and wait for two minutes before turning it back on. If a fault code appears, look up the code. If the microwave oven will not power up, locate the technical data sheet behind the panel or cabinet for diagnostics information. On some models you will need the actual service manual for the model you are working on to properly diagnose the microwave oven. The service manual will assist you in properly placing the microwave oven in the service test mode for testing the microwave oven functions.

2. **Check for external factors.** You must check for external factors not associated with the appliance. Is there electricity to the microwave oven? Is the vent system blocked? On countertop models, is there enough open space surrounding the microwave oven for proper ventilation? The voltage at the receptacle is between 108 volts and 132 volts during a load on the circuit. Do you have the correct polarity? (See Chapter 6.)

3. **Disconnect the electricity.** Before working on the microwave oven, disconnect the electricity. This can be done by pulling the plug out of the wall receptacle. Or disconnect the electricity at the fuse panel or circuit breaker panel. Turn off the electricity.

WARNING *Before you begin to service any high-voltage components within the microwave oven, you must discharge the high-voltage capacitor or inverter board first. Failure to do so will cause injury or death.*

WARNING *Some diagnostic tests will require you to test the components with the power turned on. When you disassemble the control panel or remove the outer cabinet, you can position the panel in such a way that the wiring will not make contact with metal. This will allow you to test the components without electrical mishaps.*

4. **Remove the outer cabinet or control panel to gain access.** You can gain access to the stirrer motor (see Figure 24-5) by removing the access panel or the outer cabinet on a microwave oven or combination oven. From inside the oven cavity area, remove the stirrer cover and stirrer blade first in order to remove the stirrer motor if necessary.

5. **Test the stirrer motor.** To test the stirrer motor, locate the stirrer motor leads, disconnect the wires, and attach your ohmmeter leads to the motor wires. Set your ohmmeter on the R × 1K scale. The ohmmeter scale should indicate 2.5K to 4.0K ohms. If the reading is different from this, replace the stirrer motor. On some older models, the stirrer blade is belt-driven—check for a broken belt. On other models, the stirrer blade turns with the airflow from the cooling fan motor that circulates the microwave energy. Make sure the stirrer blade spins freely.

6. **Remove the stirrer motor.** On some models, the stirrer cover and stirrer blade will have to be removed first. The stirrer motor is mounted to the oven cavity with screws—remove the screws.

7. **Install the new stirrer motor.** To install a new stirrer motor, just reverse the disassembly procedure, and reassemble. Reinstall all panels and restore the electricity to the oven. Test the microwave oven operation. Perform the microwave performance and leakage test. On electronic models, make sure to take the microwave oven out of the service test mode when the repair is completed.

Turntable Motor

The turntable motor is located on the bottom of the microwave oven (Figure 24-5 part # Z). This motor turns the glass tray so that the food will cook evenly.

The typical complaints associated with the turntable motor are:

- The microwave oven will not cook the food evenly.
- The microwave oven takes too long to cook the food.
- Erratic operation.
- Unusual display readouts and/or error codes.

To handle these problems, perform the following steps:

1. **Verify the complaint.** Verify the complaint by operating the microwave oven with a glass of water and the microwave test light bar centered on the oven floor (see Figures 24-15 and 24-16). The lights in the light bar will light up if the magnetron and high-voltage system are working correctly. At the same time, place an ammeter on one wire of the service cord; the meter should indicate approximately 15 amps at 100 percent power. Check the operation of the cooling fans and stirrer blade. It may be necessary to turn off the electricity to the appliance and wait for two minutes before turning it back on. If a fault code appears, look up the code. If the microwave oven will not power up, locate the technical data sheet behind the panel or cabinet for diagnostics information. On some models you will need the actual service manual for the model you are working on to properly diagnose the microwave oven. The service manual will assist you in properly placing the microwave oven in the service test mode for testing the microwave oven functions.

2. **Check for external factors.** You must check for external factors not associated with the appliance. Is there electricity to the microwave oven? Is the vent system blocked? On countertop models, is there enough open space surrounding the microwave oven for proper ventilation? The voltage at the receptacle is between 108 volts and 132 volts during a load on the circuit. Do you have the correct polarity? (See Chapter 6.)

3. **Disconnect the electricity.** Before working on the microwave oven, disconnect the electricity. This can be done by pulling the plug out of the wall receptacle. Or disconnect the electricity at the fuse panel or circuit breaker panel. Turn off the electricity.

WARNING *Before you begin to service any high-voltage components within the microwave oven, you must discharge the high-voltage capacitor or inverter board first. Failure to do so will cause injury or death.*

WARNING *Some diagnostic tests will require you to test the components with the power turned on. When you disassemble the control panel or remove the outer cabinet, you can position the panel in such a way that the wiring will not make contact with metal. This will allow you to test the components without electrical mishaps.*

4. **Gain access to the turntable motor.** To gain access to the turntable motor, lay the oven on its back, and remove the bottom panel. On older models, the turntable motor might be belt-driven—inspect the belt.

5. **Test the turntable motor.** To test the turntable motor, locate the motor leads, disconnect the wires, and attach your ohmmeter leads to the motor wires. Set your ohmmeter on the R × 1K scale. The ohmmeter scale should indicate 25,000 to 40,000 ohms. If the reading differs from this, replace the turntable motor.

6. **Remove the turntable motor.** To remove the turntable motor, remove the screws that secure the motor to the cavity. On some older models, you might have to remove the gearbox before removing the motor.

7. **Install the new turntable motor.** To install a new turntable motor, just reverse the disassembly procedure, and reassemble. Reinstall all panels and restore the electricity to the oven. Test the microwave oven operation. Perform the microwave performance and leakage test. On electronic models, make sure to take the microwave oven out of the service test mode when the repair is completed.

Cooling Fan Motor

The cooling fan motor serves two purposes in some models. The fan motor cools the magnetron and high-voltage transformer and circulates the air through the waveguide to the stirrer blade and out the discharge vent (Figure 24-27). Over-the-range models use two fan motors to cool the microwave oven. Other models with stirrer motors circulate the air only in the high-voltage section of the microwave oven.

The typical complaints associated with the cooling fan motor are:

- The microwave oven operates but is unable to cook the food.
- The microwave oven will shut off.
- Unusual display readouts and/or error codes.

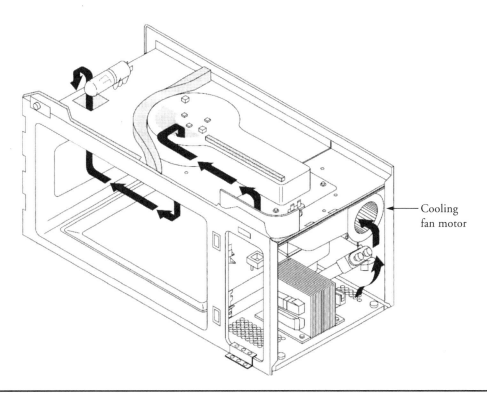

Cooling fan motor

FIGURE 24-27 A typical airflow pattern. This model shows the cooling fan circulating the air through the magnetron, waveguide, stirrer duct, and stirrer blade; through the oven cavity; and then out through the discharge vent.

1. **Verify the complaint.** Verify the complaint by operating the microwave oven with a glass of water and the microwave test light bar centered on the oven floor (see Figures 24-15 and 24-16). The lights in the light bar will light up if the magnetron and high-voltage system are working correctly. At the same time, place an ammeter on one wire of the service cord; the meter should indicate approximately 15 amps at 100 percent power. It may be necessary to turn off the electricity to the appliance and wait for two minutes before turning it back on. If a fault code appears, look up the code. If the microwave oven will not power up, locate the technical data sheet behind the panel or cabinet for diagnostics information. On some models you will need the actual service manual for the model you are working on to properly diagnose the microwave oven. The service manual will assist you in properly placing the microwave oven in the service test mode for testing the microwave oven functions.

2. **Check for external factors.** You must check for external factors not associated with the appliance. Is there electricity to the microwave oven? The voltage at the receptacle is between 108 volts and 132 volts during a load on the circuit. Do you have the correct polarity? (See Chapter 6.) Is the vent system blocked? On countertop models, is there enough open space surrounding the microwave oven for proper ventilation?

3. **Disconnect the electricity.** Before working on the microwave oven, disconnect the electricity. This can be done by pulling the plug out of the wall receptacle. Or disconnect the electricity at the fuse panel or circuit breaker panel. Turn off the electricity.

WARNING *Before you begin to service any high-voltage components within the microwave oven, you must discharge the high-voltage capacitor or inverter board first. Failure to do so will cause injury or death.*

WARNING *Some diagnostic tests will require you to test the components with the power turned on. When you disassemble the control panel or remove the outer cabinet, you can position the panel in such a way that the wiring will not make contact with metal. This will allow you to test the components without electrical mishaps.*

4. **Remove the outer cabinet or control panel to gain access.** You can gain access to the cooling fan motor by removing the access panel or the outer cabinet on a microwave oven or combination oven. If you are repairing a wall oven or an eye-level combination range, the control panel can be removed (see Figures 11-1 and 22-12) by opening the door and removing the screws that secure the panel. These might be underneath the front of the exhaust hood or just below the control panel. Some control panels are hinged; just tilt the control panel toward you for servicing. On some combination models, you might have to uninstall the appliance to gain access.

5. **Test the cooling fan motor.** To test the cooling fan motor, locate the technical data sheet for the model you are servicing, and look up the motor's operating voltage. Microwave oven manufacturers use a 120-volt AC or 12-volt DC motor. Check and see if the fan blade spins freely. Then disconnect the wires from the motor, isolating it from the rest of the circuit, and place your ohmmeter probes on the motor wire leads.

Set your ohmmeter on the R × 1 scale. For direct current (DC) motors, the ohmmeter scale should indicate 15 to 30 ohms. For alternating current (AC) motors, the ohmmeter scale should indicate 80 to 140 ohms. If the readings are different from this, replace the cooling fan motor.

6. **Remove the cooling fan motor.** To remove the cooling fan motor, disconnect the wiring, and remove the screws that secure the motor assembly to the oven frame. Depending on the model you are servicing, you may have to remove other components to access the motor.

7. **Install the new cooling fan motor.** To install a new cooling fan motor, just reverse the disassembly procedure, and reassemble. Reinstall all panels and restore the electricity to the oven. Test the microwave oven operation. Perform the microwave performance and leakage test. On electronic models, make sure to take the microwave oven out of the service test mode when the repair is completed.

Exhaust Fan Motor

The exhaust fan motor is located on the top of microwave oven (over-the-range model). See Figure 24-28. The exhaust fan will remove the heat that accumulates under the microwave oven when the consumer is cooking on the range.

The typical complaints associated with the exhaust fan motor are:

- The exhaust fan motor will not come on.
- The microwave oven overheats and shuts off.
- Unusual display readouts and/or error codes.

To handle these problems, perform the following steps:

1. **Verify the complaint.** Verify the complaint by operating the microwave oven with a glass of water and the microwave test light bar centered on the oven floor (see Figures 24-15 and 24-16). The lights in the light bar will light up if the magnetron and high-voltage system are working correctly. At the same time, place an ammeter on one wire of the service cord; the meter should indicate approximately 15 amps at 100 percent power. It may be necessary to turn off the electricity to the appliance and wait for two minutes before turning it back on. If a fault code appears, look up the code. If the microwave oven will not power up, locate the technical data sheet behind the panel or cabinet for diagnostics information. On some models you will need the actual service manual for the model you are working on to properly diagnose the microwave oven. The service manual will assist you in properly placing the microwave oven in the service test mode for testing the microwave oven functions.

2. **Check for external factors.** You must check for external factors not associated with the appliance. Is there electricity to the microwave oven? The voltage at the receptacle is between 108 volts and 132 volts during a load on the circuit. Do you have the correct polarity? (See Chapter 6.) Is the vent system blocked? On countertop models, is there enough open space surrounding the microwave oven for proper ventilation?

3. **Disconnect the electricity.** Before working on the microwave oven, disconnect the electricity. This can be done by pulling the plug out of the wall receptacle. Or disconnect the electricity at the fuse panel or circuit breaker panel. Turn off the electricity.

WARNING *Before you begin to service any high-voltage components within the microwave oven, you must discharge the high-voltage capacitor or inverter board first. Failure to do so will cause injury or death.*

WARNING *Some diagnostic tests will require you to test the components with the power turned on. When you disassemble the control panel or remove the outer cabinet, you can position the panel in such a way that the wiring will not make contact with metal. This will allow you to test the components without electrical mishaps.*

4. **Remove the outer cabinet or control panel to gain access.** You can gain access to the exhaust fan motor by removing the access panel or the outer cabinet on a microwave oven. If you are repairing a wall oven or an eye-level combination range, the control panel can be removed (see Figures 11-1 and 22-12) by opening the door and removing the screws that secure the panel. These might be underneath the front of the exhaust hood or just below the control panel. Some control panels are hinged; just tilt the control panel toward you for servicing. On some combination models or over-the-range models, you might have to uninstall the appliance to gain access (see Figure 24-28).

FIGURE 24-28
To gain access to the exhaust motor in this model, you will have to uninstall the microwave oven.

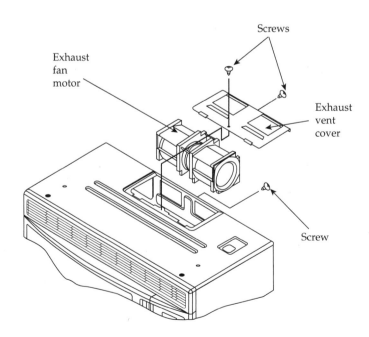

Screws

Exhaust fan motor

Exhaust vent cover

Screw

5. **Test the exhaust fan motor.** Disconnect the motor wire leads from the wiring harness. Check the motor windings for continuity. To check for a grounded winding in the motor, take the ohmmeter probes, and check from each motor wire lead terminal to the motor housing. The ohmmeter will indicate continuity if the windings are grounded.

6. **Remove the exhaust fan motor.** Remove the entire fan motor housing from the oven for easy accessibility. This is accomplished by removing the screws that secure the exhaust vent cover (Figure 24-28).

7. **Install the new exhaust fan motor.** To install a new exhaust fan motor, just reverse the disassembly procedure, and reassemble. Reinstall all panels or reinstall the oven, and restore the electricity to the oven. Test the microwave oven operation. Perform the microwave performance and leakage test. On electronic models, make sure to take the microwave oven out of the service test mode when the repair is completed.

Diagnostic Chart and Wiring Diagrams

The following diagnostic chart will help you to pinpoint the likely causes of various problems associated with microwave ovens (Figure 24-29).

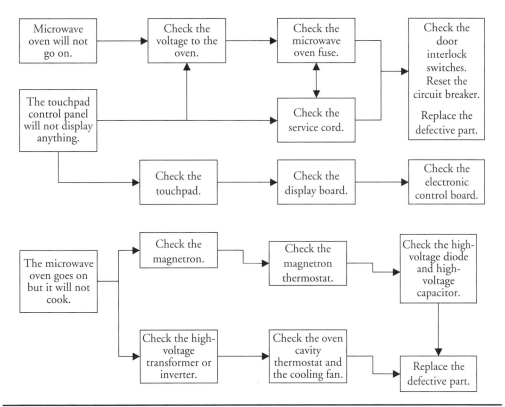

FIGURE 24-29   A diagnostic flowchart.

The wiring diagrams in this chapter are only examples. You must refer to the actual technical data sheet or the wiring diagrams for the microwave oven that you are servicing. Figure 24-30 depicts an actual wiring schematic diagram with inverter boards, and Figure 24-31 depicts an actual pictorial electrical wiring diagram of the same model. Figures 24-32 and 24-33 depict the strip circuits for the same model. The strip circuits are easier to read. Figures 24-34 and 24-35 depict a wiring schematic of a microwave oven with a high-voltage transformer, high-voltage capacitor, and high-voltage diode in the circuit.

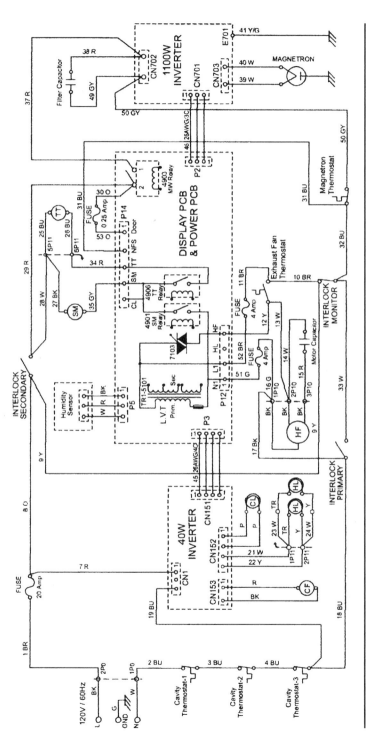

Figure 24-30 A wiring schematic.

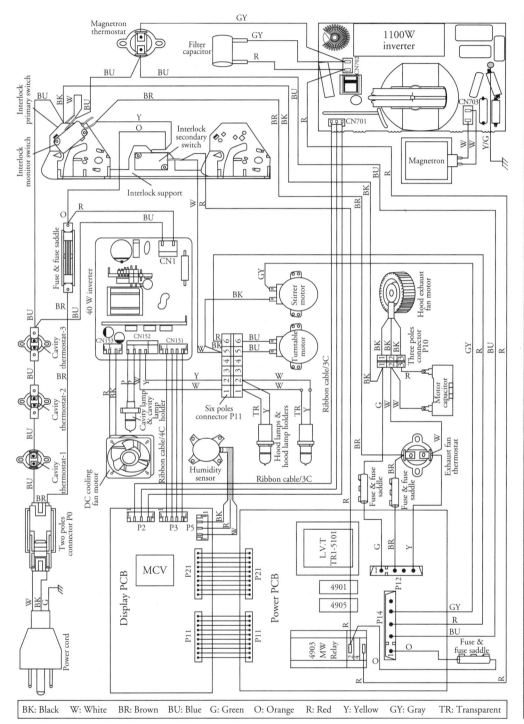

Figure 24-31 A pictorial wiring diagram.

BK: Black   W: White   BR: Brown   BU: Blue   G: Green   O: Orange   R: Red   Y: Yellow   GY: Gray   TR: Transparent

Microwave cooking

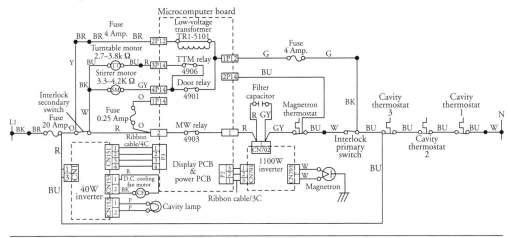

Blower fan on automatic

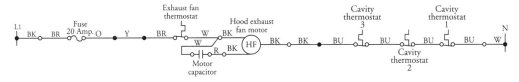

Blower fan on variable

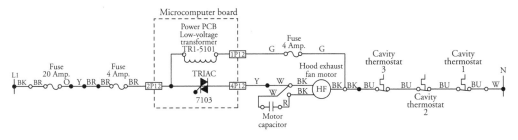

Microwave time-of-day displayed

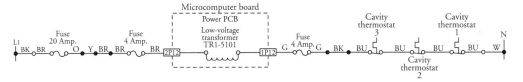

FIGURE 24-32 Microwave oven strip circuits for various functions.

Door open—cavity lamp on

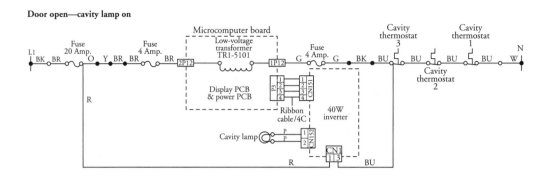

Cooktop lamp on (variable light)

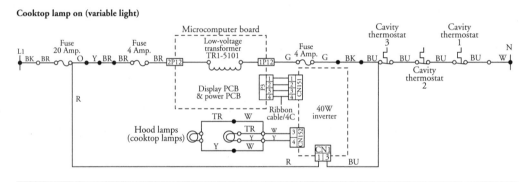

FIGURE 24-33 Microwave oven strip circuits for various functions.

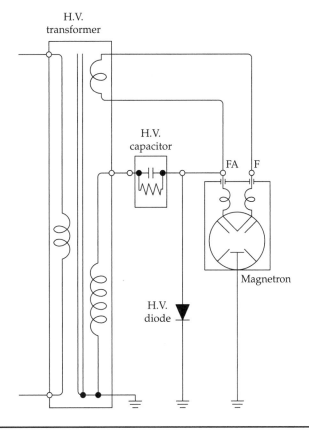

FIGURE 24-34   High-voltage circuit of microwave oven.

PART VI

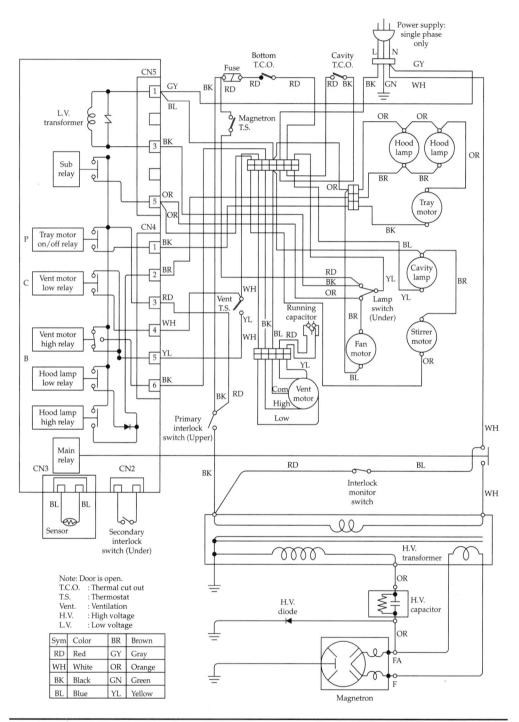

FIGURE 24-35   A wiring schematic of a microwave oven with a high-voltage transformer, high-voltage capacitor, and high-voltage diode in the circuit.

CHAPTER

Refrigerators

One of the most important applications of refrigeration (which was invented in the early 1900s) was for the preservation of food. When different types of food[1] are kept at room temperature, some of them will spoil rapidly. When foods are kept cold, they will last longer. Refrigerators prevent food spoilage by keeping the food cold.

The refrigerator consists of three parts:

- The cabinet.
- The sealed system, which consists of the evaporator coil, the condenser coil, the compressor, and the connecting tubing.
- The electrical circuitry, including fan motors and other electrical components.

This chapter covers the electrical components and how to diagnose the sealed system. The actual repair or replacement of any sealed-system component is not included in this chapter. It is recommended that you acquire refrigerant certification (or call an authorized service company) to repair or replace any sealed-system component. The refrigerant in the sealed system must be recovered properly.

The Refrigeration Cycle

The sealed system (Figure 25-1) in a refrigerator consists of a compressor, a condenser coil, an evaporator coil, a capillary tube, and a heat exchanger and its connecting tubing. This is the heart of the refrigerator that keeps the food cold inside the cabinet.

Starting at the compressor, refrigerant gas is pumped out of the compressor, through the discharge tubing, and into the condenser coil. When the gas is in the condenser coil, the temperature and pressure of the refrigerant gas greatly increase because of the capillary tube at the discharge end of the condenser coil. From the surface of the condenser coil, the heat spreads out into the room via air moving over the condenser coil. The condenser coil cools the hot refrigerant gas. As the refrigerant gas gives up the heat it obtained from inside the refrigerator cabinet, the refrigerant gas changes into a liquid. This liquid then leaves the condenser coil and enters the capillary tube.

This capillary tube is carefully made with regard to its length and inside diameter to meter out the exact amount of liquid refrigerant through the sealed system (this is designed by the manufacturer for a particular size and model). As the liquid refrigerant leaves the

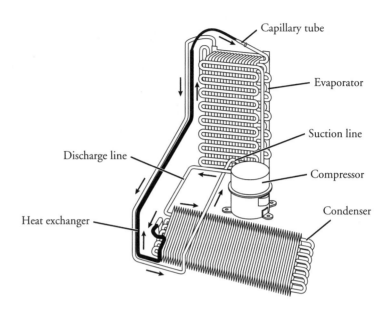

Figure 25-1
The sealed-system components in a side-by-side refrigerator.

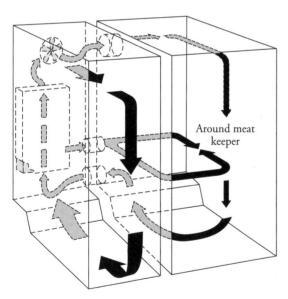

Around meat keeper

Figure 25-2   The airflow pattern in a side-by-side refrigerator.

capillary tube and enters the larger tubing of the evaporator coil, the sudden increase of tubing size causes a low-pressure area. It is here that the liquid refrigerant changes from a liquid to a mixture of liquid and gas.

As this mixture passes through the evaporator coil, the refrigerant absorbs heat from the warmer items (food) within the refrigerator cabinet, slowly changing any liquid back to gas. As the refrigerant gas leaves the evaporator coil, it returns to the compressor through the suction line.

This entire procedure is called a *cycle*. Depending on where the cold control (thermostat) is set, the thermostat can show how cold it is inside the cabinet and then control the actuation of the cooling cycle. It will determine whether to turn the system on or off to maintain the temperature within the cabinet.

Inside the cabinet, the cold air is circulated by convection and/or by means of an electrical fan. In Figure 25-2, the arrows show the airflow patterns in this type of side-by-side refrigerator. Figure 25-3 shows the air patterns in a two-door refrigerator with a top freezer.

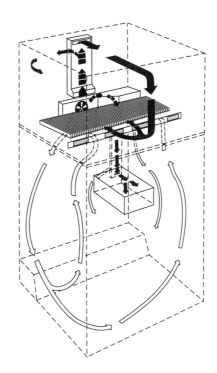

FIGURE **25-3**
The airflow pattern in a two-door refrigerator with top freezer.

Refrigerator Automatic Defrost Systems

Over the years manufacturers have developed many methods to defrost the evaporator coil. Maintaining proper moisture control within the refrigerator is crucial for proper efficient operation. Every time the consumer opens the door, moisture and heat enter the refrigerator compartment. Through either natural convection or forced-air movement, the moisture will eventually condense on the coldest spot in the refrigerator: the evaporator coil. The evaporator coil is well below the freezing temperatures, and the condensed moisture will stick to it and begin to freeze. As frost accumulates on the evaporator coil, the airflow through the evaporator coil will begin to decease and the cooling effect will also decrease.

The manual method for defrosting an evaporator coil will be time consuming, and most consumers end up destroying the evaporator coil by punching a hole in it with an ice pick or a knife. Manufacturers have developed the automatic defrost system. Most modern-day refrigerators have a timer to activate the defrost cycle. A heater element and a defrost termination thermostat have also been attached to the evaporator coil to defrost the frost buildup. When the timer calls for the defrost cycle, the sealed system is de-energized and the heater is energized. The defrost termination thermostat monitors the temperature of the evaporator coil; at around 40 to 60 degrees Fahrenheit the thermostat will turn off the heater. The defrost timer will stay in the defrost mode for up to 30 minutes, regardless of whether the frost is melted or not. The water that results from the defrost process will be directed to the evaporator drain pan and then redirected to the outside of the refrigerator to the condensate pan at the base of the refrigerator to be evaporated.

Some refrigerator models use a defrost cycle on a timed schedule. That means approximately every six to eight hours, the defrost cycle is turned on for approximately 28 to 30 minutes. This type of defrosting system runs without regard for the actual cooling demand on the sealed system. A more efficient type of defrost system developed by manufacturers is the cumulative run-time defrost system. This type of system is based on the compressor run time. The defrost timer only advances when the compressor run time has equaled a predetermined amount of run time; then the system will enter into a defrost time. One disadvantage of this type of defrost system is it does not account for the number of times the door is opened and the increased humidity enters the refrigerator compartment.

Manufacturers over time have developed a new type of automatic defrost system: the adaptive automatic defrost system. With the decreased cost of electronic components a better way has been developed to control the defrost frequency. This type of system measures the time it takes from the start of the defrost cycle until the defrost termination thermostat opens. This type of system is known as adaptive defrost.

Depending on the manufacturer, the first defrost time will occur between six to eight hours of cumulative compressor run time. The adaptive defrost control (ADC) will continually adjust the defrost intervals based on the following:

- Number of door openings.
- Compressor run time.
- The last defrost cycle.

During the defrost cycle, the ADC monitors how long the defrost termination thermostat keeps the heater energized. If the defrost termination thermostat opens in under 12 minutes, this is equal to a light frost buildup. The ADC will increase the amount of compressor run time between defrost times by two hours. If the defrost termination thermostat opens in over 12 minutes, this is equal to a heavier frost buildup. The ADC will decrease the amount of compressor run time between defrost times by two hours.

Over the course of several defrost cycles, the time between the start of the defrost cycle and the opening of the defrost termination thermostat will get closer to the ideal defrost time. The ADC has adjusted the amount of run time between defrosts to occur often enough to maintain a clean evaporator coil, but not so often as to use excessive energy. As humidity conditions change and the frost load increases or decreases, the ADC will adjust the cumulative run time to match the change in the frost load.

As mentioned earlier, not all refrigerator manufacturers use this process. They use a slightly different algorithm programmed into the ADC computer chip, but the principle of operation for most of these ADC controls is very similar. The service technician will need the wiring diagram and/or the service manual for the model they are servicing to properly diagnose and be able to turn on the defrost mode on the ADC board.

Another type of defrost system is the hot gas defrost system. This type of system is seldom used in residential refrigerators due to the additional cost of components. The hot gas defrost systems are used in commercial ice machines to aid in the harvesting process of ice. In the residential refrigerator, a bypass valve is added to the sealed system to the condenser at a point before the refrigerant begins to condense into a liquid state. When defrost is called for, the valve will open, and the hot gas enters the evaporator after the capillary tube inlet. The hot gas will thaw the frost from the evaporator coil.

A cycle defrost system is used on inexpensive refrigerators. The freezer evaporator is manual defrost, but the refrigerator evaporator is defrosted every time the compressor cycles off. In this type of system a special thermostat and two low-wattage heaters are used to defrost the evaporator.

Storage Requirements for Perishable Products

Table 25-1 lists the recommended storage temperatures, relative humidity, and the approximate storage life for perishable products. These values are used in designing commercial refrigeration systems, which house large quantities of perishable products.

Product	Storage Temp.° F	Relative Humidity %	Approximate Storage Life
Apples	30–40	90	3–8 months
Apricots	31–32	90	1–2 weeks
Artichokes	31–32	95	2 weeks
Asparagus	32–36	95	2–3 weeks
Avocados	45–55	85–90	2–4 weeks
Bananas	55–65	85–95	—
Beans (green or snap)	40–45	90–95	7–10 days
Beans, lima	32–40	90	1 week
Blackberries	31–32	95	3 days
Blueberries	31–32	90–95	2 weeks
Broccoli	32	95	10–14 days
Cabbage	32	95–100	3–4 months
Carrots	32	98–100	5–9 months
Cauliflower	32	95	2–4 weeks
Celery	32	95	1–2 months
Cherries, sour	31–32	90–95	3–7 days
Cherries, sweet	30–31	90–95	2–3 weeks
Collards	32	95	10–14 days
Corn, sweet (fresh)	32	95	4–8 days
Cranberries	36–40	90–95	2–4 months
Cucumbers	50–55	90–95	10–14 days
Dairy products			
Cheddar cheese	40	65–70	6 months
Processed cheese	40	65–70	12 months

TABLE 25-1 Storage Requirements for Perishable Products for Commercial Refrigerators and Freezers (*continued*)

PART VI

Product	Storage Temp.° F	Relative Humidity %	Approximate Storage Life
Butter	40	75–85	1 month
Cream	35–40	—	2–3 weeks
Ice cream	–20 to –15	—	3–12 months
Milk, fluid whole			
Pasteurized, grade A	32–34	—	2–4 months
Condensed, sweetened	40	—	15 months
Evaporated	40	—	24 months
Dates (dried)	0 or 32	75 or less	6–12 months
Dried fruits	32	50–60	9–12 months
Eggplant	45–50	90–95	7–10 days
Eggs, shell	29–31	80–85	5–6 months
Figs, dried	32–40	50–60	9–12 months
Figs, fresh	31–32	85–90	7–10 days
Fish, fresh	30–35	90–95	5–15 days
Haddock, cod	30–35	90–95	15 days
Salmon	30–35	90–95	15 days
Smoked	40–50	50–60	6–8 months
Shellfish, fresh	30–33	86–95	3–7 days
Tuna	30–35	90–95	15 days
Grapefruit	50–60	85–90	4–6 weeks
Grapes, American type	31–32	85–90	2–8 weeks
Grapes, European type	30–31	90–95	3–6 months
Greens, leafy	32	95	10–14 days
Guavas	45–50	90	2–3 weeks
Honey	38–50	50–60	1 year, plus
Horseradish	30–32	95–100	10–12 months
Lemons	32 or 50–58	85–90	1–6 months
Lettuce, head	32–34	95–100	2–3 weeks
Limes	48–50	85–90	6–8 weeks
Maple sugar	75–80	60–65	1 year, plus
Mangoes	55	85–90	2–3 weeks
Meat			
Bacon, cured (farm style)	60–65	85	4–6 months
Game, fresh	32	80–85	1–6 weeks

TABLE 25-1   Storage Requirements for Perishable Products for Commercial Refrigerators and Freezers (*continued*)

Product	Storage Temp.° F	Relative Humidity %	Approximate Storage Life
Beef, fresh	32–34	88–92	1–6 weeks
Hams and shoulders, fresh	32–34	85–90	7–12 days
Cured	60–65	50–60	0–3 years
Lamb, fresh	32–34	85–90	5–12 days
Livers, frozen	–10–0	90–95	3–4 months
Pork, fresh	32–34	85–90	3–7 days
Smoked sausage	40–45	85–90	6 months
Fresh	32	85–90	1–2 weeks
Veal, fresh	32–34	90–95	5–10 days
Melons, Cantaloupe	36–40	90–95	5–15 days
Honeydew and Honey Ball	45–50	90–95	3–4 weeks
Watermelons	40–50	80–90	2–3 weeks
Mushrooms	32	90	3–4 days
Milk	34–40	—	7 days
Nectarines	31–32	90	2–4 weeks
Nuts (dried)	32–50	65–75	8–12 months
Okra	45–50	90–95	7–10 days
Olives, fresh	45–50	85–90	4–6 weeks
Onions (dry) and onion sets	32	65–70	1–8 months
Oranges	32–48	85–90	3–12 weeks
Orange juice, chilled	30–35	—	3–6 weeks
Papayas	45	85–90	1–3 weeks
Parsley	32	95	1–2 months
Parsnips	32	98–100	4–6 months
Peaches	31–32	90	2–4 weeks
Pears	29–31	90–95	2–7 months
Peas, green	32	95	1–3 weeks
Peppers, sweet	45–50	90–95	2–3 weeks
Pineapples, ripe	45	85–90	2–4 weeks
Plums, including fresh prunes	31–32	90–95	2–4 weeks
Popcorn, unpopped	32–40	85	4–6 months
Potatoes, early crop	50–55	90	0–2 months
Potatoes, late crop	38–50	90	5–8 months

TABLE 25-1   Storage Requirements for Perishable Products for Commercial Refrigerators and Freezers (*continued*)

Product	Storage Temp.° F	Relative Humidity %	Approximate Storage Life
Poultry			
Fresh chicken	32	85–90	1 week
Fresh goose	32	85–90	1 week
Fresh turkey	32	85–90	1 week
Pumpkins	50–55	70–75	2–3 months
Radishes—spring, prepacked	32	95	3–4 weeks
Raisins (dried)	40	60–70	9–12 months
Rabbits, fresh	32–34	90–95	1–5 days
Raspberries, black	31–32	90–95	2–3 days
Raspberries, red	31–32	90–95	2–3 days
Rhubarb	32	95	2–4 weeks
Spinach	32	95	10–14 days
Squash, summer	32–50	85–95	5–14 days
Squash, winter	50–55	70–75	4–6 months
Strawberries, fresh	31–32	90–95	5–7 days
Sugar, maple	75–80	60–65	1 year, plus
Sweet potatoes	55–60	85–90	4–7 months
Syrup, maple	31	60–70	1 year, plus
Tangerines	32–38	85–90	2–4 weeks
Tomatoes, mature green	55–70	85–90	1–3 weeks
Tomatoes, firm ripe	45–50	85–90	4–7 days
Turnips, roots	32	95	4–5 months
Vegetables (mixed)	32–40	90–95	1–4 weeks
Yams	60	85–90	3–6 months

TABLE 25-1   Storage Requirements for Perishable Products for Commercial Refrigerators and Freezers

Large warehouses are usually equipped to store foods at those temperatures best adapted to prolonging the safe storage period for each type of food. In the domestic refrigerator, most foods are kept at 34 to 40 degrees Fahrenheit (1 to 4 degrees Celsius) with an optimum temperature of 37 degrees Fahrenheit, and the humidity is kept around 50 percent. The freezer temperature is between zero and minus 10 degrees Fahrenheit (–17 to –23 degrees Celsius).

It can be difficult to maintain these temperatures and humidity for each individual product. Therefore, refrigerator manufacturers have designed separate compartments within the refrigerated cabinet to maintain a variable temperature and humidity selected by the consumer. The storage life of various products will vary in a domestic refrigerator/freezer (Table 25-2). This period will be influenced by many factors, such as the storage

DOMESTIC REFRIGERATOR/FREEZER FOOD STORAGE TIPS			
Foods	**Refrigerator**	**Freezer**	**Storage Tips**
DAIRY PRODUCTS			
Butter	1 month	6 to 9 months	Wrap product tightly or cover.
Milk and cream	1 week	Not recommended	Check the date on the carton. Close tightly. Do not store unused portions in the original container. Do not freeze the cream unless whipped.
Cream cheese, cheese spread, and cheese food	1 to 2 weeks	Not recommended	Wrap product tightly.
Cottage cheese	3 to 5 days	Not recommended	Keep product stored in original carton. Check carton date.
Sour cream	10 days	Not recommended	Keep product stored in original carton. Check carton date.
Hard cheese (Swiss, Cheddar, and Parmesan)	1 to 2 months	4 to 6 months May become crumbly	Wrap tightly. Cut off mold.
EGGS			
Eggs in the shell	3 weeks	Not recommended	Refrigerate with the small ends facing down.
Leftover yolks or whites	2 to 4 days	9 to 12 months	For each cup of yolks to be frozen, add 1 teaspoon of sugar for use in sweet dishes, or 1 teaspoon of salt for nonsweet dishes.
FRUITS			
Apples	1 month	8 months (cooked)	Store unripe or hard apples at 60° to 70°F (16° to 21°C).
Bananas	2 to 4 days	6 months (whole peeled)	Ripen at room temperature before refrigerating. Bananas will darken when refrigerated.
Pears, plums, avocados	3 to 4 days	Not recommended	Ripen at room temperature before refrigerating. Avocados will darken when refrigerated.
Berries, cherries, apricots	2 to 3 days	6 months	Ripen at room temperature before refrigerating.
Grapes	3 to 5 days	1 month (whole)	Ripen at room temperature before refrigerating.
Citrus fruits	1 to 2 weeks	Not recommended	Store uncovered at 60° to 70°F (16° to 21°C).

TABLE 25-2 Storage Requirements for Perishable Products for a Domestic Refrigerator/Freezer (*continued*)

DOMESTIC REFRIGERATOR/FREEZER FOOD STORAGE TIPS			
Foods	**Refrigerator**	**Freezer**	**Storage Tips**
Pineapples, cut	2 to 3 days	6 to 12 months	Will not ripen after purchase. Use quickly.
VEGETABLES			
Asparagus	1 to 2 days	8 to 10 months	Do not wash before refrigerating. Store in crisper drawer.
Brussels sprouts, broccoli, cauliflower, green peas, lima beans, onions, peppers	3 to 5 days	8 to 10 months	Wrap odorous foods. Leave the peas in the pods.
Cabbage, celery	1 to 2 weeks	Not recommended	Wrap odorous foods and refrigerate in crisper drawer.
Carrots, parsnips, beets, turnips	7 to 10 days	8 to 10 months	Remove tops. Wrap odorous foods and refrigerate in crisper drawer.
Lettuce	7 to 10 days	Not recommended	
POULTRY and FISH			
Chicken and turkey, whole	1 to 2 days	12 months	Keep in original packaging for refrigeration. Place in the meat and cheese drawer. When freezing longer than two weeks, overwrap with freezer wrap paper.
Chicken and turkey, pieces	1 to 2 days	9 months	
Fish	1 to 2 days	2 to 6 months	
MEATS			
Bacon	7 days	1 month	
Beef or lamb, ground	1 to 2 days	3 to 4 months	Fresh meats can be kept in original packaging for refrigeration.
Beef or lamb, roast and steak	3 to 5 days	6 to 9 months	Place in the meat and cheese drawer. When freezing longer than two weeks, overwrap with freezer wrap paper.
Ham, fully cooked, whole	7 days	1 to 2 months	
Ham, fully cooked, half	5 days	1 to 2 months	
Ham, fully cooked, slices	3 days	1 to 2 months	
Luncheon meats	3 to 5 days	1 to 2 months	Unopened, vacuum-packed luncheon meat may be kept up to two weeks in the meat and cheese drawer.

TABLE 25-2   Storage Requirements for Perishable Products for a Domestic Refrigerator/Freezer (*continued*)

DOMESTIC REFRIGERATOR/FREEZER FOOD STORAGE TIPS			
Foods	Refrigerator	Freezer	Storage Tips
Pork, roast	3 to 5 days	4 to 6 months	
Pork, chops	3 to 5 days	4 months	
Sausage, ground	1 to 2 days	1 to 2 months	
Veal	3 to 5 days	4 to 6 months	
Frankfurters	7 days	1 month	Processed meats should be tightly wrapped and stored in the meat and cheese drawer.

TABLE 25-2   Storage Requirements for Perishable Products for a Domestic Refrigerator/Freezer

temperature, the type of container, the condition of the food, and the kind of food. For food storage tips, consult the use and care guide.

To test the temperature in a refrigerator/freezer, place a thermometer in a glass of water and place it in the center of the refrigerator compartment. After 24 hours, check the thermometer. To test the temperature in the freezer compartment, place a thermometer between two frozen packages. After 24 hours, check the thermometer. If the temperature controls needs to be readjusted, retake temperatures as listed earlier.

Safety First

Any person who cannot use basic tools or follow written instructions should *not* attempt to install, maintain, or repair any refrigerators. Any improper installation, preventive maintenance, or repairs could create a risk of personal injury or property damage.

If you do not fully understand the installation, preventive maintenance, or repair procedures in this chapter, or if you doubt your ability to complete the task on your refrigerator or freezer, please call your service manager.

Before continuing, take a moment to refresh your memory of the safety procedures in Chapter 2.

Refrigerators in General

Much of the troubleshooting information in this chapter covers refrigerators in general, rather than specific models, in order to present a broad overview of service techniques. The illustrations in this chapter are for demonstration purposes only, to clarify the description of how to service these appliances. They in no way reflect on a particular brand's reliability.

Location and Installation of Refrigerator

Thoroughly read the installation instructions that are provided with every new refrigerator. These instructions will provide you with the information you need to properly install the refrigerator. The following are some general principles that should be followed when performing the installation:

- The refrigerator must be installed on a solid floor capable of supporting the product up to 1000 pounds.

- For proper air circulation around the refrigerator, some models require a one-inch clearance at the rear and top of the cabinet and adequate clearance near the front grille at the bottom of the refrigerator.

- Do not leave the refrigerator on its side longer than necessary to remove the shipping base.

- When removing or reversing the doors on a refrigerator, always reinstall them according to the installation instructions, and remember to realign the doors properly.

- Level the refrigerator cabinet so that the doors close properly.

What's That "Different" Sound in Your Kitchen?

If you have bought or serviced a new refrigerator within the past few years, you've probably noticed that it sounds "different." Here's why: New refrigerators use only half as much electricity as the older models. In fact, a new 20.6-cubic foot refrigerator with top freezer uses no more electricity than a 75-watt light bulb. Most new refrigerators are also larger, and they have such added conveniences as automatic defrost systems, ice makers, and perhaps even a "built-in" look. These new features result in different sounds, such as:

- **High-pitched whine**   This is due to the more energy-efficient compressors that have smaller, higher-speed motors.

- **Soft hum**   This is from the evaporator fan in the freezer and/or from the condenser fan under the refrigerator.

- **Clicks**   These can be from the automatic defrost timer switching on and off, the thermostat turning the refrigerator on and off, or the water valve refilling the ice maker.

- **Boiling and/or gurgling or trickling water**   When the refrigerator stops running, the refrigerant continues to circulate within the system or the defrost water runs into the drain pan.

- **Running water and "thuds"**   These sounds occur as the ice cube tray fills and as the ice cubes drop into the storage bin.

To help mute these new sounds:

- Be sure that the refrigerator is level and that the defrost water collection pan is in position (usually reachable behind the bottom front "toe plate").

- Put a piece of carpet or a sound-absorbing ceiling tile on the wall behind the refrigerator.

- Allow enough space between the back of the refrigerator and the wall, unless it is designed as a "built-in" unit. Check your use and care manual for the needed space.

To reduce the compressor run time:

- Vacuum the condenser coils at least twice a year—more often if you have pets.
- Keep your freezer at least three-fourths full. Use partially filled water jugs to fill any large empty space.

The Major Appliance Consumer Action Panel, or MACAP, was an independent complaint-mediation group made up of professionals with expertise in textiles, equipment, consumer law, and engineering who volunteered their time. Unfortunately, MACAP went out of business in the last few years, but the information that is printed in this book is still valuable and worth reading for guidance. If you are experiencing any problems with your product or service company, contact your local Better Business Bureau (BBB).

Many of the consumers who filed "refrigerator sound" complaints with the MACAP represent one- or two-family households; they have recently moved to smaller retirement homes; or they have remodeled, and the kitchen is now open to a family living area. Sounds are more noticeable in quieter surroundings. Consumers with hearing aids are especially sensitive.

Some consumers reported to MACAP that their refrigerators are "louder" than an identical model in a friend's or a relative's house. This might be because of the number of people in the house, as well as different furnishings and room arrangements. Carpeting, drapery, upholstered furniture, and wall coverings can help to muffle refrigerator sounds.[2]

Are Refrigerators/Freezers Snowbirds?

Putting that extra refrigerator or freezer in any area in which the temperature falls below 60 degrees Fahrenheit might be a problem in colder areas when the winter months are approaching. Combination refrigerator/freezers and freezers with automatic defrost systems are sensitive to the ambient air temperature surrounding them. As this ambient temperature rises, the compressor runs more to maintain the storage temperature in the fresh food and frozen food compartments, thus wasting energy. As the ambient temperature falls, compressor operation decreases.

When temperatures fall below 60 degrees Fahrenheit, the compressor will not operate long enough to maintain low storage temperatures in the freezer compartment. This is because the fresh food compartment contains the primary sensor, and it is satisfied quickly at a low ambient temperature. The lower the temperature goes, the worse this condition becomes. At about 38 to 42 degrees Fahrenheit, the compressor will not run at all. The freezer compartment temperature will increase to the ambient room temperature, and the frozen food in the compartment will defrost and spoil.

Combination refrigerator/freezers and automatic-defrost freezers should not be operated in unheated places, like garages or porches, where room temperatures are likely to drop below 60 degrees Fahrenheit, unless they are specifically designed for operation in low temperatures. Check the manufacturer's use and care manual for the lowest safe operating temperature.

At any time the temperature will be less than 60 degrees Fahrenheit, it is best to empty the freezer compartment of the refrigerator/freezer in order to prevent defrosting and the possible spoilage of frozen foods. You might want to consider removing all of the food from the unit, turning it off, and propping the door open if you will be absent for an extended period of time.

PART VI

The door must stay open to prevent mold and odor. *Don't* do this if your refrigerator has a latch-type handle (pre-1958 model) because of the potential for child entrapment.

Manual-defrost freezers can usually be operated in an unheated garage or porch without affecting the unit or the frozen food. However, check the use and care manual to determine if your unit needs special care.[3]

"Freezer" in Single-Door Refrigerator Has Limited Function

"I just bought a single-door refrigerator and my ice cream won't harden. It's like cold soup," a consumer recently complained to MACAP. "If I turn the temperature control to a colder setting, the food in the refrigerator section freezes, but the ice cream remains soup. Something's wrong with the freezer."

Approximately 6 percent of all refrigerators sold each year are "single-door" models. That is, the model has only one outside door. Inside (usually across the top or to one side) is a small freezer compartment with its own door. Most combination refrigerator/freezers have two or more doors on the outside, providing independent access to separate freezer or refrigerated sections.

In the course of investigating this consumer's complaint, MACAP found that consumers have reason to be confused about the capabilities of "single-door" refrigerators. Many manufacturers refer to the separate frozen food compartment as a "freezer section," as a "freezer compartment," or as a "freezer" in their specification literature. MACAP found that a majority of consumers expect to be able to keep such hard-to-freeze items as ice cream and orange juice in this "freezer" compartment. These items have a high sugar content, and they freeze at lower temperatures than water.

The Association of Home Appliance Manufacturers (AHAM), in its nationally accepted standard, calls such a unit a "basic refrigerator" and specifies that it is "intended for short-term storage of foods at temperatures below 32 degrees Fahrenheit and normally above 8 degrees Fahrenheit." However, most newer models have temperatures at (or near) freezing level, which is not adequate to freeze foods. Distilled water freezes at 32 degrees Fahrenheit, but all frozen foods must be stored at a temperature lower than that to freeze. Vegetables begin to freeze at 29 to 31 degrees Fahrenheit, meats at 25 to 29 degrees Fahrenheit, and orange juice concentrate at about 8 degrees Fahrenheit. Ice cream begins to stiffen at 27 degrees Fahrenheit, but is considered at an ideal hardness for scooping at 8 degrees Fahrenheit.

If reference is made to the compartment being a "food freezer" or a "frozen food storage compartment" (as in double-door units), the consumer can expect that it will store already frozen foods for several days, or even months, without deterioration. However, a "freezer section" or "freezer compartment," as found in single-door refrigerators, might only freeze ice cubes.

MACAP recommends that consumers determine their food freezing needs and carefully read available literature before making a purchasing decision.[4]

Refrigerator Maintenance

The inside of the cabinet should be cleaned at least once a month to help prevent odors from building up. Of course, any spills that might happen should be wiped up immediately. Wash all removable parts by hand with warm water and a mild detergent; then rinse and dry the parts. The inside walls of the cabinet, the door liners, and the gaskets should also be washed using warm water and a mild detergent, rinsed, and dried. Never use cleaning waxes, concentrated detergents, bleaches, ammonia, or cleansers containing petroleum products on plastic parts. Never use cleaning products with a lemon scent; the lemon will be absorbed into the liner permanently and it may also affect the food. On the outside of the cabinet, use a sponge with warm water and a mild detergent to clean dust and dirt. Then rinse off and dry thoroughly.

At least two times a year, the outside cabinet should be waxed with an appliance wax or with a good auto paste wax. Waxing painted metal surfaces provides rust protection. The defrost pan, which is located behind the toe plate or behind the cabinet, should be cleaned out once a month. The condenser coil should also be cleaned of dust and lint at least once a month. The floor should be free of dirt and debris when the cabinet is rolled out away from the wall. After the cabinet is rolled back into place, you must check to be sure that the cabinet is level.

Food Odors and Molds

Odors in the refrigerator compartment or the freezer compartment cannot occur by themselves. The only way that odors can occur is by storing foods in an unsealed container or unwrapped. Another way that odors occur is from food spillage or from rotten or spoiled food. In new refrigerators there may be a plastic odor, but this is normal and it will dissipate in time. Here are some more tips to help in odor removal:

1. Place a box of baking soda in the refrigerator's fresh food and freezer compartments. Replace according to the instructions on the box.

2. Place some activated charcoal in a shallow metal pan inside the fresh food or the freezer compartment of the refrigerator. When the charcoal loses its effectiveness, place the metal pan in the oven and heat it on a low temperature for a couple of hours to rejuvenate it. Do not use charcoal briquettes used for grilling; it is not the same activated charcoal.

3. Place some vanilla extract in a small dish and place in the refrigerator's fresh food compartment for three weeks. Do not place in the freezer compartment; the vanilla extract will freeze and be ineffective.

On occasion the ice cubes will have a bad taste and will smell like food. Sometimes, the food odors come from the refrigerator compartment. To be able to tell which storage compartment is producing the food odors, try the following:

1. Fill ice trays with tap water and freeze them.

2. Remove the ice cubes from the tray and place in a bowl.

3. Place a bowl in the freezer compartment for a few days.

4. Taste or smell the ice cubes.

If the taste or odor is present in the ice cubes, then the odors are coming from the refrigerator or freezer compartment. The odors are present in the air and as they circulate between the two compartments, the ice cubes absorb the odors. If the bowl of ice cubes has no odors or bad taste, then the bad taste and odor are coming from the water supply that feeds the automatic ice maker and/or water dispenser. Once every two or three weeks, replace the old ice cubes with new ones.

Food molds often grow on baked goods, produce, and leftovers and dairy products. Mold is caused by microbes that attach to the food surface and causes the food to go bad. Underneath the food's surface the mold cells attack the remainder of the food, causing it to rot out.

As food is stored in the refrigerator it loses taste, texture, and nutritional value. Improper handling or storing of foods can cause food-related illness or even disease. To clean the mold from the refrigerator, throw away spoiled food, and clean the walls and shelves in the refrigerator and freezer compartments. Follow the use and care instructions that came with the refrigerator for proper cleaning instructions.

Step-by-Step Troubleshooting by Symptom Diagnosis

When servicing an appliance, don't overlook the simple things that might be causing the problem. Step-by-step troubleshooting by symptom diagnosis is based on diagnosing malfunctions, with possible causes arranged into categories relating to the operation of the refrigerator. This section is intended only to serve as a checklist to aid you in diagnosing a problem. Look at the symptom that best describes the problem you are experiencing with the refrigerator, and then correct the problem.

The Refrigerator Does Not Operate

- Check and see if the refrigerator is plugged in.
- Check voltage at receptacle.
- Check the temperature controls. Are they off?
- Check the electronic control board. Run the test mode.

Compressor Will Not Run

- Is there voltage at the wall receptacle? Check this with the voltmeter.
- Check for loose electrical connections.
- Is the condenser coil dirty? A dirty condenser coil will overheat the compressor.
- Check the condenser fan motor.
- Test the cold control for continuity.
- Test the compressor, the relay, and the overload switch.
- Check and see if the refrigerator is in the defrost mode.

Compressor Kicks Out on Overload

- Check for high or low voltage when the compressor tries to start. High voltage will overheat the compressor. Low voltage will try to run the compressor with the start winding. A compressor is designed to start and run within a 10 percent tolerance of the rated voltage.
- Test the capacitor. A shorted or open capacitor will overheat the compressor.
- Test the compressor relay.
- Test the overload for continuity.
- Test the compressor windings for a short.

Refrigerator Is Too Cold

- Check the damper control setting. Check to see if the damper is stuck open (thermostatically controlled dampers only).
- Test the cold control switch contacts for continuity. Test for stuck contacts.
- Check temperature sensors.
- Check the location of the refrigerator. If outside in the winter, the ambient temperature may be too cold.

Refrigerator Is Too Warm

- Check for restricted air circulation around the condenser coil.
- Check the location of the refrigerator.
- Check the door gaskets for proper sealing.
- Check to see if the cabinet light is staying on when the door is closed.
- Check the defrost heaters. Use a clamp-on ammeter (or wattmeter) to test the heaters if they are coming on when the refrigeration cycle is running.
- Check the cold control setting.
- Check the compressor. Is it operating properly?
- Is the evaporator fan running?
- Check the air duct for restriction.
- Check temperature sensors.
- Check for a leaking air duct.
- Check the evaporator coil for excessive frost buildup.
- Check the defrost cycle. Is it working properly?
- Check the damper control setting. Check to see if the damper is stuck closed (thermostatically controlled dampers only).

Refrigerator Is Too Noisy

- Check for loose parts.
- Check for rattling pipes.
- Check the fan assembly, the evaporator, and the condenser.
- Check the compressor.
- If these are normal operational noises, inform the consumer.
- Refrigerator not properly leveled.
- Check the floor—it may not be structurally sound.

Sweating on the Outside of the Cabinet

- Check the location of the refrigerator. If located in an area of high humidity, it will sweat.
- Check for a void in the insulation between the cabinet and the inner liner.

- Test the mullion and/or stile heaters for continuity.
- Is the energy saver switch in the "on" position?
- On older models, check for wet insulation.
- Check for suction line or any low side tubing touching the cabinet.
- Check for water leaks from the ice maker.
- Check chilled water supply lines and connections.
- Check for a kinked, misaligned, or blocked drain system.
- Check the defrost drain pan for misalignment or for leaking cracks.
- Are the doors aligned and sealing properly?

Sweating on the Inside of the Cabinet

- Check for any abnormal usage. Instruct the consumer on proper usage.
- Check the door gaskets for proper sealing.
- Check for defrost drain water leaking into the cabinet.
- Is the condensate drain blocked?
- Are the doors aligned properly?
- Inspect all access holes where tubing or wires enter the refrigerator/freezer.
- Seal with Permagum if necessary.
- Inspect cabinet outer walls and seams for any openings. Seal with Permagum if necessary.
- Are there excessive door openings on hot, humid days?
- Check for improper food storage.

Incomplete Defrosting of the Evaporator Coil or High Temperature During the Defrost Cycle

- Test the defrost thermostat.
- Check for loose wiring in the defrost electrical circuit.
- Test the defrost timer for continuity.
- Test for defective defrost heaters.

Odor in Cabinet

- Check for spoiling food in the cabinet.
- Check the defrost water drain system.
- Check the defrost heaters.

Excessive Frost Buildup on Evaporator Coil in the Freezer Section

- Check the defrost cycle.
- Check for loose wiring.
- Is the heater making contact with the evaporator coil?

- Check for proper door alignment.
- Check the door gaskets.

Freezer Section Run Time Is Too Long

- Check the thermostat setting.
- Check for excessive loading of unfrozen food.
- Check for incorrect wiring.

Temperature in Freezer Section Is Higher Than Normal

- Check the thermostat for proper temperature calibration.
- Test evaporator fan motor and blade.
- Check the defrost timer.
- Check for excessive loading of unfrozen food.
- Check door gasket for proper sealing.

Refrigerator Runs Excessively or Continuously

- Check if the interior lights are staying on continuously.
- Check condenser coil for air restriction.
- Check door gaskets.
- On models with automatic ice makers, make sure the ice maker is operating properly.

Temperature-Controlled Drawers Are Too Warm

- Check control settings.
- Check the freezer control; it may be set too low.
- Check that the drawer is not improperly positioned.

Troubleshooting Sealed-System Problems

If you suspect a sealed-system malfunction, be sure to check all external factors first. These include:

- Thermostats
- Compressor
- Relay and overload on the compressor
- Interior lights
- Evaporator and condenser fans
- Timers
- Refrigerator/freezer getting good air circulation
- Food loaded in the refrigerator/freezer properly

- Check if heat exchanger has separated
- Check the wiring harness

After eliminating all of these external factors, you will systematically check the sealed system. This is accomplished by comparing the conditions found in a normally operating refrigerator/freezer. These conditions are:

- Refrigerator/freezer storage temperature
- Wattage
- Condenser temperature
- Evaporator inlet sound (gurgle, hiss, etc.)
- Evaporator frost pattern
- High-side pressure[5]
- Low-side pressure[5]
- Pressure equalization time

One thing to keep in mind is that no single indicator is conclusive proof that a particular sealed-system problem exists. Rather, it is a combination of findings that must be used to definitely pinpoint the exact problem.

Refrigerant Leak
The following symptoms may indicate there is a refrigerant leak in the sealed system:

- Temperatures in the storage area are below normal.
- The wattage and amperage are below normal, as indicated on the model/serial plate.
- The condenser coil will be cool to the touch at the last pass or even as far as midway through the coil.
- At the evaporator coil, you will hear a gurgling noise, a hissing noise, or possibly an intermittent hissing or gurgling noise.
- When the evaporator coil cover is removed, the evaporator coil will show a receded frost pattern.
- The high- and low-side pressures will be below normal.[5]
- The pressure equalization time might be normal or shorter than normal.

Overcharged Refrigerator
If the sealed system is overcharged, the symptoms are:

- The storage temperature will be higher than normal.
- The wattage and amperage are above normal, as indicated on the model/serial plate.
- The temperature of the condenser coil will be above normal.
- At the evaporator coil, you will hear a constant gurgling noise. Generally, this is a higher sound level than normal.

- When the evaporator coil cover is removed, the evaporator coil will show a full frost pattern. If you remove the back cover, located behind the refrigerator/freezer, you will possibly see the suction line frosted back to the compressor.
- The high- and low-side pressures will be above normal.[5]
- The pressure equalization time will be normal.

Slight Restriction

The symptoms of a slight restriction in the sealed system are:

- The storage temperature will be below normal.
- The wattage and amperage are below normal, as indicated on the model/serial plate.
- The temperature of the condenser coil will be slightly below normal.
- At the evaporator coil, you will hear a constant gurgling noise and a low sound level.
- When the evaporator coil cover is removed, the evaporator coil pattern will be receded.
- The high- and low-side pressures will be below normal.[5]
- The pressure equalization time will be longer than normal.

Partial Restriction

The symptoms of a partial restriction in the sealed system are:

- The storage temperature will be higher than normal.
- The wattage and amperage are below normal, as indicated on the model/serial plate.
- The temperature of the condenser coil will be below normal more than halfway on the coil.
- At the evaporator coil, you will hear a constant gurgling noise and a considerably low sound level.
- When the evaporator coil cover is removed, the evaporator coil will be considerably receded.
- The high- and low-side pressures will be below normal.[5]
- The pressure equalization time will be longer than normal.

Complete Restriction

The symptoms of a complete restriction in the sealed system are:

- The storage temperature will be warm.
- The wattage and amperage will be considerably below normal, as indicated on the model/serial plate.
- The temperature of the condenser coil will be cool or at room temperature.
- At the evaporator coil, you will hear no sounds.
- When the evaporator coil cover is removed, the evaporator coil will not have any frost on it or the frost will be melting.
- The high-side pressure will be equal to the pressure of refrigerant at room temperature.[5]

- The low-side pressure will be in a deep vacuum.[5]
- There will be no pressure equalization time.

Moisture Restriction
The symptoms of a moisture restriction in the sealed system are:

- The storage temperature will be above normal.
- The wattage and amperage will be considerably below normal, as indicated on the model/serial plate.
- The temperature of the condenser coil will be below normal.
- At the evaporator coil, you will hear a constant gurgle, low sound level, or no sound at all.
- When the evaporator coil cover is removed, the evaporator coil might have some frost on the evaporator inlet.
- The high-side pressure will be below normal.[5]
- The low-side pressure will be below normal or in a deep vacuum.[5]
- The pressure equalization time will be longer than normal or there will be no equalization at all.

Low-Capacity Compressor
The symptoms of a low-capacity compressor in the sealed system are:

- Temperatures in the storage area will be above normal.
- The wattage and amperage will be below normal, as indicated on the model/serial plate.
- The temperature of the condenser coil will be below normal.
- At the evaporator coil, you will hear a slightly reduced gurgling noise.
- When the evaporator coil cover is removed, the evaporator coil will show a normal frost pattern.
- The high-side pressure will be below normal; the low-side pressure will be above normal.[5]
- The pressure equalization time might be normal or shorter than normal.

Repair Procedures

Each repair procedure is a complete inspection and repair process for a single refrigerator component. It contains the information you need to test the components and replace them, if necessary.

Any person who cannot use basic tools should *not* attempt to install, maintain, or repair any refrigerator. Any improper installation, preventative maintenance, or repairs will create a risk of personal injury, as well as property damage. Call the service manager if installation, preventative maintenance, or the repair procedure is not fully understood.

Electronic Control Board and Touchpad

The electronic control board and touchpad are located in the fresh food compartment as soon as you open the door on some models. On other models the electronic control board and touchpad are located on the outside of the freezer door. The touchpad allows the consumer to monitor and control temperatures within the refrigerator and freezer sections. The electronic control on the freezer door also allows the consumer to dispense water and ice from the refrigerator's freezer section. Another great feature that was added by manufacturers is the ability for technicians to enter into the service mode to run any function within the refrigerator. If any problem occurs in the refrigerator, an error/fault code appears and signals the consumer that they have a problem with the refrigerator.

The typical complaints associated with the electronic control board and touchpad are:

- Unable to program the touchpad panel functions.
- The display board will not display anything.
- Unable to control the temperatures.
- The compressor will not run.
- Unusual display readouts and/or error codes.

To prevent electrostatic discharge (ESD) from damaging expensive electronic components, follow the steps in Chapter 11.

To handle these problems, perform the following steps:

1. **Verify the complaint.** Verify the complaint by operating the refrigerator controls. Turn off the electricity to the appliance and wait for two minutes before turning it back on. If a fault code appears, look up the code. If the refrigerator will not power up, locate the technical data sheet behind the panel or cabinet for diagnostics information. On some models you will need the actual service manual for the model you are working on to properly diagnose the refrigerator. The service manual will assist you in properly placing the refrigerator in the service test mode for testing the refrigerator functions.

2. **Check for external factors.** You must check for external factors not associated with the appliance. Is there electricity to the refrigerator? Is the electrical receptacle polarized and properly grounded? The voltage at the receptacle is between 108 volts and 132 volts during a load on the circuit. Do you have the correct polarity? (See Chapter 6.)

3. **Disconnect the electricity.** Before working on the refrigerator, disconnect the electricity. This can be done by pulling the plug out of the wall receptacle. Or disconnect the electricity at the fuse panel or circuit breaker panel. Turn off the electricity.

WARNING *Some diagnostic tests will require you to test the components with the power turned on. When you disassemble the control panel, you can position it in such a way that the wiring will not make contact with metal. This will allow you to test the components without electrical mishaps.*

4. **Gain access to the electronic control board and touchpad.** You can gain access to the electronic control board and touchpad by removing the screws on the access panel. Depending on the model you are servicing, the electronic control board and touchpad can be in the refrigerator or freezer compartment, or the outside of the freezer door.

5. **Test the electronic control board and touchpad.** If you are able to run the refrigerator's diagnostic test mode, check the different functions of the refrigerator. Use the technical data sheet for the model you are servicing to locate the test points on the wiring schematic. Check all wiring connections and wiring. Using the technical data sheet, test the electronic control and display board, input voltages, and output voltages.

6. **Remove the defective component.** To remove the defective component, remove the screws that secure the PCB to the refrigerator. Disconnect the connectors from the electronic control board or display board.

7. **Install the new component.** To install a new component, just reverse the disassembly procedure, and reassemble. Reinstall all panels or the console panel, and restore the electricity to the refrigerator. Test the refrigerator operation. Make sure to take the refrigerator out of the service test mode when the repair is completed.

Door Gasket

The door gasket consists of a vinyl rubber gasket with a magnet. The magnet helps secure the door closed to keep the cold inside the box and the heat out.

The typical complaints associated with door gasket failure are:

- Sweating inside the cabinet.
- Temperatures inside the cabinet are warmer than normal.
- Ice forming on the freezer walls.
- Door gaskets not maintained properly.

To handle these problems, perform the following steps:

1. **Verify the complaint.** Verify the complaint by checking the door gasket for proper sealing and alignment. Inspect the gaskets for any damage.

2. **Check for external factors.** You must check for external factors not associated with the appliance. Is the appliance installed properly? Were the doors reinstalled correctly?

3. **Disconnect the electricity.** Before working on the refrigerator/freezer, disconnect the electricity. This can be done by pulling the plug from the receptacle. Or disconnect the electricity at the fuse panel or at the circuit breaker panel. Turn off the electricity.

4. **Gain access to the door gaskets.** To access the door gaskets, open the refrigerator/freezer door. The gaskets are located on the door.

5. **Test the door gaskets.** To test the gaskets for proper sealing, take a dollar bill and place it between the gasket and the flange of the outer cabinet (Figure 25-4). Pull on the dollar bill. When pulling on the dollar bill, you should feel some tension as the bill is gripped. This means that the gasket is making good contact with the

FIGURE 25-4
The gasket must
make full contact with
the cabinet flange.

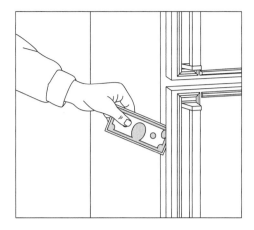

refrigerator/freezer flange. Repeat this test in other areas where you suspect problems with the gasket. If the gasket fails this test, the next step is to replace the gasket. For the doors to close and seal properly, the refrigerator should tilt backwards 1/4 of an inch. This is accomplished by raising the front legs or wheels according to the installation instructions. If you are still unable to get the doors to close properly, check the doors for sagging or warping (Figure 25-5). Also, check the floor to see if it is level under the refrigerator/freezer. Check from front to back and from side to side.

6. **Remove the door gasket.** Before you get started, remove all of the food from the door. To remove the door gasket, pull back on the gasket—this exposes the retaining strip and screws. Loosen the screws about halfway, but do not remove them (Figure 25-6). Gently remove the gasket from around the door (Figure 25-7). On some models the door gasket is held in a retainer track. Just pull on the gasket corners to remove it. This type of gasket installation uses no screws to secure it to the door.

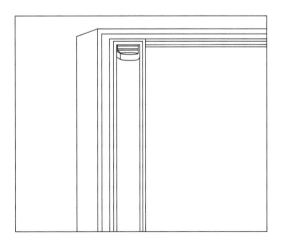
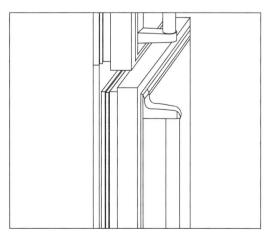

FIGURE 25-5   Warped or sagging doors must be corrected so that the refrigerator/freezer will operate properly.

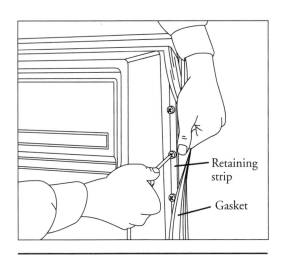

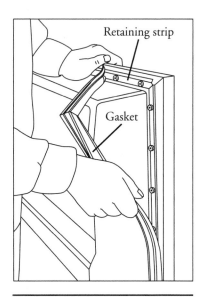

FIGURE 25-6   Peel back the gasket to gain access to the retaining strip and screws.

FIGURE 25-7   Carefully remove the gasket so as not to damage the inner door liner.

7. **Install a new door gasket.** Before installing the new gasket, soak it in warm water for about 15 to 20 minutes. This will make the gasket soft and easier to install.

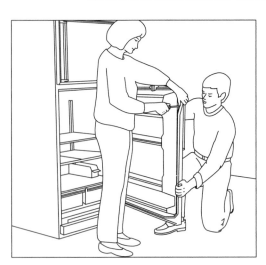

FIGURE 25-8   Have someone help you hold the door straight when tightening the screws.

Starting at either top corner, insert the flange of the gasket behind the retaining strip and/or door liner. Proceed all the way around the door. When the gasket is in place, begin to tighten the screws slightly all around the door. Now close the door; the gasket should make contact with the cabinet flange evenly and all around the door (Figure 25-8). The door gasket might be adjusted by aligning the door panel, as shown in Figure 25-9. To align the door, twist the door in the opposite direction of the warp. Close the door, and check that the gasket is sealed against the cabinet. Now that the door gasket is sealing properly, tighten the screws completely. If the gasket is distorted, or if it has wrinkles in it, use a hair dryer to heat the gasket and mold it to its original form.

Then recheck to be sure the gasket seats against the flange properly. Next, check the gap between the door and the cabinet on the hinge side. Use a penny, which is about 3/4 of an inch in diameter, to check the gap. Slide the penny from the top hinge to the bottom hinge. The door might be adjusted by moving the top hinge and by adding or removing shims to the

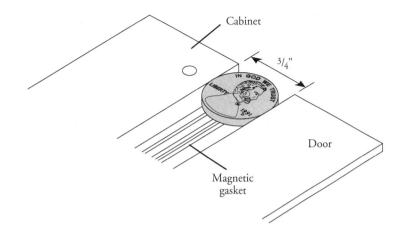

FIGURE 25-9
The correct gap between the door and cabinet will allow the door to close and seal properly. The gap should also be checked as the gasket starts rolling inward when the door is closed.

center and bottom hinges. Be sure that the doors line up evenly with the sides of the cabinet and evenly with each other. Also, check that when both doors are opened simultaneously (top-mount refrigerators only), they do not hit against each other. For other models, align the new door gasket against the door. Starting at the upper corner, push on the gasket until it seats into the retainer and it is flush to the door.

Thermostat (Cold Control)

The thermostat (cold control) is located in the fresh food compartment (Figure 25-10). The cold control maintains the temperature in the refrigerator. It turns the compressor and fans on and off at preset temperature settings.

The typical complaints associated with failure of the thermostat (cold control) are:

- The refrigerator/freezer is not cold enough.
- The refrigerator/freezer is too cold.
- The refrigerator/freezer runs all the time.
- The refrigerator/freezer doesn't run at all.

To handle these problems, perform the following steps:

1. **Verify the complaint.** Verify the complaint by checking the control setting. Turn the control off; then turn it on again and see if the refrigerator/freezer starts.
2. **Check for external factors.** You must check for external factors not associated with the appliance. Is the appliance installed properly? Explain to the user how to set the controls. The voltage at the receptacle is between 108 volts and 132 volts during a load on the circuit. Do you have the correct polarity? (See Chapter 6.)
3. **Disconnect the electricity.** Before working on the refrigerator/freezer, disconnect the electricity to the refrigerator/freezer. This can be done by pulling the plug from the receptacle. Or disconnect the electricity at the fuse panel or at the circuit breaker panel. Turn off the electricity.

WARNING *Some diagnostic tests will require you to test the components with the power turned on. On some models, when you disassemble the control panel, you can position it in such a way that the wiring will not make contact with metal. This will allow you to test the components without electrical mishaps.*

4. **Gain access to the thermostat.** To access the thermostat, open the fresh food door. Look for the control dial that has the word "off" printed on it. This is the control that turns the compressor on and off. Remove the dial (Figure 25-10). Next, remove the two screws that secure the control (Figure 25-11). Remove the wires from the terminals. On some models, the capillary tube is inserted in the air duct; on other models, the capillary tube might be attached to the evaporator coil. At this point, if the capillary tube is attached to the coil, do not remove the capillary tube yet. Test the control first.

5. **Test the thermostat.** If the capillary tube lost its charge, the refrigerator/freezer might not cool or it could freeze the food in the fresh food compartment or keep the temperatures warmer than normal. To test the thermostat, place the ohmmeter probes on the terminals (Figure 25-12). Set the range scale on R × 1, and test for continuity. With the control set in the "off" position, you should not read continuity. When the control is set to the highest position, you should read continuity. If the thermostat is good, the problem must be elsewhere.

6. **Remove the thermostat.** With the thermostat control housing already removed, the capillary tube must now be removed. Because there are many different models on the market today, Figure 25-13 represents only a few types. The capillary tube might be routed through the control housing (Figure 25-13a). The capillary tube might be secured to the evaporator by means of a clamp (Figure 25-13b). The capillary tube might be inserted into a housing that senses how cold the air is (Figure 25-13c). Whichever way the capillary tube is installed, remove it carefully so as not to damage the other components.

7. **Install a new thermostat.** To install the new thermostat, just reverse the order of disassembly, and reassemble. Then test the control. Remember to reinstall the capillary tube in the same location from which it was removed. If you do not, the refrigerator/freezer will not work properly.

Defrost Timer

The purpose of the defrost timer is to regulate the frequency of the defrost cycles and their duration. The defrost timer also limits the maximum amount of time that the defrost heater can be energized. There

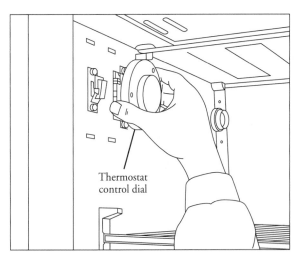

Thermostat control dial

FIGURE 25-10   Pull the dial off gently to gain access to the thermostat.

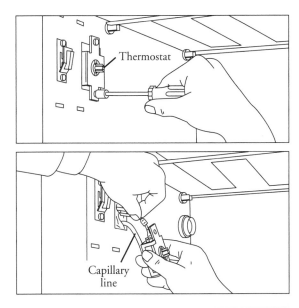

Figure 25-11 Be sure that the electricity is disconnected before attempting to remove the thermostat. Note how the capillary tube is routed and if it is secured to anything. Also note in which direction the terminal end of the thermostat is pointed when removing it. If you install the thermostat upside down, the dial indicators will be incorrect. The thermostat inside the cabinet could be set on the wrong position.

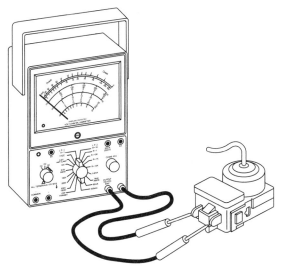

Figure 25-12 Check thermostat switch contacts for continuity and inspect the capillary tube for damage.

are two types of configurations used in a mechanical timer. They are continuous run and cumulative run. The difference between the two is the way that the timer motor is energized. The continuous run timer will be energized anytime when the refrigerator is plugged in, and the cumulative run timer is energized when the cold control is calling for cooling and the compressor is running.

The typical complaints associated with failure of the defrost timer are:

- The refrigerator/freezer does not defrost.
- The storage temperature in the cabinet is too warm.
- The compressor will not run.

To handle these problems, perform the following steps:

1. **Verify the complaint.** Verify the complaint by asking the customer to describe what the refrigerator/freezer is doing or did. On some models you will need the actual service manual for the model you are working on to properly diagnose the refrigerator.

2. **Check for external factors.** You must check for external factors not associated with the appliance. Is the appliance installed properly? Are the doors aligned properly? The voltage at the receptacle is between 108 volts and 132 volts during a load on the circuit. Do you have the correct polarity? (See Chapter 6.)

3. **Disconnect the electricity.** Before working on the refrigerator/freezer, disconnect the electricity to the appliance. This can be done by pulling the plug from the receptacle. Or disconnect the electricity at the fuse panel or at the circuit breaker panel. Turn off the electricity.

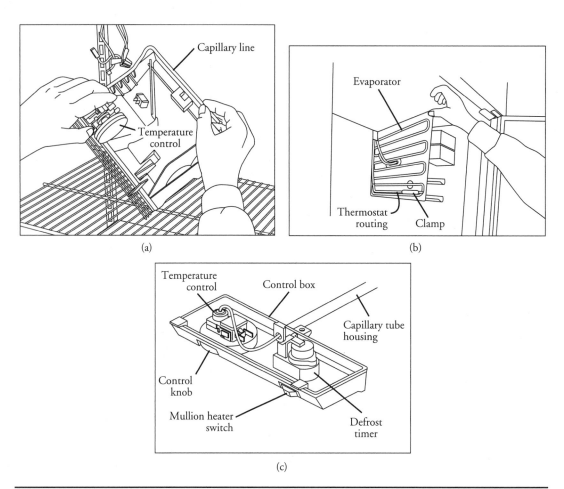

Capillary line

Temperature control

(a)

Evaporator

Thermostat routing   Clamp

(b)

Temperature control   Control box

Capillary tube housing

Control knob

Mullion heater switch   Defrost timer

(c)

FIGURE 25-13   (a) The capillary tube is attached to the control panel. It senses the airflow temperature. (b) The capillary tube is attached to the evaporator plate. Note exactly where the capillary tube is attached. If you remove it for any reason, you must reattach it in the same position. (c) Capillary tube location within the housing.

WARNING *Some diagnostic tests will require you to test the components with the power turned on. When you disassemble the control panel, you can position it in such a way that the wiring will not make contact with metal. This will allow you to test the components without electrical mishaps.*

4. **Gain access to the defrost timer.** To access the defrost timer, you must first locate it. On some models, the defrost timer is located on the bottom, behind the toe plate; or it might be behind the temperature control housing, in the fresh food section (Figure 25-14); or it might be in the back of the refrigerator, behind the rear leg (Figure 25-15).

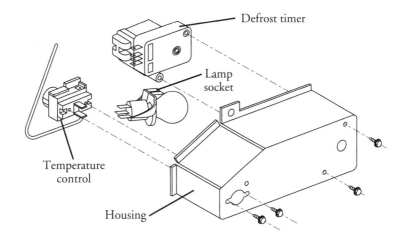

FIGURE **25-14**
Defrost timer and
thermostat located
together in one
housing.

Defrost timer

Lamp
socket

Temperature
control

Housing

5. **Remove the defrost timer.** In order to test the defrost timer, it must be removed
   from its mounting position. Remove the two mounting screws from the defrost
   timer (see Figure 25-15). Next, remove the wire harness plug from the defrost timer
   (Figure 25-16).

6. **Test the defrost timer.** To test the defrost timer, place a screwdriver in the timer cam
   slot (Figure 25-17), and turn it clockwise until you hear the first "snap." The defrost
   timer is now in the defrost cycle. At this point, you must read the wiring diagram to
   determine which numbered terminals are for the defrost circuit. For the purpose of
   demonstrating how to check for continuity of the switch contacts, Figure 25-18a
   illustrates the internal components of this sample timer. Set the ohmmeter scale on
   $R \times 1$, and place the probes on the terminals marked 2 and 3 (Figure 25-19). You
   should measure continuity. Next, rotate the timer cam until you hear the second
   "snap." The meter will show no continuity, indicating that the defrost cycle is over
   and that the refrigeration cycle begins.

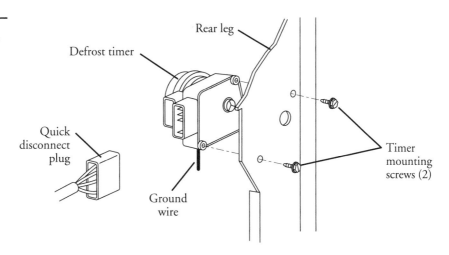

FIGURE **25-15**
Remember to always
reconnect the green
ground wire to the
defrost timer and
ground.

Rear leg

Defrost timer

Quick
disconnect
plug

Ground
wire

Timer
mounting
screws (2)

PART VI

FIGURE 25-16
Not all defrost timers
have harness plug
connectors. Some are
wired with individual
wires. If you forget
how to reconnect the
wires properly, read
the wiring diagram.

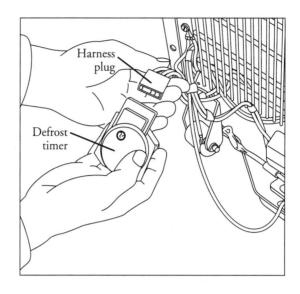

Now place the meter probes on the terminals marked 3 and 4. The ohmmeter will
show continuity, indicating the refrigeration cycle is activated. Turn the timer cam
once again, until you hear the first "snap." The meter will show no continuity.
At no time should there be continuity between terminals 2 and 4. (If so, the switch
contacts are burned and welded together and the defrost timer must be replaced.)
If the defrost timer passes this portion of the test, you must determine if the timer
mechanism is functioning. Place the ohmmeter probes on the timer motor leads and
read the resistance. The resistance can be between 800 and 4000 ohms, depending
on the type of timer used by the manufacturer. If you are unable to read resistance,
the timer motor is defective.

If the defrost timer passes this portion of the test, rotate the timer cam until you hear
the first "snap." Advance the timer cam again, counting the number of clicks until you

FIGURE 25-17
Always rotate the
timer cam clockwise.

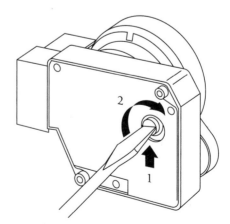

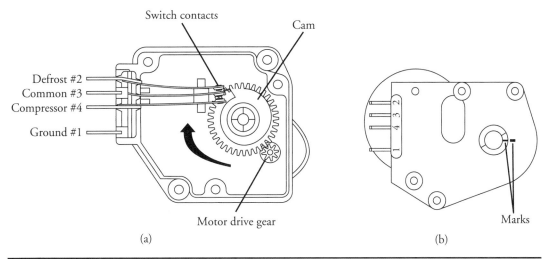

Switch contacts

Cam

Defrost #2
Common #3
Compressor #4

Ground #1

Motor drive gear

Marks

(a)

(b)

FIGURE 25-18 If you have continuity between terminals 2, 3, and 4, this indicates the switch contact points are all welded together. Symptoms in the refrigerator/freezer will be warmer temperatures than normal.

hear the second "snap." Write down the number of clicks on a piece of paper. Now rotate the timer cam again until the marks line up (Figure 25-18b), which indicates the beginning of the defrost cycle, and the "snap" is heard. Advance the timer cam and count the clicks until there is one click left before the end of the defrost cycle. Take the timer and reconnect it to the wiring harness (see Figure 25-16). Place the defrost timer on a nonmetallic surface.

Reconnect the voltage supply to the refrigerator/freezer.

FIGURE 25-19
Check switch contacts
for continuity.

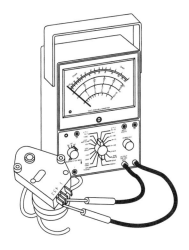

NOTE Be cautious when working with live wires. Avoid getting shocked. Place the ammeter jaws around the wire attached to the number 4 terminal. The meter should indicate no amperage. Next, place the jaws on the number 2 terminal wire. The ammeter should indicate some amperage. Wait for approximately 10 to 15 minutes: you should hear a "snap," indicating that the timer has completed the defrost cycle. At this point, the ammeter will show no amperage on number 2, but will indicate current flow at number 4. If not, replace the timer.

7. **Install a new defrost timer.** To install the new defrost timer, just reverse the order of disassembly, and resemble. Remember to reconnect the ground wire to the defrost timer.

Adaptive Defrost Control

Conventional defrost systems use electromechanical timers with a fixed defrost cycle. Adaptive defrost systems use an electronic control to determine when the defrost cycle is necessary. In order to accomplish the correct time to defrost the evaporator coil, the adaptive defrost control monitors the following refrigerator operations:

- The length of time that the refrigerator doors were open since the last defrost cycle.
- The length of time the compressor has run since the last defrost cycle.
- The amount of time the defrost heaters were on in the last defrost cycle.

The typical complaints associated with failure of the adaptive defrost control are:

- The refrigerator/freezer temperature is warm.
- The refrigerator/freezer does not defrost.
- The compressor will not run.
- Unusual display readouts and/or error codes.

To prevent electrostatic discharge (ESD) from damaging expensive electronic components, follow the steps in Chapter 11.

To handle these problems, perform the following steps:

1. **Verify the complaint.** Verify the complaint by operating the refrigerator controls. Turn off the electricity to the appliance and wait for two minutes before turning it back on. If a fault code appears, look up the code. If the refrigerator will not power up, locate the technical data sheet behind the control panel or cabinet for diagnostics information. On some models you will need the actual service manual for the model you are working on to properly diagnose the refrigerator. The service manual will assist you in properly placing the refrigerator in the service test mode for testing the refrigerator functions.

2. **Check for external factors.** You must check for external factors not associated with the appliance. Is there electricity to the refrigerator? Is the electrical receptacle polarized and properly grounded? The voltage at the receptacle is between 108 volts and 132 volts during a load on the circuit. Do you have the correct polarity? (See Chapter 6.)

3. **Disconnect the electricity.** Before working on the refrigerator, disconnect the electricity. This can be done by pulling the plug out of the wall receptacle. Or disconnect the electricity at the fuse panel or circuit breaker panel. Turn off the electricity.

WARNING *Some diagnostic tests will require you to test the components with the power turned on. When you disassemble the control panel, you can position it in such a way that the wiring will not make contact with metal. This will allow you to test the components without electrical mishaps.*

4. **Gain access to the adaptive defrost control.** You can gain access to the adaptive defrost control by removing the screws on the access panel. Depending on the model you are servicing, the adaptive defrost control can be in the rear of the refrigerator (Figure 25-20).

5. **Test the adaptive defrost control board.** If you are able to run the refrigerator's diagnostic test mode, check the different functions of the refrigerator. Use the technical data sheet for the model you are servicing to locate the test points on the wiring schematic. Check all wiring connections and wiring. Using the technical data sheet, test the adaptive defrost control board, input voltages, and output voltages.

FIGURE 25-20
On this model, the adaptive defrost control is located on the back of the refrigerator. Remove the screws that secure the cover to the back. This will expose the board for testing.

6. **Remove the defective component.** To remove the defective component, remove the screws that secure the PCB to the refrigerator. Disconnect the connectors and wires from the adaptive defrost control board.

7. **Install the new component.** To install a new component, just reverse the disassembly procedure, and reassemble. Reinstall all panels or the console panel, and restore the electricity to the refrigerator. Test the refrigerator operation. Make sure to take the refrigerator out of the service test mode when the repair is completed.

Evaporator Fan Motor

The evaporator fan motor provides air circulation over the evaporator coil located in the freezer compartment of the refrigerator. It also provides air circulation throughout the refrigerator compartments to remove the heat from within the refrigerator. There are two types of evaporator motors used in the modern refrigerator. One type is a shaded pole, single-speed motor that runs on 120 VAC, and the other evaporator motor is a PWM (pulse width modulation), three-speed motor utilizing a permanent magnet, four-pole, DC motor, which operates with the electronic control board.

The typical complaints associated with failure of the evaporator fan motor are:

- The refrigerator/freezer temperature is warm.
- The evaporator fan motor runs slower than normal.
- The evaporator fan motor does not run at all.
- The evaporator fan motor is noisy.
- Unusual display readouts and/or error codes.

To handle these problems, perform the following steps:

WARNING *Some diagnostic tests will require you to test the components with the power turned on. When you disassemble the control panel, you can position it in such a way that the wiring will not make contact with metal. This will allow you to test the components without electrical mishaps.*

FIGURE 25-21
After removing the evaporator cover in this type of refrigerator, remove the heat shield to gain access to the components.

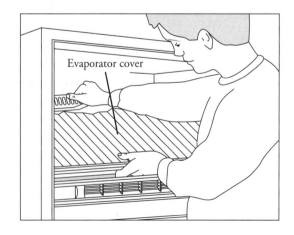

Evaporator cover

1. **Verify the complaint.** Verify the complaint by asking the customer to describe what the refrigerator/freezer is doing or did. Is the evaporator fan motor running? Is it noisy? Turn off the electricity to the appliance and wait for two minutes before turning it back on. If a fault code appears, look up the code. If the refrigerator will not power up, locate the technical data sheet behind the control panel or cabinet for diagnostics information. On some models you will need the actual service manual for the model you are working on to properly diagnose the refrigerator. The service manual will assist you in properly placing the refrigerator in the service test mode for testing the refrigerator functions.

2. **Check for external factors.** You must check for external factors not associated with the appliance. Is the appliance installed properly? Is there something hitting the fan blade? Is there electricity to the refrigerator? Is the electrical receptacle polarized and properly grounded? The voltage at the receptacle is between 108 volts and 132 volts during a load on the circuit. Do you have the correct polarity? (See Chapter 6.)

3. **Disconnect the electricity.** Before working on the refrigerator/freezer, disconnect the electricity. This can be done by pulling the plug from the receptacle. Or disconnect the electricity at the fuse panel or at the circuit breaker panel. Turn off the electricity.

4. **Gain access to the evaporator fan motor.** To access the evaporator fan motor, the evaporator cover must be removed (Figure 25-21). Remove the screws that secure the cover in place. On some models, the evaporator fan assembly is located on the rear wall of the interior freezer compartment.

5. **Test the evaporator fan motor.** The shaded pole, 120 VAC evaporator fan motor should be tested for proper resistance, as indicated on the wiring diagram. To test the evaporator shaded pole, fan motor, remove the wires from the motor terminals. Next, place the probes of the ohmmeter on the motor terminals (Figure 25-22). Set the scale on $R \times 1$. The meter should show resistance. If no reading is indicated, replace the motor. If the fan blade does not spin freely, replace the motor. If the fan motor runs and it is noisy (bad bearings), replace the motor. To test the PWM (pulse width modulation) motor, you do not use an ohmmeter. You must observe circuit polarity; otherwise, the motor or electronic control board will short out. Set your multimeter on DC volts. DC common is not AC common. Using the wiring diagram, you will verify two voltage potentials: (a) Red wire to white wire—power for internal electronic control board. (b) Yellow to white—power for the fan motor. Keep in mind that this type of motor is a DC motor. PWM motors can be run for short periods of time by using a 9-volt battery. Connect the white wire to the negative (–) battery terminal only. Next, connect the red and yellow wires to the positive (+) battery terminal. If the motor runs, disconnect the battery from the motor.

6. **Remove the evaporator fan motor.** To remove the evaporator fan motor, you must first remove the fan blade. On most models, just pull the blade off the motor shaft. Be careful not to break the blade. On other models, the fan blade is held on the motor shaft with screws. Remove the screws. Then remove the screws that secure the fan assembly to the cabinet (Figure 25-23). On some models, you must remove the fan shroud (Figure 25-24) by removing the shroud screws.

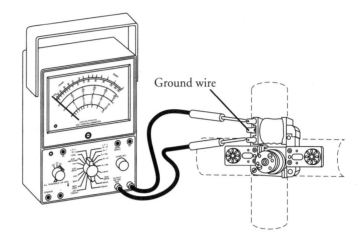

FIGURE 25-22
Check the evaporator
fan motor for
resistance. Also check
the motor for
grounded windings.

Ground wire

7. **Install a new evaporator fan motor.** To install the new evaporator fan motor, just reverse the order of disassembly, and reassemble. When reinstalling the fan blades onto the motor shaft, the fan blades should be positioned on the shaft so that one-third of its depth (approximately 1/4 inch) protrudes through the fan orifice in the direction of airflow. When reinstalling any shrouds, grilles, ducts, or gaskets, always position them correctly to ensure the proper airflow through the evaporator and within both compartments of the refrigerator/freezer. Remember to reconnect the ground wire to the motor. Reconnect the wires to the motor terminals, and test. On electronic models, make sure to take the refrigerator out of the service test mode when the repair is completed.

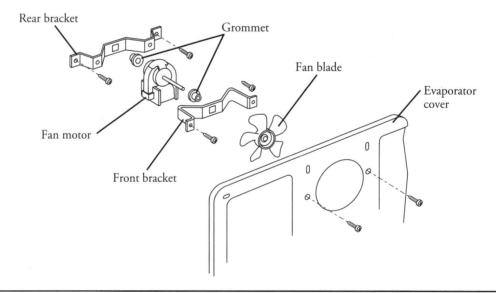

Rear bracket

Grommet

Fan blade

Evaporator cover

Fan motor

Front bracket

FIGURE 25-23   Exploded view of an evaporator fan motor assembly.

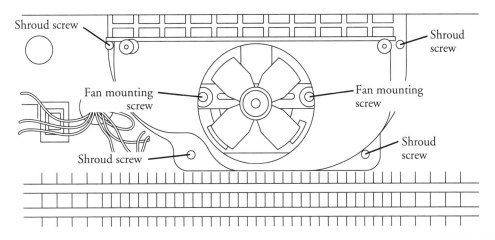

Shroud screw

Shroud screw

Fan mounting screw

Fan mounting screw

Shroud screw

Shroud screw

Shroud screw

FIGURE 25-24   Removing the fan shroud to gain access to the motor terminals.

Condenser Fan Motor

The condenser fan motor can be either a 120 VAC single-speed or a DC single-speed fan motor. You must determine which type you are servicing from the wiring diagram. *Warning: If the DC motor shorts out, you will have to replace the electronic control board also.* The condenser fan motor is located near the compressor in the machine compartment in the rear of the refrigerator on most models. When operating, the condenser fan motor will pull air across the condenser coil and then exhaust it past the compressor and out through the front of the refrigerator. The condenser fan will remove the heat from the condenser coil.

The typical complaints associated with failure of the condenser fan motor are:

- The refrigerator/freezer temperature is warm.
- The condenser fan motor runs slower than normal.
- The condenser fan motor does not run at all.
- The compressor is sometimes noisier than normal.
- Unusual display readouts and/or error codes.

To handle these problems, perform the following steps:

1. **Verify the complaint.** Verify the complaint by asking the customer to describe what the refrigerator/freezer is doing. Is the condenser fan motor running? Turn off the electricity to the appliance and wait for two minutes before turning it back on. If a fault code appears, look up the code. If the refrigerator will not power up, locate the technical data sheet behind the control panel or cabinet for diagnostics information. On some models you will need the actual service manual for the model you are working on to properly diagnose the refrigerator. The service manual will assist you in properly placing the refrigerator in the service test mode for testing the refrigerator functions.

2. **Check for external factors.** You must check for external factors not associated with the appliance. Is the appliance installed properly? Is there any foreign object blocking the condenser fan blade? Is there electricity to the refrigerator? Is the electrical receptacle polarized and properly grounded? The voltage at the receptacle is between 108 volts and 132 volts during a load on the circuit. Do you have the correct polarity? (See Chapter 6.)

3. **Disconnect the electricity.** Before working on the refrigerator/freezer, disconnect the electricity. This can be done by pulling the plug from the receptacle. Or disconnect the electricity at the fuse panel or at the circuit breaker panel. Turn off the electricity.

WARNING *Some diagnostic tests will require you to test the components with the power turned on. When you disassemble the control panel, you can position it in such a way that the wiring will not make contact with metal. This will allow you to test the components without electrical mishaps.*

4. **Gain access to the condenser fan motor.** Pull the refrigerator/freezer out and away from the wall. Remove the back panel, which is located at the bottom of the refrigerator/freezer. This will expose the compressor, the condenser fan assembly, and the condenser coil (Figure 25-25).

5. **Test the condenser fan motor.** A 120 VAC condenser fan motor should be tested for proper resistance, as indicated on the wiring diagram. Check the fan blade for obstructions. The blade should turn freely. Next, rotate the fan blade and check for bad bearings. If you hear any unusual noises coming from the motor, or if the fan blade is sluggish when spinning, replace the motor.

To test the condenser fan motor, remove the wires from the motor terminals. Next, place the probes of the ohmmeter on the motor terminals (Figure 25-26). Set the scale on R × 1. The meter should show resistance. If no reading is indicated, replace

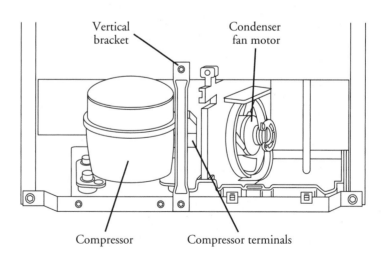

FIGURE 25-25
Removing the rear panel will expose the components that need to be serviced.

Vertical bracket

Condenser fan motor

Compressor

Compressor terminals

the motor. When testing a DC condenser fan motor, unplug the refrigerator to reset the electronic control board. At the condenser fan connector, check for 12 volts DC from the red to white wire and from the pink and white wire. If there is 12 volts DC, replace the motor. You may also have to test the electronic control board for a short. Use the technical data sheet to assist you in your final diagnosis. To test for a shorted DC condenser fan motor, place your ohmmeter leads between white and red or white and yellow wires. If you read less than 1k ohm, replace the motor.

6. **Remove the condenser fan motor.** To remove the condenser fan motor, you must first remove the fan blades. Unscrew the nut that secures the blades to the motor. Remove the blades from the motor. Then remove the motor assembly by removing the mounting bracket screws (Figure 25-27).

7. **Install a new condenser fan motor.** To install the new condenser fan motor, just reverse the order of disassembly, and reassemble. Remember to reconnect the ground wire to the motor. Reconnect the wires to the motor terminals, and test. On electronic models, make sure to take the refrigerator out of the service test mode when the repair is completed.

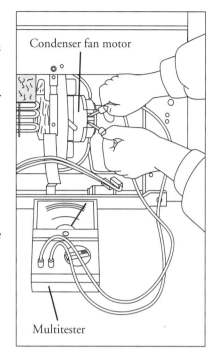

FIGURE 25-26 Remember to set the meter on the ohm scale when testing for resistance in the condenser fan motor.

Defrost Heater

Most manufacturers also use a single-calrod type, radiant heater, mounted under the evaporator coil for maximum defrosting of the evaporator coil. Other manufacturers use defrost heaters that are made with nickel-chromium wire, encased in a glass tube, having both tensile strength and high resistance to current flow, and are mounted to the evaporator coil.

The typical complaints associated with failure of the defrost heater are:

- The refrigerator temperature is warm.
- The freezer temperature is warm.
- The refrigerator/freezer does not defrost.
- Food is spoiling.
- No ice cubes.
- Unusual display readouts and/or error codes.

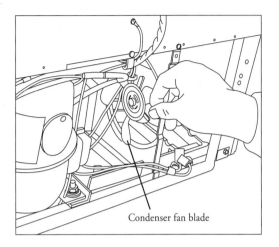

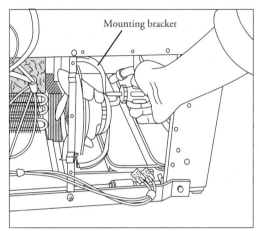

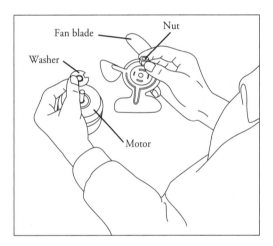

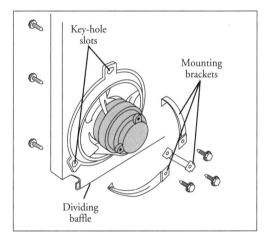

Figure 25-27 Remove the fan blade first. This will prevent the blades from bending out of shape and becoming off balance. On some models, the compressor is located within inches of the motor. Be careful! The compressor might be hot, and you could burn yourself.

To handle these problems, perform the following steps:

1. **Verify the complaint.** Verify the complaint by asking the customer to describe what the refrigerator/freezer is doing. Is food spoiling? Check the temperature in the compartments. Check for ice buildup on the evaporator cover. Turn off the electricity to the appliance and wait for two minutes before turning it back on. If a fault code appears, look up the code. If the refrigerator will not power up, locate the technical data sheet behind the control panel or cabinet for diagnostics information. On some models you will need the actual service manual for the model you are working on to properly diagnose the refrigerator. The service manual will assist you in properly placing the refrigerator in the service test mode for testing the refrigerator functions.

2. **Check for external factors.** You must check for external factors not associated with the appliance. Is the appliance installed properly? Is there electricity to the refrigerator? Is the electrical receptacle polarized and properly grounded? The voltage at the receptacle is between 108 volts and 132 volts during a load on the circuit. Do you have the correct polarity? (See Chapter 6.)

3. **Disconnect the electricity.** Before working on the refrigerator/freezer, disconnect the electricity. This can be done by pulling the plug from the receptacle. Or disconnect the electricity at the fuse panel or at the circuit breaker panel. Turn off the electricity.

WARNING *Some diagnostic tests will require you to test the components with the power turned on. When you disassemble the control panel, you can position it in such a way that the wiring will not make contact with metal. This will allow you to test the components without electrical mishaps.*

4. **Gain access to the defrost heater.** To access the defrost heater, the evaporator cover must be removed (see Figure 25-21). Remove the screws that secure the cover in place.

5. **Test the defrost heater.** A defrost heater should be tested for proper resistance, as indicated on the wiring diagram. To test the defrost heater, remove the wires from the heater terminals. Next, place the probes of the ohmmeter on the heater terminals (Figure 25-28). Set the scale on R × 1. The meter should show resistance. If no reading is indicated, replace the defrost heater.

6. **Remove the defrost heater.** To remove the defrost heater in this type of model, you must first remove the reflector shield (Figure 25-29). Bend the clip up and lift the shield. Do the same for the other end of the heater. Once the shield is removed, you can lift the defrost heater from its brackets.

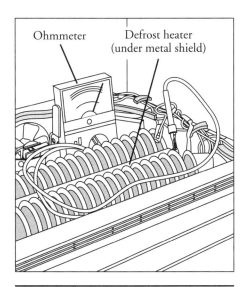

FIGURE 25-28 If the model you are repairing has glass defrost heaters in it that look black, dark smoky-gray, or burned, the heater is either defective or soon will be.

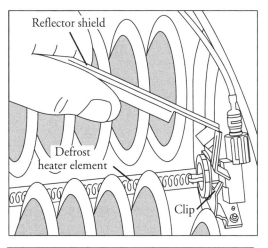

FIGURE 25-29 When removing this type of heater, be careful not to break the glass. The clip must be straight so the heater can slide out of the brackets.

7. **Install a new defrost heater.** To install the new defrost heater (Figure 25-30), just reverse the order of disassembly, and reassemble.

NOTE *On some models, do not touch the glass because it will shorten the life of the heater. Remember to reconnect the wires to the heater. When reinstalling any shrouds, grilles, ducts, or gaskets, always position them correctly to ensure the proper airflow through the evaporator and within both compartments of the refrigerator/freezer. On electronic models, make sure to take the refrigerator out of the service test mode when the repair is completed.*

Defrost Thermostat

The defrost thermostat is a bimetal switch installed on the evaporator coil, which provides over-temperature protection during defrost. The defrost heaters will defrost the evaporator coil within a given time. But, if the evaporator coil is totally defrosted before the time has expired, the defrost thermostat will open up, shutting off the defrost heater to prevent the evaporator coil area from overheating.

The typical complaints associated with failure of the defrost thermostat are:

- The refrigerator temperature is warm.
- The freezer temperature is warm.
- The refrigerator/freezer does not defrost.
- Food is spoiling.
- No ice cubes.
- Unusual display readouts and/or error codes.

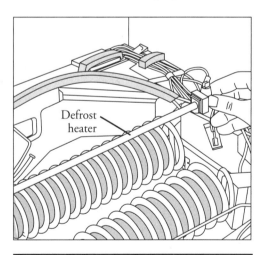

Defrost heater

FIGURE 25-30 After the heater is installed and the shield is back in place, bend back the clip on the bracket. When you reinstall the wires on the terminals, use one hand to hold the end of the heater. This will prevent the glass from cracking if you push down too hard.

To handle these problems, perform the following steps:

1. **Verify the complaint.** Verify the complaint by asking the customer to describe what the refrigerator is doing. Is food spoiling? Check the temperature in all compartments. Check for ice buildup on the evaporator cover. Turn off the electricity to the appliance and wait for two minutes before turning it back on. If a fault code appears, look up the code. If the refrigerator will not power up, locate the technical data sheet behind the control panel or cabinet for diagnostics information. On some models you will need the actual service manual for the model you are working on to properly diagnose the refrigerator. The service manual will assist you in properly placing the refrigerator in the service test mode for testing the refrigerator functions.

2. **Check for external factors.** You must check for external factors not associated with the appliance. Is the appliance installed properly? Is there electricity to the refrigerator? Is the electrical receptacle polarized and properly grounded? The voltage at the receptacle is between 108 volts and 132 volts during a load on the circuit. Do you have the correct polarity? (See Chapter 6.)

3. **Disconnect the electricity.** Before working on the refrigerator/freezer, disconnect the electricity. This can be done by pulling the plug from the receptacle. Or disconnect the electricity at the fuse panel or at the circuit breaker panel. Turn off the electricity.

WARNING *Some diagnostic tests will require you to test the components with the power turned on. When you disassemble the control panel, you can position it in such a way that the wiring will not make contact with metal. This will allow you to test the components without electrical mishaps.*

4. **Gain access to the defrost thermostat.** To access the defrost thermostat, the evaporator cover must be removed (see Figure 25-21). Remove the screws that secure the cover in place.

5. **Test the defrost thermostat.** Failure of a defrost thermostat usually results in a frost-blocked evaporator. To test the defrost thermostat, disconnect the wires to isolate the thermostat from the rest of the defrost circuit. Next, place the probes of the ohmmeter on the defrost thermostat wire leads (Figure 25-31). Set the meter scale on R × 1. The meter will show continuity when the thermostat is either frozen or very cold, indicating the defrost thermostat is good. The defrost thermostat switch contacts close when the temperature is colder than its temperature rating (Figure 25-32). If no reading is indicated, replace the defrost thermostat. At ambient temperature, you will read no continuity, which will indicate the thermostat might be good.

FIGURE 25-31
Test the defrost thermostat when it is connected to the evaporator coil.

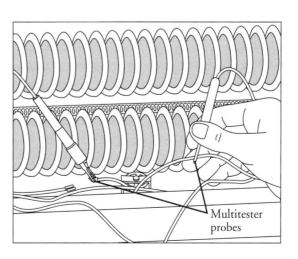

Multitester probes

Part number	Temperature setting °F	
	Open	Close
L - 45	45	25
L - 50	50	30
L - 55	55	35
L - 60	60	40
L - 70	70	50
L - 80	80	30
L - 90	90	35

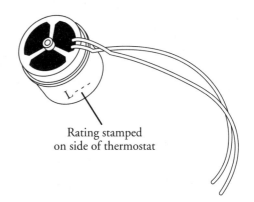

Rating stamped
on side of thermostat

FIGURE 25-32   A temperature rating chart for common defrost thermostats.

6. **Remove the defrost thermostat.** To remove the defrost thermostat, you must remove the hold-down clamp. On some models, the defrost thermostat and clamp are one assembly. On other models, the defrost thermostat clamps around the evaporator tubing. As shown in Figure 25-33, remove this type of defrost thermostat by squeezing in on the clip and lifting the thermostat up.

7. **Install a new defrost thermostat.** To install the new defrost thermostat, just reverse the order of disassembly, and reassemble. Remember to reconnect the wires to the thermostat. When reinstalling any shrouds, grilles, ducts, or gaskets, always position them correctly to ensure the proper airflow through the evaporator and within both compartments of the refrigerator/ freezer. On models that have the defrost thermostat attached to the evaporator coil, you must reinstall the defrost thermostat in the same location from which it was removed. On electronic models make sure to take the refrigerator out of the service test mode when the repair is completed.

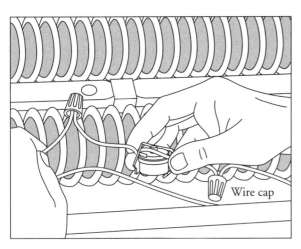

Wire cap

FIGURE 25-33   When replacing the defrost thermostat, be sure you reinstall it in the same position from which you removed it. Otherwise, the defrost cycle will not function properly.

Compressor, Relay, and Overload Protector

The compressor (reciprocating or rotary type) is the heart of the vapor compression system. It is used to circulate the refrigerant throughout the sealed system. The relay and overload are attached to the compressor. The relay starts the compressor, and the overload protects the compressor. All three components are located in the machine compartment in the rear of the refrigerator.

The relay can be either a current or a PTC (positive temperature coefficient) type device. The overload is a bimetal switch that is secured to the outer shell of the compressor.

The typical complaints associated with failure of the compressor are:

- The refrigerator temperature is warm.
- The freezer temperature is warm.
- The refrigerator does not run at all.
- Food is spoiling.
- Unusual display readouts and/or error codes.

To handle these problems, perform the following steps:

1. **Verify the complaint.** Verify the complaint by asking the customer to describe what the refrigerator is doing. Turn off the electricity to the appliance and wait for two minutes before turning it back on. If a fault code appears, look up the code. If the refrigerator will not power up, locate the technical data sheet behind the control panel or cabinet for diagnostics information. On some models you will need the actual service manual for the model you are working on to properly diagnose the refrigerator. The service manual will assist you in properly placing the refrigerator in the service test mode for testing the refrigerator functions.

2. **Check for external factors.** You must check for external factors not associated with the appliance. Is the appliance installed properly? Check for a voltage drop during refrigerator startup. Is there electricity to the refrigerator? Is the electrical receptacle polarized and properly grounded? The voltage at the receptacle is between 108 volts and 132 volts during a load on the circuit. Do you have the correct polarity? (See Chapter 6.)

3. **Disconnect the electricity.** Before working on the refrigerator, disconnect the electricity. This can be done by pulling the plug from the receptacle. Or disconnect the electricity at the fuse panel or at the circuit breaker panel. Turn off the electricity.

WARNING *Some diagnostic tests will require you to test the components with the power turned on. When you disassemble the control panel, you can position it in such a way that the wiring will not make contact with metal. This will allow you to test the components without electrical mishaps.*

4. **Gain access to the compressor.** To access the compressor, pull the refrigerator out and away from the wall. Remove the back panel, which is located at the bottom of the refrigerator. This will expose the compressor, the condenser fan assembly, and the condenser coil (see Figure 25-25). Next, remove the compressor terminal cover (Figure 25-34) by removing the retaining clip that secures the cover. Remove the terminal cover.

5. **Test the compressor relay.** To test the compressor current relay, remove the relay by pulling it off the compressor terminals without twisting it (Figure 25-35a). Remove the wires from the relay and label them. On the relay body is stamped the word TOP. Hold the relay so that TOP is in the up position.

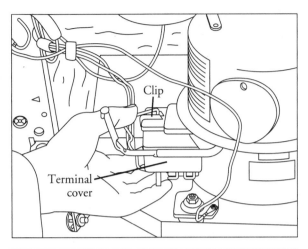

FIGURE 25-34   Removing the terminal cover to gain access to the relay and overload protector.

Next, place the probes of the ohmmeter on the relay terminals marked S and M. Set the meter scale on R × 1. The reading will show no continuity. Then remove the probe from the terminal marked M, and place it on the side terminal marked L. The reading will show no continuity. Now, move the probe from terminal S, and place it on the terminal marked M. The reading will show continuity. With the probes still attached, turn the relay upside down (Figure 25-35b), and perform the same tests. By turning the relay over, the switch contacts in the relay will close. When you retest the relay, you should get the opposite results: You should have continuity between terminals S and M and between S and L; however, the meter will not read continuity between M and L. If the relay fails this test, replace it. The elimination method is the best way to test a PTC (positive temperature coefficient) relay. You would first test the run capacitor, overload, and then run the compressor with a fused test cord. If all of these components check out okay, replace the relay.

6. **Test the overload protector.** To test the overload protector, remove the wires from the overload and compressor terminals. Then remove the overload protector from the compressor by removing the retaining clip that secures the overload protector to

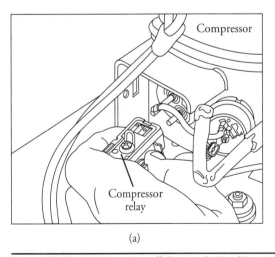

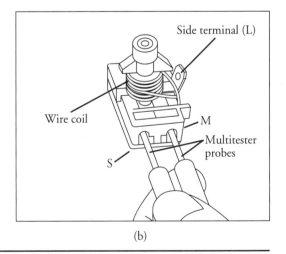

(a)   (b)

FIGURE 25-35   Pull the relay off the terminals without twisting it. This will prevent you from breaking the compressor terminals.

the compressor (Figure 25-36a). Next, place the probes of the ohmmeter on the overload terminals (Figure 25-36b). Set the meter scale on R × 1. The reading will show continuity. If not, replace the overload protector.

7. **Test the compressor.** To test the compressor, remove the relay and the overload protector. This will expose the compressor terminals. The compressor terminals are marked C, S, and R. C indicates the common winding terminal; S indicates the start winding terminal; and R indicates the run winding terminal. (Refer to the actual wiring diagram for the model that you are servicing.) Set the meter scale on R × 1, touch the probes together, and adjust the needle setting to indicate a zero reading. Next, place the probes of the ohmmeter on the terminals marked S and R (Figure 25-37a). The meter reading will show continuity. Now place the meter probes on the terminals marked C and S. The meter reading will show continuity. Finally, place the meter probes on the terminals marked C and R. The meter reading will show continuity. The total number of ohms measured between S and R is equal to the sum of C to S plus C to R. The compressor should be tested for proper resistance, as indicated on the wiring diagram.

To test the compressor for ground, place one probe on a compressor terminal, and attach the other probe to the compressor housing or any good ground (Figure 25-37b). Set the meter scale to R × 1000. The meter reading will show no continuity. Repeat this for the remaining two terminals. The meter reading will show no continuity. If you get a continuity reading from any of these terminals to ground, the compressor is grounded. Replace it.

To test compressors with inverter boards, three-phase AC compressors, and DC compressors, you must consult the service manual or technical data sheet for the model you are servicing for the correct testing procedures.

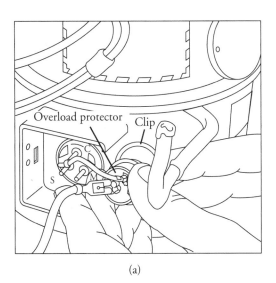

(a)

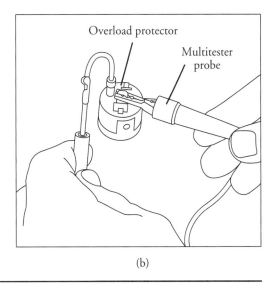

(b)

Figure 25-36   Testing the overload protector for continuity between the terminals.

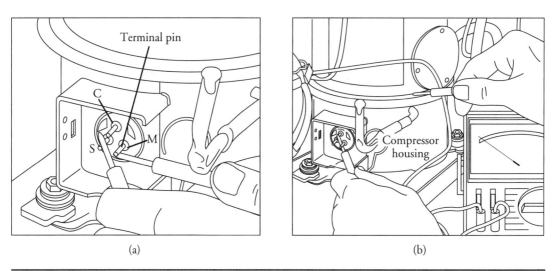

FIGURE 25-37 Testing the compressor motor windings for resistance.

8. **Replace the overload and relay.** To install the new overload or relay, just reverse the order of disassembly and reassemble. Remember to reconnect the wires to the overload or relay. On electronic models, make sure to take the refrigerator out of the service test mode when the repair is completed.

Diagnostic Charts

The following refrigerator diagnostic charts will help you to pinpoint the likely causes of the particular problem associated with these appliances (Figures 25-38 and 25-39).

Refrigerator Electrical Schematics

The wiring diagrams in this chapter are examples only. You must refer to the actual wiring diagram for the refrigerator you are servicing. Figure 25-40 depicts an actual wiring schematic for a side-by-side refrigerator, which includes an automatic ice maker and ice and water dispensers. On models with the rapid electrical diagnosis (R-E-D) feature, a technician can make a quick and accurate diagnosis of electrical faults without disassembling the refrigerator.[6] To perform this test, a special adapter is connected to the wiring harness through the multicircuit connector, which is located behind the front grill. Upon separating the multicircuit connector, the parallel circuits in the wiring harness will be isolated. This process will permit you to test all of the electrical components and the related wiring within the main wiring harness. If an R-E-D test adapter is not available, you can still check the circuits with an ohmmeter.

CAUTION *Disconnect the electricity from the refrigerator/freezer before measuring resistances.*

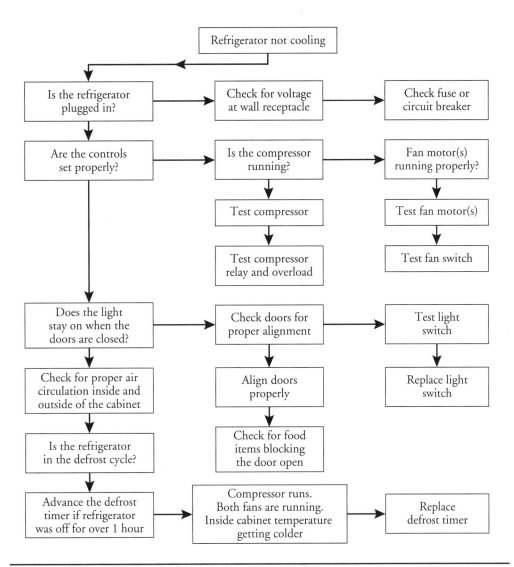

FIGURE 25-38 The diagnostic flowchart: Refrigerator not cooling.

A basic understanding of the symbols used in the schematic diagram is essential (see Figure 25-40). The numbered terminals, located in the multicircuit connector, are shown on the schematic diagram. The component circuits on the schematic diagram are indicated by an arrow and a number. The point of the arrow indicates a male terminal, and the tail of the arrow indicates a female terminal. The number identifies the terminal location in the connector (Figure 25-41).

Remember how to read a wiring schematic. Give the following examples a try.

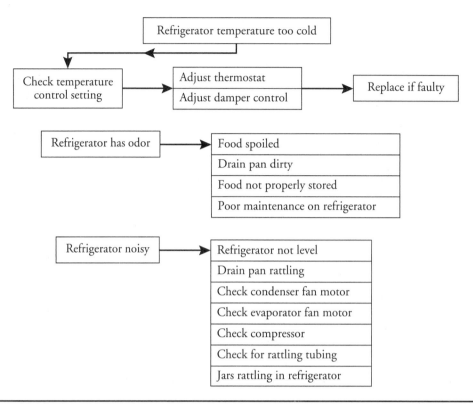

Figure 25-39   The diagnostic flowchart: Refrigerator temperature too cold.

Example #1

The customer explains that the food in the freezer is thawing and the food in the refrigerator is warm. The customer states the compressor runs constantly. You made your observation and confirmed that the food is thawing and the compressor is operating. You also noticed frost on the evaporator cover, which indicates a defrost problem. This indicates to you the possibility of three components malfunctioning: the defrost heaters, the defrost thermostat, or the defrost timer. Also, there is a possibility of a broken or loose wiring connection. Set the refrigerator in the defrost mode first. Advance the defrost timer until you hear the first "snap" sound coming from the timer.

Using the ohmmeter, set the range on R × 1, and adjust the meter scale to read zero. Separate the multicircuit connector. *Did you turn off the electricity?* To test the defrost heaters, insert one meter probe into the number 5 male connector pin and the other meter probe into the number 1 male connector pin (see Figure 25-41a). The ohmmeter should show the combined resistance of the heaters. This resistance is 16 ohms. If the test is okay, continue on to the defrost thermostat. If not, replace the heaters.

When testing the defrost thermostat, insert one meter probe into the number 1 male connector pin and the other meter probe into the number 4 male connector pin (see Figure 25-41b). The meter reading should indicate zero ohms. If not, replace the defrost thermostat.

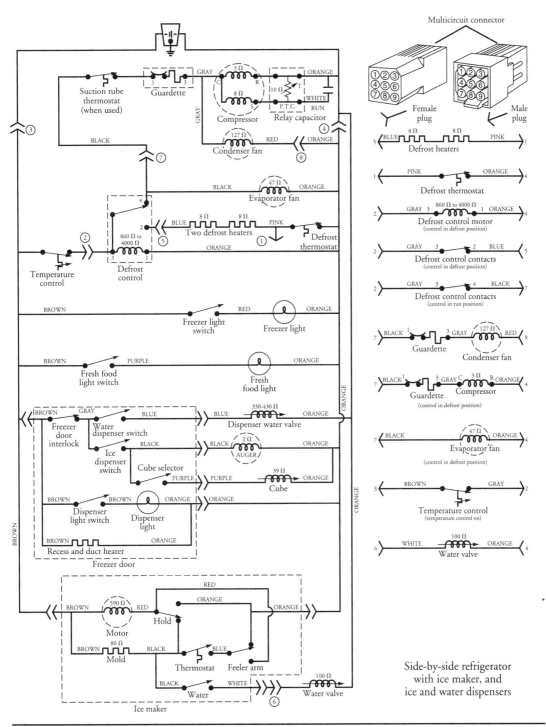

FIGURE 25-40 Side-by-side refrigerator wiring schematic.

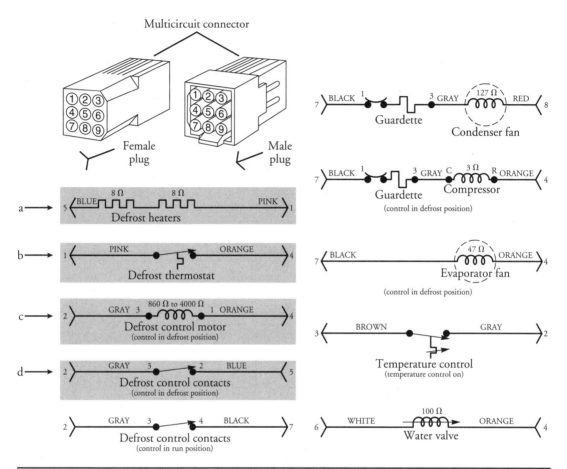

Figure 25-41 Side-by-side refrigerator wiring schematic of multicircuit connector circuitry. (a) Defrost heater circuit. (b) Defrost thermostat circuit. (c) Defrost control motor circuit. (d) Defrost control switch contacts circuit.

Now test the defrost timer motor. Insert one meter probe into the number 4 male connector pin and the other meter probe into the number 4 female connector pin (see Figure 25-41c). The meter should show between 800 and 4000 ohms. Adjust the ohmmeter range accordingly. To test the defrost timer switch contacts, insert one meter probe into the number 2 female connector pin and the other meter probe into the number 5 female connector pin (see Figure 25-41d). The meter should show zero ohms. If not, replace the defrost timer control.

Figure 25-42 depicts the active defrost circuits. If you want to check the entire defrost circuit, insert one meter probe into the number 3 male connector pin and the other meter probe into the number 4 male connector pin. The meter should read continuity. Be sure that the cold control (thermostat) is set for maximum cooling and that the defrost timer is set on defrost.

FIGURE 25-42
Illustration of the active circuits when the refrigerator is operating in the defrost mode.

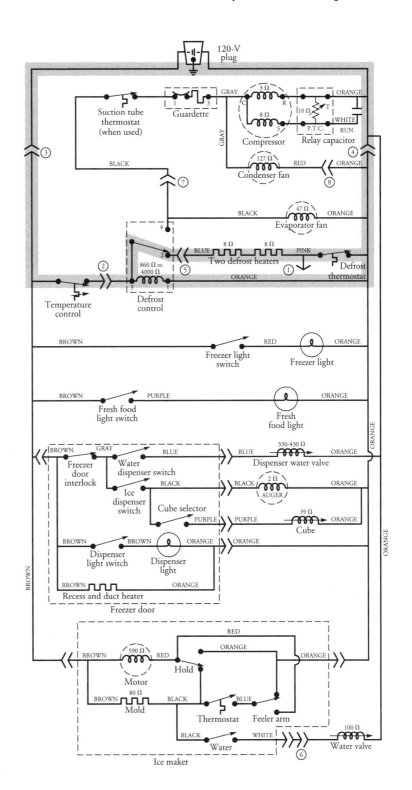

Example #2

The wiring diagrams in Figures 25-43 and 25-45 are examples only. You must refer to the actual wiring diagram for the refrigerator that you are servicing. Figure 25-43 depicts an actual wiring schematic for a cycle-defrost refrigerator.

The refrigerator door is closed, the thermostat is calling for cooling, and the compressor is running. What circuits are active? Trace the active circuits in Figure 25-43. The active components are the thermostat, overload, relay, compressor, stile heater, and the mullion heater. Check the end result with Figure 25-44.

The thermostat in Figure 25-45 has cycled off, and the switch contacts are open. Also, the compressor stopped running. What circuits are active? Trace the active circuits in Figure 25-45. The active components are the evaporator heater, drain heater, stile heater, and the mullion heater. Check the end result with Figure 25-46. Current will flow through the overload, the relay coil, and the compressor running winding, but there is not enough current to energize the start relay and to run the compressor.

FIGURE 25-43
A typical cycle-defrost refrigerator electrical schematic.

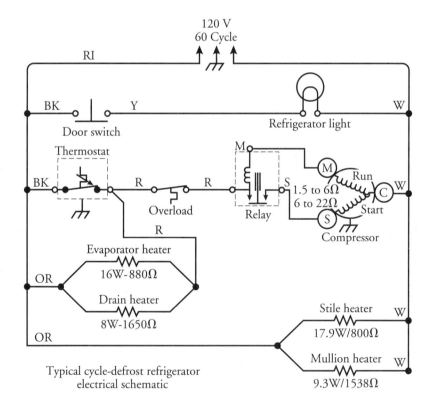

Typical cycle-defrost refrigerator
electrical schematic

FIGURE 25-44
The active circuits when the refrigerator is operating in the cooling mode.

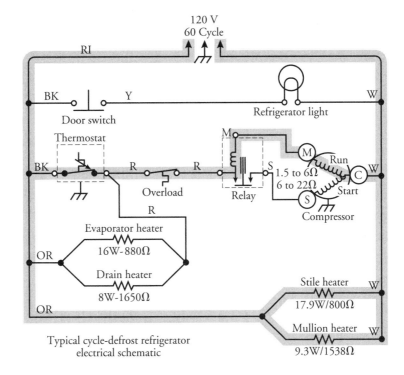

Typical cycle-defrost refrigerator electrical schematic

FIGURE 25-45
The cycle-defrost refrigerator wiring schematic, showing that the refrigerator has cycled off.

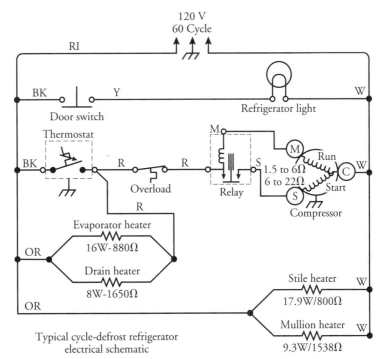

Typical cycle-defrost refrigerator electrical schematic

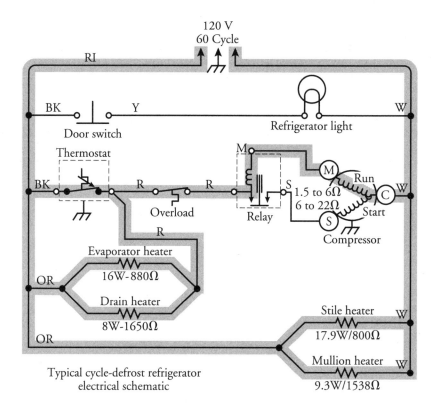

FIGURE 25-46
The active circuits when the refrigerator is operating in the defrost mode.

Typical cycle-defrost refrigerator electrical schematic

Example #3

The wiring diagram in Figure 25-47 is an example only. It depicts a no-frost refrigerator. The refrigerator is in the cooling mode, and the compressor is running. What circuits are active? Trace the active circuits. Check the end result with Figure 25-48.

Example #4

The wiring diagram in Figure 25-49 is an example only. It depicts a no-frost refrigerator with an adaptive defrost control. What is the purpose of the ADC board? How does the PTC relay work?

Example #5

The wiring diagram in Figure 25-50 is an example only. It depicts a no-frost refrigerator with an adaptive defrost control, inverter board technology, and a 230 volt, three-phase compressor. The refrigerator's main power source is 120 volt AC. The strip circuits will help the technician to diagnose the refrigerator circuit. To better understand the wiring schematic, use the strip circuits to trace out the active circuits on the wiring schematic.

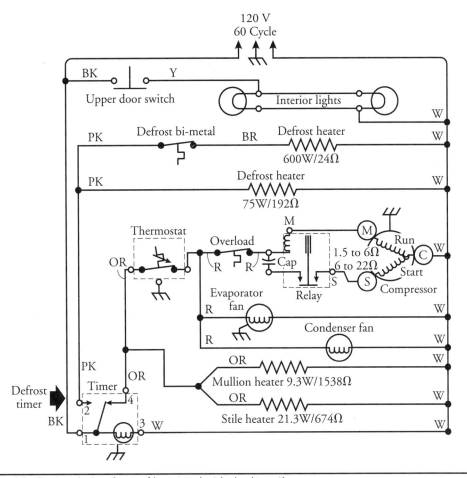

FIGURE 25-47   A typical no-frost refrigerator electrical schematic.

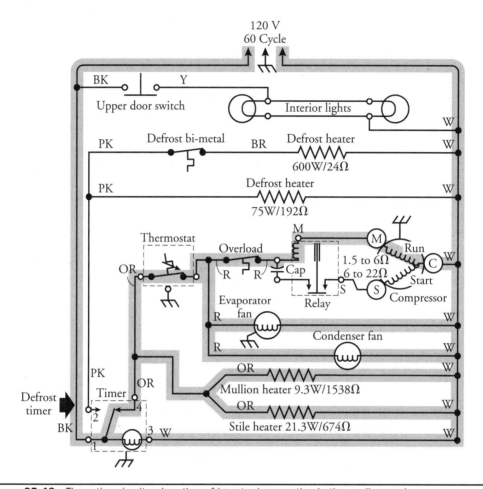

FIGURE 25-48   The active circuits when the refrigerator is operating in the cooling mode.

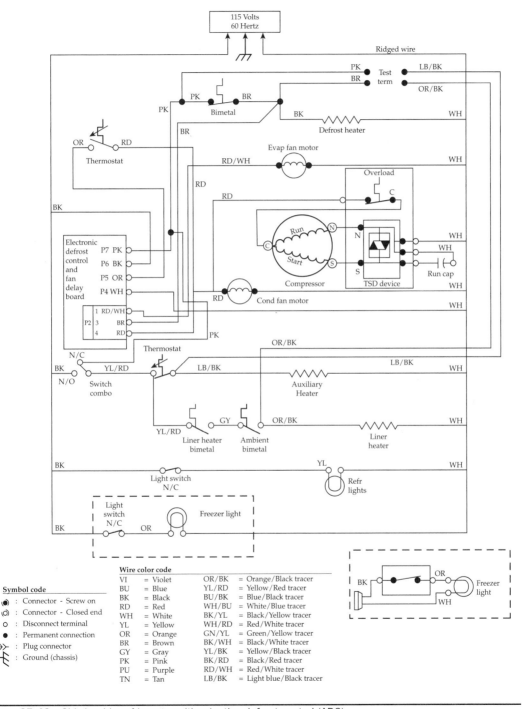

FIGURE 25-49 Side-by-side refrigerator with adaptive defrost control (ADC).

FIGURE 25-50
Side-by-side
refrigerator with
inverter board
technology. The
refrigerator has a
230-volt, three-phase
compressor and
adaptive defrost
control. (*Continued*)

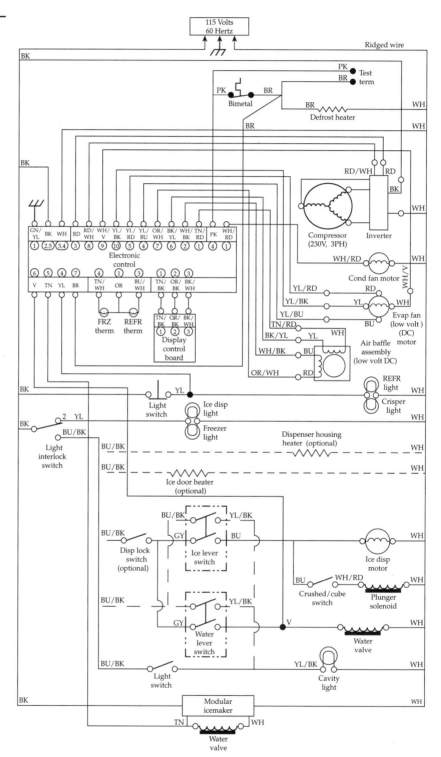

FIGURE **25-50**
Side-by-side
refrigerator with
inverter board
technology. The
refrigerator has a
230-volt, three-phase
compressor and
adaptive defrost
control.

The Cooling Circuits

The compressor

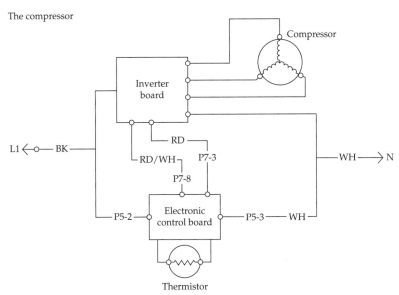

The Condenser Fan Motor

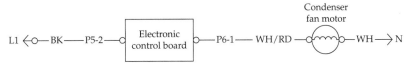

The Evaporator Fan Motor

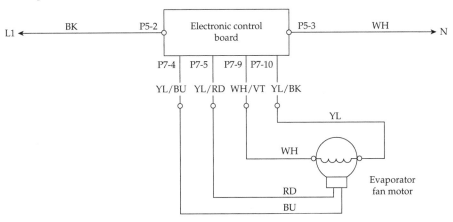

The Defrost Cycle

The defrost heater

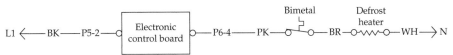

Endnotes

1. Dairy products, meats, seafood, fruits, and vegetables will all spoil rapidly if not kept cold or frozen. The colder temperatures of the refrigeration compartment should be between 35 and 40 degrees Fahrenheit to prevent the spoiling of foods. Most foods will last from three to seven days at that temperature. If the foods were frozen and packaged properly, they could last for several weeks in a domestic refrigerator with a temperature of zero degrees Fahrenheit to –10 degrees Fahrenheit.

2. Reprinted from MACAP Consumer Bulletin, Issue no. 15, December 1992.

3. Reprinted from MACAP Consumer Bulletin, Issue no. 12, November 1989.

4. Reprinted from MACAP Consumer Bulletin, Issue no. 6, February 1984.

5. If you open up the sealed system, you will void your warranty. The sealed system must be repaired by an authorized service company.

6. Robinair part # 14442, purchased at your local parts dealer.

Freezers

Freezers are conveniences for people who have very large families or for people who do not frequent the supermarket. They are especially useful in homes with smaller refrigerators or refrigerators having only an ice cube tray compartment. Home freezers come in chest and upright models. Two designs of upright models are available on the market today: manual defrost and automatic defrost. Home freezers are available with wire shelves and baskets, and with storage shelves on the doors in upright models.

The sealed system in the freezer operates the same as the refrigerator/freezer models (see Chapter 25). The only difference is temperature. The domestic freezer operates at a colder temperature. The reason for colder temperatures is to maintain food preservation for a longer period of time.

This chapter covers the electrical components only and how to diagnose the sealed system. The actual repair or replacement of any sealed-system component is not included in this chapter. It is recommended that you acquire refrigerant certification (or call an authorized service company) to repair or replace any sealed-system component. The refrigerant in the sealed system must be recovered properly according to Environmental Protection Agency (EPA) guidelines.

Upright Freezers

Upright freezers are similar to refrigerator/freezers in design and operation. They share some of the same features:

- Fan motors
- Compressor
- Automatic defrost system
- Door gasket
- Thermostat
- Interior lighting

On manual-defrost models,[1] the evaporator coils are the shelves inside the cabinet. Figure 26-1 illustrates the refrigerant flow on this type of manually defrosted upright freezer. The condenser coils are embedded between the cabinet liners and are secured to the

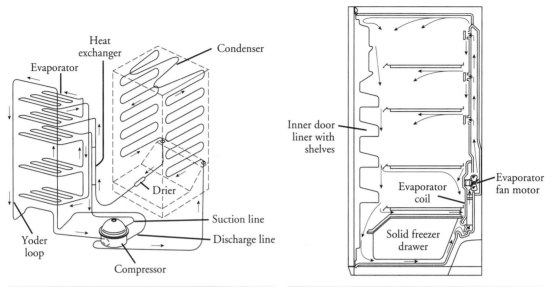

FIGURE 26-1 Refrigerant flow of an upright freezer with manual defrost.

FIGURE 26-2 The airflow pattern of an upright automatic-defrost freezer.

inside wall of the outer cabinet. This provides for even heat removal and it eliminates the need for a condenser fan motor.

Automatic-defrost models[2] use a fan motor to circulate the air inside the cabinet through air ducts. The evaporator coil is mounted on the inside back wall of the inner liner. Figure 26-2 illustrates the airflow pattern in an upright automatic-defrost freezer. Figure 26-3 illustrates the refrigerant flow in this type of upright freezer.

FIGURE 26-3
The refrigerant flow of an upright automatic-defrost freezer.

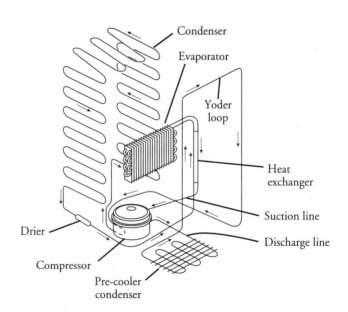

To diagnose and repair the upright freezer, consult the troubleshooting and the repair procedures sections of this chapter.

Chest Freezers

The chest freezer (Figure 26-4) has the evaporator coils and the condenser coils embedded between the inner liner and the outer cabinet. These coils are inaccessible for replacement or repair if a refrigerant leak occurs. The differences between the upright freezer and the chest freezer are:

- Door hinges.
- Gasket.
- Location and access of temperature controls.
- Location and access of the compressor, relay, and overload protector. Most models have a power indicator light. This light stays on as long as the freezer is plugged into the wall receptacle. The light alerts the consumer when the power to the freezer is off, but it does not tell you what the temperature is inside the cabinet.

Chest freezers must be defrosted once or twice a year to remove the ice buildup from the inside. To gain access to the components, remove the side access panel (Figure 26-5).

Today some manufacturers have designed chest freezers with an automatic defrost feature similar to upright freezers.

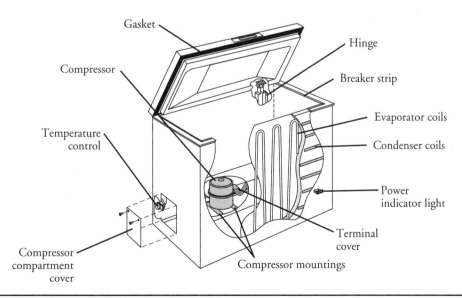

FIGURE 26-4 The component location of a chest freezer.

PART VI

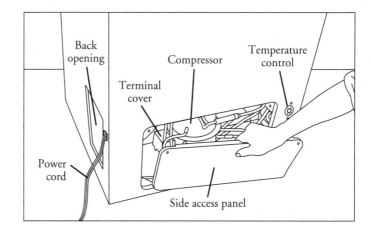

FIGURE 26-5
Removing the side access panel to gain access to the components.

Freezer Storage and Temperatures

Storing and preserving food in a freezer is a great way to maintain the quality, freshness, and nutritional value of food products. The recommended freezer temperature should be 0 degrees Fahrenheit or colder. At this temperature food should last indefinitely if packaged properly. Remember these three rules when packaging foods for freezer storage: wrap tightly, double-wrap the product again, and wrap individual portions separately. To prevent freezer burn, you must wrap food tightly to remove as much air as possible and also double-wrap the food product. When you wrap food in individual portions, you do not have to thaw out a large amount of food. You just take out what you need for that meal.

Maintaining a food's quality depends on several factors: the quality of the raw product, the procedures used during processing, the way the food is stored, and the length of storage. The recommended storage time takes these factors into consideration.

Table 26-1 depicts a storage chart for various types of food products. Also, I recommend that dates be placed on food products to protect the consumer from eating spoiled or outdated food.

Food	Storage Time at 0° F	Comments
Meat, Fish, Poultry		
Bacon	Freezing not recommended	Saltiness encourages rancidity.
Corned beef	Use within 1 month	
Frankfurters	1 to 2 months	Emulsion may be broken and the product will "weep."
Ground beef, lamb, or veal	2 to 4 months	
Ground pork	1 to 2 months	
Ham and picnic cured ham	1 to 2 months	Saltiness encourages rancidity.

TABLE 26-1   Appropriate Freezing Periods for Various Types of Foods (*continued*)

Food	Storage Time at 0° F	Comments
Luncheon meats	Use within 1 month	Emulsion may be broken and the product will "weep."
Beef roast	4 to 12 months	Freeze product in original packaging for 2 weeks. If needed to store longer, wrap in freezer wrap. For patties, separate with wax paper.
Lamb or veal roast	6 to 9 months	
Pork roast	3 to 6 months	
Sausage, dry, smoked	Use within 1 month	Freezing alters the flavor.
Beef steaks	6 to 9 months	Freeze product in original packaging for 2 weeks. If needed to store longer, wrap in freezer wrap. For patties, separate with wax paper.
Beef chops	4 to 6 months	
Lamb or veal steaks and chops	3 to 4 months	
Pork steaks and chops	2 to 3 months	
Venison or game birds	6 to 12 months	
Fish		
Lean fish, cod, flounder, haddock, and sole fillets and steaks	6 months	Freeze product in original packaging for 2 weeks. If needed to store longer, wrap in freezer wrap.
Bluefish, mackerel, perch, and salmon	2 to 3 months	
Breaded fish	3 months	
Clams	2 to 3 months	
Cooked fish or seafood	3 to 6 months	
King crab	5 months	
Lobster tails	2 to 3 months	
Oysters	2 to 4 months	
Scallops	3 to 6 months	
Shrimp, uncooked	10 months	
Poultry		
Chicken, cut up or whole	9 to 12 months	Freeze product in original packaging for 2 weeks. If needed to store longer, wrap in freezer wrap. For patties, separate with wax paper.
Chicken livers	3 to 4 months	
Cooked poultry	3 to 5 months	
Duck, turkey	6 to 9 months	
Fruit		
Berries, cherries, peaches, pears, pineapples, etc.	12 months	
Citrus fruit and juice frozen at home	6 months	
Fruit juice concentrates	8 to 12 months	

TABLE 26-1 Appropriate Freezing Periods for Various Types of Foods (*continued*)

Food	Storage Time at 0° F	Comments
Vegetables		
Home frozen	10 months	Cabbage, celery, salad greens, and tomatoes for slicing do not freeze successfully; tomatoes for soups, stews, or sauces can be frozen successfully.
Purchased frozen	8 months	
Baked Goods		
Yeast bread and rolls, baked	2 to 4 months	Freezing does not refresh the baked goods. It can only maintain whatever the quality of the food was before freezing.
Rolls, partially baked	2 to 3 months	
Bread, unbaked	1 month	
Quick bread, baked	2 to 3 months	
Cake, baked, unfrosted	2 to 3 months	
Angel food cake	2 to 6 months	
Chiffon sponge cake	2 months	Freezing does not refresh the baked goods. It can only maintain whatever the quality of the food was before freezing.
Cheesecake	2 to 3 months	
Chocolate cake	4 months	
Fruit cake	6 to 12 months	
Yellow or pound cake	6 months	
Cake, baked, frosted	3 to 4 months	
Cookies, baked	8 to 12 months	
Pie, baked	1 to 2 months	
Fruit pie, baked	6 to 8 months	
Cake, unbaked	1 month	
Main Dishes		
Meat, fish, and poultry, pies and casseroles	3 to 4 months	
TV dinners, including shrimp, ham, pork, or frankfurter	3 to 4 months	
TV dinners, including beef, turkey, chicken, or fish	6 months	
Frozen Foods – Home		
Baked muffins	6 to 12 months	
Unfrosted doughnuts	2 to 4 months	
Waffles	1 month	
Bread	3 months	
Cake	3 months	

TABLE 26-1 Appropriate Freezing Periods for Various Types of Foods (*continued*)

Food	Storage Time at 0° F	Comments
Casseroles – meat, fish, or poultry	2 to 4 months	
Cookies, baked and dough	2 to 3 months	
Nuts, salted	6 to 8 months	
Nuts, unsalted	9 to 12 months	
Pies, unbaked fruit	8 months	
Dairy Products		
Butter	6 to 9 months	
Margarine	12 months	
Whipped butter and margarine	**Do not freeze**	Emulsion will break and the product will separate.
Buttermilk, sour cream, and yogurt	**Do not freeze**	
Camembert cheese	3 months	
Cottage, farmer's cheese (dry curd only)	1 to 3 months	Do not freeze creamy cottage cheese; it will become mushy.
Neufchatel cheese	**Do not freeze**	
Cheddar cheese	6 weeks	
Edam, gouda, Swiss, brick cheeses, etc.	6 to 8 weeks	
Processed cheese food products (loaves, slices)	4 to 6 months	
Roquefort, blue cheese	3 months	
Cream – light, heavy, half and half	2 to 4 months	
Whipped cream	1 to 2 months	
Eggs in shell	**Do not freeze**	Yolks will thicken and will taste unsatisfactory in cooked products.
Whole eggs out of the shell or egg yolks	12 months	Beat thoroughly with either ½ teaspoon of salt or 2 tablespoons of sugar per cup of yolk or whole egg to control the thickening of the yolk; use in food products that ordinarily use salt or sugar as an ingredient.
Egg whites	12 months	Added salt or sugar not necessary.
Ice cream, ice milk, sherbet, frozen yogurt	2 months	
Milk	1 month	

TABLE 26-1 Appropriate Freezing Periods for Various Types of Foods

Safety First

Any person who cannot use basic tools or follow written instructions should *not* attempt to install, maintain, or repair any freezers. Any improper installation, preventive maintenance, or repairs could create a risk of personal injury or property damage.

If you do not fully understand the installation, preventive maintenance, or repair procedures in this chapter, or if you doubt your ability to complete the task on your freezer, please call your service manager.

Before continuing, take a moment to refresh your memory on the safety procedures in Chapter 2.

Freezers in General

Much of the troubleshooting information in this chapter covers freezers in general, rather than specific models, in order to present a broad overview of service techniques. The illustrations in this chapter are for demonstration purposes only, to clarify the description of how to service these appliances. They in no way reflect a particular brand's reliability.

Step-by-Step Troubleshooting by Symptom Diagnosis

When servicing an appliance, don't overlook the simple things that might be causing the problem. Step-by-step troubleshooting by symptom diagnosis is based on diagnosing malfunctions, with possible causes arranged into categories relating to the operation of the freezer. This section is intended only to serve as a checklist to aid you in diagnosing a problem. Look at the symptom that best describes the problem you are experiencing with the freezer, and then correct the problem.

The Freezer Does Not Operate

- Check and see if the freezer is plugged in.
- Check to ensure that the freezer is not plugged into a circuit that has ground fault interrupt.
- Check voltage at receptacle.
- Check circuit breakers and/or fuses.
- Check the temperature controls. Are they off?

Compressor Will Not Run

- Is there voltage at the wall receptacle? Check this with the voltmeter.
- Check for loose electrical connections.
- Is the condenser coil dirty? A dirty condenser coil will overheat the compressor.
- Check the condenser fan motor.
- Test the cold control for continuity.
- Test the compressor, the relay, and the overload switch.
- Check and see if the freezer is in the defrost mode.

Compressor Kicks Out on Overload

- Check for high or low voltage when the compressor tries to start. High voltage will overheat the compressor. Low voltage will try to run the compressor with the start winding. A compressor is designed to start and run within a 10 percent tolerance of the rated voltage.
- Test the capacitor. A shorted or open capacitor will overheat the compressor.
- Test the compressor relay.
- Test the overload for continuity.
- Test the compressor windings for a short.

Freezer Is Too Warm

- Check for restricted air circulation around the condenser coil.
- Check the location of the freezer.
- Check the door gaskets for proper sealing.
- Check the defrost heaters. Use a clamp-on ammeter (or wattmeter) to test the heaters if they are coming on when the refrigeration cycle is running.
- Check the cold control setting.
- Check the compressor. Is it operating properly?
- Is the evaporator fan running?
- Check the air duct for restriction.
- Check for a leaking air duct.
- Check the evaporator coil for excessive frost buildup.
- Check the defrost cycle. Is it working properly?

Freezer Is Too Noisy

- Check for loose parts.
- Check for rattling pipes.
- Check the fan assembly, the evaporator, and the condenser.
- Check the compressor.
- If these are normal operational noises, inform the consumer.
- Freezer not properly leveled.
- Check the floor—it may not be structurally sound.

Sweating on the Outside of the Cabinet

- Check the location of the freezer. If located in an area of high humidity, it will begin to sweat.
- Check for a void in the insulation between the cabinet and the inner liner.
- On older models, check for wet insulation.

- Check for suction line or any low-side tubing touching the cabinet.
- Check for water leaks from the ice maker.
- Check for a kinked, misaligned, or blocked drain system.
- Check the defrost drain pan for misalignment or for leaking cracks.
- Are the doors aligned and sealing properly?

Incomplete Defrosting of the Evaporator Coil or High Temperature During the Defrost Cycle

- Test the defrost thermostat.
- Check for loose wiring in the defrost electrical circuit.
- Test the defrost timer for continuity.
- Test for defective defrost heaters.

Odor in Cabinet

- Check for spoiling food in the cabinet.
- Check the defrost water drain system.
- Check the defrost heaters.

Excessive Frost Buildup on Evaporator Coil

- Check the defrost cycle.
- Check for loose wiring.
- Is the heater making contact with the evaporator coil?
- Check for proper door alignment.
- Check the door gaskets.

Freezer Run Time Is Too Long

- Check the thermostat setting.
- Check for excessive loading of unfrozen food.
- Check for incorrect wiring.
- Room or outside weather is too hot.
- Freezer has been recently disconnected for a period of time. Freezers require a minimum of four hours to cool down.
- Check for large amounts of warm water or hot food that have been stored in the freezer recently.
- Freezer lid open too long or too frequently. Advise consumer.
- Check the temperature (cold) control; it might be set too cold.
- Check that the freezer lid/door gasket is not worn, dirty, cracked, or poorly fitted. Clean or change gasket.

Temperature in the Freezer Is Higher Than Normal

- Check the thermostat for proper temperature calibration.
- Check the temperature (cold) control. Is it set too warm?
- Test evaporator fan motor and blade.
- Check the defrost timer.
- Check the lid or door. Is it open too long or too frequently?
- Check for excessive loading of unfrozen food.
- Check door gasket for proper sealing.
- Check the freezer and see if it has been disconnected for a period of time.

Temperature Inside Freezer Is Too Cold

- Check the temperature (cold) control. Reset the control to the correct temperature setting.

Freezer Runs Excessively or Continuously

- Check if the interior lights are staying on continuously.
- Check condenser coil for air restriction.
- Check door gaskets.
- On models with automatic ice makers, make sure the ice maker is operating properly.

Troubleshooting Sealed-System Problems

If you suspect a sealed-system malfunction, be sure to check all external factors first. These include:

- Thermostats
- Compressor
- Relay and overload on the compressor
- Interior lights
- Evaporator and condenser fans
- Timers
- Refrigerator getting good air circulation
- Food loaded in the freezer properly
- Check if heat exchanger has separated
- Check the wiring harness

After eliminating all of these external factors, you will systematically check the sealed system. This is accomplished by comparing the conditions found in a normally operating freezer. These conditions are:

- Freezer storage temperature
- Wattage

- Condenser temperature
- Evaporator inlet sound (gurgle, hiss, etc.)
- Evaporator frost pattern
- High-side pressure[3]
- Low-side pressure[4]
- Pressure equalization time

One thing to keep in mind is that no single indicator is conclusive proof that a particular sealed-system problem exists. Rather, it is a combination of findings that must be used to definitively pinpoint the exact problem (see Chapter 25).

Repair Procedures

Each repair procedure is a complete inspection and repair process for a single freezer component. It contains the information you need to test the components and replace them, if necessary.

Any person who cannot use basic tools should *not* attempt to install, maintain, or repair any freezer. Any improper installation, preventative maintenance, or repairs will create a risk of personal injury, as well as property damage. Call the service manager if installation, preventative maintenance, or the repair procedure is not fully understood.

Thermostat (Cold Control)

The thermostat (cold control) is located in the freezer compartment in an upright freezer or in the bottom compartment in a chest type model (Figure 26-4). The cold control maintains the temperature in the freezer. It turns the compressor and fans on and off at preset temperature settings.

The typical complaints associated with failure of the thermostat (cold control) are:

- The freezer is not cold enough.
- The freezer is too cold.
- The freezer runs all the time.
- The freezer doesn't run at all.

To handle these problems, perform the following steps:

1. **Verify the complaint.** Verify the complaint by checking the control setting. Turn the control off and on again to see if the freezer starts up. Is the power indicator light on? If the freezer will not power up, locate the technical data sheet behind the control panel or cabinet for diagnostics information. On some models you will need the actual service manual for the model you are working on to properly diagnose the freezer.

2. **Check for external factors.** You must check for external factors not associated with the appliance. Is the appliance installed properly? Explain to the user how to set the controls. The voltage at the receptacle is between 108 volts and 132 volts during a load on the circuit. Do you have the correct polarity? (See Chapter 6.)

3. **Disconnect the electricity.** Before working on the freezer, disconnect the electricity. This can be done by pulling the plug from the receptacle. Or disconnect the electricity at the fuse panel or at the circuit breaker panel. Turn off the electricity.

WARNING *Some diagnostic tests will require you to test the components with the power turned on. On some models, when you disassemble the control panel, you can position it in such a way that the wiring will not make contact with metal. This will allow you to test the components without electrical mishaps.*

4. **Gain access to the thermostat.** To access the thermostat, remove the access panel (see Figure 26-5). Next, remove the two screws that secure the control. Remove the wires from the terminals. The capillary tube is inserted into a channel. Do not remove capillary tube yet. Test the control first.

5. **Test the thermostat.** To test the thermostat, place the ohmmeter probes on the terminals (Figure 26-6). Set the range scale on R × 1, and test for continuity. With the control set in the "off" position, you should not read continuity. When the control is set to the highest position, you should read continuity. If the thermostat is good, the problem must be elsewhere.

6. **Remove the thermostat.** With the thermostat control housing already removed, the capillary tube must now be removed. Remove the capillary tube from the channel.

7. **Install a new thermostat.** To install the new thermostat, just reverse the order of disassembly, and reassemble. Then test the control. Remember to reinstall the capillary tube in the same location from which it was removed. Be careful not to kink the tube. If you do, the freezer will not work properly.

FIGURE 26-6
Testing the thermostat with a multitester with the range set on the ohms scale. If you read no continuity, replace the thermostat.

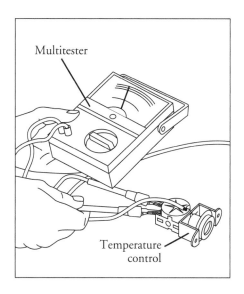

Multitester

Temperature control

PART VI

Power Indicator Light

The power indicator light lets the consumer know that the freezer is plugged into the receptacle and voltage is present.

The typical complaints associated with the failure of the power indicator light are:

- The light is not on.
- The light flickers.
- The light is dim.

To handle these problems, perform the following steps:

1. **Verify the complaint.** Verify the complaint by checking if the freezer is plugged into the wall receptacle. Is the power indicator light on? If the freezer will not power up, locate the technical data sheet behind the control panel or cabinet for diagnostics information. On some models you will need the actual service manual for the model you are working on to properly diagnose the freezer.

2. **Check for external factors.** You must check for external factors not associated with the appliance. Is the appliance installed properly? Check the voltage to the freezer. The voltage at the receptacle is between 108 volts and 132 volts during a load on the circuit. Do you have the correct polarity? (See Chapter 6.)

3. **Disconnect the electricity.** Before working on the freezer, disconnect the electricity. This can be done by pulling the plug from the receptacle. Or disconnect the electricity at the fuse panel or at the circuit breaker panel. Turn off the electricity.

WARNING *Some diagnostic tests will require you to test the components with the power turned on. On some models, when you disassemble the control panel, you can position it in such a way that the wiring will not make contact with metal. This will allow you to test the components without electrical mishaps.*

4. **Gain access to the power indicator light.** To access the power indicator light, use a screwdriver to pry out the power indicator light from the front of the freezer cabinet. Next, remove the wires from the indicator light (Figure 26-7).

5. **Test the power indicator light.** To test the power indicator light, place the ohmmeter probes on its terminals (Figure 26-8). Set the range scale on R x 1, and test for continuity. The meter should show continuity; if not, replace the component.

6. **Install a new power indicator light.** To install the new power indicator light, just reverse the order of disassembly, and reassemble (Figure 26-9). Then test the control.

Gasket

The door gasket is mounted to the lid on a chest freezer and mounted on the door of an upright model. The door gasket consists of a vinyl rubber gasket with a magnet. The magnet helps secure the door closed to keep the cold inside the box and the heat out.

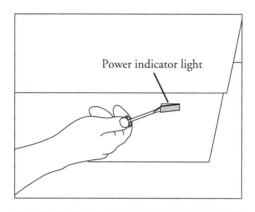

FIGURE 26-7 Pry gently to remove the power indicator light. Be careful not to scratch the cabinet.

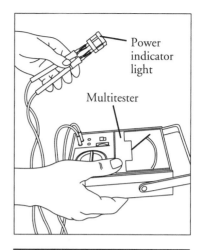

FIGURE 26-8 Testing the power indicator light.

The typical complaints associated with failure of the door gasket are:

- Sweating inside and/or outside of the cabinet.
- Temperatures inside the cabinet are warmer than normal.
- Ice is forming on the freezer walls.
- Door gasket not maintained properly.

To handle these problems, perform the following steps:

1. **Verify the complaint.** Verify the complaint by checking the door gasket for proper sealing and alignment. Inspect the gaskets for any damage.

2. **Check for external factors.** You must check for external factors not associated with the appliance. Is the appliance installed properly?

3. **Disconnect the electricity.** Before working on the refrigerator/freezer, disconnect the electricity. This can be done by pulling the plug from the receptacle. Or disconnect the electricity at the fuse panel or at the circuit breaker panel. Turn off the electricity.

4. **Test the door gasket.** To test the gasket for proper sealing, take a dollar bill and place it between the gasket and the flange of the outer cabinet. Pull on the dollar bill. When pulling on the dollar bill, you should feel some tension as the gasket and flange grip the bill. Repeat this test in other areas where you suspect problems with the gasket. If the gasket fails this test, the next step is to replace the gasket.

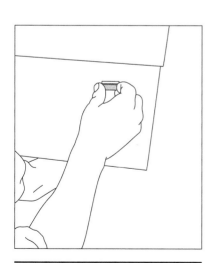

FIGURE 26-9 When installing the power indicator light, you must first reconnect the wires and then insert the light in the cabinet.

FIGURE 26-10
Removing the screws
from the hinges and
lifting the door from
the freezer.

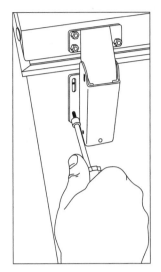
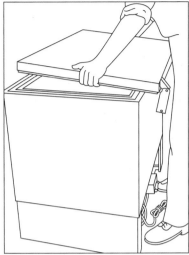

5. **Gain access to the door gasket.** To access the door gasket, the door must be removed (Figure 26-10). Turn the door over on its back.

6. **Remove the gasket.** Remove the gasket, either by prying the studs out or by removing the screws (Figure 26-11).

7. **Install a new door gasket.** Before installing the new gasket, soak it in warm water for 15 to 20 minutes. This will make the gasket soft and easier to install. Starting at either top corner, insert the flange of the gasket behind the retaining strip and/or door liner. Proceed all the way around the door. When the gasket is in place, begin to tighten the screws slightly all around the door or reinstall the studs. If the gasket is distorted, or if it has wrinkles in it, use a hair dryer to heat the gasket and mold it to its original form. Then be sure the gasket seats against the flange properly. Next, check the gap between the door and the cabinet; adjust it, if necessary (see step 4).

FIGURE 26-11
Removing the studs
(inset). On some
models, you must
remove the screws
that secure the gasket
to the door.

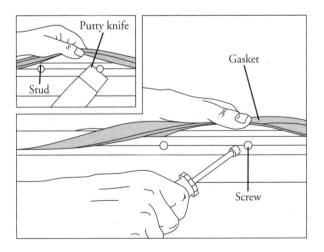

Defrost Timer (Automatic Defrost Models)

The purpose of the defrost timer is to regulate the frequency of the defrost cycles and their duration. The defrost timer also limits the maximum amount of time that the defrost heater can be energized. There are two types of configurations used in a mechanical timer. They are continuous run and cumulative run. The difference between the two is the way that the timer motor is energized. The continuous run timer will be energized anytime when the refrigerator is plugged in, and the cumulative run timer is energized when the cold control is calling for cooling and the compressor is running.

The typical complaints associated with failure of the defrost timer are:

- The freezer does not defrost.
- The storage temperature in the cabinet is too warm.
- The compressor will not run.

To handle these problems, perform the following steps:

1. **Verify the complaint.** Verify the complaint by asking the customer to describe what the freezer is doing or did. If the freezer will not power up, locate the technical data sheet behind the control panel or cabinet for diagnostics information. On some models you will need the actual service manual for the model you are working on to properly diagnose the freezer.

2. **Check for external factors.** You must check for external factors not associated with the appliance. Is the appliance installed properly? Is the door aligned properly? The voltage at the receptacle is between 108 volts and 132 volts during a load on the circuit. Do you have the correct polarity? (See Chapter 6.)

3. **Disconnect the electricity.** Before working on the freezer, disconnect the electricity to the appliance. This can be done by pulling the plug from the receptacle. Or disconnect the electricity at the fuse panel or at the circuit breaker panel. Turn off the electricity.

WARNING *Some diagnostic tests will require you to test the components with the power turned on. When you disassemble the control panel, you can position it in such a way that the wiring will not make contact with metal. This will allow you to test the components without electrical mishaps.*

4. **Gain access to the defrost timer.** To access the defrost timer, you must first locate it. On some models, the defrost timer is located on the bottom, behind the toe plate; or it might be behind the temperature control housing, in the freezer section (Figure 25-14); or it might be in the back of the freezer, behind the rear leg (Figure 25-15) or in the compressor compartment.

5. **Remove the defrost timer.** In order to test the defrost timer, it must be removed from its mounting position. Remove the two mounting screws from the defrost timer (see Figure 25-15). Next, remove the wire harness plug from the defrost timer (Figure 25-16).

6. **Test the defrost timer.** To test the defrost timer, place a screwdriver in the timer cam slot (Figure 25-17), and turn it clockwise until you hear the first "snap." The defrost timer is now in the defrost cycle. At this point, you must read the wiring diagram to determine which numbered terminals are for the defrost circuit. For the purpose of demonstrating how to check for continuity of the switch contacts, Figure 25-18a illustrates the internal components of this sample timer. Set the ohmmeter scale on R x 1, and place the probes on the terminals marked 2 and 3 (Figure 25-19). You should measure continuity. Next, rotate the timer cam until you hear the second "snap." The meter will show no continuity, indicating that the defrost cycle is over and that the refrigeration cycle begins.

Now place the meter probes on the terminals marked 3 and 4. The ohmmeter will show continuity, indicating the refrigeration cycle is activated. Turn the timer cam once again, until you hear the first "snap." The meter will show no continuity. At no time should there be continuity between terminals 2 and 4. (If so, the switch contacts are burned and welded together and the defrost timer must be replaced.) If the defrost timer passes this portion of the test, you must determine if the timer mechanism is functioning. Place the ohmmeter probes on the timer motor leads and read the resistance. The resistance can be between 800 and 4000 ohms, depending on the type of timer used by the manufacturer. If you are unable to read resistance, the timer motor is defective.

If the defrost timer passes this portion of the test, rotate the timer cam until you hear the first "snap." Advance the timer cam again, counting the number of clicks until you hear the second "snap." Write down the number of clicks on a piece of paper. Now rotate the timer cam again until the marks line up (Figure 25-18b), which indicates the beginning of the defrost cycle, and the "snap" is heard. Advance the timer cam and count the clicks until there is one click left before the end of the defrost cycle. Take the timer and reconnect it to the wiring harness (see Figure 25-16). Place the defrost timer on a nonmetallic surface.

Reconnect the voltage supply to the refrigerator/freezer. *Note:* Be cautious when working with live wires. Avoid getting shocked. Place the ammeter jaws around the wire attached to the number 4 terminal. The meter should indicate no amperage. Next, place the jaws on the number 2 terminal wire. The ammeter should indicate some amperage. Wait for approximately 10 to 15 minutes: you should hear a "snap," indicating that the timer has completed the defrost cycle. At this point, the ammeter will show no amperage on number 2, but will indicate current flow at number 4. If not, replace the timer.

7. **Install a new defrost timer.** To install the new defrost timer, just reverse the order of disassembly and reassemble. Remember to reconnect the ground wire to the defrost timer.

Defrost Heater (Automatic Defrost Models)

Manufacturers also use a single-calrod type, radiant heater mounted under the evaporator coil for maximum defrosting of the evaporator coil.

The typical complaints associated with failure of the defrost heater are:

- The freezer temperature is warm.
- The freezer does not defrost.
- Food is spoiling.
- No ice cubes.

To handle these problems, perform the following steps:

1. **Verify the complaint.** Verify the complaint by asking the customer to describe what the freezer is doing. Is food spoiling? Check the temperature in the compartment. Check for ice buildup on the evaporator cover. If the freezer will not power up, locate the technical data sheet behind the control panel or cabinet for diagnostics information. On some models you will need the actual service manual for the model you are working on to properly diagnose the freezer.

2. **Check for external factors.** You must check for external factors not associated with the appliance. Is the appliance installed properly? Is there electricity to the freezer? Is the electrical receptacle polarized and properly grounded? The voltage at the receptacle is between 108 volts and 132 volts during a load on the circuit. Do you have the correct polarity? (See Chapter 6.)

3. **Disconnect the electricity.** Before working on the freezer, disconnect the electricity. This can be done by pulling the plug from the receptacle. Or disconnect the electricity at the fuse panel or at the circuit breaker panel. Turn off the electricity.

WARNING *Some diagnostic tests will require you to test the components with the power turned on. When you disassemble the control panel, you can position it in such a way that the wiring will not make contact with metal. This will allow you to test the components without electrical mishaps.*

4. **Gain access to the defrost heater.** To access the defrost heater, the evaporator cover must be removed (see Figure 25-23). Remove the screws that secure the cover in place.

5. **Test the defrost heater.** A defrost heater should be tested for proper resistance, as indicated on the wiring diagram. To test the defrost heater, remove the wires from the heater terminals. Next, place the probes of the ohmmeter on the heater terminals (Figure 26-12). Set the scale on R × 1. The meter should show resistance. If no reading is indicated, replace the defrost heater.

6. **Remove the defrost heater.** To remove the calrod defrost heater, you must first defrost the ice from the evaporator coil. Bend the clips that secure the heater in place and then remove the defrost heater.

7. **Install a new defrost heater.** To install the new defrost heater, just reverse the order of disassembly and reassemble. When reinstalling any shrouds, grilles, ducts, or gaskets, always position them correctly to ensure the proper airflow through the evaporator and within the compartment of the freezer.

FIGURE 26-12
Frozen evaporator coil.
Test the defrost
thermostat first before
you defrost the
evaporator coil.

Defrost
Thermostat

Defrost
Heater

Defrost Thermostat (Automatic Defrost Models)

The defrost thermostat is a bimetal switch installed on the evaporator coil that provides over temperature protection during defrost. The defrost heaters will defrost the evaporator coil within a given time. But, if the evaporator coil is totally defrosted before the time has expired, the defrost thermostat will open up, shutting off the defrost heater to prevent the evaporator coil area from overheating.

The typical complaints associated with failure of the defrost thermostat are:

- The freezer temperature is warm.
- The freezer does not defrost.
- Food is spoiling.
- No ice cubes.

To handle these problems, perform the following steps:

1. **Verify the complaint.** Verify the complaint by asking the customer to describe what the freezer is doing. Is food spoiling? Check the temperature in the compartment. Check for ice buildup on the evaporator cover. If the freezer will not power up, locate the technical data sheet behind the control panel or cabinet for diagnostics information. On some models you will need the actual service manual for the model you are working on to properly diagnose the freezer.

2. **Check for external factors.** You must check for external factors not associated with the appliance. Is the appliance installed properly? Is there electricity to the freezer? Is the electrical receptacle polarized and properly grounded? The voltage at the receptacle is between 108 volts and 132 volts during a load on the circuit. Do you have the correct polarity? (See Chapter 6.)

3. **Disconnect the electricity.** Before working on the freezer, disconnect the electricity. This can be done by pulling the plug from the receptacle. Or disconnect the electricity at the fuse panel or at the circuit breaker panel. Turn off the electricity.

WARNING *Some diagnostic tests will require you to test the components with the power turned on. When you disassemble the control panel, you can position it in such a way that the wiring will not make contact with metal. This will allow you to test the components without electrical mishaps.*

4. **Gain access to the defrost thermostat.** To access the defrost thermostat, the evaporator cover must be removed (see Figure 26-12). Remove the screws that secure the cover in place.

5. **Test the defrost thermostat.** Failure of a defrost thermostat usually results in a frost-blocked evaporator. To test the defrost thermostat, disconnect the wires to isolate the thermostat from the rest of the defrost circuit. Next, place the probes of the ohmmeter on the defrost thermostat wire leads. Set the meter scale on R × 1. The meter will show continuity when the thermostat is either frozen or very cold, indicating the defrost thermostat is good. The defrost thermostat switch contacts close when the temperature is colder than its temperature rating. If no reading is indicated, replace the defrost thermostat. At ambient temperature, you will read no continuity, which will indicate the thermostat might be good.

6. **Remove the defrost thermostat.** To remove the defrost thermostat, you must remove the hold-down clamp. On some models, the defrost thermostat and clamp are one assembly. On other models, the defrost thermostat clamps around the evaporator tubing. As shown in Figure 26-12, remove this type of defrost thermostat by squeezing in on the clip and lifting the thermostat up.

7. **Install a new defrost thermostat.** To install the new defrost thermostat, just reverse the order of disassembly and reassemble. Remember to reconnect the wires to the thermostat. When reinstalling any shrouds, grilles, ducts, or gaskets, always position them correctly to ensure the proper airflow through the evaporator and within the compartment of the freezer. On models that have the defrost thermostat attached to the evaporator coil, you must reinstall the defrost thermostat in the same location from which it was removed.

Evaporator Fan Motor (Automatic Defrost Models)

The evaporator fan motor provides air circulation within the freezer cabinet and over the evaporator coil which is located in the freezer compartment of the freezer. One type of motor used in freezers is a shaded pole, single-speed motor that runs on 120 VAC.

The typical complaints associated with failure of the evaporator fan motor are:

- The freezer temperature is warm.
- The evaporator fan motor runs slower than normal.
- The evaporator fan motor does not run at all.
- The evaporator fan motor is noisy.

To handle these problems, perform the following steps:

1. **Verify the complaint.** Verify the complaint by asking the customer to describe what the freezer is doing or did. Is the evaporator fan motor running? Is it noisy? If the refrigerator will not power up, locate the technical data sheet behind the control panel

or cabinet for diagnostics information. On some models you will need the actual service manual for the model you are working on to properly diagnose the freezer.

2. **Check for external factors.** You must check for external factors not associated with the appliance. Is the appliance installed properly? Is there something hitting the fan blade? Is there electricity to the freezer? Is the electrical receptacle polarized and properly grounded? The voltage at the receptacle is between 108 volts and 132 volts during a load on the circuit. Do you have the correct polarity? (See Chapter 6.)

3. **Disconnect the electricity.** Before working on the freezer, disconnect the electricity. This can be done by pulling the plug from the receptacle. Or disconnect the electricity at the fuse panel or at the circuit breaker panel. Turn off the electricity.

WARNING *Some diagnostic tests will require you to test the components with the power turned on. When you disassemble the control panel, you can position it in such a way that the wiring will not make contact with metal. This will allow you to test the components without electrical mishaps.*

4. **Gain access to the evaporator fan motor.** To access the evaporator fan motor, the evaporator cover must be removed. Remove the screws that secure the cover in place. On some models, the evaporator fan assembly is located on the rear wall of the interior freezer compartment.

5. **Test the evaporator fan motor.** The shaded-pole, 120 VAC evaporator fan motor should be tested for proper resistance, as indicated on the wiring diagram. To test the evaporator shaded-pole fan motor, remove the wires from the motor terminals. Next, place the probes of the ohmmeter on the motor terminals (Figure 25-22). Set the scale on R × 1. The meter should show resistance. If no reading is indicated, replace the motor. If the fan blade does not spin freely, replace the motor. If the fan motor runs and it is noisy (bad bearings), replace the motor. Also check the motor for grounded windings; if grounded, replace the motor.

6. **Remove the evaporator fan motor.** To remove the evaporator fan motor, you must first remove the fan blade. On most models, just pull the blade off the motor shaft. Be careful not to break the blade. On other models, the fan blade is held on the motor shaft with screws. Remove the screws. Then remove the screws that secure the fan assembly to the cabinet (Figure 25-23). On some models, you must remove the fan shroud (Figure 25-24) by removing the shroud screws.

7. **Install a new evaporator fan motor.** To install the new evaporator fan motor, just reverse the order of disassembly and reassemble. When reinstalling the fan blades onto the motor shaft, the fan blades should be positioned on the shaft so that one-third of its depth (approximately 1/4 inch) protrudes through the fan orifice in the direction of airflow. When reinstalling any shrouds, grilles, ducts, or gaskets, always position them correctly to ensure the proper airflow through the evaporator and within the compartment of the freezer. Remember to reconnect the ground wire to the motor. Reconnect the wires to the motor terminals, and test.

Compressor, Relay, and Overload Protector

The compressor (reciprocating or rotary type) is the heart of the vapor compression system. It is used to circulate the refrigerant throughout the sealed system. The relay and overload are attached to the compressor. The relay starts the compressor, and the overload protects the compressor. All three components are located in the machine compartment in the rear or bottom of the freezer. The relay can be either a current or a PTC (positive temperature coefficient) type device. The overload is a bimetal switch that is secured to the outer shell of the compressor.

The typical complaints associated with failure of the compressor are:

- The freezer temperature is warm.

- The freezer does not run at all.

- Food is spoiling.

To handle these problems, perform the following steps:

1. **Verify the complaint.** Verify the complaint by asking the customer to describe what the refrigerator is doing. If the freezer will not power up, locate the technical data sheet behind the control panel or cabinet for diagnostics information. On some models you will need the actual service manual for the model you are working on to properly diagnose the freezer.

2. **Check for external factors.** You must check for external factors not associated with the appliance. Is the appliance installed properly? Check for a voltage drop during freezer startup. Is there electricity to the freezer? Is the electrical receptacle polarized and properly grounded? The voltage at the receptacle is between 108 volts and 132 volts during a load on the circuit. Do you have the correct polarity? (See Chapter 6.)

3. **Disconnect the electricity.** Before working on the freezer, disconnect the electricity. This can be done by pulling the plug from the receptacle. Or disconnect the electricity at the fuse panel or at the circuit breaker panel. Turn off the electricity.

WARNING *Some diagnostic tests will require you to test the components with the power turned on. When you disassemble the control panel, you can position it in such a way that the wiring will not make contact with metal. This will allow you to test the components without electrical mishaps.*

4. **Gain access to the compressor.** To access the compressor, pull the freezer out and away from the wall. Remove the back panel, which is located at the bottom of the freezer (Figure 26-5). This will expose the compressor. Next, remove the compressor terminal cover (Figure 25-34) by removing the retaining clip that secures the cover. Remove the terminal cover.

5. **Test the compressor relay.** To test the compressor current relay, remove the relay by pulling it off the compressor terminals without twisting it (Figure 25-35a). Remove the wires from the relay and label them. On the relay body is stamped the word TOP. Hold the relay so that TOP is in the up position.

Next, place the probes of the ohmmeter on the relay terminals marked S and M. Set the meter scale on R × 1. The reading will show no continuity. Then remove the probe from the terminal marked M, and place it on the side terminal marked L. The reading will show no continuity. Now, move the probe from terminal S, and place it on the terminal marked M. The reading will show continuity. With the probes still attached, turn the relay upside down (Figure 25-35b), and perform the same tests. By turning the relay over, the switch contacts in the relay will close. When you retest the relay, you should get the opposite results: You should have continuity between terminals S and M and between S and L; however, the meter will not read continuity between M and L. If the relay fails this test, replace it. The elimination method is the best way to test a PTC relay; you would first test the run capacitor, overload, and then run the compressor with a fused test cord. If all of these components check out okay, replace the relay.

6. **Test the overload protector.** To test the overload protector, remove the wires from the overload and compressor terminals. Then remove the overload protector from the compressor by removing the retaining clip that secures the overload protector to the compressor (Figure 25-36a). Next, place the probes of the ohmmeter on the overload terminals (Figure 25-36b). Set the meter scale on R × 1. The reading will show continuity. If not, replace the overload protector.

7. **Test the compressor.** To test the compressor, remove the relay and the overload protector. This will expose the compressor terminals. The compressor terminals are marked C, S, and R. C indicates the common winding terminal, S indicates the start winding terminal, and R indicates the run winding terminal. (Refer to the actual wiring diagram for the model that you are servicing.) Set the meter scale on R × 1, touch the probes together, and adjust the needle setting to indicate a zero reading. Next, place the probes of the ohmmeter on the terminals marked S and R (Figure 25-37a). The meter reading will show continuity. Now place the meter probes on the terminals marked C and S. The meter reading will show continuity. Finally, place the meter probes on the terminals marked C and R. The meter reading will show continuity. The total number of ohms measured between S and R is equal to the sum of C to S plus C to R. The compressor should be tested for proper resistance, as indicated on the wiring diagram.

To test the compressor for ground, place one probe on a compressor terminal, and attach the other probe to the compressor housing or any good ground (Figure 25-37b). Set the meter scale to R × 1000. The meter reading will show no continuity. Repeat this for the remaining two terminals. The meter reading will show no continuity. If you get a continuity reading from any of these terminals to ground, the compressor is grounded. Replace it.

8. **To replace the overload and relay.** To install the new overload or relay, just reverse the order of disassembly and reassemble. Remember to reconnect the wires to the overload or relay.

Diagnostic Charts

The following freezer diagnostic charts will help you to pinpoint the likely causes of the particular problem associated with these appliances (Figures 26-13 and 26-14).

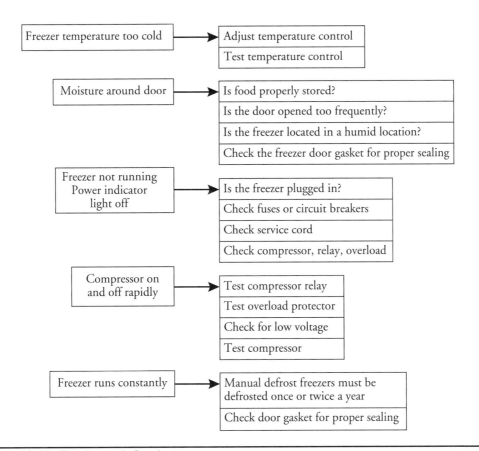

Figure 26-13 The diagnostic flowchart.

Freezer Electrical Schematics

The wiring diagrams in this chapter are examples only. You must refer to the actual wiring diagram for the freezer you are servicing. Figures 26-15, 26-16, and 26-17 show a typical wiring schematic and strip circuits for an automatic defrost and manual freezer.

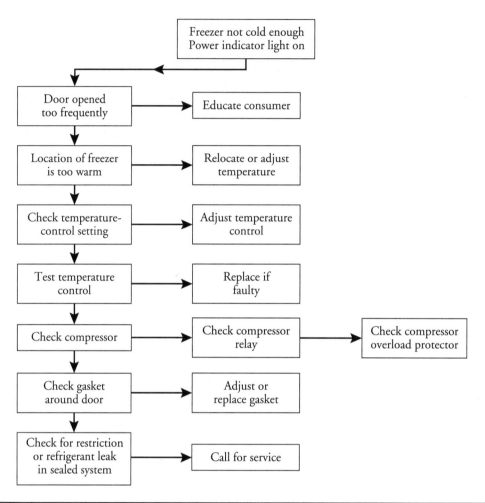

Figure 26-14   The diagnostic flowchart: Freezer not cold enough/power indicator light on.

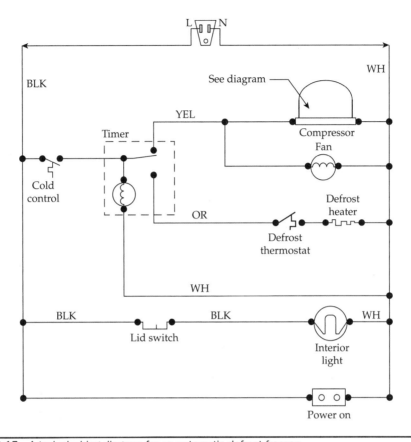

FIGURE 26-15 A typical wiring diagram for an automatic defrost freezer.

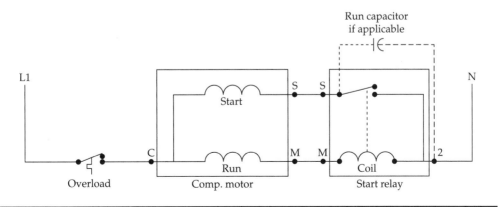

FIGURE 26-16 A typical strip circuit for a freezer.

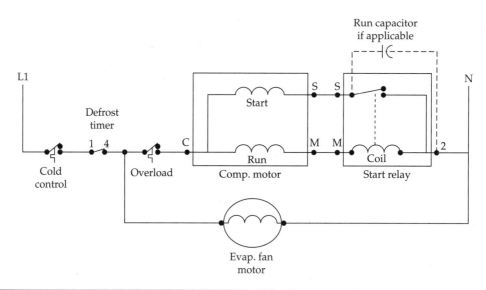

FIGURE 26-17   A typical strip circuit for an automatic defrost freezer.

Endnotes

1. Manual defrost freezers must be defrosted once or twice a year, depending upon usage.

2. Automatic defrost freezers will self-defrost the evaporator coil. The defrost components are the defrost timer, the defrost thermostat, and the defrost heater.

3. If you open up the sealed system, you will void your warranty. The sealed system must be repaired by an authorized service company.

4. If you open up the sealed system, you will void your warranty. The sealed system must be repaired by an authorized service company.

Automatic Ice Makers

The automatic ice maker is an independent appliance, installed in the freezer or in the freezer compartment of a refrigerator. The ice maker has three distinct operations: fill, freeze, and harvest the ice cubes. It is the refrigerator/freezer cooling system that allows the water to freeze into whatever shape is designed by the ice mold or tray. The amount of time it takes for the ice maker to produce and harvest ice cubes will depend on:

- Freezer temperature
- Food load conditions
- Number of door openings
- Ambient temperature

Safety First

Any person who cannot use basic tools or follow written instructions should *not* attempt to install, maintain, or repair any automatic ice makers. Any improper installation, preventive maintenance, or repairs could create a risk of personal injury or property damage.

If you do not fully understand the installation, preventive maintenance, or repair procedures in this chapter, or if you doubt your ability to complete the task on your automatic ice maker, please call your service manager.

The following precautions should also be followed:

- Never place fingers or hands on the automatic ice maker mechanism while the refrigerator/freezer is plugged in.
- Disconnect the electrical supply to the freezer or refrigerator before servicing the automatic ice maker.
- Be careful of any sharp edges on the automatic ice maker, which might result in personal injury.
- Do not attempt to operate the automatic ice maker unless it has been properly reinstalled, including the grounding and electrical connections.

Before continuing, take a moment to refresh your memory on safety procedures in Chapter 2.

Automatic Ice Makers in General

Much of the troubleshooting information in this chapter covers automatic ice makers in general, rather than specific models, in order to present a broad overview of service techniques. The illustrations that are used in this chapter are for demonstration purposes only, to clarify the description of how to service ice makers. They in no way reflect on a particular brand's reliability.

Principles of Operation for Type-1 and Type-2 Automatic Ice Makers

The automatic ice maker is an electromechanical device that, when properly installed in the freezer compartment of a refrigerator or freezer, will automatically perform the following functions:

- Detect when water is completely frozen in the ice cube molds
- Eject the ice cubes into a storage container
- Refill the ice cube maker mold with water
- Stop the operation of the ice cube maker when the storage bin is full or when the shutoff arm is in the up position

The freezer temperature determines the efficiency of the automatic ice maker. The colder the temperature in the freezer, the faster ice will freeze in the mold or tray. In order for the ice maker to harvest the ice cubes, the temperature in the freezer should be colder than 12 degrees Fahrenheit.

The water supply (Figure 27-1) is metered into the ice maker mold or tray by a water valve, usually located in the compressor compartment. The water freezes in the automatic ice maker mold or tray, as it would in a manual ice cube tray.

Type-1 Ice Maker

Type-1 ice makers, as illustrated in Figure 27-2, have a thermostat that is mounted on the ice maker mold, which will sense the temperature of the ice. When the water and ice have cooled down to 17 degrees Fahrenheit (±3 degrees), the thermostat will activate the ice maker motor and heater circuit. The motor will rotate the ejector blade, allowing it to rest on the ice cubes and, at the same time, exert pressure. The ice maker mold heater will warm up the ice cubes just enough to release them from the mold. This procedure will take approximately three to five minutes to release the ice cubes from the mold.

When the ice cubes become free from the ice maker mold, the ejector blade will continue to rotate, scooping the ice cubes out of the ice maker mold and depositing them into the ice bucket. The shutoff arm moves up and down as the ice maker cycles.

At the end of the cycle, the arm will lie on top of the ice cubes. As the cubes fill the ice bucket, the shutoff arm rises to a designated point and turns off the ice maker, halting ice production. As the ice level in the bucket falls from use, the ice maker cycle will resume. Then the water valve opens again, allowing the water to enter the ice maker mold to be frozen into ice cubes, thus beginning the cycle all over again. The fill time for the type-1 ice maker is approximately seven seconds. This type of ice maker can produce up to eight pounds of ice within a 24-hour period, providing the conditions are ideal. Some customers

A typical ice maker water supply installation. Note the direction of the water flow.

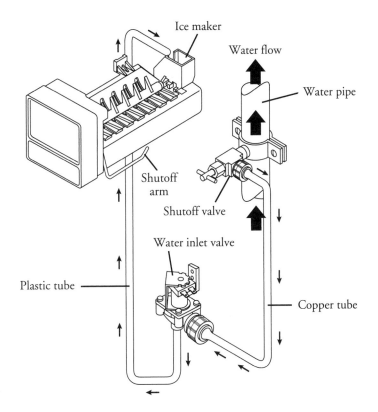

will have ice production between three to five pounds per day, depending on usage and door openings.

Type-2 Ice Maker

The type-2 ice maker (Figure 27-3) has the components in the ice maker head necessary for the ice-making operation. The temperature in the freezer compartment must be below

Type-1 ice maker.

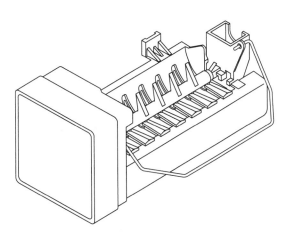

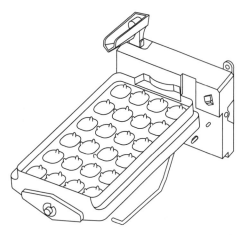

FIGURE 27-3   Type-2 ice maker.

15 degrees Fahrenheit before the ice maker will begin to operate. Inside the ice maker head assembly is a thermostat that senses the temperature in the freezer compartment. When the thermostat is satisfied, it will energize the timer motor. When the motor is energized, it will begin to turn a timing gear and the shutoff arm will also begin to move. The shutoff arm will descend into the ice bucket to determine how much ice is there. If the bucket is full, the shutoff arm will rest against the ice cubes and the ice maker will not cycle. If the shutoff arm continues to its normal position, the ice bucket is not full and the cycle will continue.

If the cycle continues, the tray will begin to rotate. When the tray has rotated about 140 degrees, one corner of the tray engages the tray stop, which prevents that part of the tray from rotating further (Figure 27-4). The rest of the tray will continue to rotate and twist the tray to about 40 degrees. At this point, the twisting of the tray will loosen the ice cubes. As the shaft continues to turn, the tray stop will begin to retract. As this happens, the tray will rapidly release the ice cubes into the ice bucket. After the ice has been released, the tray will continue to rotate until it has completed its 360-degree rotation. Near the end of the tray rotation, the water valve fill switch is energized, allowing the water to enter and fill the tray for the next cycle (Figure 27-5). The fill time for type-2 ice makers is approximately 12 seconds.

Ice Production Rate

Before a technician can repair an ice maker, he or she must test the freezer compartment temperature first. The freezer capacity must maintain a temperature of 0 degrees Fahrenheit (plus or minus 5 degrees) under loaded conditions. To ensure that the ice maker functions properly, the refrigeration system must be capable of reducing the temperature at the ice maker thermostat to approximately 17 degrees Fahrenheit (plus or minus 3 degrees). The greater the system capacity, the shorter the ice freezing cycle and the higher the ice production rate.

Airflow direction around the ice maker mold, consumer usage, and ambient temperatures will also affect the ice production rate. The only component in the ice maker that can alter the length of the ice freezing cycle is the thermostat.

The amount of water and the temperature of the water that enters the ice maker mold can affect ice production. The fill level is affected by water pressure, water line saddle valve or tee, water valve flow washer, and ice maker timing. Never install a water line to the ice maker using a self-piercing saddle valve. This type of valve will reduce the water pressure in the line to the ice maker in a short matter of time.

Under optimum conditions, ice harvests of 3.5 to 4.5 pounds in 24 hours can be expected. Realistically the consumer may have only 2.5 to 3.5 pounds in 24 hours. To obtain this amount of ice, three conditions must exist: correct airflow temperature, correct freezer temperature, and correct temperature at the ice maker mold.

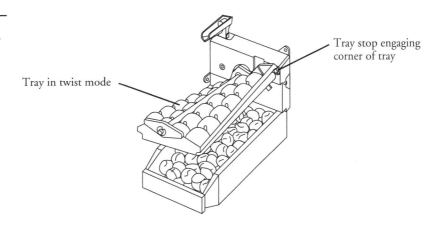

FIGURE 27-4
The ice tray will twist, releasing the ice cubes.

Tray stop engaging corner of tray

Tray in twist mode

Testing the Type-1 Ice Maker

The type-1 ice maker is designed to allow all of the components to be tested without removing the ice maker and without moving the refrigerator away from the wall to test the water valve. The ice maker in Figure 27-6a is the old-style ice maker, and it is not in production any more. Parts for this ice maker are still available, however. Figure 27-6b is the new-style modular ice maker. The components that make up this ice maker are illustrated in Figure 27-7.

The test holes that are on the ice maker head module (see Figure 27-6b) are identified as "N," "M," "H," "T," "L," and "V." These are the test points for testing this type of ice maker. The letters indicate the following:

- N = neutral side of line voltage
- M = ice maker motor connection
- H = mold heater connection
- T = Ice maker thermostat
- L = L1 side of line voltage
- V = ice maker water valve connection

This test requires the electricity to be turned on.

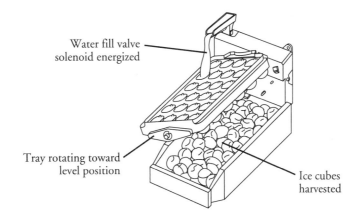

FIGURE 27-5
The ice cube tray must return to a level position after the filling cycle is complete.

Water fill valve solenoid energized

Tray rotating toward level position

Ice cubes harvested

PART VI

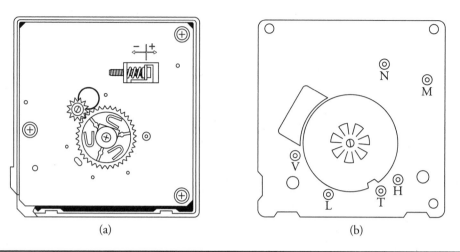

(a)   (b)

FIGURE 27-6   (a) Old-style type-1 ice maker. (b) New-style type-1 ice maker.

The following test holes, when jumped out with a jumper wire, will perform the following:

- T – H = Jumps out the thermostat and starts the ice maker cycle.
- H – N = Turns on the ice mold heater.
- M – N = Runs the ice maker motor.

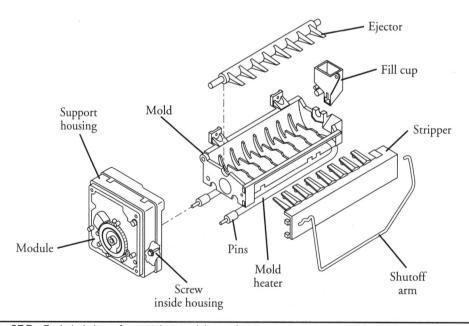

FIGURE 27-7   Exploded view of new-style type-1 ice maker.

- L – N = *Do not use a jumper wire.* It will short out the refrigerator and/or ice maker. If you place your voltmeter leads in these test holes, your meter will read 120 volts.
- L – V = The ice maker water valve will turn on.

All electrical components in a type-1 ice maker can be resistance tested from the front module with the outer cover removed and the 120 volts disconnected from the refrigerator (Figures 27-6 and 27-8).

NOTE *Caution should be used when working with live wires. Avoid getting shocked. Stay clear of live wires. Only handle the meter probes by the insulated handles.*

To test this ice maker, the unit must be installed and plugged into the freezer ice maker receptacle, the shutoff arm must be placed in the down position, and the temperature in the freezer should be colder than 12 degrees Fahrenheit. Set your multimeter for voltage, and place one probe in test point "L" and the other probe in test point "N." Be sure that the test

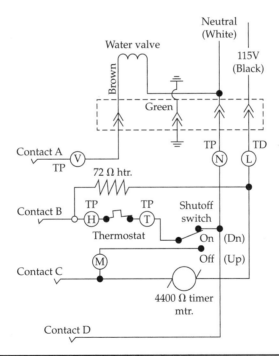

FIGURE 27-8 A resistance check can be performed on all components from the front of the ice maker module. Be safe. Unplug the refrigerator or turn off the circuit breaker to the refrigerator.
  Inserting the ohmmeter leads between test points L and H will check the resistance of the heater (72 ohms, ± 10 percent).
  Inserting the leads between L and M will check the motor resistance (approximately 4400 ohms).
  Inserting the ohmmeter leads between test points V and N will check the resistance of the water valve coil (approximately 300 ohms). This check is especially helpful since it allows you to check the continuity of the water valve without having to remove the water valve or pull the refrigerator away from the wall.

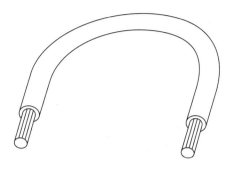

FIGURE 27-9 The jumper wire is used to short out the test points on the ice maker module.

probes go into the test points about 1/2 inch. You should have a reading of 120 volts, which indicates line voltage to the ice maker. Next, place the meter probes in test points "T" and "H." This will verify if the bimetal thermostat is open or closed. If the thermostat is open, you will read 120 volts. If it is closed, you will read no line voltage. At this point, you are going to use an insulated jumper wire (Figure 27-9) to short the test points "T" and "H."[1] This procedure will run the motor. If the motor doesn't run, replace the ice maker module assembly. If the ice maker motor runs, replace the bimetal thermostat. If you leave the jumper wire in for half of a revolution, you will feel the mold heater heating up. This means that the mold heater is working. Now remove the jumper wire, and the water valve will be energized in the last half of the revolution.

The ejector blade shown in Figure 27-10 is rotating clockwise. This illustration shows what is about to happen as the cycle begins (from the stop position) and rotates 360 degrees back again.

Another way to test the water valve without cycling the ice maker is to place one end of the jumper wire in test point "N" and the other end in test point "V." Water will immediately enter the ice maker mold, so be ready to disconnect the jumper wire. If no water enters the mold, check the water supply, the water valve, and the connecting tubing.

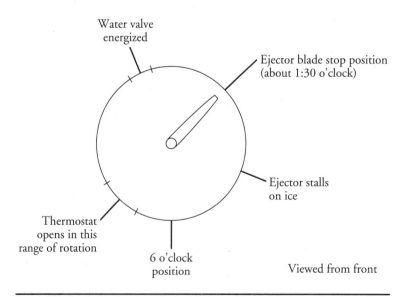

FIGURE 27-10 Indicating the ejector blade position when ice maker is cycling.

Testing the Type-1 Electronic Ice Maker

When this type of ice maker is turned on (Figure 27-11), a green LED will light on the right side of the cover. This indicator will let the consumer and the technician know that there is power (voltage) to the ice maker. The LED is also used for a fault code if something is wrong with the ice maker function. When a fault occurs, the LED will flash on and off for half a second at a time until the ice maker is repaired. To place the electronic ice maker in the service test mode, you will need to locate the technical data sheet in the refrigerator or locate the service manual for the model you are servicing. Figure 27-12 depicts an exploded view of a type-1 electronic ice maker.

Testing the Type-2 Ice Maker

In testing the operation of the type-2 ice maker head, you will have to initiate the harvest cycle. Do not remove the ice maker for this test. The shutoff arm will not operate during the

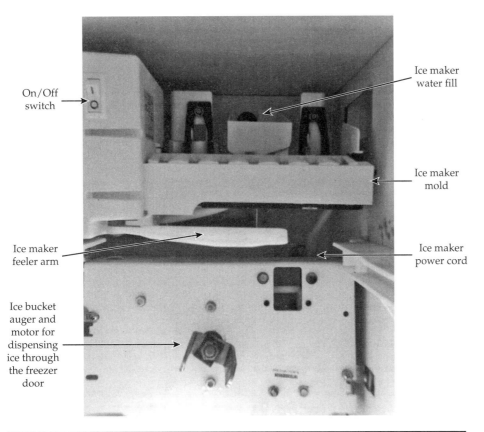

On/Off switch

Ice maker water fill

Ice maker mold

Ice maker feeler arm

Ice maker power cord

Ice bucket auger and motor for dispensing ice through the freezer door

FIGURE 27-11 Type 1 (electronic ice maker) mounted on the rear wall in a side-by-side refrigerator.

PART VI

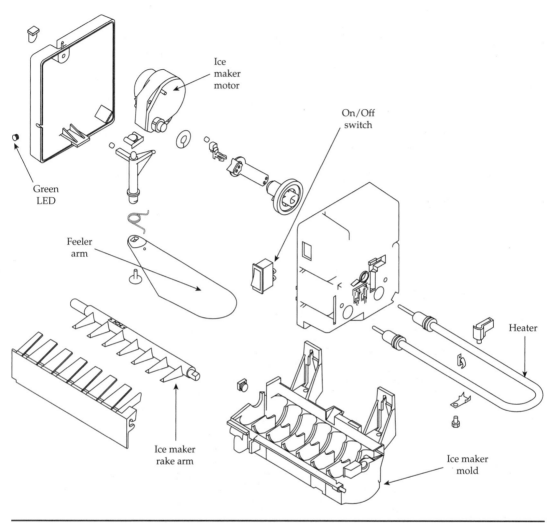

Figure 27-12   Exploded view of type-1 (electronic model) ice maker.

manual harvest. When performing the following procedures, refer to Figure 27-13. To cycle the ice maker, do the following:

1. Place the shutoff arm in the "on" position.
2. Push the switch actuator tab down, and hold it. This will activate the ice maker motor.
3. Push the tray lock tab toward the tray shaft. This will unlock the tray.
4. Twist the tray clockwise. This will start the cycle.
5. As the tray turns past 30 degrees, you can release the switch actuator tab.

FIGURE 27-13
Manually twist the ice tray clockwise. When the tray reaches 30 degrees, release the switch actuator.

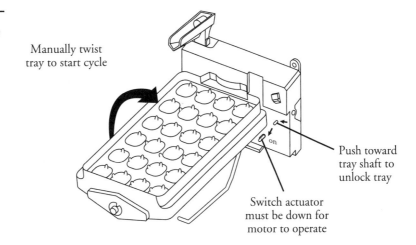

Manually twist tray to start cycle

on

Push toward tray shaft to unlock tray

Switch actuator must be down for motor to operate

6. After the harvest cycle is complete, empty the water from the ice tray. This will prevent the next automatic harvest cycle that occurs from dumping water into the ice bucket. When you perform a manual harvest, you interrupt the timing sequence only. A manual harvest will not reset the timing mechanism.

If you are unable to start the harvest cycle, check the motor shaft to see if it turns while you depress the switch actuator (Figure 27-14). Also check the temperature in the freezer. If the temperature is above 15 degrees Fahrenheit, the ice maker will not operate. If the motor shaft still does not rotate, remove the ice maker and test the unit on the work bench. Use a test cord to test the ice maker motor (Figure 27-15).[2]

FIGURE 27-14
To shut off the ice maker, lift the shutoff arm and rest it on the tab, as indicated by the broken lines.

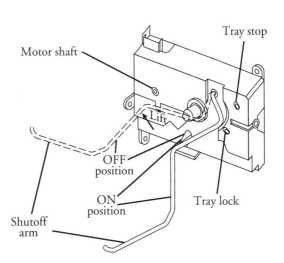

Tray stop

Motor shaft

Lift

OFF position

ON position

Tray lock

Shutoff arm

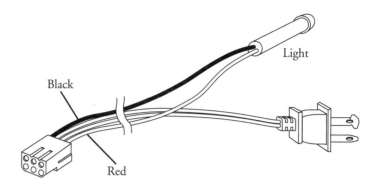

FIGURE 27-15
A test cord to test
a type-2 ice maker.

Water/Temperature Problems

The water quality can cause the ice maker to malfunction or produce poor-quality ice cubes. If the minerals in the water or sand particles become lodged in the water valve fill screen, it can restrict or stop the water from entering the ice maker mold. If sand particles bypass the fill screen in the water valve, it can possibly keep the water valve from closing completely, thus causing the water to enter the ice maker mold continuously. If this condition happens, the ice maker will produce small cubes or may even flood the ice maker compartment and freezer. Another indication of a defective water valve is that the ice maker fill tube will be completely frozen shut, possibly causing water to flow onto the floor every time the ice maker cycles. The minerals in the water can also cause lime deposits to build up on the ice maker mold. If this happens, the ice cubes will not be released easily from the mold.

Ice production can be slowed down if the temperature in the freezer compartment is above normal. To correct this problem, adjust the temperature controls to a colder position. The more the doors are opened, the colder the temperature setting must be.

Step-by-Step Troubleshooting Type-1 Ice Makers by Symptom Diagnosis

When servicing an appliance, don't overlook the simple things that might be causing the problem. Step-by-step troubleshooting by symptom diagnosis is based on diagnosing malfunctions, with possible causes arranged in categories relating to the operation of the automatic ice maker. This section is intended only to serve as a checklist to aid you in diagnosing a problem. Look at the symptom that best describes the problem you are experiencing with the automatic ice maker, and then correct the problem.

Ice Maker Producing Too Much Ice

- Check to see if the shutoff arm is not connected to the actuator in the ice maker head.
- Check the shutoff arm. Is it bent out of its original shape?
- Check the shutoff linkage in the ice maker head. Is it broken?

Ice Maker Will Not Make Ice/Low Ice Production

- Check freezer temperature.
- Is the shutoff arm in the "off" position?
- Check motor operation. Did it stall? Are the gears stripped?

- Test the voltage at the ice maker. Is the ice maker plugged into the receptacle?
- Test the thermostat for continuity. Bypass the thermostat. Will the ice maker run?
- Check the shutoff linkage in the ice maker head. Is it broken?
- Is the water supply turned on? Does water enter the ice maker mold when it cycles?
- Check the ice maker fill tube. Is it frozen?
- Check for an ice jam.
- Check for defective wiring.

Ice Cubes Are Too Small

- Check the ice maker. Is it level?
- Check the fill tube. Is it frozen?
- Check the water supply.
- Check the ice maker water valve.
- Check the water pressure. Is it between 20 and 120 psi?
- Check for a self-piercing saddle valve. Mineral deposits will restrict the opening.
- Cycle the ice maker, and catch the fill water in a glass. Measure the amount of water. Are there at least 140 cc of water in the glass?
- Test for openings in the mold heater.
- Check the ice maker thermostat. There might be insufficient thermal bond between the thermostat and the ice maker mold.

Ice Maker Is Producing Hollow Ice Cubes

- Cycle the ice maker, and catch the fill water in a glass. Measure the amount of water. Are there at least 140 cc of water in the glass?
- Check for improper airflow in the freezer compartment. Direct the airflow away from the ice maker thermostat.
- Check the ice maker thermostat.
- Check the temperature in the freezer compartment.

Ice Maker Is Flooding the Freezer Compartment or Ice Bucket

- Check the thermostat.
- Check for an ice jam when the ice maker is in the fill position.
- Check for a leaky water valve.
- Check for proper ice maker water fill. Too much water will spill over the mold, causing the ice in the bucket to freeze together.
- Check the ejector blade position. If the blade is in the 12 o'clock position, the ice maker motor has stalled.
- Check the ice maker module for contamination and/or burned switch contacts.
- Check the linkage for proper operation.

PART VI

- Is the refrigerator level? Is the ice maker level?
- Check water pressure. Ice makers fill according to time, not volume. Water pressure should be between 20 and 120 psi.
- Is the fill tube located in the fill cup?

Step-by-Step Troubleshooting Type-2 Ice Makers by Symptom Diagnosis

When servicing an appliance, don't overlook the simple things that might be causing the problem. Step-by-step troubleshooting by symptom diagnosis is based upon diagnosing malfunctions, with possible causes arranged into categories relating to the operation of the automatic ice maker. This section is intended only to serve as a checklist to aid you in diagnosing a problem. Look at the symptom that best describes the problem you are experiencing with the automatic ice maker, and then proceed to correct the problem.

Ice Maker Will Not Run

- Test the freezer temperature.
- Check the shutoff arm. Is it in the "off" position?
- Run an ice maker test. Does the ice maker motor run?
- Check for defective wiring.
- Is the ice maker plugged into the receptacle?
- Test for voltage at the ice maker receptacle.

Ice Cubes Are Stuck Together

- Test temperature of freezer.
- Check for proper fill.
- Check ice tray for mineral deposits. Mineral deposits will cause the ice to stick to the tray, which can cause the ice cubes to stick together on the next cycle.

Ice Maker Spills Water from the Tray

- Check the ice tray. When the harvest cycle is complete, the ice cube tray should return to its starting position and the tray should be level.
- Check the inlet water fill tube. Be sure that the fill tube and fill trough fit together properly.
- Check for a leaking water inlet valve.

Water Will Not Enter Ice Tray

- Check for proper water supply to ice maker.
- Check water valve strainer for restrictions.
- Check the water valve.
- Check for proper water pressure.
- Check wiring circuit.

Ice Cubes Too Small, or Some of the Ice Cube Compartments Are Empty

- Check the ice maker. Is it level?
- Check the fill tube. Is it frozen?
- Check the water supply.
- Check the ice maker water valve.
- Check the water pressure. Is it between 20 and 120 psi?
- Check for a self-piercing saddle valve. Mineral deposits will restrict the opening.
- Cycle the ice maker and catch the fill water in a glass. Measure the amount of water. Are there at least 200 cc of water in the glass? The fill time will be between 12 and 13 seconds.

Diagnostic Chart

The diagnostic chart (Figure 27-16 and 27-17) will help you to pinpoint the likely cause of the problem with an ice maker.

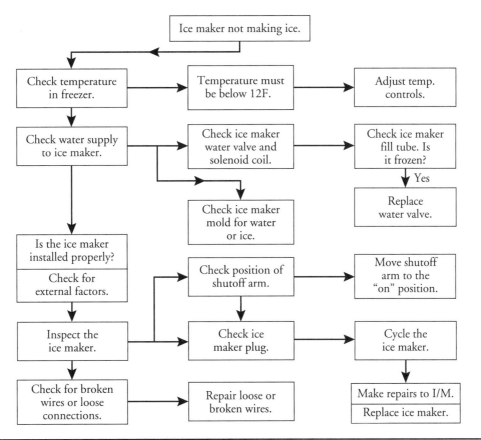

FIGURE 27-16   The diagnostic flowchart: Ice maker not making ice.

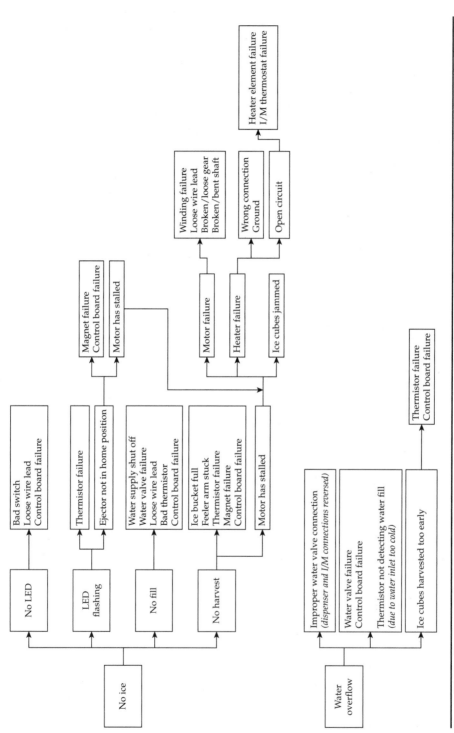

Figure 27-17   The diagnostic flowchart: Type-1 electronic ice maker.

Repair Procedures

Each repair procedure is a complete inspection and repair process for a single ice cube maker component, containing the information you need to test and replace components.

Any person who cannot use basic tools should *not* attempt to install, maintain, or repair any ice makers. Any improper installation, preventative maintenance, or repairs will create a risk of personal injury, as well as property damage. Call the service manager if installation, preventative maintenance, or the repair procedure is not fully understood.

Type-1 Ice Maker: Module

The ice maker module runs all of the operations of the ice maker. It is located on the end of the ice maker (Figure 27-8). All diagnostic tests can be performed on the ice maker module.

The typical complaints associated with failure of the ice maker module are:

- Ice maker is not making any ice.
- Ice maker stalls in the middle of cycling.

To handle these problems, perform the following steps:

1. **Verify the complaint.** Verify the complaint by checking the ice maker and the temperature in the freezer. It will be helpful if you can locate the actual service manual for the ice maker model you are working on to properly diagnose the ice maker. The service manual will assist you in properly placing the ice maker in the service test mode for testing the ice maker functions.

2. **Check for external factors.** You must check for external factors not associated with the ice maker. For example, is the ice maker installed properly? The voltage at the wall receptacle and the ice maker receptacle is between 108 volts and 132 volts during a load on the circuit. Do you have the correct polarity? (See Chapter 6.)

3. **Disconnect the electricity.** Before working on the ice maker, disconnect the electricity to the refrigerator/freezer. This can be done by pulling the plug from the receptacle. Or disconnect the electricity at the fuse panel or at the circuit breaker panel. Turn off the electricity.

WARNING *Some diagnostic tests will require you to test the components with the power turned on. When you disassemble the ice maker cover, you can position the ice maker in such a way that the wiring will not make contact with metal. This will allow you to test the ice maker components without electrical mishaps.*

4. **Gain access to the ice maker.** Open the freezer door to access the ice maker. Remove the screws that secure the ice maker to the wall of the freezer (Figure 27-18). Next, unplug the ice maker from the receptacle and remove the ice maker (Figure 27-19a). Now depress the retaining tab, and pull the wiring harness from the ice maker head (Figure 27-19b).

5. **Disassemble and remove the ice maker module.** After testing the ice maker and determining that you must replace the ice maker module, it is time to disassemble it.

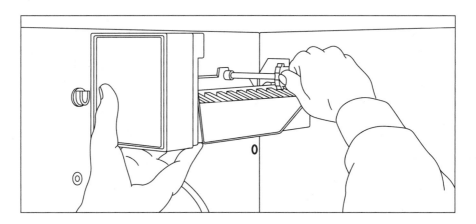

FIGURE 27-18   Support the ice maker when removing the screws.

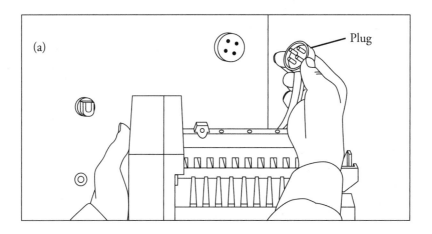

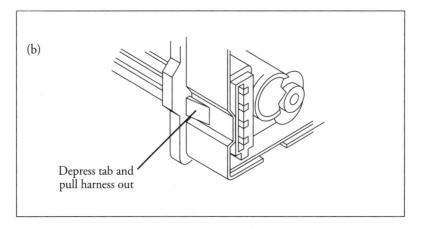

FIGURE 27-19   (a) Pull the ice maker plug out of the receptacle. (b) Remove the wire harness by depressing the tab, and then pull it.

FIGURE 27-20
Removing the shutoff
arm by prying it out of
the slot. When
installing the shutoff
arm, be sure that the
end is pressed in all
the way.

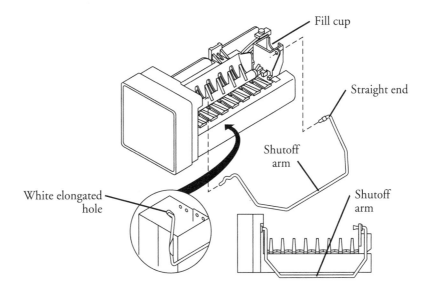

Fill cup

Straight end

Shutoff
arm

Shutoff
arm

White elongated
hole

Insert a flat-blade screwdriver between the shutoff arm and the white elongated
hole on the ice maker head (Figure 27-20). Remove the shutoff arm by prying the
arm out of the ice maker head. Pry off the ice maker cover (Figure 27-21) with a coin
or a screwdriver. This will expose the ice maker module. To remove the ice maker
module, remove the three screws (Figure 27-22), and pull the module out of the ice
maker head. Inspect the module linkage and switch contacts (Figure 27-23).

NOTE *The ground terminal is slightly longer than the other three terminals. The ground connection
is either made first or breaks last when the wiring harness is removed or plugged in.*

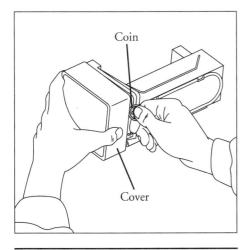

Coin

Cover

FIGURE 27-21 Gain access to the ice maker
module.

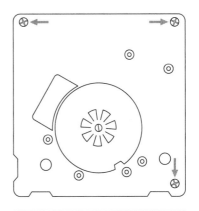

FIGURE 27-22 Removing the three
screws to remove the module.

PART VI

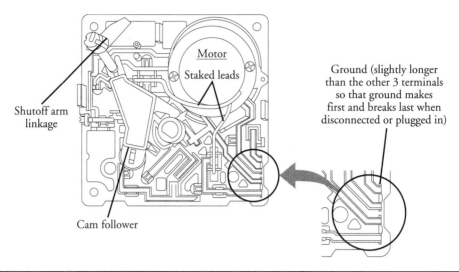

FIGURE 27-23   Inspecting the shutoff arm linkage, cam follower, and switch contacts.

6. **Install a new ice maker module.** To install the new ice maker module, just reverse the order of disassembly, and reassemble. To reinstall the shutoff arm on the ice maker (see Figure 27-20), you must first insert the straight end into the round hole in the fill cup. Be sure the flat side on the arm goes through the fill cup. This will prevent the shutoff arm from coming out of the fill cup hole. Next, insert the other end of the shutoff arm into the white elongated hole in the ice maker housing. Do not install the arm into any of the round holes in the ice maker housing. Push on the arm so that it will be completely in place and even with the ice maker surface. Reinstall the ice maker in the freezer, and test it.

Type-1 Ice Maker: Thermostat

The ice maker thermostat is located in the ice maker housing, and it is secured to the ice maker mold with two clips. Make sure the freezer temperature is colder than 12 degrees Fahrenheit. Remember that the ice maker thermostat will close between 15 and 17 degrees Fahrenheit.

The typical complaints associated with failure of the ice maker thermostat are:

- Ice maker is not making any ice cubes.
- Ice maker is producing hollow ice cubes.

To handle these problems, perform the following steps:

1. **Verify the complaint.** Verify the complaint by checking the ice maker and the temperature in the freezer. It will be helpful if you can locate the actual service manual for the ice maker model you are working on to properly diagnose the ice maker.

The service manual will assist you in properly placing the ice maker in the service test mode for testing the ice maker functions.

2. **Check for external factors.** You must check for external factors not associated with the ice maker. Is the ice maker installed properly? The voltage at the wall receptacle and the ice maker receptacle is between 108 volts and 132 volts during a load on the circuit. Do you have the correct polarity? (See Chapter 6.)

3. **Disconnect the electricity.** Before working on the ice maker, disconnect the electricity to the refrigerator/freezer. This can be done by pulling the plug from the receptacle. Or disconnect the electricity at the fuse panel or at the circuit breaker panel. Turn off the electricity.

WARNING *Some diagnostic tests will require you to test the components with the power turned on. When you disassemble the ice maker cover, you can position the ice maker in such a way that the wiring will not make contact with metal. This will allow you to test the ice maker components without electrical mishaps.*

4. **Gain access to the ice maker thermostat.** Open the freezer door to access the ice maker. Remove the screws that secure the ice maker to the wall of the freezer (see Figure 27-18). Next, unplug the ice maker from the receptacle, and remove the ice maker (see Figure 27-19a). Now press the retaining tab, and pull the wiring harness from the ice maker head (see Figure 27-19b). Insert a flat-blade screwdriver between the shutoff arm and the white elongated hole on the ice maker head (see Figure 27-20). Remove the shutoff arm by prying the arm out of the ice maker head. Pry off the ice maker cover (see Figure 27-21) with a coin or a screwdriver. This will expose the ice maker module. To remove the ice maker module, remove the three screws (see Figure 27-22), and pull the module out of the ice maker head. Inspect the linkage, the cam follower, and the switch contacts. Next, remove the two screws that secure the ice maker head to the mold (Figure 27-24). Separate the ice maker head from the mold assembly, and you will see the thermostat on the mold side of the ice maker head.

5. **Remove the thermostat.** To remove the thermostat, use needle nose pliers to pull out the retaining clips (Figure 27-25). Remove the thermostat.

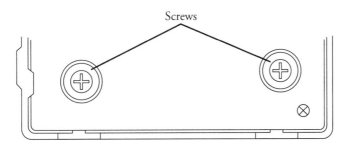

Screws

FIGURE 27-24 Removing the two screws.

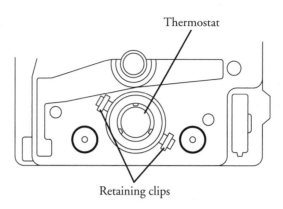

FIGURE 27-25
Removing the
retaining clips with a
pair of needle nose
pliers.

Thermostat

Retaining clips

6. **Install a new thermostat.** To install the new thermostat, just reverse the order of disassembly, and reassemble the ice maker. Be sure that you properly index the pins on the thermostat. Before you assemble the ice maker head to the mold assembly, you must apply new thermal bonding material to the thermostat. This will allow the thermostat to make better contact with the ice maker mold. Reinstall the ice maker in the freezer, and test.

Type-1 Ice Maker: Mold and Heater Assembly

The ice maker mold and heater assembly is attached to the ice maker head. This component is replaced as one unit.

The typical complaints associated with failure of the ice maker mold and heater assembly are:

- Ice maker is not making any ice.
- Ice will not come out of the mold.
- There are black pieces of mold coating material in the ice cubes.

To handle these problems, perform the following steps:

1. **Verify the complaint.** Verify the complaint by checking the ice maker and the temperature in the freezer. It will be helpful if you can locate the actual service manual for the ice maker model you are working on to properly diagnose the ice maker. The service manual will assist you in properly placing the ice maker in the service test mode for testing the ice maker functions.

2. **Check for external factors.** You must check for external factors not associated with the ice maker. Is the ice maker installed properly? The voltage at the wall receptacle and the ice maker receptacle is between 108 volts and 132 volts during a load on the circuit. Do you have the correct polarity? (See Chapter 6.)

3. **Disconnect the electricity.** Before working on the ice maker, disconnect the electricity to the refrigerator/freezer. This can be done by pulling the plug from the receptacle. Or disconnect the electricity at the fuse panel or circuit breaker panel. Turn off the electricity.

WARNING *Some diagnostic tests will require you to test the components with the power turned on. When you disassemble the ice maker cover, you can position the ice maker in such a way that the wiring will not make contact with metal. This will allow you to test the ice maker components without electrical mishaps.*

4. **Gain access to the ice maker mold assembly.** Open the freezer door to access the ice maker. Remove the screws that secure the ice maker to the wall of the freezer (see Figure 27-18). Next, unplug the ice maker from the receptacle, and remove the ice maker (see Figure 27-19a). Now press the retaining tab, and pull the wiring harness from the ice maker head (see Figure 27-19b).

   Insert a flat-blade screwdriver between the shutoff arm and the white elongated hole on the ice maker head (see Figure 27-20). Remove the shutoff arm by prying it out of the ice maker head. Pry off the ice maker cover (see Figure 27-21) with a coin or screwdriver. Insert a Phillips screwdriver into the access ports in the module (Figure 27-26) to loosen the screws. Separate the ice maker head assembly from the mold assembly. Also remove the ejector, the fill cup, the stripper, and the shutoff arm (see Figure 27-8). Do not remove the heater; the heater and mold come as a complete assembly.

5. **Install a new ice maker mold assembly.** To install a new ice maker mold assembly, just reverse the order of disassembly, and reassemble. When assembling the ice maker, apply a thin film of silicone grease to the end of the ejector that goes into the fill cup. This will prevent the ejector from freezing to the fill cup. Also apply a thin film of silicone grease to the other end of the ejector. Before installing the stripper, apply a heavy film of silicone grease to the top surface of the mold that is covered by the stripper. This will prevent the water from wicking over the ice maker mold every time it cycles. Before you assemble the ice maker head to the mold assembly, you must apply new thermal bonding material to the thermostat. This will allow the thermostat to make better contact with the ice maker mold. Reinstall the ice maker in the freezer, and test it.

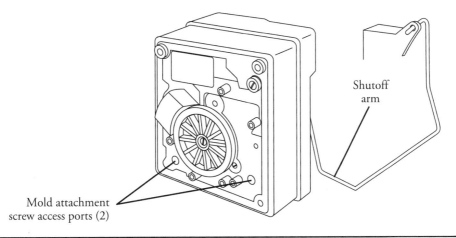

Shutoff arm

Mold attachment screw access ports (2)

FIGURE 27-26   The entire ice maker head can be removed by loosening the two screws through the access ports.

Water Valve

The water valve controls the flow of water into the ice maker, and is solenoid-operated and located in the rear of the refrigerator/freezer. When it is energized, water in the supply line will pass through the valve body and into the ice maker.

The typical complaints associated with failure of the water valve are:

- Ice maker is not making any ice.
- Ice maker fill tube is frozen.
- Ice maker mold fills with very little water.

To handle these problems, perform the following steps:

1. **Verify the complaint.** Verify the complaint by checking the ice maker and the temperature in the freezer. It will be helpful if you can locate the actual service manual for the ice maker model you are working on to properly diagnose the ice maker. The service manual will assist you in properly placing the ice maker in the service test mode for testing the ice maker functions.

2. **Check for external factors.** You must check for external factors not associated with the ice maker. Is the ice maker installed properly? Check the water supply to the water valve. The voltage at the wall receptacle and the ice maker receptacle is between 108 volts and 132 volts during a load on the circuit. Do you have the correct polarity? (See Chapter 6.)

3. **Disconnect the electricity.** Before working on the water valve, disconnect the electricity to the refrigerator/freezer. This can be done by pulling the plug from the receptacle. Or disconnect the electricity at the fuse panel or circuit breaker panel. Turn off the electricity.

WARNING *Some diagnostic tests will require you to test the components with the power turned on. When you disassemble the rear cover, you can position the ice maker in such a way that the wiring will not make contact with metal. This will allow you to test the water valve without electrical mishaps.*

4. **Gain access to the water valve.** To access the water valve, pull the refrigerator/freezer away from the wall. Remove the back access panel. Shut off the water supply to the ice maker.

5. **Remove the water valve.** Remove the water inlet tube from the water valve. Then remove the screws from the water valve bracket that secure the valve to the cabinet (Figure 27-27). Next, remove the ice maker fill line. Finally, disconnect the wiring harness from the solenoid coil of the valve.

6. **Test the water valve.** Using your ohmmeter, set the scale on R × 10, and place the probes on the solenoid coil terminals (Figure 27-28). The meter should show resistance between 200 and 500 ohms. If not, replace the water valve. If you determine that the water valve is good but there is little water flow through the valve, inspect the inlet screen. If this screen is filled with debris, it must be cleaned

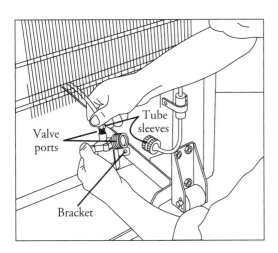

FIGURE 27-27   Disconnect the water inlet line and the outlet line. Disconnect the wiring harness from the solenoid coil.

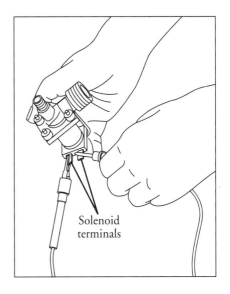

FIGURE 27-28   Testing the solenoid coil for resistance.

out. To accomplish this, use a small flat-blade screwdriver and pry out the screen (Figure 27-29). Then wash out the screen, making sure that all of the debris is removed. After cleaning out the debris, reinstall the screen, and test the water valve.

7. **Install a new water valve.** To install the new water valve, just reverse the disassembly procedure, and reassemble. Also reconnect the ground wire from the water valve bracket to the refrigerator/freezer cabinet. Secure the water supply line to the cabinet. This will prevent the supply line from leaking when the refrigerator/freezer is pushed back against the wall. Check for water leaks before you push the refrigerator/freezer back against the wall. Also inspect the fill tubing for any cracks, etc. (Figure 27-30).

Water Fill Valve Adjustment

The average water fill for a type-1 ice maker is 140 cc. This type of ice maker will produce eight cubes per harvest. If the technician needs to adjust the fill adjustment screw, which is located on the side of the ice maker head (Figure 27-31), turning the fill adjustment screw clockwise decreases the fill amount and decreases the cycle time slightly; turning the fill adjustment screw counterclockwise increases the fill amount and increases the cycle time slightly.

By adjusting the water fill to the ice maker you can increase or decrease the amount of water in the mold by:

- Quarter-turn of the adjustment screw = 2.5 cc or .08 ounces
- Half-turn of the adjustment screw = 5 cc or .16 ounces
- Three-quarter-turn of the adjustment screw = 7.5 cc or .24 ounces
- One turn of the adjustment screw = 10 cc or .36 ounces

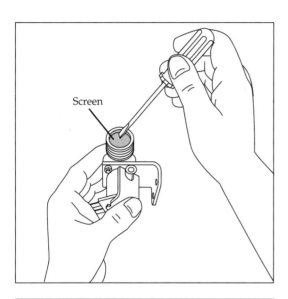

FIGURE 27-29   Removing the inlet screen. Clean it out with warm water and soap.

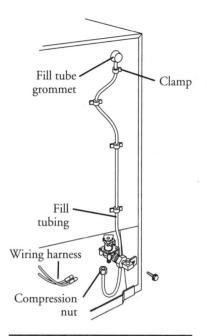

FIGURE 27-30   Checking for defective tubing. Check the clamp and compression nuts for leaks.

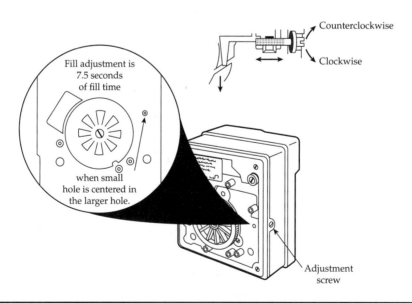

FIGURE 27-31   The illustration depicts where the water fill adjustment screw is located in the ice maker. The factory presets the adjustment. Before making adjustments check the water supply and type of shutoff valve first.

Ice maker cycle times will vary depending on the type of refrigerator or freezer in the consumer's home (side-by-side versus top-mount or bottom mount). Cycle times vary between 2 and 2½ hours and can produce up to 96 ice cubes within a 24-hour period.

Type-2 Ice Maker: Head and Tray

The typical complaints associated with failure of the type-2 ice maker head are:

- Ice maker is not making any ice.
- Ice will not come out of the mold.

To handle these problems, perform the following steps:

1. **Verify the complaint.** Verify the complaint by checking the ice maker and the temperature in the freezer. It will be helpful if you can locate the actual service manual for the ice maker model you are working on to properly diagnose the ice maker. The service manual will assist you in properly placing the ice maker in the service test mode for testing the ice maker functions.

2. **Check for external factors.** You must check for external factors not associated with the ice maker. Is the ice maker installed properly? The voltage at the wall receptacle and the ice maker receptacle is between 108 volts and 132 volts during a load on the circuit. Do you have the correct polarity? (See Chapter 6.)

3. **Disconnect the electricity.** Before working on the ice maker, disconnect the electricity to the refrigerator/freezer. This can be done by pulling the plug from the receptacle. Another way to disconnect the electricity is at the fuse panel or circuit breaker panel. Turn off the electricity.

WARNING *Some diagnostic tests will require you to test the components with the power turned on. When you disassemble the ice maker, you can position the ice maker in such a way that the wiring will not make contact with metal. This will allow you to test the ice maker without electrical mishaps.*

4. **Gain access to the ice maker.** Open the freezer door to access the ice maker. Remove the screws that secure the ice maker to the wall of the freezer. Next, unplug the ice maker from the receptacle and remove the ice maker.

5. **Disassemble and remove the ice maker head assembly.** After testing the ice maker and determining that you must replace the ice maker head assembly, it's time to disassemble it. First, remove the retaining clip from the end of the tray, and then slide the tray off the ice maker drive shaft (Figure 27-32). Next, remove the fill cup and screw. On the back of the ice maker head is a clamp that holds the tray drive shaft in place; remove it. Pull the shaft out of the ice maker head. Then place a flat-blade screwdriver behind the shutoff arm, and pry it loose (Figure 27-33).

6. **Install a new ice maker head.** To install the new ice maker head, just reverse the order of disassembly, and reassemble. When replacing a defective ice maker head, only replace it with an exact replacement with the same amount of cycle operation time. Reinstall the ice maker in the freezer, and test.

FIGURE 27-32
FIGURE 27-32
Always reinstall the ice cube tray in the same position as when it was removed. If the ice tray is installed upside down, water will flood the ice bucket every time the ice maker cycles.

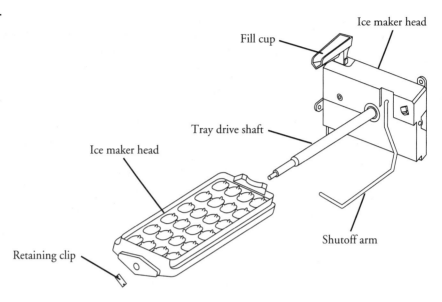

Type-2 Ice Maker Tray

The typical complaints associated with failure of the type-2 ice maker tray are:

- Ice maker is not making any ice.
- Ice will not come out of the mold.
- Water in the tray freezes into a solid block of ice.

FIGURE 27-33
Gently pry off the shutoff arm.

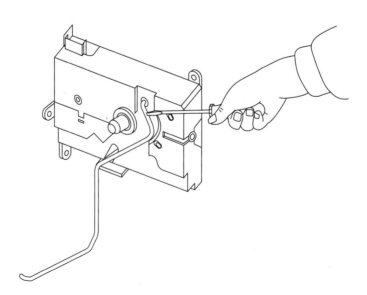

To handle these problems, perform the following steps:

1. **Verify the complaint.** Verify the complaint by checking the ice maker and the temperature in the freezer. It will be helpful if you can locate the actual service manual for the ice maker model you are working on to properly diagnose the ice maker. The service manual will assist you in properly placing the ice maker in the service test mode for testing the ice maker functions.

2. **Check for external factors.** You must check for external factors not associated with the ice maker. Is the ice maker installed properly? The voltage at the wall receptacle and the ice maker receptacle is between 108 volts and 132 volts during a load on the circuit. Do you have the correct polarity? (See Chapter 6.)

3. **Remove the ice maker tray.** To remove the ice maker tray, remove the retaining clip from the end of the tray, and then slide the tray off the ice maker drive shaft (see Figure 27-32).

4. **Install a new ice tray.** To install the new ice tray, just reverse the disassembly procedure, and reassemble. Always reinstall the tray in the same position as that from which it was removed for proper operation. Test the ice maker. Let it run through a complete cycle and harvest. Then inspect the ice maker tray.

Type-1 Ice Maker Cycle Chart and Wiring Schematic

Figure 27-34 illustrates the complete cycle of the type-1 ice maker. A wiring schematic for a type-1 ice maker (electronic model) is depicted in Figure 27-35.

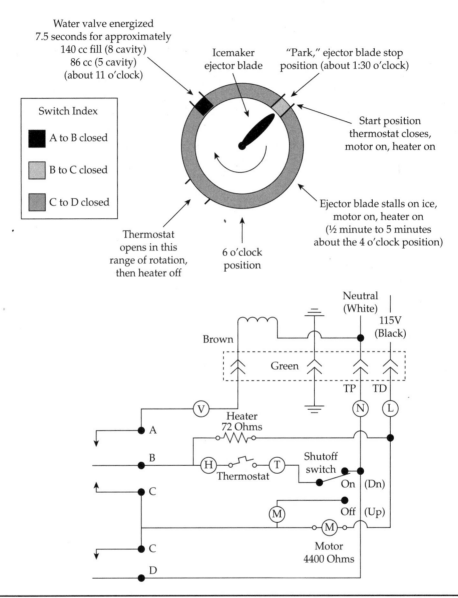

Water valve energized
7.5 seconds for approximately
140 cc fill (8 cavity)
86 cc (5 cavity)
(about 11 o'clock)

Icemaker
ejector blade

"Park," ejector blade stop
position (about 1:30 o'clock)

Start position
thermostat closes,
motor on, heater on

Switch Index

A to B closed

B to C closed

C to D closed

Thermostat
opens in this
range of rotation,
then heater off

6 o'clock
position

Ejector blade stalls on ice,
motor on, heater on
(½ minute to 5 minutes
about the 4 o'clock position)

Neutral
(White)

115V
(Black)

Brown

Green

TP   TD

V

N   L

Heater
72 Ohms

A

Shutoff
switch

B

H   T

On   (Dn)

Thermostat

C

M

Off   (Up)

M

C

Motor
4400 Ohms

D

FIGURE 27-34 One operation cycle is equal to one revolution of the ejector arm. (*continued*)

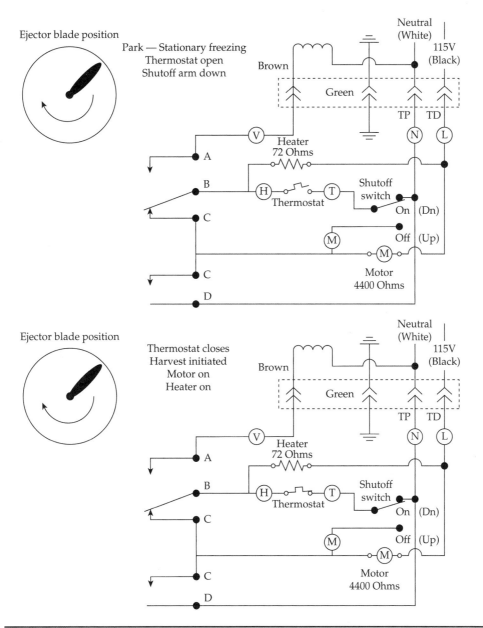

FIGURE 27-34 One operation cycle is equal to one revolution of the ejector arm. (*continued*)

PART VI

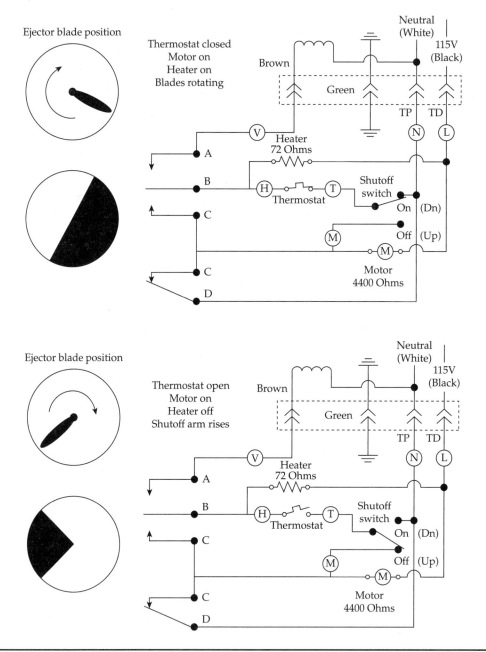

FIGURE 27-34   One operation cycle is equal to one revolution of the ejector arm. (*continued*)

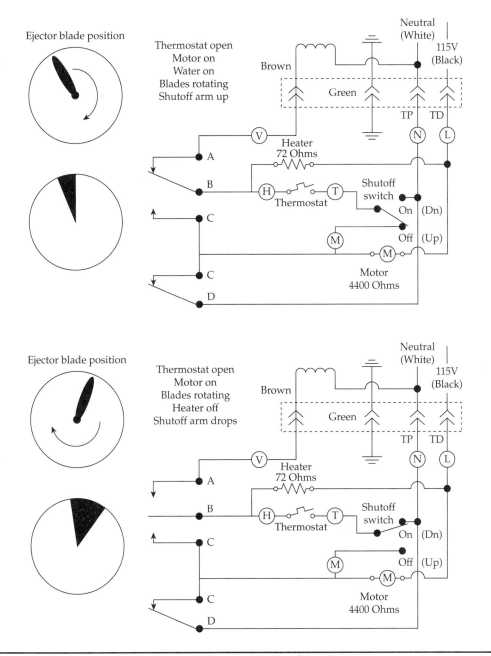

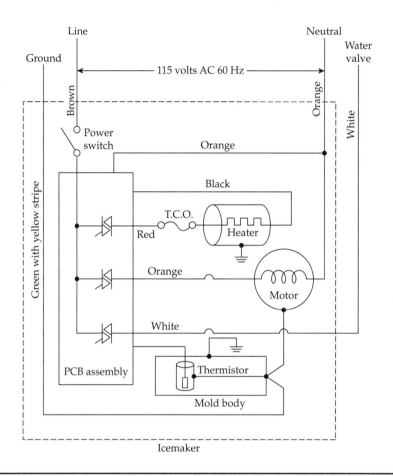

FIGURE 27-35   Wiring schematic of a type-1 ice maker (electronic model).

Endnotes

1. The jumper wire should be made of 14-gauge wire and should be approximately 8 to 10 inches long, with 5/8 inch of the insulation removed from both ends. Do not short any contacts other than those specified in this test procedure. Why? Because you can damage the ice maker, injure yourself, or both.

2. You can purchase this ice maker test cord from your parts distributor.

Residential Under-the-Counter
Ice Cube Makers

There are three operating systems in the self-contained ice cube maker:

- **The refrigeration system**   The refrigeration sealed system in the Type 1- self-contained ice cube maker operates similar to a refrigerator/freezer sealed system. The older models used R-12 and the newer models use R-134a. The ice cube maker has a hot gas solenoid valve to harvest the ice; it allows the high pressure refrigerant gas to bypass the condenser and flow through the condenser accumulator tube (Figure 28-3). When the refrigerant enters the condenser accumulator, the hot gas enters into the evaporator plate and it will evenly heat the evaporator plate so that the ice slab will release from the evaporator plate quickly and evenly.

- **The water system**   The water system provides fresh water to the ice cube maker for ice production. The water in the reservoir will recirculate as ice is produced. The water system will also flush away any impurities, minerals, and contaminates at the end of the cycle. The water system will recirculate the water in the clean cycle.

- **The electrical system**   The ice cube maker's electrical system provides voltage for the refrigeration and the water systems to operate, and controls the operating and cleaning cycles.

Safety First

Any person who cannot use basic tools or follow written instructions should *not* attempt to install, maintain, or repair any residential under-the-counter ice cube makers.

Any improper installation, preventive maintenance, or repairs could create a risk of personal injury or property damage.

If you do not fully understand the installation, preventive maintenance, or repair procedures in this chapter, or if you doubt your ability to complete the task on your automatic ice maker, please call your service manager.

The following precautions should also be followed:

- Never place fingers or hands on the automatic ice maker mechanism while the ice cube maker is plugged in.
- Disconnect the electrical supply to the ice cube maker before servicing the ice cube maker.
- Be careful of any sharp edges on the ice cube maker, which might result in personal injury.
- Do not attempt to operate the ice cube maker unless it has been properly reinstalled, including the grounding and electrical connections.

Before continuing, take a moment to refresh your memory on the safety procedures in Chapter 2.

Automatic Ice Cube Makers in General

Much of the troubleshooting information in this chapter covers ice cube makers in general, rather than specific models, in order to present a broad overview of service techniques. The illustrations that are used in this chapter are for demonstration purposes only, to clarify the description of how to service ice cube makers. They in no way reflect a particular brand's reliability.

Type 1-Self-Contained Ice Cube Maker

This type of ice cube maker is a freestanding, self-contained refrigeration appliance that produces a slab of ice, which is then cut into ice cubes (Figure 28-1). The production of ice cubes is all done automatically, and the entire mechanism is stored within the ice cube maker cabinet. The self-contained ice cube maker can also be installed under the counter. The thickness of the ice cubes can be adjusted by the thickness control, located on the control panel. This ice cube maker can produce up to 50 pounds of ice cubes in a 24-hour period. The amount of ice cubes will vary, depending on where it is installed, the room temperature, and the supply water temperature.

Principles of Operation

In the freeze cycle, water flows constantly, and it is recirculated over the evaporator freeze plate until a slab of ice is formed (Figure 28-2). When the ice slab reaches a predetermined thickness, the evaporator freeze plate temperature is sensed by the thermostat and the freeze cycle is terminated. At that point, the defrost cycle will begin to release the ice slab from the evaporator freeze plate. The ice slab will slide down onto the cutter grid, which cuts the ice slab into ice cubes. At the same time, the ice maker has automatically switched back into the freeze cycle.

When the defrost cycle begins, the remaining water in the water tank reservoir is discarded through the overflow tube. At that point, fresh water will enter through the water valve and go into the water tank reservoir.

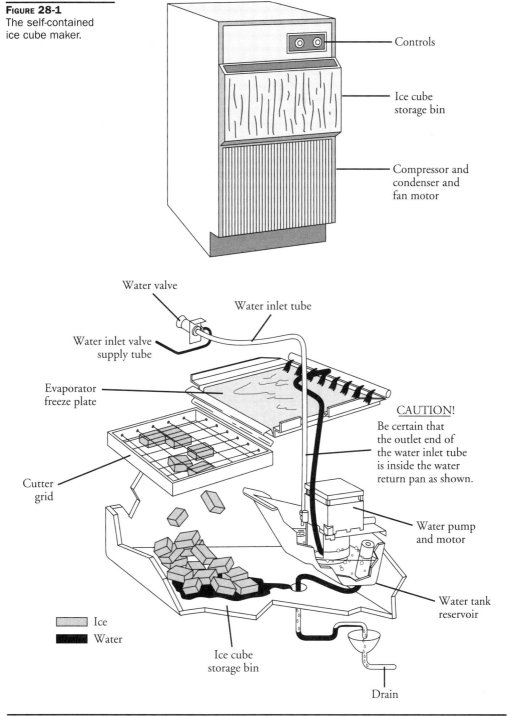

FIGURE 28-1
The self-contained ice cube maker.

Controls

Ice cube storage bin

Compressor and condenser and fan motor

Water valve

Water inlet tube

Water inlet valve supply tube

Evaporator freeze plate

CAUTION!
Be certain that the outlet end of the water inlet tube is inside the water return pan as shown.

Cutter grid

Water pump and motor

Ice

Water

Water tank reservoir

Ice cube storage bin

Drain

FIGURE 28-2 A pictorial view of the water system.

The refrigeration cycle (Figure 28-3) in this type of ice maker is similar to that in a conventional refrigerator/freezer. The compressor pumps the refrigerant into the condenser coil, which is cooled by a fan and motor. The refrigerant leaves the condenser coil as a high-pressure liquid, passes through the dryer, and enters the capillary tube. The refrigerant is next metered through the capillary tube and then enters the evaporator freeze plate. The refrigerant gas then leaves the evaporator freeze plate and returns to the compressor.

When the ice maker goes into the defrost cycle, it energizes the hot gas solenoid, which reverses the refrigeration cycle, during which the condenser fan motor and water pump will stop. The hot gas passes through the evaporator freeze plate, heating it up enough to release the ice slab. The thermostat senses the temperature of the evaporator freeze plate again and activates the freeze cycle. The hot gas solenoid valve will then close, the water valve will close, the condenser fan motor will start, the water pump will start, and the freeze cycle will begin to manufacture a new slab of ice.

FIGURE 28-3
A pictorial view of a refrigeration system.

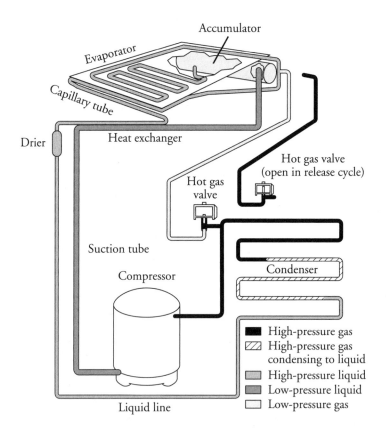

Step-by-Step Troubleshooting by Symptom Diagnosis

When servicing the type-1-self-contained ice cube maker, don't overlook the simple things that might be causing the problem. Step-by-step troubleshooting by symptom diagnosis is based on diagnosing malfunctions, with possible causes arranged into categories relating to the operation of the automatic ice maker. This section is intended only as a checklist to aid you in diagnosing a problem. Look at the symptom that best describes the problem you are experiencing with the automatic ice maker, and then correct the problem.

Compressor Will Not Run and There Is No Ice in the Storage Bin

- Is the ice cube maker located in an area where the temperature is below 55 degrees Fahrenheit?
- Test for proper voltage supply.
- Check for loose or broken wires.
- Test the compressor, relay, and overload protector.
- Check the controls for the proper setting.
- Test the bin thermostat for continuity. If contacts are open, replace the thermostat.
- Test the compressor.

Compressor Runs, but There Is No Ice in Storage Bin

- Check water supply.
- Check water valve.
- Check evaporator thermostat.
- Check the hot gas solenoid. It might be stuck "open."
- Check for sealed-system problems.
- Check for excessive use of ice cubes.
- Test cutter grid.
- Check wiring against wiring diagram.
- Is the water inlet tube from the water valve inserted in the return trough?
- Check condenser fan motor.

Ice Storage Bin Is Full of Ice and the Compressor Runs Continuously

- Check the calibration on the bin thermostat.
- Test the bin thermostat for continuity. Are the contacts stuck shut?
- Check wiring against the wiring diagram.

Low Ice Production

- Is the ice cube maker located in an area where the temperature is below 55 degrees Fahrenheit?
- Inspect the storage bin. Is water falling on the ice cubes?

- Check the calibration on the bin thermostat.
- Check the thickness control. Ice cubes produced should be between 1/2 and 5/8 inch thick.
- Check the hot gas solenoid. It might be stuck partially open.
- Check for sealed-system problems.
- Check the water supply in the reservoir. There might not be enough water circulating over the evaporator freeze plate.

Excessive Water Dripping on the Ice Cubes

- Is the water tank overflowing? Check for a blocked overflow tube.
- Is the water trough installed properly?
- Is the water inlet tube from the water valve inserted in the return trough?
- Check cutter grid for ice jam.
- Check the water deflector position.
- Check water valve for leaks.

Ice Cubes Are Too Thin

- Check thickness control setting.
- Check to see if there is enough water being circulated over the evaporator freeze plate.
- Check for restrictions in the water system.
- Check the water pump, motor, and the distributor tube.
- Check the thermostat calibration.

Ice Cubes Are Too Thick

- Check thickness control setting.
- Check the thermostat calibration.

The Condenser Fan Will Not Run During the Freeze Cycle

- Check the fan blades for binding on the shroud.
- Test the condenser motor for continuity.
- Check for open circuits against the wiring diagram.
- Check for a defective evaporator thermostat.

Water Pump Will Not Run

- Check the pump for binding in the housing.
- Check for open circuits against the wiring diagram.
- Test the pump motor for continuity.
- Check for a defective evaporator thermostat.

Water Tank Is Empty

- Water does not enter tank until the first defrost cycle is initiated.
- Check for an open circuit to the water valve solenoid.
- Check the water line for complete restriction.
- Check for a defective evaporator thermostat.
- Check the water valve; it might be stuck shut.
- Test the water valve solenoid for continuity.
- Check the water inlet tube from the water valve. It might not be directing the water into the tank.
- Check for a clogged water inlet screen in the water valve.

Treating the Water

In the freeze cycle, as the water passes over the evaporator freeze plate, the impurities in the water are rejected and only the pure water will stick to the plate. The more dissolved solids that are present in the water, the longer the freezing cycle. Bicarbonates, which are found in the water, are the most troublesome of all impurities. These impurities can cause:

- Scaling on the evaporator freeze plate
- Clogging of the water distributor head
- The water valve and many other parts in the water system to clog up

If the impurities become too concentrated in the water system, they can cause cloudy cubes and/or mushy ice.

All of the water system parts that come in direct contact with the water might become corroded if the water supply is high in acidity. The water might have to be treated in order to overcome problems with the mineral content. The most economical way to treat the water supply is with a polyphosphate feeder. This feeder is installed in the water inlet supply to prevent scale buildup. This will require less frequent cleaning of the ice maker. To install one of these feeders, follow the manufacturer's recommendations in order to treat the water satisfactorily.

Cleaning Instructions for the Type-1 Ice Cube Maker

The manufacturer of this type of ice cube maker recommends that the ice maker be cleaned occasionally to help combat lime and mineral deposit buildup.

To clean the water system parts and the evaporator freeze plate, turn off the ice maker with the cycle switch. Open the bin door, and remove the cutter grid by removing the two thumb screws. Unplug the cutter grid, and remove it from the storage bin (Figure 28-4). A drain plug is located under the water tank. Remove it to drain the water out of the tank (Figure 28-5). After all of the water has been removed, reinstall the plug. Pour half a gallon of hot water into the tank, and set the switch to "clean." The hot water will circulate through the water pump assembly and over the evaporator freeze plate, including all the

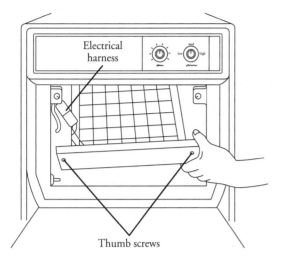

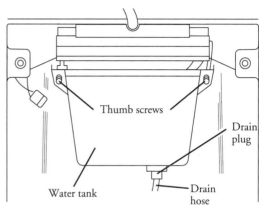

FIGURE 28-4   Removing the thumbscrews, disconnecting the electrical harness, and pulling the cutter grid out of the bin.

FIGURE 28-5   The water tank is located in the storage bin.

water system components. Let the water circulate for five minutes, and then drain the water out of the tank. Replace the plug. Mix ice machine cleaner with half a gallon of hot water, and pour it into the water tank. If you use a recognized ice machine cleaner, follow the instructions on the label for best results. If you would rather prepare your own solution, add six ounces of citric acid and phosphoric acid to half a gallon of hot water, and pour into the water tank. Turn the switch to "clean," and circulate this solution for 20 minutes or longer; then drain the water. Follow with two clean water rinses that circulate for five minutes, and then drain the water again.

Remove the splash guard, the water dispenser tube, and the plastic water pump tank. Place them in a solution of mild laundry bleach for five minutes, and then rinse. Use one ounce of bleach to one gallon of hot water. Be sure the water temperature does not exceed 145 degrees Fahrenheit—it could damage the plastic parts. Finally, sanitize the ice bin, door, ice cube scoop, grid panel, and grid with a bleach solution.

Reinstall all parts in the reverse order of disassembly, and test the ice machine operation. After the cleaning treatment, apply a release agent to the evaporator plate. This will retard any future buildup of scale and mineral deposits, and it will make the plate more slippery, which will provide for better ice slab release.

To clean the condenser coil, remove the screws that secure the front grille, and then remove the grille (Figure 28-6). Vacuum all lint and dust from the coil and from the surrounding area (Figure 28-7). Reinstall the grille. The frequency of cleaning will be determined by the surrounding conditions.

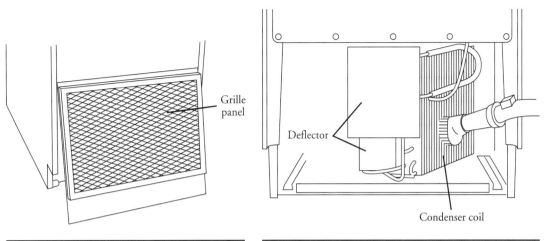

Figure 28-6 Front grille held in place with two screws located on the toe panel section.

Figure 28-7 The condenser section of the ice maker.

Prevention of Water-Utilizing System Explosions

In certain water-utilizing refrigeration systems, water can leak into the refrigerant side of the system. This can lead to an explosion of system components, including but not limited to the compressor. If such an explosion occurs, the resulting blast can kill or seriously injure anyone in the vicinity.

Systems at Risk of Explosion

Water-utilizing systems that have single-wall heat exchangers may present a risk of explosion. Such systems may include:

- Water-source heat pump/air conditioning systems
- Water cooling systems, such as ice makers, water coolers, and juice dispensers

Water-utilizing systems that have single-wall heat exchangers present a risk of explosion, unless they have one of the following:

- A high-pressure cutout that interrupts power to *all* leads to the compressor
- An external pressure-relief valve

How an Explosion Occurs

If the refrigerant tubing in the heat exchanger develops a leak, water can enter the refrigerant side of the system. This water can come in contact with live electrical connections in the compressor, causing a short circuit or a path to ground. When this occurs, extremely high temperatures can result. The heat buildup creates steam vapor that can cause excessive pressure throughout the entire system. This system pressure can lead to an explosion of the compressor or other system components.

Service Procedures

In light of the risk of explosion, be especially alert for signs of water leaking into the refrigerant side of the system. Whenever servicing or troubleshooting a water-utilizing system, always check to see if it has either a pressure-relief valve or a high-pressure cutout as previously described. If the system does not have at least one of these, *disconnect all electrical power*, and look for indications that water has leaked into the refrigerant side of the system. These indications may include:

- Observation of or a report of a blown fuse or tripped circuit breakers.
- Signs that water has leaked to the outside of the system.
- Reports that the system has made gurgling or percolating noises.
- A history of loss of refrigerant charge without a leak being found in the system.

NOTE *Common leak-detection methods will not detect a water-to-refrigerant leak in the system's heat exchanger(s).*

- Observation of or a report of the compressor giving off an unusual amount of heat.

If *any* of these indications are present, do the following checks to determine if water has leaked into the refrigerant side:

Step 1: Check for a Ground Fault (a Short to Ground)

- Check the compressor for a ground fault (also known as a short circuit to ground).
- If a ground fault does not exist, go to step 2.
- If a ground fault does exist, keep the power off. *Warning:* To avoid electric shock, electrocution, and terminal venting with ignition, do not energize a compressor that has a ground fault. Mark and red-tag the compressor to indicate that there is a ground fault. Do not reconnect the power leads. Tape and insulate each power lead separately. Proceed to step 2. Do not replace the compressor or energize the system before performing step 2.

Step 2: Check for Water in the System

Once the compressor is cool to the touch, open the system process valve slightly to see if any water comes out of the system. *Warning:* Opening the system process valve while the compressor is hot can cause severe burns from steam coming out of the valve. If any water comes out of the process valve, the entire system *must* be replaced. See the section "Replacing a Single-Wall Water-Utilizing System."

If water does not come out of the process valve, there is still a possibility that some water has leaked into the refrigerant side of the system. To address this possibility, determine if the system has a history of losing refrigerant charge without a leak being found or repaired.

If you find *any* indication of a history of losing refrigerant charge without detection of a leak, this is a sign that refrigerant has leaked into the water inside the heat exchanger. The entire system *must* be replaced. See the section "Replacing a Single-Wall Water-Utilizing System."

If you do not find any indication of a history of loss of charge without detection of a leak, you still need to install one of the following:

- A high-pressure cutout that interrupts power to *all* leads to the compressor
- An external pressure-relief valve

Also, if you found a ground fault in the compressor in step 1, replace the compressor before applying power to the system.

Replacing a Single-Wall Water-Utilizing System

When replacing a single-wall water-utilizing system, replace the system with one that has one of the following:

- A double-wall heat exchanger(s)
- A high-pressure cutout that interrupts power to *all* leads to the compressor
- An external pressure-relief valve

Repair Procedures

Each repair procedure is a complete inspection and repair process for a single ice cube maker component, containing the information you need to test and replace components. The actual repair or replacement of any sealed-system component is not included in this chapter. It is recommended that you acquire refrigerant certification (or call an authorized service company) to repair or replace any sealed-system component. The refrigerant in the sealed system must be recovered properly.

Type-1 Ice Cube Maker Compressor, Relay, and Overload Protector

The compressor (reciprocating type) is the heart of the vapor compression system. It is used to circulate the refrigerant throughout the sealed system. The relay and overload are attached to the compressor. The relay starts the compressor and the overload protects the compressor. All three components are located in the machine compartment in the rear or front of the ice maker. The relay can be either a current or a PTC (positive temperature coefficient) type device. The overload is a bimetal switch that is secured to the outer shell of the compressor.

The typical complaints associated with failure of the compressor, relay, and overload protector are:

- Ice maker does not run at all.
- No new ice production.
- Ice cubes in the storage bin are melting rapidly.
- Compressor won't run; it only hums.

To handle these problems, perform the following steps:

1. **Verify the complaint.** Verify the complaint by asking the customer to describe what the ice maker is doing. It will be helpful if you can locate the actual service manual for the ice maker model you are working on to properly diagnose the ice maker.

The service manual will assist you in properly placing the ice maker in the service test mode for testing the ice maker functions.

2. **Check for external factors.** You must check for external factors not associated with the appliance. Is the appliance installed properly? Does the ice maker have the correct voltage? The voltage at the wall receptacle is between 108 volts and 132 volts during a load on the circuit. Do you have the correct polarity? (See Chapter 6.)

3. **Disconnect the electricity.** Before working on the ice maker, disconnect the electricity. This can be done by pulling the plug from the receptacle. Or disconnect the electricity at the fuse panel or at the circuit breaker panel. Turn off the electricity.

WARNING *Some diagnostic tests will require you to test the components with the power turned on. When you disassemble the ice maker panel cover, you can position the panel in such a way that the wiring will not make contact with metal. This will allow you to test the ice maker components without electrical mishaps.*

4. **Gain access to the compressor.** Access the compressor. To access the compressor, remove the front grille. Remove the two screws in the condensing unit base, and pull the unit toward you. Be careful to not damage any refrigerant lines. Next, remove the compressor terminal cover (Figure 28-8) by removing the retaining clip that secures the cover. Remove the terminal cover.

5. **Test the compressor relay.** To test the compressor relay, remove the relay by pulling it from the compressor terminals without twisting it (Figure 28-9). On the relay body is stamped the word TOP. Hold the relay so that TOP is in the up position. Next, place the probes of the ohmmeter on the relay terminals marked S and M. Set the meter scale on R x 1. The reading will show no continuity. Next, remove the probe from the terminal marked M, and place the probe on the side terminal marked L. The reading will show no continuity. Now, move the probe from terminal S, and place it on the terminal marked M. The reading will show continuity. With the probes still attached, turn the relay upside down (see Figure 25-35b), and

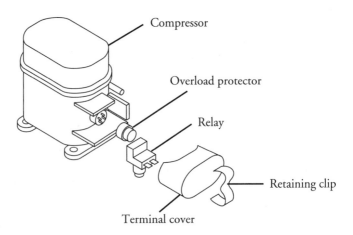

FIGURE 28-8
A pictorial view of compressor, overload, and relay.

Compressor

Overload protector

Relay

Retaining clip

Terminal cover

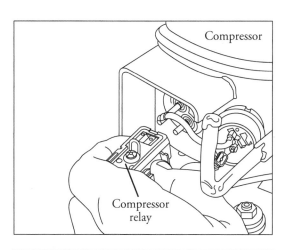

Figure 28-9   Pull the relay off the compressor. Be careful not to break the compressor pins.

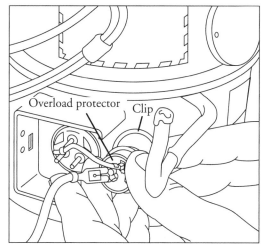

Figure 28-10   Disconnect the overload protector.

perform the same tests. By turning the relay over, the switch contacts in the relay will close. When you retest the relay, you should get the opposite results. You should have continuity between terminals S and M and between S and L. The meter will not show continuity between M and L. If the relay fails this test, replace it.

6. **Test the overload protector.** To test the overload protector, remove the wires from the overload protector and compressor terminals. Then remove the overload protector (from the compressor) by removing the retaining clip that secures the overload protector (Figure 28-10). Next, place the probes of the ohmmeter on the overload protector terminals. Set the meter scale on R × 1. The reading will show continuity. If not, replace the overload protector.

7. **Test the compressor.** To test the compressor, remove the relay and the overload protector. This will expose the compressor terminals. The compressor terminals are marked C, S, and R: C indicates the common winding terminal, S indicates the start winding terminal, and R indicates the run winding terminal. Next, place the probes of the ohmmeter on the terminals marked S and R (Figure 28-11). Set the meter scale on R × 1, and adjust the needle setting to indicate a zero reading. The meter reading will show continuity. Now place the meter probes on the terminals marked C and S. The meter reading will show continuity. Finally, place the meter probes on the terminals marked C and R. The meter reading will show continuity. The total number of ohms measured between S and R is equal to the sum of C to S plus C to R.

To test the compressor for ground, place one probe on a compressor terminal and the other probe on the compressor housing or on any good ground (Figure 28-12). Set the meter scale to R × 1000. The meter reading will show no continuity. Repeat this for the remaining two terminals. The meter reading will show no continuity. If you get a continuity reading from any of these terminals to ground, the compressor is grounded and must be replaced.

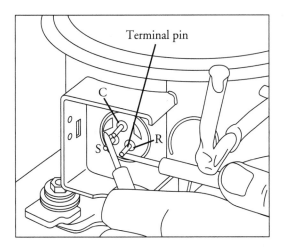

FIGURE 28-11   Testing the compressor windings.

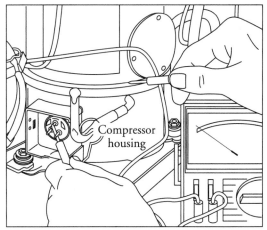

FIGURE 28-12   Testing the compressor for ground.

Bin Thermostat

The bin thermostat is located behind the control panel (Figure 28-14). The thermostat sensing bulb is located in the ice maker bin compartment. It will sense how much ice is in the storage bin.

The typical complaints associated with bin thermostat failure are:

- The ice maker runs all the time, making too much ice.
- The ice maker doesn't run at all.

To handle these problems, perform the following steps:

1. **Verify the complaint.** Verify the complaint by checking the sensing tube and bulb and the control settings. It will be helpful if you can locate the actual service manual for the ice maker model you are working on to properly diagnose the ice maker. The service manual will assist you in properly placing the ice maker in the service test mode for testing the ice maker functions.

2. **Check for external factors.** You must check for external factors not associated with the appliance. Is the appliance installed properly? Explain to the user how to set the controls. The voltage at the wall receptacle is between 108 volts and 132 volts during a load on the circuit. Do you have the correct polarity? (See Chapter 6.)

3. **Disconnect the electricity.** Before working on the ice maker, disconnect the electricity. This can be done by pulling the plug from the receptacle. Or disconnect the electricity at the fuse panel or at the circuit breaker panel. Turn off the electricity.

WARNING *Some diagnostic tests will require you to test the components with the power turned on. When you disassemble the ice maker control panel cover, you can position the panel in such a way that the wiring will not make contact with metal. This will allow you to test the ice maker components without electrical mishaps.*

4. **Gain access to the bin thermostat.** To access the bin thermostat, remove the screws from the escutcheon (Figure 28-13), and remove the panel. Next, remove the screws from the control bracket (Figure 28-14). Pull back on the control bracket, exposing the controls.

5. **Test the bin thermostat.** To test the bin thermostat, remove the wires from the thermostat terminals and place the ohmmeter probes on those terminals (Figure 28-15). Set the range scale on R × 1, and test for continuity. The meter should read continuity between the contacts if the temperature of the capillary tube is above 42 degrees Fahrenheit. The meter should not read continuity between the contacts if the temperature of the capillary tube is below 36 degrees Fahrenheit.

6. **Remove the bin thermostat.** Remove the bin thermostat from the control bracket by removing the two screws (see Figure 28-13). Remove the bin thermostat well from the left wall of the liner. Next, remove the five clips (under the gasket) from the left side of the liner (Figure 28-16). Now, bend the liner flange forward, and remove the capillary tube and thermostat control.

7. **Install a new bin thermostat.** To install the new bin thermostat, just reverse the order of disassembly, and reassemble. Then test the control. Remember to reinstall the capillary tube in the same location from which it was removed. If you do not, the ice maker will not cycle properly.

Evaporator Thermostat

The evaporator thermostat is located behind the control panel (Figure 28-14). The thermostat's capillary tube is attached to the evaporator freeze plate. The purpose of the evaporator thermostat is to end the freeze cycle and initiate the harvest cycle when the ice thickness has been reached.

FIGURE 28-13
Removing the control panel to expose the controls.

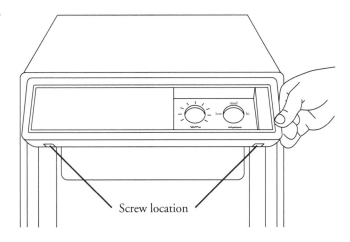

Screw location

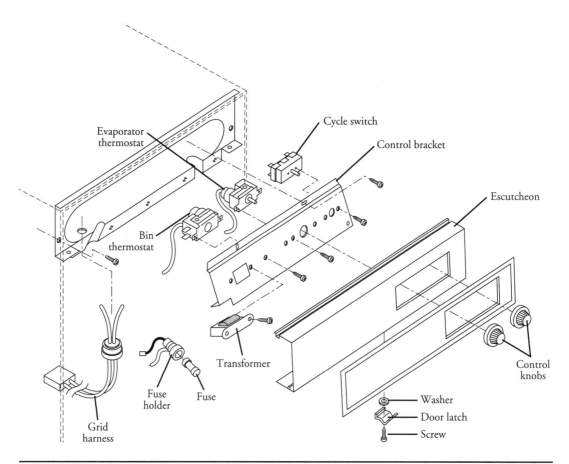

FIGURE 28-14   A pictorial view of the control panel.

The typical complaints associated with failure of the evaporator thermostat are:

- Unable to control ice cube thickness.
- Ice maker water not circulating.
- The ice maker runs, but there is no ice in the bin.

To handle these problems, perform the following steps:

1. **Verify the complaint.** Verify the complaint by checking the sensing tube and bulb and the control settings. It will be helpful if you can locate the actual service manual for the ice maker model you are working on to properly diagnose the ice maker. The service manual will assist you in properly placing the ice maker in the service test mode for testing the ice maker functions.

FIGURE **28-15**
The bin thermostat.

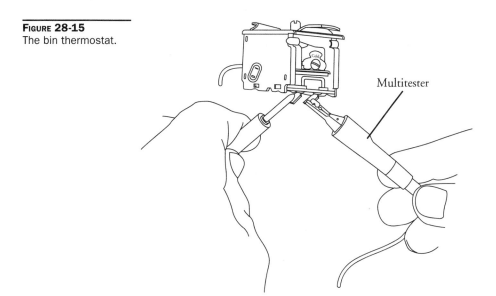

Multitester

2. **Check for external factors.** You must check for external factors not associated with the appliance. Is the appliance installed properly? Explain to the user how to set the controls. The voltage at the wall receptacle is between 108 volts and 132 volts during a load on the circuit. Do you have the correct polarity? (See Chapter 6.)

3. **Disconnect the electricity.** Before working on the ice maker, disconnect the electricity. This can be done by pulling the plug from the receptacle. Or disconnect the electricity at the fuse panel or at the circuit breaker panel. Turn off the electricity.

FIGURE **28-16**
Peel back the liner to gain access to the capillary tube.

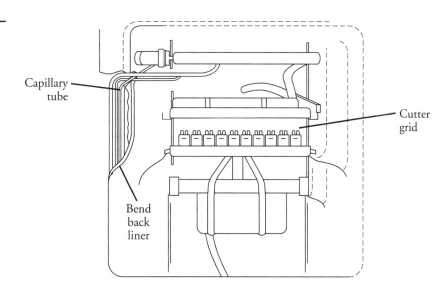

Capillary tube

Cutter grid

Bend back liner

WARNING *Some diagnostic tests will require you to test the components with the power turned on. When you disassemble the ice maker control panel cover, you can position the panel in such a way that the wiring will not make contact with metal. This will allow you to test the ice maker components without electrical mishaps.*

4. **Gain access to the evaporator thermostat.** To access the evaporator thermostat, remove the screws from the escutcheon (see Figure 28-13), and remove the panel. Next, remove the screws from the control bracket (see Figure 28-14). Pull back on the control bracket, exposing the controls.

5. **Test the evaporator thermostat.** To test the evaporator thermostat, remove the wires from the thermostat terminals, and place the ohmmeter probes on terminals 1 and 2 (Figure 28-17). Set the range scale on R × 1, and test for continuity. The meter should show continuity between the contacts if the temperature of the evaporator freeze plate is 30 degrees Fahrenheit or warmer. The meter should not read continuity between the contacts if the temperature of the evaporator freeze plate is +10 to –3 degrees Fahrenheit. By disconnecting the water pump at the terminal board and operating the ice maker without the pump, the evaporator thermostat action can be easily observed. This will cause the thermostat to cycle in a matter of a few minutes.

6. **Remove the evaporator thermostat.** Remove the cutter grid by removing the thumbscrews (see Figure 28-4). Remove the evaporator thermostat from the control bracket by removing the two screws (see Figure 28-14). Remove the clamp from underneath the evaporator freeze plate, which secures the capillary tube to the evaporator. Next, remove the five clips, under the gasket, from the left side of the liner (see Figure 28-16). Now, bend the liner flange forward, and remove the capillary tube and thermostat control.

FIGURE 28-17
The evaporator
thermostat.

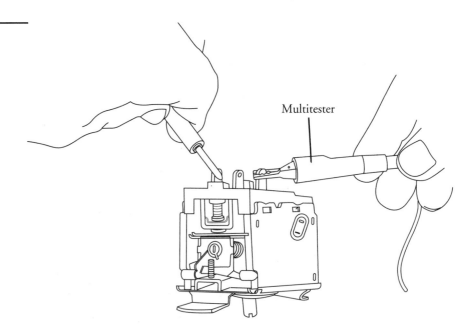

Multitester

7. **Install a new evaporator thermostat.** To install the new evaporator thermostat, just reverse the order of disassembly, and reassemble. Then test the control. Remember to reinstall the capillary tube in the same location from which it was removed. Also, the capillary tube must be taped to the hot gas restrictor tube. If you do not do these things, the ice maker will not cycle properly.

Hot Gas Solenoid Valve

The hot gas solenoid valve allows the high pressure refrigerant gas to bypass the condenser and flow through the condenser accumulator tube (Figure 28-3). When the refrigerant enters the condenser accumulator, the hot gas enters into the evaporator plate and it will evenly heat the evaporator plate so that the ice slab will release from the evaporator plate quickly and evenly. The hot gas solenoid valve is accessed through the rear of the ice maker on some older models. On newer models the hot gas valve is located behind the condenser coil.

The typical complaints associated with failure of the hot gas solenoid valve are:

- Ice maker runs, but there is no ice production.
- Evaporator freeze plate will not heat up to release ice slab.
- Ice maker runs continuously.

To handle these problems, perform the following steps:

1. **Verify the complaint.** Verify the complaint by checking the ice maker cycles. It will be helpful if you can locate the actual service manual for the ice maker model you are working on to properly diagnose the ice maker. The service manual will assist you in properly placing the ice maker in the service test mode for testing the ice maker functions.

2. **Check for external factors.** You must check for external factors not associated with the appliance. Is the appliance installed properly? Explain to the user how to set the controls. The voltage at the wall receptacle is between 108 volts and 132 volts during a load on the circuit. Do you have the correct polarity? (See Chapter 6.)

3. **Disconnect the electricity.** Before working on the ice maker, disconnect the electricity. This can be done by pulling the plug from the receptacle. Or disconnect the electricity at the fuse panel or at the circuit breaker panel. Turn off the electricity.

WARNING *Some diagnostic tests will require you to test the components with the power turned on. When you disassemble the ice maker control panel cover, you can position the panel in such a way that the wiring will not make contact with metal. This will allow you to test the ice maker components without electrical mishaps.*

4. **Gain access to the hot gas solenoid valve.** To access the hot gas solenoid valve (Figure 28-18), remove the grille (see Figure 28-6). Next, remove the deflector from the condenser (see Figure 28-7).

5. **Test the hot gas solenoid valve.** Test the hot gas solenoid valve for continuity. Remove the wires from the solenoid coil. Place the ohmmeter probes on the solenoid coil terminals (see Figures 28-18 and 28-19). Set the range scale on R × 1, and test

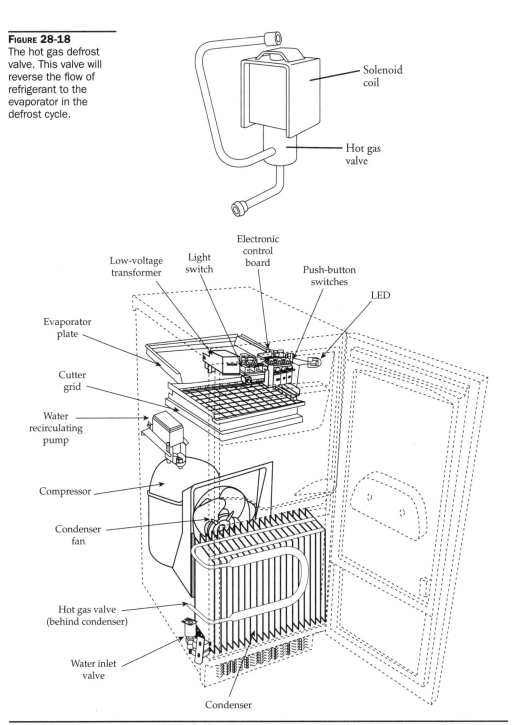

Figure 28-18
The hot gas defrost valve. This valve will reverse the flow of refrigerant to the evaporator in the defrost cycle.

Solenoid coil

Hot gas valve

Electronic control board

Low-voltage transformer

Light switch

Push-button switches

LED

Evaporator plate

Cutter grid

Water recirculating pump

Compressor

Condenser fan

Hot gas valve (behind condenser)

Water inlet valve

Condenser

Figure 28-19   Type 1 - Newer model - Self-contained ice cube maker with electronic controls and thermistors. This model has the hot gas valve located behind the condenser coil. Also, the water inlet valve is located on the left side of the condenser coil.

for continuity. To test the hot gas valve itself, connect a 120-volt fused service cord (Figure 28-20) to the solenoid coil. Listen for a click sound as the plunger rises up. Now, disconnect the service cord, and you will hear the plunger drop back. If you cannot hear a distinct click sound from the hot gas valve, it will need to be replaced by an authorized service company (the sealed system might be under warranty from the manufacturer) or by a licensed refrigerant technician.

The solenoid coil is a separate component that can be replaced without replacing the entire hot gas valve assembly. Another way to test the hot gas valve is to leave the wires off the solenoid coil and reconnect the service cord to the solenoid coil. This test requires the electricity to be turned on.

CAUTION *Tape the solenoid coil leads that were removed so that they will not touch the chassis when you plug in the ice maker for this test. Be cautious when working with live wires. Avoid getting shocked!*

With the ice maker plugged in and running, feel the hot gas defrost tube—it should feel warm or hot when the valve is energized.

6. **Remove the hot gas solenoid coil.** To remove the hot gas solenoid coil, remove the spring clip from the top of the coil, and then remove the coil (be sure the electricity is off).

7. **Install a new hot gas solenoid coil.** To install the new solenoid coil, just reverse the order of disassembly, and reassemble. Then test the valve.

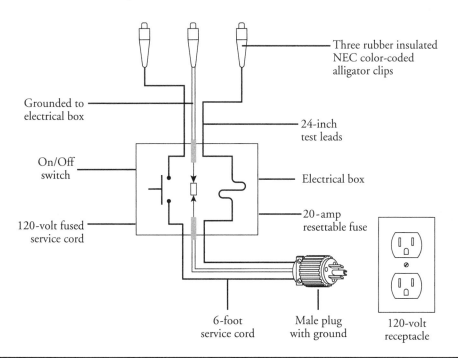

FIGURE 28-20 A 120-volt fused service cord test.

Water Valve

The water inlet valve controls the flow of water into the ice cube maker, and is solenoid-operated. When it is energized, water in the supply line will pass through the valve body and into the water reservoir.

The typical complaints associated with failure of the water valve are:

- Ice maker runs, but there is no ice production.
- No water is circulating across the evaporator freeze plate.
- Water is flooding the storage bin, causing the ice to melt.

To handle these problems, perform the following steps:

1. **Verify the complaint.** Verify the complaint by checking the ice maker cycles. It will be helpful if you can locate the actual service manual for the ice maker model you are working on to properly diagnose the ice maker. The service manual will assist you in properly placing the ice maker in the service test mode for testing the ice maker functions.

2. **Check for external factors.** You must check for external factors not associated with the appliance. Is the appliance installed properly? Explain to the user how to set the controls. Is the water supply turned on? The voltage at the wall receptacle is between 108 volts and 132 volts during a load on the circuit. Do you have the correct polarity? (See Chapter 6.)

3. **Disconnect the electricity.** Before working on the ice maker, disconnect the electricity. This can be done by pulling the plug from the receptacle. Or disconnect the electricity at the fuse panel or at the circuit breaker panel. Turn off the electricity.

WARNING *Some diagnostic tests will require you to test the components with the power turned on. When you disassemble the ice maker control panel cover, you can position the panel in such a way that the wiring will not make contact with metal. This will allow you to test the ice maker components without electrical mishaps.*

4. **Gain access to the water valve.** To access the water valve, remove the top insulated panel. The water valve is located in the upper-right front corner (older models). On newer models, the water valve is located next to the condenser (Figure 28-19).

5. **Remove and test the water valve.** In order to test the water valve solenoid coil, the water valve must be removed from the storage bin. Shut off the water supply to the ice maker. Now disconnect the water line from the valve. Next, remove the screws from the water valve bracket. Pull on the valve to release it from the receptacle in the liner. Place the ohmmeter probes on the solenoid coil terminals (Figure 28-21). Set the range scale on R × 1, and test for continuity. If there is no continuity, replace the water valve.

6. **Install a new water valve.** To install the new water valve, just reverse the order of disassembly, and reassemble. Then test the valve. Don't forget to turn on the water supply.

FIGURE 28-21
The water valve.

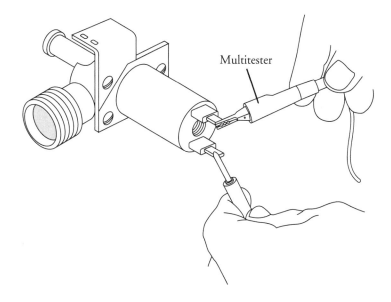

Multitester

Condenser Fan Motor

The condenser fan motor is a 120 VAC, single-speed fan motor. The condenser fan motor is located near the compressor in the machine compartment in the rear of the ice cube maker. The condenser fan motor, when operating, will pull air across the condenser coil and then exhaust it past the compressor and out through the rear of the ice cube maker. The condenser fan will remove the heat from the condenser coil.

The typical complaints associated with failure of the condenser fan motor are:

- The ice maker has stopped producing ice.
- The condenser fan motor runs slower than normal.
- The condenser fan motor does not run at all.
- The compressor is sometimes noisier than normal.

To handle these problems, perform the following steps:

1. **Verify the complaint.** Verify the complaint by asking the customer to describe what the ice maker is doing. Is the condenser fan motor running during the freeze cycle? It will be helpful if you can locate the actual service manual for the ice maker model you are working on to properly diagnose the ice maker. The service manual will assist you in properly placing the ice maker in the service test mode for testing the ice maker functions.

2. **Check for external factors.** You must check for external factors not associated with the appliance. Is the appliance installed properly? Are there any foreign objects blocking the condenser fan blades? The voltage at the wall receptacle is between 108 volts and 132 volts during a load on the circuit. Do you have the correct polarity? (See Chapter 6.)

3. **Disconnect the electricity.** Before working on the ice maker, disconnect the electricity to the ice maker. This can be done by pulling the plug from the receptacle. Or disconnect the electricity at the fuse panel or at the circuit breaker panel. Turn off the electricity.

WARNING *Some diagnostic tests will require you to test the components with the power turned on. When you disassemble the ice maker control panel cover, you can position the panel in such a way that the wiring will not make contact with metal. This will allow you to test the ice maker components without electrical mishaps.*

4. **Gain access to the condenser fan motor.** To access the condenser fan motor, remove the front grille. Remove the two screws in the condensing unit base, and pull the unit toward you. Be careful not to damage any refrigerant lines.

5. **Test the condenser fan motor.** To test the condenser fan motor, remove the wires from the motor terminals. Next, place the probes of the ohmmeter on the motor terminals (Figure 28-22). Set the meter scale on R × 1. The meter should show some resistance. If no reading is indicated, replace the motor. If the fan blades do not spin freely, replace the motor. If the bearings are worn, replace the motor.

6. **Remove the condenser fan motor.** To remove the condenser fan motor, you must first remove the fan blades. Unscrew the nut that secures the blades to the motor. Remove the blades from the motor. Then remove the motor assembly by removing the mounting bracket screws (Figure 28-23).

7. **Install a new condenser fan motor.** To install the new condenser fan motor, just reverse the order of disassembly, and reassemble. Remember to reconnect the ground wire to the motor. Reconnect the wires to the motor terminals, and test.

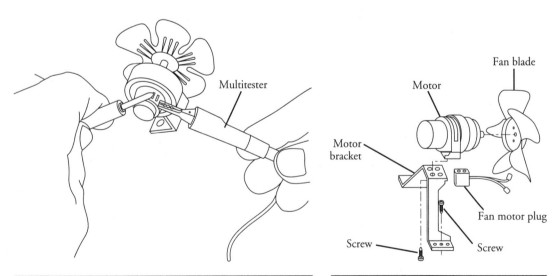

Figure 28-22   The condenser fan motor.

Figure 28-23   A pictorial view of the condenser fan motor assembly.

Water Pump

The water pump will circulate the water from the water tank across the evaporator freezing plate (Figure 28-2). It is located behind the water tank (Figure 28-5) on the right side.

The typical complaints associated with failure of the water pump are:

- Ice maker runs, but there is no ice production.

- No water circulating across the evaporator freeze plate.

To handle these problems, perform the following steps:

1. **Verify the complaint.** Verify the complaint by checking the ice maker cycles. It will be helpful if you can locate the actual service manual for the ice maker model you are working on to properly diagnose the ice maker. The service manual will assist you in properly placing the ice maker in the service test mode for testing the ice maker functions.

2. **Check for external factors.** You must check for external factors not associated with the appliance. Is the appliance installed properly? Explain to the user how to set the controls. The voltage at the wall receptacle is between 108 volts and 132 volts during a load on the circuit. Do you have the correct polarity? (See Chapter 6.)

3. **Disconnect the electricity.** Before working on the ice maker, disconnect the electricity. This can be done by pulling the plug from the receptacle. Or disconnect the electricity at the fuse panel or at the circuit breaker panel. Turn off the electricity.

WARNING *Some diagnostic tests will require you to test the components with the power turned on. When you disassemble the ice maker control panel cover, you can position the panel in such a way that the wiring will not make contact with metal. This will allow you to test the ice maker components without electrical mishaps.*

4. **Gain access to the water pump.** To access the water pump, remove the bin door, the front insulated panel, and the inner bin door. A drain plug is located under the water tank. Remove it to drain the water out of the tank (see Figure 28-5). Next, remove the thumbscrews that secure the water tank, and remove the tank.

5. **Test the water pump.** To test the water pump motor, isolate the motor, and place the probes of the ohmmeter on the motor terminals (Figure 28-24). Set the meter scale on R × 1. The meter should show some resistance. If no reading is indicated, replace the water pump. Next, check the motor with a 120-volt fused service cord (see Figure 28-20).

6. **Remove the water pump.** To remove the water pump (Figure 28-25), remove the screws from the water pump bracket that secure the pump to the liner. Disconnect the discharge hose from the pump. Remove the water pump.

7. **Install a new water pump.** To install a new water pump, just reverse the order of disassembly, and reassemble. Reconnect the wires to the motor terminals, and test.

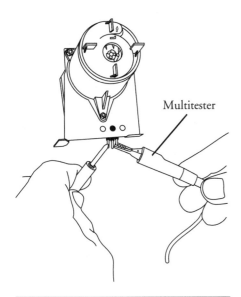

Multitester

FIGURE 28-24   A bottom view of the water pump. When checking the pump, be sure the inlet is free of debris.

Cutter Grid

The cutter grid is located inside the storage bin (Figure 28-4). It is used for cutting the ice slab into ice cubes.

The typical complaints associated with failure of the cutter grid are:

- Ice slabs lie on top of the cutter grid.
- Cutter grid is not cutting ice slab into cubes evenly.

To handle these problems, perform the following steps:

1. **Verify the complaint.** Verify the complaint by checking the ice slab and cutter grid fuse. It will be helpful if you can locate the actual service manual for the ice maker model you are working on to properly diagnose the ice maker. The service manual will assist you in properly placing the ice maker in the service test mode for testing the ice maker functions.

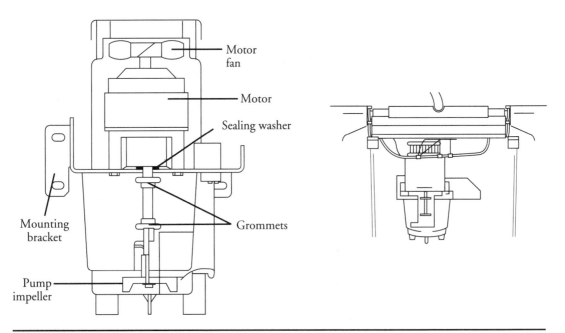

Motor fan

Motor

Sealing washer

Grommets

Mounting bracket

Pump impeller

FIGURE 28-25   A pictorial view of the water pump.

2. **Check for external factors.** You must check for external factors not associated with the appliance. Is the appliance installed properly? Explain to the user how to set the controls. The voltage at the wall receptacle is between 108 volts and 132 volts during a load on the circuit. Do you have the correct polarity? (See Chapter 6.)

3. **Disconnect the electricity.** Before working on the ice maker, disconnect the electricity. This can be done by pulling the plug from the receptacle. Or disconnect the electricity at the fuse panel or at the circuit breaker panel. Turn off the electricity.

WARNING *Some diagnostic tests will require you to test the components with the power turned on. When you disassemble the ice maker control panel cover, you can position the panel in such a way that the wiring will not make contact with metal. This will allow you to test the ice maker components without electrical mishaps.*

4. **Gain access to the cutter grid.** To access the cutter grid, open the bin door and remove the cutter grid by removing the two thumbscrews. Unplug the cutter grid, and remove it from the storage bin (see Figure 28-4).

5. **Test the cutter grid.** Examine the cutter grid for broken wires, and check the connecting pins for corrosion (Figure 28-26). As you inspect the cutter grid, look for cracked or broken insulators in the frame. Next, place the probes of the ohmmeter on the cutter grid plug terminals (Figure 28-27). Set the meter scale on R × 1. The meter should show continuity. If no reading is indicated, one or more grid wires or insulators are defective.

6. **Repair the cutter grid.** If the cutter grid frame and insulators are broken, it would be advisable to replace the entire cutter grid. Using a C-clamp, compress the spring clip to relieve the tension (Figure 28-28). Next, use a pair of pliers to compress the adjacent spring clip, and remove the buss bar. Do the same procedure for the other side of the cutter grid. The insulators, clips, and grid wires can now be removed and replaced. If any of the grid wires break, it is time to replace all of the grid wires.

7. **Install a new cutter grid.** To install a new cutter grid, just reverse the order of disassembly, and reassemble. Reconnect the wires to the cutter grid, and test it.

FIGURE 28-26
A cutter grid cuts an ice slab into strips and then into cubes.

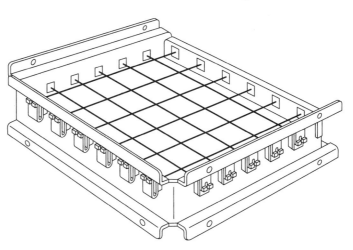

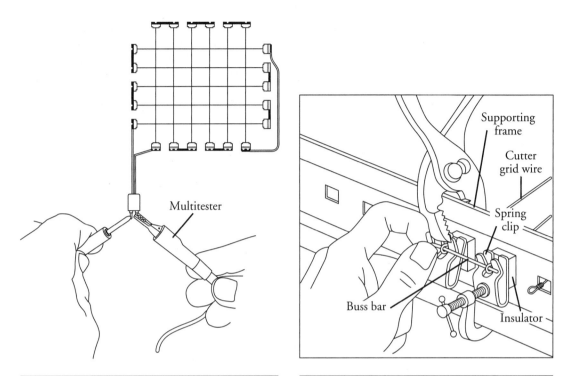

FIGURE 28-27   A pictorial view of the cutter grid circuit.

FIGURE 28-28   Side view of the cutter grid, showing the clips, buss bars, and insulators.

Cutter Grid Transformer and Fuse

The cutter grid transformer and fuse are located behind the control panel. Low voltage is used on the cutter grid to cut up the ice slab.

The typical complaints associated with failure of the grid transformer or fuse are:

- Ice slabs lie on top of the cutter grid.
- Cutter grid is not cutting ice slab into cubes evenly.

To handle these problems, perform the following steps:

1. **Verify the complaint.** Verify the complaint by checking the ice slab and cutter grid. It will be helpful if you can locate the actual service manual for the ice maker model you are working on to properly diagnose the ice maker. The service manual will assist you in properly placing the ice maker in the service test mode for testing the ice maker functions.

2. **Check for external factors.** You must check for external factors not associated with the appliance. Is the appliance installed properly? The voltage at the wall receptacle is between 108 volts and 132 volts during a load on the circuit. Do you have the correct polarity? (See Chapter 6.)

3. **Disconnect the electricity.** Before working on the ice maker, disconnect the electricity. This can be done by pulling the plug from the receptacle. Or disconnect the electricity at the fuse panel or at the circuit breaker panel. Turn off the electricity.

WARNING *Some diagnostic tests will require you to test the components with the power turned on. When you disassemble the ice maker control panel cover, you can position the panel in such a way that the wiring will not make contact with metal. This will allow you to test the ice maker components without electrical mishaps.*

4. **Gain access to the cutter grid transformer and fuse.** To access the cutter grid transformer and fuse, remove the screws from the escutcheon (see Figure 28-13), and remove the panel. The fuse is located on the control bracket on the left side. To remove the fuse, push in and twist—it will pop out of the holder (Figure 28-29). Next, remove the screws from the control bracket (see Figure 28-14). Pull back on the control bracket, exposing the controls.

5. **Test the cutter grid transformer and fuse.** To test the transformer, disconnect the wires from the transformer to isolate it from the circuit. Use a 120-volt fused service cord, and connect it to the primary side of the transformer.

NOTE *This test requires the electricity to be turned on. Be cautious when working with live wires. Avoid getting shocked!*

You might have to look at the wiring diagram for assistance in identifying the primary side and for the proper color-coding of the wires. Using the volt meter, connect the probes to the secondary side of the transformer. Plug in the 120-volt fused service cord; the meter should read 8.5 volts. Unplug the service cord.

To test for resistance, disconnect the power cord, and set the ohmmeter scale on R x 1. Place the probes on the primary wires of the transformer. The meter should show resistance. Next, place the probes on the secondary wires of the transformer (Figure 28-30). The meter should show resistance. If the transformer fails either test, replace it.

FIGURE 28-29
The cutter fuse, located behind the control panel.

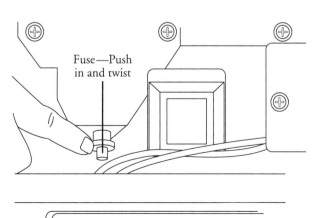

Fuse—Push
in and twist

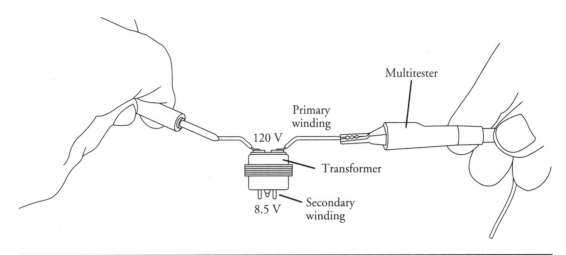

FIGURE 28-30   The cutter grid transformer.

To test the fuse, place the probes on each end of the fuse (Figure 28-31). Set the ohmmeter scale on R × 1. The meter should show continuity. If not, replace the fuse.

6. **Remove the transformer.** To remove the transformer, remove the screws that secure it to the control bracket (see Figure 28-14).

7. **Install a new transformer or fuse.** To install a new transformer or fuse, just reverse the order of disassembly, and reassemble. Reconnect the wires to the components, and test it.

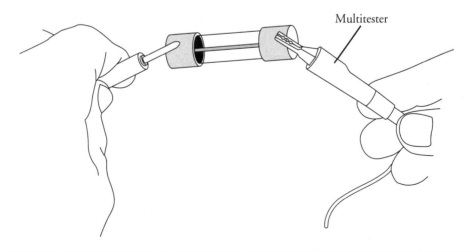

FIGURE 28-31   The cutter grid fuse. If this fuse blows, the cutter grid will not function. Check for a short.

Cycle Switch

The cycle switch is located in the control panel. This switch will start the ice cube maker cycle, and when it is positioned in the clean cycle, it will start the cleaning cycle.

The typical complaints associated with failure of the cycle switch are:

- Unable to turn on the clean cycle.
- Unable to turn on the ice maker.

To handle these problems, perform the following steps:

1. **Verify the complaint.** Verify the complaint by checking the control settings. It will be helpful if you can locate the actual service manual for the ice maker model you are working on to properly diagnose the ice maker. The service manual will assist you in properly placing the ice maker in the service test mode for testing the ice maker functions.

2. **Check for external factors.** You must check for external factors not associated with the appliance. Is the appliance installed properly? Explain to the user how to set the controls. The voltage at the wall receptacle is between 108 volts and 132 volts during a load on the circuit. Do you have the correct polarity? (See Chapter 6.)

3. **Disconnect the electricity.** Before working on the ice maker, disconnect the electricity. This can be done by pulling the plug from the receptacle. Another way to disconnect the electricity is at the fuse panel or at the circuit breaker panel. Turn off the electricity.

WARNING *Some diagnostic tests will require you to test the components with the power turned on. When you disassemble the ice maker control panel cover, you can position the panel in such a way that the wiring will not make contact with metal. This will allow you to test the ice maker components without electrical mishaps.*

4. **Gain access to the cycle switch.** To access the cycle switch, remove the screws from the escutcheon (see Figure 28-13), and remove the panel. Next, remove the screws from the control bracket (see Figure 28-14). Pull back on the control bracket, exposing the controls.

5. **Test the cycle switch.** To test the cycle switch for continuity between the switch contacts, refer to the wiring diagram for the correct switch positions (Figure 28-32). Disconnect the wires from the switch, and turn the switch to the "on" position. Place the probes of the ohmmeter on the cycle switch terminals marked 2 and 3. Set the meter scale on R × 1. The meter should show continuity. Now place the probes of the ohmmeter on the cycle switch terminals marked 1 and 6. The meter should show continuity. Next, turn the cycle switch to the "clean" position. Place the probes of the ohmmeter on the cycle switch terminals marked 5 and 6. The meter should show continuity. If the switch fails these tests, replace it.

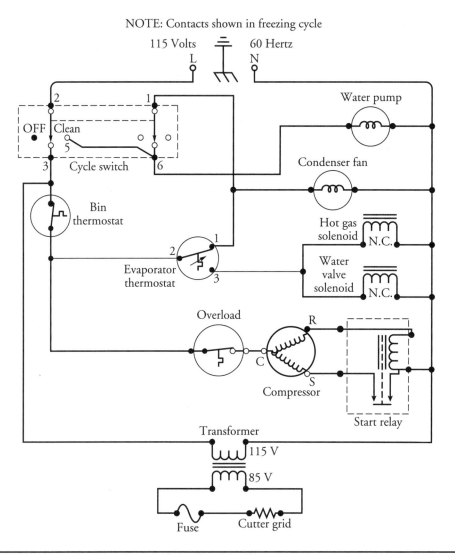

FIGURE 28-32 A sample wiring diagram of a self-contained ice maker.

6. **Remove the cycle switch.** To remove the cycle switch, remove the two screws that secure it to the control bracket (see Figure 28-14).

7. **Install a new cycle switch.** To install a new cycle switch, just reverse the order of disassembly, and reassemble. Connect the wires to the switch terminals according to the wiring diagram, and test it.

Self-Contained Ice Cube Maker Performance Data

There are a number of factors that can affect the overall performance of the ice cube maker.

Thermostat Position	Cut-In Temperature	Cut-Out Temperature
Evaporator thermostat – warm	38 degrees F ±2 degrees	11 degrees F ±4 degrees
Evaporator thermostat – coldest	38 degrees F ±2 degrees	−3 degrees F ±3 degrees
Bin thermostat	41 degrees F ±2 degrees	35 degrees F ±2 degrees

TABLE 28-1   Ice Thickness Controlled by Evaporator Thermostat

Installation

- Installation in a dusty environment will cause the ice cube maker to operate continuously and have a low ice production rate.
- Installation in a greasy atmosphere will cause the ice cube maker to operate in high temperature conditions with low to no ice production.
- Installation next to a high temperature appliance will cause the ice cube maker to operate in high temperature conditions with low to no ice production.

Water Temperature

- Higher water temperatures will increase the operating time of the ice cube maker.
- Higher water temperatures will cause decreased ice production in a 24-hour period.

Ambient Temperature

- High temperatures will decrease ice production.
- Ice storage bin melting ice above 25 percent.

Maintenance

- Restricted air movement through the condenser coil and front grille will cause low to no ice production.

Other

- If the ice cubes are more than ½ inch thick the ice production will decrease and the operating time will increase.

See Table 28-1, Table 28-2, and Table 28-3 for additional technical data.

Ambient Temperature °F	24-Hour Ice Production in Pounds			
100	36	36	35	34
90	41	40	39	37
80	46	44	43	41
70	51	48	46	44
To obtain the maximum ice production rate you will need to have the ideal conditions. The capacities that are shown above are the average production rates and they can vary depending on conditions.				

TABLE 28-2   Water Temperature

Temperature	Cycle Time in Minutes
Ambient 70°F Water 60°F	25–29
Ambient 90°F Water 60°F	27–33
Ambient 110°F Water 60°F	38–46
Ambient 70°F Water 80°F	28–34
Ambient 90°F Water 80°F	32–40
Ambient 110°F Water 80°F	40–48

TABLE 28-3   Ice Cube Operating Cycle Times

Troubleshooting Water and Its Effect on Ice Making

When making ice cubes in a self-contained ice cube maker, the quality of the ice should be defined as solid, clear, and free of taste or odor. The ice cube maker can provide this if the quality of the water is pure. The following charts (Tables 28-4 and 28-5) will show some of the ingredients that can affect ice cube production.

Type-1 Self-Contained Ice Cube Maker Wiring Diagrams

Figures 28-33 through 28-37 depict a wiring schematic and strip circuits for a type-1 self-contained ice cube maker with electronic components.

Type-2 Self-Contained Ice Cube Maker

This type of ice cube maker is a free-standing, self-contained refrigeration appliance that produces ice cubes (Figure 28-38). This self-contained ice cube maker can also be installed under the counter. This type of ice cube maker is similar in operation to some commercial ice machines. The type-2 ice cube maker is designed to make clear ice cubes from most water supplies.

Ingredient	Effect	Solution
Algae	Horrible taste and odor	1 Install a carbon filter.
Minerals: Calcium Magnesium Potassium Sodium	Cloudy ice cubes Refreezing Slow cutting of the ice slab	1 Check the water flow for a restriction (water valve or water line kinked). 2 Change the water source. 3 Install a polyphosphate feeder or water softener.

TABLE 28-4   Ingredients that Affect Ice Cube Quality

Ingredient	Effect	Solution
Chlorine Iron Manganese	Staining of ice cube maker components	1 Clean ice cube maker with ice machine cleaner and ice machine sanitizer. 2 Install a water softener and iron filter.

Permanent water hardness
    Calcium or magnesium
    Chlorides
    Nitrates
    Sulfates
	Scale will form on ice cube maker components	1 Abrasive cleaning.
		2 Polyphosphate feeder or water softener reduces or eliminates the need for abrasive cleaning.

Temporary water hardness
    Calcium or magnesium
    Carbonates
	Scale will form on ice cube maker components	1 Clean ice cube maker with ice machine cleaner and ice machine sanitizer.
		2 Polyphosphate feeder or water softener reduces the frequency of cleaning the ice cube maker by 50%.

TABLE 28-5   Ingredients that Affect the Ice Cube Maker Operation and Quantity of Ice Production

Principles of Operation

On startup, the bin thermostat closes; the water valve and the hot gas valve are energized for 3 minutes. This process will ensure the ice making process starts with fresh water and the refrigerant pressures are equalized prior to the compressor starting up. The compressor will start in about 3 minutes after the hot gas valve and the water valve de-energize. In the freeze cycle, water flows constantly over the evaporator and into each cube cell (Figure 28-39). The length of the freeze cycle is determined by a thermostat or thermistor, which monitors the refrigeration system (Figures 28-40, 28-41, and 28-42). At the end of the freeze cycle, a thermostat or thermistor will signal the beginning of the harvest cycle. In the harvest cycle (Figure 28-43) the water pump and the condenser fan motor will be de-energized and the compressor will still be running. The hot gas valve and the water valve will be energized for the remainder of the harvest cycle. The hot refrigerant gas will warm up the evaporator, releasing the ice cubes into the storage bin (Figure 28-44). At the same time, the water trough is being purged with fresh water. At the end of the harvest cycle, the freeze cycle will begin anew. As ice fills the storage bin, it will come in contact with the bin thermostat (Figure 28-45). When this happens the ice cube maker will shut off completely until the ice level drops in the storage bin. When no ice comes in contact with the bin thermostat, the freeze cycle will begin again.

    The refrigeration cycle (Figure 28-46) in this type of ice maker is similar to that in a conventional refrigerator/freezer. The compressor pumps the refrigerant into the condenser coil, which is cooled by the condenser fan motor. The refrigerant leaves the condenser coil as a high-pressure liquid, passes through the dryer, and enters the capillary tube. The refrigerant is next metered through the capillary tube and then enters the evaporator freeze plate as a low-pressure liquid. The refrigerant low-pressure gas then leaves the evaporator freeze plate and returns to the compressor.

    When the ice maker goes into the defrost (harvest) cycle, it energizes the hot gas solenoid, which reverses the refrigeration cycle, during which the condenser fan motor and

PART VI

Figure 28-33
A sample wiring diagram of a self-contained ice maker with electronic components.

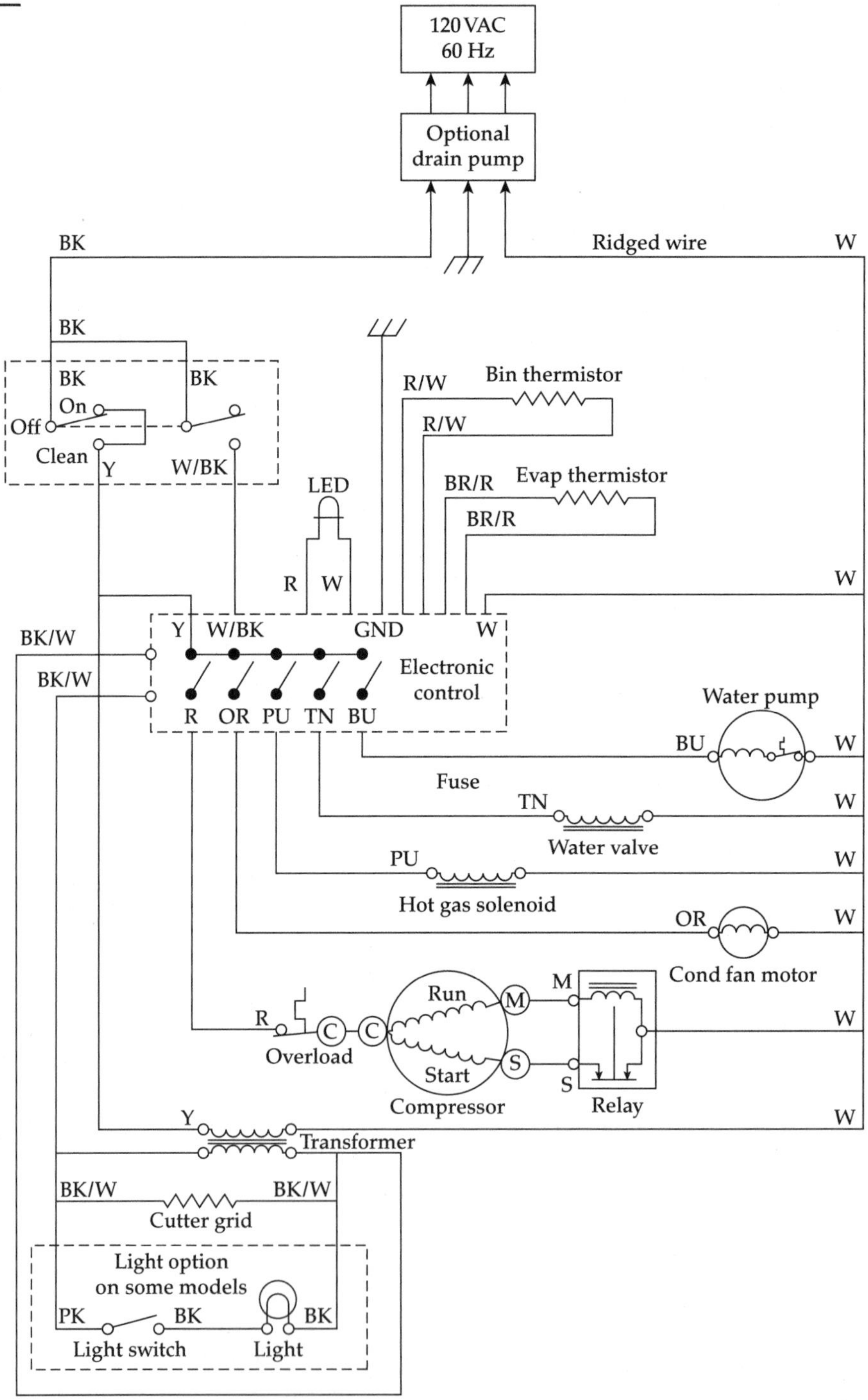

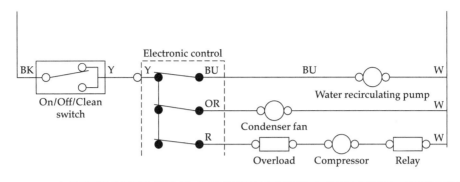

FIGURE 28-34   Strip circuit (electronic model) – Ice production mode.

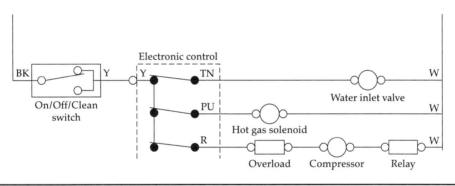

FIGURE 28-35   Strip circuit (electronic model) – Harvest mode.

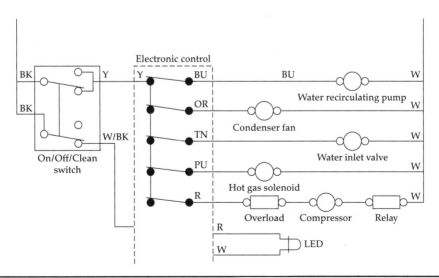

FIGURE 28-36   Strip circuit (electronic model) illustrates the first 25 seconds of the diagnostics/clean cycle.

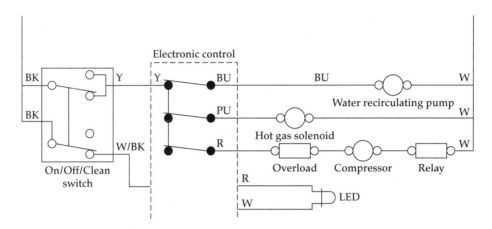

FIGURE 28-37   Strip circuit (electronic model) illustrates the last 47 minutes of the diagnostics/clean cycle.

water pump will stop. The hot gas passes through the evaporator freeze plate, heating it up enough to release the ice cubes. The thermostat senses the temperature of the evaporator freeze plate again and activates the freeze cycle. The hot gas solenoid valve will then close, the water valve will close, the condenser fan motor will start, the water pump will start, and the freeze cycle will begin to manufacture new ice cubes.

FIGURE 28-38   Type-2 self-contained ice cube maker.

FIGURE 28-39   Ice cube maker in the start-up cycle. Water is circulating over the evaporator.

FIGURE 28-40   Ice beginning to form on the evaporator.

FIGURE 28-41   The ice cubes are getting larger in the mold.

FIGURE 28-42   The ice cubes have reached their desired thickness. The water stops circulating.

FIGURE 28-43   The hot gas valve is energized. The ice cubes are defrosting off the evaporator.

FIGURE 28-44 The end of the harvest cycle. The ice cubes leave the evaporator.

FIGURE 28-45 The ice cubes enter the storage bin. When the ice cubes reach the bin thermostat, the ice production will stop.

Step-by-Step Troubleshooting by Symptom Diagnosis

When servicing the type-2 self-contained ice cube maker, don't overlook the simple things that might be causing the problem. Step-by-step troubleshooting by symptom diagnosis is based on diagnosing malfunctions, with possible causes arranged into categories relating to the operation of the automatic ice cube maker. This section is intended only as a checklist to aid you in diagnosing a problem. Look at the symptom that best describes the problem you are experiencing with the automatic ice cube maker, and then correct the problem.

No Ice Production

- Check the water supply. Is it turned on?
- Check and make sure the stand pipe is connected to the water trough.
- Check the water pump.

Low Ice Production

- Check the ambient temperature around the ice maker. Lower ice cube production is normal in higher temperatures.
- Check the evaporator for deposit buildup.
- Check ice thickness setting.

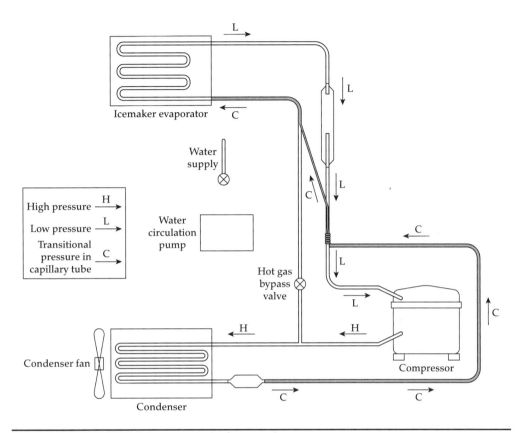

FIGURE 28-46 Type-2 ice cube maker in the freeze cycle.

- Check for a dirty condenser coil.
- Check hot gas valve.
- Check refrigeration system.

Ice Is Slow to Release from the Evaporator
- Check the evaporator for deposit buildup.
- Check the ice maker may not be level. Level ice maker.

Ice Maker Will Not Fill with Water
- Check and make sure the stand pipe is connected to the water trough.
- Check the water valve.
- Check wiring.

Continuous Ice Production

- Check the bin thermostat.
- Check bin thermostat bulb alignment.
- Check ice cube thickness.

Shallow or Incomplete Ice Cubes

- Check the water level.
- Check and make sure the stand pipe is connected to the water trough.
- Check the water pressure. The water pressure should be 20 to 120 psi.
- Is the ice maker level?
- Are you operating the ice maker without the front and back panels? This type of ice maker must have the front and back cover installed to operate properly.
- Check the thermostat or thermistor.
- On electronic models, check the control board.

Compressor Not Running, Not Making Ice; Condenser Fan Motor Running

- Check compressor relay.
- Check compressor overload.
- Check compressor.

Compressor Not Running, Not Making Ice; Condenser Fan Motor Not Running

- Check for voltage to the ice maker.
- Check circuit breaker and/or fuses.
- Check switches.
- Check wiring.

Water in the Ice Storage Bin

- Check the storage bin drain system.
- Check external drain system.
- Check drain pump.

Water Will Not Stop Filling the Trough

- Check the water valve.
- Check the electronic control board.

Ice Cube Maker Noisy

- Check the condenser fan motor.
- Check the condenser fan blade for proper alignment and for any obstruction.
- Check the refrigerant copper lines. Make sure they are not touching anything.

Poor Ice Quality (Soft or Cloudy Cubes)

- Check the quality of the water inlet.
- Evaporator may have to be cleaned.
- Check for water splashing onto the ice cubes.

No Water in the Trough

- Make sure the stand pipe is inserted into the trough.
- Check the water valve.
- Check how the water circulates over the evaporator. Make sure that the water does not splash out of the trough.
- Check the leveling of the ice maker.

Repair Procedures

Each repair procedure is a complete inspection and repair process for a single ice cube maker component, containing the information you need to test and replace components. The actual repair or replacement of any sealed-system component is not included in this chapter. It is recommended that you acquire refrigerant certification (or call an authorized service company) to repair or replace any sealed-system component. The refrigerant in the sealed system must be recovered properly.

Adjusting the Ice Cube Thickness

The typical complaints associated with size of ice cubes are:

- Water supply to ice maker inadequate.
- Ice cubes too small.
- Ice cubes too large.
- Ice cubes are being produced in a sheet.

To handle these problems, perform the following steps:

1. **Verify the complaint.** Verify the complaint by asking the customer to describe what the ice maker is doing. It will be helpful if you can locate the actual service manual for the ice maker model you are working on to properly diagnose the ice maker. The service manual will assist you in properly placing the ice maker in the service test mode for testing the ice maker functions.

2. **Check for external factors.** You must check for external factors not associated with the appliance. Is the appliance installed properly? Does the ice maker have the correct water supply? Does the ice maker have the correct voltage? The voltage at the wall receptacle is between 108 volts and 132 volts during a load on the circuit. Do you have the correct polarity? (See Chapter 6.)

3. **Disconnect the electricity.** Before working on the ice maker, disconnect the electricity. This can be done by pulling the plug from the receptacle. Or disconnect the electricity at the fuse panel or at the circuit breaker panel. Turn off the electricity.

PART VI

WARNING *Some diagnostic tests will require you to test the components with the power turned on. When you disassemble the ice maker panel cover, you can position the panel in such a way that the wiring will not make contact with metal. This will allow you to test the ice maker components without electrical mishaps.*

4. **Gain access to the electronic control board.** Open the ice cube maker door, and remove the screws from the front cover (Figure 28-47).

5. **Adjust the ice cube size dial on the electronic control board.** Look on the electronic control board for the ice cube thickness adjustment dial. If you need to make the ice cubes thicker, turn the dial clockwise to a higher number. If you need to make the ice cubes thinner, turn the dial counterclockwise to a smaller number (Figure 28-48).

6. **Test ice cube thickness.** Reinstall the front access cover and reconnect the electricity to the ice cube maker. Remove the old ice cubes from the bin and test the ice maker. If the ice cube size is not correct, then repeat steps 3 through 6. The type-2 ice cube maker produces the ice cube in a slab. When the ice slab is harvested, it will fall vertically into the ice storage bin, breaking into cubes. Figure 28-49 depicts the correct thickness.

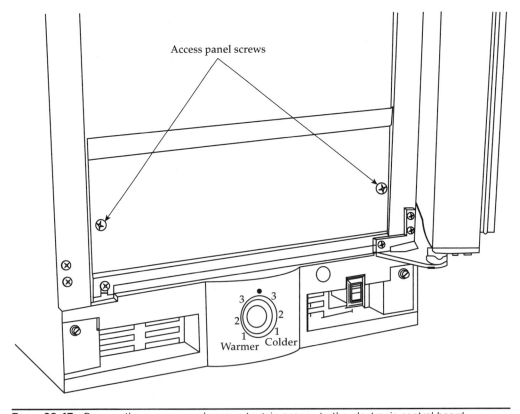

FIGURE 28-47 Remove the access panel screws to gain access to the electronic control board.

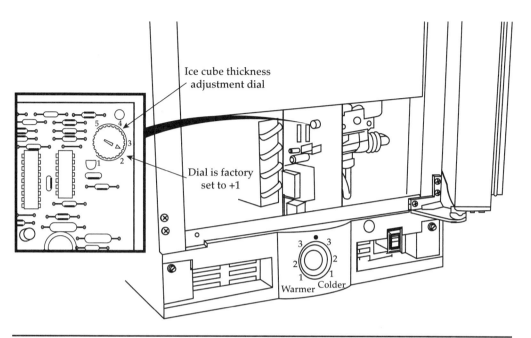

FIGURE 28-48 The ice cube thickness adjustment is located on the electronic control board. Turn adjustment dial clockwise for thicker cubes. Turn adjustment dial counterclockwise for thinner ice cubes.

Type-2 Ice Cube Maker Compressor, Relay, and Overload Protector

The compressor (reciprocating type) is the heart of the vapor compression system. It is used to circulate the refrigerant throughout the sealed system. The relay and overload are attached to the compressor. The relay starts the compressor, and the overload protects the compressor. All three components are located in the machine compartment in the rear of the ice maker. The relay is a current relay. The overload is a bimetal switch that is secured to the relay assembly, which will attach to the outer shell of the compressor.

The typical complaints associated with failure of the compressor, relay, and overload protector are:

- Ice maker does not run at all.
- No new ice production.
- Ice cubes in the storage bin are melting rapidly.
- Compressor won't run; it only hums.

To handle these problems, perform the following steps:

1. **Verify the complaint.** Verify the complaint by asking the customer to describe what the ice maker is doing. It will be helpful if you can locate the actual service manual for the ice maker model you are working on to properly diagnose the ice maker. The service manual will assist you in properly placing the ice maker in the service test mode for testing the ice maker functions.

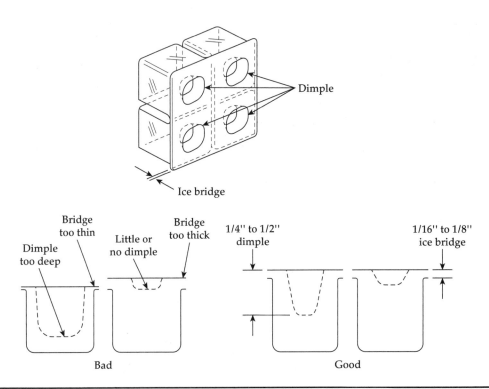

FIGURE 28-49   This type of ice cube maker produces a dimple in the ice cubes formed by the cascading water over the evaporator. Sometimes the dimples are different sizes. The ice cube adjustment dial is preset from the factory to produce an ice bridge of approximately 1/16 inch to 1/8 inch resulting in a dimple of approximately ¼ inch to ½ inch in depth. An ice cube that has less of a dimple and a thicker ice bridge will cause the ice slab to stay intact when entering the storage bin. It can also cause the storage bin to overfill.

2. **Check for external factors.** You must check for external factors not associated with the appliance. Is the appliance installed properly? Does the ice maker have the correct voltage? The voltage at the wall receptacle is between 108 volts and 132 volts during a load on the circuit. Do you have the correct polarity? (See Chapter 6.)

3. **Disconnect the electricity.** Before working on the ice maker, disconnect the electricity. This can be done by pulling the plug from the receptacle. Or disconnect the electricity at the fuse panel or at the circuit breaker panel. Turn off the electricity.

WARNING   *Some diagnostic tests will require you to test the components with the power turned on. When you disassemble the ice maker panel cover, you can position the panel in such a way that the wiring will not make contact with metal. This will allow you to test the ice maker components without electrical mishaps.*

4. **Gain access to the compressor.** Access the compressor. To access the compressor, remove the rear cover (Figure 28-50). Next, remove the compressor terminal cover (Figure 28-8) by removing the retaining clip that secures the cover. Remove the terminal cover.

5. **Test the compressor relay.** To test the compressor relay, remove the relay and overload by pulling it from the compressor terminals without twisting it (Figure 28-51). On the relay body is stamped the word TOP. Hold the relay so that TOP is in the up position. Next, place the probes of the ohmmeter on the relay terminals marked S and M. Set the meter scale on R × 1. The reading will show no continuity. Next, remove the probe from the terminal marked M, and place the probe on the side terminal marked L. The reading will show no continuity. Now, move the probe from terminal S, and place it on the terminal marked M. The reading will show continuity. With the probes still attached, turn the relay upside down (see Figure 25-35b), and perform the same tests.

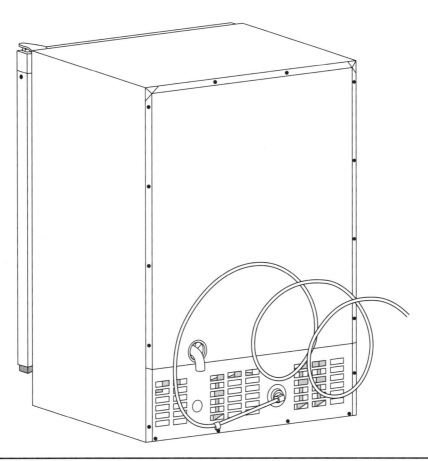

FIGURE 28-50 Remove the screws from the rear cover to gain access to the compressor, condenser fan motor, and water valve.

PART VI

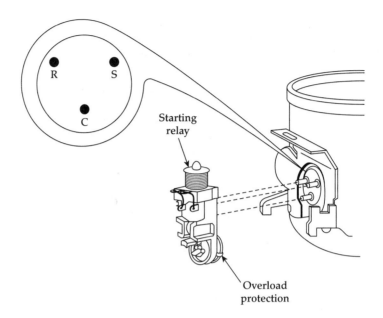

FIGURE 28-51
Disconnect the relay
and overload
assembly.

By turning the relay over, the switch contacts in the relay will close. When you retest the relay, you should get the opposite results. You should have continuity between terminals S and M and between S and L. The meter will not show continuity between M and L. If the relay fails this test, replace it.

6. **Test the overload protector.** To test the overload protector, remove the wires from the overload protector and compressor terminals. Then remove the overload protector from the compressor by removing the retaining clip that secures the overload protector (Figure 28-51). Next, place the probes of the ohmmeter on the overload protector terminals. Set the meter scale on R × 1. The reading will show continuity. If not, replace the overload protector.

7. **Test the compressor.** To test the compressor, remove the relay and the overload protector. This will expose the compressor terminals. The compressor terminals are marked C, S, and R: C indicates the common winding terminal, S indicates the start winding terminal, and R indicates the run winding terminal. Next, place the probes of the ohmmeter on the terminals marked S and R (Figure 28-51). Set the meter scale on R × 1, and adjust the needle setting to indicate a zero reading. The meter reading will show continuity. Now place the meter probes on the terminals marked C and S. The meter reading will show continuity. Finally, place the meter probes on the terminals marked C and R. The meter reading will show continuity. The total number of ohms measured between S and R is equal to the sum of C to S plus C to R.

To test the compressor for ground, place one probe on a compressor terminal and the other probe on the compressor housing or on any good ground (Figure 28-12). Set the meter scale to R × 1000. The meter reading will show no continuity. Repeat this for the remaining two terminals. The meter reading will show no continuity. If you get a continuity reading from any of these terminals to ground, the compressor is grounded and must be replaced.

Water Valve

The water inlet valve controls the flow of water into ice cube maker, and is solenoid-operated. When it is energized, water in the supply line will pass through the valve body and into the water reservoir. The water valve is located in the rear of the ice cube maker.

The typical complaints associated with failure of the water valve are:

- Ice maker runs, but there is no ice production.
- No water is circulating across the evaporator freeze plate.
- Water is flooding the storage bin, causing the ice to melt.

To handle these problems, perform the following steps:

1. **Verify the complaint.** Verify the complaint by checking the ice maker cycles. It will be helpful if you can locate the actual service manual for the ice maker model you are working on to properly diagnose the ice maker. The service manual will assist you in properly placing the ice maker in the service test mode for testing the ice maker functions.

2. **Check for external factors.** You must check for external factors not associated with the appliance. Is the appliance installed properly? Explain to the user how to set the controls. Is the water supply turned on? The voltage at the wall receptacle is between 108 volts and 132 volts during a load on the circuit. Do you have the correct polarity? (See Chapter 6.)

3. **Disconnect the electricity.** Before working on the ice maker, disconnect the electricity. This can be done by pulling the plug from the receptacle. Or disconnect the electricity at the fuse panel or at the circuit breaker panel. Turn off the electricity.

WARNING *Some diagnostic tests will require you to test the components with the power turned on. When you disassemble the ice maker control panel cover, you can position the panel in such a way that the wiring will not make contact with metal. This will allow you to test the ice maker components without electrical mishaps.*

4. **Gain access to the water valve.** To access the water valve, remove the rear panel. The water valve is located in the rear of the machine compartment.

5. **Test the water valve.** Place the ohmmeter probes on the solenoid coil terminals (Figure 28-21). Set the range scale on R × 1, and test for continuity. If there is no continuity, replace the water valve.

6. **Install a new water valve.** To install the new water valve, just reverse the order of disassembly and reassemble. Then test the valve. Don't forget to turn on the water supply.

Water Pump

The water pump will circulate the water from the water tank across the evaporator freezing plate (Figure 28-43).

The typical complaints associated with failure of the water pump are:

- Ice maker runs, but there is no ice production.
- No water circulating across the evaporator freeze plate.

To handle these problems, perform the following steps:

1. **Verify the complaint.** Verify the complaint by checking the ice maker cycles. It will be helpful if you can locate the actual service manual for the ice maker model you are working on to properly diagnose the ice maker. The service manual will assist you in properly placing the ice maker in the service test mode for testing the ice maker functions.

2. **Check for external factors.** You must check for external factors not associated with the appliance. Is the appliance installed properly? Explain to the user how to set the controls. The voltage at the wall receptacle is between 108 volts and 132 volts during a load on the circuit. Do you have the correct polarity? (See Chapter 6.)

3. **Disconnect the electricity.** Before working on the ice maker, disconnect the electricity. This can be done by pulling the plug from the receptacle. Or disconnect the electricity at the fuse panel or at the circuit breaker panel. Turn off the electricity.

WARNING *Some diagnostic tests will require you to test the components with the power turned on. When you disassemble the ice maker control panel cover, you can position the panel in such a way that the wiring will not make contact with metal. This will allow you to test the ice maker components without electrical mishaps.*

4. **Gain access to the water pump.** To access the water pump, open the ice maker door, and then remove the front cover. A drain plug is located in the water trough. Remove it to drain the water out of the trough (see Figure 28-43). Next, remove the screws that secure the water pump. Remove the water pump

5. **Test the water pump.** To test the water pump motor, isolate the motor and place the probes of the ohmmeter on the motor terminals. Set the meter scale on R × 1. The meter should show some resistance. If no reading is indicated, replace the water pump. Next, check the motor with a 120-volt fused service cord (see Figure 28-20). If there is rust on the water pump, replace it.

6. **Install a new water pump.** To install a new water pump, just reverse the order of disassembly and reassemble. Reconnect the wires to the motor terminals, and test.

Thermistor

The thermistors used in the type-2 ice cube maker plug into the electronic control board (Figure 28-52) and they monitor the ice cube maker's functions and troubleshooting capabilities. The thermistors, in conjunction with the electronic control board, determine the length of the freeze and harvest cycles.

FIGURE 28-52
Exploded view of a
type-2 ice cube maker
control panel.

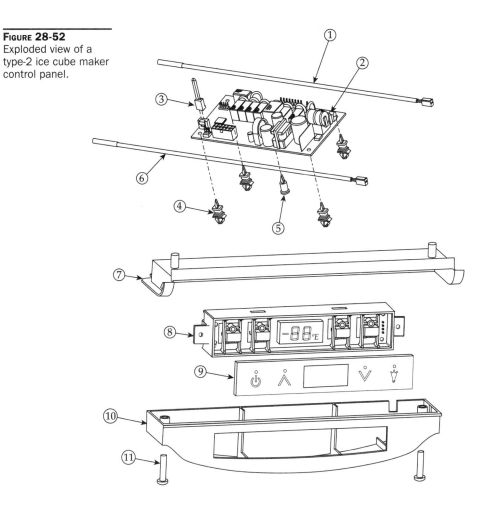

Item	Description
1	Thermistor–ice level/ice bin, senses ice cube level in the storage bin and maintains ice cube levels
2	Electronic circuit board–monitors functions of ice cube maker
3	Jumper switch
4	Electronic control board supports
5	Electronic control board support
6	Thermistor–condenser dryer inlet –liquid line, signals length of ice cube making and harvesting
7	Baseplate liner
8	Electronic display assembly—displays temperatures and service error codes
9	Display glass
10	Display housing
11	Pan screws

The typical complaints associated with failure of thermistor are:

- Ice maker runs, but there is no ice production.
- Too much ice in the storage bin.
- Ice maker not running, error code displayed.

To handle these problems, perform the following steps:

1. **Verify the complaint.** Verify the complaint by checking the ice maker cycles. It will be helpful if you can locate the actual service manual for the ice maker model you are working on to properly diagnose the ice maker. The service manual will assist you in properly placing the ice maker in the service test mode for testing the ice maker functions.

2. **Check for external factors.** You must check for external factors not associated with the appliance. Is the appliance installed properly? Explain to the user how to set the controls. The voltage at the wall receptacle is between 108 volts and 132 volts during a load on the circuit. Do you have the correct polarity? (See Chapter 6.)

3. **Disconnect the electricity.** Before working on the ice maker, disconnect the electricity. This can be done by pulling the plug from the receptacle. Or disconnect the electricity at the fuse panel or at the circuit breaker panel. Turn off the electricity.

WARNING *Some diagnostic tests will require you to test the components with the power turned on. When you disassemble the ice maker control panel cover, you can position the panel in such a way that the wiring will not make contact with metal. This will allow you to test the ice maker components without electrical mishaps.*

4. **Gain access to the thermistor.** To gain access to the thermistor, disassemble the control panel (Figure 28-52).

5. **Test the thermistor.** To test the thermistor, isolate the thermistor from the control board, and place the probes of the ohmmeter on the thermistor terminals. Set the meter scale on R $\times$ 1. The meter should show some resistance. For the exact resistance reading you will have to look it up in the service manual. If no reading is indicated, replace the thermistor. On some models, you will have to enter into the service mode to diagnose the problem with the thermistor. Thermistors generally fail due to moisture or physical damage.

6. **Install a new thermistor.** To install a new thermistor, just reverse the order of disassembly and reassemble. Reconnect the thermistor to the electronic control board, and test. Don't forget to take the ice maker out of the service mode after the repairs are made.

Electronic Control Board

The electronic control board (Figure 28-52) monitors the ice cube maker's functions and troubleshooting capabilities.

The typical complaints associated with failure of electronic control board are:

- Ice maker runs, but there is no ice production.
- Ice maker not running, error code displayed.

To handle these problems, perform the following steps:

1. **Verify the complaint.** Verify the complaint by checking the ice maker cycles. It will be helpful if you can locate the actual service manual for the ice maker model you are working on to properly diagnose the ice maker. The service manual will assist you in properly placing the ice maker in the service test mode for testing the ice maker functions.

2. **Check for external factors.** You must check for external factors not associated with the appliance. Is the appliance installed properly? Explain to the user how to set the controls. The voltage at the wall receptacle is between 108 volts and 132 volts during a load on the circuit. Do you have the correct polarity? (See Chapter 6.)

3. **Disconnect the electricity.** Before working on the ice maker, disconnect the electricity. This can be done by pulling the plug from the receptacle. Or disconnect the electricity at the fuse panel or at the circuit breaker panel. Turn off the electricity.

WARNING *Some diagnostic tests will require you to test the components with the power turned on. When you disassemble the ice maker control panel cover, you can position the panel in such a way that the wiring will not make contact with metal. This will allow you to test the ice maker components without electrical mishaps.*

4. **Gain access to the electronic control board.** To gain access to the electronic control board, disassemble the control panel (Figure 28-52).

5. **Test the electronic control board.** To test the electronic control board you will need the service manual for the model you are servicing. You will need to enter the service mode to test the electronic control board. After you enter the service mode, you can test for voltage at the relays outputs. See wiring schematic for locations (Figure 28-53).

6. **Install a new electronic control board.** To install a new electronic control board, just reverse the order of disassembly and reassemble. Reconnect all connectors on the electronic control board, and test. Don't forget to take the ice maker out of the service mode after the repairs are made.

Condenser Fan Motor

The condenser fan motor is a 120 VAC, single-speed fan motor. The condenser fan motor is located near the compressor in the machine compartment in the rear of the ice cube maker. The condenser fan motor, when operating, will pull air across the condenser coil and then exhaust it past the compressor and out through the rear of the ice cube maker. The condenser fan will remove the heat from the condenser coil.

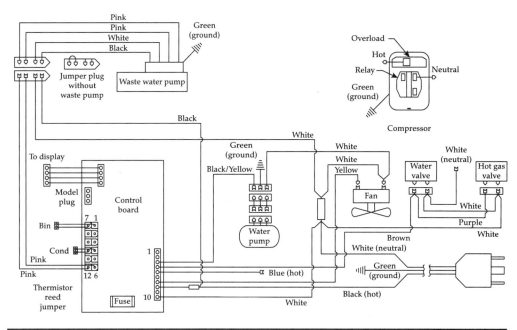

FIGURE 28-53 A typical wiring diagram for a type-2 ice cube maker with electronic controls.

The typical complaints associated with failure of the condenser fan motor are:

- The ice maker has stopped producing ice.
- The condenser fan motor runs slower than normal.
- The condenser fan motor does not run at all.
- The compressor is sometimes noisier than normal.

To handle these problems, perform the following steps:

1. **Verify the complaint.** Verify the complaint by asking the customer to describe what the ice maker is doing. Is the condenser fan motor running during the freeze cycle? It will be helpful if you can locate the actual service manual for the ice maker model you are working on to properly diagnose the ice maker. The service manual will assist you in properly placing the ice maker in the service test mode for testing the ice maker functions.

2. **Check for external factors.** You must check for external factors not associated with the appliance. Is the appliance installed properly? Are there any foreign objects blocking the condenser fan blades? The voltage at the wall receptacle is between 108 volts and 132 volts during a load on the circuit. Do you have the correct polarity? (See Chapter 6.)

3. **Disconnect the electricity.** Before working on the ice maker, disconnect the electricity to the ice maker. This can be done by pulling the plug from the receptacle. Or disconnect the electricity at the fuse panel or at the circuit breaker panel. Turn off the electricity.

WARNING *Some diagnostic tests will require you to test the components with the power turned on. When you disassemble the ice maker control panel cover, you can position the panel in such a way that the wiring will not make contact with metal. This will allow you to test the ice maker components without electrical mishaps.*

4. **Gain access to the condenser fan motor.** To access the condenser fan motor, remove the rear panel.

5. **Test the condenser fan motor.** To test the condenser fan motor, remove the wires from the motor terminals. Next, place the probes of the ohmmeter on the motor terminals (Figure 28-22). Set the meter scale on R × 1. The meter should show some resistance. If no reading is indicated, replace the motor. If the fan blades do not spin freely, replace the motor. If the bearings are worn, replace the motor.

6. **Remove the condenser fan motor.** To remove the condenser fan motor, you must first remove the fan blades. Unscrew the nut that secures the blades to the motor. Remove the blades from the motor. Then remove the motor assembly by removing the mounting bracket screws (Figure 28-23).

7. **Install a new condenser fan motor.** To install the new condenser fan motor, just reverse the order of disassembly and reassemble. Remember to reconnect the ground wire to the motor. Reconnect the wires to the motor terminals, and test.

Type-2 Ice Cube Maker Performance Data

See on the section "Self-Contained Ice Cube Maker Performance Data" in this chapter. Figure 28-54 depicts the ice cube production for the type-2 ice cube maker.

Ice Production Rates

Ambient Temp/Water Temp °F	Approximate Ice Production (lbs/day)
50/50	60
60/50	60
70/50	58
80/50	54
90/70	47
100/70	40

FIGURE 28-54 Type-2 ice cube maker ice production rates. The amount of ice cubes produced will vary depending on operating conditions, how clean the condenser coil is, type of installation, and application of the appliance.

Room Air Conditioners

Room air conditioners (RACs) are available in different sizes (rated in BTUs) and cabinet styles.[1] These units can be installed in a window or through the wall, or are available as a portable unit on wheels (Figure 1-16). Manufacturers have designed RACs with electromechanical controls or electronic controls. Consumers can purchase an RAC that is straight cool only, cool with heat (electric), heat pump (reverse cycle), or a combination of cool and heat (electric heat and reverse cycle).

The room air conditioner consists of the following:

- The base pan, fan housing, divider section (bulkhead), and outer cabinet
- The sealed system, which consists of the evaporator coil, the condenser coil, the compressor, and the connecting tubing
- The electrical circuitry, including the fan motor and other electrical components

Principles of Operation

The room air conditioner, when installed and running properly, will circulate the air in a room or area, removing the heat and humidity; some models will heat the air in the winter months. At the same time, the air filter, located behind the front grille, will filter out dust particles. Most models have a fresh air intake feature, which allows fresh outside air to enter the room when the unit is running. The thermostat will control the comfort level in the room or area, and cycle the air conditioner on and off according to the temperature setting.

Before continuing, take a moment to refresh your memory of Chapter 9. These two chapters combined will make servicing room air conditioners a breeze.

Safety First

Any person who cannot use basic tools or follow written instructions should *not* attempt to install, maintain, or repair any room air conditioners. Any improper installation, preventive maintenance, or repairs could create a risk of personal injury or property damage. If you do not fully understand the installation, preventive maintenance, or repair procedures in this chapter, or if you doubt your ability to complete the task on your room air conditioner, please call your service manager.

This chapter covers the electrical components and how to diagnose the sealed system. The actual repair or replacement of any sealed-system component is not included in this chapter. It is recommended that you acquire refrigerant certification (or call an authorized service company) to repair or replace any sealed-system component, as the refrigerant in the sealed system must be recovered properly.

Before continuing, take a moment to refresh your memory on the safety procedures in Chapter 2.

Room Air Conditioners in General

Much of the troubleshooting information in this chapter covers room air conditioners in general, rather than specific models, in order to present a broad overview of service techniques. The illustrations that are used in this chapter are for demonstration purposes only, to clarify the description of how to service these appliances. They in no way reflect on a particular brand's reliability.

Electrical Requirements

One of the most common problems that I have encountered over the years is when a consumer plugs the room air conditioner into an existing 115-volt receptacle only to find out that the circuit breaker keeps tripping. Most people do not take into account that this receptacle is connected to the other receptacles and/or lights in the same room and on the same branch circuit. This means that there is a limit to how many items can be connected to one branch circuit. Before a consumer purchases a room air conditioner, he or she needs to add up the total amperage or wattage of the items that are plugged into the branch circuit where the air conditioner is going to be plugged in. For example, on a 15-amp circuit, the total wattage on a 120-volt circuit should not exceed 1800 watts (Table 29-1). For safety

Circuit Breaker Size Rated in Amps	Volts	Watts	BTUs
15	120	1800	6138
20	120	2400	8184
25	120	3000	10,230
30	230	6900	23,529
40	230	9200	31,372
50	230	11,500	39,215
60	230	13,800	47,058
70	230	16,100	54,901
80	230	18,400	62,744
100	230	23,000	78,430
Conversion formulas: Watts = Volts x Amps; BTUs = Watts x 3.41; Amps = Watts/Volts			

TABLE 29-1   Conversion Formulas for Amps, Volts, Watts, and BTUs

FIGURE 29-1
An electrical receptacle guide illustrating the type of receptacle needed for the amperage rating.

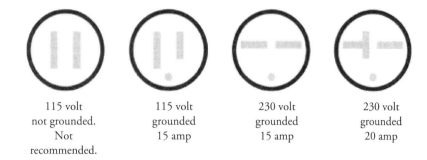

115 volt
not grounded.
Not
recommended.

115 volt
grounded
15 amp

230 volt
grounded
15 amp

230 volt
grounded
20 amp

reasons, all of the items, including the air conditioner, should stay below the amperage rating on the circuit breaker by 20 percent. The 115-volt, 15-amp circuit is reserved for lights, computers, televisions, stereos, etc. The 115-volt, 20-amp circuit is reserved for a refrigerator, dishwasher, automatic washer, or a garbage disposer only. Most 230-volt, 15- to 60-amp receptacles are on a single branch circuit. Figure 29-1 illustrates the type of plug connections needed for a room air conditioner. All room air conditioners and portable air conditioners must have a properly polarized and grounded receptacle, with the correct operating voltage for the unit purchased, preferably on its own branch circuit (Table 29-2). If the consumer only has multiple receptacles on one branch circuit, advise him or her to calculate the total amperage or wattage of each item (for example: television, computer, printer, computer monitor, lamps, microwave oven, etc.). Advise the consumer not to exceed the rated branch circuit, circuit breaker, or fuse. If the consumer does not correct the problem, this could lead to a possible fire hazard if the home wiring overheats.

Location and Installation of Room Air Conditioner

Room air conditioners and portable models can be installed in various types of windows (Figures 29-2 and 29-3) or through the wall (Figure 29-4) for permanent installation. It is the installer's responsibility to make sure that the air conditioner is installed properly according to the manufacturer's specifications and local building codes. If the air conditioner is installed in a window, the installer must make sure that the air conditioner is secured to the window so that it will not fall out, causing injury, death, or property damage. Through-the-wall models should be secured to the wall to prevent the unit from being pushed out of the

Operating Voltages	Minimum Voltages	Maximum Voltages
115	104	127
120	108	132
208	188	229
220	198	242
230	207	253
240	216	264

TABLE 29-2   The Minimum/Maximum Operating Voltages for an Air Conditioner

PART VI

Types of windows

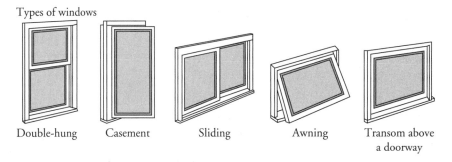

Double-hung   Casement   Sliding   Awning   Transom above
                                                                a doorway

FIGURE 29-2   Room air conditioners can be installed in these types of windows. When selecting an air conditioner to purchase, measure the window opening.

wall when the unit is installed or serviced. Portable air conditioners are much easier to install. The air intake and discharge hoses are installed through a small opening in the window. This type of unit can be moved around from room to room easily.

The following are some helpful tips when considering a location to install a window air conditioner:

- Make sure you have the correct size (BTUs) for the area being cooled.[2]

- When selecting an air conditioner, make sure that the air vents will be able to point to the center of the room for better air circulation.

FIGURE 29-3
A typical portable air conditioner installation.

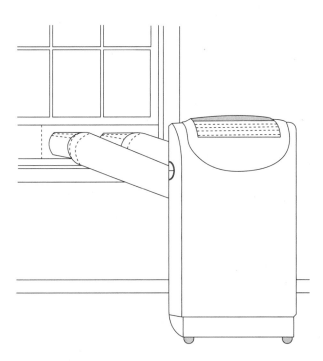

FIGURE 29-4
(a) A typical wall installation. The air conditioner must be installed with a slight pitch so that the condensate water can drain to the rear of the unit. (b) If the outside vents cannot clear the wall structure, purchase an air conditioner with a sleeve without side vents.

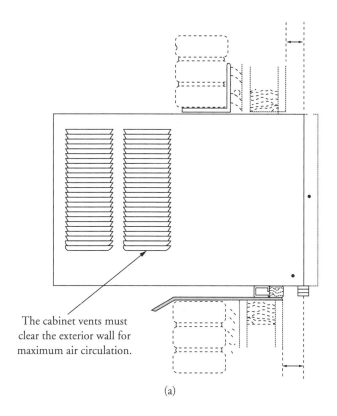

The cabinet vents must clear the exterior wall for maximum air circulation.

(a)

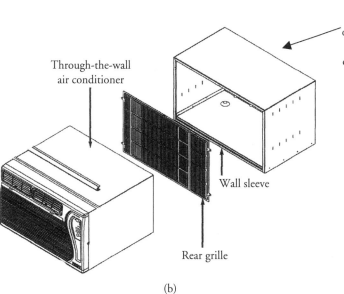

Some models have no side vents. The outside air is drawn into the rear and discharged through the rear.

Through-the-wall air conditioner

Wall sleeve

Rear grille

(b)

- When purchasing a window air conditioner, purchase one that is designed for the type of window in the home (see Figure 29-2). This will make for an easier installation.

- Make sure the unit will fit the window opening. Measure the opening of the window.

- Check the electrical outlet voltage. The voltage should match the operating voltage of the unit being installed.

- The electrical outlet should not be farther away than the length of the service cord.

- For larger air conditioners, install the brackets that come with the air conditioner. These brackets will reduce the stress on the window.

- Make sure there is adequate airflow with no obstructions on the outside of the building.

- Do not install a room air conditioner in an area where the temperature will exceed 120 degrees Fahrenheit.

The following are some helpful tips when considering a location to install a through-the-wall air conditioner:

- Make sure you have the correct size (BTUs) for the area being cooled.[2]

- When selecting an air conditioner, make sure that the air vents will be able to point to the center of the room for better air circulation.

- Measure the depth of the wall where the air conditioner will be installed. Make sure that the outside louvers will not be blocked (see Figure 29-4).

- Purchase an air conditioner with a slide-out chassis or with a wall sleeve for easier installation.

- Check the electrical outlet voltage. It should match the operating voltage of the unit being installed.

- The electrical outlet should not be farther away than the length of the service cord.

- Make sure there is adequate airflow with no obstructions on the outside of the building.

- Do not install a room air conditioner in an area where the temperature will exceed 120 degrees Fahrenheit.

Step-by-Step Troubleshooting by Symptom Diagnosis

When servicing an air conditioner, don't overlook the simple things that might be causing the problem. Step-by-step troubleshooting by symptom diagnosis is based on diagnosing malfunctions, with possible causes arranged into categories relating to the operation of the room air conditioner. This section is intended only to serve as a checklist to aid you in diagnosing a problem. Look at the symptom that best describes the problem you are experiencing with the air conditioner, and then correct the problem.

Air Conditioner Will Not Run

- Is the unit service cord plugged in?
- Check for voltage at the receptacle.
- Are the controls set properly?
- Check for a blown fuse or tripped circuit breaker. Check for total wattage on the branch circuit.
- Check the selector switch or electronic control board.
- Check all wiring connections and wiring.

Air Conditioner Will Not Cool or Cools Slightly

- Check for a dirty filter.
- Check for a dirty or restricted evaporator coil.
- Check for a dirty or restricted condenser coil.
- Check fan speed setting. Are the louvers adjusted correctly?
- Check the thermostat setting.
- Check the refrigeration cycle for leaks or an undercharge.
- Is the air conditioner the right size for the area to be cooled?[2]
- Is the fresh air intake or exhaust door open?

Compressor Will Not Run

- Check for the correct voltage at the receptacle.
- Check the control settings.
- Check the wiring connectors and wiring.
- Check the thermostat setting.
- Check the selector switch or electronic control board.
- Test the compressor for an open, short, or grounded winding.
- Test the overload.
- Test the compressor capacitor for an open, short, or ground.

Fan Motor Will Not Run

- Check the fan selector switch, main selector switch, or electronic control board.
- Check the wiring connectors and wiring.
- Test the fan capacitor for an open, short, or ground.
- Test the fan motor windings for an open, short, or ground.

Evaporator Coil Freezes Up

- Check for a dirty filter.
- Check for a dirty or restricted evaporator coil.

- Check the thermostat.
- Check fan motor.
- Check the refrigeration cycle for leaks, undercharge, or capillary tube restriction.

Condenser Coil Frozen (Heat Pump Models Only)

- Check defrost thermostat (outdoor thermostat) switch contacts and the capillary tube placement.
- Check the solenoid coil on the reversing valve.
- Check the reversing valve.
- Check the control settings.

Compressor Runs Continually and Will Not Cycle Off

- Check for excessive heat load. Is the air conditioner the right size for the area to be cooled?[2]
- Check for a partial refrigeration restriction in the line. Is the evaporator coil partially iced up?
- Check for a refrigerant leak.
- Check the running amperage against the model number identification plate on the air conditioner.
- Check the thermostat for proper operation.

Thermostat Will Not Cycle Off the Air Conditioner

- Check for stuck thermostat contacts.
- Check the thermostat setting.
- Check the wiring connections. Is the thermostat wired correctly?
- Is the air conditioner the right size for the area to be cooled?[2]

Thermostat Will Not Cycle the Air Conditioner On

- Check the thermostat bulb and capillary tube for loss of charge.
- Test the thermostat.
- Check the wiring connections and wiring.
- Check the control settings.

Thermostat Will Short-Cycle

- Is the air conditioner the right size for the area to be cooled?[2]
- Check for a dirty or restricted evaporator coil.
- Check for a dirty filter.
- Check the positioning of the thermostat bulb. Test the thermostat differential.
- Check and make sure that the plenum gasket is sealing properly.
- Check for outside air leakage into the air conditioner.

Compressor Runs and then Cycles on Overload

- Check for low voltage at the receptacle.
- Check the compressor overload.
- Did the compressor restart before the sealed system had a chance to equalize? Advise the customer to wait about three or four minutes after the air conditioner cycles off before restarting the compressor.
- Check the wiring connections and wiring.
- Check the compressor capacitor for an open, short, or ground.
- Check for a dirty or restricted condenser coil.
- Check the amperage of the air conditioner.
- Check for a kinked discharge line.

Air Conditioner Is Noisy

- Check the air conditioner installation.
- Make sure the fan blade and blower wheel are not striking the chassis.
- Check the compressor mounts and tubing.
- Inspect for loose cabinet parts.

Water Leaks Inside the Home

- Check the evaporator drain pan and drain for a blockage.
- Check weather sealing around the outer cabinet. Reseal if necessary.
- Check for water droplets on the outside of the base pan. If so, the evaporator drain pan might be cracked. Inspect the evaporator drain pan.
- Check the back side of the discharge grille. If it is wet, the evaporator coil might be dirty.
- Inspect all gaskets between the unit, outer cabinet, and window panes for air leakage.
- Check the angle of slope on the air conditioner installation.
- Check for a dirty evaporator coil.

Water Leaks on the Outside of the Home

- Check for water droplets on the outside of the base pan. If so, the evaporator drain pan might be cracked. Inspect the evaporator drain pan.
- Check for water between the condenser fan shroud and the compressor. If the water has collected around the compressor, inspect the fan shroud. Is it detached from the condenser coil?
- Check for a dirty condenser coil.
- Inspect the condenser fan blade—is the slinger ring in the correct position?

Air Conditioner Will Not Heat or There Is Not Enough Heat

- On electric heat models, check the electric heater and thermostats.
- On heat pump models, check the reversing valve, reversing valve solenoid coil, and thermostats.
- Check for a dirty filter.
- Check for an open fresh air intake or exhaust vent.
- Is the air conditioner the right size for the area?[2]
- Check for a dirty evaporator coil.
- Check the wiring connections and wiring.
- Check selector switch or electronic control board.

Heat Pump Will Not Go Into Defrost

- Check the defrost control and thermostats.
- Check the reversing valve and solenoid coil.
- Check the wiring connections and wiring.
- Check the electronic control board.

Air Conditioner Is Cooling When the Controls Are Set for Heat

- Check the wiring connections and wiring.
- Check the reversing valve and solenoid coil.
- Check the selector switch or electronic control board.
- Check the thermostats.

Room Air Conditioner Maintenance

Room air conditioners (including portable models) have air filters that need cleaning every 225 to 360 fan-hours of operation. The discharge grille area also needs vacuuming to remove the dust buildup. Twice a year, the following areas need to be inspected and cleaned:

- The evaporator coil
- The condenser coil
- The evaporator pan and base pan
- The indoor blower housing and blower wheel
- All the wiring connections and wiring
- The electrical and mechanical controls
- The voltage at the receptacle
- The inside and outside of the air conditioner
- All gaskets
- The drain system (clean it, too)

- The cabinet seal (clean the outer cabinet)
- The copper tubing

Twice a year you must inspect all control components: electrical and mechanical, electronic, as well as the power supply. The technician must use the proper testing instruments (voltmeter, ohmmeter, ammeter, wattmeter, etc.) to perform electrical tests. The technician should also use an air conditioner or refrigeration thermometer to test the room, outdoor, and coil operating temperatures. Use a sling psychrometer to measure the wet bulb temperatures indoors and outdoors.

When cleaning the air conditioner, use an approved cleaner to wash the unit. Remember to protect the electrical components and fan motor with plastic to prevent the water from damaging the components. Refer to the use and care manual that comes with every air conditioner for further maintenance instructions on the model you are servicing. Do not plug in or run the air conditioner after using water to clean the unit. Wait a few hours, allowing the air conditioner to completely dry out. To prevent electrical mishaps, the air conditioner must be totally dry before you can plug it in.

Performance Data

After you have completed the maintenance or repair on the air conditioner, perform an electrical test by checking the amperage or wattage on the unit, and compare the readings with the information on the model number data tag. At the same time, perform readings on the following:

- The room temperature and outside temperature.
- The temperature differential of the intake and discharge air through the evaporator coil. Take a reliable thermometer, place it in front of the air intake (where the air filter is located), and take a reading. Then place the thermometer in the discharge grille, and take a reading of the air blowing into the room. The difference between the two readings will be the temperature drop. This reading will vary among manufacturers and models. The temperature drop should be between 18 and 31 degrees Fahrenheit.
- The temperature differential of the intake and discharge air through the condenser coil. Use the same reliable thermometer to take the readings.
- Use a sling psychrometer to measure the indoor and the outdoor wet bulb temperatures. The sling psychrometer will measure the relative humidity in the room and outside (Table 29-3).
- Measure the operating voltage.
- Measure the startup and cycling amperage or wattage of the unit.

Take the readings and match them against the manufacturer's performance data. You can locate the air conditioner performance data on the manufacturer's Web site or in the manufacturer's service manual. The data that you accumulated should match—if it doesn't, adjustments will have to be made to bring the air conditioner up to manufacturer's standards. You might have to replace a component, clean the unit, or correct the installation.

Temperature of the Room or Outside (Dry Bulb) in Degrees Fahrenheit	The Temperature Difference Between the Dry Bulb and the Wet Bulb Temperatures in Degrees Fahrenheit								
	4	5	6	7	8	9	10	11	12
	Relative Humidity % at Pressure = 30.00 Inches								
40	68	60	52	45	37	29	22	15	7
50	74	67	61	55	49	43	38	32	27
52	75	69	63	57	51	46	40	35	29
54	76	70	64	59	53	48	42	37	32
56	76	71	65	60	55	50	44	39	34
58	77	72	66	61	56	51	46	41	37
60	78	73	68	63	58	53	48	43	39
62	79	74	69	64	59	54	50	45	41
64	79	74	70	65	60	56	51	47	43
66	80	75	71	66	61	57	53	48	44
68	80	76	71	67	62	58	54	50	46
70	81	77	72	68	64	59	55	51	48
72	82	77	73	69	65	61	57	53	49
74	82	78	74	69	65	61	58	54	50
76	82	78	74	70	66	62	59	55	51
78	83	79	75	71	67	63	60	56	53
80	83	79	75	72	68	64	61	57	54
82	84	80	76	72	69	65	61	58	55
84	84	80	76	73	69	66	62	59	56
86	84	81	77	73	70	66	63	60	57
88	85	81	77	74	70	67	64	61	57
90	85	81	78	74	71	68	65	61	58
100	86	83	80	77	73	70	68	65	62

TABLE 29-3   Psychrometric Table. A Wet Bulb Depression Chart. Formula: DB − WB = WDB

Repair Procedures

Each repair procedure is a complete inspection and repair process for a single room air conditioner component. It contains the information you need to test and replace components.

Electronic Components

The electronic components consist of the following: electronic control board, touchpad, and remote control unit.

The typical complaints associated with the electronic components are:

- Unable to program the touchpad panel functions.
- The display board will not display anything.
- Unusual display readouts.
- Unable to control the temperatures.
- The compressor will not run.
- The fan motor will not run.

To prevent electrostatic discharge (ESD) from damaging expensive electronic components, follow the steps in Chapter 11.

To handle these problems, perform the following steps:

1. **Verify the complaint.** Verify the complaint by operating the air conditioner controls. Turn off the electricity to the air conditioner, and wait for two minutes before turning it back on. If a fault code appears, look up the code. If the air conditioner will not power up, locate the technical data sheet behind the control panel for diagnostics information. It will be helpful if you can locate the actual service manual for the air conditioner model you are working on to properly diagnose the air conditioner. The service manual will assist you in properly placing the air conditioner in the service test mode for testing the air conditioner functions.

2. **Check for external factors.** You must check for external factors not associated with the air conditioner. Is there electricity to the air conditioner? The voltage at the wall receptacle must be within ±10 percent of the voltage rating on the model and serial data plate. Do you have the correct polarity? (See Chapter 6.) Is the electrical receptacle polarized and properly grounded?

3. **Disconnect the electricity.** Before working on the air conditioner, disconnect the electricity. This can be done by pulling the plug out of the wall receptacle. Or disconnect the electricity at the fuse panel or circuit breaker panel. Turn off the electricity.

WARNING *Some diagnostic tests will require you to test the components with the power turned on. When you disassemble the control panel or remove the outer cabinet, you can position the panel in such a way that the wiring will not make contact with metal. This will allow you to test the components without electrical mishaps.*

4. **Gain access to the electronic components.** You can gain access to the electronic components by removing the front grille and the screws on the control panel. Some units have either a one-piece or a two-piece grille with locking tabs and/or screws (Figure 29-5). Be careful not to break the tabs on the grille. On window models, you might have to remove the air conditioner from the window to gain access to the controls. On other models, you should be able to gain access by removing the screws that secure the control panel to the air conditioner frame. Next, tilt the control panel away from the air conditioner, making sure not to pull any of the wires off the controls.

WARNING *Do not touch the wiring or capacitor until it is discharged.*

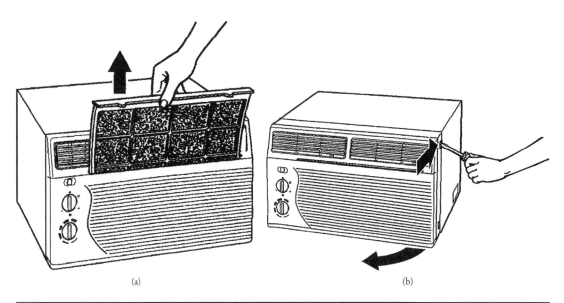

(a)   (b)

FIGURE 29-5 On some models, the air filter can be removed by sliding it out of the unit or by removing the front grille. Be careful not to damage the tabs on the grille.

WARNING *A capacitor will hold a charge indefinitely, even when it is not currently in use. A charged capacitor is extremely dangerous. Discharge all capacitors immediately any time that work is being conducted in their vicinity. Redischarge after repowering the equipment if further work must be done. Many capacitors are internally fused. If you are not sure, you can use a 20,000 ohm 2-watt resistor to discharge the capacitor. Do not use a screwdriver to short out the capacitor. By doing so, you will blow out the fuse in the capacitor and the capacitor will not work. Safely use an insulated pair of pliers to remove the wires from the capacitor and place the resistor across the capacitor terminals. When checking a dual capacitor with a capacitor analyzer or ohmmeter, you must test both sides of the capacitor.*

5. **Test the electronic components.** If you are able to run the air conditioner test mode, check the different functions of the air conditioner. Use the technical data sheet for the model you are servicing to locate the test points on the wiring schematic. Check all wiring connections and wiring. Using the technical data sheet, test the electronic control and display board, input voltages, and output voltages.

6. **Remove the defective component.** To remove the defective component, remove the screws that secure the printed circuit board to the air conditioner frame. Disconnect the connectors from the electronic control board and display.

7. **Install the new component.** To install a new component, just reverse the disassembly procedure, and reassemble. Reinstall all panels and the front grille, and restore the electricity to the air conditioner. Test the room air conditioner operation. To prevent ESD from damaging expensive electronic components, simply follow the steps in Chapter 11.

Thermistor

The thermistors used in the electronic models plug into the electronic control board, and they monitor the ambient and outdoor temperatures and troubleshooting capabilities.

The typical complaints associated with failure of the thermistor are:

- The air conditioner will not cool or heat.
- Erratic temperature control.
- The air conditioner doesn't run at all.

To handle these problems, perform the following steps:

1. **Verify the complaint.** Verify the complaint by checking the control setting. Turn the control to the lowest setting for cool or the highest setting for heat; then turn it back to a normal setting to see if the air conditioner starts cooling or heating. It will be helpful if you can locate the actual service manual for the air conditioner model you are working on to properly diagnose the air conditioner. The service manual will assist you in properly placing the air conditioner in the service test mode for testing the air conditioner functions.

2. **Check for external factors.** You must check for external factors not associated with the air conditioner. Is the air conditioner installed properly? Is the exhaust or fresh air intake vent open? Explain to the user how to set the controls. The voltage at the wall receptacle must be within ±10 percent of the voltage rating on the model and serial data plate. Do you have the correct polarity? (See Chapter 6.)

3. **Disconnect the electricity.** Before working on the air conditioner, disconnect the electricity to the unit. This can be done by pulling the plug from the receptacle. Or disconnect the electricity at the fuse panel or at the circuit breaker panel. Turn off the electricity.

WARNING *Some diagnostic tests will require you to test the components with the power turned on. When you disassemble the control panel or remove the outer cabinet, you can position the panel in such a way that the wiring will not make contact with metal. This will allow you to test the components without electrical mishaps.*

4. **Gain access to the thermistor.** To access the thermistor, remove the front grille and filter. Some units have either a one-piece or a two-piece grille with locking tabs and/or screws (see Figure 29-5). Be careful not to break the tabs on the grill. On window models, you might have to remove the air conditioner from the window to gain access to the controls. On other models, you should be able to gain access by removing the screws that secure the control panel to the air conditioner frame. Next, tilt the control panel away from the air conditioner, making sure not to pull any of the wires off the controls.

WARNING *Do not touch the wiring or capacitor until it is discharged. A capacitor will hold a charge indefinitely, even when it is not currently in use. A charged capacitor is extremely dangerous. Discharge all capacitors immediately any time that work is being conducted in their vicinity. Redischarge after repowering the equipment if further work must be done. Many capacitors are internally fused. If you are not sure, you can use a 20,000 ohm 2-watt resistor to discharge the capacitor. Do not use a screwdriver to short out the capacitor. By doing so, you will blow out the fuse in the capacitor and the capacitor will not work. Safely use an insulated pair of pliers to remove the wires from the capacitor and place the resistor across the capacitor terminals. When checking a dual capacitor with a capacitor analyzer or ohmmeter, you must test both sides of the capacitor.*

5. **Test the thermistor.** The thermistor is attached to the evaporator coil; trace the thermistor wire back to the electronic control board. Disconnect the thermistor connector from the board. Set the ohmmeter on R × 10K, and place the probes on the connector pin terminals. Measure the resistance of the thermistor. Using the technical data sheet, look for the reading and see if the results match. The reading can vary ±10 percent on the chart. Remove the thermistor from the evaporator coil, and submerge the thermistor in ice water for five minutes—the resistance will increase. As the thermistor warms up to ambient temperature, the resistance should return to the original reading. If you suspect an erratic thermistor, replace it with a duplicate of the original.

6. **Install a new thermistor.** To install the new thermistor, just reverse the order of disassembly, and reassemble. Then test the thermistor. Remember to reinstall the sensor in the same location from which it was removed. If you do not, the air conditioner will not cycle properly.

Thermostat (Operating)

The operating thermostat is located in the control panel. The operating thermostat monitors the ambient room temperature and cycles the air conditioner on and off.

The typical complaints associated with failure of the thermostat are:

- The air conditioner will not cool enough.
- The room or area is too cold.
- The air conditioner runs all the time.
- The air conditioner doesn't run at all.

To handle these problems, perform the following steps:

1. **Verify the complaint.** Verify the complaint by checking the control setting. Turn the control to the lowest setting for cool or the highest setting for heat; then turn it back to a normal setting to see if the air conditioner starts cooling or heating. It will be helpful if you can locate the actual service manual for the air conditioner model you are working on to properly diagnose the air conditioner.

2. **Check for external factors.** You must check for external factors not associated with the air conditioner. Is the air conditioner installed properly? Is the exhaust or fresh

air intake vent open? Explain to the user how to set the controls. The voltage at the wall receptacle must be within ±10 percent of the voltage rating on the model and serial data plate. Do you have the correct polarity? (See Chapter 6.)

3. **Disconnect the electricity.** Before working on the air conditioner, disconnect the electricity to the unit. This can be done by pulling the plug from the receptacle. Or disconnect the electricity at the fuse panel or at the circuit breaker panel. Turn off the electricity.

WARNING *Some diagnostic tests will require you to test the components with the power turned on. When you disassemble the control panel or remove the outer cabinet, you can position the panel in such a way that the wiring will not make contact with metal. This will allow you to test the components without electrical mishaps.*

4. **Gain access to the thermostat.** To access the thermostat, remove the front grille and filter. Some units have either a one-piece or a two-piece grille with locking tabs and/or screws (see Figure 29-5). Be careful not to break the tabs on the grille. On window models, you might have to remove the air conditioner from the window to gain access to the controls. On other models, you should be able to gain access by removing the screws that secure the control panel to the air conditioner frame. Next, tilt the control panel away from the air conditioner, making sure not to pull any of the wires off the controls.

WARNING *Do not touch the wiring or capacitor until it is discharged. A capacitor will hold a charge indefinitely, even when it is not currently in use. A charged capacitor is extremely dangerous. Discharge all capacitors immediately any time that work is being conducted in their vicinity. Redischarge after repowering the equipment if further work must be done. Many capacitors are internally fused. If you are not sure, you can use a 20,000 ohm 2-watt resistor to discharge the capacitor. Do not use a screwdriver to short out the capacitor. By doing so, you will blow out the fuse in the capacitor and the capacitor will not work. Safely use an insulated pair of pliers to remove the wires from the capacitor and place the resistor across the capacitor terminals. When checking a dual capacitor with a capacitor analyzer or ohmmeter, you must test both sides of the capacitor.*

5. **Test the thermostat.** To test the thermostat, disconnect the wires from the thermostat terminals. On two-wire thermostats, remove the wires from the control, label them, and place the ohmmeter probes on the terminals (Figure 29-6). Set the range scale on R × 1, and test for continuity. With the thermostat set in the warmest (off) position, you should not read continuity. When the thermostat is set to the coldest (highest) position, you should read continuity. Inspect the capillary tube for any damage. If the thermostat capillary tube has lost its charge, the air conditioner will not function properly. For thermostats with more than two terminals (heat models) on the control, you must refer to the wiring diagram on the technical data sheet for the correct terminals to test (Figures 29-7, 29-8, 29-9, 29-10, 29-11, 29-12, 29-13, and 29-14). If the thermostat is good, the problem must be elsewhere.

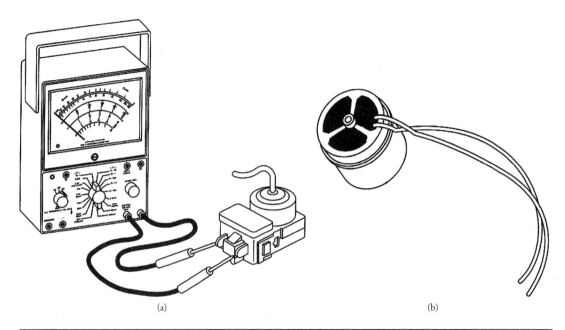

Figure 29-6   (a) Testing the thermostat. (b) Bimetal defrost thermostat.

6. **Remove the thermostat.** With the thermostat exposed and the wires already removed, the capillary tube must now be removed from the evaporator coil. The capillary tube is held in place on the coil with clips. Remove the capillary tube from the clips. If the clips come off the evaporator coil, remember where the clips go back. The placement of the clips is crucial for the air conditioner to function properly. Now remove the screws that secure the thermostat body to the control panel.

7. **Install a new thermostat.** To install the new thermostat, just reverse the order of disassembly, and reassemble. Then test the thermostat. Remember to reinstall the capillary tube in the same location from which it was removed. If you do not, the air conditioner will not cycle properly.

Thermostat (Defrost)

The defrost thermostat is only used on heat pump models. On heat pump models with electric heat, this control is a dual-purpose control that acts as an outdoor thermostat and a defrost control. When the thermostat sensing bulb, attached to the condenser coil, senses enough icing on the outdoor coil, it will shut off the compressor and turn on the electric heating element until the outdoor coil temperature reaches above 43 degrees Fahrenheit; then the electric heater will turn off and the air conditioner will resume in the reverse cycle mode (heat). When the outdoor coil temperature drops below 20 degrees Fahrenheit, the air conditioner will operate in electric heat mode continuously until the outdoor coil temperature rises above 43 degrees Fahrenheit. The fan motor will not turn off when defrost occurs, and the reversing four-way valve will not reverse. On models without electric heat,

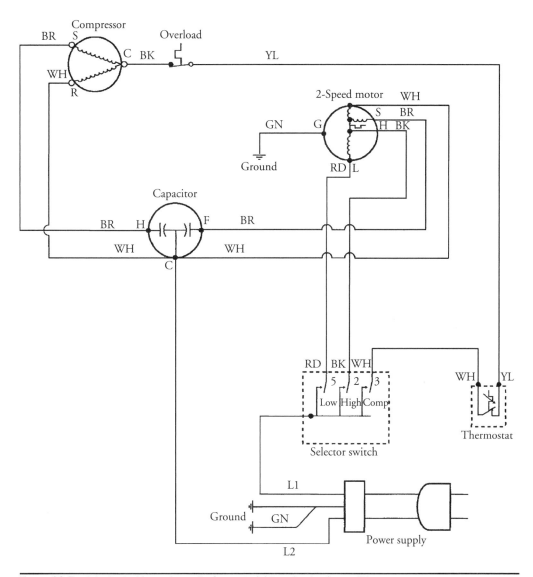

FIGURE 29-7 A typical wiring schematic for a straight-cool-only air conditioner.

the reversing four-way valve will reverse until the ice has defrosted. Some models use a bimetal type of thermostat, which initiates the defrost cycle.

The typical complaints associated with failure of the defrost thermostat (heat pump models only) are:

- Ice buildup on the outdoor coil.
- The air conditioner will not heat.

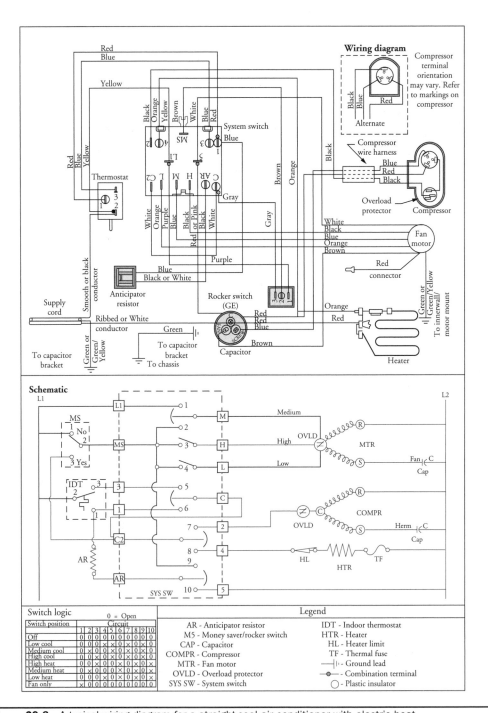

FIGURE 29-8   A typical wiring diagram for a straight-cool air conditioner with electric heat.

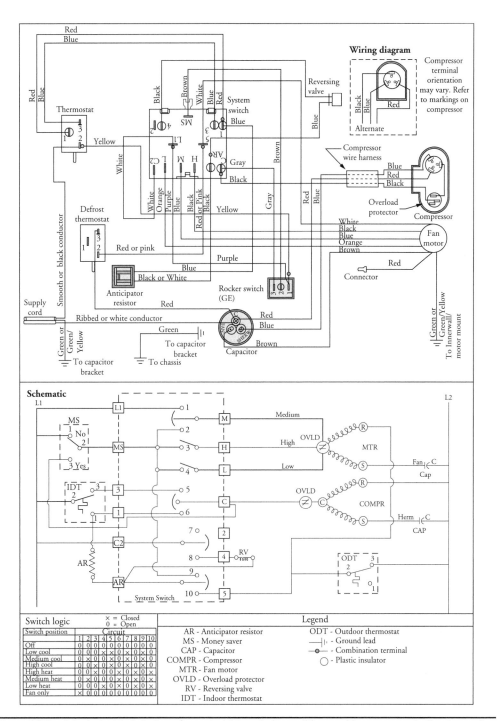

FIGURE 29-9 A typical wiring diagram for a straight-cool air conditioner with reverse cycle (heat pump).

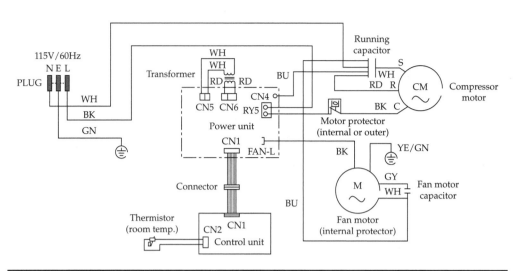

FIGURE 29-10 A typical wiring diagram, straight-cool air conditioner with electronic controls.

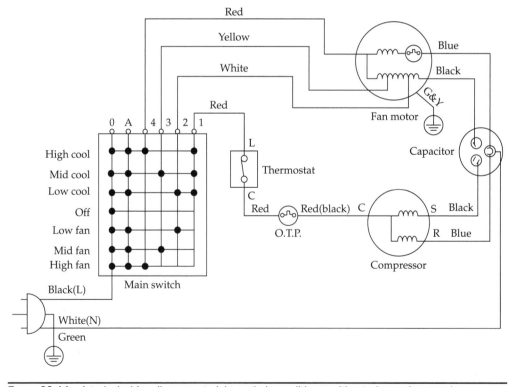

FIGURE 29-11 A typical wiring diagram, straight-cool air conditioner without electronic controls.

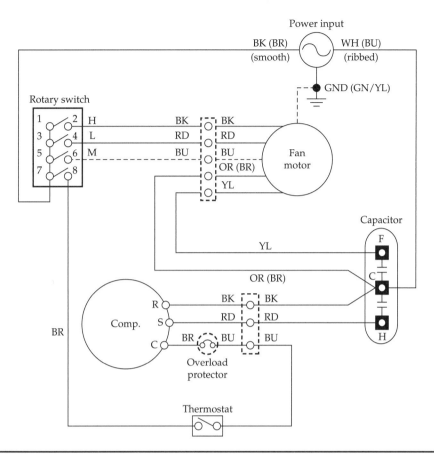

FIGURE 29-12 A typical wiring diagram, straight-cool air conditioner with rotary control.

To handle these problems, perform the following steps:

1. **Verify the complaint.** Verify the complaint by checking the control setting. Turn the control to the highest setting for heat, and then turn it back to a normal setting to see if the air conditioner starts heating. It will be helpful if you can locate the actual service manual for the air conditioner model you are working on to properly diagnose the air conditioner.

2. **Check for external factors.** You must check for external factors not associated with the air conditioner. Is the air conditioner installed properly? Is the exhaust or fresh air intake vent open? Explain to the user how to set the controls. The voltage at the wall receptacle must be within ±10 percent of the voltage rating on the model and serial data plate. Do you have the correct polarity? (See Chapter 6.)

3. **Disconnect the electricity.** Before working on the air conditioner, disconnect the electricity to the unit. This can be done by pulling the plug from the receptacle. Or disconnect the electricity at the fuse panel or at the circuit breaker panel. Turn off the electricity.

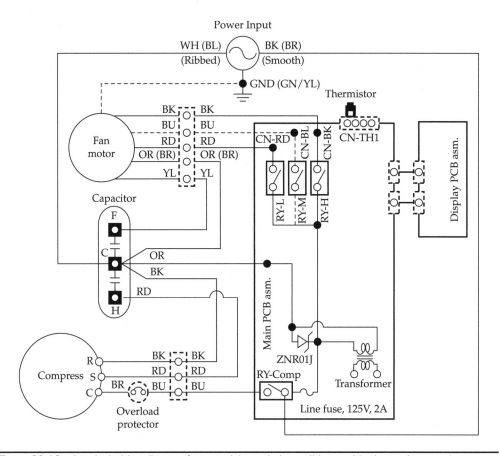

FIGURE 29-13   A typical wiring diagram for a straight-cool air conditioner with electronic controls.

WARNING *Some diagnostic tests will require you to test the components with the power turned on. When you disassemble the control panel or remove the outer cabinet, you can position the panel in such a way that the wiring will not make contact with metal. This will allow you to test the components without electrical mishaps.*

4. **Gain access to the defrost thermostat.** To access the defrost thermostat, remove the front grille and filter. Some units have either a one-piece or a two-piece grille with locking tabs and/or screws (see Figure 29-5). Be careful not to break the tabs on the grille. On window models, you might have to remove the air conditioner from the window to gain access to the defrost thermostat. On other models, you should be able to gain access by removing the screws that secure the control panel to the air conditioner frame. Next, tilt the control panel away from the air conditioner, making sure not to pull any of the wires off the controls. Some manufacturers might place the defrost thermostat control behind the bulkhead of the unit. The easiest way to locate that type of control is to follow the capillary tube from the condenser

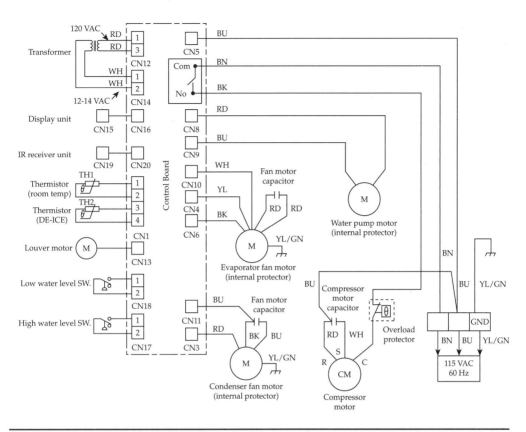

FIGURE 29-14 A typical wiring diagram for a portable air conditioner.

coil to the defrost thermostat body. The defrost thermostat is a nonadjustable thermostat without a stem or knob attached to it. On other models, the defrost thermostat might be a bimetal thermostat that is clamped to the end of the condenser coil with two wires coming out of it that lead back to the control panel.

WARNING *Do not touch the wiring or capacitor until it is discharged. A capacitor will hold a charge indefinitely, even when it is not currently in use. A charged capacitor is extremely dangerous. Discharge all capacitors immediately any time that work is being conducted in their vicinity. Redischarge after repowering the equipment if further work must be done. Many capacitors are internally fused. If you are not sure, you can use a 20,000 ohm 2-watt resistor to discharge the capacitor. Do not use a screwdriver to short out the capacitor. By doing so, you will blow out the fuse in the capacitor and the capacitor will not work. Safely use an insulated pair of pliers to remove the wires from the capacitor and place the resistor across the capacitor terminals. When checking a dual capacitor with a capacitor analyzer or ohmmeter, you must test both sides of the capacitor.*

5. **Test the defrost thermostat.** To test the defrost thermostat, disconnect the wires from the defrost thermostat terminals or from the selector control switch, label them, and place the ohmmeter probes on the terminals (see Figure 29-6a). Set the range scale on R × 1, and test the thermostat for continuity. Refer to the technical data sheet for the model you are servicing for the position of the switch contacts (open or closed position in the heat cycle) so that you can get the correct ohmmeter reading. Inspect the capillary tube for any damage. If the thermostat capillary tube has lost its charge, the air conditioner will not function properly. For defrost thermostats with more than two terminals on the control, you must refer to the wiring diagram on the technical data sheet for the correct terminals to test (see Figures 29-7, 29-8, 29-9, 29-10, 29-11, 29-12, 29-13, and 29-14). If the defrost thermostat is good, the problem must be elsewhere.

6. **Remove the thermostat.** With the thermostat exposed and the wires already removed, the capillary tube must now be removed from the condenser coil. The capillary tube is held in place on the coil with clips. Remove the capillary tube from the clips. If the clips come off the condenser coil, remember where the clips go back. The placement of the clips is crucial for the air conditioner to function properly. Now remove the screws that secure the thermostat body to the control panel. If you are servicing a model that has a bimetal clamp-on defrost thermostat (see Figure 29-6b), remove the clamp that secures the thermostat to the condenser coil. Remember where the thermostat was secured.

7. **Install a new thermostat.** To install the new defrost thermostat, just reverse the order of disassembly, and reassemble. Then test the defrost thermostat by running the heat cycle. Just remember to reinstall the capillary tube or bimetal clamp in the same location from which it was removed. If you do not, the air conditioner will not cycle properly.

Capacitor

A capacitor is a device that stores electricity to provide an electrical boost for motor starting. Most high-torque motors need a capacitor connected in series with the start winding circuit to produce the desired rotation under a heavy starting load. Some manufacturers will also add a run capacitor in the circuit for motor efficiency.

The typical complaints associated with failure of the capacitor are:

- Fuse is blown or the circuit breaker trips.
- Fan motor will not run.
- Fan motor has a burning smell.
- Motor tries to start and then shuts off on overload.

To handle these problems, perform the following steps:

1. **Verify the complaint.** Verify the complaint by operating the air conditioner. Listen carefully, and you will hear if there are any unusual noises or if the circuit breaker trips. If you smell something burning, immediately turn off the air conditioner, and pull out the plug. It will be helpful if you can locate the actual service manual for

the air conditioner model you are working on to properly diagnose the air conditioner.

2. **Check for external factors.** You must check for external factors not associated with the air conditioner. Is the air conditioner installed properly? Does it have the correct voltage? The voltage at the wall receptacle must be within ±10 percent of the voltage rating on the model and serial data plate. Do you have the correct polarity? (See Chapter 6.)

3. **Disconnect the electricity.** Before working on the air conditioner, disconnect the electricity. This can be done by pulling the plug from the electrical outlet. Be sure that you only remove the air conditioner plug. Or disconnect the electricity at the fuse panel or at the circuit breaker panel. Turn off the electricity.

WARNING *Some diagnostic tests will require you to test the components with the power turned on. When you disassemble the control panel or remove the outer cabinet, you can position the panel in such a way that the wiring will not make contact with metal. This will allow you to test the components without electrical mishaps.*

4. **Gain access to the capacitor.** Some models have the capacitor mounted on the fan motor; on other models, it is mounted behind the electrical controls. Remove the front grille and filter. Some units have either a one-piece or a two-piece grille with locking tabs and/or screws (see Figure 29-5). Be careful not to break the tabs on the grille. On window models, you might have to remove the air conditioner from the window to gain access to the controls. On other models, you should be able to gain access by removing the screws that secure the control panel to the air conditioner frame. Next, tilt the control panel away from the air conditioner, making sure not to pull any of the wires off the controls.

WARNING *Do not touch the wiring or capacitor until it is discharged. A capacitor will hold a charge indefinitely, even when it is not currently in use. A charged capacitor is extremely dangerous. Discharge all capacitors immediately any time that work is being conducted in their vicinity. Redischarge after repowering the equipment if further work must be done.*

5. **Test the capacitor.** Many capacitors are internally fused. If you are not sure, you can use a 20,000 ohm 2-watt resistor to discharge the capacitor. Do not use a screwdriver to short out the capacitor. By doing so, you will blow out the fuse in the capacitor and the capacitor will not work. Safely use an insulated pair of pliers to remove the wires from the capacitor and place the resistor across the capacitor terminals. When checking a dual capacitor with a capacitor analyzer or ohmmeter, you must test both sides of the capacitor. Set the ohmmeter on the highest scale, and then place one probe on one terminal and the other probe on the other terminal (Figure 29-15). Observe the meter action. While the capacitor is charging, the ohmmeter will read nearly zero ohms for a short period of time. Then the ohmmeter reading will slowly return toward infinity. If the ohmmeter reading deflects to zero and does not return to infinity, the capacitor is shorted and should be replaced. If the ohmmeter reading remains at infinity and does not dip toward zero, the capacitor is open and should be replaced.

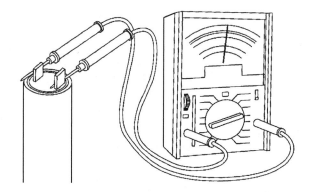

Figure 29-15
Testing a capacitor
for an open or
short circuit.

To test for a grounded capacitor, leave one meter probe on the capacitor terminal, and attach the other probe to the outer casing of the capacitor or air conditioner chassis ground. There should be no reading if the capacitor is not grounded. Next, remove the probe from the capacitor terminal, and place the probe on the other terminal. Again, there should be no reading indicated if the capacitor is not grounded.[3]

When using a capacitor analyzer to test capacitors, it will show whether the capacitor is "open" or "shorted." It will tell whether the capacitor is within its microfarads rating, and it will show whether the capacitor is operating at the proper power-factor percentage. The instrument will automatically discharge the capacitor when the test switch is released.

6. **Remove the capacitor.** Remove the capacitor from its mounting bracket.

7. **Install a new capacitor.** To install the new capacitor, just reverse the disassembly procedure, and reassemble.

NOTE *A capacitor is rated by its working voltage (WV or WVac) and by its storage capacity in microfarads (µF). Always replace a capacitor with one that has the same voltage rating and the same (or up to 10 percent greater) microfarad rating.*

Fan Motor

Most air conditioner models use a single fan motor with a double shaft for the fan blade and blower wheel. There are some models that use two fan motors, powered by AC or DC volts. Review the wiring diagram to see what type of motor(s) you are dealing with.

The typical complaints associated with failure of the fan motor are:

- No air is blowing out of the discharge grille.
- There is no cooling or heating.
- When the motor runs, there are loud noises.
- The fuse or circuit breaker trips when the air conditioner is started.

To handle these problems, perform the following steps:

1. **Verify the complaint.** Verify the complaint by operating the air conditioner. Listen carefully, and you will hear if there are any unusual noises or if the circuit breaker

trips. It will be helpful if you can locate the actual service manual for the air conditioner model you are working on to properly diagnose the air conditioner.

2. **Check for external factors.** You must check for external factors not associated with the air conditioner. Is the air conditioner installed properly? Does it have the correct voltage? The voltage at the wall receptacle must be within ±10 percent of the voltage rating on the model and serial data plate. Do you have the correct polarity? (See Chapter 6.)

3. **Disconnect the electricity.** Before working on the air conditioner, disconnect the electricity. This can be done by pulling the plug out of the wall receptacle. Be sure that you only remove the air conditioner plug. Or disconnect the electricity at the fuse panel or at the circuit breaker panel. Turn off the electricity.

WARNING *Some diagnostic tests will require you to test the components with the power turned on. When you disassemble the control panel or remove the outer cabinet, you can position the panel in such a way that the wiring will not make contact with metal. This will allow you to test the components without electrical mishaps.*

4. **Gain access to the fan motor.** Before you begin to remove the fan motor, test the motor windings. Remove the front grille and filter. Some units have either a one-piece or a two-piece grille with locking tabs and/or screws (see Figure 29-5). Be careful not to break the tabs on the grille. On window models, you might have to remove the air conditioner from the window to gain access to the controls. On other models, you should be able to gain access by removing the screws that secure the control panel to the air conditioner frame. Next, tilt the control panel away from the air conditioner, making sure not to pull any of the wires off the controls.

WARNING *Do not touch the wiring or capacitor until it is discharged. A capacitor will hold a charge indefinitely, even when it is not currently in use. A charged capacitor is extremely dangerous. Discharge all capacitors immediately any time that work is being conducted in their vicinity. Redischarge after repowering the equipment if further work must be done. Many capacitors are internally fused. If you are not sure, you can use a 20,000 ohm 2-watt resistor to discharge the capacitor. Do not use a screwdriver to short out the capacitor. By doing so, you will blow out the fuse in the capacitor and the capacitor will not work. Safely use an insulated pair of pliers to remove the wires from the capacitor and place the resistor across the capacitor terminals. When checking a dual capacitor with a capacitor analyzer or ohmmeter, you must test both sides of the capacitor.*

5. **Disconnect the motor wire leads.** Disconnect the motor wire leads from the selector switch, and label them. Check the motor windings for continuity (Figure 29-16). Check for resistance from the common wire lead (white) to the high-speed (black) wire lead (Figure 29-16a). Then check the resistance from the common wire lead (white) to the medium-speed (blue) and the common wire lead (white) to the low-speed (red) winding (Figures 29-16b and 29-16c). If the fan motor has a capacitor wire (brown), check for resistance from the brown wire lead to the common wire lead (white).[4] To check for a grounded winding in the motor, take the ohmmeter

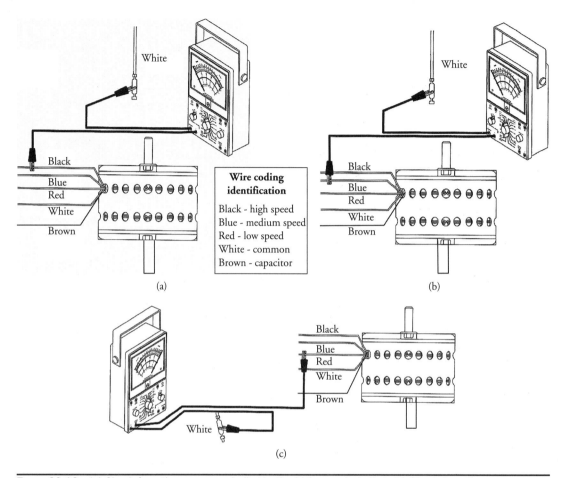

Wire coding identification

Black - high speed
Blue - medium speed
Red - low speed
White - common
Brown - capacitor

FIGURE 29-16   (a) Check from the common winding to the high-speed winding. (b) Check from the common winding to the medium-speed winding. (c) Check from the common winding to the low-speed winding.

probes and check from each motor wire lead to the motor housing (Figure 29-17). The ohmmeter will indicate continuity if the windings are grounded.

Finally, spin the motor shaft—it should turn freely. If the shaft is hard to turn, replace the fan motor. Now move the motor shaft in an up-and-down motion perpendicular to the motor body. The shaft should have virtually no movement. If there is a lot of play in the end bell bearings, replace the fan motor.[5]

6. **Remove the fan motor.** To remove the fan motor, the air conditioner will have to be uninstalled or the chassis slid out of the outer case and placed on a solid table or workbench.

WARNING   *This procedure will require two people to uninstall and move the air conditioner.*

Figure 29-17
Checking for a
grounded motor.

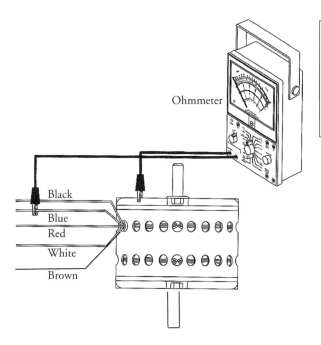

Ohmmeter

Wire coding identification
Black - high speed
Blue - medium speed
Red - low speed
White - common
Brown - capacitor

Black
Blue
Red
White
Brown

Depending on which model you are servicing, the removal of the fan motor will vary from manufacturer to manufacturer. However, the basic steps are the same for every air conditioner: The fan motor is secured to the bulkhead (Figure 29-18). In order to remove it, you must first remove the condenser fan blade. To gain access to the fan blade, remove the screws and brackets[6] that secure the condenser coil to the fan shroud. On some models, the upper housing has a cover—remove it. The condenser coil will have to be lifted up out of the base pan and gently moved out of the way to gain access to the fan blade. You only have to move the condenser coil enough to gain access to the fan blade and to provide room to remove the fan blade without damaging the condenser coil.

Warning *When moving the condenser coil away from the fan blade, do not kink or break the copper tubing. Remember that there is high-pressure refrigerant in the lines.*

Remove the setscrew or clamp from the condenser fan blade. When removing the fan blade, be careful not bend or damage it. To gain access to the evaporator blower wheel (see Figure 29-18), remove the screws from the evaporator blower housing, exposing the blower wheel. Depending on which model you are servicing, you may have to remove part of the housing to gain access. Next, remove the clamp from the blower wheel. You will not be able to remove the blower wheel at this time until the fan motor is removed from the bulkhead. With the condenser fan blade removed and the evaporator blower wheel loose on the motor shaft, you are now ready to remove the motor. Remove the fan motor wiring from the control area, making sure

PART VI

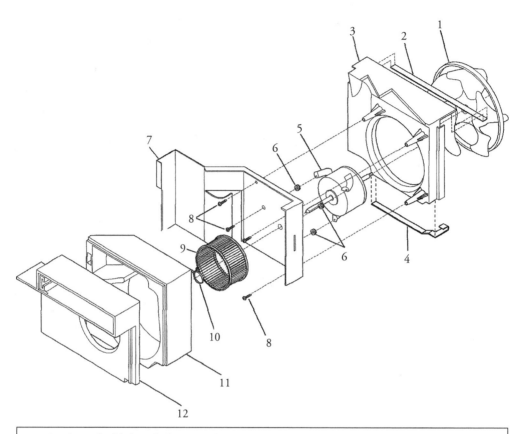

1. Condenser fan blade with clamp or set screw	7. Bulkhead housing
2. Condenser fan housing seal	8. Screws
3. Condenser coil and condenser fan blade housing	9. Evaporator blower wheel
4. Condenser fan housing seal	10. Blower wheel clamp
5. Fan motor	11. Evaporator blower housing
6. Fan motor mounting nuts	12. Evaporator blower housing

Figure 29-18   An exploded view of the fan motor, condenser fan blade, blower wheel, and the housings.

to free up the fan motor wiring harness that may be secured to the chassis. Now remove the fan motor nuts or bolts that secure the fan motor to the bulkhead. Grab hold of the fan motor, and pull it out of the air conditioner while removing the evaporator blower wheel from the motor shaft. The blower wheel will remain in the evaporator blower housing.

NOTE *Replace the fan motor with a duplicate of the original for easier installation. With the fan motor removed from the unit, it is the perfect time to chemically clean the remainder of the air conditioner. Everything will be exposed, and it will be easier to clean the unit.*

7. **Reinstall the fan motor.** To reinstall the fan motor, just reverse the instructions in step 6, and reassemble. When you reinstall the condenser fan blade, make sure to place it back on the motor shaft in the same position. Most room air conditioner condenser fan blades have a slinger ring that is attached to the paddles. The slinger ring will sit in a sump area in the base pan, allowing the fan blade to pick up the condensate water, flinging it against the condenser coil. Before reinstalling the outer cabinet or installing the air conditioner in the wall, test the air conditioner first. Make sure the fan blade and the blower wheel are not hitting against anything.

Compressor and Overload Protector

The compressor (reciprocating or rotary type) is the heart of the vapor compression refrigeration system. It is used to circulate the refrigerant throughout the sealed system. The overload is attached to the compressor housing and protects the compressor.

The typical complaints associated with failure of the compressor are:

- The air conditioner will not cool.
- The room temperature is warm.
- The air conditioner does not run at all.
- The compressor makes a humming or buzzing noise and stops.
- The circuit breaker trips when the air conditioner starts up.

To handle these problems, perform the following steps:

1. **Verify the complaint.** Verify the complaint by asking the customer to describe what the air conditioner is doing. It will be helpful if you can locate the actual service manual for the air conditioner model you are working on to properly diagnose the air conditioner.

2. **Check for external factors.** You must check for external factors not associated with the air conditioner. Is the air conditioner installed properly? Does it have the correct voltage? Check for a voltage drop during air conditioner startup. The voltage at the wall receptacle must be within ±10 percent of the voltage rating on the model and serial data plate. Do you have the correct polarity? (See Chapter 6.)

3. **Disconnect the electricity.** Before working on the air conditioner, disconnect the electricity. This can be done by pulling the plug from the receptacle. Or disconnect the electricity at the fuse panel or at the circuit breaker panel. Turn off the electricity.

WARNING *Before you begin to service any high-voltage components within the air conditioner, you must discharge the high-voltage capacitors first.*

WARNING *A capacitor will hold a charge indefinitely, even when it is not currently in use. A charged capacitor is extremely dangerous. Discharge all capacitors immediately any time that work is being conducted in their vicinity. Redischarge after repowering the equipment if further work must be done. Many capacitors are internally fused. If you are not sure, you can use a 20,000 ohm 2-watt resistor to discharge the capacitor. Do not use a screwdriver to short out the capacitor. By doing so, you will blow out the fuse in the capacitor and the capacitor will not work. Safely use an insulated pair of pliers to remove the wires from the capacitor and place the resistor across the capacitor terminals. When checking a dual capacitor with a capacitor analyzer or ohmmeter, you must test both sides of the capacitor.*

WARNING *Some diagnostic tests will require you to test the components with the power turned on. When you disassemble the control panel or remove the outer cabinet, you can position the panel in such a way that the wiring will not make contact with metal. This will allow you to test the components without electrical mishaps.*

4. **Gain access to the compressor.** To access the compressor, pull the air conditioner out and away from the window or wall. Remove the outer cabinet (on some models). This will expose the compressor, fan motor, and electrical controls (Figure 29-19). Next, remove the compressor terminal cover (Figure 29-20) by removing the retaining nut or clip that secures the cover. Remove the terminal cover and washer.

5. **Test the overload protector.** To test the overload protector, remove the wires from the overload and compressor terminals. Next, place the probes of the ohmmeter on the overload terminals (see Figure 29-20). Set the meter scale on R × 1. The reading will show continuity. If not, replace the overload protector.

6. **Test the compressor.** The compressor terminals are marked C, S, and R: C indicates the common winding terminal, S indicates the start winding terminal, and

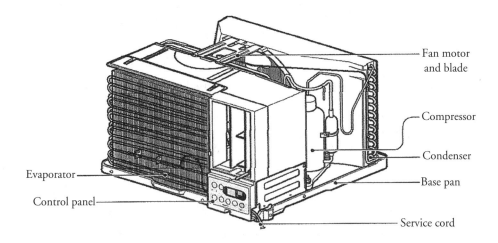

FIGURE 29-19 An illustration of the component location on a slide-out chassis wall air conditioner.

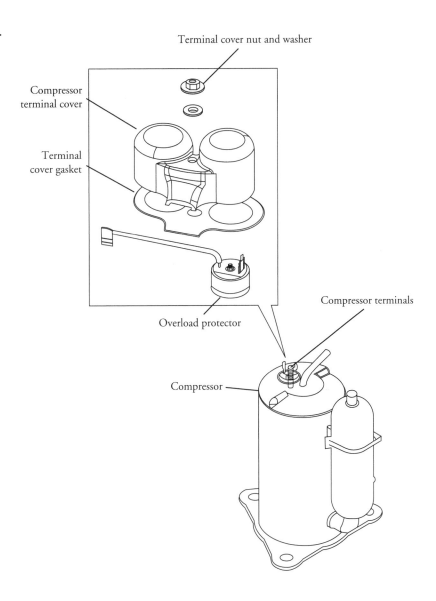

FIGURE 29-20
An exploded view of
a rotary compressor,
overload protector,
and terminal cover.

Terminal cover nut and washer

Compressor
terminal cover

Terminal
cover gasket

Overload protector

Compressor terminals

Compressor

R indicates the run winding terminal. Refer to the wiring diagram for the model that
you are servicing. Set the meter scale on R × 1, touch the probes together, and adjust
the needle setting to indicate a zero reading. Next, place the probes of the ohmmeter
on the terminals marked S and R (see Figure 25-37a). The meter reading will show
continuity. Now place the meter probes on the terminals marked C and S. The meter
reading will show continuity. Finally, place the meter probes on the terminals
marked C and R. The meter reading will show continuity. The total number of ohms
measured between S and R is equal to the sum of C to S plus C to R. The compressor
should be tested for proper resistance, as indicated on the wiring diagram.

To test the compressor for ground, place one probe on a compressor terminal, and attach the other probe to the compressor housing or to any good ground (see Figure 25-37b). Set the meter scale to R × 1000. The meter reading will show no continuity. Repeat this for the remaining two terminals. The meter reading will show no continuity. If you get a continuity reading from any of these terminals to ground, the compressor is grounded. Replace it.

WARNING *This procedure may require two people to uninstall and move the air conditioner.*

Electric Heater Element

Air conditioner heating elements are made with a nickel-chromium wire, having both tensile strength and high resistance to current flow. The resistance and voltage can be measured with a multimeter to verify if the element is functioning properly. Some manufacturers are also using calrod heaters instead of nickel-chromium wire heaters. This type of heater can be tested for resistance and voltage, too.

The typical complaint associated with failure of the electric heater element is that the air conditioner will not heat at all. To handle this problem, perform the following steps:

1. **Verify the complaint.** Verify the complaint by checking the control setting. Turn the control to the highest setting for heat; then turn it back to a normal setting to see if the air conditioner starts heating. It will be helpful if you can locate the actual service manual for the air conditioner model you are working on to properly diagnose the air conditioner.

2. **Check for external factors.** You must check for external factors not associated with the air conditioner. Is the air conditioner installed properly? Is the exhaust or fresh air intake vent open? Check for the correct voltage to the air conditioner. The voltage at the wall receptacle must be within ±10 percent of the voltage rating on the model and serial data plate. Do you have the correct polarity? (See Chapter 6.) Explain to the user how to set the controls.

3. **Disconnect the electricity.** Before working on the air conditioner, disconnect the electricity to the unit. This can be done by pulling the plug from the receptacle. Or disconnect the electricity at the fuse panel or at the circuit breaker panel. Turn off the electricity.

WARNING *Some diagnostic tests will require you to test the components with the power turned on. When you disassemble the control panel or remove the outer cabinet, you can position the panel in such a way that the wiring will not make contact with metal. This will allow you to test the components without electrical mishaps.*

4. **Gain access to the electric heater element circuitry.** To access the electric heater element circuitry, remove the front grille and filter. Some units have either a one-piece or a two-piece grill with locking tabs and/or screws (see Figure 29-5). Be careful not to break the tabs on the grill. On window models, you might have to remove the air conditioner from the window to gain access to the controls. On other models, you should be able to gain access by removing the screws that secure the control panel to the air conditioner frame. Next, tilt the control panel away from the air conditioner, making sure not to pull any of the wires off the controls.

WARNING *Do not touch the wiring or capacitor until it is discharged. A capacitor will hold a charge indefinitely, even when it is not currently in use. A charged capacitor is extremely dangerous. Discharge all capacitors immediately any time that work is being conducted in their vicinity. Redischarge after repowering the equipment if further work must be done. Many capacitors are internally fused. If you are not sure, you can use a 20,000 ohm 2-watt resistor to discharge the capacitor. Do not use a screwdriver to short out the capacitor. By doing so, you will blow out the fuse in the capacitor and the capacitor will not work. Safely use an insulated pair of pliers to remove the wires from the capacitor and place the resistor across the capacitor terminals. When checking a dual capacitor with a capacitor analyzer or ohmmeter, you must test both sides of the capacitor.*

5. **Test the electric heater element.** At this point, you do not have to disassemble the air conditioner to determine if the electric heater element is defective. Locate the wiring diagram (see Figure 29-8), and trace the electric heater element circuit. You will notice that the heater element, thermal fuse, and heater limit control (thermostat) are wired in series with the selector switch and operating thermostat. These five components make up the heating circuit. Locate the electric heater element wiring, and disconnect the wires from the terminals. For example, in the wiring diagram shown in Figure 29-8, the orange and red wires from the electric heater element go to the capacitor common terminal and to the number 4 terminal on the selector switch. If you remove those two wires from their terminals, you have just isolated the electric heater element, thermal fuse, and the heater limit control. Set your ohmmeter to the R × 1 scale, and check for continuity between those two wires.[7] If the ohmmeter reads resistance in the circuit, then the problem is elsewhere. If your ohmmeter reads an open circuit, you will have to remove the electric heater assembly for further testing.

6. **Remove the electric heater assembly.** The electric heater assembly is located in the evaporator blower housing on the discharge side. It may be behind a metal screen on older models. For newer models, the heater element is located behind the evaporator coil. To remove the heater assembly, the air conditioner will have to be uninstalled.

WARNING *This procedure may require two people to uninstall and move the air conditioner.*

Remove the screws that secure the top panel on the evaporator blower housing (see Figure 29-18). On some models, part of the evaporator housing can be removed for easy access to the heater assembly. Once you gain access to the heater assembly, remove the screws that secure the assembly to the air conditioner. Replace with a duplicate of the original.

NOTE *With the electric heater assembly removed from the unit, it is the perfect time to chemically clean the remainder of the air conditioner. Everything will be exposed, and it will be easier to clean the unit. Just remember to protect the electrical components and fan motor.*

7. **Install the heater assembly.** To install a new electric heater assembly, just reverse the disassembly procedure, and reassemble. Reinstall all panels and the front grille, and restore the electricity to the air conditioner. Test the room air conditioner heat operation.

Reversing Valve Solenoid Coil

A reversing valve solenoid is a device used to convert electrical energy into mechanical energy. When the solenoid is energized, it acts like an electromagnet and is positioned to move a predesignated metal object within the reversing valve.

The typical complaint associated with failure of the reversing valve solenoid coil is that the air conditioner will not heat. To handle this problem, perform the following steps:

1. **Verify the complaint.** Verify the complaint by checking the control setting. Turn the control to the lowest setting for cool or the highest setting for heat; then turn it back to a normal setting to see if the air conditioner starts cooling or heating. It will be helpful if you can locate the actual service manual for the air conditioner model you are working on to properly diagnose the air conditioner.

2. **Check for external factors.** You must check for external factors not associated with the air conditioner. Is the air conditioner installed properly? Is the exhaust or fresh air intake vent open? Explain to the user how to set the controls. The voltage at the wall receptacle must be within ±10 percent of the voltage rating on the model and serial data plate. Do you have the correct polarity? (See Chapter 6.)

3. **Disconnect the electricity.** Before working on the air conditioner, disconnect the electricity to the unit. This can be done by pulling the plug from the receptacle. Or disconnect the electricity at the fuse panel or at the circuit breaker panel. Turn off the electricity.

WARNING *Some diagnostic tests will require you to test the components with the power turned on. When you disassemble the control panel or remove the outer cabinet, you can position the panel in such a way that the wiring will not make contact with metal. This will allow you to test the components without electrical mishaps.*

4. **Gain access to the reversing valve solenoid coil.** To access the reversing valve solenoid coil wiring, remove the front grille and filter. Some units have either a one-piece or a two-piece grill with locking tabs and/or screws (see Figure 29-5). Be careful not to break the tabs on the grille. On window models, you might have to remove the air conditioner from the window to gain access to the controls. On other models, you should be able to gain access by removing the screws that secure the control panel to the air conditioner frame. Next, tilt the control panel away from the air conditioner, making sure not to pull any of the wires off the controls.

WARNING *Do not touch the wiring or capacitor until it is discharged. A capacitor will hold a charge indefinitely, even when it is not currently in use. A charged capacitor is extremely dangerous. Discharge all capacitors immediately any time that work is being conducted in their vicinity. Redischarge after repowering the equipment if further work must be done. Many capacitors are internally fused. If you are not sure, you can use a 20,000 ohm 2-watt resistor to discharge the capacitor. Do not use a screwdriver to short out the capacitor. By doing so, you will blow out the fuse in the capacitor and the capacitor will not work. Safely use an insulated pair of pliers to remove the wires from the capacitor and place the resistor across the capacitor terminals. When checking a dual capacitor with a capacitor analyzer or ohmmeter, you must test both sides of the capacitor.*

5. **Test the reversing valve solenoid coil.** To test the solenoid coil, disconnect the wires from the selector switch terminals. Look on the wiring diagram for the correct wires to remove. Set the ohmmeter on the R × 10K scale, and place the meter probes on the solenoid coil wire terminals. You should read resistance in the coil. If not, replace the solenoid coil with a duplicate of the original.

6. **Replace the reversing valve solenoid coil.** To access the solenoid coil, pull the air conditioner out and away from the window or wall. Remove the outer cabinet (on some models). This will expose the compressor, reversing valve, fan motor, and electrical controls (see Figure 29-19). The reversing valve is located near the compressor. Remove the retaining nut and washer from the pilot valve stem on the reversing valve, and remove the solenoid coil.

7. **Install the reversing valve solenoid coil.** To install a new solenoid coil, just reverse the disassembly procedure, and reassemble. Reinstall all panels and the front grille, and restore the electricity to the air conditioner. Test the room air conditioner heat operation.

Selector Switch

The selector switch is mounted in the control panel. Its main purpose is to select which cycle to run—for example: heat, cool, or fan only.

The typical complaints associated with failure of the selector switch are:

- The air conditioner will not cool or heat.

- The air conditioner doesn't run at all.

To handle these problems, perform the following steps:

1. **Verify the complaint.** Verify the complaint by checking the control setting to see if the air conditioner starts cooling or heating. It will be helpful if you can locate the actual service manual for the air conditioner model you are working on to properly diagnose the air conditioner.

2. **Check for external factors.** You must check for external factors not associated with the air conditioner. Is the air conditioner installed properly? Is the exhaust or fresh air intake vent open? Is the voltage correct to the air conditioner? The voltage at the wall receptacle must be within ±10 percent of the voltage rating on the model and serial data plate. Do you have the correct polarity? (See Chapter 6.) Explain to the user how to set the controls.

3. **Disconnect the electricity.** Before working on the air conditioner, disconnect the electricity to the unit. This can be done by pulling the plug from the receptacle. Or disconnect the electricity at the fuse panel or at the circuit breaker panel. Turn off the electricity.

WARNING *Some diagnostic tests will require you to test the components with the power turned on. When you disassemble the control panel or remove the outer cabinet, you can position the panel in such a way that the wiring will not make contact with metal. This will allow you to test the components without electrical mishaps.*

4. **Gain access to the selector switch.** To access the selector switch, remove the front grille and filter. Some units have either a one-piece or a two-piece grill with locking tabs and/or screws (see Figure 29-5). Be careful not to break the tabs on the grille. On window models, you might have to remove the air conditioner from the window to gain access to the controls. On other models, you should be able to gain access by removing the screws that secure the control panel to the air conditioner frame. Next, tilt the control panel away from the air conditioner, making sure not to pull any of the wires off the controls.

WARNING *Do not touch the wiring or capacitor until it is discharged. A capacitor will hold a charge indefinitely, even when it is not currently in use. A charged capacitor is extremely dangerous. Discharge all capacitors immediately any time that work is being conducted in their vicinity. Redischarge after repowering the equipment if further work must be done. Many capacitors are internally fused. If you are not sure, you can use a 20,000 ohm 2-watt resistor to discharge the capacitor. Do not use a screwdriver to short out the capacitor. By doing so, you will blow out the fuse in the capacitor and the capacitor will not work. Safely use an insulated pair of pliers to remove the wires from the capacitor and place the resistor across the capacitor terminals. When checking a dual capacitor with a capacitor analyzer or ohmmeter, you must test both sides of the capacitor.*

5. **Test the selector switch.** To test the selector switch, locate the selector switch circuit on the wiring diagram (see Figures 29-7, 29-8 29-9 29-10, 29-11, 29-12, 29-13, and 29-14). Identify the terminals that turn on the different components to be tested. Only remove those wires for the switch contacts you are testing. Set the ohmmeter on the R × 1 scale. Next, place the ohmmeter probes on those terminals. Then select the setting by either rotating the dial or by pressing the proper button on the switch (Figure 29-21). If the switch contacts are good, your meter will show continuity. Test all of the remaining component settings on the selector switch. Remember to check the wiring diagram for the correct switch contact terminals (those that correspond to the setting that you are testing).

6. **Remove the selector switch.** To remove the selector switch, remove all of the wires from the switch and label them, remove the screws that hold the switch to the control panel, and remove the switch.

7. **Install the new selector switch.** To install the new selector switch, just reverse the disassembly procedure, and reassemble. Then reattach the wires to the switch terminals according to the wiring diagram. Reassemble the control panel. When you are reassembling the control panel, make sure that the wires do not become pinched between the control panel and the air conditioner frame.

Troubleshooting Sealed-System Problems

If you suspect a sealed-system malfunction, be sure to check out all external factors first. These include:

- Thermostats
- Compressor

FIGURE 29-21
The selector switch
contact test points.

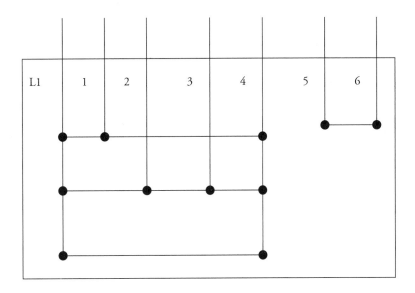

Selector switch

Switch position	Contacts closed
Off	None
Cool	L1 to 1 & 4
Heat	L1 to 2, 3, 4, 5, & 6
Fan only	L1 to 4

- Overload on the compressor
- Fan motor
- Evaporator and condenser coils getting good air circulation
- Air conditioner installation
- Make sure the heater is not on at the same time as cooling
- Make sure that the reversing valve and solenoid coil are operating properly

After eliminating all of these external factors, you will then systematically check the sealed system. This is accomplished by comparing the conditions found in a normally operating air conditioner. These conditions are:

- Room temperature
- Wattage
- Condenser temperature
- Evaporator inlet sound (gurgle, hiss, etc.)
- Evaporator cooling pattern

- High-side pressure[8]
- Low-side pressure[8]
- Pressure equalization time

One thing to keep in mind: No single indicator is conclusive proof that a particular sealed-system problem exists. Rather, a combination of findings must be used to definitively pinpoint the exact problem.

Low-Capacity Compressor

Symptoms of a low-capacity compressor in the sealed system are:

- Temperatures in the room or area will be above normal.
- The wattage and amperage will be below normal, as indicated on the model/serial plate.
- The temperature of the condenser coil will be below normal.
- At the evaporator coil, you will hear a slightly reduced gurgling noise.
- The evaporator coil will show a normal cooling pattern.
- The high-side pressure will be below normal, and the low-side pressure will be above normal.[8]
- The pressure equalization time might be normal or shorter than normal.

Refrigerant Leak

Symptoms of a refrigerant leak in the sealed system are:

- Temperatures in the room or area will be below normal.
- The wattage and amperage will be below normal, as indicated on the model/serial plate.
- The condenser coil will be cool to the touch at the last pass, or even as far as midway through the coil.
- At the evaporator coil, you will hear a gurgling noise, a hissing noise, or possibly an intermittent hissing or gurgling noise.
- The evaporator coil will show a frost pattern in the lower rungs of the coil.
- The high- and low-side pressures will be below normal.[8]
- The pressure equalization time might be normal or shorter than normal.

Overcharged Air Conditioner

If the sealed system is overcharged, the symptoms are:

- The room temperature will be higher than normal.
- The wattage and amperage will be above normal, as indicated on the model/ serial plate.
- The temperature of the condenser coil will be above normal.

- At the evaporator coil, you will hear a constant gurgling noise—generally, a higher sound level than normal.
- The evaporator coil will show a full frost pattern. If you remove the cover, you will possibly see the suction line frosted back to the compressor.
- The high- and low-side pressures will be above normal.[8]
- The pressure equalization time will be normal.

Slight Restriction

Symptoms of a slight restriction in the sealed system are:

- The room temperature will be below normal.
- The wattage and amperage will be below normal, as indicated on the model/serial plate.
- The temperature of the condenser coil will be slightly below normal.
- At the evaporator coil, you will hear a constant gurgling noise and a low sound level.
- The evaporator coil cooling pattern will be receded.
- The high- and low-side pressures will be below normal.[8]
- The pressure equalization time will be longer than normal.

Partial Restriction

Symptoms of a partial restriction in the sealed system are:

- The room temperature will be higher than normal.
- The wattage and amperage will be below normal, as indicated on the model/serial plate.
- The temperature of the condenser coil will be below normal more than halfway on the coil.
- At the evaporator coil, you will hear a constant gurgling noise and a considerably lower sound level.
- The evaporator coil cooling pattern will be considerably receded.
- The high- and low-side pressures will be below normal.[8]
- The pressure equalization time will be longer than normal.

Complete Restriction

Symptoms of a complete restriction in the sealed system are:

- The room temperature will be warm.
- The wattage and amperage will be considerably below normal, as indicated on the model/serial plate.
- The temperature of the condenser coil will be cool or at room temperature.
- At the evaporator coil, you will hear no sounds.
- The evaporator coil will not be cool.

- The high-side pressure will be equal to the pressure of refrigerant at room temperature.

- The low-side pressure will be in a deep vacuum.[8]

- There will be no pressure equalization time.

Moisture Restriction

Symptoms of a moisture restriction in the sealed system are:

- The room temperature will be above normal.

- The wattage and amperage will be considerably below normal, as indicated on the model/serial plate.

- The temperature of the condenser coil will be below normal.

- At the evaporator coil, you will hear a constant gurgle, low sound level, or no sound at all.

- The evaporator coil might have some frost on the evaporator inlet.

- The high-side pressure will be below normal.[8]

- The low-side pressure will be below normal or in a deep vacuum.[8]

- The pressure equalization time will be longer than normal or there will be no equalization at all.

Reversing Valve (Heat Pump Models)

In a straight-cool air conditioner, the refrigerant flows from the compressor discharge through the condenser coil, capillary tube, evaporator coil, and back through the suction line to the compressor. The ability of an air conditioner to reverse the direction of the refrigerant flow is achieved with a reversing valve installed in the refrigerant circuit. The reversing valve is designed as a single-port, double-throw valve. It turns the function of the evaporator coil from a heat absorption coil into a heat dissipation coil (condenser coil) (Figure 29-22). When the solenoid coil is energized, the valve will reverse the refrigerant flow. For cooling, in a heat-pump air conditioner, the refrigerant flows from the compressor discharge through the reversing valve to the outdoor coil, through the capillary tube and indoor coil, and back through the reversing valve to the suction port on the compressor. For heating, the refrigerant will flow from the compressor discharge, through the reversing valve, indoor coil, capillary tube, and outdoor coil, and then back through the reversing valve to the suction port on the compressor (Figures 29-23 and 29-24).

The reversing valve itself is a non-serviceable component and must be replaced with a duplicate of the original if it fails. It is recommended that a qualified technician with refrigerant certification replace the reversing valve. The refrigerant in the sealed system must be recovered properly.

Reversing valve–cooling

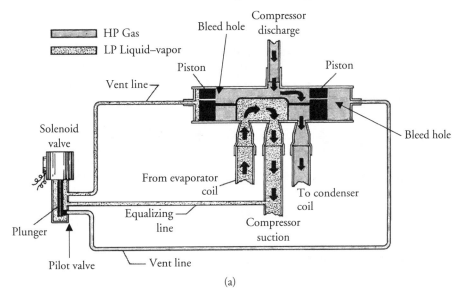

Reversing valve–heating

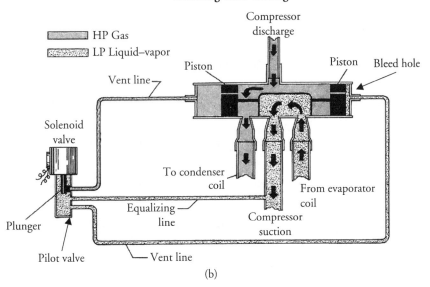

FIGURE 29-22 The refrigerant flow in a reverse-cycle valve. (a) The reversing valve solenoid coil is de-energized in the cooling mode. (b) The reversing valve solenoid coil is energized in the heating mode.

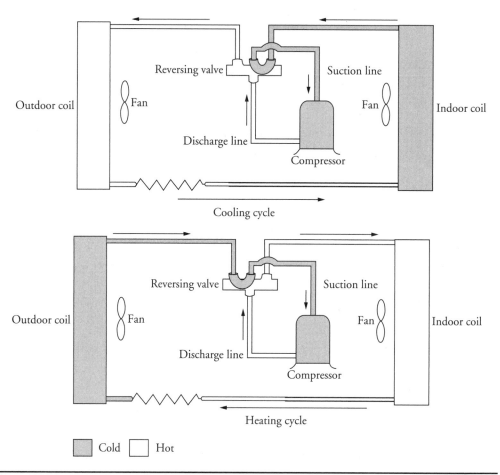

FIGURE 29-23 The refrigerant flow in a heat pump for cooling and heating.

Diagnostic Charts

The following diagnostic flowcharts, tables, and wiring diagrams will help you to pinpoint the likely causes of problems with room air conditioners (see Tables 29-4, 29-5 , and 29-6 and Figures 29-7, 29-8, 29-9, 29-10, 29-11, 29-12, 29-13, 29-14, 29-25, and 29-26).

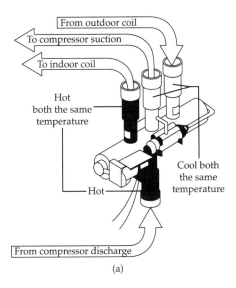

From outdoor coil

To compressor suction

To indoor coil

Hot
both the same
temperature

Cool both
the same
temperature

Hot

From compressor discharge

(a)

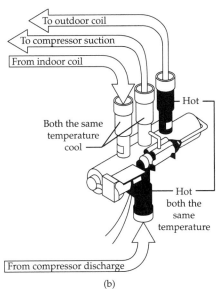

To outdoor coil

To compressor suction

From indoor coil

Hot

Both the same
temperature
cool

Hot
both the
same
temperature

From compressor discharge

(b)

FIGURE 29-24 (a) Illustrates the reversing valve in the heating mode. (b) Illustrates the reversing valve in the cooling mode. The reversing valve is also known as a four-way valve.

Reversing Valve Operating Condition	Discharge Line from the Compressor	Suction Line to the Compressor	Copper Line to the Inside Coil	Copper Line to the Outside Coil	Reversing Valve Left Pilot Capillary Tube	Reversing Valve Right Pilot Capillary Tube	Probable Cause	Solution
Normal cooling	Hot	Cool	Cool	Hot	Same temp as valve body	Same temp as valve body	Normal	None
Normal heating	Hot	Cool	Hot	Cool	Same temp as valve body	Same temp as valve body	Normal	None
Reversing valve will not shift from cooling to heating	Hot	Cool	Cool	Hot	Same temp as valve body	Hot	Bleeder hole blockage or leaking piston	Replace reversing valve
	Hot	Cool	Cool	Hot	Same temp as valve body	Same temp as valve body	Vent tube blockage	Replace reversing valve
	Hot	Cool	Cool	Hot	Hot	Hot	Both pilot valve ports are open	Replace reversing valve
	Warm	Cool	Cool	Hot	Same temp as valve body	Warm	Bad compressor	Replace compressor
Reversing valve will begin to shift and it will not complete the reversal	Hot	Warm	Warm	Hot	Same temp as valve body	Hot	Low on refrigerant or valve body damage	Check refrigerant charge or replace
	Hot	Warm	Warm	Hot	Hot	Hot	Both pilot valve ports are open	Replace reversing valve
	Hot	Hot	Hot	Hot	Same temp as valve body	Hot	Stuck valve or valve body damage	Replace reversing valve
	Hot	Hot	Hot	Hot	Hot	Hot	Both pilot valve ports are open	Replace reversing valve

TABLE 29-4   Reversing Valve Touch Test Chart (*continued*)

Reversing Valve Operating Condition	Discharge Line from the Compressor	Suction Line to the Compressor	Copper Line to the Inside Coil	Copper Line to the Outside Coil	Reversing Valve Left Pilot Capillary Tube	Reversing Valve Right Pilot Capillary Tube	Probable Cause	Solution
Fluctuating heating	Hot	Cool	Hot	Cool	Same temp as valve body	Same temp as valve body	Internal valve leaking at piston	Replace reversing valve
	Hot	Cool	Hot	Cool	Valve body warmer than usual	Valve body warmer than usual	Internal valve leaking at piston	Replace reversing valve
Reversing valve will not shift from heating to cooling	Hot	Cool	Hot	Cool	Same temp as valve body	Same temp as valve body	High refrigerant pressure differential	Restart air conditioner and recheck
	Hot	Cool	Hot	Cool	Hot	Same temp as valve body	Bleeder hole blockage	Replace reversing valve
	Hot	Cool	Hot	Cool	Hot	Same temp as valve body	Internal valve leaking at piston	Replace reversing valve
	Hot	Cool	Hot	Cool	Hot	Hot	Defective pilot valve	Replace reversing valve
	Warm	Cool	Warm	Cool	Warm	Same temp as valve body	Bad compressor	Replace compressor

Note: Before replacing the reversing valve, turn the air conditioner off for a few minutes, and restart the cycle. On some occasions, the reversing valve may return to normal operation.

TABLE 29-4 Reversing Valve Touch Test Chart

Sealed System Condition	Suction Line Pressure (Technicians Only)	Liquid Line Pressure (Technicians Only)	Suction Line to Compressor	Compressor Discharge Line	Condenser Coil	Capillary Tube	Evaporator Coil	Frost Line	Amperage or Wattage	Pressure Equalization Rate
Normal Operation	Normal pressure readings	Normal pressure readings	Slightly below room temperature	**WARNING** very hot to the touch	**WARNING** very hot to the touch	Warm	Cold coil and maintaining proper temperatures	Suction line from inside box not frozen	Normal meter readings	Normal
Sealed System Overcharged	Higher than normal pressure readings	Higher than normal pressure readings	Heavily frosted; may be very cold to the touch	Between slightly warm to hot	Between hot to warm to the touch	Cool below room temperature	Cold coil and possibly not maintaing temperatures	All the way back to the suction line	Higher than normal meter readings	Normal to slightly longer
Sealed System Undercharged	Pressure readings are lower than normal	Pressure readings are lower than normal reading	Warm to the touch; possibly near room temperature	**WARNING** hot to the touch	The entire coil feels warm	Warm	Inlet to coil feels extremely cold while the outlet from the coil will be below room temperature	Partial	Lower than normal meter readings	Normal
Partial Restriction Within Sealed System	Lower than normal pressure readings, possibly in a deep vacuum	Intermittent lower than normal reading	Warm to the touch; possibly near room temperature	**WARNING** very hot to the touch	The top passes in the coil are warm and the lower passes are cool near to room temperature	Feels like room temperature; between cool to colder	Inlet to coil feels extremely cold while the outlet from the coil will be below room temperature	Intermittent / Frost line will begin to grow in length	Lower than normal meter readings	Intermittent
Complete Restriction Within Sealed System	Pressure readings are in a deep vacuum	Ambient readings	Feels the same as room temperature	Feels the same as room temperature	Feels the same as room temperature	Feels the same as room temperature	No refrigeration or air conditioning	None	Lower than normal meter readings	No equalization
Out of Refrigerant Possible Leak in System	The pressure reading will be from 0 PSIG to 30" vacuum	Atmospheric reading	Feels the same as room temperature	Can feel like cool to hot	Feels the same as room temperature	Feels the same as room temperature	No refrigeration or air conditioning	Non existent	Lower than normal meter readings	Normal
Low Capacity Compressor	Higher than normal pressure readings	Lower than normal readings	Cool to room temperature	Cooler than normal	Low	Warm to box temperature	Partial or half of the evaporator frost pattern	Partial to non existent	Lower than normal	Quicker than normal

TABLE 29-5 Refrigeration Sealed System Diagnosis Chart

Conditions	Amperage or Wattage	Condenser Coil Temperature	Frost Line	Compressor Discharge Line Temperature	Low-Side Pressure (For Service Technicans Only)	High-Side Pressure (For Service Technicians Only)	Fresh Food Compartment Temperature	Freezer Compartment Temperature
Plugged condenser coil	Higher than normal	Higher than normal	Full	Higher than normal	Higher than normal	Higher than normal	Warmer than normal readings	Warmer than normal readings
Blocked condenser fan assembly	Higher than normal	Higher than normal	Full	Higher than normal	Higher than normal	Higher than normal	Warmer than normal readings	Warmer than normal readings
Blocked evaporator fan assembly	Lower than normal	Lower than normal	Frost back to compressor	Lower than normal	Lower than normal	Lower than normal	Warmer than normal readings	Warmer than normal readings
Evaporator coil iced up (defrost failure)	Lower than normal	Lower than normal	Frost back to compressor	Lower than normal	Lower than normal	Lower than normal	Warmer than normal readings	Warmer than normal readings
High head load	Higher than normal	Higher than normal	Full	Higher than normal	Higher than normal	Higher than normal	Warmer than normal readings	Warmer than normal readings
High ambients	Higher than normal	Higher than normal	Full	Higher than normal	Higher than normal	Higher than normal	Warmer than normal readings	Warmer than normal readings
Damper failed closed	Lower than normal	Lower than normal	Full	Lower than normal	Lower than normal	Lower than normal	Warmer than normal readings	Cooler than normal readings
Damper failed open	Slightly higher than normal	Slightly higher	Full	Normal	Slightly higher than normal	Normal	Cooler than normal	Normal to slightly warmer readings

TABLE 29-6 Conditions That Will Mimic Sealed System Failures

PART VI

FIGURE 29-25
The diagnostic flowchart: The air conditioner is cooling when the heat function is selected.

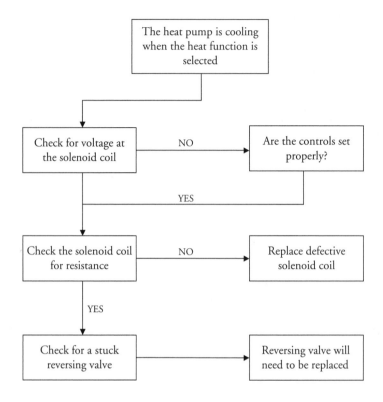

Cooling Capacity

Using the charts in Figures 29-27 and 29-8, follow the guidelines listed here to choose the correct size air conditioner.

1. Use the chart in Figure 29-27 to determine the square footage of the area to be cooled. If the room to be cooled is next to an adjacent area to be cooled, this will count as two areas. Determine the square footage of each area and then add the two totals.

2. Locate the square footage on the left or right side of the cooling capacity chart (Figure 29-28), and then draw a horizontal line across the chart. Select the correct air conditioner capacity from one of the three diagonal bands on the chart, and then go straight down the chart to the correct daytime cooling capacity BTU/hr.

The diagonal bands on the chart help compensate for variations in cooling applications. The bands indicate a range of BTU/hr capacities normally required to cool such an area.

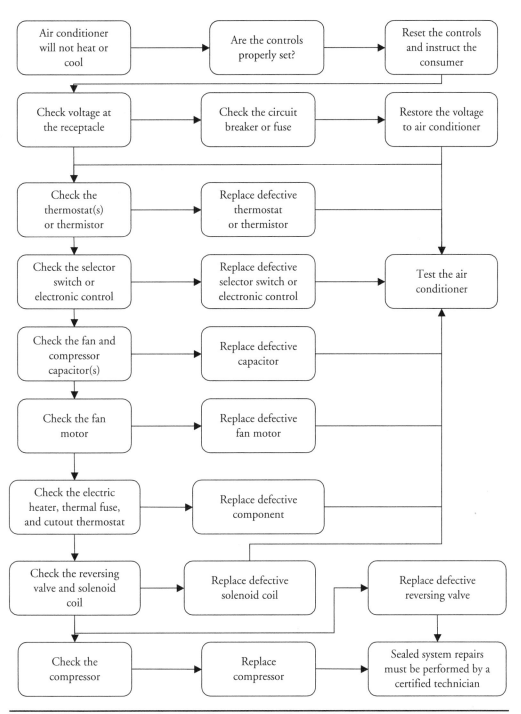

FIGURE 29-26 The diagnostic flowchart: Air conditioner will not heat or cool.

FIGURE 29-27
Floor area table.

	8'	10'	14'	18'	22'	26'	30'	34'	38'
10'	80	100	140	180	220	260	300	340	380
12'	96	120	168	216	264	312	360	408	456
14'	112	140	196	252	308	364	420	476	532
16'	128	160	224	288	352	416	480	544	608
18'	144	180	252	324	396	468	540	612	684
20'	160	200	280	360	440	520	600	680	760
22'	176	220	308	396	484	572	660	748	836
24'	192	240	336	432	528	624	720	816	912
26'	208	260	364	468	572	676	780	884	988
28'	224	280	392	504	616	728	840	952	1064
30'	240	300	420	540	660	780	900	1020	1140
32'	256	320	448	576	704	832	960	1088	1216
34'	272	340	476	612	748	884	1020	1156	1292
36'	288	360	504	648	792	936	1080	1224	1368
38'	304	380	532	684	836	988	1140	1292	1444
40'	320	400	560	720	880	1040	1200	1360	1520
42'	336	420	588	756	924	1092	1260	1428	1596
44'	352	440	616	792	968	1144	1320	1496	1672
46'	368	460	644	828	1012	1196	1380	1564	1748
48'	384	480	672	864	1056	1248	1440	1632	1824
50'	400	500	700	900	1100	1300	1500	1700	1900

FIGURE 29-28
Air conditioner sizing chart. If the area to be cooled includes a kitchen area, add 4000 BTU/hr. If more than two people occupy the area, add 600 BTU/hr per person. If only one person occupies the area, then subtract 600 BTU/hr.

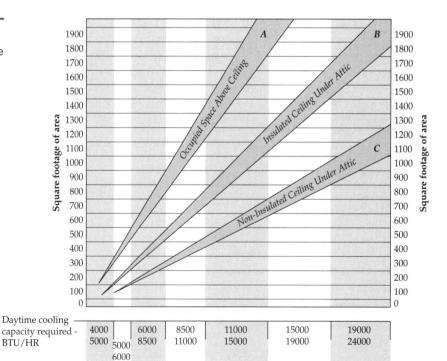

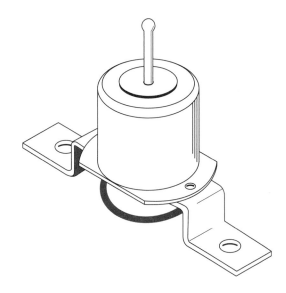

Figure 29-29
A bellows type thermostat (drain pan valve), which is located in the base pan of the air conditioner. This is a nonelectric component.

Drain Pan Valve

During the cooling mode of operation, condensate water, which collects in the base pan, is picked up by the condenser fan blade slinger ring and sprayed onto the condenser coil where it evaporates and cools the refrigerant.

During the heating mode of operation, it is necessary to remove the standing water from the base pan to prevent it from freezing during cold months. To provide a means to drain this water from the base pan, a bellows type of thermostat (drain valve) is installed in the base pan of the air conditioner (Figure 29-29). This thermostat (drain valve) is temperature sensitive (nonelectric component) and will open when the outside temperature reaches approximately 40 degrees Fahrenheit. The thermostat (drain valve) will close gradually as the outside temperature rises above 40 degrees Fahrenheit. For the valve to fully close, the outside temperature must reach approximately 60 degrees Fahrenheit.

Endnotes

1. Room air conditioners (RACs) are available with top or side discharge air grilles.
2. Refer to Chapter 1 for correct sizing.
3. Capacitors with a metal outer case can only be checked for a ground. Start capacitors are encased in plastic and they cannot be checked for a ground.
4. Some fan motor manufacturers may use different wiring colors to indicate the various speeds. Check the technical data sheet for the model you are servicing for the correct fan motor winding wiring identification and color coding.
5. When the bearings wear out in the end bell of the motor, the motor will begin to make a vibrating sound or a growling noise.

6. On some models, there may be a brace that secures the condenser fan shroud to the bulkhead. This brace must be removed to allow access to the condenser fan blade.

7. If you did not read resistance, it does not mean that the electric heater element is defective. After removing the heater assembly, test the thermal fuse and the heater limit control (thermostat mounted to heater frame) for continuity separately. The thermal fuse is a non-resetable fuse—it will open the heating circuit if the heating temperature or amperage exceeds its rating, indicating that there may be other problems with the air conditioner (for example, a defective fan motor, dirty filter, or dirty evaporator coil). The thermal fuse acts as the last line of defense if the heater limit control fails. The heater limit control will also open the circuit, limiting the heating, if there is a reduction in airflow. On some models, this device will reset automatically, allowing the electric heater to come back on again. Both of these devices are installed in the heating circuit to prevent the heating element from causing a fire in the air conditioner and home. These safety devices must remain in the circuitry. Do not bypass these safety devices. Replace the defective component with a duplicate of the original.

8. If you open up the sealed system, you will void your warranty. The sealed system must be repaired by an authorized service company and by an EPA certified technician.

Dehumidifiers

D ehumidifiers are household appliances that remove moisture (humidity) from an area. Humidity is defined as the amount of water vapor in the air and is measured in terms of relative humidity (RH %). The higher the RH percentage, the more water vapor in the air. The optimum relative humidity level in an area is generally considered to be between 30 percent and 50 percent. Humidity levels exceeding 50 percent with temperatures over 70 degrees Fahrenheit can be uncomfortable, unhealthy, and in most cases cause personal property damage. Anything above this range may promote bacteria growth or mold. In colder climates, during the heating season, humidity levels should be maintained between 30 percent and 40 percent to prevent moisture from collecting or condensing on the windows and doors. Dehumidifiers (Figure 30-1) were designed to remove unwanted moisture from the air in a given area.

Sizing a Dehumidifier

Dehumidifiers, when properly set, can maintain the desired humidity level in that area. The capacity is usually measured in pints per 24 hours and is determined by two factors: the square footage of the space that needs to be dehumidified and the conditions that exist in the space before dehumidification. Table 30-1 depicts a dehumidifier sizing chart.

Another measure that can determine the size of dehumidifier the consumer needs is the amount of air the appliance can move over a certain period of time. Dehumidifiers measure this in cubic feet per minute (CFM). The CFM is an important measure for dehumidification because the unit can only extract moisture from the air that passes through the dehumidifier. All of the air in a room must flow through the dehumidifier in order to remove moisture. To calculate the CFM you need from the dehumidifier, multiply the cubic feet of the area by the recommended air changes per hour (ACH) in Table 30-1, and then divide that number by 60 minutes. The resulting number is the CFM the dehumidifier should have.

Principals of Operation

A refrigeration system (Figure 30-2) is used to collect the moisture from the air. The dehumidifier may have either a reciprocating (split phase) or a rotary (permanent split capacitor) compressor that circulates the refrigerant in the dehumidifier. The evaporator operating temperatures are around 33 to 37 degrees Fahrenheit. When moisture-ladened air passes over the

Figure 30-1
A typical portable
dehumidifier.

Condition Without Dehumidification	500 Sq. Ft.	1,000 Sq. Ft.	1,500 Sq. Ft.	2,000 Sq. Ft.	2,500 Sq. Ft.	3,000 Sq. Ft.
In humid weather, the area is moderately damp and it has a musty odor. (60–70% RH) ACH=3	10 pints	14 pints	18 pints	22 pints	26 pints	30 pints
In very damp areas, it always feels damp and spots appear on the walls and floors. (70–80% RH) ACH=4	12 pints	17 pints	22 pints	27 pints	32 pints	37 pints
In areas that are very wet, the walls and the floor are sweating or there is seepage. (80–90% RH) ACH=5	14 pints	20 pints	26 pints	32 pints	38 pints	44 pints
In areas that are extremely wet; wet floor, laundry room, high-load conditions. (90–100% RH) ACH=6	16 pints	23 pints	30 pints	37 pints	44 pints	51 pints
The higher the humidity in an area, the more often the air should cycle through the dehumidifier. This is known as Air Changes per Hour, or ACH.						

Table 30-1   Dehumidifier Sizing Guidelines

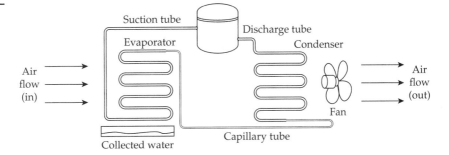

FIGURE 30-2
A typical refrigeration system used in a dehumidifier. The air is drawn in through the evaporator and expelled out through the contender coil by a fan.

evaporator, the moisture will collect on the evaporator coil and condense into a liquid. The cool air is then drawn through the condenser coil and warms the air before it is recirculated back into the room area. The water collected on the evaporator coil will run off the coil into a bucket. Many dehumidifier models include a built-in in humidistat, a device that allows the consumer to set the humidity level desired for that room. Once the room reaches the desired humidity level, the dehumidifier will cycle on and off automatically to maintain that RH level.

Safety First

Any person who cannot use basic tools or follow written instructions should *not* attempt to install, maintain, or repair any dehumidifiers. Any improper installation, preventive maintenance, or repairs could create a risk of personal injury or property damage. If you do not fully understand the installation, preventive maintenance, or repair procedures in this chapter, or if you doubt your ability to complete the task on your dehumidifier, please call your service manager.

This chapter covers the electrical components and how to diagnose the sealed system. The actual repair or replacement of any sealed-system component is not included in this chapter. It is recommended that you acquire refrigerant certification (or call an authorized service company) to repair or replace any sealed-system component, as the refrigerant in the sealed system must be recovered properly.

Before continuing, take a moment to refresh your memory on the safety procedures in Chapter 2.

Dehumidifiers in General

Much of the troubleshooting information in this chapter covers dehumidifiers in general, rather than specific models, in order to present a broad overview of service techniques. The illustrations that are used in this chapter are for demonstration purposes only to clarify the description of how to service these appliances. They in no way reflect a particular brand's reliability.

PART VI

Location and Installation of a Dehumidifier

When installing the dehumidifier in the home, you must allow 12 inches of space on all sides of the dehumidifier so that air can circulate properly (see Figure 30-3). The dehumidifier can be installed on the floor, table, or shelf. When installing the dehumidifier on a table or shelf, make sure the table or shelf can support the weight of the dehumidifier, including the weight of the bucket full of water. Next, plug the dehumidifier into a 120 volt properly grounded receptacle, properly polarized. *Safety note: Do not use an extension cord with the dehumidifier.*

Dehumidifier Maintenance

The dehumidifier's coils should be chemically cleaned at least annually to remove the dirt and slime buildup within the unit. Clean the outer case and cabinet (Figure 30-4a) with a soft cloth and a mild detergent. Next, vacuum the grille, or use a brush on the grill. Do not use abrasives or bleach on the dehumidifier. About once a month, remove the water bucket and clean it to avoid growth of mold, mildew, and bacteria. Don't forget to wash the air filter and shake off the excess water and let it air dry, and then reinstall it (Figures 30-4b and c) at least once a month also.

Step-by-Step Troubleshooting by Symptom Diagnosis

When servicing a dehumidifier, don't overlook the simple things that might be causing the problem. Step-by-step troubleshooting by symptom diagnosis is based on diagnosing malfunctions, with possible causes arranged into categories relating to the operation of the dehumidifier. This section is intended only to serve as a checklist to aid you in diagnosing a problem. Look at the symptom that best describes the problem you are experiencing with the dehumidifier, and then correct the problem.

FIGURE 30-3
Place the dehumidifier within 12 inches of any walls or other blockages that might prevent proper air circulation.

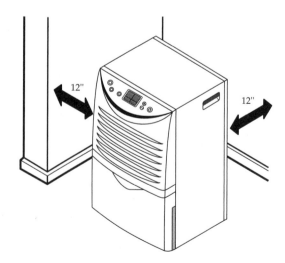

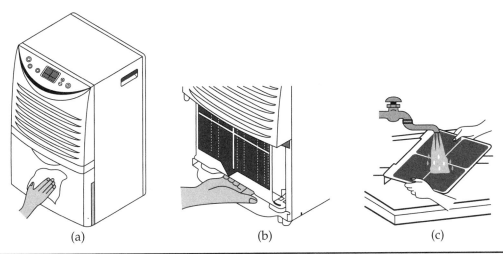

FIGURE 30-4 Performing maintenance on the dehumidifier. Clean the outside of the unit (a), front grille (b), and filter (c).

Dehumidifier Will Not Run (No Fan, No Compressor)

- Check control settings.
- Check voltage to the unit.
- Check bucket; it may be full of water.
- Check wiring.
- Run diagnostics; you may have bad humidistat.
- Check the humidity level in the room and compare that reading against the control setting.

Condenser Fan Motor Runs (Compressor Not Running)

- Check the de-icer thermostat on mechanical models.
- Check thermistor on electronic models.
- On electronic models, run diagnostics; replace electronic control board.
- Check compressor overload and relay.
- Check compressor windings.

Condenser Fan Motor Does Not Run (Compressor Is Running)

- Check voltage going to condenser fan motor.
- Test the condenser fan motor windings.
- Check condenser fan blade.

Condenser Fan Motor and Compressor Running (Evaporator Coil Frozen Up)

- Check the room temperature (nonde-icer models); the room temperature must be higher than 65 degrees Fahrenheit. Frost should disappear within 60 minutes. If not, check the sealed system.

Condenser Fan Motor and Compressor Running (No Dehumidification)

- Dehumidifier may have a sealed-system problem.
- Test the humidity in the room; might have low humidity in area.
- Check the air filter.
- Check the evaporator coil. Is it clean?

Dehumidifier Is Running (Insufficient Dehumidification)

- Check and make sure that the grilles are not obstructed.
- Check air filter.
- Check the evaporator coil. Is it clean?
- Run diagnostics on the sealed system; it might have a low refrigerant charge.
- Check the room size. The dehumidifier might be too small for the square footage.
- Check and make sure that the doors and windows are closed.
- Check the room temperature.

Dehumidifier Runs Continuously

- Check the control settings.
- Check the room humidity level; may have to adjust humidity setting.

Repair Procedures

Each repair procedure is a complete inspection and repair process for a single dehumidifier component. It contains the information you need to test and replace components.

Electronic Components

The electronic components consist of the following: electronic control board and touchpad. The typical complaints associated with the electronic components are:

- Unable to program the touchpad panel functions.
- The display board will not display anything.
- Unusual display readouts.
- Unable to control the relative humidity.
- The compressor will not run.
- The condenser fan motor will not run.

To prevent electrostatic discharge (ESD) from damaging expensive electronic components, follow the steps in Chapter 11.

To handle these problems, perform the following steps:

1. **Verify the complaint.** Verify the complaint by operating the dehumidifier controls. Turn off the electricity to the dehumidifier, and wait for two minutes before turning it back on. If a fault code appears, look up the code. If the dehumidifier will not

power up, locate the technical data sheet behind the control panel for diagnostics information. It will be helpful if you can locate the actual service manual for the dehumidifier model you are working on to properly diagnose the dehumidifier. The service manual will assist you in properly placing the dehumidifier in the service test mode for testing the dehumidifier functions.

2. **Check for external factors.** You must check for external factors not associated with the dehumidifier. Is there electricity to the dehumidifier? The voltage at the wall receptacle must be within ±10 percent of the voltage rating on the model and serial data plate. Do you have the correct polarity? (See Chapter 6.) Is the electrical receptacle polarized and properly grounded?

3. **Disconnect the electricity.** Before working on the dehumidifier, disconnect the electricity. This can be done by pulling the plug out of the wall receptacle. Or disconnect the electricity at the fuse panel or circuit breaker panel. Turn off the electricity.

WARNING *Some diagnostic tests will require you to test the components with the power turned on. When you disassemble the control panel or remove the outer cabinet, you can position the panel in such a way that the wiring will not make contact with metal. This will allow you to test the components without electrical mishaps.*

4. **Gain access to the electronic components.** You can gain access to the electronic components by removing the front grille and the screws on the control panel (Figures 30-5a, b, c, and d). Next, tilt the control panel away from the dehumidifier, making sure not to pull any of the wires off the controls.

WARNING *Do not touch the wiring or capacitor until it is discharged. A capacitor will hold a charge indefinitely, even when it is not currently in use. A charged capacitor is extremely dangerous. Discharge all capacitors immediately any time that work is being conducted in their vicinity. Redischarge after repowering the equipment if further work must be done. Many capacitors are internally fused. If you are not sure, you can use a 20,000 ohm 2-watt resistor to discharge the capacitor. Do not use a screwdriver to short out the capacitor. By doing so, you will blow out the fuse in the capacitor and the capacitor will not work. Safely use an insulated pair of pliers to remove the wires from the capacitor and place the resistor across the capacitor terminals. When checking a dual capacitor with a capacitor analyzer or ohmmeter, you must test both sides of the capacitor.*

5. **Test the electronic components.** If you are able to run the dehumidifier test mode, check the different functions of the dehumidifier. Use the technical data sheet for the model you are servicing to locate the test points on the wiring schematic. Check all wiring connections and wiring. Using the technical data sheet, test the electronic control and display board, input voltages, and output voltages.

6. **Remove the defective component.** To remove the defective component, remove the screws that secure the printed circuit board to the dehumidifier console frame. Disconnect the connectors from the electronic control board and display.

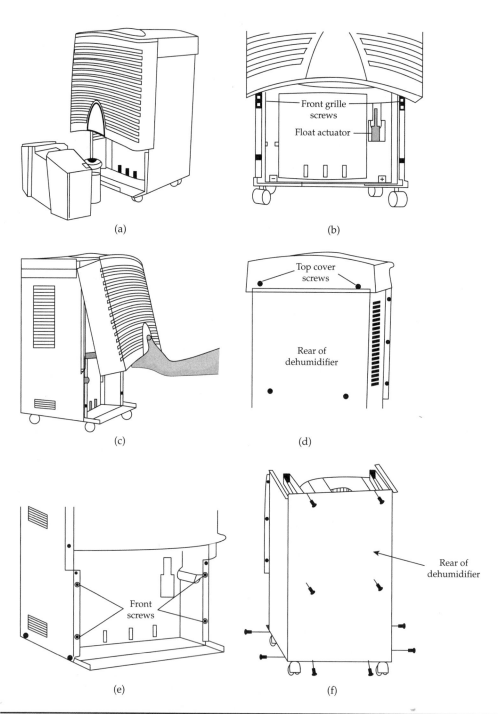

FIGURE 30-5   This figure depicts how to gain access to the components. Remove the bucket and screws as shown.

7. **Install the new component.** To install a new component, just reverse the disassembly procedure and reassemble. Reinstall all panels and the front grille and bucket. Then, restore the electricity to the dehumidifier. Test the dehumidifier operation. To prevent ESD from damaging expensive electronic components, simply follow the steps in Chapter 11.

Capacitor

A capacitor is a device that stores electricity to provide an electrical boost for motor starting. Most high-torque motors need a capacitor connected in series with the start winding circuit to produce the desired rotation under a heavy starting load. The capacitor can also be used for motor efficiency.

The typical complaints associated with failure of the capacitor are:

- Fuse is blown or the circuit breaker trips.
- Fan motor will not run.
- Fan motor has a burning smell.
- Motor or compressor tries to start and then shuts off on overload.

To handle these problems, perform the following steps:

1. **Verify the complaint.** Verify the complaint by operating the dehumidifier. Listen carefully, and you will hear if there are any unusual noises or if the circuit breaker trips. If you smell something burning, immediately turn off the dehumidifier and pull out the plug. It will be helpful if you can locate the actual service manual for the dehumidifier model you are working on to properly diagnose the dehumidifier.

2. **Check for external factors.** You must check for external factors not associated with the dehumidifier. Is the dehumidifier installed properly? Does it have the correct voltage? The voltage at the wall receptacle must be within ±10 percent of the voltage rating on the model and serial data plate. Do you have the correct polarity? (See Chapter 6.)

3. **Disconnect the electricity.** Before working on the dehumidifier, disconnect the electricity. This can be done by pulling the plug from the electrical outlet. Be sure that you only remove the dehumidifier plug. Or disconnect the electricity at the fuse panel or at the circuit breaker panel. Turn off the electricity.

WARNING *Some diagnostic tests will require you to test the components with the power turned on. When you disassemble the control panel or remove the outer cabinet, you can position the panel in such a way that the wiring will not make contact with metal. This will allow you to test the components without electrical mishaps.*

4. **Gain access to the capacitor.** Some models have the capacitor mounted on the fan motor; on other models, it is mounted behind the electrical controls. Remove the front grille and filter. Some units have either a one-piece or a two-piece grille with locking tabs and/or screws (see Figures 30-5 and 30-6). On other models, you should be able to gain access by removing the screws that secure the control panel to the dehumidifier frame.

PART VI

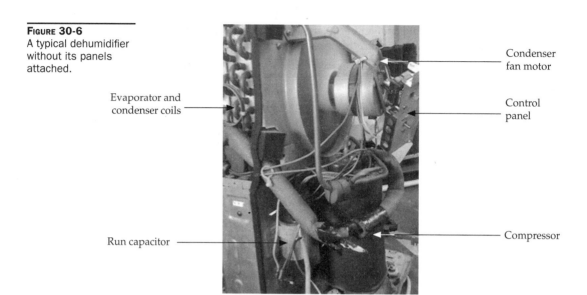

Figure 30-6
A typical dehumidifier without its panels attached.

Evaporator and condenser coils

Run capacitor

Condenser fan motor

Control panel

Compressor

WARNING *Do not touch the wiring or capacitor until it is discharged. A capacitor will hold a charge indefinitely, even when it is not currently in use. A charged capacitor is extremely dangerous. Discharge all capacitors immediately any time that work is being conducted in their vicinity. Rediscarge after repowering the equipment if further work must be done.*

5. **Test the capacitor.** Many capacitors are internally fused. If you are not sure, you can use a 20,000 ohm 2-watt resistor to discharge the capacitor. Do not use a screwdriver to short out the capacitor. By doing so, you will blow out the fuse in the capacitor and the capacitor will not work. Safely use an insulated pair of pliers to remove the wires from the capacitor and place the resistor across the capacitor terminals. When checking a dual capacitor with a capacitor analyzer or ohmmeter, you must test both sides of the capacitor. Set the ohmmeter on the highest scale and then place one probe on one terminal and the other probe on the other terminal (Figure 29-15). Observe the meter action. While the capacitor is charging, the ohmmeter will read nearly zero ohms for a short period of time. Then the ohmmeter reading will slowly return toward infinity. If the ohmmeter reading deflects to zero and does not return to infinity, the capacitor is shorted and should be replaced. If the ohmmeter reading remains at infinity and does not dip toward zero, the capacitor is open and should be replaced.

To test for a grounded capacitor, leave one meter probe on the capacitor terminal, and attach the other probe to the outer casing of the capacitor or dehumidifier chassis ground. There should be no reading if the capacitor is not grounded. Next, remove the probe from the capacitor terminal, and place the probe on the other terminal. Again, there should be no reading indicated if the capacitor is not grounded.

When using a capacitor analyzer to test capacitors, it will show whether the capacitor is "open" or "shorted." It will tell whether the capacitor is within its microfarads rating and it will show whether the capacitor is operating at the proper power-factor percentage. The instrument will automatically discharge the capacitor when the test switch is released.

6. **Remove the capacitor.** Remove the capacitor from its mounting bracket.

7. **Install a new capacitor.** To install the new capacitor, just reverse the disassembly procedure and reassemble.

NOTE *A capacitor is rated by its working voltage (WV or WVac) and by its storage capacity in microfarads (μF). Always replace a capacitor with one that has the same voltage rating and the same (or up to 10 percent greater) microfarad rating.*

Fan Motor

Dehumidifiers use a single fan motor with a single shaft for the fan blade or blower wheel. The typical complaints associated with failure of the fan motor are:

- No air is blowing out of the discharge grille.

- When the motor runs, there are loud noises.

- The fuse or circuit breaker trips when the air conditioner is started.

To handle these problems, perform the following steps:

1. **Verify the complaint.** Verify the complaint by operating the dehumidifier. Listen carefully, and you will hear if there are any unusual noises or if the circuit breaker trips. It will be helpful if you can locate the actual service manual for the dehumidifier model you are working on to properly diagnose the dehumidifier.

2. **Check for external factors.** You must check for external factors not associated with the dehumidifier. Is the dehumidifier installed properly? Does it have the correct voltage? The voltage at the wall receptacle must be within ±10 percent of the voltage rating on the model and serial data plate. Do you have the correct polarity? (See Chapter 6.)

3. **Disconnect the electricity.** Before working on the dehumidifier, disconnect the electricity. This can be done by pulling the plug out of the wall receptacle. Be sure that you only remove the dehumidifier plug. Or disconnect the electricity at the fuse panel or at the circuit breaker panel. Turn off the electricity.

WARNING *Some diagnostic tests will require you to test the components with the power turned on. When you disassemble the control panel or remove the outer cabinet, you can position the panel in such a way that the wiring will not make contact with metal. This will allow you to test the components without electrical mishaps.*

4. **Gain access to the fan motor.** Before you begin to remove the fan motor, test the motor windings. Remove the outer wrapper and the front grille and filter (Figures 30-5 and 30-6).

WARNING *Do not touch the wiring or capacitor until it is discharged. A capacitor will hold a charge indefinitely, even when it is not currently in use. A charged capacitor is extremely dangerous. Discharge all capacitors immediately any time that work is being conducted in their vicinity. Redischarge after repowering the equipment if further work must be done. Many capacitors are internally fused. If you are not sure, you can use a 20,000 ohm 2-watt resistor to discharge the capacitor. Do not use a screwdriver to short out the capacitor. By doing so, you will blow out the fuse in the capacitor and the capacitor will not work. Safely use an insulated pair of pliers to remove the wires from the capacitor and place the resistor across the capacitor terminals. When checking a dual capacitor with a capacitor analyzer or ohmmeter, you must test both sides of the capacitor.*

5. **Disconnect the motor wire leads.** Disconnect the motor wire leads from the control panel, and label them. Check the motor windings for continuity (Figure 30-6). Using the wiring diagram, check for resistance on the condenser fan motor windings. To check for a grounded winding in the motor, take the ohmmeter probes and check from each motor wire lead to the motor housing. The ohmmeter will indicate continuity if the windings are grounded.

   Finally, spin the motor shaft—it should turn freely. If the shaft is hard to turn, replace the fan motor. Now move the motor shaft in an up-and-down motion perpendicular to the motor body. The shaft should have virtually no movement. If there is a lot of play in the end bell bearings, replace the fan motor.[1]

6. **Remove the fan motor.** To remove the fan motor (Figure 30-6), remove the screws that hold the fan motor bracket in place. Then, remove the fan blade or blower wheel.

   Depending on which model you are servicing, the removal of the fan motor will vary from manufacturer to manufacturer. However, the basic steps are the same for every dehumidifier.

NOTE *Replace the fan motor with a duplicate of the original for easier installation.*

NOTE *With the fan motor removed from the unit, it is the perfect time to chemically clean the remainder of the dehumidifier. Everything will be exposed, and it will be easier to clean the unit.*

7. **Reinstall the fan motor.** To reinstall the fan motor, just reverse the instructions in step 6, and reassemble. When you reinstall the condenser fan blade, make sure to place it back on the motor shaft in the same position. Before reinstalling the outer cabinet on the dehumidifier, test the dehumidifier first. Make sure the fan blade or the blower wheel is not hitting against anything.

Compressor and Overload Protector

The compressor (reciprocating or rotary type) is the heart of the vapor compression system. It is used to circulate the refrigerant throughout the sealed system. The overload is attached to the compressor housing. The overload protects the compressor. Some models are equipped with a relay to assist the compressor in starting.

The typical complaints associated with failure of the compressor are:

- The dehumidifier does not run at all.
- The compressor makes a humming or buzzing noise and stops.
- The circuit breaker trips when the dehumidifier starts up.

To handle these problems, perform the following steps:

1. **Verify the complaint.** Verify the complaint by asking the customer to describe what the dehumidifier is doing. It will be helpful if you can locate the actual service manual for the dehumidifier model you are working on to properly diagnose the dehumidifier.

2. **Check for external factors.** You must check for external factors not associated with the dehumidifier. Is the dehumidifier installed properly? Does it have the correct voltage? Check for a voltage drop during dehumidifier startup. The voltage at the wall receptacle must be within ±10 percent of the voltage rating on the model and serial data plate. Do you have the correct polarity? (See Chapter 6.)

3. **Disconnect the electricity.** Before working on the dehumidifier, disconnect the electricity. This can be done by pulling the plug from the receptacle. Or disconnect the electricity at the fuse panel or at the circuit breaker panel. Turn off the electricity.

WARNING *Before you begin to service any high-voltage components within the dehumidifier, you must discharge the high-voltage capacitors first.*

WARNING *A capacitor will hold a charge indefinitely, even when it is not currently in use. A charged capacitor is extremely dangerous. Discharge all capacitors immediately any time that work is being conducted in their vicinity. Redischarge after repowering the equipment if further work must be done. Many capacitors are internally fused. If you are not sure, you can use a 20,000 ohm 2-watt resistor to discharge the capacitor. Do not use a screwdriver to short out the capacitor. By doing so, you will blow out the fuse in the capacitor and the capacitor will not work. Safely use an insulated pair of pliers to remove the wires from the capacitor and place the resistor across the capacitor terminals. When checking a dual capacitor with a capacitor analyzer or ohmmeter, you must test both sides of the capacitor.*

WARNING *Some diagnostic tests will require you to test the components with the power turned on. When you disassemble the control panel or remove the outer cabinet, you can position the panel in such a way that the wiring will not make contact with metal. This will allow you to test the components without electrical mishaps.*

4. **Gain access to the compressor.** To access the compressor, remove the outer cabinet (Figures 30-5 and 30-6). This will expose the compressor, fan motor, and electrical controls (Figure 30-6). Next, remove the compressor terminal cover (Figure 29-20) by removing the retaining nut or clip that secures the cover. Remove the terminal cover and washer.

5. **Test the overload protector.** To test the overload protector, remove the wires from the overload and compressor terminals. Next, place the probes of the ohmmeter on the overload terminals (see Figure 29-20). Set the meter scale on R × 1. The reading will show continuity. If not, replace the overload protector.

6. **Test the compressor.** The compressor terminals are marked C, S, and R: C indicates the common winding terminal, S indicates the start winding terminal, and R indicates the run winding terminal. Refer to the wiring diagram for the model that you are servicing. Set the meter scale on R × 1, touch the probes together, and adjust the needle setting to indicate a zero reading. Next, place the probes of the ohmmeter on the terminals marked S and R (see Figure 25-37a). The meter reading will show continuity. Now place the meter probes on the terminals marked C and S. The meter reading will show continuity. Finally, place the meter probes on the terminals marked C and R. The meter reading will show continuity. The total number of ohms measured between S and R is equal to the sum of C to S plus C to R. The compressor should be tested for proper resistance, as indicated on the wiring diagram.

To test the compressor for ground, place one probe on a compressor terminal, and attach the other probe to the compressor housing or to any good ground (see Figure 25-37b). Set the meter scale to R × 1000. The meter reading will show no continuity. Repeat this for the remaining two terminals. The meter reading will show no continuity. If you get a continuity reading from any of these terminals to ground, the compressor is grounded. Replace it.

Bucket Switch

The bucket switch is a single-pole, double-throw (SPDT) device that shuts off the dehumidifier when the bucket is full of water. The switch is mounted to the frame directly behind the bucket (Figure 30-7). The switch is activated by the float assembly.

The typical complaints associated with failure of the bucket switch are:

- The dehumidifier does not run at all.
- Water spills out of the dehumidifier.

FIGURE 30-7
Do not run the dehumidifier without the bucket. (a) The float assembly. (b) The float switch in the rear of the dehumidifier.

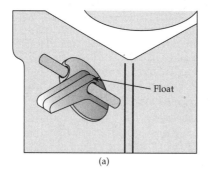

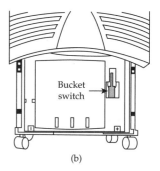

(a)

(b)

To handle these problems, perform the following steps:

1. **Verify the complaint.** Verify the complaint by asking the customer to describe what the dehumidifier is doing. It will be helpful if you can locate the actual service manual for the dehumidifier model you are working on to properly diagnose the dehumidifier.

2. **Check for external factors.** You must check for external factors not associated with the dehumidifier. Is the dehumidifier installed properly? Does it have the correct voltage? Check for a voltage drop during dehumidifier startup. The voltage at the wall receptacle must be within ±10 percent of the voltage rating on the model and serial data plate. Do you have the correct polarity? (See Chapter 6.)

3. **Disconnect the electricity.** Before working on the dehumidifier, disconnect the electricity. This can be done by pulling the plug from the receptacle. Or disconnect the electricity at the fuse panel or at the circuit breaker panel. Turn off the electricity.

WARNING *Before you begin to service any high-voltage components within the dehumidifier, you must discharge the high-voltage capacitors first.*

WARNING *A capacitor will hold a charge indefinitely, even when it is not currently in use. A charged capacitor is extremely dangerous. Discharge all capacitors immediately any time that work is being conducted in their vicinity. Redischarge after repowering the equipment if further work must be done. Many capacitors are internally fused. If you are not sure, you can use a 20,000 ohm 2-watt resistor to discharge the capacitor. Do not use a screwdriver to short out the capacitor. By doing so, you will blow out the fuse in the capacitor and the capacitor will not work. Safely use an insulated pair of pliers to remove the wires from the capacitor and place the resistor across the capacitor terminals. When checking a dual capacitor with a capacitor analyzer or ohmmeter, you must test both sides of the capacitor.*

WARNING *Some diagnostic tests will require you to test the components with the power turned on. When you disassemble the control panel or remove the outer cabinet, you can position the panel in such a way that the wiring will not make contact with metal. This will allow you to test the components without electrical mishaps.*

4. **Gain access and removal of the bucket switch and float.** Pull the bucket toward you, releasing the actuator from the float assembly (Figures 30-5a and b). Remove the outer wrapper to gain access to the bucket switch (Figure 30-6). To remove the float assembly, pull the pin (Figure 30-8), and remove the float assembly. To remove the bucket switch assembly (Figure 30-9a), squeeze the tabs to remove the bucket switch.

5. **Test the bucket switch.** Use the ohmmeter to test the resistance between the three terminals on the switch. Figure 30-9b shows testing between common and normally open terminals. In the switch rest position, there will be maximum resistance. From common to the normally closed terminal, there should be zero resistance. When the bucket switch lever is depressed, common to the normally open terminals, there should be zero resistance. From the common to the normally closed terminals, you should have a reading of maximum resistance.

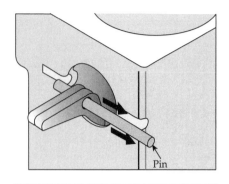

FIGURE 30-8 Pull the pin to release the float assembly.

6. **Install the new component.** To install a new component, just reverse the disassembly procedure and reassemble. Reinstall all panels and the front grille and bucket. Then, restore the electricity to the dehumidifier. Test the dehumidifier operation. To prevent ESD from damaging expensive electronic components, simply follow the steps in Chapter 11.

Defrost Thermostat

The defrost thermostat is mounted on the suction line (Figure 30-10).

The typical complaints associated with failure of the defrost thermostat (heat pump models only) are:

- Ice buildup on the outdoor coil.
- The air conditioner will not heat.

To handle these problems, perform the following steps:

1. **Verify the complaint.** Verify the complaint by checking the control setting. Turn the control to the highest setting for heat, and then turn it back to a normal setting to see if the dehumidifier starts heating. It will be helpful if you can locate the actual service manual for the dehumidifier model you are working on to properly diagnose the dehumidifier.

2. **Check for external factors.** You must check for external factors not associated with the dehumidifier. Is the dehumidifier installed properly? Explain to the user how to set the controls. The voltage at the wall receptacle must be within ±10 percent of the voltage rating on the model and serial data plate. Do you have the correct polarity? (See Chapter 6.)

3. **Disconnect the electricity.** Before working on the dehumidifier, disconnect the electricity to the unit. This can be done by pulling the plug from the receptacle. Or disconnect the electricity at the fuse panel or at the circuit breaker panel. Turn off the electricity.

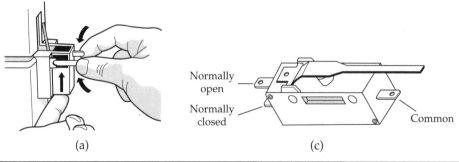

(a)   (c)

FIGURE 30-9   (a) Squeeze the tabs to release the bucket switch. (b) SPST bucket switch. Use the ohmmeter to test the switch contacts.

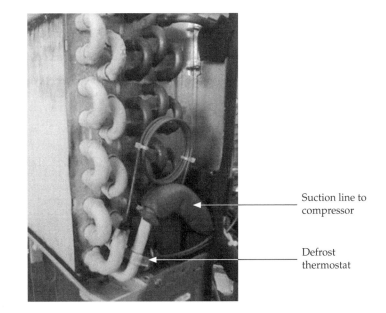

FIGURE 30-10
Defrost thermostat attached to the suction line. When the evaporator goes below 37 degrees Fahrenheit, the defrost thermostat opens, shutting off the compressor. The condenser fan motor will continue to run. When the suction line temperature rises to 59 degrees Fahrenheit, the defrost thermostat closes and the compressor starts up.

Suction line to compressor

Defrost thermostat

WARNING *Some diagnostic tests will require you to test the components with the power turned on. When you disassemble the control panel or remove the outer cabinet, you can position the panel in such a way that the wiring will not make contact with metal. This will allow you to test the components without electrical mishaps.*

4. **Gain access to the defrost thermostat.** To access the defrost thermostat, remove the outer wrapper and the front grille and filter (Figures 30-5 and 30-6). Next, tilt the control panel away from the dehumidifier, making sure not to pull any of the wires off the controls. The manufacturer placed the defrost thermostat control on the suction line (Figure 30-10). The defrost thermostat is a nonadjustable thermostat without a stem or knob attached to it. On other models, the defrost thermostat might be a bimetal thermostat that is clamped to the end of the evaporator coil with two wires coming out of it that lead back to the control panel.

WARNING *Do not touch the wiring or capacitor until it is discharged. A capacitor will hold a charge indefinitely, even when it is not currently in use. A charged capacitor is extremely dangerous. Discharge all capacitors immediately any time that work is being conducted in their vicinity. Redischarge after repowering the equipment if further work must be done. Many capacitors are internally fused. If you are not sure, you can use a 20,000 ohm 2-watt resistor to discharge the capacitor. Do not use a screwdriver to short out the capacitor. By doing so, you will blow out the fuse in the capacitor and the capacitor will not work. Safely use an insulated pair of pliers to remove the wires from the capacitor and place the resistor across the capacitor terminals. When checking a dual capacitor with a capacitor analyzer or ohmmeter, you must test both sides of the capacitor.*

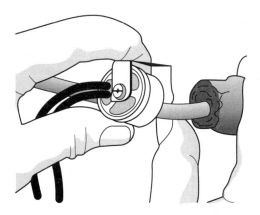

FIGURE 30-11 Remove the insulation from around the defrost thermostat so that the spring clamp is exposed.

5. **Test the defrost thermostat.** To test the defrost thermostat, disconnect the wires from the defrost thermostat terminals on the control panel, label them, and place the ohmmeter probes on the terminals (see Figure 29-6b). Set the range scale on R × 1, and test the thermostat for continuity. If the temperature is above 60 degrees Fahrenheit, the resistance reading will be zero, indicating that the contacts are closed. Now, place the thermostat in a bucket of ice. With the ohmmeter leads still attached, the contacts should open when the temperature drops below 37 degrees Fahrenheit.

6. **Remove the thermostat.** With the thermostat exposed and the wires already removed, the defrost thermostat must now be removed from the suction line. The thermostat is held in place on the suction line with a spring load clamp (Figure 30-11). Remove the thermostat clamp. If the clamp comes off the suction line, remember where the clamp goes back. The placement of the clamp and thermostat is crucial for the dehumidifier to function properly.

7. **Install a new thermostat.** To install the new defrost thermostat, just reverse the order of disassembly and reassemble. Then test the defrost thermostat by running the dehumidifier. Just remember to reinstall the bimetal clamp in the same location from which it was removed. If you do not, the dehumidifier will not cycle properly.

Wiring Diagrams and Schematics

The wiring diagrams and schematics that are used in this chapter might be helpful to diagnose a dehumidifier (see Figures 30-12, 30-13, 30-14, 30-15, and 30-16). These diagrams are not a substitution for the exact wiring schematic that might be on the inside cover of the outer wrapper.

Endnote

1. When the bearings wear out in the end bell of the motor, the motor will begin to make a vibrating sound or a growling noise.

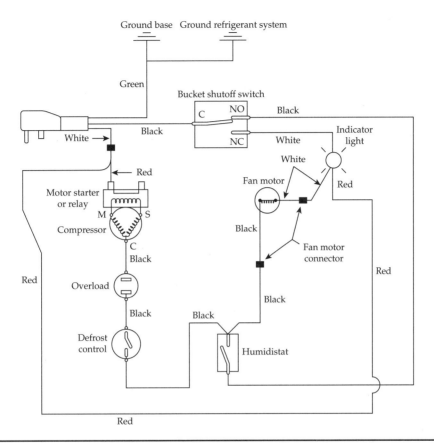

FIGURE 30-12 A typical model with a reciprocating compressor with a single-speed fan motor.

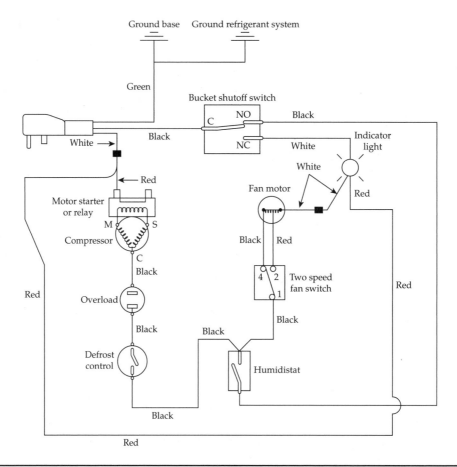

FIGURE 30-13   A typical model with a reciprocating compressor with a double-speed fan motor.

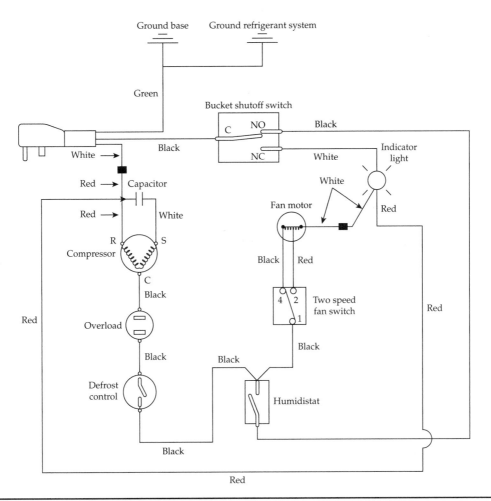

FIGURE 30-14 A typical model with a rotary compressor with a double-speed fan motor.

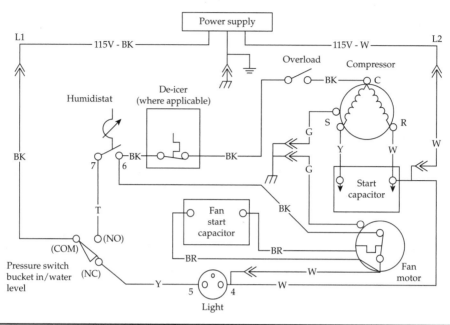

FIGURE 30-15   A typical wiring schematic for a dehumidifier with mechanical controls.

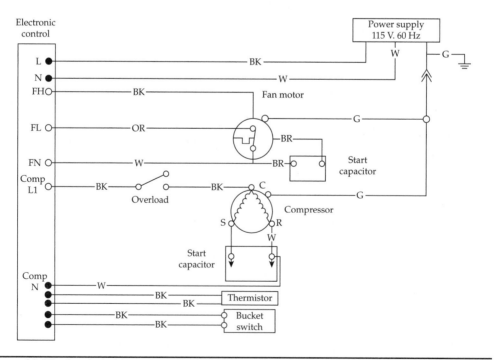

FIGURE 30-16   A typical wiring schematic for a dehumidifier with electronic controls.

Glossary

air shutter   An adjustable shutter attached to the primary air openings of a burner used to control the amount of primary air entering the burner.

aldehyde   A class of compounds that can be produced from incomplete combustion of gas fuels.

allen wrench   An L-shaped tool that is used to remove hex screws.

alternating current (AC)   Electrical current that flows in one direction and then reverses itself and flows in the opposite direction. In 60-cycle current, the direction of flow reverses every 120th second.

ambient temperature   Temperature of air that surrounds an object on all sides.

ammeter   A test instrument used to measure current.

ampere   The number of electrons passing a given point in one second.

armature   The section of a motor that turns.

automatic defrost timer   A device connected in an electrical circuit that shuts off the refrigeration cycle and turns on the electric heaters to melt the ice off the evaporator coil.

bake element   Lower heating element in an oven used for baking foods.

belt   A band of flexible material used to transfer mechanical power from one pulley to another.

bimetal (strip or thermostat)   Two dissimilar metals joined together to form one unit with a differential expansion rating. It will bend if there is a temperature change.

blower wheel   A device attached to the indoor side of the fan motor shaft used to circulate air across the evaporator coil.

bracket   A hardened structure used to support a component.

broil element   Upper heating element in an oven used for broiling foods.

BTU   (British Thermal Unit) The amount of heat required to raise the temperature of one pound of water one degree Fahrenheit.

burner   A device in which an air and gas mixture has entered into the combustion zone to be expelled and burned off.

burner head   The portion of the burner past the outlet mixing tube that contains the burner ports.

cabinet   The outer wrapper of an appliance.

cam   A rotating surface with rises and falls that open and close the switches in a timer.

capacitor   A device that stores electricity; used to start and/or run circuits on many large electric motors. The capacitor reduces line current and steadies the voltage supply while greatly improving the torque characteristics of the fan motor or compressor.

capillary tube   A metering device used to control the flow of refrigerant in a sealed system. It usually consists of several feet of tubing with a small inside diameter.

Celsius   The metric system for measuring temperature. The interval between the freezing point and the boiling point of water is divided into 100 degrees. This is also called the centigrade scale.

circuit   A path for electrical current to flow from the power source to the point of use and back to the power source.

circuit breaker   A safety device used to open a circuit if that circuit is overloaded.

closed (circuit)   An electrical circuit in which electrons are flowing.

combustion   The rapid chemical reaction of a gas with oxygen to produce heat and light.

component   An individual mechanical or electrical part of an appliance.

compressor   An apparatus similar to a pump that is used to increase the pressure of refrigerant that is to be circulated within a closed system.

condenser coil   A component of the refrigeration system that transfers the heat that comes from within a closed compartment to the outside surrounding area.

condenser fan   A fan motor and fan blades used to cool the condenser coil.

contact points   Two movable objects, or contacts, that come together to complete a circuit or that separate and break a circuit. These contact points are usually made of silver.

continuity   The ability of a completed circuit to conduct electricity.

control   A device, either automatic or manual, that is used to start, stop, or regulate the flow of liquid, gas, or electricity.

current   The flow of electrons from a negative to a positive potential.

cycle   A series of events that repeat themselves in the same order.

defective   Refers to a component that does not function properly.

defrost cycle   That part of the refrigeration cycle in which the ice is melted off the evaporator coil.

defrost thermostat (heat pump AC models)   A dual-purpose control that acts as an outdoor thermostat and defrost control.

defrost timer   A device used in an electrical circuit to turn off the refrigeration long enough to permit the ice to be melted off the evaporator coil.

diagnosis   The act of identifying a problem based on its signs and symptoms.

direct current (DC)   Electric current that flows in one direction in a circuit.

end bell   The end plates that hold a motor together.

energize   To supply electrical current for operation of an electrical component.

evaporator coil   A component of the refrigeration system that removes the heat from within a closed compartment.

evaporator fan   A motor and fan blades used to circulate the cold air.

Fahrenheit   The standard system for measuring temperature. The freezing point of water is 32 degrees Fahrenheit, and the boiling point of water is 212 degrees Fahrenheit.

flashback   Also known as extinction pop. Where the burner flames enter back into the burner head and continue to burn after the gas supply has been turned off.

fuse   A safety device used to open a circuit if that circuit is overloaded.

gasket   A flexible material (either airtight or watertight) used to seal components together.

ground   A connection to earth or to another conducting body that transmits current to earth.

ground wire   An electrical wire that will safely conduct electricity from a structure to earth.

heat anticipator   Used to provide better thermostat and room air temperature control.

Hertz (Hz)   A unit of measurement for frequency. One hertz equals one cycle per second.

hot gas defrost   A defrosting system in which hot refrigerant from the condenser coil is directed to the evaporator coil for a short period of time to melt the ice from the evaporator coil.

housing A metal or plastic casing that covers a component.

idler pulley A device that rests on or presses against a drive belt to maintain a specified tension on it.

insulation Substance used to retard or slow down the flow of heat through a substance.

insulator A material that does not conduct electricity. It is used to isolate current-carrying wires or components from other metal parts.

ladder diagram A wiring schematic in which all of the components are stacked in the form of a ladder.

lead (wire) A section of electrical wiring that is attached to a component.

lint Fine pieces of cotton fiber that have broken away from a garment.

module A self-contained device with a group of interconnecting parts designed to do a specific job.

nut driver A tool used to remove or reinstall hexagonal-head screws or nuts.

ohm A unit of measurement of electrical resistance.

ohmmeter A test instrument used for measuring resistance.

open (circuit) A break in an electrical circuit that stops the flow of current.

orifice An opening in a hood cap (orifice spud) through which a gas is discharged and where the gas is metered.

overload protector A device that is temperature-, pressure-, or current-operated that is used to open a circuit to stop the operation of that circuit should dangerous conditions arise.

parallel circuit Components that are parallel-connected across one voltage source. All of the branches are supplied with the same amount of voltage.

port An opening in the burner head where the air and gas mixture is discharged and lit by an ignition source.

pressure switch A device that is operated by pressure and turns a component on or off.

primary air The main air supply introduced into the burner which mixes with the gas before it reaches the burner head.

PTC (positive temperature coefficient) A resistor that increases resistance (ohms) with temperature increase.

pulley A wheel that is turned by or driven by a belt.

refrigerant   A chemical substance used in refrigeration and air conditioning that produces a cooling effect.

relay   A magnetic switch that uses a small amount of current in the control circuit to operate a component needing a larger amount of current in the operating circuit. A remote switch.

relief valve   A safety device that is designed to open before dangerous pressure is reached.

resistance   The opposition to current flow. The load in an electrical circuit.

run winding   Electrical winding of a motor, which has current flowing through it during normal operation of the motor.

safety thermostat   A thermostat that limits the temperature to a safe level.

schematic diagram   A line drawing that gives the electrical paths, layout of components, terminal identification, color codes of wiring, and sometimes the sequence of events of an appliance.

series circuit   A circuit in which all of the components are sequentially connected in the same line. If one component fails, all the components fail.

short circuit   An electrical condition in which part of a circuit errantly touches another part of a circuit and causes part or all of the circuit to fail (and trip the circuit breaker or blow a fuse).

slinger ring   A ring attached to the circumference of the condenser fan blade. The slinger ring is positioned close to the base pan of the air conditioner so water in the base pan is picked up by the rotating ring and dispersed into the hot condenser air stream where it evaporates.

solenoid   A cylindrical coil of insulated wire that establishes a magnetic field in the presence of current.

start winding   A winding in an electric motor used only during brief periods when the motor is starting.

surface element   The top cooking element on a range used for cooking foods.

switch   A device to turn current on or off in an electrical circuit.

temperature   A measure of heat energy or the relative lack thereof.

terminal   A connecting point in a circuit to which a wire would be attached to connect a component.

test light   A light provided with test leads that is used to test electrical circuits.

thermistor   A device that exhibits a large change in electrical resistance with a change in temperature.

thermometer   A device used to measure temperature.

thermostat   A device that senses temperature changes and that usually operates a control relay.

thermostat (operating)   A thermostat that controls the operating temperature of a component.

transformer   A device that raises or lowers the main AC supply voltage.

venturi throat   A section of the mixing tube body that narrows down and then flares out again.

voltage   The difference in potential between two points; the difference in static charges between two points.

voltmeter   A test instrument used to measure voltage.

VOM (voltohm-milliammeter)   A test instrument used to measure voltage, resistance, and amperage.

water column (WC)   A unit used for pressure. A one-inch water column equals a pressure of 0.578 ounces per square inch.

watt   A unit to measure electrical power.

wattmeter   A test instrument used to measure electrical power.

Index

H